1001 INVENTIONS
THAT CHANGED THE WORLD

1001 INVENTIONS
THAT CHANGED THE WORLD

GENERAL EDITOR JACK CHALLONER

PREFACE BY TREVOR BAYLIS

THUNDER BAY
P·R·E·S·S

San Diego, California

○ Patent applications to be processed at
the U.S. Patent Office in Washington, D.C.

Thunder Bay Press
An imprint of Printers Row Publishing Group
9717 Pacific Heights Blvd, San Diego, CA 92121
www.thunderbaybooks.com • mail@thunderbaybooks.com

Copyright © 2009, 2016, 2017, 2021 Quarto Publishing plc

Conceived, designed, and produced by
Quintessence Editions, an imprint of The Quarto Group
The Old Brewery, 6 Blundell Street
London, N7 9BH

Printers Row Publishing Group is a division of Readerlink Distribution Services, LLC.
Thunder Bay Press is a registered trademark of Readerlink Distribution Services, LLC.

Correspondence regarding the content of this book should be sent to Thunder Bay Press, Editorial Department, at the above address. Author and rights inquiries should be addressed to Quintessence Editions, The Old Brewery, 6 Blundell Street, London N7 9BH, United Kingdom.

Thunder Bay Press
Publisher: Peter Norton • Associate Publisher: Ana Parker
Acquisitions Editor: Kathryn Chipinka Dalby

QUAR.345017

Quintessence Editions
Update Editor   Ruth Patrick
Update Designer   Glenn Howard
Publisher   Samantha Warrington

Original Edition
Senior Editor   Jodie Gaudet
Editor   Frank Ritter
Editorial Assistant David Hutter
Art Director   Akihiro Nakayama
Designer   Philip Hall
Picture Researcher   Sunita Sharma-Gibson
Editorial Director   Ruth Patrick
Publisher   Philip Cooper

Library of Congress Control Number: 2021934775

ISBN: 978-1-64517-820-0

Printed in Singapore

25 24 23 22 21   1 2 3 4 5

Contents

Preface

By Trevor Baylis, OBE, inventor

If you run your eyes down the list of inventions in this book, you will be amazed that many everyday items are a lot older than you thought they were. The fishhook, for instance, first appeared 35,000 years ago, while computer programs were developed as early as 1843. And I believe there is such an invention in all of us. How many of us have a great idea but do nothing about it, assuming it has been thought of before, only to realize months or even years later that if you had got off your backside and done something about it, you could have been listed in this book.

I also believe in the great inventor Louis Pasteur's tenet that chance favors the prepared mind. In fact, it was purely by chance that I was watching a television program about the spread of HIV/AIDS in Africa one night. I was appalled when I saw naked bodies being thrown into open graves. The reporter said that the only way to control this epidemic effectively was by spreading health information and education through radio broadcasts. But there was a problem: most of Africa is without electricity, while batteries are unaffordable for many. Watching the program, I suddenly pictured myself in colonial times, wearing a pith helmet and monocle with a gin and tonic in my hand, listening to my large wind-up gramophone with His Master's Voice records blaring out of a large horn on top of a turntable. It instantly occurred to me that if you can get all that sound by dragging a needle around a piece of Bakelite, then surely there is enough power in the spring to drive a small dynamo which in turn could drive a radio. This was my "Eureka!" moment on the way towards developing the clockwork radio, and the rest went like, well, clockwork.

This story serves to illustrate that many inventions—like soap, the flushing toilet, scissors, or the fountain pen—are inspired responses to perceived needs. Yet, ironically, it is often something dramatic like war that motivates individuals to produce weapons like rockets, before people find uses for them that benefit mankind in a more agreeable way.

But once you've had that big idea, how many of us know what to do next? Remember: a picture is worth a thousand words—but a prototype is worth millions! Alas, many people don't know that if you go down to the pub and tell everyone about your idea, you cannot protect it legally any longer. The keyword is "intellectual property" because nobody pays you for a good idea, but they might pay you for a piece of paper (like a patent, design registration, or copyright document) that says you own the idea. Luckily, I knew about intellectual property when I invented the clockwork radio. But I still had to rely on the Patent Office to tell me that I should consult a patent attorney. (I like to think that that's because lawyers are the only people who can write without punctuation—something I never learned at school.)

Hence, I think it's about time for us to take inventors more seriously. Perhaps we should make the process of invention part of the National Curriculum, so when someone comes up with a good idea, they will know what steps to take to protect it, and how to get the product to the market. One of my favorite expressions is "Art is pleasure, invention is treasure," and when you look at the list of these 1,001 inventions you realize how many of them are vital to our everyday life. Moreover, inventions are crucial to the continuing growth of industry and commerce, and inventors like Frank Whittle—whose company built the first jet engine in 1937, the year I was born—have ensured that our previously gigantic planet has become a "global village" that none of our forefathers could have ever imagined.

I must congratulate the contributors of this book for compiling such an interesting and thought-provoking list of inventions. The collection invites reflection on what is important to us in our daily lives, and it is also very entertaining. The book celebrates how ordinary men and women have changed our lives—both socially and commercially—with their knowledge and creativity, and the later section in particular provides ample evidence of how quickly our inventors change the world.

Trevor Baylis, London

Introduction

By Jack Challoner, General Editor

To invent is to create something new—something that did not exist before. An invention can be an idea, a principle (such as democracy), a poem, a dance, or a piece of music, but in this book we have restricted ourselves to technological inventions. Technology is the practical application of our understanding of the world to achieve the things we need or want to do. Technology goes beyond "things" such as computers or bicycles: it includes techniques and processes, such as the alphabet, numerical systems, and the extraction of metals from their ores.

Scientific theories and discoveries are not included in this book. Science is our way of unraveling the laws of nature through theory and experiment. Science and technology are interdependent. The internal combustion engine, for example, could not have been invented without a scientific understanding of thermodynamics and the chemistry of combustion. Conversely, inventions such as the microscope, the radio telescope, and the computer have greatly aided scientific investigation.

When I began to compile this book's list of inventions, I wondered whether 1,001 would be too many. But I soon realized that countless technological innovations, large and small, have played roles in defining the way we live. Every one of the inventions in this book has changed the world in some way, but so have many others that I had to leave out.

So how did I decide upon the final list? I knew that readers would pick up the book to discover who invented the paper clip, the nonstick pan, or scissors—thus, the index contains virtually any invention of that kind. This inevitably led to a bias toward modern inventions, although the increasing pace of technological innovation dictated that this always would be the case. I also wanted the book to introduce important but less well-known technologies. You will find these by browsing or by following cross-references—or by chancing upon intriguing and eye-catching images.

1001 Inventions That Changed the World is ordered chronologically. However, there is not always a definitive date for the first appearance of an invention. The incandescent light bulb, for example, developed gradually over the entire nineteenth century. Rubber was first vulcanized in ancient Mesopotamia, but it was not really used until much later. So, for the light bulb, we have given the date of the first working incandescent bulb in a form we would recognize today; for vulcanized rubber, we cited the date of the introduction of the modern industrial process. For ancient technologies where no record exists of who invented them or when, we have based the dates on the oldest existing evidence or on other best guesses.

1001 Inventions That Changed the World is divided into eight chapters of fairly equal length. Arranging their history in this way is inevitably fairly arbitrary, but we have done our best to begin each chapter at a moment in history when society or technology were undergoing significant changes.

As chapter one reveals, stone tools are widely accepted as the first human invention, made and used by our hominid ancestors around 2.5 million years ago. The inventions of the early Stone Age (the Paleolithic)—controlled fire, clothing, the sharp blade, and the spear as well as stone tools—are perhaps the most important inventions of all time. Without these technologies, humankind might never have moved from hunting and gathering to a more settled lifestyle.

It is no coincidence that the rate of technological progress increased dramatically once people were more settled. The first great civilization arose around 7,000 years ago in Mesopotamia (in what is now Iraq). The wheel and axle, another candidate for the world's most important invention, originated in ancient Mesopotamia—as did glass and irrigation. Within the next 2,000 years or so, civilizations sprang up in the Indus Valley (in the Indian subcontinent), in Egypt, and in the Yellow River Valley in China.

The introduction of metals was a great leap forward. The new materials made better weapons and tools, as well as agricultural implements. Most metals are bound tightly with other elements in ores. In the Bronze Age and Iron Age, people began to learn how to extract metals from those ores.

Around 2,500 years ago, another great civilization was at its height around the Mediterranean. Ancient Greece is famed for its philosophers, but its mathematicians and engineers also made vital contributions to science and technology. Archimedes, for example, is credited with the earliest machines—the lever, the block and tackle, and the screw.

Chapter two takes us from the Romans to the start of the Industrial Age. In the first few hundred years CE, most technological innovation took place in the Indus Valley and in ancient China. Many of the important inventions created in the Indus Valley remained unknown to the rest of the world, but people did learn of the Chinese inventions—in particular, the "four great inventions": gunpowder, paper, the compass, and printing.

After the Roman Empire's decline in the fifth century CE, much of the West was plunged into what is often called the Dark Ages, but Chinese civilization remained stable and continued to innovate. The Islamic Empire emerged in the eighth century, and Islamic scholars passed on and added much to the work of the ancient Greek thinkers, which would have otherwise have been lost. During the European Renaissance, from the late fourteenth until the early seventeenth centuries, rational thought in an increasingly ordered and affluent society led to technologies such as the printing press, the microscope, and the telescope.

The seventeenth and eighteenth centuries were a period of rapid progress in astronomy, biology, physics, chemistry, and mathematics. This period saw the development of a bewildering array of instruments— thermometers, hygrometers, vacuum pumps, barometers, and mechanical calculators—that allowed scientists to probe and manipulate the physical world as never before. During this time, scientists also began to experiment with the two vitally important forces of electricity and magnetism.

New insights into physics and chemistry encouraged technological innovation and helped to drive the Industrial Revolution, which first arrived in Great Britain toward the end of the eighteenth century, initiated by the introduction of coke-based iron smelting. For centuries iron had been smelted using charcoal, derived from wood. The appetite for iron in the seventeenth century had been so voracious that Britain was all but deforested. By replacing charcoal with coke, derived from coal rather than wood, Britain saved its remaining forests and increased its iron production.

Extracting coal presented its own problems; biggest among them was flooding. Pumping water from mines was a challenge that led to the steam engine in the early 1700s. In the second half of the eighteenth century, James Watt improved the efficiency of the basic pumping engines and introduced more powerful engines that could produce rotary motion.

Chapter three of this book begins at the end of the eighteenth century, when Britain's Industrial Revolution was in full swing. The high-pressure steam engine led to the introduction of railroad locomotives and steam-powered ships, which quickly revolutionized transport systems across much of the world. Engineers developed smaller, more powerful steam engines, later to be followed by the internal combustion engine. At the same time, the mechanization of the textile industry encouraged a move to factory-based work; thousands left the land to work in big cities.

Chapter four begins in the mid-nineteenth century, when the punctuated technological developments of the Industrial Revolution gave way to more sustained technological progress. The new level of sophistication and progress enabled the European colonies to thrive— this was the Age of Empires. Thanks to new forms of communication and transport, the empires could coordinate their activities across continents. It was around this time, too, that the United States became the powerhouse of innovation that it continues to be in the twenty-first century. The U.S. inventor Thomas Edison alone was to hold more than 1,000 patents.

All the pieces were now in place for the development of the technologically advanced world we know today. The chemical industry was born in the 1830s, making dyes and fertilizers as well as, by the end of the century, plastics. The introduction of gas lighting and then electric lighting allowed for restless innovation and industry to continue both day and night. Telecommunications continued to advance, with the telegraph giving way to the telephone and the radio. By the late nineteenth century, where chapter five begins, many countries had extensive electrical supply systems, telephone networks, roads, and railroads. And, of course, the motorcar was beginning to make its existence felt, having been patented in Germany in 1886. Another invention of this period would have a profound effect on the twentieth century: the powered airplane. Film and television complete the list of major technologies that underpin the modern world.

War tends to provide a boost to innovation. Chapter six begins before the advent of World War II, which brought such technological innovations as DDT and the atomic bomb. This chapter also covers the postwar period, when optimism about science and technology was high. This period saw the realization of the digital computer. In the early 1950s, where chapter seven begins, the integrated circuit was invented and, most significantly, the first satellite. Our final chapter is characterized by incredible medical technologies and by the rise of consumer electronics. It begins with the dawn of the Internet, a technology that has had a profound effect on our lives, and which is likely to affect us more profoundly still.

Inventions have played a defining role in human history—transforming us from primitive hunter-gatherers into a sophisticated, settled, and determining species. Today's world contains countless technologies of varying sophistication that enrich our lives. Not all inventions are for good, however; some inventions cause harm, with or without human intention. And there is a persuasive argument that technological progress in general may threaten our survival, by allowing unrestricted growth of the human population and by upsetting the natural balance of our planet.

If necessity is the mother of invention, then ingenuity is surely the father. And since ingenuity is as innate as love, hate, kindness, and mistrust, human beings will no doubt continue to invent new things for use in peace and war, and refine those inventions already in existence.

Index of Inventions

The invention of the first stone tools more than 2 million years ago was the moment when humankind started to distinguish itself from all other species on the planet. It took our forebears another 1.2 million years before they found other ingenious ways to use their natural resources—by learning to control fire, build shelter, and make items of clothing. But having gained momentum, the ideas kept coming, and the inventions that followed resulted in full-blown civilizations.

◁ Ancient warriors forged swords using metal and fire.

The
ANCIENT WORLD

BCE

Stone Tools

(*c.* 2,600,000 BCE)

Early humankind ushers in the age of inventions.

The very first human invention consisted of sharp flints, found and used in their natural state by primitive peoples, who then went on to purposely sharpen stones. The practice reaches back to the very dawn of humankind; stone tools found in 1969 in Kenya are estimated to be 2,600,000 years old.

The principal types of tools, which appeared in the Paleolithic period, and varied in size and appearance, are known as core, flake, and blade tools. The core tools are the largest and most primitive, and were made by working on a fist-sized piece of rock or stone (core) with a similar rock (hammerstone) and knocking large flakes off one side to produce a sharp crest. This was a general-purpose implement used for hacking, pounding, or cutting. Eventually, thinner and sharper core tools were developed, which were more useful. Much later, especially during the last 10,000 years of the Stone Age, other techniques of producing stone artifacts—including pecking, grinding, sawing, and boring—came into play.

The evolution of tool making enabled early humankind to complete many tasks previously impossible or accomplished only very crudely. Animals could be skinned, defleshed, and the meat divided up with stone cutters, cleavers, and choppers. Clothing was made from animal hides cleaned with rough stone scrapers and later punctured with awls. Hunting became more efficient with spearheads fashioned from stone flakes. And with the aid of stone adzes (axes), early humankind could create shelter and begin to shape the physical world to its liking. **MF**

"The best materials ... include obsidian (a form of natural glass), chert, flint, and chalcedony."

Floyd Largent, writer

SEE ALSO: CONTROLLED FIRE, BUILT SHELTER, CLOTHING, SPEAR, FISHHOOK, DRILL, SHARP STONE BLADE

◩ *Stone Age humans became adept at chipping flakes of hard, volcanic rocks to make tools and weapons.*

Controlled Fire
(*c.* 1,420,000 BCE)

Homo erectus harnesses lightning.

Fire is an essential tool, control of which helped to start the human race on its path to civilization. The original source of fire was probably lightning, and for generation blazes ignited in this manner remained the only source of fire.

Initially Peking Man, who lived around 500,000 BCE, was believed to be the earliest user of fire, but evidence uncovered in Kenya in 1981, and in South Africa in 1988, suggests that the earliest controlled use of fire by hominids dates from about 1,420,000 years ago. Fires were kept alive permanently because of the difficulty of reigniting them, being allowed to burn by day and damped down at night. Flint struck against

---

"Seek wood already touched by fire. It is not then so very hard to set alight."

African proverb

---

pyrites or friction methods were the most widespread methods of producing fire among primitive people.

The first human beings to control fire used it to keep warm, cook their food, and ward off predators. It also enabled them to survive in regions previously too cold for human habitation. They also used it in "fire drives" to force animals or enemies out of hiding. Controlled fire was important in clearing forest for roadways, grasslands for grazing, and agricultural lands—uncontrolled, the fire destroyed the potential of the soil. Mastering fire also opened up the possibilities of smelting metals, enabling humankind to escape the limitations of the Stone Age. **MF**

SEE ALSO: OIL LAMP, CANDLE, OVEN

Built Shelter
(*c.* 400,000 BCE)

Homo heidelbergensis builds the first hut.

The earliest evidence for built shelter appears to have been constructed by *Homo heidelbergensis*, who lived in Europe between around 800,000 BCE and 200,000 BCE. Anthropologists are uncertain whether these were ancestors of *Homo sapiens* (humans) or *Homo neanderthalensis* (Neanderthals) or both.

At the French site of Terra Amata, which dates back around 400,000 years, archeologists have found what they believe to be the foundations of large oval huts. One of these shows evidence of fire in a hearth, although other archeologists postulate that natural processes could be responsible. Archeology on sites from hundreds of thousands of years ago is

---

" . . . next to agriculture, [shelter] is the most necessary to man. One must eat, one must have shelter."

Philip Johnson, architect

---

complicated. Claims of the discovery of built shelters in Japan from more than 500,000 years ago were discredited in 2000. In fact, all evidence for humans in Japan before 35,000 years ago is currently questionable.

We do know that our ancestors spent time in caves for hundreds of thousands of years. But caves are only found in certain areas. Whether they started building 100,000 or 400,000 years ago, their ability to create shelters close to food, water, and other resources provided our ancestors with protection against the elements and dangerous animals. Living close to work also gave them more time to experiment with different ways of doing things; in other words, time to invent. **ES**

SEE ALSO: DRIED BRICK, FORT, WATTLE AND DAUB,
FIRED BRICK, PLASTER

Clothing (*c.* 400,000 BCE)

Early humans cover their nakedness.

Around 400,000 years ago, *Homo sapiens* devised a solution to protect the vulnerable naked human body from the environment—clothes. Anthropologists believe the earliest clothing was made from the fur of hunted animals or leaves creatively wrapped around the body to keep out the cold, wind, and rain.

Determining the date of this invention is difficult, although sewing needles made from animal bone dating from about 30,000 BCE have been found by archeologists. However, genetic analysis of human body lice reveals that they evolved at the same time as clothing. Scientists originally thought the lice evolved 107,000 years ago, but further investigations placed their evolution a few hundred thousand years earlier.

> *"Clothes make the man. Naked people have little or no influence in society."*
>
> Mark Twain, *More Maxims of Mark* (1927)

Clothing has changed dramatically over the centuries, although its ancient role as an outward indication of the status, wealth, and beliefs of the wearer is as important as ever. During the Industrial Revolution the textile industry was the first to be mechanized, enabling increasingly elaborate designs to be made at a faster rate. In the twenty-first century, mechanization has allowed sophisticated practical clothing to be devised to protect us from dangers such as extreme weather, chemicals, insects, and outer space. Without clothes we would not have been able to explore and exploit our world and the surrounding universe to the extent that we have. **FS**

SEE ALSO: SEWING, SHOE, WOVEN CLOTH, BUTTON, HELMET, BUCKLE, SPINNING JENNY, SPINNING MULE

Spear (*c.* 400,000 BCE)

Humans learn to kill with sharpened poles.

The earliest example of a sharpened wooden pole, or spear, comes from Schöningen in Germany. There, eight spears were dated to 400,000 BCE. The ancient hominid hunters who sharpened each pole used a flint shaver to cut away the tip to form a point and then singed the tip in the fire to harden the wood, making it a more effective weapon. A similar technique was used by hunters in Lehringen near Bremen in Germany, where a complete spear was found embedded inside a mammoth skeleton, suggesting such spears were used mainly for hunting rather than warfare or self-defense. The need for food was so great that a mammoth would be attacked with only a flimsy spear, although its use would have been more to scare the mammoth in the direction of a trap or pit dug previously than to attack it directly.

Around 60,000 BCE, Neanderthals living in rock shelters and temporary hunting camps in France sharpened small pieces of flint and slotted them into the tips of their spears. Hunters in the Sahara used sharpened stones in the same way, while Central Americans used obsidian, a natural volcanic glass. Around the world, Stone Age people gradually learned how to work small stones or flints into tiny, sharpened blades known as microliths for use as spear points. The greatest advance, however, came with the development of metalworking, notably copper, in southeast Europe after 5000 BCE, followed by bronze, an alloy of copper and tin, around 2300 BCE, and then iron a millennium later. These new technologies allowed hunters and warriors to make hard, sharp, effective spear points. **SA**

SEE ALSO: STONE TOOLS, SHARP STONE BLADE, METALWORKING, ATLATL, BOW AND ARROW

→ *A spear-carrying Hittite warrior in a tenth-century BCE relief from Carchemish on the Turkey-Syria border.*

Fishhook (*c.* 35,000 BCE)

Early humans discover how to retain their caught fish.

"Opportunities are . . . everywhere and so you must always let your hook be hanging."

Augustine "Og" Mandino, writer

⬆ *Prehistoric ivory barbed harpoon hooks show evidence of meticulous carving.*

➡ *Fishing with weighted hooks in a bas-relief of the pyramid of Teti, Egypt, Sixth Dynasty, 2325–2175 BCE.*

The major problem with dating inventions earlier than the written word is that there are no first-hand accounts documenting their conception or use. Paleoarcheologists have the difficult task of piecing together the prehistory of man based on scraps of physical evidence left behind by our ancient ancestors. The fishhook is one such ingenious conception of early man and is probably more important to the success of humans than most of us would suspect.

The earliest examples of fishhooks so far found by archeologists date from around 35,000 BCE. Appearing well before the advent of metalworking, early fishhooks were fashioned from durable materials of organic origin such as bone, shells, animal horn, and wood. With the addition of a variety of baits on the hook, prehistoric man gained access, previously largely denied, to an easy source of energy loaded with protein and fat. Adding fish to his diet also ensured a healthy intake of essential fatty acids.

Over thousands of years the technology of fishhooks has evolved to optimize prey attraction, retention, and retrieval. The very earliest fishhooks of all are thought to have been made from wood, although, being more perishable than those of bone or shell, very few examples of these primitive hooks have survived. Wood might seem much too buoyant a material to be ideal for catching fish, but actually wooden hooks were used until the 1960s for catching species such as burbot.

Gaining easy access to adequate food supplies is thought to have been an essential factor in the success of early man. To fish in fecund waters requires very little energy and time, and this enabled our ancestors to pursue other activities, meaning that they were able, not just to survive, but to prosper. **JB**

SEE ALSO: STONE TOOLS, SHARP STONE BLADE, TRAWLER FISHING, MODERN HARPOON

Tally Stick (c. 35,000 BCE)

Counting makes its debut in Swaziland.

Tally sticks, or tallies, are batons of bone, ivory, wood, or stone into which notches are made as a means of recording numbers or even messages. The archeological and historical records are rich in tallies, with the Lebombo bone as the earliest example. Found in a cave in the Lebombo Mountains in Swaziland and made from a baboon's fibula, it dates back to 35,000 BCE. Its markings suggest that it is a lunar phase counter, indicating an appreciation of math far beyond simple counting.

Tally sticks became the primary accounting tool of medieval Europe, which was largely illiterate. During the 1100s King Henry I of England established the Exchequer to be responsible for the collection

> *". . . with worn-out, worm-eaten rotten bits of wood . . . a savage mode of keeping accounts . . ."*
>
> Charles Dickens, novelist

and management of revenues. To keep track of taxes owed and paid, split tally sticks were employed. Usually made of squared hazel wood, notches were made the thickness of the palm of the hand to represent £1000, the thickness of a thumb for £100, a little finger for £10, a swollen barley grain for £1, and a thin score mark for a shilling. The notches would span the stick's width, which subsequently would be split so that both halves had the same markings, to avoid forgeries. The halves differed in length; the longer half, or stock, was for the person making the payment, hence "stockholder," and the shorter half, or foil, for the recipient of the money or goods. **RP**

SEE ALSO: ABACUS, MECHANICAL CALCULATOR, POCKET CALCULATOR, DIGITAL ELECTRONIC COMPUTER

Drill (c. 35,000 BCE)

Early humans learn how to bore small holes.

It is thought that early man used a primitive drill—perhaps a modified spear—to pierce wood and animal skins. Much later, the woodworkers of ancient Egypt refined this technique by making any necessary holes with a bow drill. Adapted from the fire-stick, it had a cord wrapped round it and was held taut with a bow. Holding the drill vertically, the operator moved the bow backward and forward, pressing downward on alternate turns, with an idle return stroke. (There is also evidence of dental drilling from as long ago as 9000 BCE, accomplished by the same means.) The Romans replaced the bow drill with the auger, but the bit froze between turns. It was not until the Middle Ages that use of the carpenter's brace made continuous rotation of the drill possible.

The term "drill" may either refer to the machine supplying the rotational energy needed for penetration, or to the "drill bit," which is the part that rotates and actually cuts into the material. Various kinds of drills have evolved to meet specific needs.

Any drill has the capacity to make small holes in wood or brick, but more powerful machines are required to create pipe-sized holes in masonry or metal. Modern drills include a chuck to grip the drill bits or simple attachments. Some drills have chucks that can be unscrewed in order to receive larger attachments, such as sanding tools, wire brushes, grinding stones, and circular saws.

The tip of a drill bit is conical in shape with cutting edges. The fluted part, or body, of a drill is now usually made of hardened, high-carbon steel. The angle formed by the tapering sides of the point determines how large a chip is taken off with each rotation. The bit also has helical flutes, which affect the drill's cutting and chip-removal properties. **MF**

SEE ALSO: SEED DRILL, TWIST DRILL BIT, ELECTRIC DRILL

Sharp Stone Blade (*c.* 30,000 BCE)

Stone Age humans progress to sharpening their tools and weapons.

The use of stone instruments more than two million years ago heralded what we call the Stone Age and the very origins of humankind. While it is impossible to date when distinctly worked (rather than simply found) stone blades first appeared in the world, it seems to have occurred circa 30,000 BCE.

The technique that evolved to create sharp stones is now called lithic reduction. This involves the use of an implement (made of stone itself or of wood or bone) to strike a stone block in order to break off flakes. Such flakes will be naturally sharp and can be turned into a range of useful tools and weapons such as scrapers, scythes, knives, arrow heads, or spear points. Some early toolmakers may also have used what was left of the stone block to make axe heads.

Various kinds of stone were used to make blades, although one of the most popular was flint—leading to the term "flintknapper" to describe anyone making stone blades by lithic reduction. As the techniques of flintknapping developed, particularly the use of repetitive blows at particular angles, the craftsmen were able to gain much greater control over the size, sharpness, and type of blade.

The period after the end of the last Ice Age, 10,000 years ago, was characterized by increasingly sophisticated stone tools with multiple uses. Other tools were produced using blades made by knapped flint or obsidian, a type of naturally occurring glass. Small, sharp blades, known as microliths, became part of wooden cutting implements for use in farming, as well as barbs on arrows and spears, making them particularly effective as hunting weapons. **RBd**

SEE ALSO: STONE TOOLS, SPEAR, BOW AND ARROW, CHISEL

↗ *Used gripped in the hand, this Egyptian handaxe with knapped edges dates from 9000–2700 BCE.*

"Regardless of our ancestral heritage, we're all descended from flintknappers."

Bert Mathews, sharp stone maker

Sewing (c. 25,000 BCE)

Clothing is fitted using needle and thread.

The history of sewing is closely allied to the history of tools. The earliest needles ever discovered date from the Paleolithic era (the early Stone Age), around 25,000 BCE. Key finds from that period include needles in southwest France and near Moscow in Russia. These were made of ivory or bone, with an eyelet gouged out. Some have been found alongside the remains of foxes and hares that were used for their fur.

Sewing gave our early ancestors the opportunity to make clothing more closely tailored to the human body, improving its insulation and comfort, as well as inviting decoration. Early scraps of cloth found in France and Switzerland have included decorative seeds or animal teeth sewn on by thread, applied perhaps with the aid of fishbones or thorns. Native Americans sewed with the tips of agave leaves.

Metal needles were developed in the Bronze Age (2000–800 BCE) and initially were made of several strands of wire melted together. Needles from this era have been found in North Africa and China, where steel was introduced. The first known stitched buttonhole dates from 4200 BCE.

Embroidery—complex, decorative needlework—appeared in Bronze Age Egypt and India. In China silk was being sewn and embroidered in the same era. Protective thimbles have been used since Roman times. The famous Bayeux Tapestry, depicting the Norman invasion of England, is an example of crewelwork, a form of embroidery with loosely twisted yarn. At least four types of stitch have been identified in the tapestry. Later, the mechanization of textile production began in the sixteenth century with the stocking frame, which led to automated looms. Hand-stitching was transformed from the 1830s onward by the arrival of the sewing machine. **AC**

SEE ALSO: CLOTHING, SHOE, WOVEN CLOTH, SPINNING JENNY, SPINNING MULE, SEWING MACHINE

Atlatl (c. 23,000 BCE)

Early humans extend spear-throwing range.

When Spaniards first met the Aztecs in around 1500, the explorers were horrified when their armor was easily penetrated by the Aztec throwing darts. The Aztecs achieved this feat with the atlatl, a simple device used by many ancient peoples for long-range hunting. It probably dates from around 23,000 BCE.

The atlatl consists of a throwing board and a dart about 6 feet (180 cm) long. The board, typically about 2 feet (60 cm) long, has a spur at its end. The dart's rear is cut down the middle so that it fits onto the spur like two fingers around a card.

Gripping a handle at the front end of the throwing board, the atlatl thrower can hurl the dart with considerably more force than he could by hand.

> *"The atlatl is the tool ancient peoples used to 'bring home the bacon.'"*
>
> Robert "Atlatl Bob" Perkins, primitive technologist

During the thrower's tennis swing-like motion, the flexible dart flexes and energy builds up. The dart is weighted with a stone tip and often another counterweight to maximize the buildup of energy.

When the atlatl dart is released, the spring energy in the flexible dart is added to the forward force, accelerating the dart to speeds that can exceed 100 miles per hour (160 km/h). The atlatl was so effective at bringing down prey that some scholars speculate it may have played a significant role in the extinction of the North American woolly mammoth. Now, at least 25,000 years after its invention, the atlatl is still used by enthusiastic hobbyists. **LW**

SEE ALSO: STONE TOOLS, SPEAR, BOW AND ARROW, BOOMERANG, CROSSBOW, CATAPULT

Bow and Arrow (*c.* 20,000 BCE)

Distant targets come within deadly reach for the first time.

Evidence of the early use of bows and arrows has been found in cave paintings in Western Europe and North Africa. Its development probably arose in the Upper Paleolithic (Old Stone Age) around 20,000 BCE, when people realized that the weapon would enable hunters to kill outside their throwing range.

Bows and arrows were portable, easy to make, and the materials to make them were relatively easy to obtain. The bow consisted of a thin flexible shaft of wood; this was bent, and a length of sinew, deer gut, plant fiber, or rawhide was strung tightly between its ends. Sometimes the bowstring was twisted to make it stronger. Ash, mahogany, and yew were all used for bows. Sometimes the wood was backed with sinew to make the bow stronger and stop it breaking.

The arrow was a thin shaft of wood, sharpened at one end, with feathers attached to the other to give it aerodynamic stability. Arrowheads were made from flint or other rocks, antler, or bone.

The bow was the first machine that stored energy. Energy from the archer's muscles gradually transferred to the bow as it was drawn back; when the bow was released, it gave the projected arrow a far greater velocity than that produced by a spear-thrower. In about 1500 BCE a shorter and lighter bow was developed, the composite bow. Short and curved, it was built up from layers of materials that reacted differently under tension or compression. It was an accurate weapon to use from horseback.

Modern bows are made from fiberglass, carbon, and aluminum as well as wood, while the arrows are usually made of composite materials. **MF**

SEE ALSO: STONE TOOLS, SPEAR, CARPENTRY, ATLATL, SLING, CROSSBOW, CATAPULT

⬈ *A relief of archers from the Mortuary Temple of Ramses III, Twentieth Dynasty, 1184–1153 BCE.*

"Whose arrows are sharp, and all their bows bent, their horses' hoofs shall be counted like flint ..."

Isaiah 5:28

Boomerang (c. 18,000 BCE)

The advent of an easily retrievable weapon.

The oldest boomerang so far found was discovered in a cave in the Carpathian Mountains in southern Poland and is believed to be date from 18,000 BCE. The practice of throwing wood has also been illustrated in North African rock paintings that date from the Neolithic Age (approximately 6000 BCE). The wood thrown consists variously of a "throwing club," where the effect is concentrated at one end, or a "throwing stick," a sharpened, straight rod of hard wood that rotates, or a boomerang, which developed from these into a specialized form and has a return throw.

Ancient tribes in Europe are said to have used a throwing axe; in Egypt a special type of curved stick was used by the Pharaohs for hunting birds. The use of throwing woods is thought to have spread throughout North Africa from Egypt to the Atlantic.

Boomerangs are most commonly associated with Australian Aborigines. They have been made in various shapes and sizes depending on their geographic origins and intended function. In the past they have been used as hunting weapons, musical instruments, battle clubs, and recreational toys. The most recognizable type is the returning boomerang. Some have "turbulators" (bumps or pits on the top surface) to make the flight more predictable. A returning boomerang is an airfoil and its rapid spin makes it fly in a curve rather than a straight line.

Other types of boomerang are of the non-returning sort, and some were not thrown at all but were used in hand-to-hand combat by Aboriginal people. The throwing wood, however, was mainly used for hunting rather than as a battle weapon. **MF**

SEE ALSO: STONE TOOLS, SHARP STONE BLADE, ATLATL, CARPENTRY

◄ *A boomerang in Aboriginal art on the main gorge wall at Carnarvon Gorge, Australia.*

Braided Rope (c. 17,000 BCE)

Fibers are twisted into a valuable tool.

One of the oldest artifacts in the world, rope is still extensively used in many environments. It seems unlikely that it will be replaced for many years. Traditionally made from natural fibers such as hemp, jute, or coir, rope is now also made from synthetic materials such as nylon and even steel.

Rope is braided fiber, twisted to form a supple, strong medium. Its strength is tensile, so its main use is to link objects, one of which acts as a stable anchor for the others to hang from or pull against. The oldest evidence of man-made rope was found in the caves of Lascaux, southwest France, and date from 17,000 BCE. Rope has always been used to tie and carry prey, making it an essential hunting tool.

> *"There are nearly as many types of rope as there are fibrous materials on Earth."*
>
> Brendan McGuigan, writer

Before machinery made it possible to create long lengths of rope, essential in sailing ships, weaving fibers was done by hand—an arduous process. The ancient Egyptians developed the first tools for weaving rope, which they used to move huge stones. Machines for spinning long lengths of rope were later housed in buildings called cake-walks, or roperies, which could be up to 300 yards (275 m) long. A prime example of such a ropery exists in the former naval dockyard in Chatham, England, where rope is still produced on the premises after nearly 300 years. This ropery, 440 yards (400 m) long, was built in 1720 and at that time was the longest building in England. **JG**

SEE ALSO: WOVEN CLOTH, SAIL, NYLON

Lunar Calendar

(*c.* 15,000 BCE)

Early humans record the passing of time.

The earliest known lunar calendar is in the caves at Lascaux, southwest France, and dates from around 15,000 BCE. Various series of spots represent half of the moon's near-monthly cycle, followed by a large empty square, which perhaps indicates a clear sky.

A lunar calendar counts months (a period of 29.530588 days) and is based on the phases of the moon. Months have twenty-nine and thirty days alternately, and additional days are added every now and then to keep step with the actual moon phase.

The lunar calendar was widely used in parts of the ancient world for religious observation. Agriculturally the lunar calendar is confusing as it takes no account

> "Day is pushed out by day, and each new moon hastens to its death ..."

Horace, *Odes*, Book II

of annual seasonal variations in temperature, daylight length, plant growth, animal migration, and mating. The lunar month divides into the solar year twelve times but with 10.88 days remaining.

Meton of Athens (circa 440 BCE) noticed that nineteen solar years were equal to 234.997 lunar months. This led to the nineteen-year Metonic cycle where years three, five, eight, eleven, thirteen, sixteen, and nineteen had thirteen lunar months each, and all the other years had twelve months. **DH**

SEE ALSO: WATER CLOCK, TOWER CLOCK, GREGORIAN CALENDAR, TIME ZONES

← *A lunar calendar appears as spots above an image of a bull in the painted caves at Lascaux, France.*

Alcoholic Drink

(*c.* 10,000 BCE)

A pleasurable beverage appears.

The accidental fermentation of a mixture of water and fruit in sunlight is thought to have led to the first discovery of an alcoholic drink by a prehistoric people. Evidence of intentionally fermented beverages exists in the form of Stone Age beer jugs dated as early as the Neolithic period (10,000 BCE). Other jugs have been excavated in Southwest Asia and North Africa.

Alcoholic beverages have been an integral part of many cultures, used as a source of nutrition, in meals, for celebrations, and also in religious ceremonies. Alcohol can give a sense of wellbeing, but also acts as a depressant, lowering behavioral inhibitions.

Alcohol consumption became a status symbol for the wealthy. During the Middle Ages, concoctions were distilled to produce spirits. Alcohol has also served as a thirst quencher when water was polluted. In the 1700s, home-brewing processes were replaced by commercially made beer and wine, which became important for the economies of Europe.

Beer was the first known alcoholic beverage, but many others have been produced since then. The Chinese are thought to have produced yellow wine 4,000 years ago. In Europe the monasteries owned the best vineyards; French monks produced a sparkling wine, which was named after the Champagne region of France. Brandy is supposed to have been accidentally discovered when a Dutch trader tried boiling wine "to remove the water and save cargo space." (Brandewijn means "burnt wine" in Dutch.)

Attitudes to alcohol consumption have varied over time and different countries have limited the hours when drinking establishments are open, or even banned the sale of alcohol altogether, as Americans did in the Prohibition, between 1920 and 1933. **MF**

SEE ALSO: POTTERY, GLASS, DISTILLATION, STANDARD MEASURES, WIDGET IN-CAN SYSTEM

Pottery (*c.* 10,000 BCE)

Japanese hunter-gatherers create the first fired clay pots.

As applies to all early inventions, we do not know the name of the man or woman who invented pottery. No first potter carved his or her name or initials in the base of a pot to claim first prize. However, it has long been assumed that whoever the creative person was, he or she would have lived somewhere in the Near East of Asia. It was therefore something of an archeological shock when, in the 1960s, pots dating to around 10,000 BCE were discovered on the Far Eastern side of Asia, thousands of miles away at Nasunahara on the island of Kyusu in Japan. These pots, found in caves, were made by nomadic hunter-gatherers, rather than settled farmers or urban dwellers. Just as important, the pots were made by firing or heating the clay to harden it, suggesting that these people had knowledge of advanced technologies.

The significance of the first Japanese pots is that they predate the first pots made in the Near East by around 1,000 years. Those pots, found in Iran, were made by drying the clay in the sun in order to harden it, a far more primitive technology than firing the clay.

The Japanese pots have a round base and widen gently to a ridged top and a rounded, incised rim. They are known as Incipient Jômon, because they are the forerunners of the jômon or "cord-marked" vessels developed in Japan around 9000 BCE. These later pots had pointed bases and were made by building up coils of clay into the desired shape. The patterns of the cord-marked pots were often quite complex, suggesting that they were intended for ritual or funerary use rather than for such everyday uses as cooking and storage. **SA**

> *"Vases and shards ... the true alphabet of archeologists in every country."*
>
> Sir William Flinders Petrie, archeologist

SEE ALSO: CONTROLLED FIRE, OVEN, GRANARY, DRIED BRICK, FIRED BRICK, BLAST FURNACE

◸ *An incised pottery jar dating from the Jômon culture of Japan, circa 9000 BCE.*

Oil Lamp (*c.* 10,000 BCE)

Night is banished forever by a burning lump of fat.

The humble oil lamp may only be needed to provide light during the occasional power cut today, but for thousands of years versions of it allowed man to see by night, as well as provide decoration and symbolic power in ceremonies and festivals. Only with the invention of the Argand Lamp in 1780, and eventually electric lighting, were oil lamps all but extinguished.

Estimates suggest that crude lamps were first used around 80,000 BCE. A lamp is a vessel containing flammable oil with a slow-burning wick designed to draw up the fuel from the reserve. Early man made lamps from stone or seashell crucibles filled with animal fat, with a piece of vegetation as the wick.

The first real oil lamp appeared alongside settled agriculture around 10,000 BCE. (the Upper Paleolithic period—otherwise known as the Stone Age). With the planting of the first crops came the potential for plant oils, such as olive oil, to be used in these lamps. As well as a source of light, they were important symbols in rituals and ceremonies—the Bible and Koran both contain many references.

The Romans mass-produced clay lamps (a newly made batch was discovered buried in Pompeii by the great eruption of 79 CE). In the Middle Ages candles became popular, but these never produced a flame as bright as an oil lamp. However, in the eighteenth century the Industrial Revolution provided the pressure necessary for innovation. In 1780 the scientist Aimé Argand developed a brighter lamp with a metal casing that burned oil with a steady, smokeless flame, but with the advent of electric lighting a civilization-old technology was finally laid to rest. **SR**

SEE ALSO: CONTROLLED FIRE, CANDLE, GAS LIGHTING, LIGHTBULB, ARGAND LAMP, LAVA LAMP

↗ *An oil lamp from Port of Alexandria, Egypt, dating from the Roman Empire, first century CE.*

"I gave illuminating oil for lighting the lamps of your temple."

Inscription of Nesuhor (589–570 BCE)

Sling (*c.* 10,000 BCE)

Stones are put to a lethal purpose.

The sling is a prehistoric weapon probably dating back more than 10,000 years. The oldest known surviving slings were found in Tutankhamen's tomb, dating from 1325 BCE. And of course slings feature in the Bible, most famously in the story of David and Goliath.

A sling is used to throw a missile many times farther than is possible with the human arm alone. It consists of a cradle, or pouch, in between two lengths of cord. A stone is placed in the pouch. Both cords are held in the hand, and the thrower draws back his arm and swings the sling up and forward. One of the two cords is released and the stone is projected away.

As a weapon, the sling was a great success, being cheap to make, light to carry, easy to use, and relying

> *"David defeated Goliath with a sling and a rock. He killed him without even using a sword."*
>
> Samuel 17:50

on ammunition that was readily available. Not surprisingly, it became common all over the ancient world, except Australia where spears seem to have been preferred. A slingstone can be thrown by an expert up to 650 yards (600 m), farther than most bows could achieve, though with less accuracy. The Greeks and Romans introduced lead shot as an option for ammunition, although stones remained most popular. By the Middle Ages the sling had largely given way to more sophisticated weaponry. **RBd**

SEE ALSO: SPEAR, ATLATL, WOVEN CLOTH, CATAPULT

◄ *Advancing soldiers with slings in an Assyrian relief of 650 BCE at the Palace of Sennacherib, Iraq.*

Granary (*c.* 9500 BCE)

Early farmers build the first grain storage.

The world's first purpose-built granary was unearthed by Ian Kuijt, Associate Professor of Anthropology at the University of Notre Dame, Indiana, at Dhra' on a parched plateau next to the Dead Sea in the Jordan Valley. The structure, roughly 9 feet (2.9 m) square, was found about 2 feet (0.6 m) underground.

The granary is smaller than other nearby structures that appear to have been houses, an indication of its different use. Interestingly, the structure has two levels, an architectural feature never seen before in buildings of this age. Its significance as the world's first granary is that it belongs to one of the world's first settled farming communities, built just as people began to live in one place all year round rather than wandering from place to place in search of their food. In other words, it marks the time when early people were making the historic transition from hunter-gatherers to settled farmers.

The granary enabled people to store wheat and barley grains, nuts, and other produce harvested in the summer to see them through the winter months, or indeed through an unproductive summer. With food in storage, and thus constant supply, the population could rise, in turn spurring technological advances in agriculture and other occupations.

There were, of course, downsides to this development, as the first farmers concentrated on a reduced variety of crops, unbalancing their diets in comparison to the wide range of food foraged by their ancestors. The first farmers of Dhra' probably did little more than weed and water the crops that were already there, but they were helped by the fact that, at that time, the plateau they farmed was far wetter than it is now, giving them a wider variety of crops to tend than can grow in the region today. **SA**

SEE ALSO: BUILT SHELTER, TALLY STICK, POTTERY, IRRIGATION, SCRATCH PLOW

Metalworking (*c.* 8700 BCE)

Mesopotamians fashion objects from metals.

The use of metals to make tools, weapons, or jewelry has been one of humanity's pivotal achievements. Manipulated metals are everywhere, from kitchen utensils to high-tech weapons and tools. Even items that contain no metal are likely to owe some debt to a metal tool that was used in their construction.

As near as archeologists can tell, the love affair between humans and metals probably began around 8700 BCE, evidenced by a copper pendant found in northern Iraq. Smelting, the extraction of metal from a metal-containing rock, began around 5000 BCE when copper ores were melted to get at the metal. By 4000 BCE people were using gold and adding arsenic to copper to create arsenical bronze, probably the first

> *"Iron weapons revolutionized warfare and iron implements did the same for farming."*

Alan W. Cramb, Professor of Engineering

man-made alloy, or metal mixture. Although harder than copper, arsenical bronze took a heavy toll. Metalworking gods of several cultures are described as lame, probably the result of their long exposure to the valued but unfortunately poisonous metal.

Around 3500 BCE, tin and copper were combined to make bronze. Trade spread the hot new trend and people made bronze weapons, armor, decorations, and tools. The Bronze Age gradually came to an end as trade dwindled and tin became hard to come by. Iron and the Iron Age were the replacement. Centuries later, people mixed small amounts of carbon with their iron to make steel. **LW**

SEE ALSO: CONTROLLED FIRE, SPEAR, CHISEL, HELMET, BRAZING AND SOLDERING, CHAIN MAIL

Dugout Canoe (*c.* 7500 BCE)

Hollowed-out logs become the first boats.

Sometimes there is no real need to be clever, or complex, or even particularly sophisticated when it comes to inventions. Sometimes simple wins.

This is definitely the case with the dugout canoe. The people of 7500 BCE needed a way to travel on water, but many of the materials used in the very earliest boatbuilding still lay a long way ahead in the future. So they came up with a simple answer using the technology that was accessible to them.

The dugout canoe is, in its most basic terms, a hollowed-out log, nothing more than a tree trunk laid down on its side and its interior removed. All that was required was that the hollowed log had to be big enough for at least one person to sit inside, and the

> *"The ancient Greeks used dugouts and called them monooxylon, which means 'single tree.'"*

John Crandall, *Dugout Canoes*

wood had to be sound, not rotten. If a log fulfilled these two criteria, it was a potential canoe.

As these vessels were made before the invention of metal tools, the logs were hollowed out using a controlled fire and a sharpened rock implement (known as an adze) to scrape away the burned wood. Then, to reduce drag in the water, the front and the back of the log were fashioned into a point.

Dugout canoes have been excavated in various locations in northern Europe and are the oldest form of boat ever discovered. Before their arrival there were no other forms of water travel in existence—only swimming and clinging to driftwood. **CL**

SEE ALSO: CONTROLLED FIRE, FISHHOOK, CARPENTRY, ROWBOAT, SAIL, RUDDER, STEAMBOAT

Chisel (*c.* 7500 BCE)

The chisel becomes a standard building tool.

Chisel-like tools have been dated to the Paleolithic era, which stretches across a vast expanse of evolutionary time, from before the first *Homo sapiens* to roughly 10,000 BCE. During this time humans were making and refining stone tools, which became gradually more specialized over time. Other materials were also used, and bone chisels from around 30,000 BCE have been uncovered in Southern France, near the village of Aurignac. Although very difficult to date exactly, it is thought that by about 7500 BCE, what we would recognize today as a chisel was in fairly common use.

By the time of the Bronze Age, chisels had become quite varied and included gouges—chisels with curved blades—and tanged chisels, where the blade is connected to the handle by a collar. The Greek architect Manolis Korres believes the chisels used by the ancient Greeks were actually sharper and sturdier than today's versions. While working on restoring the Parthenon, Korres made reconstructions of various ancient tools by looking at tool marks in marble. Of course, the ancient Greeks needed specialized tools to craft their iconic temple in Acropolis.

During the medieval period, carpenters employed tools known as "former" chisels. These had a broad, flared blade, which was used to carve rough wood. A mallet could be used with stouter tools, called "firmer" chisels, to shape and finish wood. There were also other chisels for detailed and speciality work.

The chisel has changed little since medieval times, although you will probably find a less impressive selection of tools in a modern DIY store than you would have in a medieval carpenter's workshop. **HB**

SEE ALSO: STONE TOOLS, SHARP STONE BLADE, METALWORKING, CARPENTRY

↗ *A hieroglyph of a chisel (top right) on a limestone stele from Abydos, Egypt, circa 3100–2900 BCE.*

"Ancient masons . . . could carve marble at more than double the speed of today's craftsmen."

Evan Hadingon, *Smithsonian* magazine

Dried Brick (c. 7500 BCE)

Humankind builds with portable mud blocks.

Buildings erected using preformed, shaped bricks of dried mud date back to around 7500 BCE. Examples have been found by archeologists in Cayönü in the upper Tigris Valley and close to Diyarbakir in southeast Anatolia, both in modern-day Turkey.

More recent bricks, dating from between 7000 and 6400 BCE, have been found in Jericho in the Jordan Valley and in Çatalhöyük, again in Turkey. These early bricks were made of mud molded by hand and then left out to dry and harden in the sun. The bricks were then laid into walls using a simple mud mortar. Mud is an exceptionally good material for building in dry climates: it is readily available wherever agriculture is practiced, it may be dug from riverbeds, and it has good structural and thermal qualities.

Some years later, mud bricks were shaped in wooden molds, enabling a form of organized mass production to take place. This became important, as bricks were increasingly used to build not only small-scale houses, farms, granaries, and other farm structures, but whole villages and later towns and cities, including their large palaces, temples, and other state and public buildings.

Wherever stone was unavailable, or in short supply, the humble mud brick took its place. Mud bricks were used throughout the Near East, as well as by the civilizations of Egypt and the Indus Valley, where the bricks were standardized in size, in the ratio of four units long to two units wide and one unit deep. This simple but effective building material was paramount until the first kiln-fired bricks were developed in Mesopotamia during the third millennium BCE. **SA**

SEE ALSO: BUILT SHELTER, FIRED BRICK, PLASTER, FORT, WATTLE AND DAUB

⬆ *Slave workers make bricks and construct a wall on an Egyptian Eighteenth Dynasty tomb painting.*

Sledge (*c.* 7000 BCE)

Arctic peoples invent the ice-traveling vehicle.

Long before the snowmobile, our ancestors found an environmentally friendly way to get around in the snow—the sledge. In fact, the sledge (and variations on its theme) was key in many areas of ancient life.

A sled is a vehicle that moves by sliding across the ground. Sleighs are horse-drawn vehicles, with passenger seating. Sledges tend to be large vehicles consisting of a wooden base mounted on smooth runners, useful for transporting large objects. Evidence of wooden sledge usage reaches back to 7000 BCE, to peoples living in the Arctic regions of northern Europe. Initially sledges may have been pulled by humans, but with time dogs and oxen were commandeered to take the strain. Inuits have used dog-sleds since pre-Columbian times. Sledge use has extended to hotter climates, too, including the dry, dusty lands of Mesopotamia.

Exactly where and when the sledge was developed is unknown, but it is likely that it was developed independently by different communities in the world Human-pulled sledges were key in man's early expeditions to the Arctic and Antarctic. In the twentieth century, dog teams of Huskies were used to tow sleds on expeditions. More recently, kites have been used to tow sleds. They use wind power, so fewer resources need be carried on the sledges.

In today's world the sledge is used in sport and leisure. A small sled with rounded edges at the front can provide hours of fun in the form of a toboggan. Bobsledding is a sport featuring in the Winter Olympics, where teams of competitors race down tracks in a specially streamlined vehicle. **JG**

SEE ALSO: CARPENTRY, SKI, TRAVOIS, CART, SNOWMOBILE

⬆ *A mounted, animal-drawn sledge in a rock engraving from Bohuslan, Sweden, dating from 1500–1000 BCE.*

Fort (*c.* 7000 BCE)

The debut of strong, communal defenses.

Defenses have been constructed for thousands of years. Bronze and Iron Age hillforts took advantage of natural hills for defense purposes, and the Romans built the Saxon Shore Forts along the southeast coast of Britain to deter invasion.

The word *fort* is derived from the Latin *fortis* meaning "strong" and many military installations are known as forts. The term *fortification* also refers to improving other defenses such as city walls. Permanent fortifications were built of enduring materials, but field fortifications needed little preparation, using earth, timber, or sandbags.

The arrival of cannons in the fourteenth century made medieval fortifications obsolete. Later constructions included ditches and earth ramparts to absorb the energy of cannon fire. Explosive shells in the nineteenth century led to a further evolution; the profile of the fort became lower, surrounded by an open sloping area that eliminated cover for the enemy. The fort's entry point was a gatehouse in the inner face of the ditch, with access via a bridge that could be withdrawn. Most of the fort was built underground, with passages connecting blockhouses and firing points. Guns mounted in open emplacements were protected by heavy parapets. Offensive and defensive tactics became focused on mobility. In the twentieth century defending tanks were concentrated in mobile units behind the line. If an offensive was launched, reinforcements could be sent to that area.

Reinforced concrete fortifications were common during the nineteenth and early twentieth centuries, but modern warfare has made large-scale fortifications obsolete; now, only deep underground bunkers provide sufficient protection. Today, forts mostly survive as popular tourist destinations. **MF**

SEE ALSO: **BUILT SHELTER, DRIED BRICK, BATTERING RAM, CATAPULT, GUNPOWDER, CANNON**

Travois (*c.* 7000 BCE)

Native Americans invent a carrying device.

Native Americans on the Great Plains led nomadic lives. They used buffalo for almost everything, eating their meat and making clothes and tent coverings from their skins. Living on herds of animals that were always on the move, they were constantly on the move too, which meant living in tents and owning only what could be carried to the next camp.

Ideally, however, people like to carry more than can fit into one bag. On roadways and hard ground, carts are the best solution, and in the far north snow and ice lie on the ground and dragging a sled is easy because the ground is slippery. Traveling across soft soil, however, neither of these options work. The response of the Native Americans was to invent the travois.

> *"What is life? It is a flash of a firefly in the night. It is the breath of a buffalo in the wintertime ..."*
>
> Crowfoot, chief of the Blackfoot First Nation

The travois was a tall wooden "A," 6.5 feet (2 m) high, where the things to be carried sit on the crossbar and the whole thing is dragged along on the splayed poles. The dragging ends move along quietly and with little friction. Before the Spanish arrived and introduced horses into the New World, dogs were used; harnessed to the travois, they could drag up to 66 pounds (30 kg). For the horse, the travois was scaled up and carried bigger loads. Occasionally Native Americans used the travois to carry their sick or elderly, either pulled by a horse or by multiple dogs. Boy Scouts continue to be taught to use the travois for dragging wounded comrades to this day. **DK**

SEE ALSO: **CARPENTRY, SLEDGE, WHEEL AND AXLE, CART, SNOWMOBILE**

Shoe (*c.* 7000 BCE)

Native Americans insulate human feet from the ground.

From their earliest use, simply as a protective covering for the feet, to the vast fashion industry producing them today, shoes have been essential items for humans. As with any invention from antiquity, it is uncertain when shoes were first worn, and archeological evidence has continued to complicate the issue. The oldest shoes in existence are from around 7000 BCE and were discovered in America.

The earliest shoes appear to have been constructed variously from rope, leaves, and animal skins. As these are all highly perishable materials, archeological examples are rare, but some argue that there is other evidence pointing to shoe use from up to 40,000 years ago. Archeologists examining ancient bones have noticed a reduction in the size and strength of toe bones during this period, which they attribute to the feet being covered. However, this conclusion is far from proved.

The original design for most shoes is similar to that of the modern sandal, and consisted of a protective sole held onto the foot by bands or straps. While our need for shoes may seem obvious, their invention was a major development in the ability of humans to travel, work, and endure harsh conditions.

An etiquette regarding shoes has developed in many parts of the world. In large parts of Asia it is customary for people to remove their shoes when entering a home; this practice has spread to North America and Europe in many homes. In Asia, indoor shoes are often provided by hosts, but this is less common elsewhere. Muslims invariably remove their shoes before entering a mosque. **JG**

SEE ALSO: CLOTHING, SEWING, SEWING MACHINE, AIR-CUSHIONED SOLE

↗ *A woman fastens one of her sandals, painted on the neck of a Greek amphora, circa 520 BCE.*

"It is the same to him who wears a shoe, as if the whole Earth were covered with leather."

Persian proverb

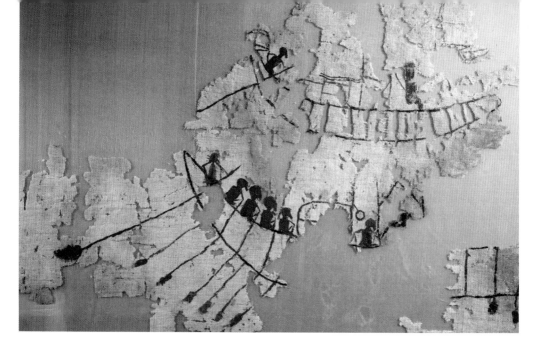

Woven Cloth (*c.* 6500 BCE)

Clothes making is revolutionized by the world's first weavers.

The first handmade material that humans created to make into clothing was felt, which was made by intermeshing animal fibers under heat and pressure. Felt lacked the necessary durability, however, and the real textile breakthrough came later with weaving.

Weaving is accomplished with a loom, a frame that holds vertical threads taut while the weaver interlaces a horizontal thread. The thread itself is obtained through spinning, in which animal or plant fibers are twisted together by hand or machine.

The earliest evidence of weaving was discovered in 1962, in the town of Çatalhöyük, Turkey. A piece of carbonized cloth, it was found to date from 6500 BCE. It is unclear whether the cloth was made from flax (a wild Mediterranean plant) or from sheep's wool. A more recent piece of linen, dating from 5000 BCE and woven from flax, was discovered in Egypt prior to this find, and it seemed likely that the Turkish cloth was made from the same material. Flax experts disagreed, however, stating that the plant was not found in this region of Turkey at that time. Wool experts concurred after discovering that the cloth is scaly; flax cloth is smooth, so they concluded that the Turkish cloth must have been made of wool. The issue was finally resolved after the cloth was dipped in an alkali solution. This would have destroyed wool, but instead it removed the cloth's black coloring to reveal a network of cross-striations consistent with flax.

It was shortly after weaving with flax that textiles made from wool and silk became available. The breakthrough led to a variety of textile products, including warmer and more durable clothing. **RB**

SEE ALSO: CLOTHING, SEWING, SPINNING WHEEL, STOCKING FRAME, SPINNING JENNY, SPINNING MULE, POWER LOOM

⬆ *Fragments of an Egyptian textile, originally decorated with images of oared boats, dated circa 4000 BCE.*

Map (*c.* 6500 BCE)

Charting of the world begins with the Babylonians.

Some of the earliest known examples of maps—in the form of Babylonian tablets—are Egyptian land drawings and paintings discovered in early tombs. However, in 1961 a town plan of Çatalhöyük in Turkey was unearthed, painted on a wall. Featuring houses and the peak of a volcano, it is around 8,500 years old.

The sixth-century tablet known as Imago Mundi shows Babylon on the Euphrates, with cities on a circular land mass, surrounded by a river. Some maps are known as T and O maps. In one, illustrating the inhabited world in Roman times, T represents the Mediterranean, dividing the continents, Asia, Europe, and Africa, and O is the surrounding Ocean. The T and O Hereford Mappa Mundi of 1300, drawn on a single sheet of vellum, includes writing in black ink and water painted green, with the Red Sea colored red.

Greek scholars developed a spherical Earth theory using astronomical observations, and in 350 BCE Aristotle produced arguments to justify this practice.

In the first century CE, Ptolemy, an astronomer and mathematician, developed a reference-line principle. His *Guide to Geography* lists 8,000 locations with their approximate latitudes and longitudes. However, Ptolemy underestimated the size of the Earth. His suggestion that India could be reached by traveling westward resulted in Columbus underestimating the distance centuries later. Cartography greatly benefited from a wealth of corrective information brought to Europe by Marco Polo in the thirteenth century.

The 1891 International Geographical Congress established specifications for a scale map of the world, and World Wars I and II brought more progress. **MF**

SEE ALSO: MAGNETIC COMPASS, ODOMETER, SEXTANT, GLOBAL POSITIONING SYSTEM (GPS)

⬆ *A drawing of the original wall map found in Çatalhöyük, Turkey, circa 6500 BCE.*

Wattle and Daub

(*c.* 6000 BCE)

Building begins with wood and mud.

The technique of wattle and daub was first pioneered by human civilizations as early as 6000 BCE as a way of weatherproofing their shelters. In its essence, wattle and daub is a way of filling in the gaps between the structural elements of wooden houses.

In a typical Tudor example, oak staves were placed vertically between structural beams and then thin twigs of a flexible hardwood, such as willow or hazel, were woven horizontally between the staves, creating a robust mesh, or "wattle." The wattle was then coated with daub—a mixture of clay or mud and animal dung, strengthened with straw or horsehair. This mixture was pressed onto the wattle by hand. The

---

"... developed when an enterprising human first daubed mud upon a branch shelter ..."

Joseph F. Kennedy, *The Art of Natural Building*

---

mud and dung helped the daub adhere to the wattle, and the fiber content prevented cracks from forming. The finished wall was sometimes burned to make it hard, like pottery, or coated with lime to make it more weatherproof. This resulted in a strong wall that kept out the wind and rain, the cold in the winter, and the heat in the summer.

Wattle and daub walls did have their disadvantages, however. If they became damp, they had a tendency to rot or become beetle-infested. And the term "breaking and entering" is thought to have originated from the ease with which criminals could enter such a building, simply by breaking through the wall. **HI**

SEE ALSO: BUILT SHELTER, CARPENTRY, DRIED BRICK,
FIRED BRICK, PLASTER

Irrigation

(*c.* 6000 BCE)

Sumerians pioneer watering channels.

It is unknown who first irrigated his crops with water brought specially from a nearby river, but archeological evidence suggests that, wherever farming began to take place, irrigation soon followed. There is evidence of irrigation from around 6000 BCE in Sumer in Mesopotamia, and also on ancient Egyptian farms near the Nile. Some 2,000 years later, irrigation occurred in Geokysur in South Russia, and in the Zana Valley in the Andes Mountains of Peru. By 3000 BCE the the same techniques were used by the Indus Valley Civilization in what is now Pakistan.

When, at around 6000 BCE, the first farmers in Mesopotamia planted their crops of barley, wheat, and other plants near the Tigris or Euphrates rivers, they relied on rain, the occasional flood, and the ability of the soil to hold water to ensure that their crops grew from seed to harvest. Water could be carried in buckets from the river, but if the rain stopped and there was a lengthy drought, the crops would die.

The problem of over-reliance on natural water supplies was solved by creating artificial means of bringing the water to the fields. The water was either diverted from a major river through canals and drainage ditches that flowed alongside the fields, or it was stored in reservoirs and ponds that were refilled in times of flood, and distributed from these. The effect of this irrigation was to extend the area of fertile land from just a narrow strip on either side of a river to a wide band that could be several miles across. Having more irrigated land resulted in more crops, and thus the ability to support a rising population. **SA**

SEE ALSO: GRANARY, SCRATCH PLOW, CANAL, DAM,
SEWAGE SYSTEM

➦ *A diagram from a Turkish manuscript of 1206 CE details an ingenious system for pumping water.*

Axe (*c.* 6000 BCE)

Early humans develop the first axe to clear areas of rain forest.

More than a million years ago, members of the species *Homo erectus* were making stone tools designed for chopping that can be described as early hand axes. They were teardrop-shaped and roughly made, flaked on either side to form a sharp cutting edge. However, not until the rise of farming during the late Stone Age did such tools come to resemble what we would now recognize as the axe. There was widespread trade in these tools around this time and stone axes have been uncovered at many Neolithic meeting places.

Axes clearly designed to be mounted (hafted) on handles have been found at a site near Mount Hagen in New Guinea. By analyzing samples of pollen from around the same era—thought be around 8,000 years ago—archeologists have concluded that they were probably employed in the opening up of the rain forest, during agricultural development, to allow light to reach crops.

By the Bronze Age in Britain, woodworkers had developed a range of axes for different cutting purposes. Archeologists have been able to suggest what these might have looked like by experimenting with their own reconstructed tools, to produce different cut marks.

Although primarily a functional tool, the axe is also a symbol of power. It is possible to identify the remains of highly ranked members of a society by looking at their grave goods, which sometimes include axes. For example, an excavation of a Bulgarian cemetery dating back to 4000 BCE uncovered a number of gold-covered axes. Their inclusion in the grave has been interpreted as signifying high levels of authority. **HB**

SEE ALSO: STONE TOOLS, SHARP STONE BLADE, CARPENTRY, SAW, CLAW HAMMER

⬆ *A worker (left) uses an adze in a painted limestone relief, Valley of the Nobles, Egypt, 1400–1390 BCE.*

Scratch Plow (*c.* 5500 BCE)

Mesopotamians invent a literally groundbreaking tool.

Around 9500 BCE, in a number of populations distant from one another, people began to select and cultivate plants for food and other purposes. These people were the first farmers. In what is now known as the Fertile Crescent in Southwest Asia, small populations engaged in small-scale farming and began to grow the eight founder crops of agriculture—emmer and einkorn wheat, hulled barley, bitter vetch, peas, chickpeas, lentils, and flax. However, it took thousands of years before the farmers developed the practices and technologies necessary to enable cultivation of the land on a larger scale.

In 5500 BCE the first plow, a tool used to prepare the soil for planting, was developed in Mesopotamia by the Indus Valley Civilization. It was known as the scratch plow and represented one of the greatest advances in agriculture. It consisted simply of a wooden stick attached to a wooden frame, but was able to aerate the soil and scratch a furrow to allow the planting of seeds. The plow was pulled by domesticated oxen and left strips of undisturbed earth between each plowed row. To increase the productivity of their fields, farmers often cross-plowed them at right angles. The squarish fields that resulted are known to archeologists as "Celtic fields."

Many different types of plow have superseded this simple device, but it is still used in many parts of the world. In certain areas, including northern Europe, the scratch plow was ineffective in dealing with sticky clay soils. However, in India farmers continue to use the primitive plow to introduce organic materials into soils that have been cultivated for up to 2,000 years. **FS**

SEE ALSO: CARPENTRY, IRRIGATION, MOLDBOARD PLOW, STEEL PLOW, CAST-IRON PLOW

⬆ *Sennedjem and his wife use a scratch plow in a wall painting of the Tomb of Sennedjem, circa 1150 BCE.*

Plaster (*c.* 5500 BCE)

Egyptians make a versatile new material.

Plaster goes by various names—plaster of Paris, partly dehydrated gypsum, or calcium sulfate hemihydrate. Gypsum is a common mineral found in a variety of crystalline forms, from the fine grain of alabaster to the large, flat blades of selenite.

Plaster was first used as a building material and for decoration in the Middle East at least 7,000 years ago. In Egypt, gypsum was burned in open fires, crushed into powder, and mixed with water to create plaster, used as a mortar between the blocks of pyramids and to provide a smooth facing for palaces. In Jericho, a cult arose where human skulls were decorated with plaster and painted to appear lifelike. The Romans brought plasterwork techniques to Europe.

> *"It is . . . poured around the stone or anything else of this kind that one wishes to fasten."*

Theophrastus, philosopher and scientist

Gypsum is found worldwide, as far east as Thailand and as far west as New Mexico, where a huge sandy deposit is used by the construction industry. The name "plaster of Paris" comes from a large deposit mined in Montmartre from the sixteenth century. The French king ordered that the wooden houses of Paris be covered in plaster as a protection against fire.

Plaster has played a key role in the fine arts as well as the building trade. The art of fresco consists of painting on a thin surface of damp plaster; and stucco is a plaster-based ornamental rendering material. In medicine, plaster was first used to support broken bones in Europe in the early nineteenth century. **AC**

SEE ALSO: BUILT SHELTER, DRIED BRICK, WATTLE AND DAUB, FIRED BRICK, REINFORCED CONCRETE

Toothpaste (*c.* 5000 BCE)

Dental care is boosted by Egyptian mixtures.

The development of pastes designed to clean teeth and freshen the breath began in Egypt as early as 5000 BCE. Myrrh, volcanic pumice, and the burned ashes of ox hooves were mixed with crushed eggshells, oyster shells, and other fine abrasives, then applied with a finger to scour teeth and help remove food and bacterial deposits.

In China around 300 BCE a nobleman named Huang-Ti claimed that toothaches could be cured by inserting pins into certain areas of a patient's gums. Huang-Ti's theories grew to become the world's first recorded and systematic approach to oral hygiene.

Generally, however, the composition of what people used as toothpaste remained an intriguing mix of practicality, myth, and superstition until well into the seventeenth century. In the first century CE, for example, it was thought that toothaches could be avoided by removing animal bones from wolves' excrement and wearing them in a band around one's neck. At the same time the Greeks and Romans were using wires to bind teeth together and began producing rudimentary instruments for tooth maintenance and extraction.

Tooth powders first became available in Europe in the late eighteenth century, although ill-conceived mixtures continued to be made available. Their highly abrasive ingredients, such as brick dust and pulverized earthenware, scoured away the protective enamel of the teeth and did more harm than good, despite the addition of glycerine to make the paste more palatable. In the 1850s chalk was added to act as a whitening agent and a new product called Crème Dentifrice saw toothpaste sold in jars for the first time. In 1873 the Colgate company began the mass production of aromatic toothpaste in jars. **BS**

SEE ALSO: SOAP, TOOTHBRUSH, DENTAL FLOSS

Carpentry (*c.* 5000 BCE)

Woodworking supplements long-established ways of working with stone.

Before the discovery of metallurgy, long before plastics, the materials that Stone Age man used were those that he found around him in nature: stone, mud, bone, and of course wood.

Wood is an extremely important material, having numerous useful properties; it floats, it burns, and it can be shaped relatively easily into a variety of different objects. The craft of shaping and using wood—carpentry—has its roots in prehistoric times.

Early woodwork consisted of the use of wood for basic tools, but there is also archeological evidence that Neanderthals were shaping wood into new forms as long ago as the middle Paleolithic (Old Stone Age, 300,000 to 30,000 years ago), using tools made from flint and stone. In this way many useful things were created from wood, including fire-hardened spears and logs hollowed out to create simple boats.

By the Neolithic (New Stone Age), basic woodworking had evolved into a more complex craft—carpentry. The largely nomadic cultures of the Paleolithic era were settling down into more agrarian societies, resulting in an increase in permanent dwellings, and these were often constructed of timber. Researched settlements in Japan and elsewhere include wooden houses of circa 5000 BCE.

The word *carpentry* actually derives from the Latin word *carpentrius*, which means maker of a carriage or wagon. Even in ancient Rome, however, carpenters were producing not only wagons but a whole array of different wooden products, from weapons (bows, spears, and large rock-throwing machines) to beautifully crafted furniture. **BG**

SEE ALSO: STONE TOOLS, BUILT SHELTER, BOW AND ARROW, SPEAR, AXE, SAW

⬈ *Carpenters at work in a painted relief of a tomb in Saqqara, Egypt, Fifth Dynasty 2450–2325 BCE.*

"Many remains of . . . stone architecture exhibit forms [that imitate] constructions in wood."

John Capotosto, writer

Rowboat (*c.* 4500 BCE)

Propelling a boat with paddles begins in Mesopotamia.

Although it is common knowledge that rowboats were used as far back as 3000 BCE in Egypt as a means of traveling and trading along the Nile River, evidence has been uncovered recently to suggest that they were in existence much earlier. In a grave uncovered in the Mesopotamian city of Eridu, archeologists found a clay model of a boat, and the grave is thought to have been dug before 4000 BCE Mesopotamia—widely cited as as "the cradle of civilization"—was the name given in the Hellenistic Period to a broad geographical area that took in what we now know as Iraq and a part of western Iran.

The model they found was of a wide boat with a shallow bottom, rather like a barge, which was designed to float on the shallow rivers of Mesopotamia. Both the Euphrates and the Tigris rivers were part of the region and, flowing from the north of the area to the south, they quickly became an integral part of the transport system set up by the emerging non-nomadic civilizations.

Since wood was in scarce supply, most of the boats in Mesopotamia were fashioned from the hollow and buoyant reeds that grew abundantly in the marshes at the mouths of the two rivers. The reeds were molded into a boat shape and held tightly in place with ropes. Bitumen was used to cover the reeds, calk the boat, and make it watertight.

Floating downstream on the current was simple enough, but going upstream was problematic. It was common practice to use animals walking alongside the water to drag the boat back but, as was discovered, often it was easier and quicker to row. **CL**

SEE ALSO: CARPENTRY, DUGOUT CANOE, SAILS, RUDDER, STEAMBOAT, SUBMARINE, MOTORBOAT, JET BOAT

"Rowing is only a magical ceremony by means of which one compels a demon to move the ship."

Friedrich Nietzsche, philosopher

◩ *A Roman mosaic of spearing octopus from a boat, originally at Dougga, Tunisia, third century CE.*

Canal (*c.* 4000 BCE)

Artificial waterways debut in the Middle East.

China's Grand Canal, completed in the thirteenth century and stretching almost 1,200 miles (1,930 km) from northern Beijing to Hangzhou in the south, is the oldest still in use today. Although the most ancient part of this waterway dates as far back as 486 BCE, canals had been in use for irrigation and transportation for centuries prior to this. The earliest evidence suggests that artificial waterways were excavated and in use across Iraq and Syria by 4000 BCE.

The first British canal, the Fossdyke, was built by the Romans, but it was not until the birth of the Industrial Revolution in the mid-eighteenth century that the construction of a canal network began in earnest, eventually totalling almost 4,000 miles (6,440 km). Canal systems also proliferated throughout Europe and the United States, with horse-drawn barges providing the principal means of cheap transportation for coal, cotton, and other commodities.

The advent of railroads in the mid-nineteenth century spelled the beginning of a decline for British canals, many of which fell into disuse for more than a hundred years until their rediscovery for boating vacations. In mainland Europe and North America, however, the distances to be traveled were much greater. Despite the arrival of the railroads, investment was warranted in wide and deep canals to admit seagoing ships into the heart of those continents; industry has reaped the benefits of canal-borne bulk transportation to this day.

Perhaps the most famous canals are those that have drastically shortened circuitous and treacherous sea voyages, including the Suez Canal of 1869, linking Europe and the East, and the Panama Canal of 1914, between the Pacific and the Atlantic, both remarkable testaments to engineering vision. **FW**

SEE ALSO: ENCLOSED HARBOR, CANAL LOCK, CANAL INCLINED PLANE

Glue (*c.* 4000 BCE)

Beeswax and saps serve as adhesives.

We may not make direct use of adhesives every day, but glue is an important component of many common manufactured items: Books, envelopes, supermarket packaging, and even cheap sneakers benefit from this invention. Although in recent decades chemists have provided us with super glues—substances so phenomenally strong that the user is warned to take extreme care—naturally occurring alternatives such as beeswax and tree sap have been in use for much longer.

In the burial sites of ancient tribes, archeologists have discovered pottery vessels whose cracks had been mended with plant saps. This tar-like glue was also applied to Babylonian statues that had

---

"Ancient Egyptians made glue by boiling animal hides and used [it] as a binder . . . for woodworking."

Joe Hurst-Wajszczuk, writer

---

eyeballs glued into their corresponding sockets. Egyptian carvings from more than 3,000 years ago portray the adhesion of veneer to sycamore, while in northern Europe 6,000-year-old clay pots have been discovered with repairs made with a glue deriving from birch bark tar.

The ancient Egyptians also developed adhesives made from animals, a technique the Romans and Greeks refined in the first five centuries BCE. The Romans subsequently made various types of glue using other natural ingredients—such as vegetables, milk, cheese, and blood—and were the first to use tar and beeswax to fill the seams of their ships. **CB**

SEE ALSO: POSTAGE STAMP, TAPE, EPOXY RESIN, SUPER GLUE, POST-IT NOTE®

Log-laid Road (*c.* 4000 BCE)

Logs open up roads in impassable terrain.

Nicknamed corduroy roads, log-laid roads consist of whole logs, or logs split down the middle, that are laid across the roadway, one tightly against the next, to create a resistant road surface over swampy or muddy land. Sand is used to cover the surface and reduce the discomfort of traveling over the corduroy-like surface.

Despite enabling easier travel through once inaccessible places, corduroy roads could be dangerous for the user. In the best of conditions the ride was already bumpy and uncomfortable, but if rain washed away the sandy cover or logs became loose or wet, the surface became highly hazardous to horses and any vehicles that were attached to them.

The first known log-laid road was constructed in

"Look well to your seat, 'tis like taking an airing/On a corduroy road, and that out of repairing."

James Russell Lowell, "A Fable for Critics"

4000 BCE. Evidence of corduroy roads, made from oak planks covering marshy areas, has been found in Glastonbury, England, dating back to 3800 BCE.

Over the centuries log-laid roads have mainly been replaced by plank roads, using flat boards instead of logs to give a smoother journey. However, both the Nazi and Soviet forces created them on the Eastern Front in World War II. More recently, corduroy roads have lost their original function and become the foundations for other surfaces after decaying very slowly in anaerobic soils. In the United States, roads such as the Alaska Highway that were built in the early twentieth century retain their log-laid foundations. **FS**

SEE ALSO: WHEEL AND AXLE, CART, SPOKED-WHEEL CHARIOT, MACADAM, MOTORCAR

Rivet (*c.* 4000 BCE)

Early Egyptians start a construction trend.

The humble rivet may be small, but is has a lot to answer for—including, quite possibly, the sinking of the *Titanic*. Rivets have been in widespread use for thousands of years but, because engineers now depend on them to secure boats, bridges, aircraft, and other more complex constructions, their reliability has become paramount.

Rivet holes have been found in Egyptian spearheads dating back to the Naqada culture of between 4400 and 3000 BCE Archeologists have also uncovered many Bronze Age swords and daggers with rivet holes where the handles would have been. The rivets themselves were essentially short rods of metal, which metalworkers hammered into a pre-drilled hole on one side and deformed on the other to hold them in place. Today, a wide variety of rivets exist, as do specialized tools for installing them.

The extensive use of rivets in modern engineering and architecture has, inevitably, increased the likelihood of the odd one or two coming unstuck. Materials scientists have blamed rivets for RMS *Titanic*'s infamous descent in 1912, killing over 1,500 people.

Jennifer McCarty and Timothy Foecke carried out an in-depth study of the sunken wreck and concluded that shoddy workmanship had sent her to the ocean floor. More specifically, a large proportion of the three million rivets driven into the ship were made with substandard iron when, McCarty and Foecke claim, they should have been made from steel. The weaker iron rivets used at the front of the *Titanic*, where it struck the iceberg, were unable to withstand the stress of an impact, as the steel rivets used in the body might have done. **HB**

SEE ALSO: WELDING, WROUGHT IRON, SCREW, NAIL, CLAW HAMMER

Wheel and Axle (*c.* 3500 BCE)

The Mesopotamian potter's wheel points the way to wheeled transport.

Most inventions do not appear out of thin air or from the ingenious brain of a brilliant scientist, but evolve from something already in existence. This is certainly true of the wheel and its attached axle, which developed from two different sources. The first was the revolving potter's wheel, invented in Mesopotamia in around 3500 BCE. Although not a tool essential to the potter's craft, the wheel did help in the faster production of better-quality pots. The second source was the sledge, a primitive but effective means of hauling large loads on parallel sleds or bars of wood. The sledge was ideal in icy and snowy conditions, and on hot sand, but not on hard, dry terrain, where great effort was required to pull it along.

Evidence that the use of the potter's wheel and the sledge came together in the invention of the wheel is found in some of the world's earliest picture-writing. Examples in Uruk, in Sumer, southern Mesopotamia, dating to around 3200 BCE, show various sledges, some with sleds, others with wheels. These first wheels were crude but effective: solid wooden discs made of two or three planks pegged together and then cut to form a wheel. When a pair of wheels was mounted on a fixed axle that enabled them to rotate together simultaneously, it was then but a short leap of the imagination to use the wheel-and-axle combination to carry a physical or human load in a cart, wagon, or chariot. Mesopotamia should not, however, claim sole deeds to this invention. Wheels were found in graves in the northern Caucasus, and wheels also appear on a clay pot from Poland; this earlier evidence dates to around 3500 BCE. **SA**

SEE ALSO: LOG-LAID ROAD, LUBRICATING GREASE, CART, LOCOMOTIVE, MOTORCAR

⬈ *Soldiers carry the king's war chariot in a relief from the Palace of Sargon II, Iraq, eighth century BCE.*

> *"[Using] wheels to reduce friction while moving objects was one of the most important inventions . . ."*
>
> Odis Hayden Griffin, engineer

Plywood

(*c*. 3500 BCE)

Egyptians learn to layer sheets of wood.

In gilding, a thin layer of gold covers an object of base material to give the appearance of a solid gold object. Plywood originated in much the same way. With fine woods in short supply in Egypt around 3500 BCE, it became necessary to find alternative solutions to the demands of high-quality furniture making. One such solution involved taking thin sheets of decorative woods and glueing them to thicker pieces of low-quality wood. This is believed to have been done purely for cosmetic and economic reasons, but the process also brings about improvements to the physical properties of the resulting hybrid wood.

Since the days of Egyptian plywood, the material has maintained its place in popular design, as illustrated by its use in the stylish furniture of Gerrit Rietveld, Marcel Breuer, and Alvar Aalto. Unlike the designers of 5,000 years ago, however, these men were aware that the plywood items they were producing would offer greater resistance to unwanted flexing. The composite wood is rigid because the grain of each successive layer is set at an angle of ninety degrees to the layer to which it is glued.

Stumbling across furniture in the middle of the night with the lights turned out can be a painful experience. Fortunately for the Egyptian carpenters, their stumbling in the dark was metaphorical and it involved stumbling across the superior strengths of plywood furniture rather than the furniture itself. In fact, the applications of plywood are by no means limited to furniture; it finds its way into building construction, the hulls of boats, shelves, and automobiles. However, it was only with the advancements in adhesives during the twentieth century that all of these uses have been practical. **CB**

SEE ALSO: CARPENTRY, CHAIR, GLUE

Lubricating Grease

(*c*. 3500 BCE)

Oils and fats speed the first wheeled vehicles.

As long as there have been wheels, there has been the need for lubrication. Any tribologist (an expert in the science of lubrication) will tell you that it serves to reduce friction. It conserves energy, reduces wear and tear, prevents overheating, and reduces noise.

The ancient civilization of Mesopotamia, and later those of Greece and Rome, used wheels in pottery for channeling water and for transportation. Olive oil was used as an axle lubricant, and an Egyptian chariot dated to 1400 BCE was found with animal fat on the axles. Fats add a crucial viscosity that water lacks.

For Roman chariot racers, wheel lubrication would have been life-saving, and a mosaic has been found in

---

"Men of former times used to employ lard … for greasing their axles."

Pliny, historian

Spain showing a man holding an amphora of oil beside the racetrack, much like the pit-stop mechanics of today. A first century BCE. bronze wheel found in Jutland had special grooves for making the axle easier to grease. More than 1,500 years later, Leonardo da Vinci invented a self-oiling axle-end, using olive oil.

Olive oil and animal fat remained the primary forms of grease even as late as the nineteenth century. Sperm oil from whales and neatsfoot oil from animal hooves were used in the British Industrial Revolution to lubricate steam engines and locomotives. In the 1850s, mineral oils, particularly petroleum oil, were developed and revolutionized industry. **AC**

SEE ALSO: WHEEL AND AXLE, CART, SPOKED-WHEEL CHARIOT, LOCOMOTIVE, MOTORCYCLE, MOTORCAR

Cart
(*c.* 3500 BCE)

Mesopotamians raise transport onto wheels.

The origins of the cart are inextricably linked to the invention of the wheel. In fact, one theory of how the wheel was invented suggests that the cart and the wheel were developed simultaneously, inspired by earlier bladed sledges that were dragged across logs.

The earliest sources of evidence for wheeled vehicles are Mesopotamian tablets. Although the dating methods used for these artifacts are not exact, the tablets are known to be from the middle of the fourth millennium BCE. Around the same time there is also evidence for wheeled vehicles in Europe, including wheel tracks at a long barrow near Kiel, Germany, and wagon pictographs found on a beaker at Bronocice, Poland. This has led archeologists to debate whether wheeled vehicles were developed in multiple places simultaneously or whether the technology quickly diffused out of Mesopotamia.

Carts have been in continuous use since their inception, evolving over time to incorporate wheels with spokes and suspension springs for added comfort. However, the invention of the automobile, and to some extent the railroad, has undoubtedly led to the cart's decline as a mode of transport.

Today carts take all sorts or shapes and functions, from traditional horse-drawn carts, through rickshaws and tuk tuks to the electric-powered golf cart. And we must not forget the ubiquitous shopping cart (or trolley), invented in 1937 in Oklahoma, United States, by Sylvan Goldman, who wanted to make it easier for his customers to buy more groceries from his chain of Piggly-Wiggly supermarkets. **RP**

SEE ALSO: LOG-LAID ROAD, WHEEL AND AXLE, LUBRICATING GREASE, SPOKED-WHEEL CHARIOT, MACADAM

⬈ *A cart transports a barrel in a Roman funerary relief from near Trier, Germany, second century CE.*

"[The customers] had a tendency to stop shopping when the baskets became too full or too heavy."

Sylvan N. Goldman, businessman

Sail (*c.* 3500 BCE)

Egyptians harness wind power on water.

For thousands of years sails have been used to harness the wind. By 3500 BCE, ancient Egyptian vessels were being blown up the Nile by the prevailing wind before returning under oars, and the Phoenicians pioneered the development of hardier vessels for sea voyages.

However, these vessels used square-rigged sails to catch the wind and carry them along with it. In order to progress into the wind the sail must instead be used as an aerofoil to produce a lifting force perpendicular to the wind passing over it. The sail can be angled toward the wind and a component of the lift force generated gives forward thrust to the vessel, thus allowing modern craft to sail within a few degrees of the very direction from which the wind is blowing.

> "Pacific island societies used an upside-down triangular sail attached to a single vertical pole."

Thomson Gale, *The World of Invention* (2006)

Sails were employed in this way in the Arabian Sea in 300 CE, but further developments were minor until the fifteenth century and the advent of the European full-rigged vessel. This bore multiple masts hung with both triangular and square sails, providing maneuverability as well as stability and power.

Commercial sailing peaked in the nineteenth century with the emergence of the Americas as competition in trade. Speed and size were paramount, characterized by clippers traveling at up to 20 knots (37 km/h) from China, North America, and Australia, and by vast full-riggers powered around Cape Horn by more than an acre (0.4 ha) of sail area. **FW**

SEE ALSO: WOVEN CLOTH, DUGOUT CANOE, ROWBOAT, RUDDER, MOTORBOAT

Oven (*c.* 3000 BCE)

Egyptians transform bread baking.

Just as the Egyptians brought the prehistoric era to an end in about 3000 BCE, they appear to have produced the first closed oven. It was invented as a way to satisfy the demand for better bread. Flatbread had been around for approximately 5,000 years, but Egyptian ovens enabled the bakers to produce bread with yeast; bread was no longer flat, it was rising.

A traditional oven is one of the simplest inventions; it traps heat within its walls in order to cook the food placed within. However, when considering the timing of the invention of the oven it is necessary to consider the agricultural advances that resulted in the need for it. After the last ice age, around 10,000 years ago, the land began to warm up

> "Heavy pottery bread molds were set in rows on a bed of embers to bake the dough … within them."

Jane Howard, *Bread in Ancient Egypt*

from its frostbitten slumber and gradually provide its inhabitants with grain and other foods from plant species that had been hibernating beneath the ice. Our predecessors may have been a little slow in responding to these new sources of nutrition but, as demonstrated by the Egyptian bakers, they eventually got there.

Open tandoor (cylindrical clay brick) ovens have been found in Mohenjo-daro, the Indus Valley city settlement also dating from 3000 BCE. However, it was the ancient Greeks who developed front-loaded and portable ovens, and used them to turn breadmaking into a profitable venture. **CB**

SEE ALSO: CONTROLLED FIRE, BUILT SHELTER, POTTERY, GAS STOVE, ELECTRIC STOVE, MICROWAVE OVEN

Flail (c. 3000 BCE)

Egyptians separate wheat from chaff with a new invention.

The flail is one of the oldest agricultural tools known to man, having been in use for more than 5,000 years. It has served as a symbol of power and even as a weapon. Despite the introduction of motor-driven harvesting machines in the nineteenth century, it is still used to this day in some parts of the world. Its primary function is for threshing—the forced separation of grain from the parent plant.

It is not clear where the flail originated, but it was certainly used in ancient Egypt. The flail is essentially a handle—called the staff—coupled at one end by a length of leather to the end of a second shorter rod. The staff is held at the free end and the rod is swung downward and from side to side. As it strikes a pile, usually spread on the ground, of harvested wheat or other grain crop, it knocks out the husks, after which the grain can be sifted out for use.

In Egypt, the flail was used as a symbol for the royal dynasties, and therefore became a mark of power. Often seen alongside a shepherd's crook, the two implements together symbolized the pharaoh's ability to provide food and look after his people, in the way that a shepherd would care for his flock. The crook and flail were also the sign of the god Osiris, lord of the Underworld, and on the coffinette of Tutankhamun, which originally contained the viscera of the dead pharaoh, he holds them crossed over his chest.

The use of flails for threshing is highly labor intensive. Today the tool has been all but replaced by modern machinery. The combine harvester can—as the name suggests—both harvest the crop and separate out the grain in a single process. **SR**

SEE ALSO: GRANARY, SCRATCH PLOW, SCYTHE, THRESHER, REAPER, COMBINE HARVESTER

⬈ *An Egyptian wall painting of 1070–945 BCE shows Osiris holding symbols of power, the flail and crook.*

"The straw was removed, and the grain along with the chaff was swept up, and placed in a basket."

John Carter, *How to Use a Flail*

Bell (*c.* 3000 BCE)

The bell tolls for the ancient Chinese.

The ancient Chinese were technologically and culturally advanced. Between 3950 and 1700 BCE, the people of the Yang-shao culture farmed pigs, grew wheat and millet, made highly specialized tools, and produced painted pottery. They also produced pottery instruments called *lings*, which became the first tuned bells. One of the earliest examples of these clay bells is a small red ling uncovered at an excavation site in the Henan Province of central China.

Later, during the Shang and Zhou Dynasties, the Chinese made bells from metal and decorated them with intricate designs. Bells came to play an important part in culture by the fifth century BCE, when sets of bronze bells were used in ritual ceremonies for musical accompaniment. Large, clapperless bells known as *zhong* were sometimes struck with mallets. It is said that these represented the sound of the Autumn Equinox, when all the crops had been harvested—in Chinese, the word *zhong* means "bell," but also "cultivated" when pronounced slightly differently.

During the Qin Dynasty, in the second century BCE, the bell became a symbol of power and authority following the installation of six large bells at the imperial court. In modern China, the bell has a different meaning: education and worship.

Today in the western world, the bell is used both functionally and symbolically. Bell chimes tell us of the time of day, but are also associated with the church and traditional celebrations such as Christmas and weddings. Hand bells are still played by members of the church community and in schools as part of music education. There are even examples of bells being used in music therapy in retirement homes and hospitals. **HB**

SEE ALSO: METALWORKING, TOWER CLOCK

Candle (*c.* 3000 BCE)

Fats, waxes, and wicks light the world.

It is difficult to attribute the invention of the candle to one society or country. The first "candles" may have been nothing more than melting lumps of animal fat set on fire. Later, these evolved into reeds dipped into animal fat, longer burning than their predecessors but still without a wick (a central slow-burning core to the candle, usually made from fiber or cord).

Archeological evidence indicates that both the Egyptians and the Greeks were using candles with wicks (not dissimilar to those we know today) as long ago as 3000 BCE. Many ancient cultures appear to have developed some variation of the candle, using materials such as beeswax or tallow or even the product of berries to make the wax. This surrounded a

---

"Native Americans burned oily fish (candlefish) wedged into a forked stick."

Bob Sherman, *Candle Making History*

---

wick made from fibers of plant material, rolled papyrus, or rolled rice paper.

Burning with a regular flame and at a constant speed, the candle remained the preferred way of producing controlled artificial light for millennia. Candles remained a cheap, efficient way of creating light throughout the Middle Ages and right up until the mid-nineteenth century, when paraffin first became commercially available and the paraffin lamp entered most homes. Since the advent of gas and then electricity, the role of candles has largely been to create a peaceful, reflective, and nostalgic atmosphere, either in a religious setting or in the home. **BG**

SEE ALSO: CONTROLLED FIRE, OIL LAMP, ARGAND LAMP, GAS LIGHTING, INCANDESCENT LIGHT BULB, LAVA LAMP

Pliers (*c.* 3000 BCE)

A tool for gripping evolves from simple tongs.

Pliers are hand tools used for gripping objects by using the principle of the lever. They utilize the hand's powerful closing grip of the fingers into the palm to apply force precisely to a small area. There have been many designs with different jaw configurations to grip, turn, pull, adjust, or cut a variety of items.

Pliers are an ancient invention and probably developed from tools used for handling hot coals when fires were used for cooking. Sticks and wooden tongs were used at first. These were replaced by metal tongs, which were effectively early pliers, around 3000 BCE when iron was being forged. A Greek Macedonian gold wreath from the fourth century BCE, also shows evidence of the use of pliers.

Modern pliers consist of three elements: a pair of PVC-sheathed handles, the pivot, and the head section with its gripping jaws or cutting edges. The pliers' jaws always meet each other at one point. Adjustable slip-joint pliers have grooved jaws, and the pivot hole (holding the rivet connecting the two halves) is elongated so that the halves can pivot in either of two positions to accommodate objects of different sizes.

Pliers may be used to grip a plumbing pipe and loosen it, repair taps, bend (round-nose pliers), and cut wire (pliers with jaws). Diagonal cutting pliers are used for cutting wire and small pins in areas that cannot be reached by larger tools. There are also smaller versions such as the mini "side-cutting" pliers used for jewelry and those in Swiss Army knives.

The basic design of pliers has changed little over the years, and they are still used in many occupations that require dexterity and precision. **MF**

SEE ALSO: CONTROLLED FIRE, METALWORKING, BRAZING AND SOLDERING

↗ *Surgical instruments, including pliers, in a bas relief at the Temple of Sobek and Haroeris, Aswan, Egypt.*

"A pair of needle nose pliers used on the Apollo 16 *lunar module sold for more than $33,000."*

Heritage Auction Galleries, Dallas, March 2008

Investment Casting

(*c.* 3000 BCE)

Humankind learns to shape metal in molds.

Investment casting is one of the oldest metalworking practices, occurring as long ago as 3000 BCE, and remains vital in producing very specific, one-piece metal designs. Today, the process is used to produce complex parts for nuclear power plants, but thousands of years ago essentially the same method was used to produce small metal ornaments and statues.

Civilizations such as the Egyptians and the Mesopotamians used the investment casting—or "lost wax"—process to create small idols or jewelry with intricate patterns. The intended shape and design of each object was first sculpted from natural beeswax, and then coated with several layers of thick

---

"For much of history, investment casting was confined to sculpture and works of art."

European Investment Casters' Federation

---

and heat-resistant plaster. This mold was then heated, the wax inside melted and drained out, and molten metal was poured into the resulting hollow space. After cooling, the plaster was removed, to reveal metal in the exact shape of the wax template.

During World War II the process was adopted extensively to produce precise components for military machinery, a trend that continued after the end of the war and expanded into other commercial industries. With the expansion came more refined ways of implementation, such as more advanced waxes, but the basic ingredients of the process have remained unchanged in millennia. **SR**

SEE ALSO: METALWORKING, SAND CASTING, DIE-CASTING

Button

(*c.* 3000 BCE)

Early humankind creates a sartorial accessory.

Buttons have been attached to clothing for around 5,000 years, but our Bronze Age ancestors used them more for ornamentation than for their potential as a fastener. In their early incarnations, buttons were simply added to clothes for decoration, while the clothes were fastened by pins and belts. The buttons were usually hand-carved from bone, wood, or horn.

It was the Greeks who first came up with the idea of using buttons to fasten clothes. The first "buttonhole" was simply a loop of thread through which a button could be passed to create a fastening.

However, buttons were not adopted in Europe until the return of the Crusaders in the thirteenth century. The introduction of this new fastening coincided with a new trend for "form-fitted" clothing and its popularity soared. By 1250, the French had established the Button Makers' Guild. In fact, the word "button" probably derives from the French *bouton* meaning "bud," or *bouter* meaning "to push."

Buttons became a status symbol, and the wealthy would wear clothing adorned with hundreds of them. By the sixteenth century the finest buttons were encrusted with precious gems and diamonds, and by the eighteenth century they were being crafted from porcelain, ivory, and glass.

The advent of London's Pearly Kings and Queens, whose costumes are covered by mother-of-pearl buttons, coincided with a huge cargo of the buttons that arrived by ship from Japan in the 1860s.

With the dawn of mass-produced buttons, their power as a status symbol diminished and so did their popularity. Most modern buttons are made of plastic, but even today highly priced clothing is often distinguished by unusual ornamental buttons. **HI**

SEE ALSO: CLOTHING, SEWING, WOVEN CLOTH, GLASS, BUCKLE

Helmet

(*c*. 3000 BCE)

Head protection debuts in Mesopotamia and Egypt.

In English, "helmet" is the generic term given to any device that protects the head, usually from impact-related damage. Today helmets can be found in a wide array of activities from sports to space exploration and are made from advanced composite materials, including plastics and Kevlar, combining maximum protection with minimum weight.

Archeological evidence suggests that helmets have been around since the third millennium BCE, being used by the ancient Mesopotamian civilizations. At this time, and for many centuries afterward, the helmet was used exclusively for the purposes of war. The ancient Egyptians were also making helmets at around the same time, taking advantage of the toughness of crocodile skin as their material.

Early arms and warfare reached a peak around the fifth century BCE with the ancient Greeks. In addition to bronze body-fitting armor and broad shields, the Greek *hoplites* (foot soldiers) also sported a bronze helmet, most often in the Corinthian style—solid metal protecting the head and neck, with a narrow aperture for the eyes, nose, and mouth. This style of helmet not only protected the head but was also fearsome for enemies to behold.

The helmet has seen many revisions over the centuries since. The Romans added hinged cheek flaps, and in medieval times visors were added, affording additional facial protection. In peacetime, protective helmets have since become highly specialized, so that, for example, a cycling helmet facilitates the passage of air to cool the head. **BG**

SEE ALSO: CLOTHING, METALWORKING, CHAIN MAIL, STANDARD DIVING SUIT

⬈ *An Assyrian helmeted winged figure, one of a pair of reliefs at the Palace of Ashurnasirpal II, Iraq, 879 BCE.*

"And he had a helmet of brass upon his head, and he was clad with a coat of mail."

Samuel 17:5, on Goliath

Ski (*c.* 3000 BCE)

Lapps use wooden skis to move on snow.

The invention of the ski has contributed greatly to society for the past 5,000 years. Unlike today, early skis were not used for fun and leisure but for work and transportation, playing a key role in both hunting and warfare. They were made of wood and were not designed for speed: They simply served the purpose of keeping the traveler on top of the snow, with walking sticks employed to keep balance.

Hunters have been using skis to chase animals in ice-covered terrain since around 3000 BCE, when the Lapps from Sapmi (a territory incorporating parts of present-day Norway, Sweden, Finland, and Russia) began to use skis extensively. However, it is not clear who invented skiing. The world's oldest surviving ski dates back to around 3000 BCE and was discovered at Kalvträsk, northern Sweden, in 1924. It is 80 inches (204 cm) long and 6 inches (15.5 cm) wide, that is, slightly longer and twice as wide as modern skis. The earliest indirect evidence for the use of skis in ancient civilizations may date back even further, with rock carvings near the White Sea and Lake Onega in Russia thought to be more than 5,000 years old. However, the most famous ancient rock carvings depicting skiers—wearing animal masks and mounted on very long skis—are located in Rødøy, Norway.

Well preserved skis have also been found under the surface of bogs in Finland and Sweden. However, predating these is the earliest ski—a Norwegian word deriving from an Old Norse term meaning "stick of wood"—which looked very different from its modern relative. It was made from the bones of big animals, and leather strips were used to attach it to the boot. **CB**

SEE ALSO: SLEDGE, ICE SKATE, CARPENTRY, SHOE, METALWORKING, SNOWMOBILE, SKI LIFT

⭱ *The world's oldest ski (displayed at the Museum of Västerbotten, Sweden) was made from pine wood.*

Ice Skate (*c.* 3000 BCE)

Finns traverse the icy terrain on bone skates.

The ice skate is believed to have been invented circa 3000 BCE in Finland. For many years scientists were not sure where exactly the skate originated, ancient models having been found throughout Scandinavia as well as Russia. However, in 2008 news emerged that people living in what is now southern Finland would have benefited the most from skating on the crude blades. This country's nickname, "the land of the thousand lakes," is an understatement as it boasts no fewer than 187,888 of them. Finland is also a cold land and therefore each winter its thousands of frozen lakes have presented serious transportation problems for the population. With neighboring villages often separated by lakes, and rowboats locked up until spring, the options were to try to navigate around the frozen water or find a way to negotiate the slippery surfaces.

The first skates consisted of the leg bones of large animals. Holes were drilled at the ends of the bones and strips of leather threaded through to tie the skates to the feet. As in skiing, skaters used thin poles to propel themselves along, and it was only with the arrival of iron runners in fourteenth-century Holland that the poles were dispensed with.

In an inadvertent homage to the skate's origins on their country's lakes, students in the Finnish city of Jyväskylä still commute to their classes by donning skates to traverse the lake that divides the city. The evolution of the skate has seen metal attached to wood and metal attached to metal, but the fundamental fact that keeping your balance enables you to glide almost effortlessly across slippery ice has ensured the skate's continued popularity. **CB**

SEE ALSO: CARPENTRY, METALWORKING, SHOE, SLEDGE, SKI, BUCKLE, ICE RINK CLEANING MACHINE

⬆ *Oxford University built these ice skates using bones similar to the ones used by the ancient Finns.*

Cuneiform (*c.* 3000 BCE)

Sumerians originate wedge-shaped writing.

About 5,000 years ago, the Sumerians of ancient Mesopotamia invented humankind's first writing system. Having already established the world's first true civilization by introducing agriculture and domesticating cattle, they decided that it was more efficient to record their economic transactions in writing rather than use tokens to represent the number of beasts and the amount of harvest they traded. Their initial use of simple pictograms (drawings representing actual things) quickly developed into a complex system of symbols where items were illustrated by one sign and their volume by another.

The Sumerians' innovation was not only used for commercial purposes, but also extended to phonetic

---

"Our earth is degenerate in these latter days. . . . The end of the world is evidently approaching."

Inscription on an Assyrian tablet

—rather than wholly pictographic—ideograms that expressed concepts such as deity and royalty as well as thoughts.

As the symbols evolved, the notes that were recorded on clay tablets became more cuneiform (wedge-shaped), owing to the wedge-tipped reed the Sumerians used as a writing utensil. They were initially drawn in vertical columns, but the writing direction soon changed to left to right in horizontal rows. Rediscovered in the nineteenth century, the cuneiform script (whose last known inscription is an astronomical text from 75 CE) carries major significance as the first means of chronicling events in writing. **DaH**

SEE ALSO: ALPHABET, INK, ABACUS, PARCHMENT, QUILL PEN, WOODBLOCK PRINTING, HINDU-ARABIC NUMERALS

Dam (*c.* 2800 BCE)

Egyptians block a river for the first time.

Dams are built for a number of purposes: to generate hydroelectric power; control flooding; safeguard water supplies for irrigation, domestic, or industrial use; provide for recreation; or ease navigation.

The earliest known dam was built by the Egyptians across the Garawi Valley in 2800 BCE and measured 370 feet (113 m) along its crest. The masonry shell was filled with earth and rubble, but as it was not sealed against water, the center of the dam was soon washed away. This failure discouraged the Egyptians from further forays into dam construction.

The Romans, armed with their knowledge of concrete, were more successful. Their constructions initially relied on sheer weight of material to resist the water, but in the first century they built the first arch-type dam at Glanum in France. The apex of the arch pointed upstream, transferring the force along the dam and into the solid bedrock of the valley sides. This design was also favored by the Mongols in fourteenth-century Iran, but it was otherwise little used until the nineteenth century, when French engineer François Zola designed his eponymous arch dam, using rational stress analysis for the first time.

In the latter half of the nineteenth century, concrete was used as the primary construction material for the first time in a gravity dam in New York and an arch dam in Queensland, Australia. More complex structures were now within reach, and multiple arch, cupola, and buttress dams sprang up around the United States.

China is home to the world's largest dam project, the Three Gorges Dam, which is expected to be fully operational in 2009. It spans the Yangtze River and has been constructed to ease flooding on the Yangtze and provide hydroelectric power for millions. **FW**

SEE ALSO: IRRIGATION, CANAL, ENCLOSED HARBOR

Chair (*c.* 2800 BCE)

Egyptians give a supporting back to a stool.

Chairs have been invented that swing, swivel, rock, roll, recline, fold, massage, and even electrocute. Before all of those, however, came the invention of the chair in its simplest form, about 4,800 years ago. More than a thousand years before that, man had invented a way of resting in a sitting position off the floor, on the simple backless seats known as stools.

Stools were raised to an art form by the ancient Egyptians. Beside creating beautiful and ornate stools, the Egyptian craftsmen also focused on function by fabricating stools that folded. Some examples have floor rails and crossing spindles with carved goose heads inlaid with ivory to resemble feathers and eyes. In the Third Dynasty (2650–2575 BCE), Egyptians were also to give stools their greatest adornment, a back to support the seated person in an upright position. By steadily increasing the height of the back from a simple lumbar support, Egyptians soon arrived at high-back chairs.

As they had with stools, the Egyptians turned chairs into art without sacrificing function for appearance. Chairs in the Middle Kingdom (2040–1640 BCE) were padded for comfort with a cushion, or they had backs of full height. These chairs were curved and fashioned from timber slats and were supported on narrow legs. Sometimes, chairs were painted to give the appearance of animal skin. In the era of the New Kingdom (1540–1070 BCE) a new feature was added to the chair: arms. Thousands of years later, with humankind becoming more sedentary than ever, retractable leg rests are a common option in the quest for the most comfortable chair ever invented. **RBk**

SEE ALSO: BUILT SHELTER, CARPENTRY, PLYWOOD, GLUE, HAMMOCK, FOLDING WHEELCHAIR

↗ *A fifth-century BCE terracotta tablet featuring a chair, from the Sanctuary of Persephone, Italy.*

"A chair is a very difficult object. A skyscraper is almost easier. That is why Chippendale is famous."

Ludwig Mies van der Rohe, architect

Soap (*c.* 2800 BCE)

Babylonians improve human hygiene.

Soap, in the form we know it today, was first produced by the Babylonians in around 2800 BCE. Clay cylinders containing a soaplike material were found during excavations of Babylon. Engraved in the side of a cylinder was a recipe for boiling fats with ashes.

Soap works by acting as an emulsifying agent. Each soap molecule consists of a long, fatty tail and an electrically charged "head." In water the soap molecules form small spheres, called micelles, where the charged heads are on the outside and the water-repelling fatty chains are in the middle. As dirt and grease are not soluble in water, they are contained within the micelles. The micelles can then be washed away, leaving behind a clean surface.

> *"Soap is . . . the first manufactured substance with which we come into contact in our lives . . ."*
>
> John A. Hunt, *A Short History of Soap*

True soap was made by boiling oil and fat with alkaline salts to form glycerin and the salts of fatty acids. The salts are solid and are like the soap we use today. Sodium salts make hard soaps, while potassium salts produce a softer product. Calcium and magnesium salts form an insoluble residue, the scum that soap produces in hard water.

The first hard white soap was produced in Spain from olive oil and the ashes of the salsola plant. However, it was only in the late nineteenth century, after processes for producing alkalis had been discovered, that there was a rapid expansion in the commercial production of soap. **HP**

SEE ALSO: TOOTHPASTE, DEODORANT, SUNTAN LOTION, ROLL-ON DEODORANT, LIQUID SOAP

Arched Bridge (*c.* 2500 BCE)

An innovation transforms architecture.

The exact date and location of the initial historic transition from simpler bridges to arch-supported types is now lost to us. The development and use of the arched bridge has been attributed variously to the Indus Valley Civilization of around 2500 BCE; the Mesopotamians, Egyptians, Sumerians, and Chinese; and the Etruscans and Romans, who built most of the surviving early arch-based architecture in Europe.

Early arches were corbeled, not really an arch as we understand the term today. A corbel is a projecting, stone supporting piece. It is a simple example of a cantilever. Such an arch is constructed by progressively corbeling from the two sides with horizontal joints until they meet at a midpoint. At the top, where the two sides meet, a capstone is placed.

The Romans, aided by their invention of a cement material to bind stone together, refined the techniques of arch construction. Arch-based Roman bridges and aqueducts may be seen today throughout many cities in Europe and the Middle East.

The basic arch design proper can be likened to a beam curved to form a semicircle and prevented from straightening and spreading by strong abutments at either end. Traditionally, the shape of a stone arch is made from wedge-shaped blocks, carefully cut to fit perfectly together.

Known as "voussoirs," these blocks gradually take the curve of the arch from a central, vertical keystone down to the outermost, horizontal footers. The weight of the bridge users pushes downward onto the keystone, and its wedge shape transfers that energy outward onto the voussoirs, thereby spreading the forces sideways and around the arch instead of straight downward. The development of the arch enabled longer and stronger bridges to be made. **MD**

SEE ALSO: FIRED BRICK, PLASTER, SUSPENSION BRIDGE, TRUSS BRIDGE, CANTILEVER BRIDGE, REINFORCED CONCRETE

Brazing and Soldering (*c.* 2500 BCE)

Metal items are joined together by another metal with a lower melting point.

Metallurgy is one of the most ancient fields of technology and also one of the most important. The use of metals has been so essential to humankind that long periods of history—the Bronze Age and the Iron Age—have been named after the metals that were used most predominantly in those times.

Being able to join pieces of metal together has always been essential in making metal artifacts. The joining can be done in a number of different ways, including welding, brazing, and soldering. Metal items to be joined by welding must themselves be partly melted before the joining can take place.

Brazing or soldering—which are sometimes called "hard" and "soft" soldering respectively, with brazing carried out at a higher temperature—are processes whereby pieces of metal are joined together by the introduction of a metal melted into liquid form. This "filler" metal acts like glue in joining the pieces of metal together. The temperature required to melt the filler metal is lower than that required to melt the metals to be joined. This factor allows metal items to be joined together without themselves ever having to undergo whole or partial melting.

Brazing was discovered before either welding or soldering. It may have occurred as early as 4000 BCE, and samples of work where brazing was used to integrate pieces of metal have been dated to 2500 BCE. The techniques of brazing and soldering have been refined over thousands of years. They continue to be important today, having applications in a variety of metallurgical fields, most notably engineering and electronics. **BG**

SEE ALSO: CONTROLLED FIRE, METALWORKING, INVESTMENT CASTING, SAND CASTING, DIE-CASTING

↗ *Metalworkers smelt over a furnace in a painted limestone panel, Egypt, second millennium BCE.*

"Zillah bore Tubal-Cain; he was the forger of all instruments of bronze and iron."

Genesis 4:22

Glass (*c.* 2500 BCE)

Egyptians develop a transparent material.

Archeological findings suggest that glass was first created during the Bronze Age in the Middle East. To the southwest, in Egypt, glass beads have been found dating back to about 2500 BCE.

Glass is made from a mixture of silica sand, calcium oxide, soda, and magnesium, which is melted in a furnace at 2,730°F (1,500°C). Most early furnaces produced insufficient heat to melt the glass properly, so glass was a luxury item that few people could afford. This situation changed in the first century BCE when the blowpipe was discovered.

Glass manufacturing spread throughout the Roman Empire in such quantities that glass was no longer a luxury. It flourished in Venice in the fifteenth century, where soda lime glass, known as *cristallo,* was developed. Venetian glass objects were said to be the most delicate and graceful in the world.

Glass is normally a clear or translucent brittle material, but it may be colored, depending on the way it has been made. The three classes of ingredients used for making glass are: alkalies, earths, and metallic oxides. Crown-glass, used for windows, uses no lead but includes black manganese oxide. Cheap bottle glass uses iron oxide, alumina, and silica.

In the 1950s Sir Alastair Pilkington introduced "float glass production," a revolutionary method still used to make glass. In this process a film of glass, which is highly viscous, is floated onto molten tin, which is fluid, and, as the two do not mix, the contact surface between them is perfectly flat.

Other developments have included safety glass, heat-resistant glass, and fiberoptics, where light pulses are sent along thin fibers of glass. Fiberoptic devices are used in telecommunications and in medicine for viewing inaccessible parts of the human body. **MF**

SEE ALSO: BLOWN GLASS, LENS, SPECTACLES, GLASS
MIRROR, OVENPROOF GLASS, FIBERGLASS, FLOAT GLASS

Welding (*c.* 2500 BCE)

Anatolians join iron pieces by hammering.

Welding is the process of joining pieces of metal with heat, pressure, or a combination of both, so that they completely fuse together.

The first instance of welding is thought to have been in the smelting of iron ore to create wrought iron, some of the earliest evidence of which was discovered in a Hattic tomb in northern Anatolia, dated to around 2500 BCE. Lumps of the iron ore were heated in furnaces until the impurities melted into a slag, trapped in pores in the still solid iron. The hot piece was then hammered to expel the liquid slag and weld together the particles of surrounding iron.

Similar methods of heating and hammering were used to join separate pieces of iron, and examples of

---

"… embracing the fracture with a pair of hot tongs and closing so tight till the weld leans out …"

Vannoccio Biringuccio, sixteenth-century writer

---

this were discovered in Tutankhamun's tomb of 1350 BCE. This type of forge welding remained the only known technique for centuries. One of the most renowned ancient examples is the Delhi Iron Pillar from the fourth or fifth century, which is a testament to the skill of the Indian metalworkers of the day.

Electricity paved the way for the development of arc welding and resistance welding, as well as the oxyacetylene torch. Welding flourished during the two world wars, and is still being developed to this day with the use of more challenging materials such as aluminum, and new technologies such as laser and electron beam welding. **FW**

SEE ALSO: METALWORKING, BRAZING AND SOLDERING,
WROUGHT IRON, HALL-HÉROULT PROCESS

Bellows (*c.* 2500 BCE)

A new invention transforms Mesopotamian metal smelting.

The ability to extract metals from their ores is one of the most significant discoveries in antiquity. Until the invention of bellows, furnace fires were stoked by breath alone. Teams of men, using blowpipes, would blow on the charcoal to supply the oxygen required to increase its temperature. The teams could achieve temperatures high enough to smelt copper and tin and melt metals such as bronze, silver, and gold.

Bellows improved this process not least because arm and leg power is considerably less exhaustible than lung power. They also enabled much larger furnaces to be used; one man with bellows could generate heat around seventy times faster than one with a blowpipe. A pan found in Talla, Mesopotamia, dated around 2500 BCE, is believed to be the earliest evidence of bellows, although they likely predate this. The pan, which held a fire, has a projection with two holes in it, thought to be where bellows were attached. Two bellows were used alternately to generate a continuous a stream of air and maintain the constant temperatures required for smelting.

Another advantage of bellows is that they use ambient air, which is higher in oxygen and lower in carbon dioxide and water vapor than exhaled breath. This enabled even higher temperatures, hot enough to smelt iron, to be achieved for the first time but, although required for ancient iron production, the presence of bellows alone does not indicate that iron was in use. It would be another millennium before bellows reached Egypt and later Europe. One thing is clear—a world of possibilities opened up once societies could extract iron from its ore. **RP**

SEE ALSO: CONTROLLED FIRE, METALWORKING, BRAZING AND SOLDERING, WELDING, WROUGHT IRON

↗ Metalworkers blow air into a furnace in a relief from Saqqara, Egypt, circa 2300 BCE.

> "He gave me a skin-bag flayed from an ox . . . and therein he bound . . . the blustering wind."

Homer, *The Odyssey*

Flush Toilet (*c.* 2500 BCE)

Sanitation arrives in the Indus Valley.

The Internet? Television? The internal combustion engine? All of these things are important, but they pale in significance next to arguably the most important invention of all time—the toilet.

Archeological research indicates that toilets flushed by water have existed since about 2500 BCE. Inhabitants of the Indus Valley developed a sophisticated system of toilets and accompanying plumbing; each house had a toilet with a seat, the waste being borne away by water in a sewer system covered with dry-clay bricks. This system was used in India for most of the existence of the Indus Valley Civilization, which ran from about 3000 to 1700 BCE.

Ancient Egypt also developed a similar system that removed waste through the use of running water. The ancient Romans were so fastidious that they constructed a toilet for use when they were traveling. Their sewerage systems were sophisticated, and public toilets were common.

After some lamentably unsanitary times from 500 to 1500 CE, the toilet saw some major innovations during the second half of the last millennium. John Harrington, godson of England's Queen Elizabeth I, had invented the water closet in 1596, but his invention was not widely adopted. The late 1700s saw development in toilet technology, with several inventors taking up Harrington's ideas and producing further refinements. One of these, developed in 1778 by Joseph Bramah, was installed in many ships.

The first all-ceramic toilet appeared in 1885, designed by china manufacturer Thomas Twyford. It incorporated in one piece the earlier innovation of the water trap, consisting of water held within a U-shaped bend in the outflow pipe that insulated the user from malodorous air in the sewage system below. **BG**

SEE ALSO: SEWAGE SYSTEM, TOILET PAPER, S-TRAP FOR TOILET, BALLCOCK

Sewage System (*c.* 2500 BCE)

Indus Valley toilets are connected to sewers.

It was probably more the need to get rid of foul smells than an understanding of the health hazards of human waste that led to the first proper sewage systems. While most early settlements grew up next to natural waterways—into which waste from latrines was readily channeled—the emergence of major cities exposed the inadequacy of this approach.

Early civilizations, like that of the Babylonians, dug cesspits below floor level in their houses and created crude drainage systems for removing storm water. But it was not until around 2500 BCE in the Indus Valley that networks of precisely made brick-lined sewage drains were constructed along the streets to convey waste from homes. Toilets in homes on the street side

---

"When the plumbers and sanitary engineers had done their work . . . diseases began to vanish."

Lewis Thomas, medical researcher and essayist

---

were connected directly to these street sewers and were flushed manually with clean water.

Centuries, later, major cities such as Rome and Constantinople built increasingly complex networked sewer systems, some of which are still in use. These days the waste is transported to industrial sewage works rather than to the sea or rivers.

After its installation, the early sewage technology of many cities in Western Europe remained in place without improvement. As recently as the late nineteenth century it was often so inadequate that fatal contagious diseases caused by foul water, such as cholera and typhoid, were still common. **RBd**

SEE ALSO: DRIED BRICK, POTTERY, FLUSH TOILET, S-TRAP FOR TOILET, BALLCOCK

Pesticide (*c.* 2500 BCE)

Sumerians use sulfur to protect their crops from rodents and insects.

Civilization was founded on agriculture. The earliest cities grew up around 9,000 years ago when nomadic hunter-gatherers settled in Mesopotamia, herding animals and growing crops for the first time. But relying on the success of an annual crop was risky. Poor weather, an infestation of insects, or crop diseases could ruin the harvest and starve a population. Humans are still unable to control the climate, but solutions to the other problems were proposed in the most ancient of times.

Early attempts to limit damage by pests were mostly physical interventions, such as crop rotation and the manual removal of grubs. The first evidence for a chemical agent comes from Sumeria in 2500 BCE, where elemental sulfur was used to ward off insects. The Sumerians had developed a sophisticated agriculture, employing irrigation and mass labor to farm barley, wheat, chickpeas, and vegetables. Sprinkling sulfur on these plantations could ward off fungi, rodents, and insects such as locusts.

Natural methods of eliminating pests dominated until World War II, when the chemical DDT (dichloro-diphenyl-trichloroethane) was first used to kill mosquitoes in a bid to reduce the diseases they spread, such as malaria and typhus. DDT was succeeded five years later by organophosphates when insects first showed immunity to DDT. Today, around 2.5 million tons of chemical pesticides are used around the world annually. However, alternative technologies, fears about toxic effects on humans, and a renewed interest in organic farming and natural pesticides are reducing our reliance on chemical pesticides. **MB**

SEE ALSO: ARTIFICIAL FERTILIZER, DDT

⌐ *Farmers inspect their fields in a wall painting from the Eighteenth Dynasty, Valley of the Nobles, Egypt.*

"[Odysseus] fumigated the hall, house and court with burning sulfur to control pests."

Homer, *The Odyssey*

Standard Measures

(*c.* 2500 BCE)

The Indus Valley facilitates fair trading.

Trade between people depends on a uniform set of weights and measures that can be used by both sides of the transaction to ensure that the amount obtained or handed over is correct. The first such standard weights were developed in the Indus Valley Civilization of southern Asia. This civilization was among the most advanced of its time—equal to any in the Near East or Egypt—and boasted large cities, such as Mohenjo-daro and Harappa. The system its merchants or accountants devised consisted of cubes of chert, a crystalline form of silica. These cubes were organized in series, doubling in weight from one unit to two units to four to eight and on to sixty-four units. The

> "We are concerned here with methodical digging for systematic information."

Sir Mortimer Wheeler, at Mohenjo-daro

next block weighed 160 units, the next 320, and then proceeded in multiples of 160. The smallest units were used by jewelers to weigh tiny amounts of gold and other precious metals and gems. The largest units were so large they were lifted with the help of a rope, and were used to weigh grain and timber.

In Mesopotamia at much the same time, natural produce such as grain was used as a comparison, but grain can vary in size and weight, making it an unreliable measure. A uniform system was thus invented, using local stones carved into the shape of a sleeping goose. The multiples of sleeping geese were surprisingly effective in regulating quantity. **SA**

SEE ALSO: METRIC SYSTEM, TIME ZONES, SI UNITS

Wrought Iron

(*c.* 2500 BCE)

The metal that gave its name to the Iron Age.

When people talk about iron, they generally mean wrought iron. This is one of three major materials whose base is iron ore—a common element that has the ability to combine with other elements and therefore occurs in many forms. In order to produce its wrought, or worked, variety, charcoal and ore are heated sufficiently to reduce iron oxide to iron without melting it. The final product contains slag and other impurities that keep it from corroding.

First produced in around 2500 B.C.E, wrought iron is the oldest form of iron and gave the Iron Age its name. Its availability increased when blast furnaces proliferated throughout Western Europe in the

> "Good iron is not hammered into nails, and good men should not be made into soldiers."

Chinese proverb

fifteenth century, before its slightly younger relative, cast iron (the malleable form of which is nowadays used in pipes as well as machine and car parts), became more popular.

Today, wrought iron is most commonly used in the restoration of historic ironwork and the construction of high-quality commissions. Steel, the third type of iron, has a higher carbon content and greater hardness. The mild steel developed by Henry Bessemer in the nineteenth century was not only stronger but cheaper to make. The introduction of steel initiated the gradual demise of what was once an indispensable material. **DaH**

SEE ALSO: METALWORKING, WELDING, BELLOWS, BLAST FURNACE, BESSEMER PROCESS

Ink

(*c.* 2500 BCE)

The Chinese introduce permanent dyes to highlight carved lettering.

An ink consists of a liquid base and a pigment, or dye. The pigment provides a colored residue that sticks to a surface when the liquid dries. The first inks were invented by the Chinese some 4,500 years ago, made from a mixture of soot, lamp oil, gelatin (from animal skins), and musk (to counteract the smell of the oil). The ink was used to blacken the raised surfaces of stone carvings to emphasize shapes and letters. Later, in China and elsewhere, more reliable inks were developed using powdered minerals, plant extracts, and berry juices as pigments.

With the advent of writing, and of papyrus and then paper, new types and colors of ink were required for use with writing implements designed for detailed and permanent texts. Some 2,500 years ago, the Chinese developed a solid ink to be stored as a stick; such inks are still in use today. When required, ink is simply scraped off the stick and mixed with water

Other early ink recipes included metal dyes, seed husks, and the ink of cuttlefish (yielding a deep brown ink known as sepia). One enduring recipe, invented some 1,600 years ago, consists of iron salts, tannin (from tree galls), and thickener. This ink is a blue-black color when first used but fades to brown over time.

With the arrival of the printing press in the fifteenth century a different sort of ink was required to stick to printing blocks. A thick, oily ink made from soot, turpentine, and walnut oil was developed specifically for printing. Modern inks are complex fluids, consisting of varying amounts of solvents, pigments, dyes, resins, lubricants, and other materials. **RBd**

SEE ALSO: PAPER, QUILL PEN, LITHOGRAPHY, FOUNTAIN PEN, BALLPOINT PEN, INKJET PRINTER

⬈ *Inked hieroglyphs of an Egyptian treatise on papyrus, Greco-Roman Period, first century BCE.*

"The palest ink will always be much better than even the sharpest memory."

Chinese proverb

Enclosed Harbor (*c. 2500* BCE)

The first man-made dock is built on the Indian coast.

The world's first enclosed harbor, or tidal dock, is believed to have been constructed thousands of years ago during the Harappan or Indus Valley Civilization. It is located at Lothal, in the present-day Mangroul harbor, on India's Gujarat coast, bordering the Indian Ocean.

The dock was discovered in 1955 and is believed to have been constructed around 2500 BCE. It was trapezoid in shape and its walls were constructed from burned brick. It measured 40 yards (37 m) from east to west and 24 yards (22 m) from north to south. Inlet channels allowed excess water to escape and prevented erosion of the banks. On its northern side the structure was connected with the estuary of the Sabarmati River, and lock gates on that side ensured that ships remained afloat in the dockyard.

The entrance to the dock was able to accommodate two ships at a time, and the dock had facilities for loading and unloading cargo from the merchants' boats that constantly plied the harbor.

Ships coming to and from the dock at Lothal probably traveled north as far as the Tigris and Euphrates river deltas. Sumerian goods transported to Lothal included cotton fabrics, beaded jewelry, and foodstuffs.

A major flood occurred in 2200 BCE, and by 1900 BCE the dock at Lothal was buried in sand and silt. These natural events initiated a period of decline in the area that lasted hundreds of years. Excavations that began in the 1950s have provided archeologists with evidence of the activities of this port town. While some researchers question the structure's intended use, the experts are agreed that it is an excellent example of ancient maritime architecture. **RH**

SEE ALSO: CANAL, DAM, DUGOUT CANOE, ROWBOAT, SAIL, RUDDER, CANAL LOCK, CANAL INCLINED PLANE

⬆ *Boats entered the trapezoid harbor at Lothal via a narrow channel that leads to the Sabarmati River.*

Tunnel (*c.* 2180 BCE)

Babylonians excavate the first enclosed roadway.

The Babylonians are said to have built a tunnel under the Euphrates River in circa 2180 BCE using what is now known as the cut-and-cover method. The river was diverted, a wide trench was dug across the riverbed, and a brick tube was constructed in the trench. The riverbed was filled in over the tube and the river allowed to resume its normal course. However, there is no firm proof of this tunnel's existence, so we need to look to the more recent past. Many tombs of the Egyptian New Kingdom pharaohs buried between 1481 and 1069 BCE in the Valley of the Kings were approached by tunnels dug in the solid rock, but these are as much entrances as tunnels.

The first real tunnel—that is, one that was dug through solid rock from both ends, to meet in the middle—was Hezekiah's Tunnel (the Siloam Tunnel) in Jerusalem. This tunnel was dug through solid rock to act as an aqueduct and bring water into the city during an imminent siege by the Assyrians. The two opposing teams of excavators made several directional errors during construction, resulting in a 1,757-foot (535 m) curving tunnel that gently slopes from the Gihon Spring down to the Pool of Siloam by the city walls; as a straight distance, it covers only 1,104 feet (309 m). More famously, and more accurately, the Greek engineer Eupalinos dug a straight tunnel through Mount Kastro on Samos to supply its capital with water. The 3,399-foot (1,036 m) tunnel, dug sometime between 550 and 530 BCE, was perfectly constructed: the two teams of excavators met in the middle with a vertical difference between the two tunnels of only 1.5 inches (3 cm). **SA**

SEE ALSO: ARCHED BRIDGE

⬆ *Hezekiah's Tunnel is an underground water channel that may have incorporated an existing cave.*

Water Filter

(*c.* 2000 BCE)

The Indus Valley purifies drinking water.

The human quest for clean, drinkable water has been going on for thousands of years, and methods of purifying water have undergone countless incarnations over this time. According to the evidence of Sanskrit writings dating to approximately 2000 BCE, water filtration appears to have been developed in the Indus Valley, located in current day Pakistan and western India. The *Sus'ruta Samhita*, ancient Sanskrit medical writings, include instructions on purifying water: "Impure water should be purified by being boiled over a fire, or being heated in the sun, or by dipping a heated iron into it, or it may be purified by filtration through sand and coarse gravel and then allowed to cool." Early purification methods were focused on the aesthetic qualities of water, such as taste and appearance, rather than hygiene.

The ancient Egyptians were also concerned with the appearance of their drinking water. As early as 1500 BCE they were using alum to settle out particles clouding their drinking water. Hundreds of years later, Hippocrates invented what is known as the "Hippocrates sleeve," a cloth sack for filtering water after it had been boiled.

In the eighteenth century modern sand filtration methods were introduced, which led to water filtration in large cities. But it was not until the nineteenth century that the link between health and water quality was established. Until this point all purifying methods were still based on the notion that pure water was simply water that looked clear and tasted good. When a cholera outbreak in London in 1855 was traced to a contaminated water source, the public finally came to realize that invisible contaminants in water could cause major health problems. **RH**

SEE ALSO: DISTILLATION, COFFEE FILTER

Mechanical Lock

(*c.* 2000 BCE)

Egyptian locksmiths learn to deter thieves.

The Egyptians, and possibly other ancient peoples around the same time, invented the first mechanical locks some 4,000 years ago. The locks were a development of the simple wooden crossbeam that slides horizontally across the back of a door to bar entry. To hold the beam, or bolt, in place, a set of movable pins were located on the back of the door which dropped by gravity into recipient holes on the bolt as it moved into place. To unlock the door from the outside, a wooden key with matching pegs or prongs was inserted through a hole; the key raised the pins above the bolt, allowing it to be pulled back by a handle. Such keys could be up to 2 feet (0.6 m) long.

> *"Later locks were so beautifully fashioned that the artist obscured the mechanical intention."*

F. J. Butter, *Locks and Builders' Hardware*

The introduction of metal locks around a thousand years ago provided smaller, stronger, and more precise locking mechanisms. "Wards," solid obstructions within the lock to counteract tampering, were introduced by the Romans. Portable "travel" locks, or padlocks, were particularly useful to merchants on the trade routes of Europe and Asia.

In Renaissance Europe the locksmith became a master craftsman. Bespoke ornamental locks were commissioned by the rich as a symbol of taste as well as prudence. Since that time, the age-old and continuous battle between locksmith and lock picker has led to many ingenious variations. **RBd**

SEE ALSO: METALWORKING, COMBINATION LOCK, PADLOCK, TUMBLER LOCK, SAFETY LOCK, TIME-LOCK SAFE, YALE LOCK

Anesthesia

(*c.* 2000 BCE)

Egyptians find ways to limit pain under surgery.

Many breakthroughs made in modern medicine, such as open heart surgery or joint replacements, would never have been possible in a world without pain control. But how did anesthesia develop?

As it turns out, early physicians never, to the best of our knowledge, resorted to knocking people out prior to performing surgery. Ancient Egyptian and Assyrian physicians compressed both carotid arteries at the same time, limiting blood flow to the brain and so inducing loss of consciousness in patients for the purpose of conducting a procedure. In addition, the Egyptians discovered that opium could help to ease pain, and the Assyrians used their own painkilling mixtures of belladonna, cannabis, and mandrake root. The Greeks and Romans copied and developed these techniques, and medieval Arabs even developed a form of inhalational anesthesia.

The advent of modern anesthesia can be traced to the latter half of the eighteenth century, when Joseph Priestley isolated nitrous oxide. Sir Humphrey Davy realized that it had anesthetic and soporific qualities, but it was considered more of an amusing way to pass the afternoon than a medical breakthrough. All that changed when a U.S. dentist started using it to perform dental extractions painlessly. A few years later, diethyl ether became the anesthetic drug of choice, first for dental procedures, and subsequently for other operations. Chloroform, which had the benefit of being less flammable, but the caveat of being much more likely to cause complications, was used in lieu of ether in some areas. **BMcC**

SEE ALSO: NITROUS OXIDE ANESTHESIA, ETHER ANESTHETIC, CHLOROFORM ANESTHETIC, MODERN GENERAL ANESTHETIC

↗ *The young Pharaoh Ramses II, in the shape of Osiris, offers relief to his father in the form of an ointment.*

"... and [Adam] slept: and [God] took one of his ribs, and closed up the flesh instead thereof."

Genesis 2:21

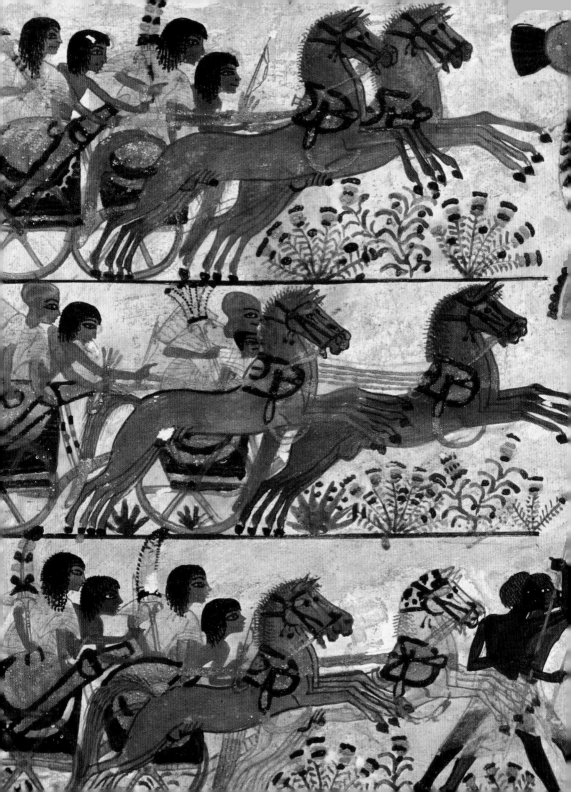

Spoked-wheel Chariot (*c.* 2000 BCE)

Egyptians develop a fast fighting platform.

The development of the spoked-wheel chariot circa 2000 BCE revolutionized warfare. Bronze tools allowed carpenters to discard the solid, heavy, planked wheel in favor of a lighter, spoked wheel. This was made by placing a set of same-length spokes around a central hub and then fixing them within a wooden, circular rim, itself held together by an outer bronze band.

Spoked wheels were larger and lighter than their predecessors and ran better over uneven ground. Used on a two-wheeled chariot that was pulled by a single horse and driven by a charioteer, with room for a warrior alongside, the charioteer could now easily outpace the foot soldier while the warrior—with the

"The Good God, Golden [Horus], Shining in the chariot, like the rising of the Sun . . ."

Tablet of victory of Amenhotep III (1391–1353 BCE)

advantage of speed and maneuverability—attacked him with spear, lance, or bow. The use of such chariots soon spread throughout the Near East. The Hyksos people introduced them to Egypt in 1600 BCE and by 1000 BCE they were in use across Europe. Independently, the Chinese began to use these new chariots around 1300 BCE. The subsequent use of iron rather than bronze made them even more effective, increasing both the speed and strength of these fearsome war machines. **SA**

SEE ALSO: CARPENTRY, METALWORKING, LOG-LAID ROAD, WHEEL AND AXLE, LUBRICATING GREASE, CART, MOTORCAR

⬅ *Chariots on a painted wooden chest from the Valley of the Kings, Egypt, circa 1330 BCE.*

Fired Brick (*c.* 2000 BCE)

Fired bricks are developed in the Middle East.

In ancient times, brick houses were made first by compacting together wet mud and clay into slabs and leaving them to dry in the sun. Once solid, the bricks were piled up to fashion a basic building. However, the major problem with sun-dried bricks is that rainy weather can revert them to wet mud. It took brick makers a long time to arrive at a solution—buildings were constructed from dried mud blocks for more than 5,000 years before the fired brick appeared.

Using a combination of clay, sand, and water, brick makers in the Middle East formed a pliable mass of matter called a clot. The clot was shaped in a wooden mold to create what is known as a "green" (that is,

"And they said to one another, 'Come, let us make bricks, and burn them thoroughly.'"

Genesis 11:3

unfired) brick. This was placed in a kiln and baked at nearly 3,600°F (2,000°C) , before being allowed to cool down into a permanently hard, more durable brick.

The fired brick enabled the construction of the first truly permanent structures—buildings much more resilient than those of mud bricks to harsh climates, changes in temperature, and weathering.

Fired bricks have been refined since they were first invented, the chemical ingredients of the clot mixture having been altered and optimized. Advances in technology have also made mass production possible. However, the premise behind the brick-making process remains exactly the same. **CL**

SEE ALSO: BUILT SHELTER, POTTERY, OVEN, DRIED BRICK, PLASTER, REINFORCED CONCRETE

Saw (*c.* 2000 BCE)

Egyptians introduce the metal-toothed saw.

The saw evolved from Neolithic tools. Archeologists have found metal-toothed Egyptian saws dating back to 2000 BCE, but China claims that the saw was invented by Lu Ban in the fifth century BCE. Early blades were of copper; the Romans then used iron and reinforced the blade at the top, holding it in a wooden frame. In the nineteenth century in Europe a rigid blade of steel with a pistol-grip handle was introduced to produce a more accurate cut.

The cutting edge of a saw blade may be either serrated or abrasive. A handsaw with a stiff serrated blade can cut on both the push and pull strokes, but flexible blades allow cutting on the pull stroke only. Each tooth is bent to a precise angle, called the "set," which is determined by the saw's intended use. Some teeth are usually splayed to each side, so that the blade does not stick, or "bind," in the cut. An abrasive saw uses an abrasive disc or band for cutting.

A number of different categories of hand-powered saws exist, designed either to be pushed forward or pulled backward, or both, and used by one or two people. These were followed by mechanically powered saws, using steam, water, petrol, or electricity, but they all had the same purpose of cutting large pieces of material into smaller ones. Later designs of saw include the circular saw (a rotating metal disc with saw teeth around its edge) and the chain saw (the blade is a chain carrying small cutting teeth).

Samuel Miller's invention of the circular saw in 1777 only came into use when mills became steam-powered. In 1813, Tabitha Babbitt, a Massachusetts Shaker spinner, invented a circular saw as an improvement for lumber production. An early chain saw was developed in 1830 by the German orthopedist Bernard Heine for cutting bone. **MF**

SEE ALSO: SHARP STONE BLADE, METALWORKING, AXE, CIRCULAR SAW, BAND SAW, CHAIN SAW

Alphabet (*c.* 2000 BCE)

First phonetic alphabet originated in Egypt.

In 1999, Yale Egyptologist John Darnell revealed to the world that the 4,000-year-old graffiti he had discovered at Wadi el Hol in Egypt's western desert represented humankind's oldest phonetic alphabet. Incorporating elements of earlier hieroglyphs and later Semitic letters, Darnell's discovery contradicted the long-held belief that alphabetic writing originated in the area of Canaan (modern-day Israel and the West Bank) midway through the second millennium BCE.

Nevertheless, the writings—carved into soft limestone cliff—are thought to be the work of Canaanites, or rather Semitic-speaking mercenaries serving in the Egyptian army during the early Middle Kingdom (*c.* 2050 BCE–*c.* 1780 BCE). Presumably

> *"All the learnin' my father paid for was a bit o' birch at one end and an alphabet at the other."*
>
> George Eliot, author

developed as a simplified version of Egyptian hieroglyphs, the alphabet enabled those soldiers—as well as ordinary people in general—to record their thoughts and to read those of others. Many of the words are thought to be the names of people—the desire to record them stemming from the belief that your afterlife would improve if people read out your name after your death.

Today, the impact of the first phonetic writing system is still felt all over the world, since all subsequent alphabets (with the exception of the Korean Hangul) have either directly, or indirectly, descended from it. **DaH**

SEE ALSO: CUNEIFORM, SEMAPHORE, BRAILLE

Umbrella (*c.* 2000 BCE)

The Chinese invent a collapsible shade.

It was either the Chinese or the ancient Egyptians who first invented the umbrella. Early records from both cultures indicate that umbrellas were used to screen monarchs and people of high standing from the sun. The job of hoisting an umbrella above the emperor was often reserved for the servant of highest rank. The Chinese developed the technology furthest, waxing their paper parasols to provide protection from rain. Around 4,000 years ago, the Chinese also made their umbrellas collapsible, and since then the overall design has changed very little.

Making its way to Rome and Greece, the umbrella was used to shade women and even effeminate men from the sun while attending the open-air theater. These umbrellas were made from leather or skins. The umbrella reached England during the reign of Queen Anne, at the start of the eighteenth century, and were used only by women for protection from the rain. These umbrellas were made from waxed or oiled silk, which became difficult to open or close when wet. But umbrella use was discouraged by the religious, who saw it as interfering with God's intention to wet the faithful, and later on by carriage drivers, who lost business from people who could walk comfortably in inclement weather.

The umbrella's association with femininity was finally shaken off in the mid-eighteenth century when a writer and hospital founder named Jonas Hanway began to carry one. He was a man of poor health, who for thirty years carried an umbrella to ward off heat and cold. Gradually the umbrella came to be accepted by both sexes equally. **DK**

SEE ALSO: WOVEN CLOTH, PAPER, METALWORKING, CARPENTRY

🡕 *In this nineteenth-century Burmese image, Buddha stands on a lotus flower underneath an umbrella.*

"The American people never carry an umbrella. They prepare to walk in eternal sunshine."

Alfred E. Smith, U.S. politician

Quernstone (*c.* 2000 BCE)

Stones developed for the grinding of grain.

As humankind ceased to live as nomadic hunter-gatherers and began to settle down and raise crops, a different style of tool became necessary. People were now able to grow grain. However, grain had to be ground into flour in order to make bread. To accomplish this task an early form of mill, called a quernstone, eventually emerged.

Approximately 4,000 years ago, humans worked out that they could place one rough stone on top of another and use the two of them to grind grain into small particles. Early versions consisted of a rough rock base, or quern, and a smaller rock that could be ground over the top of it, often referred to as a rubbing stone.

> *"Be he 'live, or be he dead I'll grind his bones to make my bread."*

Jack and the Beanstalk, *English Fairy Tales* (1890)

A major advance occurred when the top stone was made to turn on the stationary bottom stone rather than move parallel to the long axis of the stone. These so-called "rotary querns" eventually evolved to feature a central hole in the upper stone that would allow grain to be poured in from the top and flour to work its way out from between the two stones. Later societies experimented with using different types of stone—the Romans favoring types of lava for their rough and sharp surfaces.

The quernstone evolved into larger water- and wind-powered mills, but is still in use in societies where grain is ground by hand. **BMcC**

SEE ALSO: GRANARY, WINDMILL, WATERMILL, TIDAL MILL, AUTOMATIC FLOUR MILL

Aqueduct (*c.* 2000 BCE)

Water conduits invented by the Minoans.

An aqueduct is any artificial conduit for the delivery of water, though the term is often misunderstood to refer only to the arches sometimes used to enable these channels to span low ground.

Ancient civilizations on the Tigris, Euphrates, and Nile diverted water from these great rivers for irrigation, but the paucity of supply in Minoan Crete encouraged the development of complex storage and distribution systems for the first time in the second millennium BCE.

It is the Romans who are best known for their innovative water supply systems. Between 312 BCE and 226 CE the Romans constructed eleven major aqueducts to provide Rome with water.

> *"It is a wretched business to be digging a well just as thirst is mastering you."*

Titus Maccius Plautus, playwright

Aqueducts did not become commonplace again until the late nineteenth century, when rising populations in the United Kingdom outgrew local water sources, and engineers developed systems of aqueducts to provide a clean and reliable supply.

The United States followed suit in the twentieth century with the construction of vast aqueducts to supply its cities, and these, including the 444-mile-long (715 kilometer) California Aqueduct, remain among the largest and longest in the world. **FW**

SEE ALSO: CANAL, ARCHED BRIDGE, DAM, TUNNEL, SEWAGE SYSTEM

➡ *The first-century CE Pont du Gard near Nimes, France, took the Romans fifteen years to build.*

Rubber Ball (*c.* 1600 BCE)

Ancient Mesoamericans were the first people to invent rubber balls.

While other ancient civilizations were playing with balls made of stitched-up cloth or cow bladders, the people of Mesoamerica (present-day Mexico and Central America) were playing a game of life and death using balls made from a processed rubber. By adding the juice of the morning glory vine to latex (raw liquid rubber) harvested from the native rubber tree (*Castilla elastica*), they created balls that had great bounce.

As early as 1600 BCE, the Mesoamericans used this method to make resilient rubber balls that defied the natural brittleness of solid latex. Their amalgamation could be shaped into any conceivable form, but would harden within minutes, making it impossible to reshape the object afterward. They used this process for a variety of artifacts and produced balls of different sizes, the biggest being larger than a volleyball and weighing up to eight pounds (3.6 kg). These were then used in ritual ball games that had great political and religious significance.

While modern followers of sports refer to matches as "a matter of life and death," this was actually the case for the contestants on Central America's fields and ball courts. For the Mesoamericans, the games epitomized their worldview of life as a struggle between good and evil. Winners were showered with riches, whereas the leader of the "evil" losers was sacrificed in the belief that this was the only way to keep the sun shining and the crops growing.

The Mesoamericans' rubber ball was therefore a potentially life-changing device long before Charles Goodyear's heat- and sulfur-treated gum of 1839 added a new facet to leisure activities. **DaH**

SEE ALSO: SYNTHETIC RUBBER, SILICONE RUBBER

⬆ *These rubber balls, found in the seventeenth century in Peru, were the kind used in ritual ball games.*

Shadow Clock (*c*. 1500 BCE)

Egyptians harness the light of the sun to tell the time.

The sun travels across the sky at the rate of 15 degrees per hour (reappearing at a given point after one day) and the shadow that it casts moves at a similar rate. In sunny climes the shadow has been used as a clock. The most ancient clock was the vertical obelisk. This tapering column, rather like Cleopatra's Needle in London, cast a shadow that varied in its length and orientation as the day progressed.

The Egyptians had a small, portable shadow clock. It consisted of a T-shaped bar that lay on the ground, except that close to the shorter crossbar was a 90-degree bend that lifted the crossbar above the long horizontal stem so that its shadow would fall on the stem. The long stem was pointed directly toward the west point on the horizon in the morning. At noon, it was pointed in the reverse direction, toward the east. There were five variably spaced markers on the bar, the one directly under the "T" indicating where the shadow would be at noon and the

subsequent ones for the five hours between noon and sunset (or sunrise and noon if it was being used in the morning). The Egyptians divided the period when the sun was above the horizon into ten "hours." There were two more "hours" for the twilight dawn and dusk periods, and the night was divided into twelve "hours," making the twenty-four-hour day.

The use of this shadow clock required the Egyptians to have an accurate knowledge of the direction of their cardinal points. North, south, east, and west were very important to them, as is demonstrated by the sides of their pyramids, which are aligned very accurately in these directions. **DH**

SEE ALSO: **WATER CLOCK, TOWER CLOCK, PENDULUM CLOCK, CLOCKWORK MECHANISM, ATOMIC CLOCK, QUARTZ WATCH**

⬆ *This Egyptian shadow clock must be aligned to the east or west to provide accurate readings of the time.*

Scissors (*c.* 1500 BCE)

Egyptians create the first fabric cutters.

Spring-type scissors probably date from the Bronze Age. Consisting of blades connected by a C-shaped spring at the handle end, they were used in Egypt from 1500 BCE to cut silhouettes for artwork.

Pivoted scissors used in ancient Rome and parts of Asia were made of bronze and iron, as were sixteenth-century European ones. Scissors and other implements became more widely used as their quality improved with better methods of metal forging, but cast steel was not used until 1761, when Robert Hinchliffe manufactured scissors in Sheffield. Many were hand-forged with elaborate handles, but the styles were simplified in the nineteenth century to facilitate large-scale mechanical production.

> *"Scissors ... evolved, step by step, [with] many other tools destined to cut, separate, and pierce."*
>
> Massimiliano Mandel, *Scissors*

The steel used in scissors contains varying amounts of carbon, depending on the quality of scissors. Drop hammers form the rough shape of the blade from blanks made from red-hot steel bars. The blades are then trimmed and hardened. The steel may contain from 0.55 to 1.03 percent carbon, with the higher carbon percentages providing a harder cutting edge for certain applications.

Surgical and other specialized scissors are made of stainless steel; cheaper scissors are made with softer steel that is cold pressed. Shears used for sheet-metal work, called tin snips, have high-leverage handles but are constructed in the same way as scissors. **MF**

SEE ALSO: METALWORKING, WELDING, STEEL

Water Clock (*c.* 1500 BCE)

Egyptians free timekeeping from the sun.

For millennia, humankind has kept track of the progress of time by observing natural bodies, most notably the sun and the stars. In cloudy periods, however, these cannot be seen. The water clock, or *clepsydra* in Greek, is a timekeeper that works by measuring a regulated, uniform flow of water out of, or into, a vessel. With sufficient water, and a large enough vessel, this timekeeper can "run" for a day or two without needing to be refilled, or emptied.

Imagine a cylindrical water container with a hole in the bottom. The rate at which water drips out of the container is a function of the pressure exerted by the water that it contains; so the more water in the vessel (that is, the greater the "head" of water) the faster is the flow rate. When the container is full, the water level goes down quickly, but the flow is slower when it is nearly empty. Around 1500 BCE the Egyptians realized that if the sides of the bucket tapered parabolically, the water level would go down at a uniform rate, and this would make a reasonable basis for a clock. Others introduced multiple cistern systems that ensured that the head of water remained constant. The dripping of the *clepsydra* in ancient times has found an echo in the ticking of a modern clock.

Later on, the Romans and Chinese constructed complicated float systems that followed the changing water level and moved an hour hand on a circular dial, or rang bells at specific times.

Simple forms of the *clepsydra* were also used to measure specific time intervals, for the regulation of religious services or political debates, in the way that sand glasses are now used as egg timers. **DH**

SEE ALSO: SHADOW CLOCK, TOWER CLOCK, CLOCKWORK
MECHANISM, PENDULUM CLOCK, QUARTZ WATCH

➥ *A funnel-like water clock (center) in a tomb painting, Valley of Kings, Egypt, circa twelfth century BCE.*

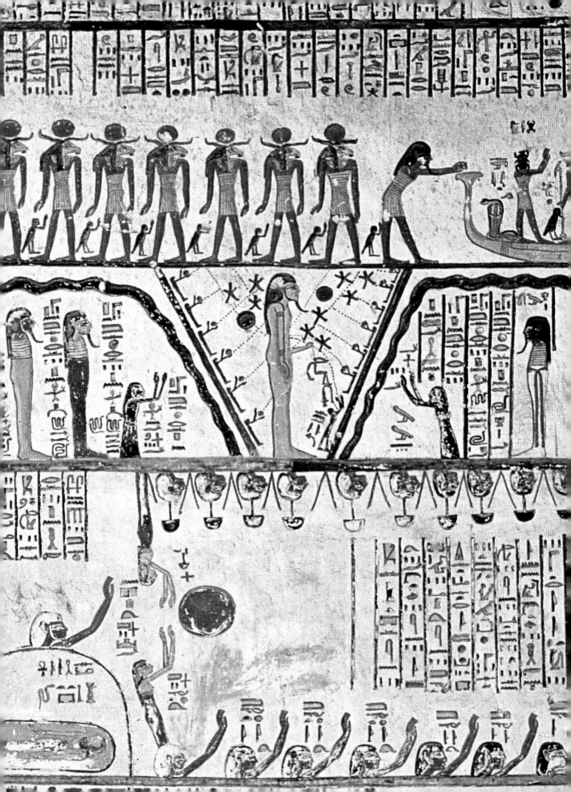

Steel (*c.* 1500 BCE)

East Africans harden iron with carbon.

Steel was first produced in carbon furnaces in sub-Saharan East Africa, around 1500 BCE. Steel is an alloy of iron and 0.2–2.4 percent carbon. It can also contain trace elements such as vanadium, manganese, or tungsten. The carbon acts as a hardening agent and prevents the lattices of iron crystals from sliding past each other. The more carbon present in steel, the harder it is, but this is at the expense of increased brittleness. By controlling the exact ratio of iron to carbon and other elements, the properties of the steel can be tuned to those needed for a specific function.

Damascan steel (also known as Wootz steel) was famed for its strength and ability to keep an edge. It actually originated from India around 300 BCE before

> *"The Iron Age itself came very early to Africa, probably around the sixth century BCE..."*
>
> Richard Hooker, historian

being widely exported; it was identified by its banded appearance. Recent studies of blades made with Wootz steel have found that they contain carbon nanotubes that contributed to their legendary properties. Unfortunately, the process for making the steel died out in the eighteenth century after the necessary ores were depleted.

Modern steel making took off in Europe in the late 1850s with the invention of the Bessemer process. The key element of this process was the removal of impurities via oxidation, achieved by blowing air through the molten iron. For the first time this allowed cheap production of steel on an industrial scale. **HP**

SEE ALSO: METALWORKING, BESSEMER PROCESS, STAINLESS STEEL, SCISSORS, SWORD, MOTORCAR

Sword (*c.* 1500 BCE)

Iron smelting brings sword-length weapons.

A sword consists of a blade and a handle, which is itself made up of a hilt, or grip, and a pommel, or counterweight. A sword blade has one or two edges for striking and cutting, and a point for thrusting. The word "sword" comes from the Old English *sweord*, meaning to wound or hurt.

Humans developed weapons from sharpened flint tools, and in the Bronze Age short-bladed weapons such as daggers were used. It was then impractical to make bronze swords more than 3 feet (90 cm) long, but with the development of smelting technology and stronger alloys, longer iron swords became possible from about 1500 BCE.

The Chinese single-edged steel sword appeared in the third century BCE. By Roman times the hilt was distinct from the short, flat blade, and by the European Middle Ages the sword had acquired its main basic shape and a variety of designs were devised to fulfill different functions. Medieval swords had a double-edged blade, a large hilt, and protective guard and were designed to be gripped in both hands. A curved blade for cutting, used in Asia, was introduced into Europe by the Turks in the sixteenth century, and in the West was modified into the cavalry saber.

Hunting swords and the naval cutlass developed from the sixteenth-century "hanger," with its convex cutting edge, as did the bayonet, developed in the seventeenth century for use with firearms. During the seventeenth and eighteenth centuries, the shorter "smallsword" became a fashion accessory. The smallsword and the rapier remained popular dueling swords well into the eighteenth century. **MF**

SEE ALSO: STONE TOOLS, SHARP STONE BLADE, SPEAR, METALWORKING, WELDING, HELMET, CHAIN MAIL, BUCKLE

→ *In a scene painted onto an ancient Greek vase, a warrior armed with a sword attacks a spear carrier.*

Rudder *(c. 1420 BCE)*

Egyptians learn to direct their watercraft from the stern.

Egyptian tomb paintings from around 1420 BCE depict a ship fitted with steering oars on either side of the stern and are thought to be the earliest evidence of the use of the rudder principle, by which water flowing past the boat's hull is redirected. The same technique was long used on Mediterranean cargo ships, but the Vikings preferred a single oar, mounted to the starboard side of the stern of their longboats. The oar could be easily lifted in shallow water but was not always effective in heavy seas, when it could be raised out of the water by the waves.

A rudder is most efficient when mounted along the vessel's centerline, and in accordance with this Chinese vessels have been designed with hinged rudders on the stern since the first century BCE. There is no evidence of such a practice in Europe until some eleven centuries later, and centerline rudders did not become widespread until the thirteenth century CE. It is not certain whether this change came about through

independent development in northern Europe or by a transfer of knowledge from China. Either way, the rudder was a key enabler for the subsequent rise of Western fleets to naval superpowers.

In the early 1900s the Wright brothers used vertical rudders behind the tailfins of their pioneering gliders to steer their first powered aircraft during its first flight of 1903. Modern aircraft use similar rudder systems to control side-to-side yawing motion. The rudder, albeit with a host of variations and specializations, is still the means by which we steer not only our ships but a multitude of other craft both in the water and the sky. **FW**

SEE ALSO: DUGOUT CANOE, ROWBOAT, CANAL, SAIL, ENCLOSED HARBOR, METAL ANCHOR, SEXTANT

⊡ *A sailing boat equipped with side rudder from a tomb in the Valley of the Nobles, Egypt, circa 1400 BCE.*

Abacus (c. 1000 BCE)

Mesopotamians usher in the age of computing with the first calculator.

The early calculating instrument we know as the abacus—consisting of a wooden frame supporting wires or rods on which wooden beads slide from side to side—was developed in Mesopotamia from a flat, sand-covered, stone counting board on which pebbles were moved. This aid to calculation was in use long before the adoption of the Hindu-Arabic numeral system and can be adapted to any numeral base. The abacus has a huge advantage over counting on the fingers of the hand, simply because it can be used to record very large numbers accurately.

The easiest type of abacus to understand is the modern Western version that uses a base of ten. Here each wire carries ten beads and represents a decadal unit, that is one, ten, 100, 1,000 and so on. A number, say 617,483, can be represented by positioning the respective number of beads, on each wire, against one side of the abacus. It is then a relatively easy task to add or subtract another number from the first one.

After a calculation, the whole abacus can also be reset for further computing by a simple shake.

Abaci were widely used throughout the ancient world and are still important as a teaching aid in pre-school. The movement of the beads helps children to understand the groupings of ten that are the foundation of our present number system.

Other abaci were produced with an interior dividing bar. The Chinese *suan-pad*, introduced about 1200 CE, had five beads on one side of the bar and two above. The Japanese *soroban* originally had a five-to-one bead distribution. The Russian *schety* followed the European pattern with ten beads and no bar. **DH**

SEE ALSO: TALLY STICK, METRIC SYSTEM, MECHANICAL CALCULATOR, POCKET CALCULATOR, MECHANICAL COMPUTER

↑ *A mosaic of c. 212 CE shows Aristotle surprised by a Roman soldier while calculating with an abacus.*

Pontoon Bridge (*c.* 1000 BCE)

Chinese make a temporary water crossing.

In 1000 BCE it was recorded that King Wen of the Zhou Dynasty, ancient China, designed the first pontoon bridge. The invention was to be incorporated into his elaborate wedding ceremony, allowing the wedding procession to cross the Weihe River.

King Wen's design was of a floating bamboo deck structure supported by boatlike pontoons to allow a water crossing. Since their invention, the floating bridges have become much more than just a decorative water crossing—they have become a military weapon. One of the earliest recorded pontoon bridges to be used in combat was built in 974 CE by the Song Army of ancient China, who constructed it in fewer than three days. However, such bridges take a lot less time to destroy or dismantle—a necessary practice to prevent the enemy from following.

King Wen's design is still being used by the military to this day. In 2003 the U.S. Army's "Assault Float Ribbon Bridge" was used to cross the Euphrates River near Al Musayib during "Operation Iraqi Freedom."

Not all pontoon bridges are temporary structures. Some are established across rivers where it is uneconomical to build a suspension bridge. These bridges include an elevated section to allow boats to pass. One of the longest of these bridges spans Lake Washington in Washington State; the Lacey V. Murrow Memorial Bridge, 6,620 feet (2,019 m) in length, was completed in 1940 and cost $10 million less than an orthodox bridge to build.

Cheap though they are, pontoon bridges are not particularly safe. They are especially vulnerable to bad weather and have been known to be destroyed by strong winds. In fact, part of the Lacey V. Murrow Memorial Bridge sank in 1990 after a heavy storm and was subsequently rebuilt. **FS**

SEE ALSO: ARCHED BRIDGE, SUSPENSION BRIDGE, TRUSS BRIDGE, CANTILEVER BRIDGE

Battering Ram (*c.* 1000 BCE)

Assyrians build a new siege weapon.

The battering ram has none of the subtleties of the Trojan horse, but the results are the same; an uninvited entry. Principally weapons of war, early battering rams were heavy wooden beams, sometimes with a metal-covered end that was on occasion shaped as a ram's head (hence the name), whose sole purpose was to breach the fortifications of towns and castles.

In its simplest mode of operation, the battering ram was carried by several people who would run with the ram and thrust it at the target with as much force as they could muster. The key to success was speed, however, and later rams were wheeled.

Battering rams became increasingly sophisticated. One important example was the siege engine of the

> *"Forty-six of [Hezekiah's] . . . towns and innumerable smaller villages [I] besieged and conquered."*

King Sennacherib of Assyria

Assyrians of circa 1000 BCE. Their ram was suspended from a covered wooden frame so that it could be continuously swung at the target, while the frame provided protection for the soldiers within. Wet hides or earth were used to defend against flaming arrows. Mounted on wheels, this ram was easily maneuvered.

Despite changes in warfare, battering rams still have their place today, attached to military vehicles. One person-operated metal rams are also used by today's law enforcement agencies. **RP**

SEE ALSO: BOW AND ARROW, FORT, WHEEL AND AXLE, CART, SPOKED-WHEEL CHARIOT

➡ *Assyrians use a wheeled battering ram to attack a fortress in a relief from Nimrud, northern Iraq.*

Sandwich (c. 1000 BCE)

Hittites serve meat between slices of bread.

The invention of the sandwich is popularly credited to John Montague, Fourth Earl of Sandwich. Its origins go much further back than this, however. Another common belief comes from the Jewish tradition—that the sandwich was invented by Hillel the Elder in the first century BCE. During Passover, Hillel the Elder's invention is commemorated in the text: "This is what Hillel did when the Temple existed: he used to enwrap the Paschal lamb, the matzo, and the bitter herbs and eat them as one." At this point of the remembrance service, the participants do likewise.

Evidence suggests that the sandwich may go back even further than this, to the days of the Hittite Empire, hundreds of years before. There are records of soldiers of the empire being issued with meat between slices of bread as their rations.

Today's sandwich comes in a multitude of varieties from international cuisines. Although there has been some controversy over what constitutes a sandwich (resulting in one court ruling in the United States), it is generally understood to be a meal made from two slices of bread and a filling.

The Earl of Sandwich's part in the story, apart from giving the meal its name, lay in popularizing the sandwich in England in the eighteenth century. It is believed that Montague liked the sandwich because he could eat it without getting grease on his fingers from meat, making it a suitable snack to eat while playing cards. Whether or not this is true is debatable, but certainly after this time the sandwich became the dominant lunchtime meal in England. Being easy to prepare, portable, and adaptable to limitless variations, the sandwich has never lost its popularity and can now be bought from thousands of dedicated outlets and chains across the world. **JG**

SEE ALSO: BREAKFAST CEREAL, POWDERED MILK, CANNED FOOD, POP-UP TOASTER, AUTOMATIC BREAD SLICER

Inoculation (c. 1000 BCE)

Chinese monk pioneers smallpox protection.

Smallpox is believed to have first appeared around 10,000 BCE. Ramses V died suddenly in 1157 BCE; his mummy bears scars that have a striking resemblance to those left by that scourge. Smallpox killed about a third of its victims and left many survivors scarred. But it was noted that survivors never got smallpox again.

After the eldest son of China's Prime Minister Wang Dan died around 1000 BCE of smallpox, Wang Dan sought a cure for it. A Daoist monk introduced the technique of variolation, a type of inoculation. Scab-coated pustules taken from survivors were ground up and blown into the nose like snuff.

Reports of inoculation reached Europe in the 1700s. In London, in 1721, Lady Wortley Montague and

---

"The English are fools ... they give their children the smallpox to prevent their catching it."

Voltaire, *Letters on the English* (c. 1778)

---

the Princess of Wales urged that four condemned prisoners be inoculated. Several months later the men were exposed to smallpox and all survived. Now that variolation was deemed safe, the royal family underwent it and the procedure became fashionable.

Exposing a person to smallpox to prevent smallpox seems madness, but as the scabs were taken from survivors, the virus had been weakened. Mortality from variolation was about 1 percent; mortality from smallpox was 20 to 40 percent. In 1774 Benjamin Jesty inoculated his wife with the cowpox virus. In 1796 Edward Jenner did the same with young James Phipps and the process of vaccination was started. **SS**

SEE ALSO: VACCINATION, CHOLERA VACCINE, RABIES VACCINE, BCG VACCINE, POLIO VACCINE, RUBELLA VACCINE

Kite (*c.* 1000 BCE)

Unknown Chinese establish a long and colorful tradition.

The kite was first invented in China about 3,000 years ago. The first recorded construction of a kite was by the Chinese philosopher Mo Zi (*c.* 470–391 BCE) who spent three years building it from wood. Materials ideal for kite building, such as silk for the sail material and bamboo for a strong, light frame, were plentiful in China, and kites were soon used for many purposes. Stories and records from ancient China mention kites that were used to measure distances, to test the wind, and to communicate during military maneuvers. The earliest Chinese kites were often fitted with musical instruments to create sounds as they were flown; they were decorated with mythical symbols.

The first kites were flat and rectangular in shape, but kites are now designed in a variety of forms, including boxes and other three-dimensional assemblies. Kites flown as a hobby are particularly popular in Asia, where kite flying is a ritual incorporated into the national festivals of many countries. The Chinese people believe that kites are lucky and they fly them to ward off evil spirits.

The kite has been used in important scientific research, including Benjamin Franklin's famous experiment to prove that lightening is electricity. The Wright brothers constructed a 5-foot (1.5 m) box kite in the shape of a biplane when they were experimenting with the principles of controlling an airplane in flight. This research helped the brothers achieve their dream of making the world's first controlled, heavier-than-air, human flight in 1903. Modern kites have been used to pull sledges over snow-covered terrain in the Antarctic. **CA**

SEE ALSO: PUBLIC ELECTRICITY SUPPLY, POWERED AIRPLANE, SUPERSONIC AIRPLANE

↗ *Kites constructed in butterfly designs have long been prominent in the Chinese kite-flying tradition.*

"Tie the [handkerchief] corners to the extremities of the cross, so you have the body of a kite ..."

Benjamin Franklin, to Peter Collinson, 1752

Pulley (*c.* 750 BCE)

Assyrians revolutionize the lifting of weights.

A pulley is one of the simplest machines, essentially a circular lever in the form of a wheel or fixed curved block, with a groove around it to accommodate a rope or belt. The earliest evidence for the existence of the pulley comes from Assyria in the eighth century BCE; a painting of a battle scene shows a warrior using a simple pulley to lift a bucket over a wall.

Pulleys are mainly used to move or lift a load. A single fixed pulley can be used to change the direction in which a force is applied, as it may be easier to pull on a rope than to drag or push the load. When the rope is fixed at one end and another pulley is added, the system effectively halves the required force, as each part of the rope carries an equal share of the load. This does not reduce the mechanical work required: work is the multiple of force and distance and the rope—now doubled in length—will need to be pulled twice as far. More pulleys can be added to make a "compound pulley" system, further multiplying the effectiveness of the force applied.

Pulleys have been in use throughout the world for many centuries. Although the earliest hard evidence of their use dates from the eighth century BCE, it is highly likely that the principle was in use long before that. Early humans most likely created pulleys by throwing a rope or a fibrous vine over the branch of a tree to hoist up a heavy weight.

It is highly probable that pulleys were used by the builders of early massive constructions, such as the ziggurats of Mesopotamia (built as early as the fourth millennium BCE), Stonehenge (built in 2200 BCE), and the pyramids of Egypt (third millennium BCE). **EH**

SEE ALSO: BRAIDED ROPE, WHEEL AND AXLE, WINCH, CRANE, LEVER, COMPOUND PULLEY

⬆ *This remnant of a simple bronze pulley without wheel was found at Gezer, in Israel.*

Buckle (*c.* 700 BCE)

The ancient Greeks introduce the buckle fastener.

The buckle originated in circa 700 BCE. Many examples survive from ancient Greece and Rome, and indeed from all over Europe into the Middle Ages. The word *buckle* comes from the Latin word *bucca* meaning "cheek." Due to its ease of use and manufacture, the buckle continues as a solution to the many fastening problems posed by clothing and equipment.

Early buckles were manufactured from bone, ivory, and metal and were used on military gear, harnesses, and armor, being favored mainly because of their durability. The use of the buckle was not restricted to these areas though; they were commonly used as fasteners on boots and shoes and, prior to the invention of the zip, on clothing.

The addition of decorative ornamentation lifted the buckle out of its utilitarian realm. Buckles made of silver and bronze and inlaid with precious stones have been found in graves and tombs such as that of Childeric I, king of the Franks, who died in 481 CE.

Jeweled shoe buckles were in vogue during the reign of Louis XIV. During the nineteenth century, the British Navy was at the height of its power, but it had one intractable problem; the sailors' clothes were fastened with laces and eyelets. In cold, wet conditions the fastenings became fiddly to use, and waterlogged clothes were poorly supported by the laces. One clever seaman allegedly had the idea of fastening a buckle to a leather strap and using this to hold up his trousers; it worked and was easy to use, even with freezing fingers.

Today, buckles remain a fashion accessory though lack the status of jewelry. **BG**

SEE ALSO: CLOTHING, METALWORKING, SHOE, BUTTON
FASTENING, HELMET, CHAIN MAIL, STANDARD DIVING SUIT

⬆ *A gold belt buckle with granulated decoration from the Orientalizing Period of ancient Greece.*

Metal Anchor (*c.* 650 BCE)

Greeks use metal to weigh down anchors.

The need to moor ships and boats is as old as the vessels themselves. The ancient world—Mesopotamia, Egypt, Greece, and Rome—used whatever came to hand for the task, from a basketful of stones to a sackful of sand, lowered by rope. Single large stones with a rope-hole became common.

Use of metal crept in gradually during classical times, as ships increased in size and varied anchor designs were developed for different situations and vessel types. In 500 BCE, bronze anchors were cast in Malta. Some crude wooden anchors had pieces of lead or other heavy metals added for weight, while a popular wooden hook design gradually became fashioned entirely in iron instead. Iron anchors have been recovered from Roman merchant vessels. Soon the classic form developed. This featured flukes (the pointed ends of arms at the anchor base) and a stock (a horizontal bar that upends the anchor to ensure that one fluke becomes well embedded).

The invention of the anchor has been credited to the legendary King Midas (700 BCE), but other tales relate how, around 650 BCE, Greek sailors first added hooks to a stone anchor and established the basic future design. Anchor design changed little over the centuries. By 1700, the prevalent anchor design was the kedge type made from iron, which featured a long shank. By 1901, a stockless anchor made from cast steel had been patented. In 2003, the world's oldest wooden anchor, with a metal crown, was found at the former ancient Greek colony of Klazomenai in modern Turkey. Dating back to 600 BCE, this may have been close to the birth of the fully metal anchor. **AK**

SEE ALSO: METALWORKING, SAIL, RUDDER, ROWBOAT, STEAMBOAT, SUBMARINE, MOTORBOAT, ENCLOSED HARBOR

⤒ *A Minoan cast-metal anchor, decorated with an octopus; the hole is for easy attachment to a rope.*

Crossbow (*c. 550* BCE)

Chinese pioneer the longbow's smaller rival.

The crossbow originated in ancient China circa 550 BCE and is thought to have been developed from the horizontal bow trap, which was used to kill game. For use as a weapon, the Chinese developed many different designs of crossbows and drawstrings. Some had stirrups attached to them to hold the bow down when the bowmen were rearming. Later crossbow-cannons had winches to pull back the strings, because people would not have been strong enough to do this unaided. The Chinese also invented grid sights in 100 CE and a machine-gun type of crossbow, which had a magazine of bolts fitted above the arrow groove; as one bolt was fired another dropped into its place. Poisoned crossbow bolts were also used.

Knowledge of the crossbow was probably transmitted from China to Europe via the Greeks and Romans. The weapon could be used by an untrained soldier to injure or kill a knight in armor. The crossbow was adopted in Europe in the tenth century CE and used throughout the Middle Ages. William the Conqueror brought the medieval crossbow to England in 1066, but the Welsh longbow supplanted the crossbow during the reign of Edward I (1272–1307). Many viewed it as an inhuman weapon, requiring no skill and having no honor, and in 1139 the Pope, through the Second Lateran Council, condemned the use of crossbows against anyone except infidels.

Today, crossbows are mostly used for target shooting in modern archery. In some countries crossbow hunting is still allowed, such as a few states in the United States, Asia, Australia, and Africa. **MF**

SEE ALSO: CARPENTRY, METALWORKING, BOW AND ARROW, HELMET, CHAIN MAIL, CATAPULT

⊡ *A bronze crossbow trigger mechanism, from the grave of the Chinese king Zhao Mo (r. 137–122 BCE).*

Crane (*c.* 550 BCE)

Greeks use cranes for heavy construction.

The extent to which human beings extend their natural capabilities through the use of machines is something that distinguishes us from other members of the animal kingdom. Cranes are an especially relevant example of this; the ability to raise and maneuver weights vastly greater than those that people could lift and move unaided has played a defining role in the development of human society.

The crane is a system of pulleys and cords or wires attached to a framework that enables the movement of heavy weights both vertically and horizontally through the use of mechanical advantage. The earliest cranes have been dated to approximately 550 BCE, although there are Greek architectural constructions still in existence that predate this by several hundred years, and that undoubtedly would have required some sort of supporting pulley mechanism. Cranes were used in ancient Greece for a variety of purposes. They were integral to Greek construction and were also used for pulling heavy loads. The Claw of Archimedes was a wall-based crane used to hoist invading ships to a great height, only to drop them to their destruction.

Cranes continued to be used extensively in ancient Rome, where a "treadmill crane" was used to help in building projects. In reconstructions, single blocks of stone weighing as much as 100 tons have been lifted considerable heights above the ground using this technology. Cranes fell out of use for a while, only to reappear in the late Middle Ages. They have continued to see heavy usage through to the present day. **BG**

SEE ALSO: BRAIDED ROPE, PULLEY, WINCH, LEVER, COMPOUND PULLEY

"Archimedes had stated that given the force, any given weight might be moved ..."

Plutarch, *Life of Marcellus*

◤ *This pivoting crane, with its weight block lifted, is displayed at the Musée Romain in Nyon, Switzerland.*

Artificial Limb (c. 550 BCE)

A Persian reportedly creates the first replacement body part.

The earliest written reference to an artificial limb occurs in an epic Indian poem, the "Rig-Veda," which was compiled between 3500 and 1800 BCE. Written in Sanskrit, the poem includes a description of the amputation of the warrior Queen Vishpla's leg during battle. Later fitted with an iron prosthesis by the Ashvins (celestial physicians), she returned to combat.

Most authorities doubt the story of Queen Vishpla, and turn to the *Histories* of Herodotus for the first plausible reference to a prosthetic limb. Herodotus describes how, in the mid sixth century BCE, Hegesistratus of Elis, a Persian soldier and seer imprisoned by the Spartans, was sentenced to death, and cut off part of his foot to escape from the stocks. Hegesistratus fashioned a wooden prosthesis to help walk the 30 miles (48 km) to Tregea, but unfortunately was captured by Zaccynthius and beheaded.

In the first century BCE, Pliny the Elder wrote in his *Natural History* of Marcus Sergius, a Roman general who led his legion against Carthage in the Second Punic War (218 to 210 BCE). The general sustained twenty-three injuries, necessitating the amputation of his right arm. An iron hand was fashioned to hold his shield, and he returned to battle. He fought four more battles, and had two horses killed from beneath him.

The oldest known prosthesis was discovered in a tomb in Capua, Italy, in 1858. Made of copper and wood, it dates to 300 BCE, during the period of the Samnite Wars. Regrettably, the Capua leg did not survive another war—it was destroyed in 1941 when the Museum of the Royal College of Surgeons was seriously damaged in an air raid. **SS**

SEE ALSO: CARPENTRY, ANESTHESIA, SPECTACLES, JOINTED ARTIFICIAL LIMB, ARTIFICIAL HEART

⤴ *A copy of a brass and plaster artificial leg found in a Roman grave in Capua, Italy, circa 300 BCE.*

> *"Demeter, goddess of agriculture, ate Pelops' shoulder, but [made him] a prosthetic ivory shoulder."*
>
> Ampulove, *History of Prosthetics*

Winch (*c. 500* BCE)

Persians use the winch in bridge building.

The first known reference to a winch is made in the writings of Herodotus of Halicarnassus on the Persian Wars in 480 BCE, in which wooden winches were used to tighten cables used in a bridge that crossed the Hellespont. The idea caught on quickly and within a hundred or so years the winch had reached Greek construction sites, though evidence suggests that it was invented by the Assyrians in the fifth century BCE.

A simple winch is used to wind rope or cable, but the tool has many more applications when fitted with a cleat to maintain tension and prevent the rope or cable from unwinding. Cleated winches have long served on boats and harborsides to keep ships and boats closely moored to docksides. They are

> *"They stretched the cables by twisting them taut with wooden windlasses."*
>
> Herodotus of Halicarnassus, historian

important for lifting work on construction sites, enabling workers to complete tall projects in a fraction of the time otherwise required.

In medieval times a cleated winch was an important component of the rack, a bedlike torture device designed to stretch a person by infinitely painful degrees. Cleated winches are used to raise flags on flagpoles and keep them aloft, and mini winches also feature on the reels of fishing rods, allowing anglers to maintain or release line tension when playing a fish. Winches are used by tow-trucks, and they play a vital role in helicopter rescues, safely extracting people from dangerous situations. **CB**

SEE ALSO: CARPENTRY, METALWORKING, PULLEY, CRANE, LEVER, COMPOUND PULLEY

Hammock (*c. 425* BCE)

Alcibiades creates a new type of bed.

The precise year of the hammock's invention is impossible to tell, but estimates of 1000 BCE are considered reasonable, with the Mayan Indians most often credited with the invention. However, there is no evidence for this, and the hammock's creation is often attributed to a later inventor. In Greece, Alcibiades (*c. 450–404* BCE) was a student of Socrates, and some sources attribute its invention to him.

Western European society was first introduced to the hammock in 1492 when Christopher Columbus returned from the Bahamas where he had found the native people resting and sleeping in them. He took some hammocks back to Europe and within a century or so they were standard issue for European sailors. In the cramped ships of the time their value was obvious, as they could be stowed away or hooked up for use almost instantly. More than any other form of bed, they allowed sailors to sleep in paradoxical harmony with the rocking movement of their ship, hanging downward under the pull of gravity while the ship rolled, pitched, and yawed around them.

Arriving in the wake of Columbus in 1500, the Portuguese explorer Pero Vaz de Caminha saw a Caribbean native asleep in a suspended fishing net. He called this innovation a *rede de dormir*, or "net for sleeping," and *rede* remains the Portuguese name for a hammock. At much the same time the Spanish conquistadors were also encountering hammocks used by Caribbean Indians. The term they used was *homoca*, itself derived from *hamaca*, the Indian name for the hamak, or hammok, tree whose bark supplied the fibers from which the hammocks were woven.

Aside from twenty-first-century materials being adopted, the hammock's design has remained largely unchanged over the centuries. **CB**

SEE ALSO: BRAIDED ROPE, WOVEN CLOTH, CHAIR

Chain Mail (*c.* 400 BCE)

Romanians make the first known protective metal shirt.

Chain mail was originally called just mail or chain in England and *maille* in France (the French word *maillé* means "meshy" or "netted"). It was not until the 1700s that chain mail became its common English name.

Mail is constructed from a series of links made from wire. These are bent into circles around a forming cylinder, and the finished links are welded or riveted into the form of a shirt. The result is a sturdy piece of armor that affords very effective protection from most cutting blows while at the same time being relatively lightweight and flexible.

Chain mail alone could not protect against crushing injuries, however, and warriors therefore combined it with a gambeson, which was worn underneath the mail. This was a padded jacket made from layers of wool and other materials that provided effective resistance to impact injuries.

The first mail shirt on record is from a Romanian Celtic chieftain's burial chamber and dates back to the fourth century BCE. Chain mail saw extensive use throughout the first half of the last millennium, being employed throughout Europe and Asia, but it was not until the thirteenth century CE that mail armor really came into its own. Extending over the whole of a knight's body, the basic mail shirt (or *hauberk*) was joined by individual mail pieces for the legs, arms, and head, providing more complete protection.

Knights did not wear this type of full armor for long, however. Items of plate armor were increasingly added to the mail, and these grew increasingly more sophisticated. Foot soldiers continued to wear chain mail until late medieval times. **BG**

SEE ALSO: CLOTHING, METALWORKING, WOVEN CLOTH, SEWING, SPEAR, SWORD, HELMET, PIKE

↗ *Christian knights wear chain mail in a fresco at the Palacio Berenguer de Aguilar in Barcelona, Spain.*

"... chain mail makers, slowly going mad [while] they clipped together chain mail rings ..."

Ursula K. Le Guin, *Tehanu* (1990)

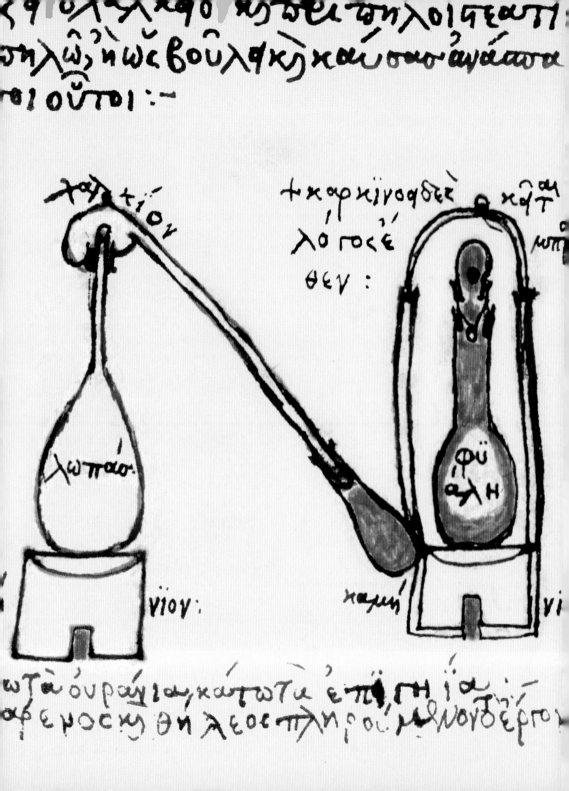

Distillation (*c.* 400 BCE)

Alcohol distillation precedes that of water.

Distillation is not a process confined to spirit production; it is a method of separating chemical substances by their volatility. Chemicals are separated from solutions by heating them until they boil and turn into gas. The gas is then collected and cooled, when it condenses into a liquid. As different chemicals boil at different temperatures, it is possible to separate them by controlling the heating temperatures.

There is evidence of the distillation of alcohol dating back to the second millennium BCE although recent evidence from Pakistan demonstrates that it was not until 400 BCE that the process was well understood. The idea of boiling water and collecting it as steam, which separates out dirt, salts, and bacteria,

---

"Reserve your right to think, for even to think wrongly is better than not to think at all."

Hypatia of Alexandria, distillery inventor

---

seems to have come 800 years later, when Hypatia of Alexandria (*c.* 350–415 CE) invented the first apparatus for distilling water. However, it was not until the eighth century CE that pure chemical substances were obtained by distillation. The alembic still was invented by Persian chemist Jabir Ibn Hayyan. Later, in the ninth century, petroleum was distilled to produce kerosene by another Muslim chemist, al-Razi, and the extraction of essential oils by steam distillation was invented by Avicenna in the eleventh century. **RP**

SEE ALSO: CONTROLLED FIRE, ALCOHOLIC DRINK, WATER FILTER, INSTANT COFFEE

← *This copy of a late third-century treatise by Zosimos of Panoplis shows two alembics and their receivers.*

Pike (*c.* 400 BCE)

Macedonians deploy the pike in the phalanx.

A major weapon advancement in the complicated history of ancient warfare, the invention of the pike in 400 BCE is credited for the Macedonian takeover of Greece, Egypt, and parts of Asia.

Philip II of Macedonia (382–336 BCE), father of the famous Alexander the Great, is credited with adopting the pike (also called the sarissa), as well as the Macedonian phalanx, a type of infantry formation of soldiers. The pike was around 20 feet (6 m) long, and this great length enabled soldiers to strike while they were themselves out of range of shorter weapons. The phalanx consisted of a tight formation of soldiers and pikes. The men in the front of the phalanx would hold their pikes straight out, creating an equivalent depth of about five rows of men.

Before Philip's invention, the Macedonian army was considered ill-equipped and ill-trained. The combination of the pike and the phalanx formation ensured that the soldiers were well defended—the phalanx arrangement only failed if the formation was broken or outflanked, which happened rarely. The pike was an effective weapon only when used in the phalanx, and was essentially useless outside of it. Away from the phalanx formation, Philip's men used javelins. They were adept in the use of both types of weapons, an impressive military feat since the skills required to use each of them are quite different.

Alexander the Great inherited Philip's military tactics with the pike and phalanx and used them to conquer Egypt, Persia, and what is now northern India. Versions of the pike were still being used in military operations up until the eighteenth century. While the phalanx may seem an unwieldy fighting unit today, it was able to act like a modern-day tank, breaking away to crash into enemy ranks with impunity. **RH**

SEE ALSO: SPEAR, METALWORKING, SWORD, HELMET, CHAIN MAIL

Magnetic Compass

(*c.* 400 BCE)

An invention fails to find its ideal application.

The Chinese discovered the orientating effect of magnetite, a magnetic ore known as lodestone (or leading stone), as early as the fourth century BCE and the earliest compasses were used for quasi-magical purposes. They consisted of a piece of lodestone floating on a stick in a bowl of water, which swung around so that it always pointed in a consistent direction. It was another thousand years before they were used for navigation. Previously navigators in the northern hemisphere had used the North Star to indicate direction, and followed earlier maps, but the compass, which aligned with the North Star, was more useful because it could be used in all conditions.

> *"Magnus magnes ipse*
> *est globus terrestris.*
> *[The whole Earth is a magnet.]"*

William Gilbert, physician and natural philosopher

Magnetic compasses work in this way because molten iron in the center of the Earth acts as a magnetic core, as if it were a giant bar magnet, and causes the needle to set parallel to the north–south axis of the globe. It was only realized later that the directions of the magnetic north and geographical north (Earth's axis) were not parallel to each other and varied by about 12 degrees.

It was later discovered that iron or steel needles stroked by a lodestone became magnetized and also lined up in a north–south direction. In 1745 Gowin Knight, an English inventor, developed a method of magnetizing steel permanently. **MF**

SEE ALSO: MAP, STEEL, SEXTANT, GYROCOMPASS, GLOBAL POSITIONING SYSTEM (GPS)

Blast Furnace

(*c.* 400 BCE)

Chinese smelting begins with bronze.

The oldest known blast furnaces were built during the Han Dynasty of China in the fourth century BCE. Early blast furnace production of cast iron evolved from furnaces used to melt bronze. Iron was essential to military success by the time the State of Qin had unified China (221 BCE). By the eleventh century CE, the Song Dynasty Chinese iron industry switched from using charcoal to coal for casting iron and steel, saving thousands of acres of woodland.

In a blast furnace, fuel and ore are supplied through the top of the furnace, while air is blown into the bottom of the chamber. The chemical reaction takes place as the material moves downward, producing molten metal and slag at the bottom, with flue gases exiting from the top of the furnace.

The oldest known blast furnaces in the West were built in Dürstel in Switzerland, the Märkische Sauerland in Germany, and Lapphyttan in Sweden, where they were active between 1150 and 1350 CE. There have also been traces of blast furnaces dated as early as 1100 found in Noraskog, also in Sweden. These furnaces were very inefficient compared to those used today.

French Cistercian monks, who are known to have been skilled metallurgists, passed on their knowledge of technological advances regarding blast furnaces between the thirteenth and the seventeenth centuries. Iron ore deposits were often donated to them and the monasteries sold their surplus iron as well as the phosphate-rich slag from their furnaces, which was used as an agricultural fertilizer.

In 1709 Abraham Darby, a Quaker iron founder in Shropshire, England, used coke instead of charcoal to smelt iron ore in his improved blast furnace. He also processed cast iron into wrought iron and steel. **MF**

SEE ALSO: CONTROLLED FIRE, OVEN, BELLOWS, STEEL, WROUGHT IRON, ELECTRIC ARC FURNACE

Catapult
(*c.* 400 BCE)

Sicilians introduce one of the first great war machines.

The word *catapult* came from two Greek words: *kata*, meaning "downward," and *pultos*, which refers to a small circular shield. *Katapultos* was taken to mean "shield piercer." The weapon was said to have been invented in 399 BCE in the Sicilian city of Syracuse and, according to Archimedes, was derived from a composite bow, which was similar to the crossbow.

Early catapults had a central lever with a counterweight at the opposite end to the projectile basket. Torsion-powered catapults entered into common use in Greece and Macedon around 330 BCE Alexander the Great used them to provide cover on the battlefield as well as during sieges.

The Chinese, Greeks, and Romans used various types of catapults. The ballista, built for Philip of Macedon, was similar to a giant crossbow and, using tension provided by twisted skeins of rope, it could aim heavy bolts, darts, or spears. The trebuchet consisted of a lever and a sling and could be used to hurl large stones. The mangonel, credited to the Romans, fired heavy objects from a bowl-shaped bucket at the end of its single giant arm.

Catapults used as siege weapons were usually constructed on the spot because they were too cumbersome to move around. Sometimes beehives or carcasses of dead animals were catapulted over castle walls to infect those inside. The weapon reached Europe during medieval times and the French used them during their siege of Dover Castle in 1216 CE Cannons replaced catapults as the standard European siege weapon in the fourteenth century CE. **MF**

SEE ALSO: SPEAR, SLING, BOW AND ARROW, FORT, LEVER, CROSSBOW, CANNON, BALLISTIC MISSILE

↗ *In a stylized medieval illustration, a basic catapult is brought into play during the course of a siege.*

"The first stone . . . fell with such weight and force upon a building that a great part . . . was destroyed."

Marco Polo, *The Travels of Marco Polo* (*c.* 1298)

Stirrup

(*c*. 300 BCE)

The Chinese gain an important advantage in mounted warfare.

The oldest recorded account of a single metal mounting stirrup is a depiction found on a pottery shard uncovered from a tomb in western China belonging to the Jin Dynasty and dating to around 300 BCE. The stirrup was at first used primarily as a tool to assist the rider in mounting his horse.

China at that time was constantly plagued by threats of mounted warfare from its northern nomadic neighbors. Considering that the Chinese developed the harness and horse collar a thousand years before their arrival in Europe, and that they had an established expertise in metal casting, it is not surprising that stirrups appeared among China's elite mounted cavalry. Their use of a single stirrup for mounting soon evolved toward using stirrups in pairs to provide a stable foundation for riding and fighting in war.

With the arrival of stirrups, the cavalry became a dominant military instrument; they fundamentally altered the approach to mounted warfare. A pottery horse dating to 322 BCE excavated from a tomb near the town of Nanjing in China's eastern Jiangsu Province is the earliest known evidence of stirrups deliberately forged and used as a pair.

Considering man's long dependence on the horse for transportation and communication, and its strategic importance in warfare, the invention of the stirrup came relatively late in history, although circumstantial evidence points to its appearance in the Middle East as early as 850 BCE. Its invention preceded a huge leap forward in the communication between and migration of cultures. **BS**

> "*The stirrup . . . enabled the horseman to become a better archer and swordsman.*"

Professor Albert Dien, historian

SEE ALSO: METALWORKING, BUCKLE, SADDLE, HORSESHOE, HORSE COLLAR

◤ *A Chinese bronze stirrup of the seventh century CE, an elegant realization of the important invention.*

Saddle

(*c.* 300 BCE)

Horsemen end the age of bareback riding.

It is unclear when humans first began to domesticate and ride horses—evidence from cave paintings in France suggests that horses might have been bridled as long ago as 15,000 BCE. But while early riders had the use of bits, bridles, and harnesses to control their mounts, they sat uncomfortably on little more than folded blankets or cloth, or rode bareback. Asian horsemen created a felt and wood saddle around 300 BCE, but it was not until around 100 CE that riders gained a saddle that offered genuine comfort.

The first padded, framed saddles were developed in Han Chin sometime between 25 and 220 CE. They consisted of a wooden frame covered in a stiff material

> *"Every occasion will catch the senses of the vain man and with that . . . saddle you may ride him."*
>
> Sir Philip Sidney, politician

such as leather, padded with cloth and shaped for comfort. To ensure a good ride, the pommel, or front, and the cantle, or rear, of the saddle were raised above the seat. What began as a simple but effective means of sitting on a horse soon became a status symbol, as riders decorated the leather of their saddles with inscribed designs and personal emblems and fashioned them with intricate ivory and other inlays.

Although the saddle had a great effect on horsemanship, its full effect was not at first realized, for the rider still remained insecure perched on his seat. It was not until the invention of the stirrup shortly afterward that the saddle truly came into its own. **SA**

SEE ALSO: METALWORKING, BUCKLE, STIRRUP, HORSESHOE, HORSE COLLAR

Moldboard Plow

(*c.* 300 BCE)

The Chinese transform a farm implement.

The simple moldboard plow was one of the most significant developments in history, but the name of its inventor is lost in time.

When humans first began tilling their fields, they would simply drag a stick or a hoe through the soil. The resulting furrows were perfect for planting seeds for cultivation. Once humans had domesticated the ox, around 7000 BCE, they were able to harness its pulling power to increase plowing efficiency. The oxen pulled a hoe contained by a wooden frame.

But the real breakthrough occurred in the third century BCE when the Chinese designed the *kuan*, or moldboard plow. This consisted of a hitch, to attach it

> *"The moldboard plow . . . buries almost all the old crop stubble, straw, and residue . . . "*
>
> Rick Kubik, farm safety expert

to an animal, and an asymmetric moldboard blade, which cut through the earth horizontally, with the added benefit of slicing through the roots of weeds. Once the earth had been cut horizontally, the forward motion of the curved plowshare pushed the soil against the blade, which turned the soil upside down before depositing it back on the ground, to the side of the new furrow. This aerated the soil, but it was the inverting of the earth that brought new advantages. Any surviving weeds were buried by the inverted earth and, especially in dry soil, nutrients and moisture were brought back to the surface. Now much larger areas could be farmed more efficiently. **DHk**

SEE ALSO: SCRATCH PLOW, STEEL PLOW, CAST-IRON PLOW

Lever (*c*. 260 BCE)

Archimedes explains how leverage works.

The lever was first described in 260 BCE by Archimedes (*c*. 287–212 BCE), but probably came into play in prehistoric times. A lever can be used to raise a weight or overcome resistance. It consists of a bar, pivoted at a fixed point known as the fulcrum. Extra power can be gained for the same effort if the position of the fulcrum is changed.

Levers may be divided into classes. First-class levers have the fulcrum in between the applied force and load, which are at opposite ends, such as with the seesaw. Second-class levers have the fulcrum at one end, and the applied force at the other, such as with a bottle opener. Finally, third-class levers have the effort in between the fulcrum and the load; for example,

---

"Give me a lever long enough and a fulcrum on which to place it, and I shall move the world."

Archimedes, mathematician and physician

---

tweezers have two class three levers that are pressed together to do the work for which they are designed.

The Egyptians used a lever in 5000 BCE for weighing, pivoting a bar at its center to balance weights and the objects to be weighed. Ramps and levers were also used to move stones higher up a structure, adapting the principle of the *shaduf*, which was developed in Egypt in 1500 BCE. This machine had a lever pivoted near one end with a water container hanging from the short arm and counterweights attached to the long arm. Several times a person's weight could be lifted by pulling down the long arm. **MF**

SEE ALSO: PULLEY, CRANE, WINCH, ARCHIMEDES SCREW, COMPOUND PULLEY, MECHANICAL LOCK, REPEATING RIFLE

Nail (*c*. 250 BCE)

Handmade nails are individually forged.

Nails were among the first metal objects made by hand. In Roman times, any sizable fortress would have a workshop where workmen fashioned the metal items required by the army. Here, workmen called "slitters" cut up iron bars for the attention of "nailers," who gave them a head and a point.

Early nails were usually square in section, and the head of each was formed simply by turning over one end to make an L-shape. Such nails were expensive to produce, and they were so valued that people sometimes burned their houses when moving in order to retrieve nails from the ashes for reuse.

In 1590 water-powered slitting mills were introduced into England. After rolling the hot iron into

---

"It is a great advantage that every honest employment is deemed honorable. I am … a nail maker."

Thomas Jefferson, U.S. president, in a letter

---

sheets, each sheet was slit into long, narrow, square-sectioned bars by rollers that cut like shears. The flat, headless, machined bars continued to be finished off as nails and spikes by hand, often by blacksmiths producing them to order. This was the procedure until the advent of the nail-making machine at the end of the eighteenth century; by the end of the nineteenth century, the handmade nail industry was extinct.

Nails are made in a variety of forms. Most common, dating from the late nineteenth century, is the "wire nail," as distinct from "stamped nails," "pins," "tacks," or "brads." Nails are now available in many different sizes and shapes, with a variety of heads. **MF**

SEE ALSO: RIVET, SCREW, CLAW HAMMER, NAIL-MAKING MACHINE, STEAM HAMMER, PNEUMATIC HAMMER

Archimedes Screw (c. 250 BCE)

Archimedes raises water for irrigation by an ingenious means.

The Archimedes screw was first mentioned in the writings of Athenaeus of Naucratis in 200 BCE. He described the use of a screw mechanism to extract bilgewater from a ship named *Syracusia*, and attributed its invention to Archimedes.

Archimedes (c. 287–212 BCE) himself lived in Syracuse, Sicily, and was devoted to the exploration of mathematics and science. The polymath is thought to have spent time studying in Egypt, and the screw named for him is used in the Nile delta to this day, more than 2,000 years later, as a means to raise water from rivers for irrigation purposes.

The Archimedes screw consists of a helix within a hollow tube, the lower end of which is placed in a fluid. The screw is then rotated and the fluid is lifted up through the spiral chamber to the top of the tube. In ancient times this tool was applied throughout the Mediterranean for irrigation. It was especially used by the Romans in their water supply systems, and as a means of extracting water from their mines in Spain.

Technology took a backward step in the Dark Ages, but the Archimedes screw reappeared in the fourteenth century as a means of supplying public fountains with water. It was then largely superseded by reciprocating pumps, but came into its own in the 1600s for the reclamation of land from the sea, particularly in the Netherlands. Powered by windmills, they raised water from low-lying land up into canals.

The Archimedes screw is still used for drainage and flood control today, but also has many modern applications in oil pumping, sewage treatment, agriculture, and even cardiac medicine. **FW**

SEE ALSO: IRRIGATION, SEWAGE SYSTEM, SCREW, LEVER, COMPOUND PULLEY, WINDMILL

⬈ *A terracotta figurine of an Egyptian worker using an Archimedes screw for irrigation, first century BCE.*

"An enormous amount of water is thrown out . . . by means of a trifling amount of labor."

Diodorus Siculus, historian

Compound Pulley

(*c.* 250 BCE)

Archimedes introduces the block and tackle.

The simple pulley enables the user to lift a load more easily by changing the direction from which the force is applied. When the rope is fixed at one end and another pulley is added, the system provides a mechanical advantage by multiplying the applied force, making it possible to lift heavy loads. More pulleys can be added to the system, now known as a "compound pulley" system, further multiplying the effectiveness of the force applied.

As an indication of the benefit of the system, the addition of a second pulley to a one-pulley lifting mechanism halves the amount of force required to make the lift. A third pulley, properly rigged, reduces the amount of force required to a quarter.

In 250 BCE, the Greek scientist and inventor Archimedes (*c.* 287–212 BCE) adopted this principle by mounting several pulleys on the same axle to create a "block" that was much more convenient to use than a series of separate pulleys. A single rope—called the "tackle"—fixed at one end, can be threaded between a fixed block and around each of the pulleys ("sheaves") in a movable block so that the load on the system is shared between the ropes under tension.

According to Plutarch, whose account provides us with the earliest record of the block and tackle, Archimedes used his new invention single-handedly to move a whole warship. Whether this is true we cannot know, but the block and tackle has certainly been an essential tool for lifting and moving heavy loads ever since. It can be found today in the cranes and lifting gear used in construction, engineering, freight loading, and warehousing as well as on almost every yacht and sailing boat, including small dinghies with the simplest of rigging. **EH**

SEE ALSO: BRAIDED ROPE, PULLEY, WINCH, LEVER, CRANE, ARCHIMEDES SCREW

Pipe Organ

(*c.* 240 BCE)

Ktesibios reinvents a primitive instrument.

Long before the development of the pipe organ, its essential musical element—a set of pipes of different sizes that resonate at different pitches when air is passed through them—existed in the form of the syrinx. This simple instrument was used widely throughout the eastern Mediterranean region.

However, in 240 BCE, Ktesibios (*c.* 285–222 BCE), a Greek engineer, developed a way of supplying a steady flow of air to the pipes. He attached them to a closed box into which air was pumped using pressurized water, creating and maintaining steady air pressure in the box. The pipes were opened or closed to the air source by a simple switch system operated

> *"All one has to do is hit the right keys at the right time and the instrument plays itself."*

J. S. Bach, composer and organist (attributed)

from a keyboard. This device was originally called *hydraulis*, later *organum*, and produced loud sounds, clearly audible outdoors and ideal for use at games and processions. The instrument appears in paintings, mosaics, and writings throughout Byzantium.

The pipe organ is the oldest musical instrument still used in classical music. It was developed in the Byzantine Empire, and was adopted by the Christian church in the first century CE. The hydraulic system was replaced by bellows in the following century. **EH**

SEE ALSO: METALWORKING, ELECTRIC GUITAR, SPEECH SYNTHESIS, ELECTRONIC SYNTHESIZER

⮕ *Bellows supply air to an organ in a thirteenth-century psalter illustration of musicians playing together.*

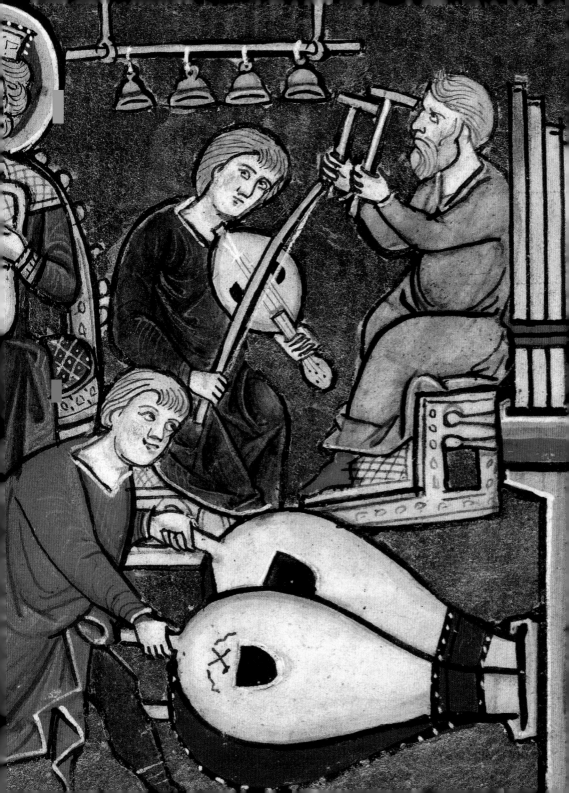

Dry Dock (*c. 200* BCE)

A Phoenician finds a new way to launch ships.

The dry dock was invented in Egypt by a Phoenician, some years after the death of Ptolemy IV Philopator, who reigned from 221 to 204 BCE. His method of launching a ship consisted of digging a trench under it close to the harbor, then making a channel from the sea to fill the excavated space with water.

Dry docks continued to be used throughout antiquity. In Europe the first dry dock was commissioned in 1495 by King Henry VIII at Portsmouth, England. Dry docks are mainly used for the maintenance and repair of ships, and more rarely for their construction because the time required to build a ship is so long. While early dry docks were often used for launching ships, slipways are more frequently used in modern times.

There are two types of dry docks: graving docks, where "graving" is the term for scouring a ship's bottom, and floating dry docks. The graving dry dock consists of a water-filled narrow basin, usually made of concrete, with gates that can be opened and closed, into which a vessel may be floated. The water is then pumped out, leaving the vessel supported on blocks, so that the ship can be serviced. When the work is finished, water is let back into the dock and the ship refloated. Earlier dry docks were built in the same shape as the ships that were to be docked there, but more recently, graving docks have been built in a box-shape, to conform to boxier ship designs.

A floating dry dock is usually built of hollow steel. The dock is first submerged, the ship is brought into its channel, and the dock is then floated by removing ballast from the hollow floor and walls. The fully drained dock supports the craft on blocks attached to the dock floor. Floating dry docks are usually operated in sheltered harbors to avoid wave damage. **MF**

SEE ALSO: ENCLOSED HARBOR, STEEL, REINFORCED CONCRETE, AUTOMATIC FLOUR MILL

Windmill (*c. 200* BCE)

China and Persia harness wind power.

The early history of the windmill is much contested, and it is not known for sure when or where it first appeared. Some date it as far back as Babylonia in the seventeenth century BCE, while others claim it was not until 200 BCE that wind power was used to pump water in China and mill cereal in Persia. It is, however, reliably documented that windmills were widespread in Persia by the seventh century CE. They ground grain between millstones rotated by wind blowing on woven reed sails mounted around a vertical axis.

The earliest European windmills were built in France and England in the twelfth century and are thought by some to have been the result of a transfer of knowledge from the returning Crusaders. However,

> *"There are, indeed, few merrier spectacles than that of many windmills bickering together…"*
>
> Robert Louis Stevenson, writer

the horizontal-axis mills of northern Europe bore scant resemblance to their Persian counterparts and were probably an entirely separate development.

Windmills became increasingly practical with the advent of fantails to turn them automatically into the wind, and improved sails for efficiency and control. Until the dawn of the Industrial Revolution, windmills sprang up in their thousands throughout Europe and North America for grinding foods, sawing wood, and pumping water for drainage, irrigation, or desalination. Since the late twentieth century, wind-driven technology has gained economic importance in the form of wind turbines for electricity generation. **FW**

SEE ALSO: SAIL, QUERNSTONE, WATERMILL, TIDAL MILL

Astrolabe (*c.* 150 BCE)

Hipparchus develops a calculator of astronomical positions.

An astrolabe is a device with which astronomers solved problems relating to time and the position of the sun and stars in the sky. Its main element is a two-dimensional circular stereographic projection of the hemispherical sky. The projection was most probably formalized by the Greek astronomer Hipparchus (190–120 BCE), who worked on the island of Rhodes.

The astrolabe was suspended vertically and a cross-arm was used to measure the altitude above the horizon of the sun (in the day) and bright stars (at night). The rim of the astrolabe is marked off in months, days, and hours, and most astrolabes have a series of longitude-specific circular main plates each marked off with lines of constant altitudes, azimuths, declinations, and right ascensions. Fitting over the plate is a cutaway fretwork (a "rete") that delineates that portion of the celestial sphere that can be seen above the horizon at any specific time at a specific latitude. The rete contains pointers that mark the positions of about twelve of the brightest stars.

By noting the elevation of the sun, or these bright stars, the traveler can tell the time of day or night. By noting how well the observed star positions correspond to a specific plate, the travelers can estimate their latitude. The stellar positions also enable the accurate establishment of north on the horizon.

Claudius Ptolemy (*c.* 85–165 CE) wrote about the stereographic projection and probably owned an astrolabe. The astrolabe was popular in the Islamic world because it enabled Muslims to ascertain prayer times and the direction of Mecca. The oldest existing instruments date from the tenth century CE. **DH**

SEE ALSO: MAP, MAGNETIC COMPASS, GYROCOMPASS, GLOBAL POSITIONING SYSTEM (GPS)

↗ *This astrolabe was used by Egyptian astronomers to measure the altitude of stars and planets.*

" . . . when I seek out the massed wheeling circles of the stars, my feet no longer touch the Earth . . ."

Claudius Ptolemy, mathematician and astronomer

Antikythera Mechanism (*c.* 150 BCE)

A controversial astronomical calculator is attributed to the Greeks.

One of the most remarkable inventions of the ancient world came to light in 1900 when a Greek sponge diver discovered the wreck of an ancient Greek or Roman cargo ship that had sunk off the Greek island of Antikythera around 80 BCE. Among the objects recovered from the wreck was a geared mechanism that, from the shape of its inscribed Greek letters, dated to between 150 and 100 BCE.

The mechanism has more than thirty gearwheels and three main dials. When reassembled, it formed a scientific instrument that could be used to calculate the astronomical positions of the sun, moon, and the five planets then known. When a date was entered via a crank, now lost, the mechanism calculated the position of the sun, moon, or planet. The Antikythera mechanism, as it is known, is the first known geared device and thus the first known clockwork mechanism, and the oldest known scientific instrument.

The concept of differential gearing was not rediscovered until the sixteenth century, while the complexity and miniaturization of its parts are comparable to the finest eighteenth-century clocks. Crucially, while the mechanism is based on theories of astronomy and mathematics developed by Greek scientists of the day, it has a heliocentric rather than the then current geocentric view of the solar system and presumes a theory of planetary motion and a knowledge of the laws of the gravity that were not known at the time. Recent research has revealed it also organized the calendar in the four-year cycle of ancient Greek games. Whoever built this mechanism would have been astonishingly ahead of his time. **SA**

"This device is . . . the only thing of its kind. The design is beautiful, the astronomy is exactly right."

Professor Michael Edmunds, Cardiff University

SEE ALSO: WATER CLOCK, SHADOW CLOCK, TOWER CLOCK, PENDULUM CLOCK, ATOMIC CLOCK, DIGITAL CLOCK

◩ *Fragments from the Antikythera Mechanism found by divers off the coast of Antikythera, Greece.*

Parchment (*c.* 150 BCE)

Animal skin becomes a writing material.

According to Pliny the Elder, parchment was developed in the city of Pergamum (now Bergama, Turkey) because a king of Egypt, fearing that Pergamum's great library might overshadow that of Alexandria, stopped exporting papyrus to the city.

It seems more likely that parchment already existed and was refined at Pergamum. Also, this was not the first time animal skin had been written on. Leather had been used occasionally, possibly dating back to circa 2000 BCE. However, previous attempts involved tanning the leather and produced documents that were slightly hairy, stiff, and one-sided. Parchment, on the other hand, was made from the skins of sheep, calves, and goats that were cleaned and, crucially,

---

" … when … Ptolemy suppressed the export of paper, parchment was invented at Pergamum …"

Pliny the Elder, *Natural History*, Book 13

---

scraped thoroughly. Both sides of the smooth, flexible surface were ideal for writing and ultimately allowed sheets to be sewn together into "books" that were far easier to read than papyrus scrolls. Although papyrus was cheaper than parchment, it was Europe's favored surface up until the fourteenth century's advances in paper making, especially for medieval illuminated manuscripts, such as the stunning *Très Riches Heures* of the Duc de Berry of the early 1400s.

The finest parchments, especially those made from the skins of very young, or even unborn, animals, were called vellum. The term is often used today for any kind of high-quality special paper. **AK**

SEE ALSO: PAPER, INK, CARBON PAPER, KRAFT PROCESS, LIQUID PAPER

Belt Drive (*c.* 100 BCE)

Romans connect a belt drive to a treadmill.

The belt drive is a vital component of most modern machines. In it, a ring of a flexible material is wound around two or more shafts. As one shaft rotates, the belt moves, causing other shafts to rotate as well. This simple pulley device has long been a versatile and reliable means of transferring power.

In 100 BCE, while constructing Haterii's Tomb in Rome, workers used a treadmill-powered crane to lift heavy material. This was a historic moment for mechanics. In 1203, French innovators replaced the human workers who had been powering belt-driven technology with a team of donkeys.

Introducing animal power was far from the final stop for the belt drive. Water-powered mills used belt drives to harness water power, and Industrial Revolution-era factories employed belt drives, called line shafts, to transfer power throughout the factory.

Belt drives are also commonly used in engine designs. Belt drives can be found in most mechanical movers, from motorcycles to helicopters. Automobile engines usually contain belt-drive systems called V-belts and serpentine belts. These systems redirect and disperse engine power to accessories.

V-belts, named for their triangular "V" profile, are generally used to power the vehicle's air-conditioning compressor, alternator, power-steering pump, and water pump. They are frequently called fan belts.

A serpentine belt, an alternative to using a combination of several V-belts, has a longer life than the combined V-belts. A serpentine belt system uses a single, long belt to power the same number of accessories as numerous V-belts. Its name comes from its complex, snakelike path around multiple shafts. A spring-loaded pulley is used to keep the serpentine belt under optimum tension. **LW**

SEE ALSO: BRAIDED ROPE, PULLEY, WATERMILL, MOTORCAR

Watermill (*c.* 100 BCE)

Greeks use the energy of running water to grind grain.

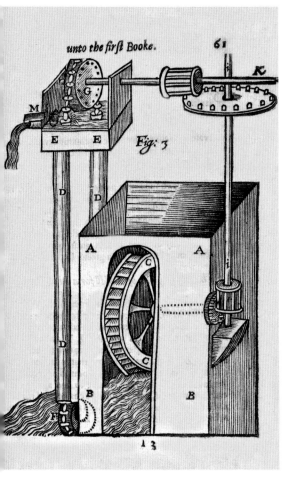

The earliest reference to watermills is found in the writings of Antipater of Thessalonica, describing their use for the grinding of grain in the first century BCE. These ancient Greek devices consisted of a millstone mounted on a vertical axis and rotated against a stationary stone bed by a horizontal paddle wheel spinning in a fast-flowing stream. This type of watermill has also been discovered throughout Ireland, Scandinavia, and China.

The Romans were the first to devise a more efficient and versatile machine with a horizontal axis, which may have been inspired by ancient Eastern waterwheels, originally used for lifting water. The medieval Islamic nations embraced the watermill from the seventh century, building mills in bridges and on the sides of moored ships, or channeling water from dams to supply them. They were used to make pulp for paper, saw wood, grind cereal, and crush sugar cane and mineral ores.

The nineteenth century brought a surge in the need for industrial power in northern Europe and North America, and new cast-iron waterwheels fed the demand. These were eventually replaced by steam engines, fueled by coal brought on the canals.

In northern England, 1880, water was used for the first time to generate electricity through a new type of turbine initially conceived in 1826 by Benoit Fourneyron. More plants followed, and in the 1920s 40 percent of electricity in the United States came from hydropower. Other countries, including Norway and Brazil, now meet almost all their electricity needs by harnessing the energy of flowing water. **FW**

SEE ALSO: **WINDMILL, TIDAL MILL, AUTOMATIC FLOUR MILL, CANAL, DAM, STEAM TURBINE, WIND TURBINE GENERATOR**

"Rest your mill-turning hand ... the nymphs [now carry out] the chores your hands performed."

Antipater of Thessalonica, epigrammatist

⬚ *In this 1635 engraving, the waterwheel powering this mill is driven by the ebb and flow of tide water.*

Blown Glass (*c.* 100 BCE)

Syrians first shape glass with the aid of a blowtube.

It was the Syrians who first learned to blow molten glass through a hollow metal tube and shape it into intricate forms. Although the technique for producing glass had existed for about two and a half millennia, it was only in approximately 100 BCE that the hazardous art of glassblowing—using glass melting at a few thousand degrees Fahrenheit—was mastered.

Glassblowing is the process for forming glass into a desirable shape, and this ability to form iconic, practical, and elegant shapes out of glass has been of incalculable value and practical benefit to society. Glassblowing machines have now largely replaced the Syrian specialists, but the science behind the technique remains the same. Molten glass is first introduced to the end of a hollow tube. A bubble of air is then blown through the tube, and as the bubble passes out of the tube a covering of molten glass forms around the sphere of air. This glass-covered bubble, still attached to the tube, is either placed within a mold of the required form and enlarged through further blowing, or sculpted with tools into the desired shape. The glass is then allowed to cool slowly to complete the process. It is the fact that glass has no set melting or freezing points that makes glassblowing possible; as the temperature rises or falls, the state of the glass gradually changes.

The rise of the Roman Empire at around the same time as the beginnings of glassblowing greatly facilitated the proliferation of the art. By adding manganese oxide to the mix, the Romans also discovered clear glass in around 100 CE, which was used for architectural purposes. **CB**

SEE ALSO: **GLASS, GLASS MIRROR, SPECTACLES, MICROSCOPE, TELESCOPE, CONTACT LENSES**

↗ *French glass blowers from the fifteenth century illustrated in* Tractatus de Herbis *by Dioscorides.*

"Blowing allowed for previously unparalleled versatility and speed of manufacture."

Rosemarie Trentinella, Metropolitan Museum of Art

In the two centuries after Julius Caesar's assassination in 44 BCE, the Roman Empire reached its peak. Also during this period and until the end of the Middle Ages, the Chinese made great contributions to the development of human culture, inventing indispensable items such as paper, the wheel, and the toothbrush. By the time of the French Revolution, the Dark Ages were over, the steam engine had been created, and technology roared into a new era of strength and power.

⊞ The need for water gives rise
to mills and aqueducts.

From ROME *to*
REVOLUTION

Odometer (*c.* 27 BCE)

Vitruvius makes measuring distances simpler.

Measuring the distance between two places is a basic task in cartography. The earliest method was to walk and count the number of times a specific foot hit the ground—a thousand right steps, for example, made a mile (from the Latin "mille," meaning one thousand).

The Roman architect and engineer, Vitruvius (*c.* 75 BCE–*c.* 15 BCE), mechanized the process. Around 27 BCE he devised a wheelbarrow-type device that dropped a pebble into a container every time its large wheel of known circumference rotated once. At first this was pushed along by hand, but it was soon incorporated into a chariot, the standard chariot wheel being 4 feet (1.2 m) in diameter. This wheel turned 400 times in a Roman mile. Needless to say, the smoothness of the road was important. The device was described by Hero of Alexander in chapter thirty-four of his book *Dioptra*.

Around 300 CE the Chinese—some sources suggest Chang Heng—devised a similar, but more musical, instrument. Every time the road wheel of a special coach rotated once, a pin moved a tooth on an internal cog wheel. Every complete rotation of the cog wheel activated a stick that banged a drum. Every tenth drum beat was replaced by a sounding gong. Distances between towns could be easily measured in this way to an accuracy of a tenth of a mile.

Early motor cars had odometers (or mileometers) fitted to one of the road wheels, these having separate gears that registered distances of 1, 10, 100, 1000 miles, and so on. Measured distances were a function of the tire pressure. Since 1980, cars have had odometers that indicate the number of miles traveled up to 999,999.

Simple hand-pushed odometers are still used today by city surveyors, and these are sometimes called waywisers or perambulators. **DH**

SEE ALSO: SURVEYOR'S PERAMBULATOR WHEEL, WHEEL AND AXLE, SPOKED-WHEEL CHARIOT, WHEELBARROW, MOTORCAR

Ball Bearing (*c.* 40)

The Romans reduce rotational friction.

Ball bearings are a low-cost method of allowing different parts of a mechanism to rotate past each other without much energy loss from friction. They have many uses, including in bicycles, gyroscopes, electric motors, and turbines. They did not come into general use until the Industrial Revolution, but the concept has been around for more than 2,000 years.

Roman Emperor Caligula (12–41 CE) had two large ships built at Lake Nemi. When the remains of these ships were recovered in the early 1930s, marine archeologists found the earliest known ball bearings. There were two types found—bronze spheres and wooden balls. The wooden ball bearings supported a rotating table, similar to a lazy Susan or dumbwaiter.

" . . . the balls . . . will touch in one point only between the load and its resistance. . . . "

Leonardo da Vinci, *The Madrid Codices* (*c.* 1490)

Prior to this discovery, historians had believed that Leonardo da Vinci invented ball bearings.

Ball bearings became so widely used in factories, vehicles, and other machinery, that during World War II, the Allies made a concerted effort to bomb German ball bearing plants in order to disrupt the German war effort. The Germans had astutely stockpiled millions of bearings and were able to continue supplying factories despite ball bearing production being halved.

These days, for applications with high load, speed, and/or precision requirements, ball bearings are increasingly being replaced by fluid bearings, which use a layer of gas or liquid to support the load. **ES**

SEE ALSO: SELF-ALIGNED BALL BEARING, ELECTRIC MOTOR, MOTORCAR, SAFETY BICYCLE, MOTORCYCLE

Claw Hammer (c. 79)

The Romans devise a tool used to pound in, and remove, nails.

Hammers—tools for striking or pounding—have been around for millions of years in the form of specially shaped stones used to break or shape other stones, bones, or wood. They are most commonly associated with woodworking. But after the invention of the nail, someone realized it would be very useful to be able to insert and remove nails with the same tool. Nails were valuable, and a carpenter who hammered one at the wrong angle would have rescued and reused it. Thus, the claw hammer was born.

A claw hammer has a two-sided head attached to a handle and can be said to be roughly T-shaped. One side of the head is the striking surface and is usually flat. The other side is a rounded or angled wedge and is used for removing nails. Archeologists found an iron claw hammer at Pompeii that was buried in the eruption of Mount Vesuvius in 79 CE. First-century Romans were skilled at making nails; archeologists found almost 900,000 at the Roman fortress of Inchtuthil in Scotland that the garrison had abandoned in the late 80s CE.

Various types of claw hammers were patented between 1867 and 1941. For many years, the hammer was the principal tool for carpenters and builders. However, since the invention of the nail gun in the 1950s, builders have increasingly been relying on them instead of hammers for their nailing needs because they are easier, faster, and more fun to use. The fact that they provide inexperienced home contractors with the capability to leave a plethora of nails in our walls is, of course, merely a side effect of the new technology. **ES**

SEE ALSO: NAIL, NAIL MAKING MACHINE, STEAM HAMMER, PNEUMATIC HAMMER

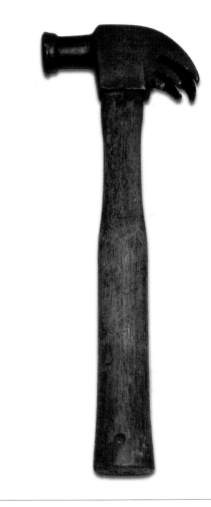

↗ *This triple claw hammer is displayed at the world's only Hammer Museum in Haines, Alaska.*

"A worker may be the hammer's master, but the hammer still prevails . . . ."

Milan Kundera, writer

Horseshoe (*c.* 100)

The Romans invent the "hipposandal."

Horses have played central roles in the histories of various powerful empires, and their employment was boosted by the invention of the horseshoe. Protecting horses' hooves from wear and tear on hard or rough surfaces allowed for longer journeys when the horse was the common mode of transport and a domestic working animal. It also made them more effective when used in the cavalry as part of a military campaign.

The precise date of its invention is unknown, but the Roman poet Catullus mentions a mule losing its shoe in the first century BCE. Evidence from Roman regions to the north of the Alps suggests that horses from what is now Germany may have been the first to use horseshoes regularly, from around 100 CE.

"Horseshoing, very likely, was invented by different nations at about the same period. . . . "

Scientific American (1891)

Over the years horseshoe design has improved from the "hipposandal" used by the Romans—which had a solid bottom and was strapped to a horse's hoof—to the U-shaped metal plate used today. The earliest known mention of the iron horseshoe is in 910. The weight and shape of early horseshoes varies on their provenance, and the climate and terrain in which the horses had to move. Blacksmiths and farriers made and fitted horseshoes using nails, and their skills helped develop metallurgy during medieval times. Today, horseshoes are often made from steel and aluminum, but also come in copper, titanium, rubber, or plastic depending on what the horse is used for. **CB**

SEE ALSO: METALWORKING, NAIL, SADDLE, STIRRUP, HORSE COLLAR

Dome (*c.* 100)

Roman engineers solve an architectural puzzle.

Domes, like arches, present problems for architects and engineers: They remain unstable until the final stone is put in place and have to support their own weight without collapsing. It was the Romans who in circa 100 first solved these technical problems when they managed to build a true dome, that is, an unsupported half-sphere.

The greatest of these is the Pantheon in Rome, which was erected in circa 123. Roman engineers tackled the problem as if it was a series of circular barrel vaults, or arches, arranged in a circle across a central point, and used concrete as a building material, in this case a mixture of lime, pumice, pieces of rock, and volcanic ash. A template for each arch was erected

". . . my own opinion is that the [Pantheon's] name is due to its round shape, like the sky."

Cassius Dio, *History of Rome* (*c.* 220)

on scaffolding and served as the mold for the poured concrete. The dome was built up in sections, with heavier concrete being used at the thicker base and lighter concrete toward the thinner top. The weight of the dome is concentrated on a ring at the top of the structure surrounding an opening that lets in the light, and reduces the weight at the top. The result is to push the weight of the dome toward its base and down to the floor below. Once the dome was completed and the concrete dry, the supports could be removed. **SA**

SEE ALSO: CARPENTRY, ARCHED BRIDGE, FIRED BRICK, REINFORCED CONCRETE

➡ *The ground plan and facade of the Pantheon, Rome, showing the immense unsupported dome.*

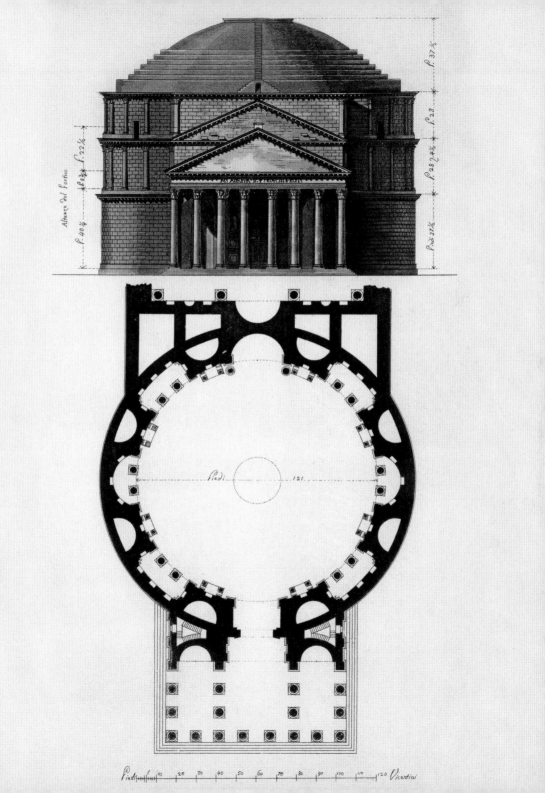

Altezze del Portico

Piedi 121

Vicentini

Piedi 10 20 30 40 50 60 70 80 90 100 110 120

Suspension Bridge (*c.* 100)

The Chinese suspend a bridge with chains.

Primitive suspension bridges, in the form of vines and fiber ropes, have been used for many thousands of years throughout Asia, Africa, and South America, such as those by the Incas. It is thought that the iron chain first replaced these frailer materials in China in *c.* 100.

The walkways of early catenary bridges were directly fixed onto the chains that spanned a valley, but in fourth-century India the road deck was instead suspended from the main cables to create a horizontal pathway that was more easily negotiable than its sloping predecessors.

Basic suspension bridges were used in military campaigns in Europe, but the first permanent example in the West was a primitive catenary bridge built over

"I have no quarrel with you, good Sir Knight, but I must cross this bridge."

King Arthur, *Monty Python and the Holy Grail* (1975)

England's River Tees circa 1741. The idea caught on in the United States, and in 1801 James Finley built the first modern suspended-deck suspension bridge across Jacob's Creek in Pennsylvania.

Wire cable eventually replaced chains, and the first permanent suspension bridge to use wire cable was the St. Antoine Bridge built in Geneva, Switzerland, in 1823. Steel was used in early constructions but the cost was prohibitively high until the Bessemer steel-making process was developed in 1855. John A. Roebling made great improvements in the reliability of bridges, and his Brooklyn Bridge of 1883 remains an iconic suspension bridge. **FW**

SEE ALSO: BRAIDED ROPE, ARCHED BRIDGE, PONTOON BRIDGE, TRUSS BRIDGE, CANTILEVER BRIDGE, BESSEMER PROCESS

Paper (105)

Ts'ai Lun initiates papermaking.

In 105 when Ts'ai Lun (50–121), a courtier in the Chinese Imperial court, invented paper, little did he realize that he was opening one of the most epoch-making chapters in history. He refined and popularized the process of mixing tree fibers and wheat stalks with the bark of a mulberry tree, then pounding them together and pouring the mixture onto a woven cloth to create a lightweight writing surface. His blended, fibrous sheets were an improvement over bamboo and wood, which were awkward and heavy, and silk, which was expensive. Successive Chinese dynasties conspired to keep his invention secret, and it was not until the start of the seventh century that papermaking techniques began to appear in Japan and Korea.

"Paper can convey a private warning, a public threat, secret temptation, open defiance . . . ."

Eric Frank Russell, *Wasp* (1957)

With the capture of Chinese paper merchants by Arab soldiers during the Battle of Talas in 751, the knowledge of papermaking soon spread across the Arab world, and first appeared in Europe in Moorish Spain early in the twelfth century.

In Europe, paper began a centuries-long battle for prominence with parchment until the invention of movable type in the fifteenth century led to a steep rise in literacy and a demand for the production of books that parchment could no longer satisfy. The eighteenth century saw paper made from linen and cotton rags that were replaced by wood and other vegetable pulps in the early nineteenth century. **BS**

SEE ALSO: INK, PARCHMENT, KRAFT PROCESS (MODERN PAPERMAKING), LIQUID PAPER, ELECTRONIC PAPER

Wheelbarrow (c. 231)

Zhuge Liang's invention makes light of moving heavy loads.

The wheelbarrow is reputed to have been invented by a Chinese chancellor, Zhuge Liang (181–234) during the Han Dynasty, who used the device in military campaigns to transport supplies for injured soldiers. It was said to have been kept secret because of the advantages it gave the Chinese over their enemies.

It was also used for early Chinese agriculture, which was said to have been thirty times more efficient than that in Europe. Designed to transport heavy loads, wheelbarrows are now used in the construction industry as well as in gardening.

A wheelbarrow is a small cart, with one or two wheels, designed to be pushed by one person using two handles at the rear. Chinese wheelbarrows often had two wheels, and the Chinese sometimes attached sails to them so that the wind could take part of the load, as is recorded by European travelers of the sixteenth century.

Wheelbarrows were seen in Europe in the twelfth century, as evidenced by a stained-glass window at Chartres Cathedral in France, dating around 1220. This is believed to be the earliest image of a wheelbarrow in Western Europe. A manuscript illumination of 1286 also shows the European wheelbarrow. This design differed from the Chinese one in that the wheel was moved from the center to the front of the box, and the propelling power was at the rear.

Designs of wheelbarrow have been developed in recent years, such as British inventor James Dyson's Ballbarrow of 1974 that uses a spherical plastic wheel, and is easier to steer than those models based on the conventional wheel design. **MF**

SEE ALSO: **WHEEL AND AXLE, SAIL, CART, LUBRICATING GREASE, BALL BEARING**

⬈ *In China the single-wheeled wheelbarrow was used to convey wealthy individuals as well as goods.*

"To carry stones and rakings of garden to places, appointed to receive 'em or, to carry earth. . . ."

Francis Gentil, *The Solitary Gardener* (1706)

Trebuchet (*c.* 500)

The Chinese invent a rock-hurling machine.

"The trebuchet was the dominant siege weapon ... lasting 100 years after the [arrival] of gunpowder."

University of Arkansas website

⤴ *A small wooden trebuchet, erected at the fortified village of Castelnaud in the Perigord region, France.*

➡ *An illustration, derived from a fifteenth-century engraving, shows the trebuchet in action.*

The trebuchet was an ancient form of artillery that first appeared in China in the fifth century BCE. The weapon of mass destruction of its time, it was an improvement on the catapult. Unlike other catapults, such as the mangonel that uses twisted rope to provide power, the trebuchet uses a counterweight to provide its force. The trebuchet dominated long-range artillery until the sixteenth century.

The main arm of a trebuchet is attached to a fulcrum in such a way that the end holding the counterweight is much closer to the fulcrum than the end that holds the projectile. As the counterweight is released, the short end of the arm drops downward rapidly. Because of the longer length on the other side of the fulcrum, the end holding the projectile is flung upward in an arc at a much faster rate, giving considerable velocity to any missile thrown from it. This effect is usually increased by adding a sling to the firing end that also swings outward while firing, effectively increasing the power of the trebuchet.

Trebuchets were far more powerful and accurate than earlier catapults and, by the time they appeared in the Mediterranean region, around 1100, they had evolved into terrifying machines of war that could be easily maneuvered to lay siege to castles with strong fortifications. They were primarily used to batter down stone walls by repeatedly firing at points of weakness. Boulders weighing as much as 300 pounds (140 kg) were pounded into fortified walls.

The trebuchet also served as an early form of biological warfare in that ordure, diseased animals, and rotting corpses were launched into besieged towns and forts. Not only did these spread disease, but having former comrades raining down from the sky had a powerfully negative impact on morale. **DHk**

SEE ALSO: **SLING, CATAPULT, BOOMERANG, ATLATL, FORT, GUNPOWDER, ROCKET, CANNON, BALLISTIC MISSILE**

Horse Collar (*c*. 500)

A Chinese device increases horse power.

Most inventions are remarkable because they appear to be ahead of their time, revelatory events that transform the world into which they appear. The horse collar seems to be somewhat the opposite, for it is difficult to see why it was not invented earlier.

The problem to be solved is obvious: A horse wearing a simple harness can pull a load weighing about 135 pounds (60 kg), but any heavier load forces the harness on to the horse's windpipe, and restricts its ability to breathe. Therefore, while horses had been domesticated, mounted, saddled, and harnessed by around 100—and so could be ridden for pleasure, work, or warfare—their role as a beast of burden was necessarily limited, and was to remain so for another 400 years.

It was not until circa 500 that a Chinese camel driver had the bright idea of devising a padded collar, which was quickly used also on horses. It took the form of a rigid construction that sat low on the horse's chest and rose round its neck to rest on its withers or shoulders. The top of the collar supported a pair of curved metal or wood hames, to which the harness was attached. The collar reduced pressure on the windpipe and allowed the horse to use its full strength, thus allowing it to push forward with its hindquarters into the collar rather than pull the weight with its shoulders.

This new design of collar reached Europe circa 920, and soon revolutionized agriculture. Horses replaced oxen as the main beasts of burden, pulling plows, harrows, harvesting machines, and other agricultural implements, as well as farm carts and wagons. **SA**

SEE ALSO: SCRATCH PLOW, CAST-IRON PLOW, STEEL PLOW, SADDLE, STIRRUP, HORSESHOE, CART

⬆ *This carved wooden horse collar, dating from around 1300, is in the History Museum, Stockholm, Sweden.*

Quill Pen (*c. 580*)

Spanish record the "pen that wrote history."

The first specific reference to a quill pen is found in the writings of St. Isidore of Seville around 580, although pens made of bird feathers are likely to have been used even earlier. The quill pen was the main writing tool in the Western world until the invention of the fountain pen in the nineteenth century. The quill's development was assisted by the rise of Christianity because its fine script was suitable for the promulgation of religion, as well as lending itself to other documents in increasingly dense text.

Although the outer wing feathers of many birds could be used, those of the goose and crow were preferred. A slit would be made in the base of the quill to allow ink to flow to the nib, with goose quills especially adept at holding the ink. The composition and size of goose quills also allowed the nib to be sliced to a broad edge, and then sharpened to an extremely fine point. Quill pens quickly became blunt, and needed to be recut frequently in order to maintain their edge. The feathers were taken only from live birds—those taken from the left wing best suited the right-handed scribe because they curved outward to the right side. Each bird could supply around ten good quills. The quill was of such importance that goose farms were prevalent across Europe.

The United States Supreme Court began a tradition of using quill pens in 1801, and its Chief Justices continued to write with them until well into the 1920s. Today white quills are still placed on attorneys' tables when the court is in session. By 1850, quill pen usage was starting to decline as the quality of steel nibs improved. **BS**

SEE ALSO: PARCHMENT, INK, PAPER, FOUNTAIN PEN, GRAPHITE PENCIL, BALLPOINT PEN

⬆ *The quill pen was the principal writing instrument from the sixth century to the mid-nineteenth century.*

Toilet Paper (*c. 589*)

The Chinese revolutionize personal hygiene.

The earliest recorded use of toilet paper comes from China in the sixth century when government official and scholar Yan Zhitui warned against using paper printed with philosophical utterances for the wiping of bottoms. By the end of the fourteenth century, when the rest of the world was using water, the Chinese were producing more than 700,000 sheets of aromatic toilet paper a year for the Imperial court.

Prior to the advent of the first commercially packaged, premoistened toilet paper by Joseph Gayetty of New York City in 1857, how people used to clean themselves depended to a large degree on where and how they lived, and their standing in society. Coconut shells were used widely throughout

> *"Paper on which there are … the names of sages, I dare not use for toilet purposes."*
>
> Yan Zhitui, *The Family Instructions of Master Yan* (589)

an egalitarian Hawaii, while lace and hemp proved popular within the French aristocracy. In ancient Rome combining rosewater with wool was common among the upper classes; colonial Americans used corncobs and old almanacs; while the ends of old anchor cables proved an unpopular but not uncommon option for Spanish and Portuguese maritime sailors of the seventeenth and eighteenth centuries.

Toilet paper was manufactured on a roll for the first time in the United States by the Scott Paper Company in 1890. In 1935 the Northern Tissue Company began to advertise "splinter-free" paper, and a softer two-ply toilet paper made its debut in 1942. **BS**

SEE ALSO: FLUSH TOILET, S-TRAP FOR TOILET, COMPOSTING TOILET, SEWAGE SYSTEM, PAPER, PAPER TISSUES

Spinning Wheel (*c. 700*)

Indians accelerate the spinning process.

The origins of the spinning wheel remain unsure, but the machine is thought to have been invented around 700 in India, where it was used to turn fibers into thread or yarn that were then woven into cloth. Earlier hand-spinning methods were superseded by mounting the spindle horizontally and rotating it by slowly turning a large wheel with the right hand. The fiber was held at an angle in the operator's left hand to produce the necessary twist.

The spinning wheel reached Europe in the Middle Ages, becoming part of a cottage industry that used simple hand-operated tools. It persisted in this context until the eighteenth century.

In Britain the new cotton industry was modeled on the old woolen cloth industry. The most complicated apparatus was the loom, worked by a single weaver and normally kept in an upstairs room where a window provided natural light. The weavers were usually men who used yarn produced by a number of women, known as "spinsters," who did the spinning. With the advent of waterpower, this early cottage industry eventually grew into a large-scale factory operation. In 1733 John Kay invented the flying shuttle that enabled the weaving process to become faster, and the Industrial Revolution saw the process become increasingly mechanized.

Modern spinning wheels use electrical and mechanical means to rotate the spindle, automatic methods to draw out fibers, and devices to work many spindles together at high speeds. Other technologies that offer faster yarn production include friction spinning and air jets. **MF**

SEE ALSO: WOVEN CLOTH, SEWING, WHEEL AND AXLE, SPINNING JENNY, SPINNING MULE, POWERED LOOM

➡ *In a miniature Indian watercolor of the nineteenth century, a woman uses a wheel to spin cotton.*

Tidal Mill (*c.* 787)

Irish monks harness tidal waterpower.

In a tidal mill, incoming water enters and fills the millpond through sluices and is then channeled out at low tide to turn the mill wheel, thus powering the millstones, and crushing the grain to flour. The first known tidal mill, dating around 787, was built on Strangford Lough in northern Ireland, and was used by monks to grind corn for the nearby monastery.

Tidal mills were used along Europe's Atlantic coast in the Middle Ages, where the high tidal ranges ensured generous payback for the millers. The number of suitable sites was limited, and the mills could only operate for a certain period after each high tide, but their output was predictable compared to weather-dependent windmills and traditional watermills.

> "*Much water goeth by the mill that the miller knoweth not of.*"

Proverb

The popularity of tidal mills waned with the arrival of the steam engine, until in 1966 French engineers harnessed the tides to generate electricity. The Rance estuary in northern France was dammed by a barrage housing twenty-four turbines capable of producing power from both the ebb and the flow of the tide. Other schemes have been developed since, but the specific site requirements and the environmental considerations around the tidal basin are limiting.

Modern alternatives to the barrage approach are to construct underwater turbines, either on a riverbed or seabed, or to hang them from tethered surface buoys that rotate in the tidal stream. **FW**

SEE ALSO: GRANARY, QUERNSTONE, WINDMILL, WATERMILL, AUTOMATIC FLOUR MILL

Gunpowder (*c.* 800)

The Chinese trigger the race toward firearms.

Few inventions can have instigated as much misery to humankind as that of the seemingly innocuous gunpowder. Created by Chinese alchemists in the ninth century, gunpowder consists of a mixture of ground saltpeter (potassium nitrate), charcoal, and sulfur in approximate proportions of 75, 15, and 10 by weight. Known as "black powder," its exposure to an open flame produces an explosion that can propel an object great distances when contained in a tube closed at one end. The Chinese experimented with different levels of saltpeter content to design rockets. Arab chemists acquired knowledge of gunpowder in the thirteenth century, rapidly employing it for military purposes, including the production of a gun made

> "*We owe to the Middle Ages the two worst inventions of humanity —romantic love and gunpowder.*"

André Maurois, novelist and writer

from a bamboo tube reinforced with iron. The spread of information arrived in Europe, where gunpowder was manufactured in larger grains of uniform size to control the speed of burning, and advances in metallurgy witnessed the rise of the cannon and handheld firearms.

By the late nineteenth century, "black powder" had been replaced by nitrocellulose, resulting in a smoother, more powerful explosion from a firearm with little smoke deposit. Smokeless powder accounts for most gunpowder produced today, usually in single-base powder (nitrocellulose) or double-base powders (nitrocellulose and nitroglycerin). **SG**

SEE ALSO: FIREWORKS, GUN, MUSKET, CANNON, ARQUEBUS, FLINTLOCK MECHANISM, BREECH-LOADING GUN, REVOLVER

Banknote (*c. 806*)

The Chinese Tang Dynasty persuades people to accept paper money.

When Marco Polo traveled to China in the late thirteenth century he was astonished to see the locals use paper money instead of coins. Prompted by a copper shortage, the Tang Dynasty (618–907) introduced this new monetary system in 806, more than 800 years before the first European banknotes.

While commodity money (the trading of goods that have an intrinsic value, such as gold and cattle) has been around since the dawn of civilization, the first standardized coinage is thought to have appeared in Lydia (western Asia Minor) in the seventh century BCE. This was the first time that the nominal value of money was higher than the worth of its inherent material. Ancient Greek historian Herodotus criticized the "gross commercialism" that this system induced.

Swedish bank Stockholms Banco issued the first European banknotes in 1660. Once again, a shortage of copper was to blame. Customers had loaned copper to the bank, but demanded it back when their coins' copper content was reduced, making the copper more valuable than the coins. Unable to respond to the demand, the institution came up with promissory banknotes to solve their liquidity problem.

The Chinese were the first to discover the risks of using money without inherent worth. By 1020 the cost of imported goods, coupled with bribes given to potential invaders, had forced the government to issue more and more notes, thereby fueling inflation. By 1455 paper money had become so devalued that the Ming Dynasty (1368–1644) decided to get rid of it altogether, leaving the world without circulating banknotes for another 200 years. **DaH**

SEE ALSO: CASH REGISTER, CREDIT CARD, CASH MACHINE (ATM)

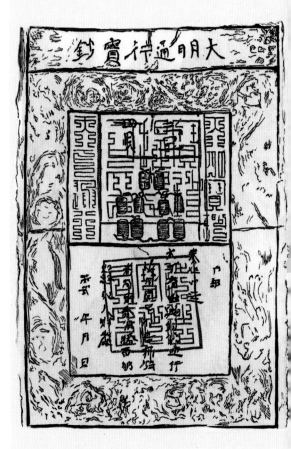

↗ *A banknote from China's Ming Dynasty period, one of the world's earliest surviving examples.*

"My notion of a wife at 40 is that a man should be able to change her, like a banknote, for two 20s."

Warren Beatty, actor, producer, and director

Woodblock Printing (*c. 868*)

A Chinese woodblock-printed masterpiece speaks of an earlier tradition.

Woodblock printing first appeared in China during the Tang Dynasty in the ninth century and was initially used in the production of textiles and Buddhist texts and amulets. A text or image was transferred to a thin layer of paper that was then glued face down onto a wooden surface using rice paste. The lines would then be cut out by the block maker. Only those portions of a page or pattern to be inked were left untouched on the block's surface, with the remainder carved away along the grain using a fist knife known as a *quan dao*. Dense hardwoods such as birch, pear, or jujube were used because they withstood moisture and insects yet their regular, fine grains lent themselves to easy engraving and printing. Once a block was completed, the ink was rubbed onto the surface of the raised lines, and a skilled artisan could produce 1,000 or more sheets a day, thus ushering in the era of mass-produced books and manuscripts.

The world's oldest extant woodblock-printed book that carries the date of its production is the *Diamond Sutra*, an Indian Buddhist Sanskrit text originally dating to 400, and later translated into Chinese and block-printed in ink on paper by an artisan, Wang Jie, on May 11, 868. It was discovered in a sealed cave in northwestern China by the archeologist Sir Marc Aurel Stein in 1907. However, it is by no means the oldest example of block printing. The degree of technical perfection in the *Diamond Sutra* suggests that the practice of woodblock printing had been long established by the time of its production.

Woodblock printing continued to be popular in East Asia well into the nineteenth century. **BS**

> *"... good woodblock printing rests upon the perfection of drawing and painting, of color and line."*
>
> Hiroshi Yoshida, *Japanese Woodblock Printing* (1939)

SEE ALSO: INK, MOVABLE TYPE, PRINTING PRESS, LITHOGRAPHY, LINOTYPE MACHINE, MONOTYPE MACHINE

⬉ *In this Japanese illustration, a craftsman cuts woodblocks while a woman uses them to print.*

Rocket (*c.* 904)

Chinese "fire arrows" herald rocketry.

In 904 at the siege of Yuzhang, in southeastern China, attacking troops were ordered to launch "flying fire" on the city gates, burning them down and allowing their army to enter and capture the city. This is the first use of "fire arrows," a term that originally meant an arrow carrying a tub of gunpowder that would explode when the arrow impacted.

The *Compendium of Important Military Techniques* (1044), written by Tseng Kung-Liang, gives details of how to launch fire arrows by gunpowder rather than using bows. By 1232, when the Chinese were fighting the Mongols, a much more recognizable rocket was being made using the exploding tubes to propel the arrows. The tubes were capped at the top, but open at

> *"At the campaign of Yuzhang, he ordered his troops to propel the 'flying fire' on the besieged city. . . ."*
>
> *Compendium of Important Military Techniques* (1044)

the bottom, and tied to the top of an arrow that, when lit, would ignite the powder and produce thrust. Whether the rockets themselves did much physical damage in the war is unclear, but the psychological effect was formidable. After seeing it used against them, the Mongols quickly developed their own versions that they used throughout their empire, and this spread the technology across the Middle East and on to Europe. By the twelfth century rockets arrived in European arsenals, reaching Italy by 1500, and then Germany, and later England. The use of the Iron Rocket against the British in India in the eighteenth century led to the development of the technology. **DK**

SEE ALSO: BOW AND ARROW, GUNPOWDER, FIREWORKS,
BALLISTIC MISSILE

Canal Lock (*c.* 984)

Qiao Weiyo devises the pound lock.

Locks interrupt a canal or river with stepped stretches of still water, thus reducing currents in the waterway and conserving deep water for passage. The forerunner of today's lock was the flash lock, already in use by the first century BCE in China, whereby part of a dam would be temporarily opened to allow passage of a vessel. Those traveling downstream were carried on the resulting surge of water, whereas those sailing in the opposite direction hauled the vessel against the torrent. Such an arrangement was dangerous and resulted in the loss of large quantities of water downstream for every vessel passing, a circumstance not appreciated by mill owners reliant on the supply.

In 984, during the construction of China's Grand

> *"To see barges waiting . . . at a lock affords a fine lesson in how easily the world may be taken."*
>
> Robert Louis Stevenson, *An Inland Voyage* (1878)

Canal, engineer Qiao Weiyo noted that in placing two flash locks 750 feet (229 m) apart, he had created an intermediate stretch of water that could be held at the level of either the upper or lower reach of the waterway, and thus the pound lock, or chamber lock, was born. Following that breakthrough, a significant improvement was the development of mitered lock gates in sixteenth-century Italy, perhaps based on the designs of Leonardo da Vinci. The miter uses the pressure of the high water on the upper side of the gate to create a secure seal until the water levels have equalized. This allowed the constructions required to withstand pent-up water to be less massive. **FW**

SEE ALSO: CANAL, CANAL INCLINED PLANE, DAM, DRY DOCK,
TIDAL DOCK

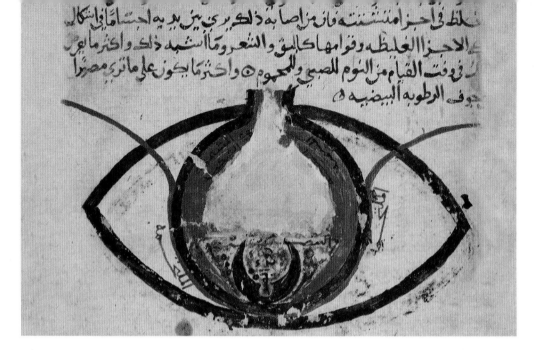

Lens (c. 984)

Ibn al-Haytham's treatise establishes optical science.

The earliest lenses were made of circular pieces of rock crystal or semiprecious stone, such as beryl and quartz, which were ground and polished so that they produced a magnified image when looked through. The oldest known lens artifact was one made of rock crystal dating from around 640 BCE and excavated in Nineveh, near the modern city of Mosul, Iraq. The most common form was circular and thicker in the middle than around the edge, and having both its front and back surfaces the same shape.

The modern convex lens developed from the ancient Greek burning glass. Here a spherical vase of water would be used to concentrate the rays of the sun onto a small area, which heated up. The heat was used to ignite fires in temples or to cauterize wounds.

The Iraqi mathematician and optics engineer Ibn Sahl (c. 940–1000) wrote the treatise On Burning Mirrors and Lenses (984) in which he set out his understanding of how curved mirrors and lenses bend and focus

light, using what is now known as Snell's law to calculate the shape of lenses. But the Iraqi Ibn al-Haytham (965–1039), also known as Alhazen, is regarded as "the father of optics" for his treatise, the Book of Optics, (1011–1021), in which he proved that rays of light travel in straight lines, explained how the lens in the human eye forms an image on the retina, and described experiments with a pinhole camera.

In the thirteenth century convex lenses were used in spectacles to correct farsightedness. The use of concave lenses, which disperse the light as opposed to concentrating it, to correct for nearsightedness, came in the early fifteenth century. **DH**

SEE ALSO: GLASS, TELESCOPE, MICROSCOPE, SPECTACLES, BIFOCALS, EYE TEST, SPECTROSCOPE, CONTACT LENSES

⬆ *A twelfth-century anatomy of the eye, from an Egyptian book on eye diseases by Al-Mutadibi.*

Sextant (c. 994)

Abu-Mahmud al-Khujandi measures the altitude of the sun above the horizon.

Iranian astronomical observer and instrument designer Abu-Mahmud al-Khujandi (circa 940–1000) constructed the first known mural sextant, with a radius of 66 feet (20 m), on an accurate north–south facing wall in Ray, near modern Tehran, Iran. The name "sextant" refers to the fact that the instrument had an angular scale that was 60 degrees in length, one sixth of a circle. (When measuring latitude, one minute is equal to one sixtieth of a degree.)

The instrument was designed to measure the altitude of the sun above the horizon at noon on the days of both the summer and winter solstice, the two dates in the year when this angle has its maximum and minimum value. From the average of these two angles, an observer could determine his or her latitude—the angular distance between the equator and the observation site.

The height of the sun in the sky was measured by looking at the shadow it cast on an accurate scale. The Al-Khujandi scale was so accurate that the latitude that he obtained was correct to a tiny fraction of a degree. Other famous mural sextants followed, including the Fakhri sextant with a radius of approximately 118 feet (36 m) constructed by Iranian Ulugh Beg in Samarkand, Uzbekistan, in around 1420. More modern astronomical sextants are smaller and pivoted at the balance point. They can be moved to measure the angular separations of stars and planets.

Handheld nautical sextants have become common in the last three centuries. They are fitted with adjustable mirrors and are used to measure the altitudes of celestial bodies. **DH**

SEE ALSO: MAP, GLASS MIRROR, ASTROLABE, HUBBLE SPACE TELESCOPE

⤒ *Arabian scholars employ devices, including a sextant (right), in a sixteenth-century German woodcut.*

Movable Type (1041)

Bi Sheng advances printing by creating characters on movable clay tablets.

In 1041, in the best traditions of China's inventive and technologically vibrant Song Dynasty (960–1280 CE), an alchemist named Bi Sheng shaped a series of reusable, moistened clay tablets, inscribed an individual Chinese character upon the surface of each one, fired them to harden and make them permanent, and in the process invented movable type. Printers then took the characters and laid them within an iron frame coated with a mix of resin, turpentine, wax, and paper ash, arranging the characters to reflect what was to become the printed page.

Unlike the Western alphabet that requires the generation and arrangement of only twenty-six characters, Bi Sheng worked in a language with over 5,000 distinct characters. Many of these needed several pieces of type to complete, and all of them required the making of multiple copies. The copies were wrapped in paper and stored within wooden framed cases when not in use, ordered according to the first syllable of their pronunciation. Like Johannes Gutenberg hundreds of years later, Bi Sheng failed to receive recognition for his invention until long after his death, despite his efforts being recorded by the great Chinese scientist Shen Kuo in his series of *Dream Pool Essays*.

The multiplicity of Chinese characters and symbols was one reason for the failure of Bi Sheng's invention to impact significantly upon Chinese society, in contrast to Gutenberg's press in fifteenth-century Europe. Another problem was that clay tablets were manifestly unsuitable for large-scale printing and were not at all durable. The limitations of clay led eventually to the invention in Korea of metal movable type, which accelerated the spread of printing in Asia in the early thirteenth century. **BS**

"... for printing hundreds or thousands of copies, it was marvelously quick."

Shen Kuo, *Dream Pool Essays* (1088)

↑ *Chinese inventor Bi Sheng, who in 1041 devised movable type using clay blocks.*

➡ *A nineteenth-century engraving of Johannes Gutenberg who introduced printing to Europe.*

SEE ALSO: INK, PAPER, PRINTING PRESS WITH MOVABLE METAL TYPE, STENOTYPE MACHINE, TYPEWRITER, LINOTYPE MACHINE

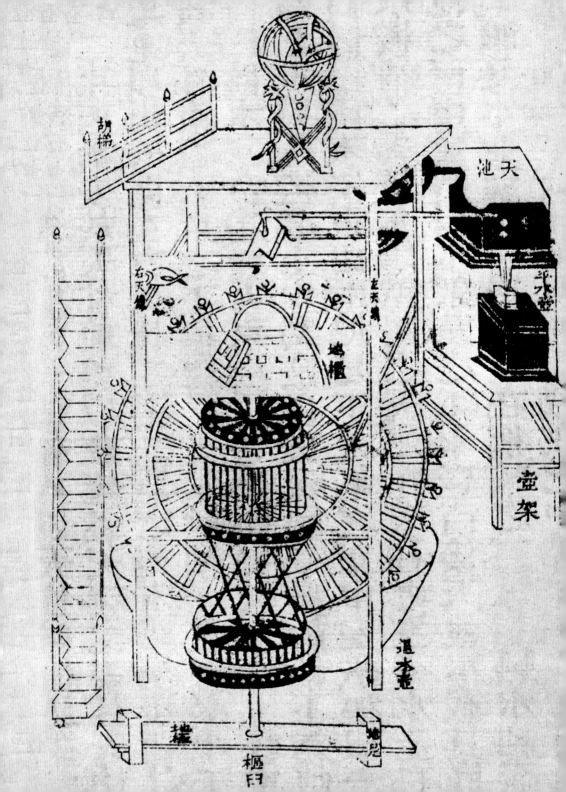

Tower Clock (1094)

Su Song builds a technological marvel.

The world's first water-driven astronomical clock tower was by far the most advanced astronomical instrument of its day. Its designer was Su Song (1020–1101), who oversaw the project with the aid of mathematician Han Gong-lian. The elaborate, 40-foot (12 m), water-powered, mechanically driven clock had bronze castings, precision gears, gear rings, and pinions. A bronze armillary sphere with a celestial globe mounted below allowed the sun, moon, and selected stars to be seen through a sighting tube.

The tower was a three-level, pagoda-like structure powered by thirty-six buckets attached to a central wheel, each of which would trip a lever and tilt forward at a predetermined point to engage the clock's

> *"… the comparison of the rotary movements … will show no discrepancy or contradiction. …"*
>
> Su Song, *The Rotation of an Armillary Sphere* (1092)

complex system of gears and counterweights. Song's greatest achievement, however, was an ingenious escapement mechanism that converted this energy discreetly from a pendulum to the gears in a concept vital in the construction of clocks, and a technology unknown to Europeans until late in the thirteenth century. This technology was a precursor to the mechanical escapement, which enabled the manufacture of all-mechanical clocks that could tell time with far greater precision. **BS**

SEE ALSO: WATER CLOCK, SHADOW CLOCK, CLOCKWORK MECHANISM, PENDULUM CLOCK, ATOMIC CLOCK

◄ *The central wheel had thirty-six buckets, and water filled each in turn to power the clock mechanism.*

Chain Drive (1094)

Su Song's clock hides a brilliant innovation.

Chinese government official Su Song (1020–1101) was also a naturalist, cartographer, astronomer, horologist, and engineer. His greatest legacy was the clock tower he built in Kaifeng. In 1086 the emperor had ordered the construction of an "armillary clock" to keep time and track celestial bodies. A finished structure was completed in 1094, and consisted of three levels. The upper level contained a rotating armillary sphere that allowed astrological observations through sighting tubes; the middle level had a bronze celestial globe; and the lower level had mechanically timed manikins that would exit doors at fixed times of the day.

Perhaps most significant, however, was the clock's innovatory driving system. At the heart of the clock tower was the *tian ti*, or "celestial ladder." This is the oldest known endless power-transmitting chain drive. The chain transmitted the power from a water wheel to turn the armillary sphere and power the clock.

Drive belts had been present in China for approximately 1,000 years before Song's chain drive. However, such belts were primitive, haphazard affairs and lacked the precision necessary to drive a clock and armillary sphere. The chain links fitted sprockets and were not subject to stretching or slippage. Chain drive is now used in a wide range of mechanical devices, especially in vehicles such as bicycles, but it was not until several centuries after Song's innovation that Europeans independently discovered the technology.

The clock was captured and dismantled by invading Jurchens from Manchuria who overran Kaifeng in 1127. It is one of the great lost artifacts of the medieval age. However, Song's 1092 treatise (*Essentials of a New Method for Mechanizing the Rotation of an Armillary Sphere and a Celestial Globe*) survived, with illustrations and descriptions of the clock. **SS**

SEE ALSO: WATER CLOCK, TOWER CLOCK, INTERNAL COMBUSTION ENGINE, CAMSHAFT, SAFETY BICYCLE

Cannon (*c.* 1128)

The Chinese develop bronze cannons from primitive bamboo forerunners.

During the Chinese Song Dynasty (circa 960–1279), artillery engineering exploded, as it were, with the development of the ancestor of the cannon: flame-throwing "fire lances" made of bamboo. When gunpowder at one end was ignited, it forced sand, lead pellets, or shards of pottery at the enemy.

When metal later replaced bamboo, probably in the early 1100s, these lances became "fire tubes" or "eruptors." The oldest record of them is a painting, dated to 1128. The early Chinese cannons could throw a ball about 50 yards (45 m). A century later they had become powerful enough to breach city walls, and were made of bronze. According to the historian of Chinese technology Joseph Needham, cannon warfare took a great step forward with the development of cannonballs that fitted the tube's bore precisely, enabling more control. Later cannons were made of cast iron, and some were wheeled.

Cannon technology spread and was developed in Europe; the Scots defended Stirling Castle with cannons in 1341, and three cannons were used at the battle of Crécy-en-Ponthieu in 1346 by England's King Edward III, who had more than a hundred of them in London. Such early European cannons were small, but by the end of the Hundred Years' War (1337–1453) giant ones were available, known as "bombards." Mons Meg, a bombard built in 1457, survives at Edinburgh Castle in Scotland; with a 22-inch (56 cm) caliber barrel, it is capable of firing gunstones weighing 331 pounds (150 kg) nearly 2 miles (3.2 km).

From the sixteenth century, lighter cannons capable of more accurate fire were developed, and these gradually evolved into modern artillery pieces, such as the howitzer used to great effect in the American Civil War (1861–1865). **AC**

SEE ALSO: **GUNPOWDER, ROCKET, GUN, MUSKET, MACHINE GUN, AUTOMATIC MACHINE GUN, BALLISTIC MISSILE**

> *"Shells are made of cast iron . . . and are sent flying toward the enemy camp from an eruptor."*
>
> Jiao Yu and Liu Ji, *Fire Dragon Manual* (*c.* 1368–1398)

⊡ *This naval cannon is integrated with its wheeled carriage for ready maneuverability aboard ship.*

⊡ *Centuries after the Chinese invention of the cannon, Chinese artillerymen fire a German one, circa 1900.*

Fireworks (*c.* 1150)

The Chinese invent explosive entertainment.

Fireworks, familiar now in sound-and-light shows on dark evenings, to celebrate festivals and to entertain, were invented in China around 1,000 years ago, following the invention of gunpowder in the first century CE. Bamboo tubes, filled with gunpowder, were thrown onto fires to create explosions at religious festivals, perhaps in the belief that the noise they made would scare off evil spirits. It is highly likely that some of these little bombs shot like rockets out of the fire, propelled by the gases they produced.

The next step seems likely to have been to attach such charged bamboo tubes to sticks and fire them with bows. The earliest evidence of devices that could be described as firework rockets comes from a written report of the battle of Kai-Keng in 1232 during the war between China and Mongolia, in which the Chinese attacked with "arrows of flying fire." After Kai-Keng, the Mongols began to make rockets as well, and probably took these with them on their travels to Europe.

There is documentary evidence for experiments involving rockets in Europe throughout the thirteenth to fifteenth centuries. English Franciscan friar and philosopher Roger Bacon (circa 1214–1294) reported on his experiments to improve gunpowder and increase the range of rockets, and, in France, Jean Froissart (circa 1337–1405) commented that rockets could be fired more accurately if launched from tubes.

For the first 700 years of their existence, fireworks were available only in one color (yellow), until the French chemist Claude Berthollet discovered potassium chlorate in around 1800 and began the development of new colors for fireworks. **SC**

SEE ALSO: GUNPOWDER, MUSKET, CANNON, ARQUEBUS, FLINTLOCK MECHANISM, BREECH-LOADING GUN, REVOLVER

← *Members of a Chinese family light a firework to honor a kitchen god as part of their New Year celebrations.*

Arabic Numerals (*c.* 1202)

Fibonacci promotes Hindu-Arabic numbers.

In 1202 Leonardo Pisano (*c.*1170–*c.*1240), known as Fibonacci, published his seminal work *Liber abaci* ("Book of the Abacus") and thus popularized Hindu-Arabic numbers in Europe. Although born in Italy, Fibonacci grew up in what is modern-day Beja, Algeria. Taught by Arab teachers, Fibonacci came in contact with our modern numeral system, which was devised in ancient India yet virtually unknown in Europe.

Until then the Roman numeral system had been prevalent throughout the continent. The system had been an improvement on the first recorded numbers found in Egypt—simple representative strokes for each digit, and a special symbol for ten—as well as the Greek (Attic) method of recording the first letters of the

---

"The idea seems so simple that its significance and importance is no longer appreciated."

Pierre-Simon Laplace (1749–1827), mathematician

---

numeral names. In the Hindu-Arabic base ten system, on the other hand, the single digits were represented by symbols whose value depended on placement (i.e., 2 in 200 being ten times greater than 2 in 20).

The first known inscriptions of these numbers date from the third century BCE, although whether they were used in a place-value system isn't clear. As early as the seventh century CE the system had reached the Arab world, recorded by mathematicians such as Al-Khwarizmi. The advent of the printing press in the fifteenth century accelerated the proliferation of Hindu-Arabic numerals, which over the centuries has become the closest thing to a universal human language. **DaH**

SEE ALSO: TALLY STICK, ABACUS, CUNEIFORM, SLIDE RULE, METRIC SYSTEM, MECHANICAL CALCULATOR

Crankshaft (c. 1206)

Al-Jazari sets rotational energy to work.

Islamic scholar Al-Jazari (1150–1220) lived in what was northern Mesopotamia—today's northeastern Syria and Iraq. A brilliant inventor, he made one of the most significant contributions to human engineering in 1206 by devising the world's first crankshaft. This conceptually simple device converts rotary into reciprocating motion, and vice versa, but it is now used in a huge number of modern machines, including automobiles. Put simply, the Industrial Revolution could not have happened without it.

The crankshaft is perhaps most commonly recognized today in modern motor engines, and is among Al-Jazari's more famous inventions. In today's cars it receives force from firing pistons that move in a

> *"Types of machines . . . came to my notice offering possibilities for types of marvelous control."*
>
> Al-Jazari, *Book of Knowledge* (1206)

linear fashion back and forth and, through movable bearings, rotates around its axis, thus converting the pistons' energy into rotation output for the wheels.

Al-Jazari's original invention was designed to pump water from wells for irrigation, with cattle employed to provide the energy—quite a contrast from the modern interpretation of the technology. Despite the difference in complexity, the principle behind the crankshaft remains the same more than 800 years after Al-Jazari first thought of it. **SR**

SEE ALSO: CAMSHAFT, LOCOMOTIVE, STEAM ENGINE, COMPOUND STEAM ENGINE, MOTORCAR

← *As an entertainment, Al-Jazari created a boat of automaton musicians; it moved via a crankshaft.*

Camshaft (c. 1206)

Al-Jazari transforms rotational power.

When the great Islamic scholar Al-Jazari (1150–1220), published in 1206 his *Kitáb fí ma'rifat al-hiyal al-handasiyya* (*Book of Knowledge of Ingenious Mechanical Devices*), it included a description of a device that could change rotational motion into reciprocating motion: the camshaft. This invention consists of a shaft that has oval-shaped lobes attached to it, which turn with the shaft itself. Because of their noncircular shape, these "cams" appear to oscillate when the shaft spins on its axis. If a cam is positioned next to a valve, as it is in the example of the internal-combustion engine, then, as the camshaft turns, the longest end of the cam will depress, and hence open, the valve each time the shaft makes a turn. Before that, the camshaft played an important role in many medieval technologies. In windmills and waterwheels, for example, camshafts transformed rotational power into the energy and modes of action needed to mill corn, saw wood, or hammer metal.

In modern times the camshaft is best known as an integral part of the internal-combustion engine. When fuel within each combustion chamber of an engine explodes, the mixture of combusting fuel and air expands and drives a piston that is connected to a crankshaft, thus translating the piston's motion into the energy needed to propel a car, lift a weight, or perform some other form of work. A camshaft—or two camshafts, depending on the engine's design—controls the valves that allow new fuel into each combustion chamber, as well as the valves that release exhaust gases from the previous fuel ignitions.

In an engine the camshaft is connected to the crankshaft by a timing belt that allows precise coordination of the firing of the combustion chambers with the intake of fuel and output of exhaust. **DHk**

SEE ALSO: WINDMILL, CRANKSHAFT, MOTORCAR

Automaton (*c.* 1206)

Al-Jazari creates the automaton that anticipates today's industrial robots.

Most people think of self-operating machines as twentieth-century inventions. Although Isaac Asimov coined the word "robotics" in 1942, and Grey Walter built the first electronic autonomous robots in 1948, the first automaton for which we have good evidence was a boat with four mechanical musicians. It was built more than eight hundred years ago by Islamic scholar Al-Jazari (1150–1220).

Al-Jazari, considered by some to be the father of robotics, wrote his *Kitáb fí ma'rifat al-hiyal al-handasiyya* (*Book of Knowledge of Ingenious Mechanical Devices*) in about 1206, while he was the palace chief engineer in Diyarbakir (located in the southeast of present-day Turkey). The book describes a boat he constructed that floated on the palace lake and entertained guests at parties with music from a flute, a harp, and two drums played by automatons. The drummers contained rotating cylinders with movable pegs. As the cylinder rotated, the pegs would strike levers that caused the drums to be played. Changing the number and location of the pegs produced different rhythms, and so the automaton was entirely programmable.

Automatons created in subsequent centuries, mainly for entertainment purposes, continued to play musical instruments, along with other activities that could be recreated in a sufficiently realistic manner.

Today, factories increasingly use robots—essentially automatons powered by electricity—for jobs that require speed, precision, strength, and/or endurance. Robots build cars, package goods, manufacture circuit boards, and perform many other tasks. Almost a million robots were in operation around the world in 2007, and the International Federation of Robotics expects this number to reach 1.2 million by the end of 2010. **ES**

> *"Obedience . . . / Makes slaves of men, and, of the human frame / A mechanized automaton."*
>
> Percy Bysshe Shelley, "Queen Mab" (1813)

⬆ *An automaton of a Spanish monk, metal with a wooden head and hands, and its separate habit.*

➡ *Al-Jazari created this water-pouring automaton for the amusement of Urtugid Dynasty princes.*

SEE ALSO: ROBOT, INDUSTRIAL ROBOT, BIPEDAL ROBOT, SURGICAL ROBOT

لتتحرك طرف الشطبة بين الرقة والزبق وأمثل صورة الغلام وجميع ما ظاهره وباطنه

Spectacles (*c.* 1250)

Venetians create the first European glasses.

In the first century the Roman philosopher and dramatist Seneca used a glass sphere full of water resting on his reading material to magnify the letters, and this method was certainly used by farsighted monks a millennium later. Glass blowers in Venice produced lenses that were used as magnifying glasses, and in Europe in the late thirteenth century these were being used in pairs, one for each eye, the holding frame being made of wood or horn.

Salvino D'Armate of Pisa (1258–1312) and the friar Alessandro da Spina (*d.* 1313) of Florence are often given the credit for the invention of spectacles, in the year 1284, but Marco Polo, in 1270, saw elderly Chinese using spectacles and, when asked, they credited the

> *". . . this [invention] enables good sight and is one of the most useful of arts . . . the world possesses."*
>
> Fra Giordano da Rivalto, sermon (1305)

invention to Arabs in the eleventh century. The Chinese were also using smoky quartz as simple non-magnifying sunglasses at that time.

The first spectacles used convex lenses and corrected for presbyopia (farsightedness). In 1451 the German cardinal Nicholas of Cusa introduced spectacles with concave lenses that corrected for myopia (nearsightedness). The explanation as to why the lenses worked was given by Johannes Kepler in his 1604 treatise on optics. By around 1730 a London optician, Edward Scarlett, perfected spectacle arms that hook behind the ears, although it was some time before this became the preferred design. **DH**

SEE ALSO: GLASS, LENS, BLOWN GLASS, EYE TEST (SNELLEN CHART), BIFOCALS, CONTACT LENSES

Land Mine (*c.* 1277)

The Chinese pioneer a concealed weapon.

Land mines—explosive weapons triggered by pressure or proximity—have been in use for many hundreds of years, and as such their exact history is somewhat clouded. There is evidence to suggest, however, that the first self-contained land mines for military purposes were used in China in 1277 against Mongol invaders. Today their use has prompted great controversy, due to the high civilian casualties that can be caused many years after conflict has ended.

The name "mine" is derived from their original use in Europe in the Middle Ages as "tunnel mines." During the siege of a castle or fort, tunnels would be dug under the walls and explosives detonated just beneath them in a bid to cause them to collapse. True anti-

> *"The fuse starts from the bottom . . . (black powder) is compressed into it to form an explosive mine."*
>
> Jiao Yu and Liu Ji, *Fire Dragon Manual* (c. 1368–1398)

personnel land mines were introduced in China by the Middle Ages. The mines were detonated either by the pressure of someone walking over them, or by operation from afar, and appeared in a variety of designs. However, their use in warfare was restricted by a scarcity of ammunition.

The mine emerged at the beginning of the sixteenth century in Europe in the form of a shallow pit of gunpowder, with a "charge" made from flint that would ignite the explosive. In later centuries mines were used to slow advancing troops, especially in the Boer War and World War I. The twentieth century saw a great proliferation of their use. **SR**

SEE ALSO: GUNPOWDER, TORPEDO

Scythe (*c.* 1300)

This European invention increased productivity and liberated workers from the sickle.

The scythe is often ranked among the world's most significant advances in agricultural implements of the past thousand years. Its appearance on the farms of Europe in the latter half of the thirteenth century was to profoundly revolutionize agricultural production. Initially used as a grass cutter to gather hay, it was later used to harvest grain. Consisting of a curved blade, sharpened on the inside of the curve, and a long wooden handle (called a snath), the scythe allowed the reaper to stand upright while cutting grass—a vast improvement over the short-bladed sickle, which required the user to stoop uncomfortably as he cut.

A worker using a sickle, which was essentially unaltered in design since its emergence in around 5000 BCE, could harvest at best only three-quarters of an acre (0.3 ha) per day, while the use of a scythe increased the production of grain harvesting to more than an acre (0.4 ha) per day. Its ergonomic design and its way of operating as a kind of lever gave it more power than the backbreaking sickle, and by the sixteenth century the scythe had all but replaced the sickle as the preferred tool for harvesting—as well as the weapon of choice in many a peasant rebellion.

By the late eighteenth century a series of wooden pegs had been added along the snath, which allowed grain to be cut and their stalks gathered in one motion for bundling into sheaves, an innovation that effectively doubled the farmer's productivity. The scythe remained the dominant tool for grain harvesting until the invention of the mechanical horsedrawn reaper in 1831, which could do as much reaping in a day as ten men with scythes. **BS**

SEE ALSO: METALWORKING, FLAIL, THRESHER, REAPER, LAWN MOWER, COMBINE HARVESTER

↗ *A peasant cuts neat rows of grass with a scythe in an illustration from the* Bedford Hours, *1423.*

"The scythe was faster ... to use than the ancient, short handled, curved, serrated blade sickle."

The Countryside Museum website

Musket (c. 1300)

Using gunpowder, Chinese gunsmiths invent a deadly rival for the longbow.

"The impact of Chinese firearms in terms of warfare and territorial expansion was profound."

Sun Laichen, Asia Research Institute

⬆ *This ornamental musket and rest, both intricately inlaid with ivory, date from around 1630.*

➡ *An English illustration shows the procedure for setting up, firing, and shouldering a musket to move onward.*

A musket is a smoothbore firearm loaded from the muzzle and fired from shoulder-level. It is larger than an arquebus and often fired from a rest on which it can be pivoted from side to side. It is difficult to pin down with certainty when the musket was invented, although according to ancient Chinese texts it was some time in the fourteenth century. Basic cannons had been fashioned by the Chinese, and Chinese weapons experts were the first to produce a device that was recognizable as a musket, but it was when this technology met the greater metallurgical prowess of the Ottoman Empire (and later that of the European powers) that a revolution in warfare took place.

It took a long time for muskets to become established; early muskets in particular took a long time to reload, could not reliably pierce armor, and were expensive and unreliable. Crucially, a cheap and deadly accurate alternative already existed: A longbow and arrows could be constructed relatively easily from locally available materials. Yet from the sixteenth century onward, Europeans, and in particular the Portuguese, were producing muskets and cannons and exporting them to Asia.

By the seventeenth century, the musket's reliability and ease of use had improved, and it enjoyed one distinct advantage over the longbow: It did not require a highly trained soldier or nobleman to use it. Despite initial resistance to this "cheapening" of "noble" warfare, the aristocracy quickly realized that enemy forces were utilizing firearms to deadly effect, and also that the vast majority of casualties were now not people of their own class, but the more expendable common man. This pattern of warfare continued and intensified through the centuries, reaching a bloody peak in World War I. **JB**

SEE ALSO: GUNPOWDER, CANNON, GUN, ARQUEBUS, BREECH-LOADING GUN, FLINTLOCK MECHANISM, REVOLVER

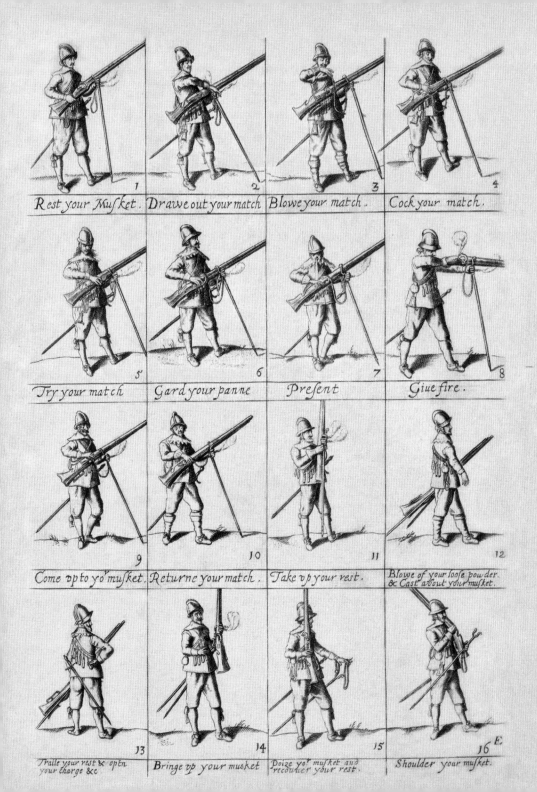

Rest your Musket.	*Drawe out your match*	*Blowe your match.*	*Cock your match.*
Try your match	*Gard your panne*	*Present*	*Giue fire.*
Come vp to yor musket.	*Returne your match.*	*Take vp your rest.*	*Blowe of your loose powder. & Cast about your musket.*
Traile your rest & open your charge &c	*Bringe vp your musket*	*Poize yor musket and recouuer your rest.*	*Shoulder your musket.*

ARSENAL

Production Line (c. 1320)

The Venetians initiate mass production.

The division of labor into many discrete tasks in a production line increased efficiency and output and enabled the mass production of high quality goods. Nowhere was this better exemplified than in the Venice Arsenal factory, where standardized parts and specialized tasks gave the Venetians a speed advantage in building warships, firearms, and, as a result, their empire.

In 1574, King Henry III of England had the privilege of seeing this formidable first factory construct complete galley ships in less than a day. With their efficient and organized approach, the Arsenal's thousands of workers manufactured complicated war tools 600 years before the start of the Industrial

---

"[The Arsenal of Venice's work force] continually produced new innovations for the Venetian fleet."

Gregory Sheridan, *The Imperial Age of Venice* (1970)

Revolution. The Arsenal was expanded on several occasions throughout its lifetime but was originally built as a shipyard in 1104. By 1320, the workers carried out specialized tasks in the production of ships. At its peak, the Arsenal employed around 16,000 people, the largest workforce in Europe. After Venice was annexed by Italy in 1875–1878, the Arsenal underwent further expansion. During the late 1800s it was the site of construction for several of the Italian Navy's most powerful ships. **LW**

SEE ALSO: POWERED LOOM, ROBOT, INDUSTRIAL ROBOT

← *A seventeenth-century perspective map locates the Arsenal, the logistical center of the Venetian navy.*

Rain Gauge (1441)

Jang Yeong-Sil measures rainfall.

In the fifteenth century Korea was a drought-plagued realm, and King Sejong (1397–1450) wished to levy land taxes based on an assessment of each farmer's potential harvest. To this end a nationwide network of rain gauges was established and the local magistrates of every village were commanded to report the rainfall to the central government.

In 1441 each village was provided with a standard cylindrical container, 17 inches (43 cm) high and 7 inches (17 cm) wide, that was mounted on a stone stand; a special ruler was used to measure the depth of rainwater that entered the gauge over a specific time. Its inventor was a civil-servant scientist, Jang Yeong-Sil. Needless to say, the method of rain measurement was rather labor-intensive. The Chinese, meanwhile, had used a similar technique to measure snowfall in 1247 CE.

In 1662 Sir Christopher Wren (1632–1723) invented the "pluviometer," a mechanical, self-emptying, tipping-bucket rain gauge. It consisted of two small, well-balanced buckets that collected rain sequentially. When a specific amount of water had fallen into one of the buckets—usually about 0.03 inches (0.1 cm)—it tipped over, emptied itself, and produced a mechanical signal. Also, a hole was punched into a slowly moving paper tape as the bucket tipped. The second bucket then started to collect rain, only to tip over when full. The more signals recorded, the greater the rainfall.

Care had to be taken in positioning Wren's gauge. It had to be sufficiently high off the ground to avoid splashes and animals, and be well away from trees, fences, and buildings to prevent shadowing and an unrepresentatively low reading. Heated gauges are now used when hail and snow are expected. **DH**

SEE ALSO: ANEMOMETER, BAROMETER, ANEROID BAROMETER, WEATHER RADAR

Coiled Spring (*c.* 1450)

Henlein perfects the heart of clockwork.

Often in history one critical invention leads to another that overshadows the first. This was true of the German locksmith Peter Henlein, the inventor of the portable or pocket watch. He created this sometime between 1504 and 1508, and it could operate for up to forty hours before it needed rewinding.

Henlein's work was made possible by the invention, more than fifty years earlier, of a single piece of metal that, when in a certain shape, would use the metal's natural elasticity to both absorb and release a force applied to it: the humble coiled spring. History does not recall who first created this most useful invention, but a small number of examples of spring-driven clocks have survived from the early fifteenth century.

> *"... a piece of work which excites the admiration of the most learned mathematicians."*
>
> *Cosmographica Pomponii Melae* (1511)

The coiled spring's ability to store energy was what made it perfect to power clockwork devices. Winding up a clock, or watch, costs physical energy as the spring becomes compressed. Clockwork gears then allow that energy to be gradually released while supplying a steady, if tiny, flow of power to a clock.

Henlein, and others, had to overcome various problems in their early models. A good spring is created from a single ribbon of steel of uniform thickness that can be squeezed without breaking. The delicate preheating required to temper coiled springs was difficult, and perfecting the formula was an important step in metallurgy. **DHk**

SEE ALSO: WATER CLOCK, SHADOW CLOCK, CLOCKWORK MECHANISM, PENDULUM CLOCK, POCKET WATCH

Arquebus (*c.* 1450)

The Spanish create the handheld cannon.

In the middle of the second millennium CE the battlefield was dominated by armored cavalry and the romantic concept of the chivalrous knight in armor. However, a technological innovation was about to take place that would completely change warfare. The invention in question was called the "hackenbushce," or arquebus, probably a derivative of the Dutch word *haakbus*, meaning "hook-gun." The arquebus was one of the first effective examples of a handheld firearm.

By this time, using gunpowder to fire projectiles was not a new idea. Cannons had existed since the early 1300s, and smaller "hand-cannons" had developed to complement these. These early firearms were basically small cannons mounted on poles or on crossbow stocks, and were fired by touching the vent-hole with a match, which would ignite the powder and fire the projectile.

The particular innovation of the arquebus, which cannot be attributed to one person but is probably of Spanish origin, lay in the use of a pivoting mechanism with the slow-burning match at one end that allowed the arquebusier, or shooter, to hold and aim the weapon with both hands, and then simply pull the other end of the pivoting lever to touch the match to the powder and fire the weapon. This mechanism was, of course, an early forerunner of the trigger, ubiquitous on modern guns, and by freeing both hands allowed the weapon to be fired with greater accuracy.

Despite this advantage, the arquebus was still grossly inaccurate when compared to the longbow or crossbow, but it still caught on, probably due to the fact that use of the arquebus was relatively easy to learn; where the longbow took years of study and considerable physical strength to master, the gun could be fired by almost anyone. **BG**

SEE ALSO: GUNPOWDER, MUSKET, FLINTLOCK MECHANISM, BREECH-LOADING GUN, REPEATING RIFLE, REVOLVER

Anemometer (c. 1450)

Alberti measures wind speed with an effective new instrument.

Devised by Leon Battista Alberti (1404–1472), the anemometer was a simple instrument to measure wind speed. It had a rectangular metal plate attached to a horizontal axis with a hinge, so that in the wind the metal plate lifted, giving an indication of relative wind speed that could be measured crudely on a curved scale bar below the plate. In light winds, the plate would move slightly on its hinge; in stronger winds, the plate would lift further. Alberti describes and illustrates this device in his book, *The Pleasure of Mathematics* (1450). The well-educated son of a wealthy merchant, Alberti was an accomplished artist, athlete, horserider, musician, mathematician, cryptographer (inventing the cipher disc), classicist, writer, cleric, and architect. He was a true Renaissance polymath, created by the intellectual culture prevailing in the Italian cities at the time.

As an artist and an architect, Alberti was inspired by Filipppo Brunelleschi's use of linear perspective and his great design for the dome of Florence Cathedral, where Alberti was a canon. In turn, Alberti's work, including the anemometer, inspired Leonardo da Vinci, who made drawings of it and saw its value in his designs for flying machines.

Alberti's simple design served its users for more than 200 years, until a British scientist, Robert Hooke (1635–1703), reinvented it in 1664, placing the moving plate beneath the curved scale bar to ensure more accurate measurements. Almost two centuries passed before the four-cup windmill anemometer was invented, in 1846, by the Irish astronomer John Thomas Romney Robinson (1792–1842). **EH**

SEE ALSO: RAIN GAUGE, BAROMETER, ANEROID BAROMETER, WEATHER RADAR

↗ *This nineteenth-century anemometer with two eight-bladed vanes was devised by G. T. Kington.*

"None of our modern craftsmen [except Alberti] has known how to write about these subjects ..."

Giorgio Vasari, *Lives of the Artists* (1550)

Printing Press with Movable Metal Type *(c. 1450)*

Gutenberg's innovatory press vastly increases the dissemination of information.

Writing was an important step in the advancement of civilization, but few books were produced and they reached a limited number of people. Such books were usually of a religious nature, handwritten in Latin, and copied by clerks for the clergy and the nobility. It was only when the printing press was developed that knowledge and ideas were spread more widely.

The earliest form of press incorporated a wooden block with raised letters on one side. Such blocks were arranged in a frame and inked, so that when pressed onto paper an impression of the letters was produced. Unfortunately the blocks disintegrated with use, and so could not produce many copies; it was very time-consuming for craftsmen to produce new blocks for letters and illustrations.

In 1450 Johannes Gutenberg (circa 1400–1468), a German printer, developed a technique where letters produced from molds of metal alloy were arranged into words, then locked together using a template to produce a whole page of typeface. This was sufficiently robust to print many hundreds of identical pages and hence the production of books became widespread, enabling more people to learn to read and increasing the demand for reading material.

Gutenberg's press would have had limited usefulness without appropriate inks. Before his time, simpler printing methods made use of water-based inks, and Gutenberg himself introduced more robust oil-based inks, including colored inks that were printed experimentally in some copies of his bible.

Flatbed-printing methods were eventually replaced by rotary presses, and by the end of the twentieth century computer to plate (CTP) technology was used, in which material was sent directly from a computer to a printing plate. **MF**

"It shall scatter the darkness of ignorance, and cause a light . . . to shine amongst men."

Johannes Gutenberg, printer

↑ *The impact on society of Gutenberg's press was comparable to that of the Internet today.*

→ *The Gutenberg Bible (1456) combined text printed with moveable type and handpainted elements.*

SEE ALSO: PAPER, INK, WOODBLOCK PRINTING, MOVABLE TYPE, ETCHING, ROTARY PRINTING PRESS, LINOTYPE MACHINE

Incipit prologus sancti Jeronimi pres-
biteri in parabolas salomonis.

Iungat epistola quos iungit sacerdoti-
um: immo carta non diuidat: quos
xpi nectit amor. Comentarios in osee-
amos: z zachariã malachiã. quoqз
posuis. Scripsisse: si licuisset pre uali-
tudine. Mittitis solacia sumptuum:
notarios nros et librarios sustenta-
tis: ut vobis potissimu nrm desuder
ingeniu. Et ecce ex latere frequens turba
diuisa poscetiu: quasi aut equu sit me
vobis esurietibз alijs laborare: aut
in ratione dati et accepti: cuiqз preter
vos obnoxi9 sim. Itaqз longa egrota
tione fractus: ne penitus hoc anno re-
ticerē: z apud vos mutus essem: tridui
opus nomini vro consecraui: interp
tatione videlicet triu salomonis vo-
luminu: massloth qd hebrei pabolas.
vulgata editio pubia vocat: coeleth
que grece ecclesiasten: latine ocionatore
possum9 dicere: sirasirim qd i linguã
nrã vertit canticu canticoз. Fertur et
panaretos: iħu filij sirach liber: z ali9
pseudographus qui sapientia salomonis inscribit. Quoз priore hebrai-
cum reperi: nõ ecclesiasticu ut apud la-
tinos: sed pabolas pnotatu. Cui iuncti
erant ecclesiastes: et canticu canticoз: ut
similitudine salomonis: nõ solu nu-
mero libroru: sed etiã materiaз gene-
re coequaret. Secundus apud hebreos
nusqз est: quia et ipse stilus greca
eloquetiã redolet: et nõnulli scriptoз
veres: huc esse iudei filonis affirmãt.
Sicut ergo iudith z thobie z machabe-
oз libros: legit quide eos ecclesia: sed
inter canoicas scripturas nõ recipit:
sic z hec duo volumina legat ad edi-
ficatione plebis: nõ ad auctoritatem
ecclesiasticoз dogmatu afirmandam.

Si tui sane septuagita interpretum
magis editio placet: habet eã a nobis
olim emedata. Neqз eni noua sic cu-
dim9: ut vetera destruam9. Et tamen cu
diligentissime legerit: sciat magis nra
scripta intelligi: que nõ in tertiu vas
trãsfusa coacuerit: sed statim de prelo
purissime emedata teste: suu saporē ser-
uauerit. Incipit parabole salomois.

Arabole salomonis
filij dauid regis isrl:
ad sciendã sapienti
am z disciplinã: ad
intelligendã verba
prudentie et suscipi
endã eruditione doctrine: iusticiã
et iudiciu z equitate: ut detur paruulis
astutia: et adolescenti scientia et intel-
lectus. Audiens sapiens sapiention erit: z
intelliges gubernacla possidebit. Ani-
aduertet parabolam et interpretatio-
nem: verba sapientiu z enigmata eoз.
Timor dñi principiu sapientie. Sapien-
tiam atqз doctrinam stulti despiciut.
Audi fili mi disciplinã pris tui et ne
dimittas legem mris tue: ut addatur
gracia capiti tuo: z torques collo tuo.
Fili mi si te lactauerint pcores: ne ac-
quiescas eis. Si dixerit veni nobiscu:
insidiemur sanguini: abscodam9 tedi-
culas cõtra insontem frustra: degluti-
mus eu sicud infernus viuentē z inte-
grum: quasi descendentē in lacu: omne
pciosã substantiã reperiem9: implebim9
domus nras spolijs. sortem mitte no-
biscum: marsupiu sit unum omniu
nrm: fili mi ne ambules cu eis. Pro-
hibe pedem tuu a semitis eoз. Pedes
eni illoз ad malu currut: z festinãt ut
effundant sãguinem. Frustra autem
iacit rete ante oculos pênatoз. Ipi qз
contra sanguine suu insidiantur: et

Toothbrush (c. 1498)

The Chinese introduce a dental savior.

The forerunner of the modern bristle toothbrush is generally believed to have originated in fifteenth-century China. A Chinese encyclopedia dating to 1498 describes the short, coarse bristles from the neck of a Siberian wild boar being embedded in a handle made from animal bone, which was then used to clean the teeth. In the seventeenth century, Chinese traders took the brush to Europe, where its popularity flourished despite boar hairs being considered too rough for sensitive European gums. Softer horsehair bristles were seen as an alternative, although boar bristles remained the most common fiber.

The toothbrush was not humankind's first attempt at dental hygiene. "Toothsticks" dating back to 3000 BCE have been uncovered during excavations of Pharaonic tombs in Egypt. These are lengths of frayed twigs or fibrous wood from shrubs, used to clean between the teeth and freshen the breath. "Chewing sticks" made from aromatic shrubs for oral hygiene and to freshen the breath were also used by the Chinese in the sixteenth century.

The first mass-produced toothbrush was designed and marketed by the English inventor William Addis in 1780 using boar hairs and swine bristles attached to the end of a downsized cow's thigh bone. More geometric designs began to appear in the mid-1840s when bristles first began to be aligned in rows. Natural bristles continued to be used up until the invention of nylon by Dupont de Nemours in 1938. The world's first electric toothbrush appeared in 1939.

The toothbrush is one of the oldest implements still used by humanity, regularly finishing above more fancied inventions—such as the car and the personal computer—on people's lists of the things they simply could not do without. **BS**

SEE ALSO: TOOTHPASTE, DENTURES, DENTAL FLOSS, ELECTRIC DENTIST'S DRILL

Glass Mirror (c. 1505)

The Venetians transform glass reflectivity.

Primitive peoples would have found their reflections in the surface of still ponds, while mirrors used in early Greek and Roman civilizations and in Europe during the Middle Ages were highly polished pieces of metal that reflected light off their surfaces. However, the real leap forward for vanity occurred during the early sixteenth century, when the Venetians developed a method of backing a plate of flat glass with a thin layer of reflecting metal that was an amalgam of tin and mercury, much increasing the clarity of the reflection.

The earliest mirrors were hand mirrors used for personal grooming, and later as objects of household decoration with frames made of ivory, silver, or carved wood. The chemical process of coating a glass surface

> *"The world has become uglier since it began to look into a mirror every day."*
>
> Karl Kraus, journalist, poet, and playwright

with metallic silver, from which modern techniques of mirror-making were developed, was discovered by Justus von Liebig in 1835. English brothers Robert and James Adam designed large and elaborate fireplaces that used mirrors to great effect. In the nineteenth century, mirrors were incorporated into pieces of furniture, such as wardrobes and sideboards.

In modern times, mirrors are used in scientific apparatus such as telescopes, cameras, and lasers and in industrial applications. Mirrors designed for electromagnetic radiation with wavelengths other than that of visible light are also used in manufacturing, especially of optical instruments. **MF**

SEE ALSO: METALWORKING, GLASS, LENS, TELESCOPE, COMPOUND CAMERA LENS, LASER, HUBBLE SPACE TELESCOPE

Pocket Watch (c. 1508)

Henlein makes the first pocket-sized timepiece.

The world's first portable timepiece, which came to be known as a "pocket-clocke," was made in Nuremberg, Germany, between 1504 and 1508 by the former apprentice locksmith and clockmaker Peter Henlein (c. 1479–1542). The miniaturization of timepieces began with the invention of the coiled spring in Italy in the late 1450s along with the development of various escapement mechanisms. However, the real breakthrough proved to be Henlein's invention of the balance spring, which greatly improved the precision of a watch's spring-driven interior.

Henlein's new pocket watch measured only a few inches in diameter, chimed on the hour, and could run for up to forty hours before it required rewinding. It was driven by scaled-down steel wheels and hand-forged springs that, despite representing a huge technological leap forward, were nonetheless persistently inaccurate. The coiled springs unwound at a varying speed, causing the watch to slow as the mainspring unwound. Early springs also had a maddening tendency to break when tightly wound, although later refinements such as cam springs were added to compensate for inherent irregularities.

It took Henlein ten years to develop his first pocket watch (this and his subsequent timepieces would go on to become known as his "Nuremberg eggs"). The high cost of the innovation saw the pocket watch become largely a status symbol of the upper classes, yet the popularity of the pocket watch was to continue for 400 years, not least as a formal gift of readily perceived worth, until the invention of the wristwatch in the early twentieth century. **BS**

SEE ALSO: CLOCKWORK MECHANISM, COILED SPRING, DIGITAL WATCH, ATOMIC CLOCK, DIGITAL CLOCK, QUARTZ WATCH

↗ *This 1650 pocket watch, like those of Henlein, indicates the time with only a single hand.*

"Peter Henlein . . . makes from a little iron a pocket clock with a lot of wheels."

Johannes Cocleus, historian

Etching for Print (c. 1510)

Hopfer simplifies printmaking.

As a decorative technique, etching had been in use for many years before the birth of Daniel Hopfer (1470–1536), possibly since antiquity. His innovation was to apply the method to printmaking. The etching process begins by covering a metal plate with a waxy material called a ground. Lines are then scratched into the ground with a needle to expose bare metal where the artist wants lines to appear on the print. The plate is then washed with (or dipped into) acid, which cuts into the exposed metal, leaving lines etched in the plate.

The longer the plate is submerged, the deeper the incision, and the darker the lines will appear on the print. For a more sophisticated finished print, the process can be repeated to allow for different tones

> "He is praised as an innovator, yet he is also maligned as derivative and labeled a mere craftsman."

Freyda Spira, historian

within the piece. Once ready, the plate is covered in ink that is then wiped away, leaving ink only in the incisions and rough areas. The plate is covered by a sheet of wet paper and passed through a press. The pressure of the press forces the paper into the incisions, leaving a mirror image of the plate on the paper.

Hopfer's technique was influential because it was easy. Another way of making prints is to engrave, but this requires metalwork skills. The only prerequisite for etching is to be able to "draw" into the ground. **JG**

SEE ALSO: METALWORKING, INK, PAPER, MOVABLE TYPE, WOODBLOCK PRINTING, LITHOGRAPHY

◄ The Village Dance, one of the earliest etchings made by Daniel Hopfer in the sixteenth century.

Sand Casting (c. 1540)

Biringuccio revives ancient skills.

Sand casting with molten metal ranks as one of the oldest of the manufacturing technologies. For many years it was a dark art and its mysteries were known only to a select few. A sixteenth-century Italian metallurgist and arms maker, Vannoccio Biringuccio (1480–c. 1539), would change this with his seminal work, De la pirotechnia (1540). Published in Venice a year after Biringuccio's death, the book is a veritable encyclopedia of metallurgical knowledge and constitutes some of the earliest printed information on sand casting and foundry techniques in general.

Born in Siena, Biringuccio, under the patronage of an Italian merchant politician and part-time tyrant, Pandolfo Petrucci, traveled widely throughout Italy and Germany, accumulating the information and experiences that he would summarize in his book.

During a typical sand-casting operation, a model or "pattern" of the item to be cast is positioned in a frame. Sand, moistened to bind it together, is then pressed closely around the pattern. This moist sand mix is a critical feature of the process and through the years various binding agents have been adopted, some of them quite unusual. Biringuccio himself recommended the use of human urine and the dregs from beer vats for the purpose.

When the frame is full and the sand rammed down, it is turned over and another frame temporarily attached. The process is then repeated, the frames separated, and the model carefully removed, leaving a cavity in each of the sand boxes that represents one half of the item to be cast. Channels ("sprues" and "runners") are cut into the sand to allow molten metal to be poured into the mold, and air to escape during the pouring. The frames are now securely reassembled together and the metal is poured. **MD**

SEE ALSO: METALWORKING, BLAST FURNACE, ELECTRIC ARC FURNACE

Combination Lock

(*c.* 1550)

Cardano devises a lock that is opened with a memorized code.

In 1206 an Arab scholar, Al-Jazari, published a book called *Kitáb fí ma'rifat al-hiyal al-handasiyya* (*Book of Knowledge of Ingenious Mechanical Devices*). The little-known academic described numerous devices that had not yet been invented, including the waterwheel, the crankshaft, and the combination lock.

In 1550 an eccentric Italian mathematician expanded on Al-Jazari's early ideas and produced what is now recognized as the first combination lock. Gerolamo Cardano (1501–1576), notoriously short of money, kept himself financially afloat through successful gambling and chess playing. But he was also a notable inventor and produced a practical product to help keep possessions secure.

Cardano's design used a number of rotating discs with notches cut into them. The lock could be secured by a pin with several teeth that hooked into the rotating discs. When the notches in the discs came into alignment with the teeth on the pin, the lock could be opened.

Today, Cardano's combination lock is considered to be one of the least secure types because, if not machined precisely, it can be opened without knowledge of the correct combination. An open lock can be achieved in very little time simply by rotating the discs and listening for the clicks indicating that the tooth has settled into its corresponding notch.

Cardano's original design is still used for low-security bicycle locks and briefcases, but now it must compete with sophisticated electronic combination locks whose codes are not cracked so easily. **FS**

SEE ALSO: MECHANICAL LOCK, PADLOCK, TUMBLER LOCK, SAFETY LOCK, TIME-LOCK SAFE, YALE LOCK

"We ride nonpolluting bicycles to save mankind, but we lock them because we can't trust mankind."

Unknown

◥ *This eighteenth-century English combination lock could accommodate any combination of four letters.*

Condom

(*c*. 1560)

Falloppio invents the prophylactic sheath.

The Italian anatomist Gabriele Falloppio (1523–1562) posthumously published the first description of the condom in *De Morbo Gallico* (1564), a treatise on syphilis. To help counter the spread of the sexually transmitted disease, Falloppio invented a linen sheath that, when dipped in a solution of salt, formed a protective barrier during intercourse. To attract the ladies, the condoms were secured by pink ribbons. Falloppio claimed that none of the 1,100 men who used the device became infected with syphilis.

This is not to say Falloppio's condom was the first. Cave paintings from Combarelles in France and drawings from ancient Egypt have been found

---

"The condom is an armor against enjoyment and a spider web against danger."

Madame de Sévigné, writer

---

depicting men wearing condoms. Over the years condoms have been made from oiled paper, thin leather, fish bladders, and even tortoiseshell.

By 1844 Charles Goodyear (of tire fame) had patented a process for the vulcanization of rubber, where intense heat transformed rubber into a strong, elastic material. The first condoms to be made of vulcanized rubber were as thick as bicycle tires with seams running down their sides. In the 1880s an updated manufacturing process led to condoms being produced by dipping glass molds into liquid latex. This process removed the seam, thus making condoms an altogether more practical prospect. **JF**

SEE ALSO: RUBBER BAND, SYNTHETIC RUBBER, SILICONE RUBBER, BIRTH CONTROL PILL, VIAGRA

Jointed Artificial Limb

(*c*. 1564)

Paré initiates modern prosthetics.

The jointed prosthetic limb originated in the 1500s and steadily improved in the next five centuries. Credit for the invention goes to Ambroise Paré (circa 1510–1590), a French barber-surgeon better known for some of his earlier achievements. For example, during the siege of Turin (1536–37) he realized that gunshot wounds were not poisonous and did not need to be cauterized with boiling oil. In his book of 1545, *La Méthod de traicter les playes faites par les arquebuses et aultres bastons à feu* (*The Method of Treating Wounds Made by Arquebuses and Other Guns*), Paré recommended simple dressings and ointments. The Frenchman also promoted the tying off of blood vessels to prevent hemorrhage during surgery (ligaturation), which had been practiced for more than a thousand years but had fallen into disuse.

Paré then invented a prosthetic limb for above-the-knee amputees, to be fitted to the thigh. It incorporated a kneeling peg leg with a knee joint that could be released by a thong running to the hip. Paré went on to devise a hand for a French Army captain to use in battle. Called "Le Petit Lorrain," its thumb was fixed but the fingers operated by springs and catches.

A similar hand had previously been devised for Gotz von Berlichingen, a knight who had garnered a romanticized reputation as the "German Robin Hood" for his kidnapping of nobles and attacking merchant convoys for booty. Later he led a section of rebels during the Peasant's War of 1525, but he quit before their ultimate defeat. In 1504 he lost his right arm when "friendly" cannon fire during the siege of Landshut struck his sword and it fell and severed his arm. He lost his hand as a result but gained another moniker: "The Knight of the Iron Hand." **SS**

SEE ALSO: ANESTHESIA, ARTIFICIAL LIMB, ARTIFICIAL HEART

Graphite Pencil (*c.* 1564)

The English invent an erasable marker.

The graphite pencil was invented in England in 1564 following the discovery of an extensive deposit of pure graphite at Seathwaite Fell near Borrowdale in Cumbria. The Borrowdale deposit was so pure it could be cut into sheets and subsequently into tiny square-profile lengths. The material left a darker mark than other less pure graphite composites, possessed a greasy texture, was extremely brittle, and quickly dirtied the hands of the user, thus requiring some form of protective sheath. However, the fact that it could be erased made it a popular alternative to ink.

The first known account of a graphite pencil was written in 1565 by the German scientist Konrad von Gesner. He described a rudimentary lead pencil

---

"A short pencil is more reliable than the longest memory."

Proverb

"enclosed in a wood holder," and it was not until the 1660s that a Keswick joiner hollowed out a piece of wood to create the forerunner of today's graphite-rod pencil. Graphite was known as *blacklead* or *plumbago* (Latin for "lead ore") until the new name was coined by the Swedish chemist K. W. Scheele as a footnote to his *Treatise on Fossils* (1779).

The Seathwaite Fell deposit remains the purest deposit of graphite ever found. It was so valuable and easily extracted that in 1752 the British parliament passed a law making the theft of graphite punishable by imprisonment. The first commercial production of pencils began in Nuremberg, Germany, in 1761. **BS**

SEE ALSO: PAPER, INK, QUILL PEN, FOUNTAIN PEN, PENCIL SHARPENER, CRAYON, BALLPOINT PEN

Theodolite (*c.* 1571)

Digges facilitates the triangulation process.

An instrument designed for measuring horizontal and vertical angles, the theodolite is of critical importance to the surveying profession. Comprising a telescopic lens that can tilt on both horizontal and vertical axes, the theodolite is essential in "triangulation," a long-established technique for surveying tracts of land. The distance between two points is measured, to become the baseline of a triangle. The theodolite is placed at one end of the line and used to determine the angle to a predefined distant point (which makes the third corner of the triangle). The instrument is then moved to other end of the baseline, where a second angle is measured to that same point. Simple trigonometry is then employed to calculate accurately the length of the other two sides.

The word *theodolite* first appears in a surveying textbook, *A Geometric Practice Named Pantometria* (1571) by mathematician and scientist Leonard Digges. Although tools for measuring angles had existed for centuries, it was Digges who introduced an "altazimuth" instrument—one that could be moved on two perpendicular axes. The inventor himself died penniless twelve years before the publication of his book. His historical standing can be attributed to his son, Thomas, who worked tirelessly to continue his father's work, and who would become a significant figure in his own right, popularizing the principles of modern science.

There have since been many technological refinements, namely Jesse Ramsden's Great Theodolite of 1797, used in the Principal Triangulation of Great Britain—the original Ordnance Survey. **TB**

SEE ALSO: VERNIER SCALE, SCREW MICROMETER

➡ *Ramsden's highly accurate Great Theodolite was first commissioned for the survey of southern Britain.*

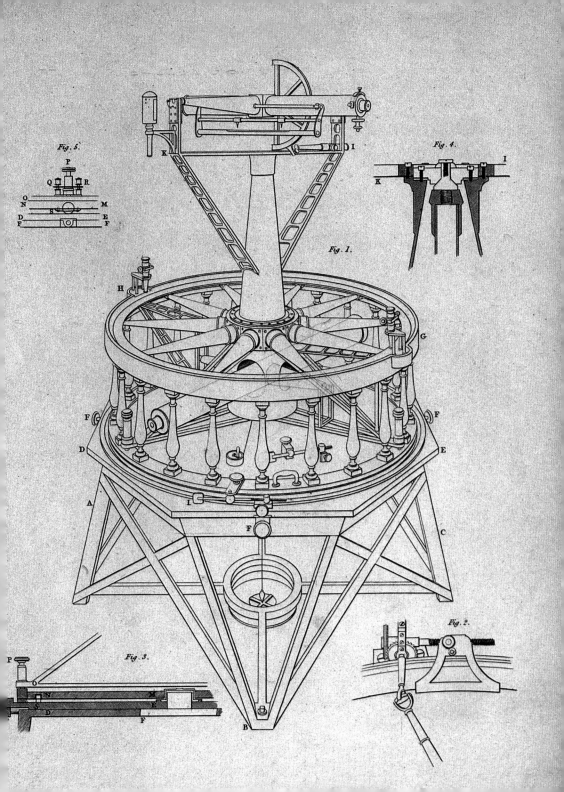

Fig. 5.

Fig. 4.

Fig. 1.

Fig. 2.

Fig. 3.

Top continuation

Day		March	Day		April
13	b	Euphrasia	13	e	Hermene
14	c	Matthildis	14	f	Tihurti,
15	d	Longin,	15	g	Basilissa,
16	e	Heribert,	16	A	Callist,
17	f	Patrici,	17	b	Anicet,
18	g	Narciss,	18	c	Eleutheri,
19	A	Ioseph	19	d	Thimon
20	b	Ioachim	20	e	Sulpiti,
21	c	Benedict,	21	f	Anselm,
22	d	Basili, M.	22	g	Soter
23	e	Victorian,	23	A	Leonti,
24	f	Gabriel	24	b	Georgi,
25	g	Mariæ Ver.	25	c	Marc,
26	A	Ludgeri,	26	d	Clet,
27	b	Rupert,	27	e	Anthim,
28	c	Guntram,	28	f	Vitalis
29	d	Eustachi,	29	g	Petrus M.
30	e	Ioann Cl.	30	A	Cath. Se.
31	f	Amos			

Day		January	Day		February
12	e	Arcadi,	13	A	Lucaia
13	f	Hilari,	13	b	Gregori,
14	g	Felix,	14	c	Valentin,
15	A	Paul. E.	15	d	Faustin,
16	b	Marcell,	16	e	Iuliana
17	c	Antoni, A	17	f	Polychron,
18	d	Pet. St. R.	18	g	Simeon,
19	e	Marius	19	A	Gabin,
20	f	Fab. Seb.	20	b	Euchari,
21	g	Agnes	21	c	Maximian,
22	A	Vincenti,	22	d	Pet. St. A.
23	b	Emerenti,	23	e	Siren,
24	c	Thimoth.	24	f	Mathias
25	d	Pauli Be.	25	g	Victor
26	e	Polycarp,	26	A	Alexand,
27	f	Ioh. Chry.	27	b	Roman,
28	g	Carol, M.	28	c	Macari, M.
29	A	Francisc. S			Im Schalt.Iahr
30	b	Martina I.			hat Febr. 29 Tag,
91	c	Pet. Nol.			Matth. ist den 25.

MAIUS. IUNIUS.

MAIUS. Hat 31. Tag. ⊙
in ♊ h. ♄. Or. 4.
20. Oc. 7.40 T.14.
20. N. 8.40 Z.21.

IUNIUS. Hat 30. Tag. ⊙
in ♋ h. ♃. Or. 4.
0. Oc. 8. 0 T. 16.
0. N. 8. 0 Z.21.

Day		May	Day		June
1	b	Philippi Iac.	1	e	Inventi,
2	c	Athanas,	2	f	Erasm,
3	d	Creutz Erf.	3	g	Lucian,
4	e	Monica	4	A	Quirin,
5	f	Gotthard,	5	b	Bonifaci,
6	g	Ioh. Port.	6	c	Norbert,
7	A	Stanisla,	7	d	Rrobert,
8	b	Mich. Ersch.	8	e	Medard,
9	c	Greg. Naz.	9	f	Prim,
10	d	Gordian,	10	g	Getuli,
11	e	Mamert,	11	A	Barnabas
12	f	Pancrati,	12	b	Basilides
13	g	Servati,	13	c	Anth. Pad.
14	A	Bonifaci,	14	d	Basili,
15	b	Robert,	15	e	Vitus
16	c	Vbald,	16	f	Benno
17	d	Bruno B.	17	g	Nicander
18	e	Venanti,	18	A	Marc, M.
19	f	Pet. Cölest.	19	b	Geruasi,
20	g	Bern. Sen.	20	c	Silueri,
21	A	Hospiti,	21	d	Alban,
22	b	Cassi,	22	e	Achati,
23	c	Desideri,	23	f	Edeltraud
24	d	Iohanna	24	g	Iohann T.
25	e	Vrban,	25	A	Gallican,
26	f	Phil. Ner.	26	b	Ioh. Paul
27	g	Ioh. Pabst	27	c	7. Schläf.
28	A	German,	28	d	Leo 2. P.
29	b	Maximin,	29	e	Pet. Paul
30	c	Felix,	30	f	Pauli Ged.
31	d	Petronilla			

MARTIUS APRILIS

MARTIUS. Hat 31. Tag. ⊙.
in ♈ h. ♂. Or. 6.
0. Oc. 6. 0 T. 12.
0. N. 12. 0 Z. 20.

APRILIS. Hat 30. Tag. ⊙.
in ♉ h. ♀. Or. 5.
4. Oc. 6. 56 T. 13.
52. N. 10. 8. Z. 19.

Day		March	Day		April
1	d	Suidbert,	1	g	Theodora
2	e	Simplici,	2	A	Franc. d. P.
3	f	Cunegund,	3	b	Nicetas
4	A	Casimir,	4	c	Isidor,
5	A	Gerasim,	5	d	Zeno M.
6	b	Euagri,	6	e	Celestin,
7	c	Thom. Aq.	7	f	Hegesipp,
8	d	Philemon,	8	g	Amand,
9	e	Francisca	9	A	Mariæ Cle.
10	f	40. Mart.	10	b	Ezechiel
11	A	Firmini,	11	c	Leo 1. P.
12	A	Gregori, P.	12	d	Zeno M.
13	b	Euphrasia	13	e	Hermene
14	c	Matthildis	14	f	Tihurti,
15	d	Longin,	15	g	Basilissa,
16	e	Heribert,	16	A	Callist,
17	f	Patrici,	17	b	Anicet,
18	g	Narciss,	18	c	Eleutheri,
19	A	Ioseph	19	d	Thimon
20	b	Ioachim	20	e	Sulpiti,
21	c	Benedict,	21	f	Anselm,
22	d	Basili, M.	22	g	Soter
23	e	Victorian,	23	A	Leonti,
24	f	Gabriel	24	b	Georgi,
25	g	Mariæ Ver.	25	c	Marc,
26	A	Ludgeri,	26	d	Clet,
27	b	Rupert,	27	e	Anthim,
28	c	Guntram,	28	f	Vitalis
29	d	Eustachi,	29	g	Petrus M.
30	e	Ioann Cl.	30	A	Cath. Se.
31	f	Amos			

Gregorian Calendar (1582)

Lilio resets the Catholic clock.

The Julian year (introduced by Julius Caesar in 45 BCE) contained exactly 365.25 days and had a leap year every fourth year (when the year number was divisible by four). But the actual year is 365.24219879 days long and thus the Roman calendar gradually became out of step with reality. By the sixteenth century, the calendar was ten days adrift from the seasons.

In 1576 Pope Gregory XIII assembled a commission of astronomers, mathematicians, and clergy, and this advisory body eventually adopted a plan suggested by the Calabrian physician Luigi Lilio, who was also known as Aloysius Lilius. On February 24, 1582, the pope declared that Thursday, October 4, 1582, was to be followed by Friday, October 15. From that date, century leap years would only be allowed if the year number was exactly divisible by 400. Thus, 2000 was a leap year, to be followed by 2400, 2800, and so on.

Roman Catholic countries immediately adopted the Gregorian Calendar, but the Protestant and Greek Orthodox countries would have none of it. Norway and Denmark finally changed their minds in 1700. Great Britain and its colonies, including the United States, eventually followed, in 1752. In that year, in those territories the day after September 2 was September 14. Japan changed to the Gregorian calendar in 1873, Russia adopted it in 1918 after the Revolution, and Greece waited until 1923.

There are still minor problems. By 13000 the calendar will be ten days out of step again. And the fact that the Earth's spin rate is slowly decreasing in a complicated fashion means that the system should be subtracting a day about every 2,000 years. **DH**

SEE ALSO: **LUNAR CALENDAR, ATOMIC CLOCK**

← *Two sections of the blade of a German sword, circa 1686, engraved with the Gregorian calendar.*

Stocking Frame (1589)

Lee mechanizes knitting.

William Lee (circa 1550–1610), a clergyman from Nottinghamshire, England, invented the stocking frame in 1589. One story suggests that he invented it to relieve his mother and sisters of the burden of knitting; another has it that a girl was showing more interest in her knitting than in him.

Knitted fabrics are constructed by the interlocking of a series of loops, with each row of loops caught into the previous row. The stocking frame allowed production of a complete row of loops, held by a long bar similar to a knitting needle; a second bar opposed it, and each loop, picked up by a piece of wire, was transferred to the first bar.

Lee's first machine, which produced coarse wool

> *"The privilege of making stockings for everyone is too important to grant to any individual."*

Queen Elizabeth I to William Lee

---

stockings, was refused a patent by Queen Elizabeth I. An improved machine produced silk stockings of finer texture, but again she refused, saying she was concerned for the livelihoods of hand knitters.

King Henry IV of France encouraged new industries, so Lee and his brother James moved nine hand frames to Rouen. They prospered until Henry IV's assassination in 1610, after which Louis XIII imposed restrictions on foreign industries. The Lees returned to England, where James set up workshops in London.

William Lee died in Paris in 1610. Today's knitting industry, while employing machinery driven by computers, still incorporates many of his ideas. **MF**

SEE ALSO: **CLOTHING, SEWING, WOVEN CLOTH, SPINNING WHEEL, SPINNING JENNY, SPINNING MULE, POWERED LOOM**

Microscope (*c.* 1590)

Hans and Sacharias Jensen combine lenses in the first compound microscope.

The earliest microscope was no more than a single small lens that magnified between six and ten times. Sacharias Jansen and his father, Hans, a lens maker, experimented with combinations of lenses and realized that greater magnification could be obtained by an inversion of the telescope. Their compound microscope combined a magnifying objective lens (the one closest to the object being investigated) with an eye lens at the opposite end of a tube. A focusing device was added by the Italian Galileo Galilei.

The circulation of blood through capillaries was observed by the Italian physiologist Marcello Malpighi (1624–1694). The popularity of microscopes was greatly enhanced by the publication of *Micrographia* (1655) by English scientist Robert Hooke.The Dutchman Anthoni van Leeuwenhoek (1632–1723) used a microscope to count the number of threads in woven cloth, and his refined instrument could magnify 270 times. Van Leeuwenhoek's microscope only had a single lens with a radius of curvature of roughly 0.7 millimeters. He was the first to see microorganisms and blood cells. Of the 500 microscopes manufactured by van Leeuwenhoek, about ten still survive.

In the eighteenth century, improved glass, coupled with multiple objective lenses with smaller focal lengths that could see much finer detail, led to much better microscopes. Stages were added so that samples under investigation could be held securely. The nineteenth century brought the familiar microscope form with its firmly mounted, vibration-free optical tube. **DH**

"Nature composes some of her loveliest poems for the microscope and the telescope."

Theodore Roszak, academic and historian

SEE ALSO: GLASS, LENS, TELESCOPE, ELECTRON MICROSCOPE, SCANNING TUNNELING MICROSCOPE

◩ *Associated with Robert Hooke, this microscope circa 1675 was part of King George III's collection.*

Newspaper (1605)

Carolus publishes the world's earliest printed news.

In 1605 Johann Carolus (1575–1634) published the first printed issue of *Relation aller Fürnemmen und gedenckwürdigen Historien* in Strasbourg, France, thereby giving the world its first newspaper.

Similar concepts had been around for more than 1,500 years. Julius Caesar established the *Acta Diurna*—a newsletter carved on stone or metal—for the citizens of Rome, and, almost 800 years later in 713, the Chinese Tang Dynasty published the *Kaiyuan Za Bao*, a news bulletin handwritten on silk.

Initially Carolus copied his newsletters by hand and sold them to rich subscribers. But in order to make his publication affordable to more people, and thus increase his revenue, he bought a printing shop in 1604. Despite his modern approach, *Relation* did not survive, so today the Dutch daily *Haarlems Dagblad* (after merging with the *Oprechte Haarlemsche Courant* from 1656) is the world's oldest existing newspaper.

Initially the medium was viewed skeptically by some. Benjamin Harris found out as much when he tried to establish the United States's first newspaper, *Publick Occurrences, Both Forreign and Domestick*, in 1690. The paper was meant to be "furnished once a moneth (or if any Glut of Occurrences happen, oftener)." But he was forced to abandon this plan after only one issue when outraged government officials decided that his publication contained "reflections of a very high order" and had been printed "Without the least Privity or Countenance of Authority."

While the Internet now poses a danger to many newspapers, more than a billion people worldwide still read a daily newspaper in print every day. **DaH**

SEE ALSO: INK, PAPER, MOVABLE TYPE, PRINTING PRESS WITH MOVABLE METAL TYPE

↗ *This 1609 edition of* Relation *is the oldest surviving newspaper, owned by Heidelberg University.*

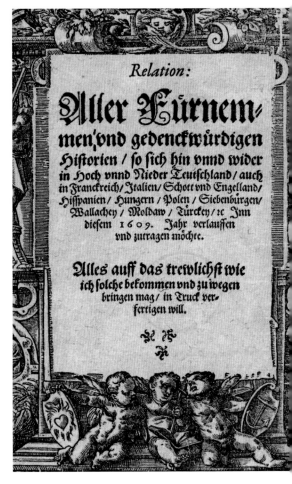

"A newspaper consists of just the same number of words, whether there be any news in it or not."

Henry Fielding, novelist

Telescope (*c.* 1609)

Lippershey develops an instrument for examining the heavens.

The legend goes that, playing one day in their father's spectacle shop, two Dutch children realized that if they looked through both a concave lens close to their eye and a concave lens held at arm's length, the local church tower was greatly magnified. Their father, Hans Lippershey (circa 1570–1619), then mounted the two lenses in a tube and tried to sell the device to the Dutch Army. Whether the credit for this invention should go to Lippershey or to, for example, Zacharius Janssen or Jacob Metius, or even the Englishman Leonard Digges, has become a matter of considerable debate. At the very least, Lippershey is generally credited with popularizing the device, and creating and disseminating designs for the first practical telescope. Soon similar instruments, known as "Dutch Trunks," were appearing all over Europe.

The Italian astronomer and physicist Galileo Galilei heard about the new device when he was in Venice in May 1609. Returning to his university in nearby Padua, he made a telescope that magnified by about twenty times and had a field of view of about one-tenth of a degree. Using this, he discovered that the sun had spots, Jupiter was accompanied by four satellites, Venus had phases, and the moon was mountainous. These results he published in March 1610 in his work, *Siderius Nuncius* (*The Sidereal Messenger*).

Telescopic astronomy never looked back. By 1611, the German astronomer Johannes Kepler was using a telescope consisting of two convex lenses, an instrument that gave greater magnification but an inverted image. In 1668 the English genius Sir Isaac Newton invented the reflecting telescope, which uses a curved mirror rather than a large lens to collect and focus light, thus eliminating the problem of severe chromatic aberration. **DH**

"[Reaching] the dim boundary . . . we measure shadows, and we search among ghostly errors . . ."

Edwin Hubble, *The Realm of the Nebulae* (1936)

⤒ *The telescope with which Galileo explored the solar system is held at the Museo della Scienza, Florence.*

⮕ *These observation-based sepia wash studies of phases of the moon were executed by Galileo in 1609.*

SEE ALSO: GLASS, GLASS MIRROR, LENS, RADIO TELESCOPE, X-RAY TELESCOPE, HUBBLE SPACE TELESCOPE

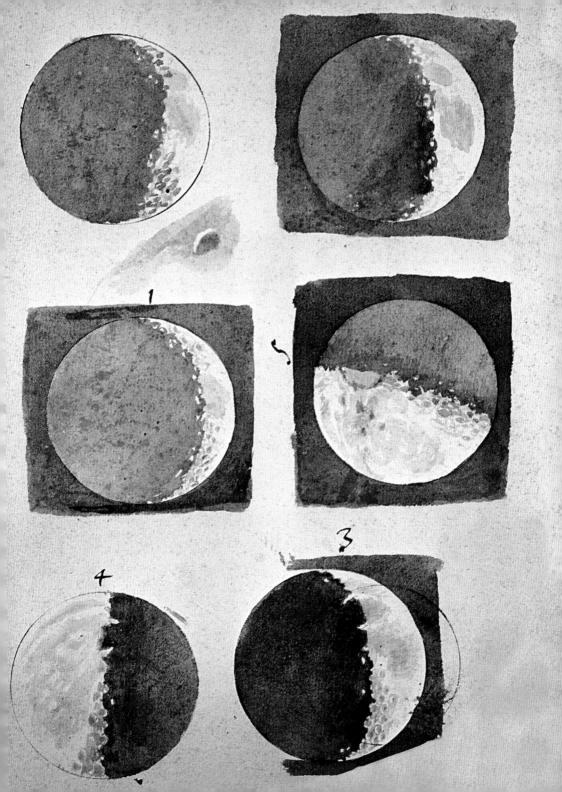

Flintlock (*c.* 1612)

An unknown French gunsmith devises a mechanism for igniting gunpowder.

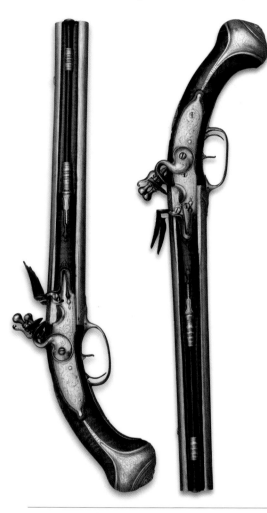

The basic mechanism from which the traditional flintlock originated is thought to have first appeared on a firearm made for King Louis XIII of France. The name of the French courtier Marin Le Bourgeoys appears on the flintlock, and it is thought to have been made in about 1612.

The flintlock works as follows. First, the hammer of the gun, which holds a piece of flint, is pulled back or rotated to the half-cock position. Gunpowder is poured into the barrel, followed by ammunition—often a steel ball—and both are pressed into position with a ramrod. A small amount of finely ground gunpowder is then placed in a compartment below the hammer, known as the flashpan. The hammer is then pulled back or rotated to the full-cock position, and the gun is ready to be fired. When the trigger is pulled, the hammer springs forward and its flint hits the frizzen, a curving piece of steel. The frizzen moves aside to expose the gunpowder in the flashpan, and sparks from the flint ignite the gunpowder in the flash pan; this, in turn, explodes the main gunpowder charge, which projects the ball out of the barrel.

Anyone wishing to discharge a round of ammunition successfully anytime in the seventeenth or eighteenth centuries was obliged to master the sequence of actions required by the flintlock mechanism. Even the fastest experts took about fifteen seconds to load and fire a flintlock weapon.

The flintlock is responsible for a number of phrases in the English language still in use today. "Lock, stock, and barrel" and "going off half-cocked" both have their origins in flintlock operation. **CL**

"Political power . . . grows out of the barrel of a gun. . . ."

Mao Zedong, political leader

SEE ALSO: GUNPOWDER, GUN, MUSKET, ARQUEBUS, BREECH-LOADING GUN, GUN SILENCER

◩ *This pair of English double-barreled flintlock pistols was made by Andrew Dolep (1648–1713).*

Cigarette (1614)

Beggars in Seville find a new way to smoke.

When Christopher Columbus arrived in the Americas in 1492 he was struck by the locals' indulgence in an unfamiliar habit. The Mayans had been smoking dried tobacco leaves since the first century BCE, and by the time the Spanish sailors discovered the New World the custom had spread throughout the continent. Possibly thinking their foreign visitors divine, the indigenous Arawaks offered Columbus and his men some of the leaves—who immediately threw them away.

One member of the crew, Rodrigo de Jerez, was not as skeptical, though, and very soon he also "drank" the dried tobacco leaves wrapped in palm or maize, thus becoming the first European smoker. Back home, his newly acquired habit frightened his compatriots so much that the Inquisition put him in jail.

Over the next few centuries the practice gradually spread all over the world, but to a mixed reception. Initially European doctors praised its medicinal properties—the French ambassador to Portugal, Jean Nicot de Villemain (who gave nicotine its name) even described it as "a panacea." Soon, however, people were beginning to realize its dangers and ban it. Mexico was the first country to outlaw smoking in places of worship, in 1575, and Turkey, Russia, and China temporarily declared the habit to be a crime punishable by execution in the 1630s.

A few years before that, in 1614, Seville in southern Spain had become the center of cigar making. It was here, in the same year, that beggars created the first cigarettes by taking leftover tobacco from cigars and rolling it in paper. However, snuff, cigars, and pipes remained more popular than cigarettes in the West for another 250 years, but then British soldiers fighting in the Crimean War (1853–1856) were won over by the cigarettes smoked by their Turkish allies. **DaH**

SEE ALSO: CONTROLLED FIRE, FRICTION MATCH, SAFETY MATCH, FIRE-LIGHTING LAMP

Slide Rule (c. 1622)

Oughtred creates a ready-reckoner.

The slide rule is a mechanical device used to carry out complicated mathematical functions. It is based on two logarithmic scales that move parallel to each other and are aligned according to the desired calculation. To multiply two numbers, for example, the logs are added and raised to the power ten; to divide, the logs are subtracted.

In 1620 Edmund Gunter (1581–1626), an English clergyman and Gresham College Professor of Astronomy, produced a logarithmic scale and used dividers to take off specific distances to do the calculations. William Oughtred (1574–1660), mathematician and rector of Albury, did away with the dividers by using two sliding Gunter rules side by side

> *"A computer who must make many difficult calculations usually has a slide rule close at hand."*
>
> Pickett manual

in circa 1622 and described his circular slide rule in *Circles of Proportion and the Horizontal Instrument* (1632). Sliding different distances multiplied and divided by different quantities. Seth Partridge (1617–1689) invented the modern slide rule in which the inner scale (the slide) is held by, and moves within, the outer scales, known as the stock or body of the rule.

In 1775 John Robertson added a cursor (an etched line in a transparent sliding attached plate) so that settings could be noted and transferred to any of a series of parallel scales. By 1815, P. M. Roget had added log-log scales, enabling powers, exponentials, and roots to be assessed easily. **DH**

SEE ALSO: STANDARD WEIGHTS AND MEASURES, POCKET CALCULATOR, DIGITAL ELECTRONIC COMPUTER, LAPTOP

Mechanical Calculator (*c.* 1623)

Schickard automates the manipulation of figures.

Early inventions to speed up calculations focused on manual solutions such as Napier's bones, which consisted of multiplication tables inscribed onto bones for calculating sums. Seeing John Napier's work, the German polymath Wilhelm Schickard (1592–1635) created a mechanical calculator that automated the process of calculation and incorporated Napier's bones. In 1623, he designed and built the "calculating clock." At around the size of a typewriter, it could handle numbers of up to six digits in length.

The calculator used a direct gear drive and rotating wheels to add and subtract. When a wheel made a complete turn, the wheel adjacent rotated one-tenth of a turn. Dials on the lower part of the machine were turned one way to perform addition, and the opposite way to perform subtraction. These dials were joined by teeth-bearing internal wheels that carried one digit every time the wheel passed from nine to zero. The upper part of the machine used Napier's bones to multiply and divide. The machine was fitted with a bell that rang when a calculation produced a result of more than six digits (and was thus too long to display).

Schickard began building a replica of his calculating clock for astronomer Johannes Kepler but it was never completed because a fire engulfed his workshop. He gave Kepler detailed instructions on how to build the calculator, but then Schickard and his family died of the plague in the 1630s and the prototype was lost. It was not until the 1950s that a sketch of the calculating clock was discovered among Kepler's papers in Russia, proving that Schickard was the originator of the mechanical calculator. **RB**

"[It] computes the given numbers automatically; adds, subtracts, multiplies, and divides."

Wilhelm Schickard

SEE ALSO: TALLY STICK, ABACUS, POCKET CALCULATOR, MECHANICAL COMPUTER, DIGITAL ELECTRONIC COMPUTER

◹ *A seventeenth-century calculator by Blaise Pascal with figures in slots above a row of metal wheels.*

Vernier Scale (1631)

Vernier refines micro measurement.

Vernier callipers are a sliding, adjustable-jaw device for measuring distances of a few centimeters to an accuracy of 0.01 centimeters. A main (ruler) scale is marked off with 0.1 centimeter divisions. Sliding parallel and alongside the main scale is a much smaller vernier scale on which ten divisions are equally spaced over 0.9 centimeters of the main scale. To subdivide the 0.1 centimeter division on the main scale into ten, the user has to select the nearest vernier division that is in line with one of the main scale divisions.

The scale was invented by French scientist and engineer Pierre Vernier (1580–1637), and the details were published in his 1631 book, *La Construction, l'usage, et les propriétés du quadrant nouveau de*

> *"There are ... fixed boundaries, beyond and about which that which is right cannot exist."*

Horace, *Satires*, Book 1 (35 BCE)

mathématiques (The Construction, Uses, and Properties of a New Mathematical Quadrant), published in Brussels. Vernier was interested in cartography and surveying, and his vernier was first used on the circular scale of a quadrant theodolite. The scale enabled angles to be measured with ease to an accuracy of one minute of arc (one-sixtieth of a degree).

Unfortunately, dividing a circular scale into an equal number of uniform degrees was a matter of considerable complexity and vernier scales did not become a common adjunct to angle-measuring devices such as telescopes and theodolites until the early nineteenth century. **DH**

SEE ALSO: SEXTANT, TELESCOPE, THEODOLITE, SCREW
MICROMETER, SPRING TAPE MEASURE

Screw Micrometer (1635)

Gascoigne improves precise measurements.

William Gascoigne (1612–44) was an English mathematician and astronomer, renowned for making scientific instruments. He was intrigued by the vernier scale and saw its potential for measuring the angular distances between stars.

In circa 1635, while working on precision optics, he noticed that a thread from a spider's web had become trapped at the exact focal point of two lenses, and that he could therefore see it sharply. This inspired him to create a thin marker that could be placed at the focal point of a lens. He added a second linear marker so that, when he looked through his telescope lens, two parallel lines could be seen within the field of view.

One of the markers he linked to a very fine screw thread, which could be used to adjust the distance between the two markers—the other of which would remain fixed. As the angular size of the field of view of a telescope was known, Gascoigne's invention enabled precise measurements to be made of the positions of astronomical objects in the sky.

Gascoigne's micrometer revolutionized accurate measurements in astronomy, but there were other uses to be discovered for this device, beside acting as a finely calibrated telescope sight.

British engineer James Watt, best known for his invention of the steam engine, adapted Gascoigne's idea in 1776 to produce a handheld micrometer screw gauge that measured actual sizes of small objects. By replacing the markers with callipers, and by knowing the size of the threads on the screw which adjusted them, he was able to add measuring wheels to the head of the screw. This allowed minute adjustments and measurements to be made. Watts's instrument significantly advanced the ability of manufacturers to make machine parts of precise dimensions. **DHk**

SEE ALSO: SCREW, ARCHIMEDES SCREW, LENS, MICROSCOPE,
TELESCOPE, VERNIER SCALE

Barometer (1643)

Torricelli researches atmospheric pressure.

Interest in atmospheric pressure arose when miners and well-diggers realized that pumps and siphons would only raise water to a maximum distance of about 33 feet (10 m). Hearing that the Grand Duke of Tuscany had a suction pump that could not raise water as far as he wanted, the Italian physicist Evangelista Torricelli (1608–1647) investigated the problem in 1643, creating what is known as the Torricelli tube.

Imagine that you have such a tube of straight glass, 40 inches (100 cm) long, sealed at one end and filled with mercury, and you carefully invert this tube, keeping the open end dipped in a reservoir of mercury. The mercury will retreat down the tube leaving a vacuum at the top. The height of mercury above the reservoir level will be about 30 inches (75 cm), and the weight of the mercury in the tube will be supported by the pressure exerted by the Earth's atmosphere as it presses down on the mercury in the reservoir.

At the time, many natural philosophers were interested in the properties of the vacuum, and the Torricelli tube was a common demonstration experiment at their meetings. Blaise Pascal (1623–1662) fitted the tube with a graduated scale around the 1660s, and what was a one-off demonstration device developed into an important instrument for measuring the variations of atmospheric pressure.

Pascal asked his brother-in-law to carry a mercury barometer up the nearby Puy-de Dôme mountain and learned that atmospheric pressure decreases with height. Edmund Halley quantified the decrease as being exponential. People at the time noted how the pressure changes with the weather, and the barometer, in various forms, has been an essential tool in weather forecasting since the nineteenth century. **DH**

SEE ALSO: RAIN GAUGE, ANEMOMETER, ANEROID BAROMETER, WEATHER RADAR

Vacuum Pump (1650)

Guericke exploits the vacuum.

A vacuum is an empty space containing nothing, not even air. Anything containing a vacuum has a much lower pressure on its inside than its outside, and this creates a tremendous force. Otto von Guericke (1602–1686), a German scientist, was the first to experiment with the power of the vacuum. In his experiments, he filled containers with water and then used a suction pump to remove the water while trying to avoid letting in any air. Wood was useless for this as it leaked air, so he used glass or metal containers. To minimize air intake, Guericke put his container in another layer of water as it was easier to stop water leakage than air leakage. The inward pressure on the containers was often so great that they would collapse. Further trials

> *"A vacuum is a hell of a lot better than some of the stuff that nature replaces it with."*
>
> Tennessee Williams, *Cat on a Hot Tin Roof* (1955)

led Guericke to the conclusion that spherical containers were optimal because their smooth shape avoided weak points in the structure. To prove the power of his discovery, Guericke demonstrated to Emperor Ferdinand III that neither fifty men nor teams of horses could pull apart two copper hemispheres that contained a vacuum.

In science vacuums are most commonly used for their ability to create a truly empty space, enabling the study of particles without the confusion of air. They are essential in many machines used in industry for pumping liquids and other materials, moving objects, and powering heavy machinery. **LS**

SEE ALSO: MICROWAVE OVEN, TELEVISION, VACUUM-SEALED JARS, VACUUM FLASK, ELECTRIC VACUUM CLEANER

Pendulum Clock (1656)

Huygens uses the pendulum to improve timekeeping accuracy.

Around 1602 Galileo Galilei noticed that the swing period of a pendulum was nearly independent of the amplitude of the oscillation, and this became the most important discovery in the history of horology. In 1656 Dutch mathematician and astronomer Christiaan Huygens (1629–1695) was the first to use a pendulum as a regulating oscillator in a clock.

The swing period of a pendulum is only a function of its length and the local gravitational field, unlike the verge and balance (foliot) oscillator, which it replaced, which had an oscillation period that depended on the force exerted by the driving spring.

Within years of Huygens's discovery, weight-driven pendulum clocks were appearing all over Europe. To provide a sufficient distance for the weights to fall, and to accommodate a reasonably long pendulum—a two-second tick-tock requires a pendulum 3 feet (1 m) long—these clocks were put in long floor-standing cases. These "grandfather" clocks were reliable to an impressive (in those days) twenty seconds a day. Around 1670 the invention of the anchor escapement led to improvements in timekeeping by enabling the amplitude of the pendulum oscillation to be reduced.

In 1676 the more fragile dead-beat escapement was introduced to high-accuracy regulating clocks. This escapement gave the pendulum a "push" only when it was near its vertical position. Coupled with a pendulum made of bars of different metals (usually brass and steel), it ensured that the length did not change as the temperature changed. The accuracy improved to about one second per day, an important aid in the work of astronomical observatories. **DH**

SEE ALSO: WATER CLOCK, SHADOW CLOCK, CLOCKWORK MECHANISM, TOWER CLOCK, DIGITAL CLOCK, POCKET WATCH

↗ *Conceived by Galileo just before his death, the clock was built as a model in the nineteenth century.*

"While fancy, like the finger of a clock, runs the great circuit, and is still at home."

William Cowper, "The Winter Evening" (1785)

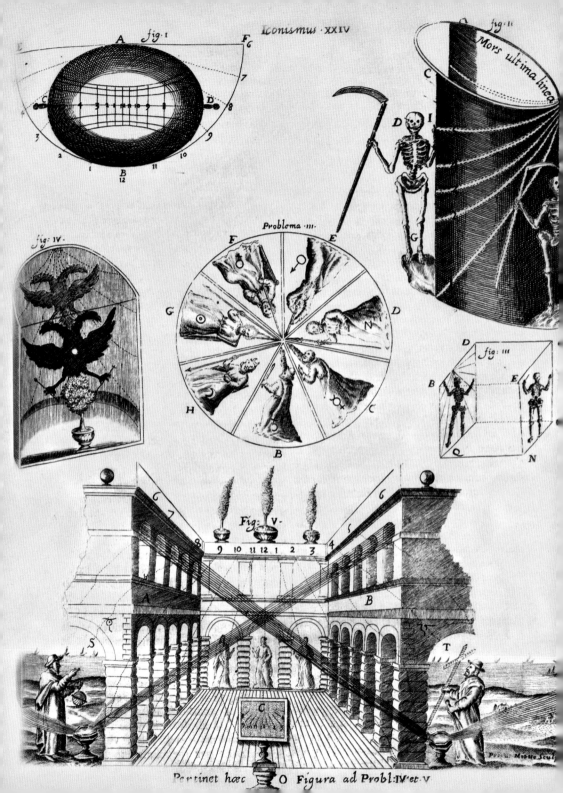

Iconismus · XXIV

fig. I

Mors ultima linea

fig. II

fig. IV

Problema · III

fig. III

Fig: V

Pertinet hæc O Figura ad Probl: IV et V

Magic Lantern (*c. 1659*)

Huygens makes the first image projector.

The "magic" lantern was used to project still images onto a wall or sheet and was an early version of the slide projector. The idea has been understood for many centuries. Light, shining through a translucent picture, will project the image onto a light-colored flat surface. The earliest reference to the use of a lantern to project images is in *Liber Instrumentorum* by Giovanni de Fontana, written around 1420.

Optics developed rapidly in Europe during the seventeenth century. As early as 1659, Dutch scientist Christiaan Huygens (1629–1695) had made a lantern with a lens to focus the light and produce a sharp image. Danish mathematician Thomas Walgensten (1627–1681) traveled throughout Europe in the 1660s, selling the

> "*. . . a lanthorn, with pictures on glass, to make strange things appear on a wall, very pretty.*"
>
> Samuel Pepys, diarist

lanterna magica. Numerous designs survive from this period and, in 1663, optician John Reeves of London was making and selling lanterns. The diarist Samuel Pepys (1633–1703) bought one from Reeves in 1666.

During the eighteenth century, improvements in lenses, light sources, and mirrors transformed the magic lantern into a powerful projector. Showmen traveled throughout Europe putting on elaborate "phantasmagoria" shows, using magic lanterns to present ghost and horror stories. **EH**

SEE ALSO: DAGUERREOTYPE, CELLULOID, KINETOSCOPE, FILM CAMERA/PROJECTOR

◄ *Athanasius Kircher's* Ars Magna Lucis Et Umbrae *(1646) led to the development of the magic lantern.*

Hygrometer (1664)

Folli measures air humidity.

The most sticky, hot, and humid places in the world tend to be found in Southeast Asia, near coastal regions around the equator. Anyone who is not used to the heavy, damp, often motionless air can find them to be very uncomfortable places to live.

Humidity, the moisture content of the air, tends to be high in these places because the heat of the sun causes the air to absorb increased moisture from the surrounding seas and oceans—the air in cold latitudes is relatively dry. But it was not until the 1600s that people were able to measure air humidity.

Technically, Leonardo da Vinci designed the first crude hygrometer in the 1440s, but in 1664 the first practical hygrometer, used to measure the moisture content of air, was invented by the Italian scientist Francesco Folli (1624–1685). Folli's invention was a finely decorated device, made of brass, that contained a mounted paper ribbon acting as a hygroscopic (moisture-absorbing) indicator. When the ribbon changed in length as a result of changes in its water volume, a simple mechanical system moved a pointer on a central brass dial marked with a graduated scale. The pointer indicated variations in humidity.

Some modern hygrometers still use principles very similar to those of Folli's original design. One commonly seen improvement of the original is that blond human hair, rather than paper ribbon, is the medium used to expand and contract in response to variations in atmospheric moisture.

However, there are now many different types of hygrometer. The most common is the dry and wet-bulb psychrometer, which compares readings of dry and water-immersed thermometers. Others use semiconductors to measure changes in electrical resistance, which is affected by humidity. **FS**

SEE ALSO: RAIN GAUGE, BAROMETER, ANEROID BAROMETER, WEATHER RADAR

Pressure Cooker (1679)

Papin's "steam digester" prefigures the modern cooking vessel.

There is a story that when French scientist and inventor Denis Papin (1647–1712) first demonstrated his wonderfully named "digester" to London's Royal Society in 1679, the device exploded. So another invention swiftly came into being: Papin's safety valve, which went on to have other applications.

By 1682, a refined version of the steam digester proved excellent at cooking food and making nutritious bones soft and tasty. After a demonstration dinner at the Royal Society in that year, one guest, leading horticulturalist John Evelyn, noted in his diary that food served up from the digester was among "the most delicious that I have ever seen or tasted."

Papin was an interesting character of diverse scientific interests. Trained in medicine as a young man, he had long been interested in food preservation. His tightly sealed digester vessel showed how atmospheric pressure affected boiling points. Under high pressure, water in the vessel produced steam that cooked food quickly at temperatures far higher than those possible in a saucepan. The cooked food was meltingly soft, its nutrients and flavor were preserved, and the cooker used little fuel. Papin quickly saw that the impoverished were among those who would benefit greatly from his device.

Papin went on to experiment with similar principles in various important early steam-engine prototypes that he developed. Meanwhile his digester also informed the history of the autoclave (whose uses include sterilizing medical instruments) and became the modern pressure cooker, which still works very much to his template. **AK**

> " He [Papin] doth not think . . . that any thing better can be made for such things, as must be stew'd. . . . "
>
> Denis Papin, *Philosophical Transactions* (1683–1775)

SEE ALSO: OVEN, STEAM ENGINE, STEAM ENGINE WITH SEPARATE CONDENSER, ELECTRIC OVEN, MICROWAVE OVEN

↖ *Mass uptake of Papin's idea of cooking food under high pressure had to wait until the twentieth century.*

Universal Joint (1676)

Hooke joins rotating shafts together.

The name of Robert Hooke (1635–1703) pops up frequently in the late seventeenth century. This was a time when a small number of scientists led the whole world in new discoveries across various scientific fields, and of this distinguished group Hooke was one of the most accomplished.

The English polymath discovered the laws of physics that govern elasticity and now bear his name. He was the first person to use the word "cell" to describe the basic building blocks that made up living things. In addition, Hooke was also a top architect—even collaborating on projects with Sir Christopher Wren. But among all of his achievements it was his often overlooked invention of the universal joint that opened up whole new possibilities to the world of applied mechanics.

Like many inventions, the universal joint evolved as the solution to a problem that the inventor had encountered personally. Hooke was a serious astronomer and recognized that the best way to improve knowledge of the universe was by building better and more accurate equipment. But some of his projects, which involved turning small screws at angles to gears with teeth, were beyond the contemporary level of manufacturing.

In 1676, while working on a way to operate an adjusting arm for his helioscope, he created the first working model of a joint that allowed power to be transmitted from one rotating shaft to another. Critically, his joint allowed for the two shafts to be at angles to each other, and maintain the angle while rotating. This made it possible, for the first time, for a rotating shaft essentially to be able to go around corners, opening up a new world of possibilities for machine designs of all types. **DHk**

SEE ALSO: CAMSHAFT, CRANKSHAFT, MOTORCAR

Centrifugal Pump (1689)

Papin improves mine ventilation.

The centrifugal pump works by drawing in a fluid (a liquid or gas) at the center of a cylindrical chamber that contains a rotating impeller with vanes. This forces the fluid to rotate outward toward the wall of the cylinder before flowing into an outlet pipe. The rotation of the fluid causes the liquid to leave with a higher velocity and pressure than when it entered.

The centrifugal pump was invented by French scientist Denis Papin in 1689 as he attempted to solve the problem of ventilating mines. Papin's device was used to pump air through mines and was also applied to furnaces, where it was known as the Hessian bellows. The basic centrifugal pump was improved by John Appold, who carried out an exhaustive study on

> *"Today, centrifugal pumps and compressors have reached efficiency levels above 90 percent."*
>
> Abraham Engeda, Michigan State University

the effect of blade shape on pump efficiency. He found that curved vanes on the impeller drastically increased the pump's efficiency. At the Great Exhibition of 1851, Appold showed his improved design, which was nearly three times more efficient than that of its nearest rival. The new design propelled the development of the centrifugal impeller, which found applications in compressors as well as pumps.

Centrifugal pumps are currently used in areas of power generation, water supply, and general industry. They are widely used in the petroleum and chemical industries because they are relatively inexpensive and can handle large volumes of fluid. **HP**

SEE ALSO: BELLOWS, STEAM PUMP, WATER PILLAR PUMP, INTERNAL COMBUSTION ENGINE

Metronome (1696)

Loulié's device sounds out musical tempo.

The tempo of a piece of music, that is, the number of beats per minute, can be established using a metronome, a type of compact, adjustable, loud clock. The most common type is powered by simple clockwork and has a vertical metal rod that swings from side to side making a loud clicking sound at every swing. The rate of swing can be adjusted by moving a small weight up or down the swinging bar. Up decreases the tempo, and down increases it. This helps musicians not only establish the intended beat, but also maintain it throughout a musical piece.

The first metronome was made in 1696 by the Parisian Étienne Loulié (1654–1702). This required a single-weighted pendulum, similar to that of a grandfather clock. It had no clock escapement to maintain the pendulum in motion, so it only gave the musician the beat for a limited time. In 1812 Dietrich Nikolaus Winkel invented a version in Amsterdam. His breakthrough was the realization that a short, 8-inch (20 cm) metal pendulum, weighted both above and below the pivot, could be made to sound out a low tempo of forty to sixty beats per minute. Johann Mälzel patented the well-known small portable metronome in 1816 using Winkel's basic design. The first composer to mark his music with the expected metronome-regulated tempo was Ludwig van Beethoven, in around 1817.

Needless to say, electronic metronomes now vie with the mechanical version. These include sophistications such as additional sounds, and can sound out complicated time signatures, such as 5/4, that are beyond the range of their predecessors. **DH**

"Tempoi parendum.
[One should be compliant
with the times.]"

Maxim of Theodosius II

SEE ALSO: WATER CLOCK, TOWER CLOCK, COILED SPRING, PENDULUM CLOCK, DIGITAL CLOCK, QUARTZ WATCH

◥ *This nineteenth-century metronome reflects the triangular design established by Johann Mälzel.*

Steam Pump (1698)

Savery uses steam to drain floodwater from mine shafts.

Coal mining is difficult and risky work, and one of the dangers in the mine shafts is flooding. While this is something modern equipment can easily handle, the best remedy for flooding in the late seventeenth century was baling with a bucket. The problem caught the attention of English military engineer Thomas Savery (c. 1650–1715), who set out to make draining faster and easier. Savery's solution was to fight fire with fire, or in this case, fight water with steam.

Steam's power had been revealed by French physicist Denis Papin and his pressure cooker in 1679. Papin had observed that bottled-up steam lifted the cooker's lid, and he envisioned steam doing the same to a piston in an engine. Papin's work inspired Savery to put steam to work in the mines. In 1698 Savery patented "The Miner's Friend," a rudimentary steam engine for pumping water from mine shafts.

Savery's device used pressurized steam to force water up a drainage pipe placed with one end in the flooded shaft. Savery's pumping system had dozens of parts—drainage pipes, valves, boilers, connector pipes, steam delivery pipes, condensers, furnaces—and one big limitation: distance. Floodwater would only travel as far as it was forced by the pressure of the steam. Savery's pump had a limit of about 25 feet (7.6 m), which curtailed its use in underground mining.

The distance limitation of Savery's pump was solved by Thomas Newcomen's atmospheric steam engine, but Savery's patent barred Newcomen from manufacturing his machine. The inventor went into business with Savery to avoid legal difficulties, and soon their engines were highly sought after. **RBk**

SEE ALSO: PRESSURE COOKER, STEAM ENGINE, STEAM ENGINE WITH SEPARATE CONDENSER

↗ *An eighteenth-century model of Savery's pump, which was dangerous to use as well as inefficient.*

" . . . such an engine may be made large enough to do the work [of] ten, fifteen, or twenty horses . . ."

Thomas Savery, *The Miner's Friend* (1702)

Seed Drill

(1701)

Tull achieves eightfold productivity by transforming how seeds are sown.

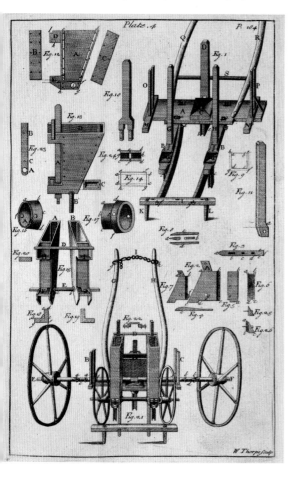

English farmer Jethro Tull (1674–1741) despaired at the waste of seeds that resulted from sowing them by scattering. Seeds would fall too close together, or onto stony ground, lie at differing depths, and plants would grow with no soil between them from which the crop could be weeded, tended, and harvested.

Tull's horsedrawn wooden seed drill improved on this situation and resulted in crop yields of up to eight times those where the seeds had been scattered. A shaped wooden drill dug an even groove of the right depth into the soil and seeds from the hopper mounted above it trickled into the groove, evenly spaced by the forward movement of the horse. Tull mounted three drills alongside each other in the machine, and so could plant three rows of seeds at a time, leaving space between these triple rows.

As a young man, Tull traveled in continental Europe in what were the early years of the Age of Enlightenment and was inspired by some of the scientific ideas he encountered. Although he is best known for inventing the seed drill, he also introduced the use of workhorses instead of cattle, invented a horsedrawn hoe, and developed the design of the plow in ways that are still in use today.

Some of his ideas proved controversial at the time, and his observation that crop nutrients are released from the soil by pulverization turned out to be misguided, but much of what he achieved established the foundations of modern agriculture in Britain, some decades before the start of the industrial and agricultural revolutions. **EH**

"When tillage begins, other arts follow. The farmers therefore are the founders of human civilization."

Daniel Webster, American statesman, 1840

SEE ALSO: DRILL, SCRATCH PLOW, CART, HORSE COLLAR

◪ *Details of Tull's seed drill, illustrated in* The Horse Hoeing Husbandry, *the book he published in 1733.*

Coke-based Iron Smelting (1709)

Darby revolutionizes iron making.

Before the introduction of plastics, iron was one of the most multipurpose materials, used to make almost everything. However, the only pure iron on Earth fell from space as meteorites, and that is far too rare to rely on. Most iron has been pushed up to the Earth's crust by activity in the planet's core, but this has reacted with many other elements, resulting in iron ore, rather than pure elemental iron. The process of separating iron from ore is called smelting: The ore is heated to a temperature at which it becomes a liquid, and then the metal is separated from the waste.

Charcoal is one of the few materials that burns hot enough to melt iron. In Britain the iron industry originally moved around the country, burning forests and then moving on, but by the seventeenth century the industry was running out of trees and wood was becoming much more expensive. Also, charcoal is soft, which means that the furnaces had to be small and iron could never be mass-produced. An alternative was needed.

Coal was no good because elements from it get into the iron and make it weak, but in the same ways that charcoal can be made from wood, coal can produce a material called coke. Coke was cleaner than the alternatives and, in 1709, ironmaster Abraham Darby I (1678–1717) built the first coke-fired blast furnace. He was the first of three generations of Abraham Darbys to perform pioneering works in the iron industry. The purity of the iron made it stronger and the use of coke allowed bigger furnaces. Soon large quantities of iron were available cheaply in Britain, playing an important role in the Industrial Revolution and in making Britain one of the dominant world powers at that time. **DK**

SEE ALSO: IRON ROCKET, CAST-IRON PLOW, IRON LUNG, BLAST FURNACE, ELECTRIC-ARC FURNACE

Atmospheric Steam Engine (1712)

Newcomen improves mine drainage.

Thomas Newcomen (1663–1729), a Devonshire blacksmith, developed the first successful steam engine in the world and used it to pump water from mines. His engine was a development of the thermic syphon built by Thomas Savery, whose surface condensation patents blocked his own designs.

Newcomen's engine allowed steam to condense inside a water-cooled cylinder, the vacuum produced by this condensation being used to draw down a tightly fitting piston that was connected by chains to one end of a huge, wooden, centrally pivoted beam. The other end of the beam was attached by chains to a pump at the bottom of the mine. The whole system

> *"Those who admire modern civilization usually identify it with the steam engine…"*

George Bernard Shaw, playwright and writer

was run safely at near atmospheric pressure, the weight of the atmosphere being used to depress the piston into the evacuated cylinder.

Newcomen's first atmospheric steam engine worked at Conygree in the West Midlands of England. Many more were built in the next seventy years, the initial brass cylinders being replaced by larger cast iron ones, some up to 6 feet (1.8 m) in diameter. The engine was relatively inefficient, and in areas where coal was not plentiful was eventually replaced by double-acting engines designed by James Watt (1736–1819). These used both sides of the cylinders for power strokes and usually had separate condensers. **DH**

SEE ALSO: PRESSURE COOKER, STEAM PUMP, STEAM ENGINE WITH SEPARATE CONDENSER, COMPOUND STEAM ENGINE

Mercury Thermometer (1714)

Fahrenheit initiates the standardized measurement of temperature.

In a mercury thermometer, mercury in a small glass bulb expands into an evacuated, linear, uniform cross-section glass tube; the amount of expansion is used to measure the temperature of the bulb. Dante Gabriel Fahrenheit (1686–1736) left Gdansk, Poland, and eventually became a glassblower and scientific instrument maker in the Netherlands. His first glass thermometer (1709) used alcohol as the expanding fluid, but this has a limited temperature difference between its freezing and boiling points. In 1714 Fahrenheit turned to mercury, a liquid metal that expands uniformly over normal temperature ranges.

Fahrenheit insisted that thermometer results should be universally reproducible, and similar temperatures should be represented by the same number. To this end he introduced, in 1724, three "fixed" points and eight graduations on his thermometer tube. Zero degrees was the lowest temperature that he could obtain in the laboratory, the temperature of a mixture of water, water ice, and ammonium chloride. Thirty-two was the temperature of an ice/pure water mixture, and ninety-six degrees was the normal temperature of a human body. From 1717 Fahrenheit was selling thermometers from a base in Amsterdam, and these, and his temperature scale, became widely used throughout Britain, the Netherlands, and Germany.

More recently the Fahrenheit scale has been defined using the freezing and boiling points of pure water, at normal atmospheric pressure, as 32 and 212 degrees (0 and 100°C). Here our typical body temperature becomes 98.6°F (37°C). **DH**

"[Scientists] should return to the plainness . . . of Observations on material and obvious things."

Robert Hooke, *Micrographia* (1664)

SEE ALSO: GLASS, BLOWN GLASS, BIMETALLIC STRIP, THERMOCOUPLE

◤ *This eighteenth-century model, made by George Adams, has graduations from -20 to 210° Fahrenheit.*

Padlock (1720)

Polhem devises a portable lock that effectively resists tampering.

Primitive padlocks have been around since medieval times, but their design left them prone to force or picking. In 1720, Swedish inventor Christopher Polhem (1661–1751) conjured up a lock that was much more resistant to the dexterous hands of lockpickers.

Polhem was one of the most gifted mechanical engineers of his day. After studying mathematics, physics, and engineering at Uppsala University, he set up as a clock repairer. His ingenuity was soon spotted by important patrons, including King Charles XI of Sweden. Polhem went on to design many intricate devices both small (watch mechanisms) and large (industrial machinery). Perhaps his most enduring invention, however, was the padlock.

His basic design comprises an elliptical cast iron body containing a series of rotating disks. When locked, the disks fit into grooves on the shackle (the U-shaped bar on top of the padlock), preventing its release from the body. Notches on the discs can be aligned with those on the shackle by rotating the correct key, thus releasing the shackle and allowing the lock to be opened.

The Swedish inventor's device became known as the Polhem lock, or Scandinavian lock, and Polhem started a factory in Stjärnsund to produce it. The Scandinavian lock came to dominate the market. The design was later strengthened by the American locksmith Harry Soref, who founded the Master Lock Company in 1921. A modified version of this design is still in use today. For his many achievements, Polhem has been honored by appearing on the back of the 500 Swedish kronor note. **MB**

SEE ALSO: MECHANICAL LOCK, COMBINATION LOCK, TUMBLER LOCK, SAFETY LOCK, TIMELOCK SAFE, YALE LOCK

↗ *An English brass padlock, constructed according to the design principles established by Polhem.*

> *"Lock-and-key, n. The distinguishing device of civilization and enlightenment."*
>
> Ambrose Bierce, writer and journalist

Pitot Tube (1732)

Pitot measures air and liquid flow rates.

The Pitot (pronounced *pea-tow*) tube is an eighteenth-century invention still flying high amid twenty-first-century technology. Designed by French astronomer, engineer, and mathematician Henri Pitot (1695–1771), this deceptively simple device is essentially a differential pressure gauge and can be used for a variety of flow-rate or speed-measuring purposes.

Pitot's pet interest was water flow, and his personal research led him to conclude that much of the accepted wisdom of the day was incorrect. He would not accept, for example, the prevailing theory that, other things being equal, the speed of flowing water increased with depth. His tube, demonstrated at the French Academy of Sciences in 1732, would show that he was right: it does not.

As well as being used in a fixed position to determine the flow rate of a liquid or gas, the L-shaped tube may be attached to a boat or airplane to measure the craft's forward speed. In all cases, the tube functions by registering the difference between the ambient pressure surrounding it and the pressure created by the flow into it (the "impact" pressure), which will increase with speed. The resulting comparative measurements can be displayed via suitable instrumentation. Various improvements to the basic design have been made over the years, although Henri Darcy's 1858 design is more or less the one still in use today.

Today, variations of, and uses for, Pitot's amazing little tube continue to grow; multiport versions exist to enable the measurement of the impact and static pressures at different points. They can be seen on Formula 1 racing cars and spacecraft. They have even been configured to perform an additional role as an aircraft antenna. **MD**

SEE ALSO: BAROMETER, ANEMOMETER, POWERED AIRPLANE

Baby Carriage (1733)

Kent creates a child's amusement.

The first known design for a baby carriage was produced in 1733 by William Kent (c. 1685–1748), the renowned English landscape-garden designer. Today the baby carriage is an essential tool for any family with children, but it was originally intended as an entertainment. Kent, who as a designer could turn his hand from furniture to ladies clothes as well as gardens, was commissioned by the third Duke of Devonshire to design something to amuse his children. He produced a shell-shaped vehicle in which a baby could sit, with an attached harness designed to fit a small pony, a dog, or a goat.

Baby carriages quickly became popular among the wealthy as fashionable toys. Gradually changes were

> *"Nobody outside of a baby carriage . . . believes in an unprejudiced point of view."*
>
> **Lillian Hellman, playwright**

made to their design, with one of the most significant being the addition of handles, which allowed a person to push the vehicle. The carriages became more popular in the 1840s when Queen Victoria bought three of the new push-style versions from Hitchings Baby Store of Ludgate Hill.

The next breakthrough came in 1889 with a new design created by William H. Richardson, who devised a special joint that enabled the bassinet to be turned to face the handles, as seen in many modern designs. He also improved the axles enabling the wheels to turn individually allowing for great maneuverability. Many of his design features are still used today. **TP**

SEE ALSO: WHEEL AND AXLE, CART, COLLAPSIBLE STROLLER, FOLDING WHEELCHAIR

Flying Shuttle (1733)

Kay vastly speeds up the weaving process.

Archeologists have found a model of a loom in an Egyptian tomb from 4,000 years ago. Yet the development of loom technology was slow until 1733, when John Kay (1704–1780) invented the flying shuttle.

Looms interlace two sets of yarn or threads together to form cloth. The first set of threads is placed lengthwise along the loom and is called the warp. The second set of threads is called the weft. The weft is carried between the warp threads by a shuttle. In traditional looms, weavers passed the shuttle through the warp by hand, and it was a slow process. Kay's flying shuttle moved on wheels in a track through the warp when the weaver pulled a cord. This was much faster than hand weaving, and could also be used to create much wider fabrics than previously possible.

Kay did not receive much benefit from his invention because weavers saw the flying shuttle as a threat to their livelihoods. They believed—incorrectly as it turned out—that the demand for cloth was constant, so if looms were more efficient, fewer weavers would be needed. Although manufacturers were glad to use Kay's invention, they did not pay him any royalties. Kay died a poor man in 1780.

The flying shuttle created a huge demand for yarn. At the time, yarn spinning was a slow process done by hand. Over the next fifty-five years, inventors worked on machines to increase the productivity of spinners. These included the spinning jenny, the water frame, and the spinning mule. All these inventions made cotton items affordable to many more people. **ES**

SEE ALSO: CLOTHING, WOVEN CLOTH, SEWING, SPINNING WHEEL, SPINNING JENNY, SPINNING MULE, POWERED LOOM

↗ *Eighteenth-century weavers' shuttles had rollers and were iron-tipped to reduce friction.*

➔ *Shuttles sped through the looms at remarkably high speeds, each with its own colored yarn.*

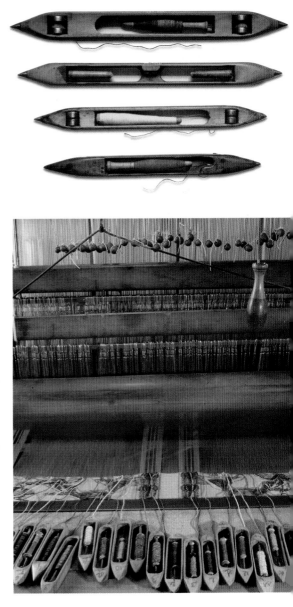

Franklin Stove (Circulating Fireplace) (1742)

Franklin invents a safer and more efficient way of heating wooden buildings.

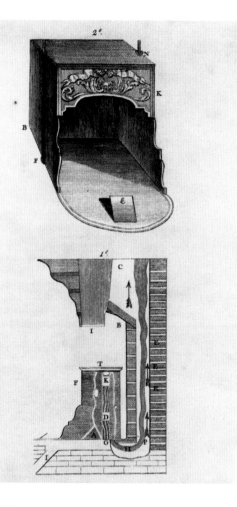

Before inventing the lightning rod and bifocal lenses, American statesman and polymath Benjamin Franklin (1706–1790) had turned his attention to keeping peoples' homes warm and safe. In the eighteenth century many homes in the United States, built of wood and heated by open hearths, were at great risk of fire. This had concerned Franklin since at least 1735, when he organized the first volunteer fire department in his adopted home town of Philadelphia, Pennsylvania. He also called for building regulations to include minimum safety standards in fireplace design—modern standards are still based on them.

In 1742 he designed a new stove that he called the "Pennsylvania fireplace"; it was later called the "Franklin stove" or "circulating fireplace." The stove was a box lined with metal that stood away from the wall, improving efficiency compared to the standard fireplace where much of the heat was lost to the wall behind it. He also added flat plates or "baffles" at the stove's rear to improve the flow of air. The only flaw, later fixed by Franklin, was that smoke had to escape through the base of the stove, and it filled the room.

The stove was first manufactured by Franklin's friend Robert Grace, but the inventor refused to patent the device in order to keep the technology freely available—a story that is often cited as an early example of open-source development.

The freedom to tinker with Franklin's design was explored in the 1780s by David Rittenhouse, who added an L-shaped chimney. History has preferred to call it the Franklin stove nonetheless, and there are many in use to this day. **AC**

SEE ALSO: CONTROLLED FIRE, OVEN, FIRED BRICK, ELECTRIC FIRE ALARM, AUTOMATIC FIRE SPRINKLER

"The use of these fireplaces in very many houses . . . is a great saving of wood to the inhabitants."

Benjamin Franklin, statesman and scientist

◩ *A 1773 engraving by François Nicolas Martinet details the function of the Franklin stove.*

Leyden Jar (1745)

Van Musschenbroek demonstrates that electricity can be stored and then discharged.

In 1745 the Dutch physicist Pieter van Musschenbroek (1692–1791) took a sealed glass vial partially filled with water, passed a conducting wire through a cork at one end and attached it to a nearby Wimshurst friction machine, which generated a static charge. The glass jar, called a Leyden Jar in honor of the inventor's home town and university, absorbed the charge, demonstrating for the first time that electricity could be produced and stored successfully and then discharged through the exposed wire to any grounded object. Musschenbroek tested the device by holding the jar in one hand and touching the charged, exposed wire with the other. He received such a shock that he swore not even a promise of the entire French nation could persuade him to do so again.

The Leyden jar created a sensation within the worldwide scientific community. The American inventor Benjamin Franklin called it "Musschenbroek's wonderful bottle." A year later the English physician William Watson, using a modified Leyden jar, successfully transmitted an electric spark along a wire stretched across the River Thames. The precise nature and makeup of electricity proved elusive until the discovery of the electron by J. J. Thompson in 1897.

Though cumbersome and grossly inefficient by modern standards, this forerunner of the modern capacitor represented the eighteenth century's single most significant advance in the understanding and harnessing of electricity. It facilitated a greater understanding of the nature of conductivity and led to a more mathematical approach in the study of the attraction of electrified bodies. **BS**

SEE ALSO: GLASS, ELECTRIC MOTOR, ELECTRICAL GENERATOR, PUBLIC ELECTRICITY SUPPLY, AC ELECTRIC POWER SYSTEM

⬀ *The original Leyden jar of 1745 was made of glass and had metal foil coatings inside and out.*

"My whole body was shaken as though by a thunderbolt."

Pieter Van Musschenbroek, physicist

Electroscope (1748)

Nollet's device detects and measures electric charge.

Jean-Antoine Nollet (1700–1770) was the first professor of experimental physics at the University of Paris. At that time electrostatics was a topic of great interest.

Nollet's electroscope was designed to detect and crudely measure electric charge. An insulated charge sensor protruded into a cylindrical container, the ends of which were closed by two flat glass windows. The bottom of the sensor (the part in the cylinder) was fitted with two leaves of metal foil (usually gold). If the sensor's opposite, extruding end was brought into contact with a negatively charged body, electrons were repelled into the two leaves and they separated. The degree of separation was a function of the size of the charge. A second negatively charged body contacting the extruding end would cause the leaves to separate more, whereas a positively charged body subsequently applied to the same end would cause the separation to decrease.

Earlier electroscopes were mainly used to investigate the amount of charge that could be produced by hand-cranked or foot-treadled frictional electrostatic machines. Here, a globe of glass was charged by rubbing it with a soft material.

Nollet was interested in the causes of electrostatic repulsion and attraction and the speed with which electric current flowed. He once estimated the latter by discharging the charge contained in a large Leyden jar through a line of 200 Carthusian monks who were connected to each other by iron wires 25 feet (7.6 m) long. As all the monks leaped with shock at the same time, he concluded that electricity traveled speedily, and could travel a long distance. **DH**

"The electron . . . crystallizes out of Schrödinger's mist like a genie emerging from his bottle."

Sir Arthur Eddington, *Nature of the Physical World*

SEE ALSO: LEYDEN JAR, ELECTRICAL GENERATOR, PUBLIC ELECTRICITY SUPPLY, AC ELECTRIC POWER SYSTEM

◰ *A negative charge passing down the electroscope's rod causes the leaves at the end to repel each other.*

Water Pillar Pump (1749)

Hell drains mines with water power.

During a period spanning the sixteenth and nineteenth centuries, the Slovakian mining town of Banská Štiavnica rose to fame both as a major source of gold and silver and as a center of excellence in the technologies needed to extract those precious metals. The area became synonymous with advances in ore extraction and processing.

Fundamental among the many problems that mining engineers had to overcome at that time was the removal of water from shafts several hundred yards deep. Human and animal power played its part in no small measure; by the end of the seventeenth century, close to a thousand men and hundreds of horses were toiling around the clock to keep the existing pumping systems working. Steam-powered systems were put to the test. However, it was asked that if an abundance of water was the problem, why not work with water, exploit its energy potential, and turn it into the solution? Jozef Hell, senior mining engineer and son of the equally talented Matej Kornel Hell, would become preeminent in the successful implementation of this revolution with his "water pillar" or "water column" pump designs.

Hell's first pump, constructed in 1749 at Banská Štiavnica's Leopold mine shaft, was a positive displacement "engine," utilizing a system of pistons and valves quite similar to a steam engine. Instead of steam, however, hydraulic pressure, developed by a column of water obtained from a surface reservoir, was used. It comprised a single, vertical cylinder with a control system using simple rods and hammers that directly operated two-way valves. A second pump could be installed beneath the first, so that the same primary water supply could be used to drive them both and pump water from an even lower level. **MD**

SEE ALSO: CENTRIFUGAL PUMP, STEAM PUMP, VACUUM PUMP, STEAM (ATMOSPHERIC) ENGINE, INSULIN PUMP

Lightning Rod (1752)

Franklin proves lightning is electricity.

American statesman and inventor Benjamin Franklin (1706–1790) was particularly interested in electricity and set up a small laboratory in his house to investigate its properties. His interest soon switched from electricity to lightning after he noticed similarities between the two. Many scientists had previously noticed a link, but none had managed to prove it.

On a stormy night in 1752, he conducted a life-threatening experiment to demonstrate that lightning is the result of an electrical buildup. He constructed a kite that carried a metal spike and flew it into the thunderstorm. The kite had a key attached near the bottom of the ribbon and Franklin noticed that it sparked as he brought his knuckles close to it. Franklin

> *"Electrical fire would ... be drawn out of a cloud silently, before it could come near enough to strike."*
>
> Benjamin Franklin, statesman and scientist

had shown that lightning was a form of electricity and he went on to use this knowledge to design a lightning rod to protect buildings. The iron rod was between 6 and 10 feet (2 and 3 m) in length, and provided a path of least resistance for the lightning, channeling it safely to the ground. He later showed that sharp rods were better than blunt ones for the purpose.

Recently it was suggested that the kite experiment was a hoax and that Franklin would have been killed if he had actually been struck by lightning. Some suggest the experiment was real but the sparks observed were actually from an electrical field and not a lightning strike. **RB**

SEE ALSO: LEYDEN JAR, ELECTROSCOPE, ELECTRICAL GENERATOR, ELECTROPLATING, ELECTROMECHANICAL RELAY

Linnaean Taxonomy
(1753)

Linnaeus classifies all plant life.

Linnaean taxonomy is the system of classification of living organisms that is used throughout the biological sciences. Its inventor, Carolus Linnaeus (1707–1778), spent most of his career in Uppsala, Sweden. Starting with the plant kingdom, Linnaeus created a hierarchy in which plants are grouped, according to similarities in their appearance, into twenty-five phyla, and then each phylum into classes, and these in turn into orders, families, genera, and species. The first description of this system was published by Linnaeus in 1753, in a two-volume work, *Species Plantarum*. He later applied the same principles to animals and minerals.

The most important feature of Linnaean taxonomy is a system known as binomial (or two-name) nomenclature. The first name identifies the genus to which the organism belongs; the second name, its unique species: for example, the common daisy is *Bellis perennis*. If necessary, the family, order, and phylum to which a genus belongs can be looked up in a floral taxonomy reference book.

Linnaeus collected, studied, and classified plants and animals, publishing his findings in successive editions of *Systema Naturae*. The first edition, published in 1735, was just eleven pages long; the tenth edition, published in 1758, detailed 4,400 species of animals and 7,700 plant species. Linnaean taxonomy, although developed a hundred years before Darwin's theory of evolution, proved to be robust and effective even as scientists have explored the evolutionary relationships between organisms. More recently, comparisons of the genetic codes of individual species have led to some reclassification of plants and animals, but the essential concepts of Linnaean taxonomy remain entirely valid today. **EH**

SEE ALSO: GENETICALLY MODIFIED ORGANISM, GENE THERAPY, AUTOMATED DNA SEQUENCER

Bimetallic Strip
(1755)

Harrison utilizes metal expansion.

The bimetallic bar was invented by the Yorkshire clockmaker John Harrison (1693–1776), and was used in his third and fifth chronometers to cancel out thermally induced variations in the balance springs.

Imagine two straight strips of metal bar—steel and brass, say—riveted, brazed, or welded together along their length. Brass expands by nineteen parts in a million for every increase in temperature of 1.8°F (1°C), and steel by thirteen parts. Heating the bar will make one metal expand more than the other and cause the bar to bend. In the above example, the brass will be on the outer side of the curve. Cooling the bar will cause it to curve the other way, with the steel on the outside.

"Some people change their ways when they see the light, others when they feel the heat."

Caroline Schroeder, pianist

A spiral bimetallic strip unwinds or tightens as a function of temperature; a cheap, robust thermometer may be made by attaching a pointer. Strips are also used in clocks, forming the circular rim of the balance wheel, where the size of the wheel changes in a way that compensates for temperature variations in the strength of the controlling spring. But by far the most common usage is as a temperature-sensitive contact breaker in thermostatically controlled devices such as refrigerators, ovens, and irons. The fact that the metals conduct electricity means that the moving end of the strip can be used to open and close an electrical switch that is connected to a heater or cooler. **DH**

SEE ALSO: MERCURY THERMOMETER, ELECTRIC STOVE, ELECTRIC IRON, REFRIGERATOR

Marine Chronometer
(1761)

Harrison's invention enables mariners to check their longitude at sea.

One way of calculating the difference between a longitude at sea and a known longitude (of Greenwich, say) was to ascertain the mean solar time on the ship, by astronomical observations, and compare it with the time at Greenwich. To this end a clock was needed that accurately kept Greenwich time despite being rocked back and forth by the ship. In 1714 the British government offered a £20,000 prize (about £1,000,000 today) to anyone who could find longitude at sea to an accuracy of 0.5 degrees.

Yorkshireman John Harrison (1693–1776) decided that an accurate clock was the answer. He built his first marine chronometer in 1735. This spring-driven clock was regulated by two connected balances that oscillated in opposite directions, thus eliminating all the effects of the ship's motion. Intentional variations in the lengths of the balance springs also compensated for temperature changes.

Harrison's third chronometer (1759) had a bimetallic temperature compensator and a remontoire to ensure that the escapement driving force was constant. Harrison's fourth chronometer (1761) embodied all his improvements into a large "pocket" watch, about 5 inches (13 cm) in diameter.

This watch was carried to Jamaica on board HMS *Deptford* in 1761. The clock error over the journey was about five seconds, equivalent to a longitude error of about one sixtieth of a degree. The ship's position was thus known to an accuracy of 1.5 miles (2 km). Harrison finally got all his prize money in 1773, and soon every ship was carrying his instrument. **DH**

SEE ALSO: ASTROLABE, CLOCKWORK MECHANISM, SEXTANT, POCKET WATCH, DIGITAL CLOCK

↗ *Harrison's marine chronometers were extremely well made, as shown by this 1770 example.*

"... every great captain ... became lost at sea despite the best available charts and compasses."

Dava Sobel, *Longitude* (1995)

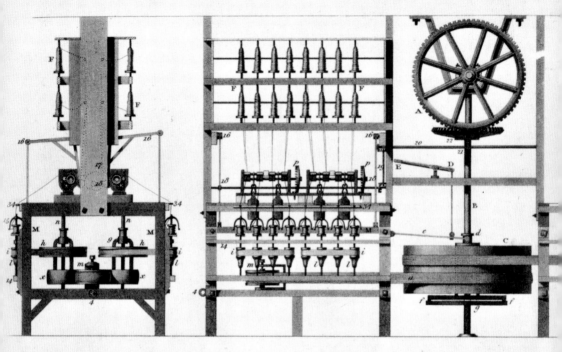

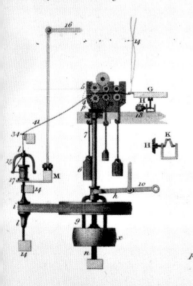

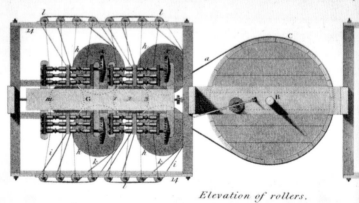

Elevation of rollers.

End View.

Fig.6.

Spinning Jenny (1764)

Hargreaves transforms cotton spinning.

Spinner and carpenter James Hargreaves (1720–1778) invented this multispool spinning wheel as a way of increasing the productivity of his cotton factory in Lancashire. The "spinning jenny" had eight spindles, all of which could be operated by a single person, who rolled a beam back and forth over the yarn until it was the correct thickness. The machine increased the production of spun yarn eightfold. One story has it that a daughter of Hargreaves knocked over a spinning wheel, and he noticed that it continued to work perfectly well. This led him to consider a machine with multiple spindles, all in an upright position.

The machine was so successful that yarn prices fell,

> *"[The] Industrial Revolution . . . opened an age of . . . production for the needs of the masses."*
>
> Ludwig von Mises, economist

upsetting the spinning community in the area. Several spinners broke into Hargreaves's house and destroyed his machines, causing him to move to Nottingham where, in 1767, he began making spinning jennies. Three years later he applied for a patent for his invention but failed in his attempt to sue local manufacturers who were using copies of his machine. Hargreaves died in 1778, the year Samuel Crompton invented an even more efficient spinning machine that was dubbed the "spinning mule." **JL**

SEE ALSO: CLOTHING, SEWING, WOVEN CLOTH, STEAM ENGINE, SPINNING WHEEL, SPINNING MULE, POWERED LOOM

🔁 *Technical drawings of Hargreaves's revolutionary waterpowered frame for spinning cotton.*

Surveyor's Perambulator Wheel (1765)

Fenn facilitates land surveying.

By the time Isaac Fenn was granted a patent, in 1765, for his distance-measuring hodometer, the device had existed, in various forms and with various names, for centuries. In Roman times it was called an odometer and comprised little more than a wheel that could be pushed along, coupled to a mechanical system for counting the number of revolutions the wheel made, and thus the distance it had traveled.

The eighteenth and nineteenth century saw the mapping of India and the division of vast tracts of land into farms in regions such as the United States and Australia. Reasonably accurate surveying and distance measurement became important. The surveyor's perambulator wheel (the "waywiser" or trundle wheel) was in everyday use. The accuracy of this device was good on a smooth surface such as a pavement or macadamed road. On rough terrain, such as farmland, wheel bounce and slippage became a problem and surveyors had to apply a series of corrections to the readings. For very accurate work, the surveyor had to resort to a tape or chain measure

A typical eighteenth-century waywiser would have a wheel diameter of around 31½ inches (80 cm), equating to a circumference of around 8¼ feet (2.5 m). This meant that two revolutions of the wheel would equate to one pole (an old English measure of length and area). A central dial with two hands, much like a clock, was attached to the unit. The bigger hand made one sweep every 320 poles—this marking 1 mile (1.6 km). The shorter hand indicated the total number of miles traversed.

Today, professional trundle wheels are extremely accurate and are likely to sport an LCD display and onboard digital storage/manipulation of data. **MD**

SEE ALSO: WHEEL AND AXLE, MAP, ODOMETER, WHEELBARROW

Steam Engine with Separate Condenser (1765)

Watt drives steam power forward with a technological breakthrough.

Scottish engineer James Watt (1736–1819) was responsible for some of the most important advances in steam-engine technology. Steam engines had been in use since the 1710s, mainly to pump water from mines. These machines depended upon steam condensing inside a large cylinder after the cylinder was cooled with cold water. As the steam condensed, it took up less space, allowing atmospheric pressure to push down on a movable piston inside the cylinder.

In 1765 Watt made the first working model of his most important contribution to the development of steam power; he patented it in 1769. His innovation was an engine in which steam condensed outside the main cylinder in a separate condenser; the cylinder remained at working temperature at all times. Watt made several other technological improvements to increase the power and efficiency of his engines. For example, he realized that, within a closed cylinder, low-pressure steam could push the piston instead of atmospheric air. It took only a short mental leap for Watt to design a double-acting engine in which steam pushed the piston first one way, then the other, increasing efficiency still further.

Watt's influence in the history of steam-engine technology owes as much to his business partner, Matthew Boulton (1728–1809), as it does to his own ingenuity. The two men formed a partnership in 1775, and Boulton poured huge amounts of money into Watt's innovations. From 1781, Boulton and Watt began making and selling steam engines that produced rotary motion; all previous engines had been restricted to a vertical, pumping action. Rotary steam engines were soon the most common source of power for factories, becoming a major driving force behind Britain's Industrial Revolution. **JC**

"I have now made an engine that shall not waste a particle of steam. It shall be boiling hot."

James Watt to his friend John Robison

⬆ *Watt's separate steam condenser allowed the main cylinder to remain hot, bringing great fuel economy.*

➡ *The top, side, and end elevation drawings of James Watt's steam engine with separate condenser.*

SEE ALSO: ATMOSPHERIC STEAM ENGINE, HIGH-PRESSURE STEAM ENGINE, COMPOUND STEAM ENGINE

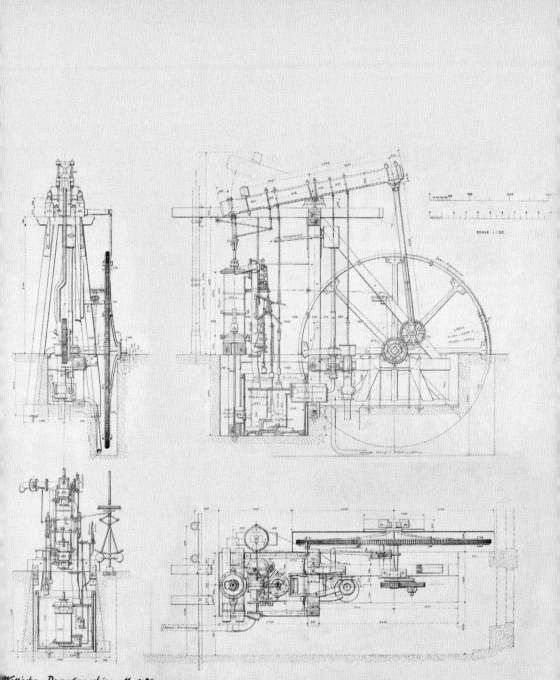

Watt'sche Dampfmaschine. M. ·1·20.   Disposition für die Aufstellung der Nachbildung in der Hauptwerkstätte München

SCALE 1 : 20

Water Frame (1769)

Arkwright's innovation accelerates the rate of textile mass production.

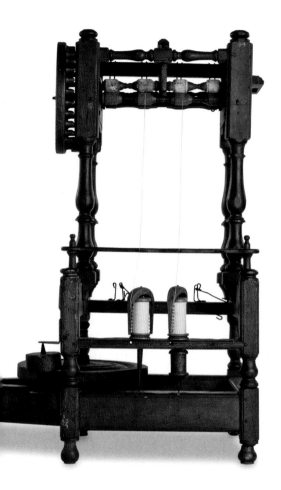

Preston wigmaker Sir Richard Arkwright (1732–1792) patented a new cotton-spinning machine in 1769. Around that time people were racing to find a fast and inexpensive way to produce good-quality cotton for the textiles industry. James Hargreaves had created the "spinning jenny" between 1764 and 1767, but this simply mimicked the action of a hand-turned spinning wheel and could not produce high-quality cotton thread. Arkwright's "water frame" would become a major catalyst for the Industrial Revolution.

While working as a wigmaker, Arkwright became interested in the spinning of cotton. He enlisted the help of clockmaker John Kay, who had worked previously with Thomas Highs on another spinning machine that had been halted by a lack of funds. Together, Arkwright and Kay built a prototype horsepowered spinning frame, which they patented in 1769. The frame drastically sped up the process of spinning cotton, producing both weft (filling yarn) and high-quality waft suitable for hosiery.

Realizing the potential for large-scale production, Arkwright perfected a model powered by waterwheel, one that became known as the "water frame." Although Arkwright is heralded as the inventor of the water frame, many believe that it was actually Highs who came up with the original design. It is thought that, while working with Kay, Arkwright obtained the secret to Highs's design and went on to use it for his water frame. This model could not be operated within workers' homes, so in 1771 Arkwright built the first textiles factory in Cromford, Derbyshire, marking the beginning of textile mass-production. **HI**

> *"One machine [produced] in an hour what had previously taken hundreds of person-hours."*
>
> Robert Clark, University of East Anglia

SEE ALSO: SPINNING WHEEL, SPINNING JENNY, SPINNING MULE, POWERED LOOM

◤ *Arkwright's water frame revolutionized cloth production in eighteenth-century Britain.*

Venetian Blind (1769)

Beran patents a window covering.

Precisely where or when the very first venetian blinds appeared has long been a point of some conjecture. Slatted blinds made from various timbers were popular throughout northern Italy in the mid to late-1700s; these consisted of slats held together by strips of fabric rather than corded cloth, and the angle of the slats could be adjusted with the use of a tilting device, not unlike what is in use today. When freed Venetian slaves took venetian blinds to France in the 1790s, the window coverings soon became known to the French as *les Persiennes* (the Persians).

In fact, by the 1700s Italy had already enjoyed a long association with the venetian blind. Archeologists have uncovered slatted window coverings amid the ruins of Pompeii, with the individual slats being fashioned from marble. Farther to the east, the earliest window coverings unearthed in modern Iran (previously known as Persia) have been dated to 4000 BCE; these were made from clay tiles. To the south, on the Mediterranean island of Crete, shutters made from an amalgam of alabaster and marble have been found among ruins attributed to the ancient Minoan civilization (2600–1100 BCE).

In England, venetian blinds were patented by Edward Beran in London on December 11, 1769. A century later they became highly popular with the Victorians as an alternative to curtains, which had become heavy, cumbersome, and unfashionable.

The modern method of adjusting the angle of the slats while keeping them synchronized and parallel was invented in 1841 by John Hampson of New Orleans, Louisiana. Blinds made from wood continued to remain popular until the development by Joe Hunter and Henry Sonnenberg of the 2-inch (50 mm) aluminum slat in the mid-1940s. **BS**

SEE ALSO: BUILT SHELTER, CARPENTRY, WOVEN CLOTH, GLASS, HALL-HÉROULT PROCESS (ALUMINUM MANUFACTURING)

Caterpillar Tracks (1770)

Edgeworth patents continuous tracks.

When you want to navigate areas where the terrain is uneven and muddy, what better solution than to take the road with you? This was precisely the conclusion of Englishman Richard Lovell Edgeworth (1744–1817) when he invented the "portable railway," the earliest incarnation of a full-track vehicle. Although his 1770 patent is open to interpretation, it could describe anything from a vehicle with shoed wheels to a system similar to that seen today where continuous tracks run between front and rear wheels.

The nineteenth century saw a glut of patents filed for vehicles sporting tracks. However, they suffered from problems such as poor steering and a lack of materials capable of taking the stresses and strains

> *"They come and pushed me off. They come with the cats … the Caterpillar tractors."*
>
> The Grapes of Wrath, Nunally Johnson screenplay

exerted by the system. But perhaps the biggest stumbling block was insufficient propulsive power, a problem overcome only with the advent of the internal combustion engine. Despite this, full-track, steam-powered vehicles had their uses; the Western Alliance for example used them during the Crimean War (1853–1856).

It is believed that the name "caterpillar track" was shrewdly trademarked in the early 1900s by Benjamin Holt, founder of Holt Manufacturing, later Caterpillar Inc., after hearing a British soldier quip how the tracked vehicles crawled like a caterpillar. The caterpillar track now appears in designs created for many terrains. **RP**

SEE ALSO: TRACTOR, BULLDOZER, MILITARY TANK

Soda Water (1771)

Priestley mixes carbon dioxide and water.

Joseph Priestley (1733–1804) grew up near a brewery in Yorkshire, England, and as a teenager saw carbon dioxide gas "floating" above deposits of fermenting grain. In 1771 this clergyman, philosopher, and chemist began to inject carbon dioxide, what he called "fixed air," into small containers of water uncontaminated by the surrounding air. By agitating the mixture for thirty minutes he was able to cause the water to absorb its own volume of carbon dioxide, and so he created the world's first drinkable glass of carbonated water. In 1772 Priestley wrote a book detailing in part how he thought carbonated water could be used to retard food spoilage and reduce the incidence of scurvy on long ocean voyages. He also wrote a paper entitled

"Let us have wine and women, mirth and laughter / Sermons and soda-water the day after."

Lord Byron, *Don Juan*

Directions for Impregnating Water with Fixed Air. Priestley's considerable legacy includes writings on the nature of electricity, ethics, religious freedom, and extensive work on the nature of gases, which led to his discovery of oxygen in 1774. He never found the time, nor likely possessed the inclination, to pursue the commercial potential of his carbonated water.

Artificially carbonated water mimicked the bubbles found in many natural springs. It was not until the invention of the soda fountain or carbonated drink dispenser by Samuel Fahnestock in 1819, however, that carbonated water achieved the kind of popularity it has today as the foundation of many soft drinks. **BS**

SEE ALSO: ALCOHOLIC DRINKS, SODA SYPHON, CROWN BOTTLE TOP

S-trap for Toilet (1775)

Cummings introduces the indoor toilet.

Without the toilet system, disease would be widespread and water undrinkable. It is an invention that is taken for granted in the modern world, but where would we be without it? Even though they began only as holes in the ground, toilets in various forms have been used since Babylonian times.

A defining step in the long and complex history of the toilet was the S-trap system developed by Alexander Cummings, a watchmaker by trade. (Thomas Crapper is often given credit for the invention of the modern toilet; although he was involved in toilet production, it was Cummings who held the patent.) Cummings's design incorporated an S-shaped bend in the drainage pipe that created a water seal between

"And you shall have an implement [and] you shall dig with it and turn and cover your refuse."

Deuteronomy 23:1 on camp sanitation

flushes. This meant that foul odors were trapped below the water and could not escape into the air.

Life improved immeasurably with the combined benefits of the flushing toilet and the closed sewer system. By blocking the odors of the sewer, Cummings made it possible to bring the toilet inside the house, and so he made the toilet desirable. Soon everyone who could afford one was happily flushing away.

The S-trap design is still used in toilets as an effective way to deal with odor. However, toilet design has since addressed other aspects of use—in Japan, standard toilets are fitted with seat warmers, jets of cleansing water, and fully automated flushers. **LS**

SEE ALSO: FLUSHING TOILET, COMPOSTING TOILET, SEWAGE SYSTEM, TOILET PAPER, MUNICIPAL WATER TREATMENT

Submersible Craft (1775)

Bushnell builds the first underwater vessel to be used for military purposes.

In 1775 Britain's North American colonies rebelled against British rule, precipitating a War of Independence. An enthusiastic American patriot, David Bushnell (1742–1824) of Saybrook, Connecticut, devised a secret weapon to counter the might of Britain's Royal Navy. He designed and built a submersible vessel to attack warships in harbor.

Bushnell's *Turtle* was an oval-shaped vessel of wood and brass, just large enough to hold one person. It had ballast water tanks that were filled to make it dive, then emptied with a hand pump to return to the surface. Two screw propellers, operated by foot pedals and a handle, allowed the operative to maneuver the vessel laterally and vertically underwater. Ingeniously, the inside of the submersible was lined with naturally luminescent wood to provide light for reading the instruments—a compass and a depth meter.

Turtle's weapon was an underwater gunpowder charge with a timer, in effect the first sea mine, although Bushnell called it a "torpedo," after a stinging crampfish. A drill was provided for attaching the charge to the hull of a ship at anchor.

Turtle was first sent into action on September 7, 1776. With an army volunteer, Sergeant Ezra Lee, at the controls, it was launched into New York harbor to attack the British flagship, HMS *Eagle*. Lee successfully brought *Turtle* up against the underside of *Eagle*'s hull, but failed to attach the charge. Getting the drill to penetrate the ship's copper-sheathed hull while maintaining position in strong currents was beyond Lee's powers. The subsequent fate of *Turtle* is obscure, but it never achieved a successful attack. **RG**

SEE ALSO: SUBMARINE, BATHYSPHERE

⬈ *A cutaway model of* Turtle; *the two propellers controlled vertical and horizontal movement.*

> *"I was obliged to rise up every few minutes to see that I sailed in the right direction."*
>
> Sergeant Ezra Lee, pilot of *Turtle*

Boring Machine (1775)

Wilkinson creates a vital tool for precise engineering in iron.

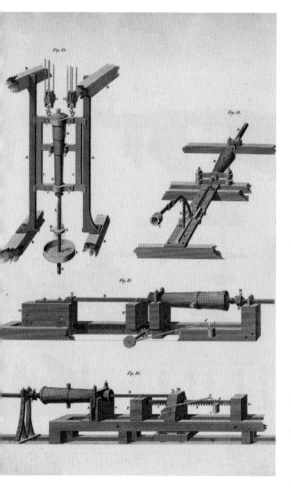

The boring machine, designed by John Wilkinson (1728–1808), was one of the foundations of the Industrial Revolution. The idea of mechanical boring was not new, but Wilkinson used it to bore better cannon with a greater degree of accuracy. More importantly, however, boring machines could be used to make precisely engineered cylinders for steam engines. Thanks to his collaboration with James Watt, inventor of the commercial steam engine, Wilkinson enjoyed a monopoly on the engine for several years, and they both became very wealthy men.

As Wilkinson's wealth grew, so did his eccentricities. He was a volatile character and was often criticized by his fellow industrialists, thanks to some shady business dealings and the implication that some of his ideas may have already been suggested by others. Some of his family relationships were strained too—he was estranged from his father, and his spectacular fallout with his brother in the late 1780s caused the collapse of his steam-engine monopoly.

In his later years Wilkinson become increasingly obsessed with iron. He arranged an enormous and impractical iron coffin in which to be buried, and prophesied his supernatural return to his beloved furnaces seven years after his death. However, extraordinarily for his time, he gave pensions to old workers who had served him well, and he was held in high regard by his employees. After his death in 1808 he was commemorated in folksong by his ironworkers and, after seven years had passed, thousands turned up at his furnaces to see whether "Iron Mad" Wilkinson would "live up to" his prophecy. **JB**

"One of the most hard-hearted, malevolent old scoundrels now living in Britain."

Lord Dundonald on John Wilkinson

SEE ALSO: DRILL, CANNON, NAIL-MAKING MACHINE, SCREW-CUTTING LATHE, COKE-BASED IRON SMELTING

▣ Rees's Cyclopaedia, *1808, showed how Wilkinson's technology could be used to bore a cannon.*

Steamboat (1776)

De Jouffroy d'Abbans powers a watercraft.

It is not uncommon for the American Robert Fulton (1765–1815) to be heralded as the inventor of the steamboat, but in actuality the true creative force behind its invention was a young French aristocrat, Claude-François-Dorothée, Marquis de Jouffroy d'Abbans (1751–1832). De Jouffroy d'Abbans, according to legend, was wild and unruly, resulting in his incarceration in a military prison on the Isle of St. Marguerite. While he was there he studied the boats passing by and developed an interest in engineering.

On his release he went to Paris and studied with the Perier brothers, examining the Watt steam engine and devising methods in which it could be applied to propelling a vessel. He began work on an experimental

> *"[The glory] belongs to the author of the experiments made on the River Saône at Lyons in 1783."*
>
> Robert Fulton on De Jouffroy's steamboat

boat, a steamship called the *Palmipède*, which he ran along the Doubs River in June and July 1776. The boat was not entirely successful, and he continued his experimental work, this time moving from Paris to Lyons. In 1783 his new model, the paddle steamer *Pyroscaphe*, was ready and ran for fifteen minutes along the Saône against the current and to a crowd of scientists and spectators. The boat was a success and ran for sixteen months, but the French Academy of Sciences in Paris refused to acknowledge it and denied de Jouffroy d'Abbans his license. Finally, embittered and impoverished, the inventor retired to the Hôtel des Invalides, where he died from cholera. **TP**

SEE ALSO: STEAM TURBINE, STEAM PUMP, STEAM ENGINE, STEAM ENGINE WITH SEPARATE CONDENSER

Circular Saw (1777)

Miller patents an innovatory form of saw.

In 1777 Samuel Miller of Southampton, England, received the first patent for a circular saw. His wind-powered machine's usefulness was limited, however, for want of a more powerful energy source.

Thirty-six years later Tabitha Babbit, a Shaker woman from the Harvard Shaker village, invented a circular saw of her own. Her religious beliefs prevented her from seeking a patent, but the new invention became popular in her community. Babbit's saw was initially human-powered, but waterwheels and steam were soon harnessed for added convenience and efficiency. Sawmills adopted the circular saw, and the tool was soon at the heart of the lumber industry.

The circular saw is a relatively simple device that dramatically improves on the efficiency of a standard handsaw, where half of each stroke is wasted effort. Circular saws cut by spinning circular serrated blades at high speeds into the timber passed through them.

The U.S. military, bolstering its technology for World War II, enlisted the saw manufacturer Skilsaw to develop a specialized saw for military use. Skilsaw's answer, the PS-12 military circular saw, could function in all conditions, including underwater, and came with a rugged camouflage paint job. The U.S. Navy put the saw into camouflaged boxes and began floating secret circular-saw units to predefined landing areas.

Nowadays the circular saw is an industrial staple. Portable circular saws allow users to saw small jobs, miter saws permit all angles of cutting, and table saws provide the backbone of most woodworking shops. Numerous blades have also been created to optimize saws for particular cutting purposes. Specialized blades, with intriguing names such as "ripping," "dado," and "thin kerf," are designed to cut through materials as varied as brick, steel, and glass. **LW**

SEE ALSO: METALWORKING, SAW, BAND SAW, CHAIN SAW

Tumbler Lock (1778)

Barron patents a thief-resistant lock.

People depend on their locks and keys a lot more than they would like to admit. Without having to stand guard over their possessions from morning to night, they are free to pursue their lives away from their homes and businesses. Locks and keys existed before Robert Barron patented his tumbler lock in 1778, but the sheer number of people now carrying keys to tumbler locks testifies to the success of his invention.

Barron's lock, which offered considerably improved security over any previous locks, was called a double-acting tumbler and was very similar to many modern models. A tumbler, essentially a lever inside the lock, prevents the bolt of the lock from being opened unless it is raised to a certain height. Barron's lock

---

"The lever tumbler lock . . . could still be picked. It merely required more skill and time."

Jock Dempsey, blacksmith

---

employed two tumblers and these needed to be lifted to different heights for the bolt to be released.

Most focused thieves could still pick the Barron lock if they had enough time, so in 1818 Jeremiah Chubb added a detecting feature in hopes of winning a reward of £100 offered by England's Portsmouth Dockyard. The detector consisted of either a spring or a specialized lever that would catch any tumbler that was raised too high. If an unlucky picker raised any tumbler higher than the detection point, the lock would jam shut, vanquishing the thief. The jammed lock would also alert the lock's owner, who could reset the lock simply by using the original key. **LW**

SEE ALSO: MECHANICAL LOCK, PADLOCK, COMBINATION LOCK, TIME-LOCK SAFE, SAFETY LOCK, YALE LOCK

Speech Synthesis (1779)

Kratzenstein reproduces vowel sounds.

The earliest speech synthesizer was created by a Russian professor, Christian Kratzenstein (1723–1795). Between 1773 and 1779 Kratzenstein made acoustic resonators and produced vowel sounds by connecting them to organ pipes.

A contemporary in Vienna, Wolfgang von Kempelen, produced a more advanced machine in 1791. His "acoustic mechanical speech machine" was able to produce single sounds and even words or short phrases. He is best known for an earlier invention, a chess-playing machine named "The Turk." This consisted of a cabinet, housing (apparently) just cogs and wheels, and a manikin with movable arms. One could not see the legless human chess player concealed inside. Once this hoax was exposed, his legitimate speech machine was discredited as well.

Alexander Graham Bell became interested in speech synthesis after he saw a replica of one of Von Kempelen's speech machines. When young, Bell had taught his pet terrier to stand between his legs and growl while he manipulated the dog's vocal tract by hand. He was eventually able to produce "How are you, Grandmamma?"

Joseph Farber improved on Von Kempelen's machine by adding a mechanical tongue and a pharyngeal cavity that could be manipulated as well. It was powered by bellows and controlled by a keyboard and could sing as well as produce speech.

The first electrical speech synthesizer was the VODER, developed by Homer Dudley and presented at the 1939 World's Fair. It saw more utility as the VOCODER, which reduced ordinary speech into a facsimile to reduce the bandwidth necessary for the telephonic transmission. This allowed a larger number of telephone calls to be transmitted over a line. **SS**

SEE ALSO: PIPE ORGAN, TELEPHONE, MICROPHONE

Spinning Mule (1779)

Crompton refines further the efficiency of mass-produced spun yarn.

The textile industry was one of the cornerstones of Britain's Industrial Revolution. The production process had changed little in centuries: yarn was spun skillfully on a human-powered wheel. Invariably performed by women and young children, it was hard work and provided little in the way of reward.

By the middle of the eighteenth century, as the demand for textiles for export grew, labor-saving devices enabling yarn to be spun at greater speed began to emerge. The two most significant developments were the water frame and the spinning jenny. The water frame used the principles of the water wheel to power the spinning frame, thus dramatically reducing the amount of human effort required; the spinning jenny, a multispool spinning wheel, boosted output by enabling a single worker to operate up to eight spools at once. In 1779 the inventor Samuel Crompton (1753–1827) combined the main features of both, creating the spinning mule. A multispooled, waterpowered spinning wheel, the mule could create a strong, thin yarn, high both in quality and consistency, suitable for any kind of textile. And it could do so at considerable speed.

The Industrial Revolution was a period of endless technical innovation, and as steam gradually became the ascendant form of power, the mule was duly converted. This gave rise to the widespread mass production of textiles, and the gradual appearance of the mighty factories that would come to dominate the Lancashire and Yorkshire landscapes. **TB**

SEE ALSO: CLOTHING, WOVEN CLOTH, SPINNING WHEEL, FLYING SHUTTLE, SPINNING JENNY, POWERED LOOM

↗ *This wood and metal replica of the spinning mule is found in the Science Museum, London.*
➡ *Workers make adjustments inside a spinning shed containing belt-driven Crompton's mules.*

Argand Lamp
(1780)

Argand revolutionizes the oil lamp, starting a worldwide hunt for the sperm whale.

Fuel-burning lamps had been used for hundreds of years without significant improvement. Then, in 1780, Swiss scientist Aimé Argand (1750–1803) invented a lamp that would revolutionize the lives of two species—*Homo sapiens* and *Physeter macrocephalus*.

Argand studied chemistry under the French chemist Antoine-Laurent de Lavoisier, who discovered that oxygen was required for burning. Argand's lamp used a hollow wick to draw more air to the inside of the flame and had a glass cylinder around the wick to increase the flow of air outside the flame. Argand also provided a way to lower or raise the wick to decrease or increase the size of the flame and thus the amount of light the lamp produced. Because of the additional oxygen, the flame burned at a higher temperature, which produced much more light. It also burned most of the carbon particles that had dirtied and dimmed older oil lamps. The glass protected the flame from air currents, which kept the amount of light steady.

Argand found that sperm whale oil produced the best flame, up to ten times brighter than candles. Since Argand lamps provided a reliable source of light after nightfall, the demand for whale oil rocketed.

In 1794, during the French Revolution, Lavoisier was executed and Argand's patent was taken away, allowing anyone to make Argand lamps. He died in London in 1803, after having spent the rest of his life experimenting on bones, coffin wood, and graveyard plants in an attempt to find the elixir of long life. Until kerosene lamps arrived in the 1850s, the sperm whale continued to be killed in large numbers for its oil. **FS**

"In the Argand ... the air and the gas were brought into contact by means of numerous small orifices."

The Mechanics' Magazine (1854)

SEE ALSO: OIL LAMP, CANDLE, GAS LIGHTING, ARC LAMP, INCANDESCENT LIGHT BULB, LAVA LAMP

◩ *This French Argand lamp has a small fuel tank and would have required frequent refilling with oil.*

Iron Rocket
(1780)

The Indians extend the range of war rockets.

The use of rockets in warfare began with the Chinese, who first developed the technology around the thirteenth century. Their new "fire arrows" were successfully deployed against the Mongols, and it was not long before the rest of the world began to experiment with them.

During the eighteenth century, the British and the French were fighting over India, each keen to possess its riches. Unfortunately for them, they discovered that the inhabitants were not always happy to hand over their land. Tipu Sultan of Mysore in southern India fought the British with a tactic, developed by him and his father, of using rocket brigades against the British infantry. The Mysoreans perfected the use of the rockets in the battlefield, developing the technology so that they could fire them over much greater distances than British weapons could achieve.

The European rockets were wooden, so they could only survive so much thrust before breaking apart. Tipu Sultan's rockets were constructed from a tube of iron, making them much stronger than wooden rockets. This extra strength meant they could withstand more thrust and fly much farther, giving the Mysoreans a tactical advantage in the field. The sheer numbers of rockets deployed, not to mention their noise and drama, disoriented the British infantry, and the rockets aimed directly at the infantry caused significant casualties.

Impressed by these rockets, the British took hundreds back to reverse engineer them. New British rockets were used at Boulogne, Copenhagen, and against the Americans at Fort Washington, with the words "rockets' red glare" eventually being included in the first verse of the American national anthem. **DK**

SEE ALSO: GUNPOWDER, FIREWORKS, ROCKET, SHRAPNEL
SHELL, BALLISTIC MISSILE

Compound Steam Engine (1781)

Hornblower introduces compound cylinders.

The steam in early steam engines was used only once; after it had pushed back the piston it was discharged into the atmosphere, A more efficient process allowed the steam to expand in two or more stages. These "compound" engines had two or more cylinders. After the steam had been expanded in the high-pressure cylinder the exhaust steam was then used to push back the piston of a following, larger-circumference, low-pressure cylinder. The two pistons were connected with cranks that enabled them to work at the required different phases. With correct size scaling, the power output per cylinder could be equalized, and the engine ran smoothly. As these systems were

> *"[Engine control] requires an intelligent man, an honest man, a sober man, a steady man . . ."*
>
> Isambard Kingdom Brunel, engineer

rather complicated, they were mainly used in industrial and marine engines. Some compound railway locomotives were built, but the tough operating conditions made them difficult to maintain.

Jonathan Hornblower (1753–1815) was originally an employee of Boulton and Watt and designed the first compound steam engine in 1781. Unfortunately, the early compounds were no more economical than simple single-cylinder engines. The concept was then revised by the Cornish engineer Arthur Wolf, who obtained a patent in 1805. Problems with the high-pressure cylinder meant that few such engines were used until the mid-nineteenth century. **DH**

SEE ALSO: STEAM ENGINE, STEAM ENGINE WITH SEPARATE
CONDENSER, HIGH-PRESSURE STEAM ENGINE

Hot-air Balloon (1783)

The Montgolfier brothers use fire to send humankind aloft.

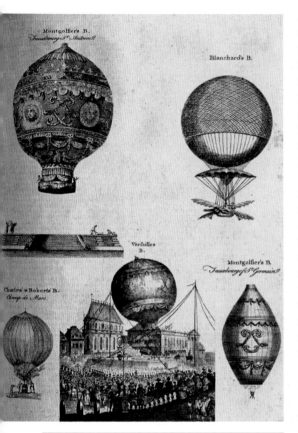

> *"Get in a supply of taffeta . . . and you will see one of the most astonishing sights in the world."*
>
> Joseph Montgolfier in a letter to his brother

⤴ *Made in 1784, soon after the Montgolfier flight, this engraving shows the proliferation of balloon designs.*

➡ *In an early test, a Montgolfier balloon lifts from its heat source, restrained by men on the ground.*

At 1:45 P.M., November 21, 1783, in the courtyard of the Château de la Muette on the outskirts of Paris, the balloon constructed by brothers Joseph-Michael (1740–1810) and Jacques-Étienne de Montgolfier (1745–1799) made its first manned flight. Piloted by a young scientist named Jean-François Pilâtre de Rozier and an army officer, the Marquis d'Adlandes, the flight high over the city lasted for twenty-five minutes and came to the ground on the Butte-aux-Cailles, 10 miles (16 km) from where it had started.

The Montgolfiers' inspiration came from watching a fire burning and speculating as to the "force" that caused the sparks, smoke, and embers to rise. They constructed a large envelope from taffeta, lit a fire beneath the opening, and watched as it rose to the ceiling. Their experiments grew grander in scale until on September 19, 1782, they were called to conduct a royal demonstration in Versailles. King Louis XVI and Queen Marie Antoinette witnessed the first passenger-carrying air flight—albeit with a sheep, a duck, and a rooster in the basket of the balloon. The eight-minute flight saw the Montgolfiers' creation rise to a height of 1,500 feet (460 m) before running out of fuel and landing safely 2 miles (3.2 km) distant.

In spite of their success, the brothers had still failed to understand the science that underpinned their work. They believed it was smoke from the burning wool and straw that caused their balloon to rise. In fact, the balloon rose because the property of air changes when its temperature is altered: The heated air inside the balloon became lighter than the colder air on the outside, causing it to ascend.

The two brothers were honored by the French Académie des Sciences and both were credited with further important inventions in their careers. **TB**

SEE ALSO: CONTROLLED FIRE, KITE, GLIDER, POWERED AIRPLANE, AUTOGYRO, JET ENGINE, SUPERSONIC AIRPLANE

VUE ET PERSPECTIVE DU JARDIN DE M^r. REVEILLON FABRIQUANT DE PAPIER,

g S^t. Antoine, à l'ancien Hôtel de Titon, où se font faites les expériences de la Machine Aérostatique de MM. Montgolfier freres, dans le

courant de l'Eté, en l'année 1783 à la satisfaction d'un concours immense d'amateurs.

DEDIÉE A M^{rs}. LES PHYSICIENS.

A Paris, chez Esnauts et Rapilly, rue S^t. Jacques, à la Ville de Coutances.

Parachute (1783)

Lenormand makes the first descent slowed by a fabric canopy.

On December 26, 1783, before a large public gathering at the base of the Montpellier Observatory in Paris, the French scientist and physicist Louis-Sébastien Lenormand jumped from the observatory's tower clinging to a 14-foot (4.2 m) parachute attached to an improvised wooden frame. Lenormand's leap of faith was the first ever documented use of a parachute and followed on from an earlier attempt at a slowed descent when he leaped from a tree holding on to nothing more than two modified parasols.

Lenormand's inspiration likely came from the popular writings of a former French ambassador to China whose memoirs included an account of Chinese acrobats floating to earth using umbrellas. Chinese legends dating to 90 BCE also tell of a group of prisoners who cheated death by leaping from a tower and slowing their descent with the aid of conical straw hats. Leonardo da Vinci sketched his famous pyramid-shaped sealed linen parachute in 1485, but there is no

evidence that his idea ever progressed beyond a sketch. Recently uncovered Renaissance manuscripts from 1470 depict a parachute not unlike Da Vinci's version, and yet predate his drawings by fifteen years.

The first descent from a great height was performed by the French balloonist André-Jacques Garnerin, who in 1797 leaped from the basket of a hot-air balloon beneath a silk parachute designed, like Da Vinci's, without an aperture, descending 2,230 feet (680 m) to Parc Monceau in Paris and landing entirely without injury before a large crowd. Garnier later adapted his design by adding an aperture to reduce the parachute's oscillations during descent. **BS**

SEE ALSO: WOVEN CLOTH, HOT-AIR BALLOON, POWERED AIRPLANE, KNAPSACK PARACHUTE

⬆ *Carried upward by a balloon, Garnerin released his parachute and descended safely to the ground.*

Bifocals (1784)

Franklin publicizes "double spectacles" combining two functions in one pair.

The ability to focus clearly on nearby objects—accommodation—decreases with age. This is known as presbyopia. It explains why older people tend to hold books at arm's length. Bifocals are glasses incorporating two lenses for each eye. The lower section of each glass corrects for presbyopia, bringing nearby objects into focus.

American statesman Benjamin Franklin (1706–1790) is normally credited with the invention of bifocals—although no one knows for sure who devised them, or when they did so. Franklin, and a few others, may have been wearing bifocals since the 1760s, but the first mention he makes of his "double spectacles" is in a letter dated August 21, 1784. Franklin certainly did much to publicize bifocals and, as a man of science—and of poor eyesight—he certainly understood the principles behind their function.

Each "lens" of Franklin's bifocal glasses consisted of two distinct pieces of glass, but toward the end of the nineteenth century Louis de Wecker (1832–1906) devised a way to fuse the two lenses into one. The term "bifocals" was actually introduced by John Isaac Hawkins (1772–1854) in 1826 to distinguish them from glasses he introduced that incorporated three distinct lenses. Hawkins's "trifocals" overcame a problem faced by older people who wear bifocals: while distant and nearby objects are in focus, vision in the middle distance range suffers.

Today, glasses are available with a much wider range of focal lengths, not just two or three. These "varifocals," first seen in the late 1950s, offer a smooth transition from one focal length to another. **JC**

SEE ALSO: GLASS, LENS, SPECTACLES, EYE TEST (SNELLEN CHART), CONTACT LENSES

⬆ *This pair of bifocals includes two separate lenses for each eye; also of interest are the sliding side arms.*

Shrapnel Shell (1784)

Shrapnel combines existing technologies in a deadly new weapon.

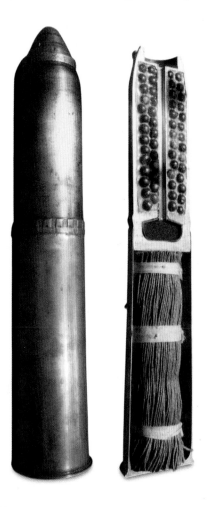

In everyday language, the word *shrapnel* is normally used to describe fragments of metal from an explosion—often those lodged into human flesh. However, the word has a more specific meaning; it is the name of an important artillery weapon invented in 1784 by Henry Shrapnel (1761–1842), then a lieutenant in the British Army.

In its original design, the shrapnel shell was a hollow iron sphere filled with gunpowder and about 200 musket balls. Protruding into the shell was a timed explosive fuse. As the sphere flew through the air, the fuse exploded, causing the shell to blast open and release the musket balls. The balls continued to travel in the same direction, at the speed at which the shell had been traveling—fast enough to cause death and destruction on a devastating scale.

Shrapnel's idea was a combination of two existing battlefield technologies. At close range, soldiers used a metal canister filled with lead balls, which sprayed out like shot from a shotgun. It was very effective. At longer ranges, they fired hollow iron shells filled with gunpowder. When the gunpowder exploded, it would send fragments of iron in all directions. This type of shell made a loud noise, but it was not very effective.

From 1787 onward, Shrapnel demonstrated his invention to the British Army; eventually, in 1803, it was accepted. The shells were first used, with the desired outcome, in Surinam in 1804. They went on to make an important contribution in many other battles during the nineteenth century. Shrapnel himself called his invention a "spherical case," but artillerymen always referred to it as the shrapnel shell. **JC**

SEE ALSO: GUNPOWDER, CANNON, MUSKET

"[The committee] do not take upon themselves to decide upon the policy of introducing it."

British Army's rejection of the shrapnel shell, 1801

◨ *The exterior and interior of a shell (the tied cord at the base represents the propellant, frequently cordite).*

Safety Lock (1784)

Bramah devises a nearly unpickable lock.

The British inventor Joseph Bramah (1748–1814) patented eighteen new ideas during his lifetime, including the fountain pen, fire engine, and beer hand-pump, but remarkably he is most famous for his safety locks. His lock design had ingeniously included notched sliders that made it nearly impossible to pick.

Bramah was so sure of his locks that he offered a cash prize to anyone who could pick it open. The Challenge Lock, as it became known, caught media attention and boosted the profile of his company as the lock resisted all effort for at least fifty years; the money remained unclaimed during Bramah's lifetime. The lock was eventually opened by A. C. Hobbs, an American, although it took him sixteen days to do so.

> "Billy . . . identified [the lock] as a Chubb rather than a Bramah, which is reputably unpickable."
>
> Frederick Forsyth, *The Fourth Protocol* (1984)

The original of the famous lock can now be found in the Science Museum in London.

Bramah's locks were very difficult to make by hand and later Bramah took on a young apprentice called Henry Maudslay. With Bramah's bright ideas and Maudslay's practical talent they created machinery that could make the locks at an economic price. Maudslay went on to invent the screw cutting lathe, which revolutionized manufacturing.

Bramah's company survives in London and is run by a descendant of Joseph. They continue to make locks based on his original design, which still exceeds British and European standards. **LS**

SEE ALSO: MECHANICAL LOCK, COMBINATION LOCK, PADLOCK, TUMBLER LOCK, TIME-LOCK SAFE, YALE LOCK

Threshing Machine (1784)

Meikle's machine separates grain from stalks.

In 1784 Scottish millwright and inventor Andrew Meikle invented the threshing machine, probably drawing inspiration from a design Michael Menzies had patented fifty years earlier. Once grain plants are harvested it is necessary to separate the grain from the plant. After a failed first attempt in 1778 Meikle built a machine that could complete this process in a fraction of the time that had previously been required. As the machines slowly spread across Britain, they were greeted with a wave of hostility from discontented villagers. Threshing had previously offered an opportunity for laborers to supplement their income during the winter period, and their livelihood was now under threat. Understandably, they revolted.

Threshing machines are not particularly safe to operate; they beat and thrash at whatever is fed to them, be that sheaves of grain plants or arms and legs. Meikle's final design used a strong drum with fixed beaters. Thus, he avoided making the mistake he'd previously made, that is, building a device that rubbed the grain rather than beating it.

The first peasants to man them were often poorly trained and unfamiliar with the dangers of automated machinery. However, what made things far worse—in the late eighteenth century at least—was that peasants drank beer rather than water, which was of unreliable cleanliness. It was common for thirsty farm workers to drink several pints throughout the course of the day, keeping relatively hydrated but all the time drifting continuously and calamitously toward inebriation. The combination of unfamiliarity with the machines, fatigue, and inebriation resulted in a great many deaths and injuries.

The invention went on to form part of the combine harvesters that further revolutionized agriculture. **CB**

SEE ALSO: FLAIL, SCYTHE, REAPER, COMBINE HARVESTER

Powered Loom
(1785)

Cartwright revolutionizes weaving and capitalizes on breakthroughs in spinning.

The dramatic rise of the textiles industry in Britain was one of the most important aspects of the world's first Industrial Revolution. Machines had been spinning yarn rapidly and effectively since the 1760s, but the mechanization of weaving did not take hold properly until after 1810. It began when Edmund Cartwright (1743–1823) patented the first powered loom in 1785.

Handlooms had become commonplace by the eighteenth century. They are relatively slow, and each one requires at least one person's full attention. In its basic form, a powered loom is simply a mechanized, automated version of a handloom. It can produce cloth faster than a handloom, and a single person can watch several or many machines.

Power for Cartwright's looms was originally supplied by waterwheels, via a drive shaft and belts and gears. But steam power increasingly took the place of waterpower during the nineteenth century, because steam-powered factories could be built anywhere, not only next to a river.

The weaving process is repetitive, lending itself well to mechanization. But many things can go wrong, as Cartwright discovered after he opened his own textiles factory in Doncaster. He quickly found solutions to many of the problems and was granted several patents between 1785 and 1792. Many other inventors made improvements, too, and by the middle of the nineteenth century, there were nearly 300,000 powered looms in use in Britain alone. **JC**

SEE ALSO: CLOTHING, WOVEN CLOTH, SPINNING WHEEL, FLYING SHUTTLE, SPINNING JENNY, SPINNING MULE

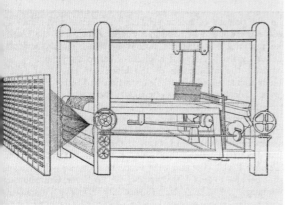

↖ *A Robert Fulton portrait of Cartwright, who became the agricultural experimenter of the Dukes of Bedford.*
← *Cartwright's steam-powered loom brought faster weaving but also cost many weavers their jobs.*

Automatic Flour Mill
(1785)

Evans transforms the flour-milling process.

In 1782, Oliver Evans (1755–1819), along with his brother Joseph, opened a village store in Maryland. There, dealing with the local milling community, Evans discovered how cumbersome the milling process was. The stone or log mills were quite primitive, requiring hours of hard labor, and the flour produced by them was often contaminated with dirt from the floor. Evans suspected that there was a better way to make flour, and he began to design an automatic flour mill.

The site of Evans's flour mill was in Red Clay Creek, where an old stone mill had been built in 1742. By 1785 the automatic flour mill was in operation. It consisted of a bucket elevator, a belt conveyor, a horizontal

> *"[Red Clay Creek] demonstrated for the first time the fully integrated automatic factory."*
>
> Eugene Ferguson, historian of technology

conveyor, and a mechanical hopper-boy, which was a rake used for spreading and cooling the flour. The entire process was mechanized and run on either waterpower or gravity. Millers were slow to warm to Evans' invention, but it slowly became obvious that his process would revolutionize flour making.

Evans received patents for his invention in Delaware, Pennsylvania, Maryland, and New Hampshire. A national patent law was passed in 1790, and the third awarded national patent went to the automatic flour mill. Both George Washington and Thomas Jefferson later commissioned Evans to build automatic flour mills at their estates. **RH**

SEE ALSO: GRANARY, QUERNSTONE, WINDMILL, WATERMILL, TIDAL MILL

Nail-making Machine
(1790)

Perkins automates nail manufacture.

It has been estimated that the average house is built using around 20,000 nails. For this we have American inventor and engineer Jacob Perkins (1766–1849) to thank—not for inventing the nail, which appeared thousands of years before he was born, but for inventing a machine that could produce upward of 200,000 nails per day, introducing mass-produced nails for the first time.

Perkins was born in Massachusetts, and worked for a goldsmith during his teenage years. He became known for creating a multitude of different inventions, of which his nail-making machine was one of his first and most famous. Nails were traditionally handmade by beating out sheets of iron into the required shape, but this was made easier with the invention of mechanical cutting processes by the mid 1700s.

In 1790 Perkins devised his nail-cutting invention, followed five years later by a patent for "a machine to cut and head nails at a single operation." Perkins set up a nail-making factory in his hometown of Newburyport, although eventually others took up his idea and built similar machines. A watermill was used to power the machine, which could cut out and head the nail in a single process. The output was an estimated 200,000 nails per day, swamping previous levels of nail production. The effect was dramatic—nail prices fell by around 70 percent within a few decades, reducing the overall cost of construction. This became increasingly important in the ensuing industrial revolution, when the demand for nails rose.

Perkins continued to innovate, and upon moving to England in 1819 he pioneered the use of steel instead of copper for printing banknotes, as well as producing many useful contrivances. **SR**

SEE ALSO: NAIL, HAMMER, WATERMILL

Guillotine (1791)

Guillotin proposes a machine for humane decapitation.

In 1789, at the start of the French Revolution, Joseph-Ignace Guillotin (1738–1814), a medical doctor of progressive views, proposed a thorough-going reform of the French penal system. Inspired by the humane and rational principles of the Enlightenment, Guillotin's proposals included a single method of execution to replace the messy horrors of breaking on the wheel and hanging by the neck. Guillotin's mechanism would prevent suffering, while making capital punishment more democratic; beheading was traditionally the punishment reserved for aristocrats—an efficient decapitation machine would spread that privilege to all classes.

In 1791 the French National Assembly appointed a committee to push the project through. Although Guillotin was involved, the prime mover was Dr. Antoine Louis, Royal Physician and Secretary of the Academy of Surgery. The basic design adopted, with a blade hauled to the top of a high frame and then released, was not novel—such machines had been in use since the Middle Ages. Its advance upon earlier models lay in the beveled-edged triangular blade.

Originally dubbed the "louison" or "louisette," after Dr. Louis, the machine was soon being called the "guillotine." It became the symbol of revolutionary extremism, as the Terror was unleashed upon alleged "enemies of the people." King Louis XVI was an early victim, executed on January 21, 1793. Despite its bloodthirsty revolutionary associations, the guillotine remained in use in France until 1981. **RG**

"The mechanism falls like thunder; the head flies off, blood spurts, the man is no more."

Joseph-Ignace Guillotin, 1789

SEE ALSO: CARPENTRY, METALWORKING, SWORD, STEEL, ELECTRIC CHAIR

◸ *An eighteenth-century model of Guillotin's humane but nevertheless horrifying killing machine.*
▢ *A lurid depiction of the execution by guillotine of Louis XVI by the caricaturist James Gillray (1757–1815).*

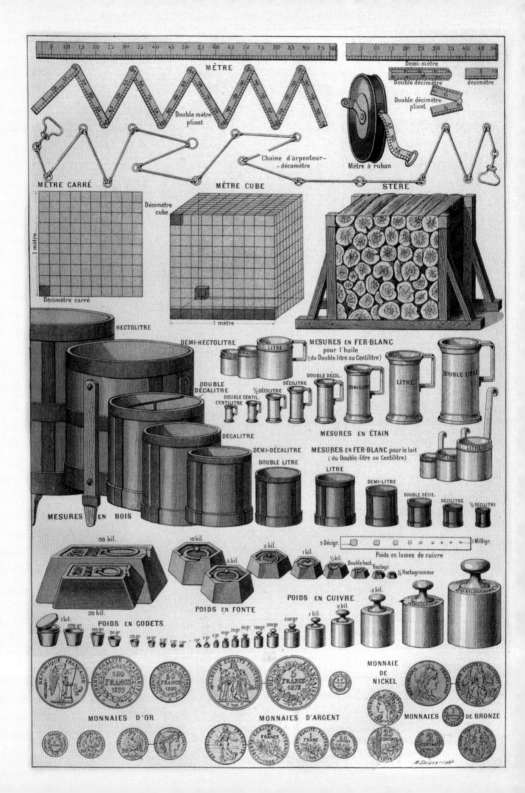

Metric System (1791)

The French initiate metric measurement.

In the nineteenth century there was a confusing multiplicity of units of measurement. In England, for example, length was measured in inches, feet, yards, furlongs, rods, chains, poles, perches, miles, and more.

In 1791 the French National Assembly instructed the Academy of Sciences to design a simple decimal system. In 1793 the unit of length, the meter, was chosen to be one ten-millionth of the distance between the north pole of Earth and the equator, the specific meridian that was chosen being the one that passes though Paris. Unfortunately, the length of the meridian had not been measured at the time and this job was carried out by Jean-Baptiste-Joseph Delambre (1749–1822) and Pierre Méchain (1744–1804).

A platinum bar engraved with two marks separated by the new "meter" was then placed in the International Bureau of Weights and Measures in Paris. Decimal divisions, such as the centimeter and kilometer were then used, the circumference of the Earth being about 40,000 kilometers (25,000 miles). The word "about" has to be used because the French did not take into account the fact that the Earth is a slightly flattened sphere, not a perfect one.

The meter also led to the metric definition of the unit of mass, the kilogram. The reforms instituted by the French Revolution defined the kilogram as being the mass of a cube, each side 10 centimeters (3.9 in) square, of pure, air-free water at the temperature of 4°C (39.2°F), the temperature at which the density of liquid water is at its maximum.

The British, being somewhat suspicious of the French, did not adopt the system. **DH**

SEE ALSO: STANDARD WEIGHTS AND MEASURES, SLIDE RULE, SPRING TAPE MEASURE

⬅ *A page from a French reference work includes the relative sizes of units of metric measurement.*

Gas Turbine (1791)

Barber initiates gas-powered technology.

In 1791 the English inventor John Barber (1734–1801) patented "an Engine for using Inflammable Air for the Purpose of procuring Motion." We now know his invention as the gas turbine, a mechanism that was simultaneously 150 years before its time and based on an idea more than 1,700 years old.

A turbine is a machine that turns the energy of moving gas or liquid into rotational energy. During the first century, the Greek inventor Hero of Alexandria developed the first steam turbine. His device drove steam from boiling water through curved nozzles to rotate a cylinder. John Barber's engine, while based on similar principles, included features not present in its ancient predecessor.

A gas turbine engine such as Barber's has three main components: a compressor to increase the pressure of the air, a combustion chamber in which the air is combined with fuel to produce an explosion, and a turbine wheel that is spun by the expanding combustion products.

Barber's design was sound, but the metallurgy required to support the high temperatures required by gas turbines was not available to him. Although many people worked on gas turbines in the nineteenth century, the first successful gas turbine was probably the eleven-horsepower engine built by Aegidius Elling in 1903.

Two uses of the gas turbine stand out: power generation (in which most of the energy goes into spinning the turbine) and jet propulsion (in which most of the energy goes into high-velocity exhaust gases). Both applications saw their first practical embodiments in 1939: A commercial power plant with gas turbines began operating in Switzerland, and the first jet aircraft flew in Germany. **ES**

SEE ALSO: GAS LIGHTING, GAS STOVE, GAS-FIRED ENGINE, STEAM TURBINE, FRANCIS TURBINE

Dentures (1791)

De Chémant secures the English patent for teeth made from porcelain.

The history of dentures stretches back far before 1791, when a dentist, Nicholas Dubois de Chémant (1753–1824), obtained the patent for them. There are records from around 700 BCE of Etruscans using dentures made from human and animal teeth. By the fifteenth century, ivory or bone dentures were in use in Europe, attached in the mouth by wire to surviving teeth. All these early forms of dentures would have been highly uncomfortable to wear, deteriorated rapidly, and contributed to the malodor of the mouth.

In 1774 Alexis Duchâteau (1714–1792), a French chemist who was dissatisfied with his own set of dentures, produced a new design that used porcelain teeth. He was helped in this endeavor by de Chémant. However, Duchâteau was unable to promote his new dentures properly and his idea stalled. De Chémant continued the experimentation and by 1787 had perfected new dentures. He applied for the patent in France, and Duchâteau, believing the idea to be his, tried unsuccessfully to sue his former friend.

During the French Revolution, de Chémant fled to England, where in 1791 he was granted the patent to his "mineral paste" dentures. For some years the famous English company Wedgewood provided de Chémant with the porcelain paste that he needed to make the false teeth; by the early 1800s, he was also producing single false teeth. De Chémant's dentures remained in use through much of the nineteenth century, until improvements were made to fit, structure, and materials. Gradually, vulcanite came into use to replace the porcelain paste, and then acrylic resins and other plastics were introduced. **TP**

SEE ALSO: TOOTHPASTE, TOOTHBRUSH, DENTAL FLOSS, ELECTRIC DENTIST'S DRILL

⬆ *Hippopotamus-ivory dentures in a smart porcelain holder, made for the Prince of Wales around 1800.*

Gas Lighting (1792)

Murdock ignites gas from burning coal as a source of lighting.

William Murdock (1754–1839) was a consummate and prolific inventor, but the invention that he is most remembered for today was the development of gas lighting, which took over from the oil and tallow system. His experiments began around 1792 when he realized that gases released from burning coal could be lit and used as a steady source of light. He is said to have burned coal in his mother's old kettle, lighting the gas that came out of the spout. By 1794, however, the kettle had been replaced by a specially built retort in which the coal was burned; gas from the burning coal was funneled through a long attached tube, to be ignited at the tube's end.

Murdock first used his system of gas lighting in his own home in Redruth, Cornwall, and continued to develop methods for producing, storing, and igniting the gas more efficiently and practically. In 1798 he moved back to Birmingham from Cornwall to work at the Boulton and Watt factory (run by Matthew Boulton, the renowned engineer, and James Watt, of steam engine fame), in which he installed his new gas lighting. In 1802 he illuminated part of the factory's exterior, to great public delight. The following year his gas lighting was installed in the Philips and Lee cotton mill in Manchester.

Why Murdock failed to obtain a patent remains a mystery, although it is possible he was discouraged from doing so by his employers, Boulton and Watt. By the mid-1800s most large towns in England were illuminated by gas lighting and had their own gas works, all based on Murdock's original invention, but for this he received little benefit. **TP**

SEE ALSO: OIL LAMP, CANDLE, ARGAND LAMP, ARC LAMP, PUBLIC ELECTRICITY SUPPLY, INCANDESCENT LIGHT BULB

⬆ *A cartoon by Thomas Rowlandson (1756–1827) shows reactions to gas lighting installed in London in 1807.*

Ambulance (1792)

Larrey introduces the medical-care vehicle.

Ambulances first began to appear on the Napoleonic battlefields of France in 1792. Their inventor, surgeon Dominique-Jean Larrey (1766–1842), had grown frustrated with regulations requiring him to stay to the rear. After observing how the mobility of the French artillery helped it to quickly disengage from an advancing enemy, Larrey proposed to the military hierarchy what he called an *ambulance volante*, or "flying ambulance," that would follow the artillery into battle and tend to the wounded where they fell.

Larrey devised a horsedrawn wheeled carriage with a central compartment able to transport two patients comfortably on leather-covered horsehair mattresses; windows on either side provided good

> *"Before ... the flying ambulance, we seldom saw men who had lost both legs and arms ..."*
>
> Dominique-Jean Larrey, surgeon

ventilation. Inside, patients could be moved in and out easily on floors set on rollers. Recessed areas contained medicines and medical equipment, and ramps at the rear doubled as emergency operating tables.

Larrey and his team tended the wounded on the basis of their injuries rather than rank or status. Those likely to die from their wounds were set aside in favor of treating those who were deemed not to be mortally wounded—this brought the French word *triage*, meaning "sorting," into the surgeon's lexicon. **BS**

SEE ALSO: SLEDGE, TRAVOIS, WHEEL AND AXLE, CART, HORSE COLLAR

◄ *Before the wheeled ambulance, Larrey (top left) used camels and side panniers to transport wounded men.*

Cotton Gin (1793)

Whitney transforms cotton processing.

The decision of Yale graduate Eli Whitney (1765–1825) to leave his Massachusetts home in 1792 and seek employment in the southern state of Georgia would radically alter the course of American history. While working on a plantation, Whitney learned of the financial need to make cotton-picking more efficient than was possible by the labor-intensive method of manually removing seeds from the cotton bolls. Within months he had constructed a device that rapidly separated the cotton from the seeds by pulling it through hundreds of short wire hooks mounted on a revolving cylinder. This method allowed only the fiber to pass through narrow slots in the iron breastwork; the seeds were left behind.

The beauty of the cotton gin (the name derived from the Southern pronunciation of *engine*) lay in its simplicity of use, whether powered by man, animal, or water. Aware of the huge demand from English textile factories, Whitney recognized the potentially great financial gains to be made from the cotton gin and secured a patent for the machine in 1794. In partnership with Phineas Miller he manufactured and serviced gins around the South, only to find that planters resented paying for something they could pirate themselves. Lawsuits brought against the planters were easily defeated and the partners were forced out of business by 1797. As Whitney slumped into debt, Southern planters profited handsomely.

The cotton gin revolutionized the cotton industry in the South and by the mid-nineteenth century the United States were supplying three-quarters of all cotton throughout the world. The invention would also leave a dark legacy. As demand intensified, so did the growth of slavery in the southern states precipitating eventual war with the North. **SG**

SEE ALSO: CLOTHING, WOVEN CLOTH, SPINNING WHEEL, SPINNING JENNY, SPINNING MULE, POWERED LOOM

Canal Inclined Plane (1794)

Fulton transfers canal boats between levels.

Use of the inclined plane for the transfer of boats between water levels dates back at least to the sixth century BCE, when ships were transported across the Isthmus of Corinth in wheeled cradles. Early Chinese engineers also made use of the principle, employing double slipway constructions in their canals to haul vessels between levels.

Modern inclined planes were pioneered by Italian architect Daviso de Arcort in Northern Ireland in the 1770s. Coal barges were raised or lowered in stages through a total of 190 feet (58 m) onto the Coalisland Canal, drawn on sloped rails by a combination of counterweighting and horsepower. The ambitious

> *"This is simply two single inclined planes in conjunction, expanding from hill to knoll . . ."*
>
> Robert Fulton, engineer

project was fraught with problems, and closed in 1787.

In 1778 William Reynolds constructed England's first inclined plane in Shropshire, followed by a steam-driven version on the Shrewsbury Canal, but despite these successes it was American Robert Fulton who claimed the British patent for the canal inclined plane in 1794. Arriving in England in 1786 as a young artist, Fulton soon recast himself as an engineer and proved adept at developing the fledgling ideas of others. Although not regarded as a true inventor, his 1796 *Treatise on the Improvement of Canal Navigation* did disseminate the principle of the inclined plane throughout France and the United States. **FW**

SEE ALSO: CANAL, CANAL LOCK, DAM

Internal Combustion Engine (1794)

Street creates a new form of power plant.

The term *internal combustion engine* tends to refer to reciprocating piston engines in which combustion is intermittent, although continuous combustion engines, such as jet engines, rockets, and gas turbines are also internal combustion engines.

In the seventeenth century, Sir Samuel Morland, an English inventor, used gunpowder to drive water pumps, creating the first rudimentary internal combustion engine, but it was not until 1794 that Robert Street actually built a compressionless engine. In 1879 Karl Benz designed and built the four-stroke engine that powered the first automobiles.

The most significant distinction between modern internal combustion engines and the early designs is the use of in-cylinder compression. The first internal combustion engines did not have compression, but ran on an air/fuel mixture sucked in during the first part of the intake stroke.

Within the internal combustion engine, the combustion of fuel and an oxidizer (normally air) occurs in confined spaces, the combustion chambers. This exothermic reaction produces gases at high temperatures and pressures, which expand, resulting in the movement of pistons that slide within the combustion chambers. Steam engines, in contrast, use external combustion chambers to heat a separate working fluid, which in turn moves the pistons.

The most common internal combustion engines are the four-stroke, gasoline-powered, spark-ignition engines, which are also used in the aeronautics industry. Internal combustion engines, generally using petroleum, are used for mobile propulsion in automobiles, trucks, motorcycles, boats, and in a wide variety of aircraft and locomotives. **MF**

SEE ALSO: STIRLING ENGINE, GAS-FIRED ENGINE, FOUR-STROKE CYCLE, TWO-STROKE ENGINE, BOURKE ENGINE

Semaphore
(1794)

Chappe pioneers a long-range, rapid, visual signaling system.

In Paris in 1791, a little-known engineer and inventor named Claude Chappe (1763–1805) began to experiment with an optical signaling system or visual telegraph. His ambition was to send complex messages via a succession of towers using a combination of signaling arms. Three years later, in 1794, working with the aid of his four brothers, Chappe demonstrated his first optical semaphore. His string of fifteen towers placed within sight of each other was able to transmit a message 120 miles (190 km) from Paris to Lille in only nine minutes. The project's burdensome costs were borne by a French leadership who were recently at war with Austria and eager for any strategic advantage in communication.

Each tower was topped by a 30-foot (9 m) mast to which a rotating arm was attached with smaller, counterbalanced wooden arms at its ends. These could move through the horizontal and vertical and in a series of seven 45-degree positions to produce 196 combinations of letters, numbers, and selected words. The semaphore's indicators seemed almost to mimic the outstretched arms of a human, and were painted black to provide maximum contrast against blue or cloudy skies. Telescopes were fitted to each tower so that its operators could relay messages on. Because spies could easily see the messages, a code book consisting of secret phrases and words was introduced for emergencies. Chappe's semaphore proved such a success that new lines of towers were soon constructed, radiating out from the French capital to cities such as Dunkirk, Brussels, and Antwerp. **BS**

SEE ALSO: ELECTROMAGNETIC TELEGRAPH, PRINTING
TELEGRAPH, DUPLEX TELEGRAPH, TELEPHONE

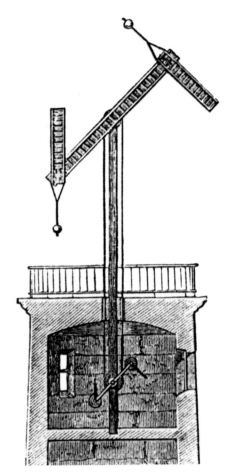

↗ *The arrangements of the mast's three wooden arms were controlled from a room in the tower below.*

"[Electricity] had to be abandoned when no adequate insulators could be found for the wires."

Abraham Chappe on his brother's experiments

Corkscrew (1795)

Henshall devises an elegant way to remove a cork from a wine bottle.

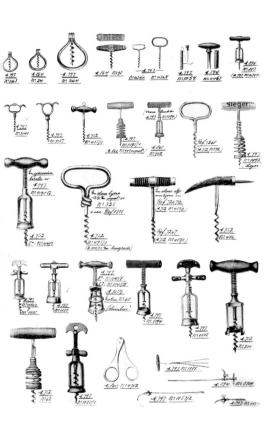

In 1795 the Reverend Samuell Henshall of Oxford, England, took the end of a gun worm—a steel, helix-tipped tool used for removing wadding and unspent charges from musket muzzles—and attached it to a wooden handle: he had invented the modern corkscrew. Between the handle and worm he added a unique concave button, designed to compress the cork. A series of ribbed protrusions on the underside of the button engaged the cork, breaking its adhesion to the bottle and preventing the cork from fraying. This so-called "Henshall button" also prevented the helical gun worm from penetrating too far into the cork. The corkscrew was manufactured in Birmingham by Michael Boulton, with the Latin phrase *obstando promoves soho patent*, meaning "by standing firm one makes advancement," inscribed on each button.

The latter half of the eighteenth century had seen refinements in glassblowing techniques and the bottle had evolved. Rudimentary, squat designs, with tapered necks and protruding corks that were pulled out by hand, had given way to mass-produced bottles with long cylindrical necks, ready to be stacked on their sides for easy storage and transportation. To prevent leakage, tighter seals had become necessary. Corks were compressed prior to being inserted into the bottle and no longer protruded from the neck, and so the need arose for a device to remove the cork.

Although it is conceded that Henshall's design came decades after other attempts by gunsmiths and blacksmiths to make gun-worm "bottlescrews," these were far less sophisticated, were never patented, and all lacked Henshall's ingenious "button." **BS**

SEE ALSO: **ALCOHOLIC DRINKS, GLASS, SCREW, BLOWN GLASS, SODA SYPHON, MUSKET**

> *"Somebody forgot the corkscrew and for several days we had to live on nothing but food and water."*
>
> W. C. Fields, comedian and actor

◩ *By the end of the nineteenth century, many inventors had added refinements to Henshall's corkscrew.*

Hydraulic Press (1795)

Bramah harnesses the force of liquid under pressure in a powerful machine.

Joseph Bramah (1748–1814), an inventor and locksmith born in Yorkshire, England, developed and patented the hydraulic press in 1795. He also invented a beer engine (1797), a papermaking machine (1805), a machine for printing bank notes with sequential serial numbers (1806), and a fountain pen (1809).

Hydraulic presses are widely used in industry for tasks that require a large force. Their capacity can range from 1 ton, or less, to more than 10,000 tons. The machine depends on Pascal's principle, which is that pressure throughout a closed system is constant. Typically it has two cylinders and pistons of differing cross-sectional areas joined by a length of small-diameter tubing. A fluid, such as oil, is displaced when either piston is pushed inward. The small piston displaces a smaller volume of fluid than the large piston, for an equal distance of movement, so any force exerted on the smaller piston is translated into a larger force on the larger piston, because the force is magnified by the increased area of the pistons.

The science of hydraulics deals with the flow of liquids in pipes, rivers, and channels and their confinement by dams and tanks. Its principles may also apply to gases. Nowadays the scope of hydraulics extends to such mechanical devices as fans and gas turbines and to pneumatic control systems, as well as hydraulic disc brakes and hydraulic garage lifts.

Bramah worked with William George Armstrong on the hydraulic press, which found numerous industrial applications. The hydraulic press, which is often named for him as the Bramah Press, was his most important invention. **MF**

SEE ALSO: **DAM, FOUNTAIN PEN, HYDRAULIC CRANE, HYDRAULIC BRAKE, HYDRAULIC JACK**

⬚ *An 1812 engraving shows in section the Bramah hydraulic press and its working components.*

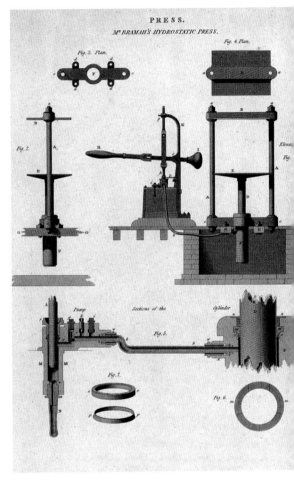

"As an accumulator of power, [the] press surpasses anything that has yet been invented …"

Scientific American (January 1864)

Vaccination (1796)

Jenner engineers immunity to smallpox.

As Edward Jenner (1749–1823) was growing up in England, smallpox had again become prevalent and was ravaging London and the countryside. Jenner became a doctor, practicing in Gloucestershire where he became interested in the link between cowpox and smallpox. Milkmaids who contracted the non-deadly cowpox seemed immune to smallpox, and Jenner, intrigued by this, began to investigate the link.

In May 1796, milkmaid Sarah Nelmes contracted cowpox from her cow, and pus-filled blisters covered her hands and arms. She visited Dr. Jenner. Realizing his opportunity to test the protective properties of cowpox on someone who had not contracted smallpox, Jenner took some pus from Sarah and

> *"[I was] … the instrument destined to take away from the world one of its greatest calamities…"*

Edward Jenner, scientist and doctor

applied it to scratches made on the arm of a young boy, James Phipps. Some days later Phipps came down with a mild form of cowpox, proving that the disease was transferable between people. Jenner then injected Phipps with smallpox, and although the boy became ill, he recovered quickly and completely.

Many more experiments confirmed the results, and in 1798 Jenner published his findings. At first there was reluctance amongst the medical industry to use his methods, but the results were too conclusive to ignore. In 1853 an Act of Parliament made vaccination with cowpox compulsory, and the numbers of people dying from smallpox dropped dramatically. **TP**

SEE ALSO: INOCULATION, CHOLERA VACCINE, RABIES VACCINE, ANTHRAX VACCINE, BCG VACCINE, POLIO VACCINE

Cast Iron Plow (1797)

Newbold casts a one-piece plow.

For almost a millennium the farming practices of the Saxons remained unchanged. To plow the ground, farmers used the primitive scratch plow developed in Mesopotamia in 5500 BCE. The work was laborious.

In 1797 Charles Newbold (1764–1835), a blacksmith from New Jersey, patented a practical plow in which the three main parts were made as one solid piece of cast iron. These were the moldboard (a curved iron plate), the share (the cutting edge of the plow attached to the moldboard), and the landslide (the stabilizing mechanism that counteracts the sideways motion of turning over the soil). Newbold also integrated a runner that allowed directional control and therefore straighter furrows.

> *"The development involved more than $30,000 in experimentation and marketing."*

Lloyd E. Griscom, *The Historic County of Burlington*

Farmers were initially wary of Newbold's plow, believing that the cast iron would poison their fields and ruin their crops. Eventually the plow gained acceptance, but this came too late for Newbold to recover his investment—$30,000, a tremendous sum in those days—because competitors had begun to develop similar plows. However, in 1807 Newbold successfully sued David Peacock, who had been issued a patent for a nearly identical plow. Newbold won $1,500 for patent infringement.

Later, Newbold's design was improved upon by producing the three molded parts separately, so that each could be removed and replaced if broken. **FS**

SEE ALSO: SCRATCH PLOW, MOLDBOARD PLOW, STEEL PLOW

Screw-cutting Lathe (1797)

Maudslay standardizes screw manufacture.

Before the days of British engineer Henry Maudslay, (1771–1831), screws were handmade and depended on the skill of the craftsman. Consequently, no two screws were alike or interchangeable.

In 1797 Maudslay created precision machinery that enabled identical screws to be produced. Without Maudslay's standardization, tasks such as building flatpack furniture would be extremely difficult.

Maudslay was the skilled apprentice of the lockmaker Joseph Bramah. Their working partnership failed when Maudslay and Bramah fell out over pay, causing Maudslay to set up his own shop in another part of London. In a quest for precision, he devised a screw-cutting lathe capable of cutting down reliably to a ten-thousandth of an inch.

To existing lathe designs Maudslay introduced gears and a lead screw that changed the pitch (distance between a complete turn) of the screw. This allowed him to cut a range of thread pitches from the same machine, rapidly and with an accuracy and control previously unseen. Maudslay's identical screws were immediately put to good use in the manufacturing of steam engines. One of his apprentices, Armstrong Whitworth, later standardized screw thread, and today's U.K. imperial unit for thread sizes is the British Standard Whitworth (BSW).

Precision screws led to the start of another Maudslay endeavor, the machine tool, a tool that makes other tools, which was to play a key role during the English Industrial Revolution. **LS**

SEE ALSO: SCREW, ARCHIMEDES SCREW, SCREW MICROMETER, STANDARDIZED SCREW SYSTEM, SQUARE-HEADED SCREW

↗ *Maudslay's lathe combined a lead screw and change wheels for reproducing screw threads.*

↪ *British engineer Henry Maudslay, who standardized manufacture of both ordinary and specialist screws.*

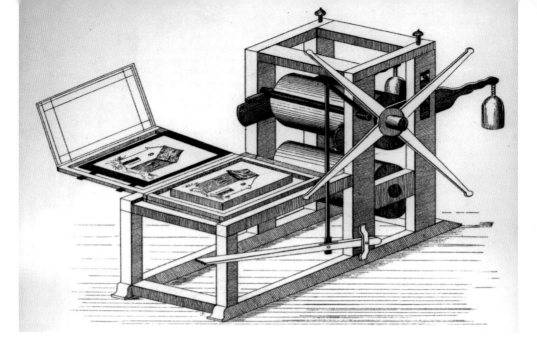

Lithography (1798)

Senefelder devises an entirely new printing process using limestone.

In 1798 Austrian actor and playwright Alois Senefelder (1771–1834) created a print by using a press to copy an image onto paper from the smooth surface of a section of limestone. Senefelder erroneously referred to his process as chemical printing. It would go on to become the most significant innovation in printing since relief printing in the fifteenth century.

Although the precise details of his discovery are vague at best, the most commonly accepted story is that, when asked by his mother to prepare a laundry list, he was unable to find a suitable piece of paper, so he used a grease pencil to write the list on the flat surface of a piece of dense Solenhofen limestone. Senefelder then at some point must have observed how the greasy residue left by the pencil became absorbed and embedded into the porous limestone, retaining its ink even after having its surface washed. Washing the stone then caused the remaining surface areas to repel ink, leaving only the drawn image and

thus eliminating the need for etching. Senefelder had discovered a new way of setting off inked and non-inked surfaces, though it was to take him another four years to fully actualize the process.

The great advantage of lithography was its ability to endlessly reproduce an image without any parts of the process wearing down. By 1800 Senefelder had refined the process and established a lithographic press in London to produce sheet music. In 1802 he set up a press in Paris, where the French ignored his rather cumbersome name for the process—Chemical Printing for Bavaria and the Electorate—and referred to it as *lithographie*, meaning "writing on stone." **BS**

SEE ALSO: INK, PAPER, WOODBLOCK PRINTING, PRINTING PRESS WITH MOVABLE METAL TYPE, ETCHING FOR PRINT

⬆ *A drawing shows how the press produces an image on paper by pressing it onto prepared limestone.*

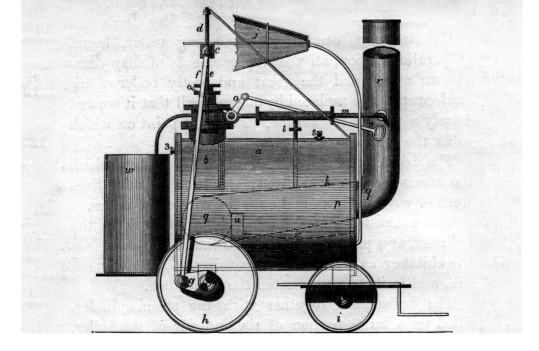

High-pressure Steam Engine (1799)

Trevithick builds the first engine powered by steam under high pressure.

By the age of nineteen, Cornishman Richard Trevithick (1771–1833) worked for the the Cornish mining industry as a consultant engineer. The mine owners were attempting to skirt around the patents owned by James Watt, inventor of the steam engine, because the royalties were costing them a fortune.

William Murdoch had developed a model steam carriage, starting in 1784, and demonstrated it to Trevithick in 1794. Trevithick thus knew that recent improvements in the manufacturing of boilers meant that they could now cope with much higher steam pressures than before. Using steam at a higher pressure, Trevithick could eliminate the need for a separate condenser, which was integral to the patents held by Watt, as well as peripherals such as the air pump. Further, Watt's low-pressure engines required large buildings to house them. By using high-pressure steam in his experimental engines, Trevithick was able to make them smaller, lighter, and more manageable.

Trevithick constructed high-pressure working models of both stationary and locomotive engines that were so successful that in 1799 he built a full-scale, high-pressure engine for hoisting ore. The "used" steam was vented out through a chimney into the atmosphere, bypassing Watt's patents. Later, he built a full-size locomotive that he called *Puffing Devil*. On December 24, 1801, this bizarre-looking machine successfully carried several passengers on a journey up Camborne Hill in Cornwall. Despite objections from Watt and others about the dangers of high-pressure steam, Trevithick's work ushered in a new era of mechanical power and transport. **DHk**

SEE ALSO: STEAM ENGINE, STEAM ENGINE WITH SEPARATE CONDENSER, COMPOUND STEAM ENGINE, LOCOMOTIVE

⬆ *Trevithick's* Puffing Devil *was the first locomotive to carry passengers, but on the road rather than rail.*

Battery (1799)

Volta creates the first device to produce its own electricity.

A battery, sometimes called a cell, is a device that converts chemical energy into electrical energy. When two or more cells are joined together in such a way that the currents produced from each flow in the same direction, they are known as a battery of cells. There are two basic types of batteries: the primary, nonrechargeable battery, where the electricity stops when the chemicals are used up, and the secondary (or storage) battery, which can be recharged.

The battery originated with Alessandro Volta (1745–1827) who, in 1799, invented the "voltaic pile," a pile of silver and zinc disks, separated by pieces of fabric saturated with sea water, that supplied an electric current when connected by a wire. His work was based on that of Luigi Galvani, who had noticed that a dead frog's legs twitched when they came into contact with two different types of metal.

Each cell had two terminals, or electrodes—a positive one, the anode, and a negative one, the cathode—suspended in a liquid known as the electrolyte. Over the next few years, a number of inventors developed other combinations of metals and electrolyte to produce more efficient batteries. In the 1880s a solid electrolyte was used and the contents were encased in covers and known as dry cells. The first portable safe device, known as the "Flashlight," was produced in 1896.

In 1859 Gaston Planté, a French physicist, produced a secondary (or storage) battery, which could be recharged and is similar to the battery, or accumulator, that is used in today's automobiles. In this battery, lead plates were immersed in sulfuric acid. **MF**

SEE ALSO: ELECTROPLATING, ARC LAMP, RECHARGEABLE STORAGE BATTERY, CAR BATTERY

⬆ *Volta's wet pile (or voltaic pile), in the* Philosophical Transactions of the Royal Society, *London, 1800.*

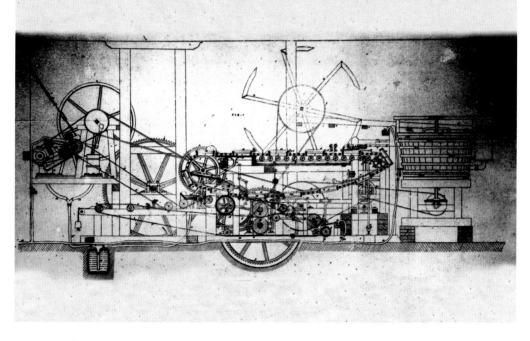

Fourdrinier Machine (1799)

Robert patents a papermaking machine that later goes on sale as the "Fourdrinier."

Before the Fourdrinier machine, paper was made one sheet at a time using a screen-bottomed frame and a mold, or vat, of wet pulp. Lifting the frame through the pulp allowed the water to drain, leaving pulp on the screen. The pulp layer was then pressed and dried. The size of a single sheet was restricted to how large a frame could be handled manually.

Paper production was a skilled affair undertaken by craftsmen, often working in guilds. But by the eighteenth century, an increased demand for paper, and a desire to circumvent the paper makers' guild, prompted Frenchman Nicholas-Louis Robert (1761–1828) to design a machine that would automate the process and produce a seamless length of paper, via a continuous belt of cloth-covered, wire-mesh screen.

After much experimentation and testing, Robert's machine received a French patent in January 1799, but the design still needed development. The political situation in France, and disagreements with Robert's original sponsors, led, in 1801, to Robert and his brother-in-law, John Gamble, obtaining a patent for the machine in England, the rights being shared with their new financial backers, Henry and Sealy Fourdrinier. Ultimately, the Fourdriniers' engineer, Bryan Donkin, would construct an improved machine—the "Fourdrinier." It produced high-quality paper and, following yet more improvements, was marketed in 1807. By 1812 such machines were operating commercially. Fearful that automation would cost them their jobs, handmade-paper workers rioted and, ignoring posted-up warning notices, attempted to destroy such equipment. **MD**

SEE ALSO: PARCHMENT, PAPER, KRAFT PROCESS (MODERN PAPER MAKING)

⬆ *A technical drawing of a papermaking machine named for its financiers rather than its inventor.*

In the first half of the nineteenth century, the use of steam and electricity as power sources had a great impact on various aspects of human life, from cooking and transport to medicine and warfare. Photography added a new dimension to leisure activities and helped police in their search for criminals. Meanwhile, ostensibly modern creations such as computer programs and fax machines already saw the light of day, and the first constant electric light paved the way for Thomas Edison's modern light bulb.

◁ The steam locomotive signals the start of a new era.

The INDUSTRIAL AGE

1800 to 1859

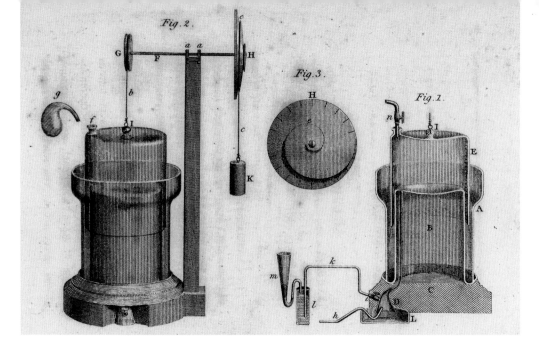

Nitrous Oxide Anesthetic (1800)

Davy's "laughing gas" proves an effective form of pain relief.

Humphry Davy (1778–1829) first noted the anesthetic effects of nitrous oxide—a colorless, almost odorless gas—while experimenting at the Pneumatic Institute in Bristol, England. Davy (best known for inventing the miner's lamp) realized that nitrous oxide both made him want to laugh (coining the term "laughing gas") and relieved his toothache. In 1800 he published a book stating that the gas might "be used with advantage during surgical operations." After Davy's observations, nitrous oxide became popular at laughing parties and fairground shows, but it was not used in surgery for another forty years.

At one fair in the United States, Horace Wells, a Connecticut dentist, observed a man who gashed his leg while under the influence of nitrous oxide. He seemed to be pain-free, and Wells immediately had one of his own teeth removed while breathing in the gas. In January 1845, Wells demonstrated the use of nitrous oxide in a dental extraction at the Harvard Medical School in Massachusetts. Unfortunately, insufficient gas was applied and the patient cried out in pain. The public humiliation resulted in Wells's loss of reputation as a dentist and tragically to his suicide three years later. The next year dentist William Morton successfully used the gas while a surgeon removed a tumor from a man's neck and use of nitrous oxide in surgery then quickly caught on in London and Paris.

Today, nitrous oxide continues to be used during childbirth and in dentistry to allay anxiety and offer some degree of analgesia. As an anesthetic it has survived chloroform, which proved too toxic, and ether, which posed too high a risk as an explosive. **JF**

SEE ALSO: ANESTHESIA, ETHER ANESTHETIC, CHLOROFORM ANESTHETIC, MODERN GENERAL ANESTHETIC

⬆ *An 1800 illustration of the mercurial air holder and breathing machine used by Davy in his researches.*

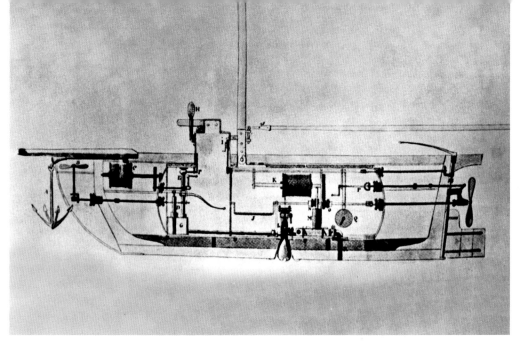

Submarine (1800)

Fulton makes a major advance in the design of underwater boats.

In 1797, Robert Fulton (1765–1815), a Pennsylvania-born artist and inventor, offered to build a submarine for the French, who were then engaged in a bitter war with Britain. He proposed to use the vessel to sink Royal Navy warships blockading French ports. The French government hesitated to become involved in what they regarded as a dishonorable style of warfare. Undeterred, Fulton went ahead with developing his machine, gaining French financial backing after Napoleon made himself First Consul in 1799.

Fulton's "mechanical Nautilus" was launched on the River Seine at Rouen in 1800. Its inventor described it as "six and a half meters long and two meters wide . . . built entirely of wood." It had a mast and sail for propulsion on the surface, which were lowered when it submerged by filling its water tanks. The crew of two to four could stay underwater for around four hours thanks to a supply of compressed air. Submerged, the vessel was driven by a hand-cranked propeller and

could maneuver using vertical and horizontal rudders. The interior was lit by candles and a glass cupola allowed the crew to see around them when semisubmerged. Like Bushnell's *Turtle* of 1775, the *Nautilus* depended on underwater explosives—primitive sea mines—to attack enemy ships.

After successful tests in the Seine, the *Nautilus* showed its ability to sink a target ship at Brest in July 1801. However, the French refused permission for Fulton to attack British warships and withdrew their support. Fulton tried to interest Britain in his invention, but as the dominant naval power they had no motive to pursue a revolutionary form of warfare. **RG**

SEE ALSO: SUBMERSIBLE CRAFT, STIRLING ENGINE, BATHYSPHERE

⬆ *An original 1806 drawing of Fulton's second, 35-foot (10.7 m) submarine, which was never built.*

Punched Card (1801)

Jacquard automates production and invents an early method of storing encoded data.

Building on a concept proposed by Jacques de Vaucanson in 1745, Joseph-Marie Jacquard (1752–1834), namesake of the famous loom, perfected a more practicable interpretation of his fellow Frenchman's idea—automation using punched cards.

Jacquard knew from experience that silk weaving, although a skillful art, was extremely repetitive. It was this aspect of the process he attacked and so he set out to control the weaving process by linking the actions of the loom to the pattern of holes on the cards. Each card had the same number of rows and columns, the presence or absence of a hole being detected mechanically and thereby determining the loom's movements. Ultimately, many such cards would be connected in sequence, enabling the loom to weave complex designs time and time again.

In 1803 Jacquard was summoned to Paris to demonstrate his invention and to work for the Conservatoire des Arts et Métiers. Improvements to his loom mechanism followed, and by 1806 the loom was declared "public property." The invention was violently opposed by weavers who felt it would cost them their jobs, and riots ensued.

From the late nineteenth century onward, the potential to store data on such cards led to their widespread use in the fields of data collection and processing. They were adopted by Charles Babbage for his analytical engine and by Herman Hollerith for tabulating the U.S. census in 1890. The computer age saw the development of cards in formats suited to the needs of programmers and for use with computer languages such as COBOL or FORTRAN. **MD**

"The real danger is the gradual erosion of individual liberties through automation . . ."

U.S. Privacy Protection Study Commission, 1977

SEE ALSO: POWER LOOM, MECHANICAL COMPUTER, COMPUTER PROGRAM, PUNCHCARD ACCOUNTING

◩ *Jacquard's punched-card system was used for the loom he created in 1804 to weave figured cloth.*

Gas Stove (1802)

Winzler's innovation brings gas to the kitchen.

Humankind has been preparing food in countless ways for many thousands of years, but perhaps the most important innovation was deciding to cook food in the first place. Open flames worked for a time, but as humans became civilized, so did their cooking. The Chinese and Japanese had closed stoves from the second and third centuries BCE respectively—long before the rest of the world. By the fifteenth century, Europe had moved toward a modern stove, but the whole world relied on wood, charcoal, coal, or oil to fuel their cooking until the ninteenth century.

Gas cooking was introduced by Zächaus Winzler (1750–c. 1830) in 1802. Winzler, a Moravian chemical manufacturer living in Austria, began hosting dinner parties where the food was cooked using a small gas cooker complete with four burners and an oven. However, gas stoves did not appear in other kitchens for another thirty years.

One kitchen got a gas stove in 1826, years before they were commercially available. James Sharp, a manager with England's Northampton Gas Company, used his home kitchen as a testing facility and installed a gas cooker of his own design.

Gas cooking eventually caught on, and by 1834 Sharp, with the financial help of Earl Spencer, started producing cookers for sale. Business was sluggish until the 1850s when gas pipelines brought the fuel within the average household's reach. By the 1880s, gas cooking had become all the rage. Since then, stoves have gone electric and microwave, but at dinner parties all over the world, you will still find gas stoves helping hosts serve hungry guests. **RBk**

SEE ALSO: CONTROLLED FIRE, OVEN, GAS LIGHTING, ELECTRIC STOVE, MICROWAVE OVEN

↗ *An early gas stove of 1837, shown with its two doors open and its upper and lower gas rings ignited.*

"Cooking is one of the oldest arts and one which has rendered us the most important service in civic life."

Jean-Anthelme Brillat-Savarin, gastronome

Powdered Milk (1802)

Krichevsky invents a versatile food substitute.

Russian doctor Osip Krichevsky first produced powdered milk in 1802. It is made by drying or dehydrating milk until it forms a fine white powder. This can be achieved either by spraying a fine mist of milk into a heated chamber or by adding the milk in a thin layer to a heated surface, from which the dried milk solids can be scraped off. Freeze-drying is now used because it conserves more nutrients and the milk can be fortified to improve its nutritional value. The resulting powder can then be stored for long periods, because the dry environment means it is less prone to bacterial contamination that would spoil fresh milk.

As well as its potential for long-term storage, powdered milk has several practical advantages over

"Things are seldom what they seem / Skim milk masquerades as cream."

Sir William Schwenck Gilbert, *H.M.S. Pinafore*

fresh milk. In the developing world, its relative light weight and the fact that it does not need refrigeration mean that it is easy to transport over long distances, without the need for expensive refrigerated trucks.

Powdered milk has found an additional use in modern science in a technique for separating proteins called Western Blotting. In this process, the protein-rich milk is used to block inappropriate binding of the antibodies used, and therefore produce a much clearer result in experiments. **SB**

SEE ALSO: INSTANT COFFEE, CANNED FOOD, VACUUM-SEALED JAR, MILK/CREAM SEPARATOR

← *A late nineteenth-century sign advertises Nestle's powdered milk, one of the company's first products.*

Leaf Spring (1804)

Elliot cushions the blows of road travel.

Londoner Obadiah Elliot invented the leaf spring in 1804 when he piled steel plates on top of each other, pinned them together, and attached them to the end of a carriage. His design remains a key component in supporting heavy goods vehicles, however, he was not the first person to add a bit of spring to transportation. Ancient Roman vehicles were suspended on elastic wooden poles, which work on the same principle.

The spring is formed by stacking several layers of steel in the shape of an arc with an axel in the center and the edges tied to the vehicle. What has made the leaf spring such a popular invention is that varying the number of leaves (steel plates) or the curvature of their configuration alters the performance and weight

"Business opportunities are like buses, there's always another one coming."

Richard Branson, British entrepreneur

capacity of the spring. They are also cheap to produce and reliable to use. Another source of their popularity, however, comes from their ability to minimize the nauseating vibrations of eighteenth-century road travel. A lack of adequate suspension meant that travel sickness was a regular occurrence. Elliot could not even out the roads, but he could cushion their blows

A rapid growth in passenger travel followed Elliot's invention, which in turn fueled other industries. Coaches required horses, and horses needed food for fuel, leading to a growth in the corn and feed industry. Passengers themselves fueled the inn and hotel boom that grew up concurrently along the way. **CB**

SEE ALSO: WHEEL AND AXLE, CART, STEEL, HORSE COLLAR

Locomotive (1804)

Trevithick demonstrates that a steam engine with wheels can pull passenger wagons.

> *"Steam is no stronger now than it was a hundred years ago but it is put to better use."*

Ralph Waldo Emerson, writer

⬆ *A portrait of the Cornish engineer Richard Trevithick, dating from 1816.*

➡ *An illustration of the locomotive Penydarren, built by Trevithick in 1803 and demonstrated in 1804.*

In the late eighteenth century, there were many wagonways and tramways in Europe. These had iron rails and horsedrawn wagons fitted with flanged wheels. The first steam locomotive to run on rails was built by Richard Trevithick (1771–1833) of Cornwall, England. Trevithick was encouraged to develop an engine that was more efficient and cheaper to run than the low-pressure Watt and Newcomen type; he was the first to harness high-pressure steam.

Trevithick's *Puffing Devil* (1801) and *London Steam Carriage* (1803) were demonstration steam vehicles, but on February 21, 1804, his *Penydarren* locomotive pulled five wagons, seventy passengers, and 10 tons of iron down an iron railway between Merthyr Tydfil and Abercynnon in south Wales. This reasonably reliable and robust machine proved that heavy trucks could be hauled along low-gradient, smooth railway lines by a smooth-wheeled, heavy locomotive.

By 1825 fellow Englishman George Stephenson had built a better railway engine—called the *Locomotion*. It pulled six coal wagons and 450 passengers in twenty-one coaches down the 9 miles (14.5 km) of railway between Stockton and Darlington, a journey that took about one hour.

Improvements came at a great pace. In October 1829, at the Rainhill Trials—a competition to select a locomotive for the first intercity railway line, between Liverpool and Manchester—George and Robert Stephenson's *Rocket* covered 50 miles (80 km) at a speed of 12 miles per hour (19 kph). This engine had a multitubular boiler that increased steam production.

Soon four-wheeled locomotives with highly inclined cylinders were being replaced by more stable locomotives with longer boilers, horizontal inside-frame cylinders, and more and larger wheels. **DH**

SEE ALSO: WHEEL AND AXLE, STEAM ENGINE, HIGH-PRESSURE STEAM ENGINE, STEAMBOAT

Electroplating (1805)

Brugnatelli's metal coating has many uses.

Electroplating, sometimes called electrodeposition, is the process by which an electric current, provided by an external supply of direct current such as a battery, is passed through a solution resulting in the chemical breakdown of this electrolyte. This results in metal being transferred from one electrode—the anode—via metal ions in the solution, to the other—the cathode—which has the effect of the target object being coated with a thin layer of the metal that had formed the anode.

The Italian chemist Luigi Brugnatelli (1761–1818) is credited with inventing electroplating in 1805. He used Alessandro Volta's earlier invention of a battery, the voltaic pile, to facilitate the first electrodeposition.

Electroplating is used in many industries for decorative as well as functional purposes. It can increase the value or improve the appearance of an object. For example, jewelry is often gold-plated, and silverware may be made of cheaper metal coated with silver. The technique is also used to silver-plate table cutlery. Coatings such as zinc and tin also provide protection against corrosion and certain objects, such as steel car bumpers, are weatherproofed by being electroplated with first nickel and then chromium. Hard chromium is used to decrease frictional wear in moving machinery by electroplating the surfaces of hydraulic pistons and camshaft-bearing diameters.

Electroplating is also used to silver-plate copper or brass electrical connectors, because silver tarnishes much more slowly and has a higher conductivity than those metals, resulting in more efficient electrical connections. Similarly, connectors used in computers and other electronic devices may be plated with gold or palladium over a barrier layer of nickel to improve electrical conduction. **MF**

SEE ALSO: METALWORKING, BATTERY

Carbon Paper (1806)

Wedgwood simplifies the copying process.

In 1806 the potter and inventor Ralph Wedgwood (1766–1837) was issued a patent for what he called his "stylographic manifold writer," a device that assisted the blind to write by employing a metal stylus rather than simply drawing by hand with the dominant writing implement of the day, a quill pen.

His writing machine was a board crisscrossed with metal wires that helped guide the hand of the blind as they wrote. Wedgwood then took a sheet of paper, saturated it in printer's ink, dried it, and placed it between a sheet of tissue paper and a second sheet within the stylographic's writing frame. Its metal stylus was then used to transfer the ink from the carbonized paper to the sheet below, eliminating any concern about keeping quill pens filled with ink.

> "Carbon paper was initially manufactured entirely by hand and on a craft basis."

Bruce Arnold, writer

Carbon paper, created almost as a by-product of the stylographic writer, was slow to gain acceptance in the business world as people were suspicious that it could lead to forgeries. It was not widely used until the advent of the first commercially successful typewriter in 1867.

Carbon paper altered during the 1860s when Lebbeus Rogers began brushing on a mix of soot, naphtha, and oil, a process that remained essentially unaltered well into the twentieth century. With the rise in electronic communication, carbon paper may well become obsolete in the future, although "Cc" (short for carbon copy) remains a feature of e-mail. **BS**

SEE ALSO: PAPER, INK, QUILL PEN, TYPEWRITER

Band Saw (1808)

Newberry's new kind of saw mechanizes carpentry.

"The band saw created a special era in American architecture . . . the gingerbread house was born."

200 Years of Woodworking (1976)

⬆ *A technical drawing shows how the band saw's continuous blade passes over its upper wheel.*

➡ *A hefty piece of timber is passed through a powerful band saw in a nineteenth-century engraving.*

Greek legend suggests that the first saw was made by Perdix, the nephew of the inventor Daedalus. He was inspired to create a cutting tool on observing the ridges on a fish's backbone. However, saws were probably in use well before this; ancient Egyptians used serrated copper saw blades.

The handsaw is limited by the need for people to power it. The circular saw, patented in the eighteenth century, helped alleviate that problem but also had its limitations, namely that it could cut no deeper than the radius of its disc. The band saw—a fast-moving cutting "strip" mounted in a machine—potentially solved both problems and has many advantages over the handsaw and the circular saw.

William Newberry of London, England, was granted a patent for the first band saw in 1808. It was a strip of flexible steel welded together to form a circle or band. When rotated rapidly, this had a vigorous cutting action and, in principle, could make light work of slicing a trunk into planks. The drawback of Newberry's invention was that, although technically sound, the metals from which saws were made at the time were not robust enough for the task. There were also problems with the join in the piece of metal, and it was not until some years later that the band saw actually began to look like a viable option.

In 1846 a Frenchwoman, Mademoiselle Crépin, patented a technique that allowed the two ends of the saw to be fixed together with a much stronger join, allowing practical band saws to be made. Further developments in the quality of steel also helped, and band saws became an essential tool. The practical results of the band saw range from an elaborate new style in architecture (American Carpenter Renaissance) to the creation of the first 3D jigsaw puzzles. **BG**

SEE ALSO: CARPENTRY, SAW, CIRCULAR SAW, CHAIN SAW

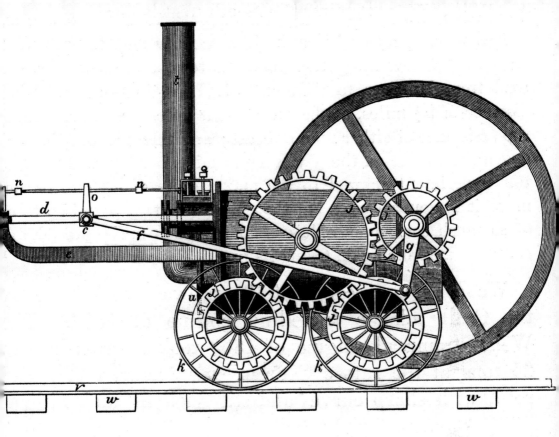

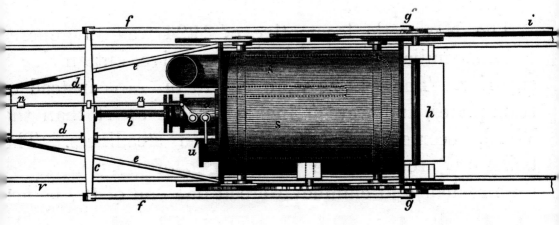

Municipal Water Treatment (1804)

Thom brings clean water to the masses.

The need for drinking water is as old as life on earth. For much of human evolution, springs, streams, rivers, and lakes provided a ready source of water, and early towns and cities eventually grew up around such important supplies. But as towns and cities got bigger, the difficulty of ensuring a clean supply increased because of greater pollution and more thirsty mouths to sate.

Methods of purifying water have been around for centuries. Sand filtration was first described as a means to remove impurities in the seventeenth century and used a method borrowed from nature. The technique was adopted by individuals, but not set up as a municipal service until 1804.

In that year, civil engineer Robert Thom (1774–1847) created the first water treatment plant in Paisley, Scotland. The water was slowly filtered through sand and gravel before being transported by horse and cart. Three years later, this rather inefficient system was improved when Glasgow began piping water to residents. (With the increasing popularity of bottled water today, transported by road again, it is curious to reflect how we appear to have come full circle.)

Thom, who came from South Ayrshire, Scotland, went on to bigger things. He was commissioned to build an improved water supply for Greenock, near Glasgow, a town that was thriving on the manufacturing trades of the Industrial Revolution. Thom designed a 5½-mile (9 km) aqueduct—known as "The Cut"—to bring water from a nearby lake to power the mills. The people of Greenock still enjoy water pumped through his system to this day, and the water source is known as Loch Thom in honor of the engineer. **MB**

SEE ALSO: AQUEDUCT, FLUSH TOILET, WATER FILTER, SEWAGE SYSTEM

Glider (1804)

Cayley boosts the quest for a flying machine.

Born into a wealthy family in Yorkshire, northern England, George Cayley (1773–1857) was a prolific inventor with an interest in human flight. He devised a heavier-than-air flying machine, with a wing to provide lift, a fuselage in which a pilot could sit, and a cruciform tail for balance and control. In 1804 he built a glider based on this design, with a kite for a wing and a pole some 5 feet (1.5 m) long as the fuselage. This seems to have flown down slopes unmanned, with varying weights of ballast onboard, although Cayley recorded that in later experiments with similar but larger gliders a man running into "a gentle breeze" had found himself lifted off the ground "for several yards."

> *"We shall be able to transport ourselves and families ... more securely by air than by water."*
>
> George Cayley, "On Aerial Navigation 1809–10"

The originality of Cayley's design lay in abandoning flapping as a means of propulsion. It had previously been assumed that a human would fly like a bird. But Cayley, having defined flight in terms of lift, drag, and thrust, confined the wing to providing lift. But having given up flapping, Cayley lacked an alternative power source to provide thrust. He later built a glider, which, in 1853, led to the first sustained manned glider flight. Powered flight, however, had to wait another half century for the petrol engine and the Wright brothers. **RG**

SEE ALSO: KITE, HOT-AIR BALLOON, POWERED AIRPLANE, JET ENGINE, SUPERSONIC AIRPLANE

→ *This reconstruction of the man-carrying glider Cayley flew in 1853 was built in 1973 by Anglia TV in England.*

Vacuum-sealed Jar
(1809)

Appert's technique stops food spoilage.

Nicolas Appert (*c.* 1750–1841) was a humble Parisian candymaker when he responded to a 12,000-franc challenge issued by Napoleon Bonaparte in 1795 to anyone who could provide the means to keep his vast armies supplied with fresh food. After experimentation, Appert realized that heating foodstuffs at boiling point for long periods of time helped prevent spoilage. After cooking his food in open kettles, it was then placed inside glass jars and heated until a seal using pitch, cork, and sealing wax was made. As the jar cooled, a vacuum was often created inside the jar removing the air necessary for bacteria to grow, and the lids were then fastened with a metallic thread.

"C'est la soupe qui fait le soldat. [An army marches on its stomach]."

Attributed to Napoleon Bonaparte

In 1806 the French Navy successfully tested Appert's preserves, which included milk, fruit, and vegetables. In 1809 he was awarded the 12,000-franc prize, and his process came to be known as "appertization." In 1810 Appert published his findings and with his prize money opened the world's first commercial cannery called the House of Appert, in Massy, south of Paris, in 1812.

Unfortunately, Appert was unable to comprehend the nature of sterilization, much less communicate to others the scientific reasons why he had succeeded. It was not until nearly fifty years later that Louis Pasteur proved his conjecture that heat destroys bacteria. **BS**

SEE ALSO: GLASS, CANNED GOODS, SELF-HEATING FOOD CAN

Breech-loading Rifle
(*c.* 1810)

Hall influences the evolution of firearms.

Prior to the invention of the breech-loader, firearms had been loaded from the end of the barrel (muzzle-loaders), an operation that took considerable time to complete. The first breech-loading rifle to see action in combat was developed by Major Patrick Ferguson and was actually a breech-loading version of a flintlock rifle. Despite being superior to conventional flintlocks in terms of rate of fire, the weapon was notorious for breaking and did not achieve widespread use.

The rifle of John H. Hall (1781–1841), the M1819, was a single-shot, breech-loader; the powder and ball were still loaded separately, but instead of being inserted into the end of the long barrel and rammed

"I have succeeded in establishing methods for fabricating arms exactly alike, and with economy."

John H. Hall

into the chamber using a rod, the rifle "broke" in the middle and the ball and powder could be loaded directly into a shortened chamber, before the barrel was clicked back into place for firing. This procedure radically shortened the time it took to reload, compared to muzzle-loaders, making the M1819 considerably more efficient. Hall received an order for 1,000 breech-loading rifles from the U.S. military.

Hall also patented devices that were employed in the construction of the M1819, including machines for drilling and straight cutting. In this way, Hall's rifle helped pioneer techniques and devices that would later become commonplace in mass production. **BG**

SEE ALSO: GUN, ARQUEBUS, FLINTLOCK MECHANISM, REVOLVER, MACHINE GUN, PORTABLE AUTOMATIC MACHINE GUN

Canned Goods

(c. 1810)

Durand makes a landmark advance in the history of food preservation.

Canning is the process by which food is preserved by sealing it into a robust and airtight container, and then sterilizing the sealed can. Cans are heated (sometimes under pressure, to achieve a higher temperature than boiling) to destroy bacteria.

Englishman Peter Durand patented a process using a tin-lined, wrought-iron "canister" in London in 1810. (Nicolas Appert had recently developed his food preservation process in France, whereby hot food was sealed with wax into glass jars.) At first, the strong metal cans were made and sealed by hand and cooked for six hours, making it an expensive process. At this stage canned food was used only by the armed forces and explorers. The can opener had not yet been invented and hungry diners had to cut or break open the tins using brute force.

In 1846, Englishman Henry Evans developed a die-cast process that increased can production from six to sixty an hour and, in the United States in 1847, Allen Taylor patented a machine-stamped tin can. Further developments in methods of lining, sealing, heating, and opening cans continued apace following Durand's early models, improving manufacturing efficiency and making canned food and drinks progressively safer and more convenient.

The rapidly increasing urban populations of Europe and the United States from the late nineteenth and early twentieth centuries onward saw a huge rise in demand for safe, transportable, and inexpensive foods. The market for canned food and drink has not looked back since. **EH**

SEE ALSO: VACUUM-SEALED JAR, CAN OPENER, SELF-HEATING
FOOD CAN

↗ *Canned foods are supplied to troops in distant wars, as these relics of the Boer War (1899–1902) testify.*

"We may find in the long run that tinned food is a deadlier weapon than the machine-gun."

George Orwell, novelist

Miner's Safety Lamp (*c.* 1815)

Davy's invention brings light to the mines and makes the dangerous work of mining safer.

*"The heroic is for hereafter . . .
for labors of pick and spade
by Davy lamp down below."*

George Meredith, *The Amazing Marriage* (1895)

⬆ *The Davy lamp consists of a cylinder of wire gauze containing a lighted wick attached to an oil reservoir.*

➡ *Miners rest in Powell Duffryn Colliery, South Wales, in 1931; their later-model safety lamps are at their feet.*

Miners' safety lamps are still referred to generically as "Davy lamps," after Sir Humphry Davy (1778–1829) who pioneered their design. He was not, however, alone in this endeavor. At the same time as Davy was developing his lamp, railway pioneer George Stephenson and Dr. William R. Clanny were also designing a lamp for miners.

The invention of the miner's lamp allowed for penetration to deeper mining seams, by providing light, albeit dim. More importantly, the lamp also gave an indication of the presence of flammable gases such as methane, by its flame burning suddenly more brightly with a blue tinge. It also indicated areas where oxygen was low, by the flame simply extinguishing, and so also functioned as a safety device.

The greatest design problem was to create a light that would not cause an explosion in the proximity of flammable gases. Through experiments Davy realized that flammable gases would only explode if they were heated to their igniting temperature, so he devised a wire mesh to enclose the flame in his lamp. The wire absorbed the heat from the flame, spreading it over a large area and so keeping the temperature below the level at which the gases would ignite. Although Davy's lamp did not give off much light due to the surrounding wire mesh, it did cut down greatly on the number of explosions, and his wire mesh system was used effectively for a hundred years or more.

By around 1900 electric lamps had started to be used, and by the 1930s they had all but replaced the original open-flame lamp, providing more light and being much safer. Despite this, the original Davy lamps continued to be used to determine the presence of gases and levels of oxygen, rather than for their light-giving properties. **TP**

SEE ALSO: GAS LIGHTING, ARC LAMP, INCANDESCENT LIGHT BULB

Arc Lamp (1809)

Davy's discovery marks an important step on the road to electric light.

> *"The most important of my discoveries have been suggested to me by my failures."*

Sir Humphrey Davy

⬆ *In the Jandus arc lamp, the flow of air was restricted by a glass cover, so the carbon rod lasted longer.*

➡ *A worker uses a mercury arc lamp to process a photograph at a General Electric factory, c. 1909.*

The name of Sir Humphry Davy (1778–1829) will forever be associated with the famous safety lamp that he developed for miners, but his demonstration of the arc lamp in many ways was more significant.

Considered one of the greatest British scientists, Davy became renowned for his mesmerizing public lectures, including his demonstration of the effects of laughing gas. In 1801 the twenty-two-year-old Davy was appointed as the director of the laboratory at the new Royal Institution in London, where he began his work in electrochemistry.

It was here that he first discovered the principles behind what would eventually become the arc lamp. He used two sticks of carbon in the form of charcoal and connected each of them by wire to opposite terminals of a battery. When he held the two carbon "electrodes" a few inches apart, an electric arc jumped between them and completed the circuit.

The side effect was that, as the arc bridged between them, the tips of the carbon electrodes were heated to incandescence and glowed with a light that can only be described as dazzling. But it did not last long. In Davy's experiments, the spark connecting the electrodes formed a curved shape, due to the movement of air currents, and so he named his creation the "arch lamp." Eventually, however, it became known as the arc lamp.

Although spectacular, in some respects Davy's discovery was premature. The battery itself was still new to science, with the very first one having been demonstrated only a few years previously by its inventor, Alessandro Volta. Davy had to use 2,000 of them to power his lamp. Because of this lack of commonly available electricity, it was not until the 1870s that the arc lamp was used as lighting. **DHk**

SEE ALSO: OIL LAMP, CANDLE, ARGAND LAMP, BATTERY, PUBLIC ELECTRICITY SUPPLY, INCANDESCENT LIGHT BULB

Dental Floss (1815)

Spear Parmly enhances dental hygiene.

Appalled by the terrible condition of his patients' teeth, U.S. dentist Levi Spear Parmly (1790–1859) was spurred into promoting the use of what is now called dental floss. He encouraged people to clean between their teeth using a thin piece of waxen silk thread, and even went so far as to state that flossing combined with the use of a toothbrush and dentrifice would eliminate bacteria from the gums. While this might have been a slight exaggeration, the use of dental floss is today widely recognized as one of the most important, and least utilized, forms of oral hygiene.

The practice of cleaning between the teeth with thread or small picks is one that predates Spear Parmly, but he reinvented the procedure and the tools to do it,

"With this apparatus regularly and daily used, the teeth and gums will be preserved free from disease."

Levi Spear Parmly

and publicized the beneficial effects of flossing. His influential book, *A Practical Guide to the Management of the Teeth*, was published in 1819. In it, he advised flossing after every meal.

Despite Spear Parmly's sound words, dental floss was slow to take off and was not mass-produced until Codman and Shurtleft patented their floss in 1874. But it was the Johnson & Johnson Corporation that was responsible for the widest distribution of dental floss, created from the same silk thread used in surgical suture production. In around 1900 the Johnson & Johnson Corporation bought Codman and Shurtleft, which remains an active division of the company. **TP**

SEE ALSO: TOOTHPASTE, TOOTHBRUSH, DENTURES, ELECTRIC DENTIST'S DRILL

Stethoscope (1816)

Laënnec revolutionizes chest medicine.

Although physician René Laënnec (1781–1826) is credited with inventing the stethoscope, doctors in ancient Greece had practiced the art of auscultation, or listening. The Frenchman's flash of inspiration came in 1816 when he was confronted by a plump, young female patient with a heart condition. Overcome by embarrassment at the thought of having to press his ear to her ample bosom, Laennec recalled having seen children tapping a log while listening at the far end. This inspired him to roll up a sheath of papers into a cylinder and apply it to her chest, with the result that he could clearly hear her heartbeat.

From this idea Laënnec developed the first true stethoscope, which consisted of a hollow wooden

"No other symbol so strongly identifies the doctor than a stethoscope."

Ariel Rogun, MD, PhD

tube around 9 inches (22 cm) long and 1 inch (2.5 cm) in diameter, known as a "Monaural" stethoscope due to having only one earpiece. Laënnec spent the next three years perfecting his design. In 1819 he published a groundbreaking book in which he showed that by giving access to the internal sounds of breathing and blood flow, stethoscopes allowed pathology to be performed for the first time on living things.

In 1829 Nicholas Comins, a Scottish doctor, devised the first flexible stethoscope, and in 1852 New York physician George P. Cammann added two earpieces for binaural sounds to Laënnec's design. In 1878 a microphone was added to amplify sounds. **JF**

SEE ALSO: MICROSCOPE, HEART-LUNG MACHINE, ARTIFICIAL HEART, ENDOSCOPE, LAPAROSCOPE

Stirling Engine (1816)

Stirling creates a quiet-running alternative to steam.

Scottish clergyman Reverend Robert Stirling (1790–1878) began work on a new type of engine in the hope of replacing steam engines in the workplace. At the time steam engines were unstable, dangerous, and prone to explode, frequently causing horrific accidents. Spurred on by the number of people attending his parish who were in danger, he developed what would subsequently be called the Stirling or hot-air engine.

The Stirling engine needs an external heat source that can be almost anything—solar, chemical, or nuclear energy. The engine is then powered by the heating and cooling of a gas contained in a cylinder. As the Stirling engine does not rely on explosions it is quiet in operation and at the time was much safer than the steam engine. Stirling's design added an economizer that increased the efficiency greatly, and in 1850 Sadi Carnot worked out the thermodynamics of engines, showing that Stirling's could theoretically be very close to perfect.

For the hot-air engine to produce energy efficiently, it needed to run at very high temperatures, and the materials of the day could not cope. It was almost entirely overshadowed by the steam engine throughout the Industrial Revolution. By the time the materials had caught up with Stirling's idea, electric motors had taken over.

These days you are most likely to find a hot-air engine in a submarine, where quiet operation is essential for stealth. However, a possible application for the future might be to use them for converting solar energy into electricity. **LS**

SEE ALSO: STEAM ENGINE, INTERNAL COMBUSTION ENGINE, GAS-FIRED ENGINE, TWO-STROKE ENGINE, BOURKE ENGINE

⬈ The cylinder is heated and the air inside it expands to push the piston; the cylinder is then cooled.

"The Stirling engine is fuel independent, it doesn't even need any fuel—the sun is enough!"

Lund Institute of Technology, Sweden

Cardboard Box (1817)

Thornhill introduces strong, light packaging.

The cardboard box is one of the most widely used methods of packaging and storing goods. Although in the last few decades its use has been threatened by new materials, the current environmental climate has seen a move back toward card as a sustainable material that can be recycled.

Before the invention of the cardboard box, the most common packing material for goods was wood, which was heavy and expensive, and unsuitable for small or light goods. In 1817 Sir Malcolm Thornhill produced the first commercial cardboard box. His invention gained in popularity, but it was not until the Kellogg brothers used it to package their cereals after the turn of the century, that the product truly took off.

> *"Inside a big cardboard box, a child is transported to a world of his or her own."*

National Toy Hall of Fame

As card can be printed on, manufacturers recognized the possibilities of making attractive packaging at low cost, and also being able to advertise on the boxes. The corrugated cardboard box was later developed in Sweden. A reinforced version of the original box, this was much stronger and useful for a wider range of applications, including both house-removals and shelters for thousands of homeless people.

A measure of the cardboard box's influence in Victorian society can be seen in the Sherlock Holmes story, "The Adventure of the Cardboard Box." It credits the box in the title, when its contents of a severed ear probably deserved more attention. **JG**

SEE ALSO: PAPER, PLYWOOD, FLAT-BOTTOM PAPER BAG, BREAKFAST CEREAL

Draisine (1817)

Drais invents an early form of the bicycle.

In 1815 the Tambora volcano in Indonesia erupted, starting a chain of events that led to the invention of the bicycle. The eruption was the largest in recorded history and dumped tons of ash into Earth's atmosphere. This caused the global temperature to drop, and did terrible things to crops.

Three years earlier, following some bad weather in 1812, the price of oats was climbing, and the German inventor Karl Drais (1785–1851) was looking for something to replace hungry horses. He designed a four-wheeled vehicle, powered by a servant sitting in the back pedaling, while the master steered from the front with a tiller. It did not catch on and Drais decided to focus instead on surveying equipment.

After the 1815 volcanic eruption, even worse weather caused oat prices to climb higher still. The need for horseless transport was even more pressing, and so Drais tried again. He switched from four wheels to two and got rid of the pedals completely. What he invented in 1817 was called the draisine or the *draisienne*, depending on where you lived.

The design was simple: a light wooden frame supported the rider between the wheels, the rider used his or her feet to push, and the machine could reach speeds of 12 miles per hour (19 kph). The machine quickly became popular, despite being quite uncomfortable to ride, and was copied all over the world, but some of the cheaper imitations lacked brakes, making them particularly dangerous. After being banned from sidewalks as a hazard to pedestrians, they went out of fashion. But they led the way for the bicycles that would follow. **DK**

SEE ALSO: WHEEL AND AXLE, SAFETY BICYCLE, MOTORCYCLE

➡ *The draisine first appeared in England in 1818, where it was also known as the "hobby horse."*

Blood Transfusion (1818)

Blundell pioneers a lifesaving medical procedure.

James Blundell (1790–1878), an obstetrician and gynecologist at Guy's Hospital, London, realized that blood transfusions offered a possible solution for women who suffered severe hemorrhage following childbirth. Blundell was familiar with the work of John Leacock, who, in 1816, had reported experiments in cats and dogs, establishing that the donor and recipient had to be of the same species.

The first human transfusion supervised by Blundell did not in fact involve an obstetric case, but a thirty-five-year-old man with stomach cancer. On December 22, 1818, the man was transfused with about fourteen ounces of blood administered by a syringe in small amounts from several human donors. Despite a slight improvement, the patient died fifty-six hours later.

To facilitate transfusions Blundell devised an apparatus, known as Blundell's Impellor, consisting of a funnel and pump for the collection of donor blood for indirect transfusion into the veins of patients. Between 1818 and 1829, Blundell and his colleagues performed ten transfusions, of which only four were successful, and transfusion remained controversial.

In 1829 writing in *The Lancet*, Blundell himself reported patients who "suffered fever, backache, headache, and passed dark urine." With hindsight scientists now know he was describing ABO incompatibility. All became clear in 1900, when Karl Landsteiner, a doctor from Vienna, reported that the serum of some people agglutinated the red cells of others. From these experiments Landsteiner identified three groups, A, B, and C (later renamed O). A fourth much rarer blood group, AB, was discovered the next year. Landsteiner suggested his findings might be applied to blood transfusions, but the idea was not taken up for more than a decade. **JF**

"Those who drag in the use of human blood for internal remedies of diseases ... sin gravely."

Thomas Bartholin, professor of anatomy, 1616–1680

⬆ *The Moncoq-Mathieu apparatus of c. 1874 was one of many developed to make blood transfusion safer.*

➡ *J. S. Elsholtz, in his* Clysmatica nova *(1667), proposed transfusions, even between humans and animals.*

SEE ALSO: VACCINATION, HYPODERMIC SYRINGE, BLOOD BANK, KIDNEY DIALYSIS

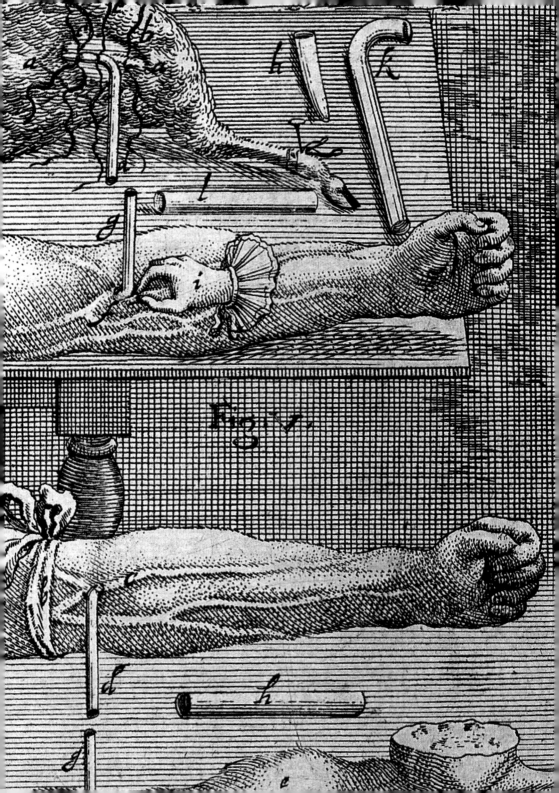

Fig: V.

Fire Extinguisher (1818)

Manby's invention saves lives in fires.

Captain George Manby (1765–1854) is perhaps most famous for his invention of the Manby Mortar, which was a device to help rescue people from shoreline shipwrecks. However, he is also heralded as the father of the modern fire extinguisher, which in itself has helped to save many thousands of lives.

Fire extinguishers in one form or another predate Manby's invention, and there is some debate over the design of the first extinguisher, although one of the earliest recorded ones was designed and used in 1723 by Ambrose Godfrey. But Godfrey's device consisted of a receptacle containing a fire-extinguishing liquid and a chamber of gunpowder with a series of fuses attached. When the fuses were lit, the gunpowder exploded and scattered the liquid. They were not widely used, although there is an account of them being used to put a fire out in London in 1729.

Captain Manby's 1818 invention was more efficient. He was inspired to invent a portable fire extinguisher after witnessing a house fire in Edinburgh and seeing the firemen's inability to fight the fire on the top floors. He designed a copper cask containing three or four gallons of potassium carbonate with the remaining space taken up by compressed air. A stopcock on the top of the extinguisher could be opened to allow the air to rush out, spreading the potassium carbonate over quite a range. The system could also be used with water, and was easily portable, allowing firemen to reach otherwise inaccessible areas in peril.

Manby's invention was soon replaced by a more efficient model. However, his use of compressed air to force the retardant out and over a large area formed the premise for the development of new prototypes, with modern extinguishers using carbon dioxide as the pressurizing agent in a similar way. **TP**

SEE ALSO: CONTROLLED FIRE, GUNPOWDER, AUTOMATIC
FIRE SPRINKLER, ELECTRIC FIRE ALARM

Macadam (1820)

McAdam lays the basis for the road system.

In the early nineteenth century, road maintenance was carried out by road gangs, who were often more interested in profit than in keeping roads in good condition. Scot John Loudon McAdam (1756–1836) was a self-taught engineer, who had been appointed trustee of a turnpike. His neighbor suffered rickets and found traveling on potholed roads extremely painful, so McAdam began to experiment with different methods of road construction.

Since the days of the Roman Empire, roads had been constructed with a bottom layer of heavy stone, topped by several layers of finer stone and a top layer of gravel. McAdam took this method, but used a compressed heavy roller to compact the layers of

> *"The ground was covered by some dark stuff that silenced all the wheels. . . . It was like magic."*

Laura Ingalls Wilder, children's writer

rocks together. McAdam's roads had graded layers of broken chippings, with a center higher than the edges and drains on either side. This meant that roads were not only more stable but needed less maintenance.

McAdam's method greatly increased the speed of road travel, but macadam had to be adapted with the arrival of the motor car. To reduce dust caused by the vacuum created under fast-moving vehicles, E. Purnell Hooley used hot tar to bond the broken stones together and patented tarmac in 1901. **SD**

SEE ALSO: LOG-LAID ROAD, MAP, MOTORCAR

➡ *Workmen equipped with a large tar-boiling machine repair the road in Ludgate Circus, London, c. 1900.*

Electromagnet (*c.* 1820)

Oersted and Ampère link magnetic fields and electricity.

The fact that the current passing through a wire conductor produces a magnetic field around the wire, and that two current-carrying wires could attract or repel each other depending on the direction of the current, was emphasized in 1820 by the independent writings of Hans Oersted and André-Marie Ampère (after whom the SI unit of measurement of electric current, the ampere, is named). It was, however, William Sturgeon (1783–1850), a physicist working at the Royal Academy, Woolwich, London, who recognized the significance of the phenomenon. He converted electromagnetic devices from toys into practical weightlifting machines.

A horseshoe of iron around which is wound a loose current-carrying coil becomes a strong metal-lifting device when the current is switched on and, just as important, the force disappears when the current is switched off. The action of the device can be speedily controlled by electricity. Electricity can flow down miles of wire, so throwing a switch in one place can activate a distant magnet to ring a bell or move a lever. Joseph Henry (1797–1878), while a professor of mathematics and physics at the Albany Academy in the United States, improved Sturgeon's electromagnet by insulating the wire and increasing the number of turns. His electromagnet lifted 2,300 pounds (1,040 kg), a world record at the time. When Henry moved to Princeton, in 1832, he used an electromagnet as an electrical relay device to link his lab to his home. This was the first practical telegraph system. Henry, S. F. B. Morse, and Charles Wheatstone all insisted that they should be credited as the telegraph pioneer. **DH**

SEE ALSO: ELECTRIC MOTOR, INSULATED WIRE, ELECTRICAL GENERATOR, MULTIPLE COIL MAGNET, BATTERY

⬆ *This horseshoe-shaped electromagnet belonged to the English chemist and physicist Michael Faraday.*

Electric Motor (1821)

Faraday shows how electricity and magnets can be made to produce rotation.

Electric motors are used in a vast number of household items, and are essential for industry. Michael Faraday (1791–1867) introduced this revolutionary technology in 1821 when he successfully demonstrated the first electricity-driven motor, a discovery that led to a golden age in the application of electrical technology.

Faraday, son of a blacksmith, started his career as a chemist but soon became involved in many aspects of science, most notably electromagnetism. In 1813 he worked under chemist Humphry Davy at the Royal Institution, where he focused on the principle of electromagnetic rotation that would later be developed into the first electric motor. Following the recent discovery of the property of electromagnetism by the Dane Hans Oersted, Faraday showed that mechanical action could be derived from electromagnetic energy in 1821. He passed an alternating current (AC) through a wire suspended in a small cup of mercury with a magnet at the bottom.

The wire swung around the magnet at speed in a circular path, proving that the electrical current had generated a magnetic field around the wire, which interacted with the magnet, resulting in the wire moving.

Faraday presented his ideas to the Royal Society at the end of 1831 and start of 1832. His findings greatly improved scientists' knowledge and understanding of electricity, magnetism, and how they interact. Faraday continued to explore the electromagnetic effect through experiments with batteries and electrolysis, but his work also inspired other scientists to explore the science behind electricity, ultimately culminating in the modern electric motor. **SR**

SEE ALSO: ELECTROMAGNET, INSULATED WIRE, ELECTRICAL GENERATOR, BATTERY, INDUCTION MOTOR

⬆ *A model replica of an early electric motor based on the original principles set down by Faraday.*

Truss Bridge (1821)

Town's lucrative invention carries the railroads over wide spans.

The concept of a bridge built from trusses—frameworks of straight parts connected to form a pattern of triangles—was first described by the Italian father of Western architecture, Andrew Palladio, in 1570. However, trussed frames had already been in use by the ancient Egyptians and Greeks, and simple timber trusses are believed to have been used in bridge construction in Europe by the time Palladio was writing.

However, the heyday of the truss bridge came in the early nineteenth century in the United States. The impetus came from the development of rail transport and the need to convey heavy rail vehicles safely across spans of varying length. Trussed frameworks offer an adaptable solution, rather like a toy construction set on a grand scale. The nineteenth-century explosion in truss designs was kick-started by Ithiel Town (1784–1844), who filed a patent for his wooden lattice truss bridge in 1821. Town was a renowned New England architect, whose design allowed unskilled workers to construct bridges quickly without the need to transport huge sections.

Truss bridges were often covered by a roof, and two bridges made to Town's design survive to this day in Connecticut, his home state. Part of the design's success can be attributed to Town's skill at marketing, and his control of the patent made him a rich man.

From the 1870s until the 1930s, iron replaced wood in truss bridge construction, and steel replaced iron thereafter. Many different designs, like the Pratt, Warren, and Howe trusses, were developed. The world's longest truss bridge, at 2.3 miles (3.7 km), connects Kansai International Airport to Osaka in Japan. But steel trusses are less suitable for long spans, for which the suspension bridge has now become the usual choice. **AC**

"[I'm looking for] improvements in the taste and science of architecture."

Ithiel Town

⬆ *A Clun Castle steam engine crossing the Royal Albert Bridge, built by I. K. Brunel in the 1850s in England.*

➡ *The viaduct above the Upper Genesee Falls at Portageville, New York, exemplifies the truss bridge.*

SEE ALSO: STEEL, ARCHED BRIDGE, PONTOON BRIDGE, SUSPENSION BRIDGE, CANTILEVER BRIDGE, AQUEDUCT

Fresnel Lens (1821)

Fresnel improves lens efficiency.

"If you cannot saw with a file or file with a saw, then you will be no good as an experimentalist."

Augustin-Jean Fresnel

French physicist Augustin-Jean Fresnel (1788–1827) was extremely interested in the properties of light. He published a number of papers on optical phenomena such as aberration and diffraction, but he is best remembered today for an invention that turned his love of physics into a practical and revolutionary device: the Fresnel lens.

Essentially he created a new type of lens that performed in the same way as a traditional one. Fresnel's design, however, was much lighter, which made it a practical option for making the large lenses needed for lighthouses. Some of the lighthouse lenses were more than 12 feet (3.6 m) tall and looked like giant beehives. The greatly improved efficiency with which these lenses could capture light meant that up to 83 percent of the light from a powerful bulb could be focused into a beam that could be seen by ships as much as 20 miles (32 km) away at sea.

Fresnel came up with his idea whilst working as a commissioner of lighthouses in Paris. The principle is simple. If you imagine a large magnifying glass, lenticular (with two convex sides) in shape, and then slice it into dozens or hundreds of concentric rings, each ring would have a different thickness, as well as a different size. Fresnel's rings were all flat on one side and had the same thickness. In order to maintain the original function of the lens— that is, its ability to focus light—he modified the individual lens rings so that the angle of the face of each ring was different. Stacking the rings back together gives you a Fresnel lens.

Today, Fresnel lenses are used in the back of some cars and vans. **DHk**

SEE ALSO: GLASS, LENS, SPECTACLES, GLASS MIRROR, MICROSCOPE, TELESCOPE

◩ *The Fresnel lens magnifies and concentrates light into a powerful beam that can travel to the horizon.*

Waterproof Raincoat (1823)

Macintosh patents the Mackintosh, a waterproof fabric perfect for use in rainwear.

It was a Scottish chemist, Charles Macintosh, who gave us one of the most widely recognized names, the Mackintosh, the eponymous and essential waterproof coat. He invented, not the coat, but the waterproof material from which such garments are made.

Macintosh's experiments began with waste products from the process of creating gas from coal. Initially he extracted ammonia from the waste products to make a violet-red dye. This process left a further waste product, called coal-tar naphtha. Macintosh began to experiment with this as a solvent, quickly realizing its waterproof qualities. He began to coat a thin material with it, but encountered two problems: the rubber was sticky, and it had a terrible odor. He combated the first problem by pressing two sheets of the fabric together, with the rubber in the middle, which created a waterproof and usable material, and this he patented in 1823.

Unfortunately, the problem of the rubbery smell was never truly overcome, and even today traditional waterproof jackets tend to have a distinct odor. On account of the smell and the slightly unwieldy nature of his material, garments were at first slow to take off, although the armed forces in particular were an early advocate of his waterproof material. Later the British inventor Thomas Hancock (1786–1865) took out a license to produce the double-layer waterproof material and improved upon it, adding a higher percentage of rubber and making it more efficient and less malodorous. Macintosh recognized Hancock's developments and in 1831 Hancock was made a partner in the firm Chas. Macintosh & Co. **TP**

SEE ALSO: CLOTHING, WOVEN CLOTH, UMBRELLA, RUBBER WELLINGTON BOOTS

↗ *A boy in the 1930s demonstrates to the camera some applications for the Mackintosh waterproofed fabric.*

"I'm an optimist, but an optimist who carries a raincoat. . . ."

Harold Wilson, former British Prime Minister

Toy Balloon (1824)

Faraday devises a rubber gas-holder that is rapidly taken up as a public amusement.

Michael Faraday (1791–1867), one of Britain's foremost physicists, is most famous for discovering principles relating to magnetic induction and the relationship between magnetism and electricity. But Faraday was also an inventive lecturer and actually initiated the tradition of the Christmas lectures at the Royal Institution (the lectures still go on to this day and are televised in the United Kingdom). Faraday even has the honor of having a unit of measurement named after him; the farad is a unit of electrical capacitance. However, an achievement of this scientist that much less often finds its way into the history books is that he also invented the toy balloon.

Faraday's career began not in physics but in chemistry. He was appointed as a chemicals assistant at the Royal Institution and performed experiments in this capacity, investigating chlorine and the nature of gases. It was while researching the properties of hydrogen that Faraday made the first rubber balloon.

Balloons had been made for many hundreds of years from intestines. These smelly amusements were touted around court in the European Middle Ages on the ends of sticks by jesters, who fashioned them into amusing shapes. Faraday's balloons were more sanitary, being made from two sheets of rubber "welded" together to form a bag. When filled with hydrogen, the balloons ascended, defying the force of gravity and enabling Faraday to make observations about the properties of the gas.

Balloons became available commercially just a year later, when rubber manufacturer Thomas Hancock began producing a balloon-making "kit." **BG**

SEE ALSO: RUBBER BALL, HOT-AIR BALLOON, TOY ELECTRIC TRAIN SET, LEGO®

⬆ *A tinted illustration of 1880 family fashions includes the suitably pink accessory of a gas-filled balloon.*

Thaumatrope (1824)

Paris inadvertently lays the foundations of cinematography and animation.

A distinguished English physician, John Ayrton Paris (1785–1856), was the inventor of the thaumatrope, a popular children's toy of Victorian Britain. It was conceived, not as a form of entertainment, but to demonstrate an optical phenomenon.

The thaumatrope is a simple device: a disc with a picture on each side and two pieces of attached string. The disc is rotated by hand to coil the attached strings, and then the strings are quickly pulled apart. This causes the disc to rotate rapidly, enabling the pictures on either side to be visible in quick succession; the two images are interpreted by the human eye as a single image. This phenomenon is known as persistence of vision and was demonstrated to a meeting of the Royal College of Physicians in 1824. One of Paris's exhibits featured a bird on one side of the disk and an empty birdcage on the other; when it was rotated quickly, the disk presented to the eye the image of a bird inside a cage.

The thaumatrope was certainly an important antecedent of both cinematography and animation. When we watch a film at the cinema we are experiencing the same persistence of vision that Paris demonstrated with the thaumatrope, since what we are seeing is a series of still images passing before our eyes at such a speed that they appear to be moving. Cartoon animation works on the same principle.

Paris himself showed little interest in the nonscientific applications of his invention, and continued to devote his efforts to medical research. He would later write some of the most important medical books of his generation. **TB**

SEE ALSO: MAGIC LANTERN, FILM CAMERA/PROJECTOR, FILM SOUND

⬆ *The brain interprets two images that are rapidly superimposed, one on the other, as a single image.*

Braille (1824)

Braille enables blind people to read and write with a system based on touch.

"Learning to read music in braille and play by ear helped me develop a damn good memory."

Ray Charles, musician

⬆ *The author Helen Keller, who lost her sight aged nineteen months, reads braille text in her youth.*

➡ *A German sheet of 1952 shows Braille's alphabet and commemorates the 100th anniversary of his death.*

Louis Braille (1809–1852) was fifteen years old when he devised a system of raised dots to reproduce the alphabet so that it could be read by touch. Braille became blind at the age of four as a result of an accident while playing in his father's saddlery workshop and was educated at the National Institute for Blind Children in Paris from the age of ten. He learned to read using raised wooden letters designed by the school's founder, Valentin Haüy.

In 1821, Charles Barbier, a former captain in the French Army, visited the school. Braille was introduced to the letter code that Barbier had developed to enable soldiers to communicate without sound or light at night. Barbier's system used large symbols to represent sounds, each using dots and dashes raised on paper. The reader had to move his finger around to recognize the sounds.

Braille realized that compact symbols readable by one touch of the finger would be quicker and more reliable and would make it possible to write without sight. By 1824 he had developed a system of identically shaped cells, each with the space for six raised dots arranged in a regular rectangle. Each letter and digit is represented by a unique pattern of dots using this grid—sixty-four combinations are possible—and the sequence of patterns is systematic to facilitate learning. Braille refined and tested the system, developed patterns for mathematical and musical notation, and published his *Method of Writing Words, Music and Plain Song by Means of Dots* in 1829.

Braille, who became a teacher at the Institute for Blind Children, later developed the system further so that sighted people could recognize the letters. He also collaborated on the development of a machine to speed up the writing process. **EH**

SEE ALSO: SPECTACLES, PARCHMENT, PAPER, BIFOCALS, LASER EYE SURGERY, VISUAL PROSTHETIC (BIONIC EYE)

LOUIS BRAILLE

GEDENKBLATT
ZU SEINEM EINHUNDERTSTEN TODESTAG

geboren am 4. Januar 1809 · gestorben am 6. Januar 1852

DEM SCHÖPFER DER PUNKTSCHRIFT UND DAMIT BEGRÜNDER
DER BLINDENBILDUNG DER WELT

Arbeitsausschuß für Blindenfragen in der Deutschen Demokratischen Republik

Blindenschrift - Alphabet

a	b	c	d	e	f	g	h	i	j
k	l	m	n	o	p	q	r	s	t
u	v	x	y	z			ss	ß	st
au	eu	ei	ch	sch			ü	ö	w
äu	ä	ie							

Grundform

, : „ . ? ! ( ) _ * - ' -

Grundform

1 2 3 4 5 6 7 8 9 0   1 9 5 2

Zahlenzeichen

Aus den 6 Punkten werden gebildet: Vollschrift, Kurzschrift, Stenografie, Mathematik-, Chemie- und Notenschrift

Verlag: Deutsche Zentralbücherei für Blinde zu Leipzig

LISTOWEL & BALLYBUNION RLY. ENGINE.

Monorail (1825)

Palmer builds a special-purpose single-track railway.

A monorail replaces the more usual two-rail transport system with a single track, and its vehicles either straddle the rail or hang from it. In the first type, the pillars supporting the rail can be of different heights, and so rough terrain can be crossed cheaply. In the second type the rail can be suspended above canals and rivers, taking up little valuable land.

In June 1825 Henry Robinson Palmer (1795–1844) opened a suspended monorail in Cheshunt, near London. Although designed to carry bricks, it could also carry passengers. For the United States Centennial Exposition of 1876, General Le-Roy Stone built a demonstration pillar monorail in Philadelphia. Other pillar monorail systems were built to carry agricultural products and mineral ores.

The most famous passenger-carrying pillar monorail was the Irish 9-mile (14.5 km) Listowel and Ballybunion railway that ran from 1888 to 1924. A German suspended monorail in Wuppertal started service in 1901 and is still in regular passenger-carrying use. In 1903 Louis Brenan patented a ground-level monorail with vehicles that were maintained upright by two large spinning gyroscopes. Fear of gyroscopic failure made this system short-lived.

One problem with a conventional two-rail system is that at high speed the train tends to oscillate from side to side, a phenomenon known as hunting. This does not happen on a monorail, allowing faster train speeds. A modern version of the pillar monorail was designed by the Swedish industrialist Dr. Alex Lennart Wenner-Gren (ALWEG). ALWEG monorails have been constructed in Florida, Japan, and Australia. **DH**

SEE ALSO: LOCOMOTIVE, CABLE CAR, ELECTRIC TRAM, MAGLEV TRAIN

⬆ *The 1888 Listowel and Ballybunion Railway in Ireland was the first monorailway of its type in the world.*

Omnibus (1826)

Baudry stumbles on a huge popular demand for public transport.

Stanislas Baudry (1780–1830) was the proprietor of a Parisian bathhouse; he was also a bus operator who provided clients with transportation to his spa. Baudry did not invent the horsedrawn bus—horsedrawn carriages of various kinds had long been in existence—but in 1826 he instigated the use of the term "omnibus," and he is therefore credited with inventing the omnibus as a concept. His inspiration came from the name of a hat-maker's shop (Omnes Omnibus, Latin for Everything for Everyone) that he regularly passed on his way to the baths.

Baudry's concept of a means of transport available to all, regardless of social class, spread across the globe and evolved into various forms with varying numbers of horses, alternative seating configurations within the vehicle, and, no doubt, different tariffs depending on their clientele. The buses survived the test of time and eventually the horses were replaced by diesel engines. Latter-day concerns for the environment have seen petroleum-based engines phased out in some countries in favor of engines run on fermented sugar. Buses powered by this technology are already in use in Italy, Poland, Spain, and Sweden.

As the success of Baudry's concept showed (it proved much more popular than the baths it was intended to promote), public transport is a vital economic need for any society. It increases trade and the exchange of knowledge, bringing consumers to companies and experts to employers, enabling a country to maximize its intellectual and commercial potential. Baudry unintentionally identified an important factor in the evolution of society. **CB**

SEE ALSO: WHEEL AND AXLE, HORSE COLLAR, LOCOMOTIVE, MONORAIL, MOTORCAR

⬆ *Shillibeer's omnibus service, begun in 1829, was the first in London, England, and plied a single route.*

Photography (1826)

Niépce uses light-sensitive chemicals to take the first photograph.

Ever since the invention by Alhazen (965–1040) of the pinhole camera, which projected an image onto a surface, people sought a way of "fixing," and thus recording, that image. By accident, in 1727, the German chemist Johann Schulze discovered that a mixture of chalk, nitric acid, and silver darkened when exposed to sunlight, and the rate of darkening increased if more silver was added. By 1777 the Swedish chemist Carl Scheele had been able to fix, or make permanent, the results of this change using ammonia.

Joseph Nicéphore Niépce (1765–1833) produced the first permanent photographic image in 1826. He first used a flat pewter plate covered with bitumen, but then quickly moved on to silver compounds. Louis Daguerre produced silvered images that were very delicate and could not be copied. Although the exposure time was about ten minutes, he managed to produce daguerreotypes of famous people such as Abraham Lincoln (1846) and Edgar Allen Poe (1848).

Around this time William Henry Fox Talbot produced the calotype, in which an intermediate stage of the process resulted in a paper "negative" that could be used to make a multitude of final "positive" prints. Unfortunately, the paper and glass plates were wet.

By 1884 George Eastman invented a dry process. Cameras became simple to use and all the complex chemistry could be left for later processing.

1907 heralded the commercial color photographic plate, and 1925 saw the introduction of 35-millimeter film. In the late 1980s, the manufacture of relatively inexpensive megapixel charge-coupled devices threatened the demise of film-based technology. **DH**

SEE ALSO: PHOTOGRAPHIC FILM, SELF-DEVELOPING FILM CAMERA, SINGLE-LENS REFLEX CAMERA, DIGITAL CAMERA

⬆ *Niépce's earliest surviving photograph is this view taken from a window of his home in 1826.*

Friction Match (1827)

Walker makes fire lighting a simple matter.

A match consists of a wooden or paper stick with a coating of chemicals at one end that ignites when struck against an appropriate surface. The ignition results from the heat generated by friction as the two surfaces are rubbed together. The match enabled people to overcome the limitations of damp tinder and bad weather and create a flame at will.

English chemist John Walker (1781–1859) created the first friction match in 1827. He owned a pharmacy in Stockton-on-Tees and manufactured an explosive chemical mixture for use in percussion caps, a component of firearms. He accidentally discovered that this mixture, made of equal quantities of antimony sulfide and potassium chlorate, ignited when rubbed along a rough surface. Walker had previously used chlorate matches, which ignited after being dipped in a bottle of sulfuric acid, but began manufacturing them using the new mixture instead. Walker called the matches "congreves." The chemical composition of

the friction match was refined, and an analysis in the 1820s showed them to be composed of five parts potassium chlorate, five parts antimony sulfide, three parts gum Arabic, and one part iron oxide.

The friction match was, however, patented by Samuel Jones and sold under the name of Lucifer matches. The ignition of these matches was violent and often occurred with a loud bang. The flame at the match head was unsteady, did not always result in ignition of the stick, and had an unpleasant odor. White phosphorus was added to remove these flaws, but then the matches had to be kept in an airtight box to prevent them from spontaneously combusting. **CA**

SEE ALSO: **CONTROLLED FIRE, FIRE-LIGHTING LAMP, SAFETY MATCH, SMOKE DETECTOR**

⬆ *Walker's matches were tipped with chlorate of potash, sulfide of antimony, gum arabic, and water.*

Insulated Wire (1827)

Henry produces stronger electromagnets.

When American Joseph Henry (1797–1878) first became interested in science at the age of sixteen, he found electricity fascinating and began experimenting with electromagnets.

Wires carrying electricity produce weak magnetic fields around them. If, however, the wire is coiled many times around a metal such as iron, the effect is magnified and the resulting magnetic field is much stronger. The first electromagnets were coiled very loosely to prevent the current-carrying wires touching and causing an electrical short circuit. Henry was the first to use an insulating cover for the wires so that they could be wrapped more tightly and in many layers, multiplying the effect. Henry's first insulation for wires

---

"As a physical philosopher he has no superior in our country [at least] among the young men."

Geologist Benjamen Silliman on Joseph Henry

---

was tediously made from strips of his wife's petticoats.

The material of the insulator has to resist the flow of electricity. This is done by ensuring that the constituent atoms have tightly bound electrons. The first transatlantic telegraph cables were insulated using gutta-percha, a latex sap from a species of tropical tree. This unfortunately deteriorated when exposed to air, so early power cables were wrapped in jute and placed inside bitumen-filled pipes. By the late 1890s wire insulators made of rubber or oil-impregnated paper became common. By World War II insulation made of synthetic rubber and polyethylene was being introduced. **LS**

SEE ALSO: BATTERY, COAXIAL CABLE, OPTICAL FIBER, HIGH-TEMPERATURE SUPERCONDUCTOR

Screw Propeller (1827)

Ressel patents a powerful force to drive ships.

With the onset of the Industrial Revolution in the late eighteenth century, engineers began to harness steam power for many vehicles, including ships. Treadmill-driven paddles had been used since antiquity, and prototype paddle steamers began to appear in the 1780s. But a parallel technology was about to emerge, or perhaps submerge, in the shape of the screw propeller. This acted entirely underwater, as opposed to the paddle which was only partly below the surface—an obvious advantage in a naval battle, with cannonballs flying around. There is still controversy over who can claim credit for the marine propeller. James Watt proposed steam-driven propellers in 1784, but never built them. Many others took out patents on the concept, although none led to a practical device.

The first truly useful screw propeller, and the first to place the screw between the ship's helm and the stern, was developed by Bohemian engineer Josef Ressel (1793–1857). The first test of the technology took place in Trieste harbor. With forty passengers onboard the steam-driven *Civeta*, Ressel managed a speed of six knots before the engine exploded. Ressel was banned from further testing. Undeterred, he continued development over the next few decades. But, Ressel lost out to foreign rivals, and there is some suspicion that the invention was sold to Britain where, in 1836, a design was tested by Francis Petit Smith.

The screw was improved by Swedish engineer John Ericsson, who in 1839 crossed the Atlantic in forty days using this means of locomotion. The technology went on to power the great steamships of the Victorian era, such as Brunel's *Great Eastern*. **MB**

SEE ALSO: RUDDER, SCREW, STEAMBOAT

➡ *One of four massive, 23-ton screw propellers made for the giant French liner* Normandie, *launched in 1932.*

Fountain Pen (1827)

Poenaru's fountain pen shortcomings are overcome at last in a breakthrough design.

"None of us can have as many virtues as the fountain pen, or half its cussedness."

Mark Twain

⬆ This Mont Blanc fountain pen has ink cartridges rather than a permanent rubber ink reservoir.

➡ This 1919 ad refers to the fact that the Treaty of Versailles was signed with a Waterman pen.

The invention of the modern fountain pen is really more a story of perfection than invention. In 1883, more than fifty years after the fountain pen was first invented, a New York insurance broker, Lewis Waterman, was set to sign an important contract and decided to honor the occasion by using the standard ink-filled pen of the day. However, fountain pens were notoriously unreliable, especially in their capacity to regulate their ink flow, so when the pen spilled ink across the contract so that it could not be signed, Waterman decided to do something about it.

Within a year Lewis Waterman had designed the world's first practical, usable, and virtually leakproof fountain pen. To regulate the flow of ink he successfully applied the principle of capillary action, with the inclusion of a tiny air hole in the nib of the pen along with grooves in the feeder mechanism to control the flow of ink from his new leakproof reservoir to the nib. Although Waterman deserves credit for the invention of the modern fountain pen we know today, he nonetheless stood on the shoulders of many who had gone before.

As early as the beginning of the eighteenth century, the chief instrument-maker to the king of France, M. Bion, crafted fountain pens with nibs, five of which survive to this day. The first steel pen point was manufactured in 1828, thought to be invented by Petrache Poenaru, and in the 1830s the inventor James Perry had several unsuccessful attempts at designing nibs that employed the principle of capillary action. But it was Lewis Waterman who overcame every obstacle and crafted a successful pen. It was so successful that by 1901, two years after Waterman's death, more than 350,000 pens of his design were sold worldwide. **BS**

SEE ALSO: INK, PAPER, STEEL, QUILL PEN, GRAPHITE PENCIL, TYPEWRITER, BALLPOINT PEN, FELT-TIP PEN

rte Plume "Ideal"
Waterman
l'Arme de la paix
1919
TRAITÉ
de
PAIX

Pencil Sharpener
(1828)

Lassimone frees pencils from the penknife.

The French painter and army officer Nicolas-Jacques Conté patented the process to make pencils in 1795, but it was another Frenchman, Bernard Lassimone, a mathematician, who filed the first patent for a pencil sharpener, in 1828. Up until that point pencils were sharpened using a penknife, which itself derived its name from its use in sharpening quills into pens.

It was not until 1847, however, that the pencil sharpener really took off in its modern form, and this was with the invention of a new sharpener by Therry des Estwaux. Perhaps, though, the most significant advance in pencil sharpener design was that by the African-American inventor John Lee Love. He designed the portable pencil sharpener, known as the Love Sharpener, which remains in wide use today, especially by artists. The pencil is placed in the opening of the sharpener and rotated by hand, with the shavings being collected in a compartment that can then be emptied. Love patented his original design on November 23, 1897, in the United States, submitting a detailed drawing. He later explained that although his patent drawing depicted a plain, utilitarian device, he also envisaged it as highly decorated, to the extent that it could be used as a desk ornament or paperweight.

The first electric pencil sharpener is often attributed to the industrial designer Raymond Loewy in the early 1940s, and sold through the Hammacher Schlemmer Company of New York. **TP**

SEE ALSO: GRAPHITE PENCIL, PAPER, STEEL

▧ *This twentieth-century sharpener includes spring-loaded jaws to hold the pencil while sharpening.*

◩ *Five pencils may be loaded into this machine; the sharpener's blade is rotated by turning the wheel.*

Multi-tube Boiler Engine (1828)

Seguin increases the efficiency of boilers.

The boiler of a steam engine uses the heat from both fire and the hot gases produced by fire to boil water and produce steam. The efficiency of this process can be greatly improved if the contact area between the hot gases and the water vessel is very large. Instead of just having a kettle-type boiler sitting on a fire, a multiple-tube boiler passes the fire gases through the boiler along a series of narrow tubes. Cornish boilers had one tube, Lancashire boilers two, and marine and locomotive boilers had many tubes, which sometimes were passed back and forward through the boiler.

The first two-pass multi-tube boiler was invented in 1828 by the French engineer Mark Seguin (1786–1875). This quickly led to a considerable improvement in the power and speed of early railway engines and was a major factor in the success of George Stephenson's *Rocket* in the Rainhill Trials on the Liverpool and Manchester railway, in October 1829.

Multiple-tube boilers such as these are known as fire-tube boilers. They are relatively compact and enable a working steam pressure to be built up quickly. As well as in locomotives, multiple-tube boilers generated steam power for anything from stationary industrial engines to steam boats.

In the modern era, boilers working at higher pressures incorporate a multiple water-tube system. Here, narrow tubes carry water and steam directly through the hot furnace, and forced convection is often used to speed the transit of the fluids in the tubes. The steam can also be superheated for use in driving turbines. Multiple water-tube systems are safer than fire-tube boilers. They are less likely to suffer catastrophic failures, and thus have an important role in the generation of nuclear power. **DH**

SEE ALSO: STEAM ENGINE, HIGH-PRESSURE STEAM ENGINE, LOCOMOTIVE, STEAMBOAT

Thermocouple (1829)

Nobili finds a way to measure high temperatures.

In 1821 the Estonian physicist Thomas J. Seebeck made an accidental discovery: Not only does a potential (that is, voltage) difference exist between the two ends of an electrical conductor if these ends are at different temperatures, but also the voltage is a direct function of the temperature difference. If a circuit is made of a uniform material, the net loop voltage is zero. If, however, two different metallic conductors, such as platinum and palladium, are connected, a positive voltage is produced.

The thermocouple, after being calibrated using the melting points of certain pure substances such as lead (621.68°F/327.6°C), silver (1,762°F/961°C), and nickel (2,647°F/1,453°C), can measure temperatures nearly up to the melting points of its two components.

The Italian physicist Leopoldo Nobili (1784–1845) was slightly less ambitious and used a series of antimony bismuth bars to construct a thermopile in 1829. This was used to investigate infrared radiation and was connected to an astatic galvanometer that measured the voltage.

Later, the French industrial chemist Henry Le Chatelier was interested in measuring the temperatures greater than 930°F (500°C) commonly encountered in the cement industry. Using pure platinum and an alloy of platinum and rhodium in his thermocouple thermometer, he quickly realized that the purity and homogeneity of the two materials was of great importance.

By the 1890s, accurate, sturdy, and more reliable thermocouple thermometers were in widespread use in such places as steelworks. This enabled great strides to be made in the study of the effects of alloying elements in the properties of metals. **DH**

SEE ALSO: METALWORKING, STEEL, MERCURY THERMOMETER, THERMOSTAT, PYROMETER

Stenotype Machine (1830)

Drais devises a spoken-word recorder.

The German inventor Karl Drais (1785–1851) is most famous for inventing the draisine or "running machine," one of the earliest forms of mechanized transport and a precursor of the modern bicycle. But in 1830 he also invented a keyboard system for recording speech that developed into what we now call a stenotype or shorthand machine.

A stenotype (also called shorthand) machine consists of a keyboard of twenty-two letters and numbers that the operator, or stenographer, can press simultaneously to spell out whole syllables, phrases, or words in one action—like playing chords on a piano. Stenographers spell out syllables phonetically, that is by their sound rather than spelling. Broadly speaking,

---

"[Their] only movement is the dancing of their fingers over the keys of their curious machines ..."

Sarah Campbell, *The Times* (February 8, 2007)

---

the left-hand fingers are used to produce the initial consonant, the right hand produces the final consonant, and the thumbs generating the vowel sounds in between.

Stenography is able to transcribe spoken words rapidly and became popular in courts of law where it is still the main recording method. A court reporter must write accurately at speeds of around 225 words per minute, although some users can reach up to 300 words a minute (five words per second). Stenographers develop their own conventions in using their machines, so that one may not even be able to read another's output. **RBd**

SEE ALSO: TYPEWRITER, MECHANICAL COMPUTER, DIGITAL
ELECTRONIC COMPUTER, PERSONAL COMPUTER

Thermostat (1830)

Ure improves temperature regulation.

Thermostats control the temperature of a system—such as an oven, car engine, or room—so that it remains close to a preset value. The control is achieved by means of a temperature-sensitive switch that operates heating or cooling devices. Many of these switches are activated by monitoring the expansion of metals, waxes, or gases. More recently, thermostatic devices have relied on thermistors in electrical circuits. The thermistor is usually a ceramic or polymer electrical resistance that changes its value significantly as a function of temperature. Thermistors were patented by Samuel Ruben in 1930.

Dr. Andrew Ure (1778–1857) was a Scottish medic and chemist who was also greatly interested in the factory system, free trade, and steam-driven machines. Realizing that textile mills needed constant temperatures to ensure uniformity in product manufacture, Ure patented a thermostat for this purpose. This was not an entirely new device. The Dutchman Cornelis Drebbel (1572–1633) had used the expansion and contraction of a vessel of mercury to control the airflow, and thus the temperature, inside a chicken incubator.

Animals have very efficient biological thermostats. The human body uses the hypothalamus to both regulate temperature and detect temperature. The typical human temperature is about 99.32°F (37.4°C) when we are awake and working, and 97.34°F (36.3°C) when we are asleep. When the skin temperature increases above 98.6°F (37°C), sweating begins, and body heat loss is enhanced by the evaporation of perspiration. At temperatures below 98.6°F (37°C), heat flow to the skin is reduced by vasoconstriction. Shivering starts, hairs become erect to increase insulation, and heat production is increased. **DH**

SEE ALSO: GAS OVEN, ELECTRIC STOVE, ELECTRIC IRON,
REFRIGERATOR, MERCURY THERMOMETER

Lawn Mower (1830)

Budding invents a machine that has come to symbolize suburban living.

Before the invention of the lawn mower, fine lawns were the exclusive preserve of those who could employ gardeners to scythe the grass. English engineer Edwin Beard Budding (1795–1846) invented the cylinder mower to maintain the grass of sports fields and the gardens of the wealthy. Patented in England in 1830, Budding's machine used a heavy roller to drive the cylinder, requiring two gardeners to push and pull it. The Regent's Park Zoological Garden in London was an early adopter, using one in 1831. By 1860 lawn mowers were being manufactured in eight roller sizes, up to 36 inches (900 cm), and in 1859 Thomas Green produced the first chain-driven mower. Grass boxes were added during the 1860s.

A steam-powered mower fueled by petrol or kerosene was patented in 1893 by James Sumner, who co-founded the Lancashire Steam Motor Company (later the Leyland Motor Company) in 1896 to produce steam-powered vehicles. This technology paved the way for the birth of the British motor industry. A lightweight cylinder mower was developed and patented by Amariah Hills in the United States in 1899, and, around 1900, Ransomes in England made the first ride-on mowing machine. Mowing technology has continued to develop, with the later invention of rotary and hover mowers and even the robotic mower.

The invention of the mower made lawns affordable and helped popularize sports that rely on grass—such as tennis, cricket, rugby, and football—as lawns and pitches became more widely accessible. **EH**

SEE ALSO: WHEEL AND AXLE, STEAM ENGINE, ELECTRIC
MOTOR, MOTORCAR

↗ *Budding's lawnmower was based on the shears used for trimming woollen cloth in textile factories.*
➡ *An advertisement for the "Buckeye Mower and Self-Raking Reaper" features a ride-on lawnmower.*

Pyrometer

(1830)

Daniell measures high temperatures using the expansion of heated platinum.

A pyrometer is a device for measuring high temperatures, specifically those above 673°F (356°C), the boiling point of mercury. John Frederic Daniell (1790–1845), the first professor of chemistry at King's College, London, invented an instrument known as a register pyrometer in 1830. This used the expansion of platinum to indicate, for example, the temperature of liquid silver. The platinum bar was placed in a hollow cylinder of plumbago, and the expansion was registered using a lever system and scale. Daniell also went on to invent an electrical battery that became known as the Daniell cell, and a dew-point hygrometer.

As the nineteenth century progressed, the accurate measurement of temperature became ever more important in manufacturing processes involving such things as pottery kilns and steel furnaces. Devices were also required that would measure temperatures up to around 5,400°F (3,000°C). Since these devices had to be distant from the high-temperature object, the mirror pyrometer invented by Frenchman Charles Féry used a reflecting telescope system to focus emitted radiation onto a detector.

Optical pyrometers were first patented in 1899, and these compared the color of the incandescent body to that of a hot current-carrying wire. This instrument relied on the fact that the color of the radiation emitted by a heated body changes from red to yellow to white as the temperature increases. **DH**

SEE ALSO: MERCURY THERMOMETER, THERMOSTAT, THERMOCOUPLE

🡔 *This pyrometer was made by Hippolyte Pixii of Paris and listed in his catalog of instruments for 1849.*
🡐 *Workers in a factory in Detroit, Michigan regulating oven temperatures using a pyrometer, in 1917.*

Electromagnetic Telegraph (c. 1830)

Schilling devises new way of communication.

Telegraphing is a way of sending messages using wires and an electric current. At one end, the sender taps out a word with a switch. Each tap completes the circuit and allows electricity to flow. The electricity flows down wires to the receiver, where it powers an indicator dial or pointer, enabling the operator to observe the message coming in.

Early forms of the telegraph were based on electrolysis, in which electricity passes through a liquid to produce a visual effect. Samuel Thomas von Sömmering's early electrolysis telegraph consisted of thirty wires immersed in acidic water, one for each letter of the German alphabet. As the letters were tapped and the circuit completed, an electrochemical reaction produced a flow of hydrogen bubbles. The message was easily deciphered by watching which wire produced the bubbles.

Baron Pavel Schilling (c. 1780–1836) decided to collaborate with von Sömmering to invent a more practical device. Around 1830, Schilling found that coiling electrical wire around a magnetized needle would make it swing one way or the other. The effect was based on electromagnetism, in which electricity flowing through a wire creates a magnetic field. If the wire is coiled, the field is amplified and is strong enough to swing the needle. He incorporated this phenomenon into the telegraph by using horizontally mounted needles to indicate letters.

The electromagnetic telegraph, first demonstrated in 1832, represented the first major use of electricity for communication. Four years later, Charles Wheatstone and William Fothergill Cooke patented their electrical telegraph, and in May 1844 Samuel Morse's first single-wire telegraph line was opened. **RB**

SEE ALSO: MORSE CODE, PRINTING TELEGRAPH, DUPLEX TELEGRAPH, ELECTROMECHANICAL RELAY, TELEPHONE

Reaper (1831)

McCormick automates crop harvesting.

Before American Cyrus McCormick (1809–1884) invented the reaper, crops were tediously gathered by hand, usually with the aid of a hand-swung scythe. Often landowners were limited by what they could reap in the fall rather than by the size of their land or the amount of seed that could be sown in the spring.

McCormick's father had started to work on the design of a "reaper," a horsedrawn machine that could automatically cut and bundle corn, but he was unsuccessful and passed his researches over to his son. Cyrus McCormick's reaper design of 1831 had a frame that he would place over himself and his horse to stabilize the contraption while its large, mechanical, armlike cutters would cut the crop.

McCormick was not happy with his first design and delayed obtaining a patent until 1834, when a rival reaper inventor appeared in Maryland. However, the reaper did not catch on, even when McCormick had improved its weight, efficiency, and reliability, and even added an ability to bundle. Farmers were wary of change, and there were almost no sales in a decade.

Eventually, however, the demand became so great that manufacturing had to move from McCormick's blacksmith shop to a factory in Chicago. The inventor grasped the international market by winning the most prestigious prize at the industrial exhibition at Crystal Palace, London, in 1851. At first the British laughed at his clumsy machine, but when they saw how efficiently it could cut crops, many rushed to place orders with the McCormick Harvesting Machine Company.

Many decades later, improvements to the reaper and the addition of first steam power and then an engine led to the modern combine harvester, which revolutionized farming in the twentieth century. **LS**

SEE ALSO: FLAIL, SCYTHE, THRESHER, COMBINE HARVESTER

Multiple Coil Magnet (1831)

Henry develops the work potential of electromagnetism.

Englishman William Sturgeon invented the simple electromagnet in 1825; five years later, American Joseph Henry (1797–1878) successfully improved its magnetic field by a factor of about 400.

The most important characteristic of modern electromagnets is the speed with which their magnetic fields can be manipulated. Basically an electromagnet is just a corkscrewlike coil of current-carrying wire—a solenoid. If a direct current is passed through the wire, the strength of the magnetic field is directly proportional to the amplitude of the current. A solenoid has only air in its core. If a stronger field is required, the core is filled with soft iron or any other suitable paramagnetic of ferromagnetic material. A disadvantage with this system is that remnant magnetism can persist after the electric current has been switched off. This has to be removed by applying a decreasing alternating current to the coil.

Direct-current electromagnets are used to pick up iron objects or hold switches in one position, on or off. Alternating-current electromagnets are used in scanning devices such as television tubes. Television tube yokes have two sets of electromagnets mounted perpendicular to each other. Current flowing through the coils controls the beam of electrons that are focused on the screen. This beam can be made to trace out a raster (scan pattern) of horizontal lines that are scanned from the top to the bottom of the screen.

Loudspeakers also rely on electromagnets. Here the central iron core is attached to the paper cone of the speaker. Alternating current through the coil causes the speaker to vibrate and emit sound. **DK**

> *"[Henry's magnetic force] totally eclipses every other in the whole annals of magnetism."*
>
> William Sturgeon on Joseph Henry

SEE ALSO: ELECTROMAGNET, ELECTRIC MOTOR, INSULATED WIRE, ELECTRICAL GENERATOR, BATTERY

◳ *Henry's electromagnet, shown here with a weight it lifted, is owned by the Smithsonian Institution.*

Dynamo (1831)

Faraday and Henry convert movement to electricity.

A dynamo is a device that converts mechanical energy into electrical energy. For example, power for bicycle lamps used to be provided by a type of dynamo in which a ribbed cylinder resting against the bike tire was made to rotate as the cyclist pedaled along.

The two main components of a dynamo are a system for producing a magnetic field (a stator) and a coil of conducting wire (an armature) that rotates in such a way that the wires continually cut through the magnetic field lines. The end product is an alternating current flowing through the wires. A third component, a commutator (a set of contacts mounted around the machine's rotating shaft) is often used to convert the alternating current into a direct current.

There is a close relationship between dynamos and electric motors. One converts movement into electric current; the other, electric current into movement. In fact, traction motors on modern electric locomotives are often used for both tasks.

English chemist and physicist Michael Faraday (1791–1867) was responsible for the invention of a whole range of electromagnetic rotary devices, and he produced the first dynamo, in 1831. American inventor Joseph Henry (1797–1878) produced a similar prototype around the same time. These demonstration devices soon gave way, in 1832, to a more effective machine devised by a French instrument maker, Antoine-Hippolyte Pixii (1808–1835). Working according to Faraday's, Pixii rotated a permanent magnet so that its poles passed by a piece of iron wrapped with wire; with each turn of the magnet, electricity passed along the wire. **RBk**

SEE ALSO: ELECTROMAGNET, ELECTRIC MOTOR, INSULATED WIRE, ELECTRICAL GENERATOR, MULTIPLE COIL MAGNET

⬈ *Pixii's 1832 magneto-electric machine was the first practical mechanical generator of electrical current.*

"Seeds of great discoveries . . . only take root in minds well prepared to receive them."

Joseph Henry

Time-lock Safe (1831)

Rutherford improves bank-vault security.

It is thought that the first time-lock safe was patented by Scotsman Williams Rutherford in 1831. His work was part of a race between banks wanting to keep their vaults secure and robbers trying to break into them.

The Romans were the first to invent locks made out of metal, and these had special notches and grooves that made them more difficult to pick. In 1784 Joseph Bramah invented the first "unpickable" lock, although one locksmith did succeed, albeit after more than fifty-one hours—more time than robbers normally have at their disposal. However, during the 1800s the incidence of bank robberies was fast increasing.

The solution to the problem was the time-lock, a clockwork device that prevented the bolt of a lock from

> *"Bankers were at first reluctant to adopt a lock that barred friend as well as foe."*

John Erroll and David Erroll, writers

being opened until a defined moment in time. Even with the correct key (or combination in the case of a combination lock), a time-lock cannot be opened until a predetermined point in time has arrived. For example, banks locking their vaults over the weekend would be unable to open them again until Monday morning.

The design that formed the basis for modern time-locks was patented more than forty years later, in 1873, by James Sargent, an employee of Yale. This model was reliable, secure, and easy to use. After its introduction in banks, robbers were forced to come up with more elaborate ways of gaining entry, such as using explosives to blow open the doors. **HI**

SEE ALSO: MECHANICAL LOCK, COMBINATION LOCK, PADLOCK, TUMBLER LOCK, SAFETY LOCK, YALE LOCK

Fire-lighting Lamp (1832)

Döbereiner brings fire instantly to hand.

Prior to the seventeenth century, anyone intent on lighting a fire would first have had to find another fire. It was either that or return to antiquity and try to coax a flame by means of one of a variety of methods involving sticks or stones. By the eighteenth century fires were being started with convex burning glasses, but these were of no use if a fire had to be lit in the dark, or when there was no direct sunlight.

The first pneumatic fire-lighter was invented in 1770. This device relied on rapid compression of gases to produce heat and the all-important flame. But it was not until the mid 1830s that the first practical fire-lighting lamp for use in the home became available. Just five years after German chemist Johann Döbereiner came up with his initial idea in 1832, 20,000 fire-lighting lamps were in use in England and Germany. They were to remain popular until safety matches, followed by modern cigarette lighters, appeared on the scene.

Döbereiner realized, through his experiments with platinum, that a fine platinum powder would produce a flame in the presence of hydrogen gas. Actually the reaction had been noted before, but Döbereiner was the first to put it to any practical use. Within days of his rediscovery of the phenomenon, he had created a lamp, the *Döbereinersche Feuerzeug* which brought hydrogen into contact with platinum to act as a lighter. The only drawback of the device was that it had to be filled with acid to produce the hydrogen.

Modern cigarette lighters employ a slightly different mechanism, with a mixture of metals substituting for platinum as the flint. When the lighter's button is depressed, a wheel rasps against the flint, releasing particles that burst into flame in the air and ignite the gas inside the lighter. **HB**

SEE ALSO: CONTROLLED FIRE, FRICTION MATCH, SAFETY MATCH

Magnetometer (1833)

Gauss gauges the energy of magnetic fields.

A simple compass indicates the direction of a magnetic field, but the magnetometer, invented by the German mathematician and scientist C. F. Gauss (1777–1855), could also measure its absolute strength. Before Gauss's time, people had compared the fields at two different spots on Earth by observing how long it took a suspended magnetized needle to make a certain number of oscillations. Gauss, and his physics professor friend Wilhelm Weber (1804–1891), gauged the field strength of the magnetized needle by measuring by how much it twisted two fibers from which it was suspended, when held at a right angle to Earth's field.

Gauss and Weber founded the *Magnetischer Verein* (Magnetic Club), whose members measured not only magnetic fields all over Earth but also their variations with time, caused by slow changes in Earth's liquid iron core and by changes in the electron clouds that surround Earth. Electron clouds are influenced by particles emitted spasmodically by the solar wind.

Modern flux-gate magnetometers are much handier. They measure an ambient magnetic field by noting the way in which it influences the induced electrical currents in a wire coil wrapped round a magnetic core, the field in the core being produced by a second AC driven coil. Magnetometers have been fitted to most interplanetary spacecraft. They also play an important role in geology, especially in prospecting for iron ore deposits, and in archeology, where they can reveal metallic objects buried in the ground. **DH**

SEE ALSO: MAGNETIC COMPASS, GYROCOMPASS, ELECTROMAGNETIC TELEGRAPH

↗ *The Kew magnetometer was designed by Thomas Jones in c. 1836 for the Kew Observatory in England.*

↘ *An engraving showing experiments with magnets includes Lampadius's magnetometer (plate 13).*

Combine Harvester (1834)

Moore speeds up grain production.

Traditional grain harvesting was a laborious process, requiring separate cutting, binding, and threshing operations. Although the mechanical reaper, invented in 1831 by Cyrus McCormick, did the cutting, farmers still had to follow the machine and bind the sheaves of grain by hand. Hiram Moore created the first successful combine harvester in 1834 with the aim of speeding up the production of grain from the vast wheatlands of America. Corn and wheat spelled big money in the 1800s, but farmers had to employ dozens of farmhands in order to reap the benefits of their harvests, and this was a costly business.

Moore's invention, developed in the farmlands of Michigan, succeeded in combining the two separate processes of cropping and threshing grain into one simplified, mechanically powered step. This creation, paradoxically, was both a blessing and a curse for farmworkers—while it saved their backs it also cost many of them their livelihoods.

The machine essentially works in a two-step process; first it harvests the corn, cutting the crop at the base with a sharp multibladed cylinder, then passing it inside the machine where it is threshed. Threshing separates the grain into wheat and chaff. The machine retains the saleable grain while leaving behind the chaff and straw, which can be used elsewhere on the farm for livestock feed and bedding.

It took as many as sixteen horses to drag the heavy machinery of early combine harvesters across the fields, but the effort was worthwhile since it combined the work of several men over many hours into one simplified movement. Steam engines were later used to ease the task, and more recently engine-powered combine harvesters have been developed, making the process of harvesting grain even more efficient. **KB**

SEE ALSO: FLAIL, REAPER, SCYTHE, THRESHER

Sewing Machine (1834)

Hunt paves the way for mass-produced clothes.

The history of the sewing machine begins in 1790, with a patent by British inventor Thomas Saint for a device (never built) to puncture leather and repeatedly pass a thread through the holes. In 1830 Frenchman Barthélemy Thimonier successfully built a machine that also used this "chain stitch" method. Within a dozen years he had built eighty machines, but they were destroyed by a mob of angry tailors.

The first person to develop the sewing machine as we know it was American Walter Hunt (1796–1859), in 1834. His crucial innovation was to use two spools of thread (on an upper spindle and a lower bobbin) and an eyed needle to create "lockstitch"—the two threads lock together when they pass through the hole.

> *"… the sewing machine was as awe-inspiring as the space capsule [was to twentieth-century people]."*
>
> Grace Rogers Cooper, writer

Hunt, also the inventor of the safety pin, failed to patent his sewing machine. This made it easy for fellow American Elias Howe to patent a lockstitch machine twelve years later. Howe defended his rights against many rivals, most notably Isaac Singer, who had tried to patent a new device based on Hunt's machine. Howe won a court case and Singer was forced to pay him royalties—but it was Singer who pioneered mass production and whose name, to this day, remains synonymous with the sewing machine. **AC**

SEE ALSO: CLOTHES, SEWING, WOVEN CLOTH, SAFETY PIN

➡ *Patented in 1846, Howe's machine proved impossible to sell until 1849. It later made him a multimillionaire.*

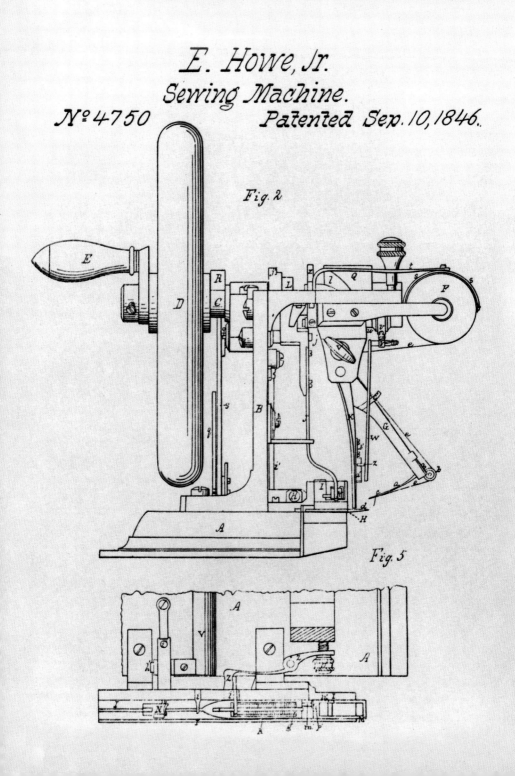

E. Howe, Jr.
Sewing Machine.
№ 4750   Patented Sep. 10, 1846.

Fig. 2

Fig. 5

Calotype Process (1835)

Talbot develops the first negative/positive process for taking photographs.

The calotype process was created by William Henry Fox Talbot (1800–1877) in the very early days of photography. Athough Talbot was not the originator of photography, his calotype formed the basis of the photographic process for more than 150 years.

The calotype, or Talbotype, process will be recognizable to anyone who is familiar with developing photographs. *Calotype* is derived from the Greek, meaning literally "good impression." Talbot used high-quality writing paper coated with light-sensitive silver chloride. The paper had to be prepared before exposure to light, and then needed ten to fifteen minutes of exposure time in bright light, usually sunlight. The paper was exposed to the light in a wooden box functionally similar to a modern camera.

Talbot also realized that short exposure of the light-sensitive paper and then chemical fixing could produce a useful negative. The fixing process removed the light-sensitive silver, making the negative easier to handle. The negative also allowed multiple "positive" images to be printed, so Talbot's invention was the direct ancestor of the developing, fixing, and printing processes still used today.

Simultaneous to Talbot's work, Louis Daguerre was developing the daguerreotype process. This process was more akin to a Polaroid than a traditional photograph. The daguerreotype was available ahead of the calotype and was initially very popular, but the lack of a reusable negative meant that the calotype process soon came to dominate photography. **JB**

SEE ALSO: CAMERA, DAGUERREOTYPE, DIGITAL CAMERA, PHOTOGRAPHIC FILM, PHOTOGRAPHY, SELF-DEVELOPING FILM

◩ *The original camera made by William Fox Talbot in 1830 for his experiments with photography.*

◩ *Photograph of Laycock Abbey in Wiltshire, England, taken in 1835 from the earliest camera negative.*

Incandescent Light Bulb (1835)

Lindsay's demonstrations makes constant electric light a reality.

Decades before Thomas Edison filed a patent for his electric lamp, Scotsman James Bowman Lindsay (1799–1862) produced constant electric light in what became a prototype of the modern light bulb.

Building on Humphry Davy's successful yet impractical platinum incandescent light, which he developed in 1802, Lindsay managed to create a more usable form of the light bulb. Having secured a position as a lecturer at the Watt Institution in Dundee, Scotland, in 1829, Lindsay began experimenting with constant electric light. He demonstrated his invention in 1835 at a public meeting in Dundee.

The light from an incandescent bulb is produced from a filament through which an electrical current is passed. Lindsay claimed that, with his light, he could "read a book at a distance of one and a half foot." This was an improvement on Davy's light, which did not last as long, was not as bright, and used platinum—an expensive material. The public was impressed by Lindsay's light, which produced no smell or smoke, did not explode, and could be kept on a tabletop.

Lindsay went on to give public lectures about his invention over the following years, but did little with it after that. He neither pursued nor established a patent for his device, and the modern light bulb as we know it would later be developed by others, such as Joseph Swan and Thomas Edison. Even so, Lindsay's invention is considered an important event in the history of lighting, and was one of the first prototypes of today's incandescent light bulb. **RH**

SEE ALSO: ARC LAMP, ARGAND LAMP, CANDLE, FLUORESCENT LAMP, GAS LIGHTING, OIL LAMP, ANGLEPOISE® LAMP

↗ *An early incandescent light bulb with the coils giving electrical contact to the filament inside the glass.*

⇥ *Engraving from 1883 of Edison's incandescent lamps showing the different kinds of carbon filament.*

Electromechanical Relay (1835)

Henry communicates with electricity.

In 1835, shortly after the American Joseph Henry (1797–1878) became a professor at Princeton University, he passed crude on/off messages from his laboratory to his nearby home on campus using an electromagnetic relay and a current-carrying wire.

In the early relays, switching an electrical current on or off magnetized the relay's electromagnet core, and this magnetic field attracted a pivoting iron armature, which itself operated a set of contacts that made or broke an electrical circuit. One of the advantages of such relays was that the current through the electromagnet could be very small, and the switched-circuit current could be much larger.

Relays formed the basis of the telegraph system and were at the heart of telephone exchanges. In modern industry, electromagnetic relays are used to control electric motors and are a vital component of automated systems of machine manufacture.

Great advances have been made in relay design, mainly to guard against the potential damage caused by the voltage increase that is produced by the collapsing magnetic field of the coil when it is deactivated. Initially this was done by placing a capacitance and resistor in series, but in modern relays a simple diode is wired across the coil.

The power consumption of relays has been greatly reduced by introducing latching relays. These only consume power when they are switched on. By using a ratchet and cam system, these relays use one pulse to turn themselves on and a subsequent pulse to turn themselves off. Also, much attention has been paid to the relay contacts. Some are placed in vacuum tubes to prevent corrosion. Others are wetted with mercury to improve switching speeds. **RH**

SEE ALSO: ELECTROMAGNET, ELECTROMAGNETIC TELEGRAPH

Mechanical Computer (1835)

Babbage pioneers the computing machine.

An accomplished mathematician and mechanical engineer, Charles Babbage (1791–1871) combined these two disciplines to create a "difference engine" capable of solving polynomial functions without having to use unreliable, hand-calculated tables.

Despite generous government funding, Babbage sadly never fully finished his difference engine and the project was abandoned in 1834. This did not stop Babbage from thinking about computing, however. In 1835 he released designs for his "analytical engine," a device similar to the difference engine but which, by using programmable punched cards, had many more potential functions than just calculating polynomials.

"To err is human, but to really foul things up requires a computer."

Farmers' Almanac

The analytical engine was never built, although Babbage produced thousands of detailed diagrams.

Using the lessons learned from the analytical engine, Babbage created a more efficient and smaller difference engine in 1849. Difference Engine No. 2 was not built until 1991, when the London Science Museum completed the machine from the original blueprints and found that it worked perfectly. Babbage's reputation as a man years ahead of his time was then restored to its rightful place. **JB**

SEE ALSO: DIGITAL-ELECTRONIC COMPUTER, LAPTOP, PALMTOP COMPUTER, PERSONAL COMPUTER, SUPERCOMPUTER

➡ *This completed section of Babbage's unfinished Difference Engine No. 1 still works perfectly today.*

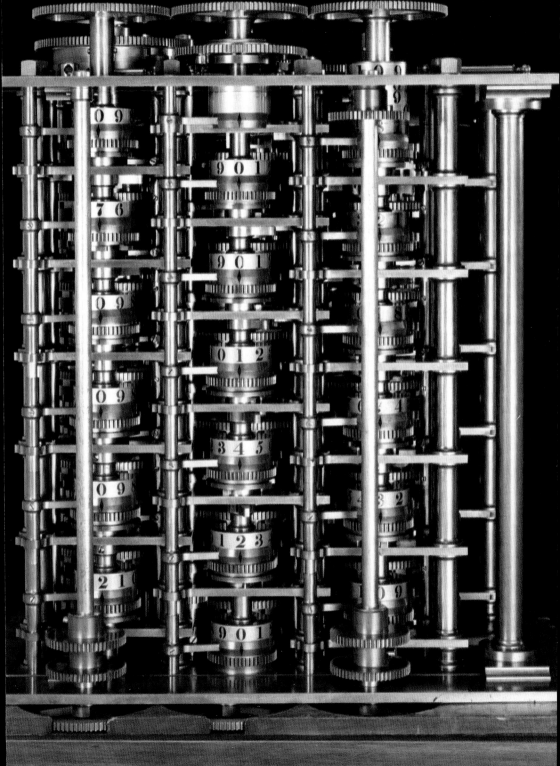

Revolver (1835)

Colt's firearm enables repeat firing without reloading.

Samuel Colt (1814–1862) produced the first revolver capable of firing five or six times without reloading. It was equipped with a cylinder containing chambers into which bullets were loaded. The cylinder would revolve with each cock of the hammer, enabling each chamber to lock into position behind the barrel and discharge the bullet with a pull of the trigger.

In anticipation of demand for the revolver, Colt opened a manufacturing plant, but poor sales forced its closure. Colt's fortunes changed after favorable reports from U.S. troops fighting in Florida and Texas instigated a government order for 1,000 pistols. With the assistance of Eli Whitney Jr., Colt opened the world's biggest armory in Hartford, Connecticut, developing a production line of handguns and interchangeable parts. By 1856 the company was gaining a reputation around the world for exceptional quality and workmanship, and Colt soon became one of the wealthiest businessmen in the United States.

Other companies, such as Smith & Wesson, emerged with their own revolver designs to challenge Colt's market monopoly. In 1873 Colt manufacturers responded with the six-shot, single-action, .45-caliber Peacemaker model, which proved immensely popular in the American "Wild West."

Further innovations in firearms witnessed the double-action revolver, which allowed a pistol to be fired with a simple pull of the trigger, and swing-out cylinders for easy loading. The revolver continued to be the weapon of choice for the military until the emergence of the semiautomatic pistol at the turn of the twentieth century. **SG**

"Abe Lincoln may have freed all men, but Sam Colt made them equal."

Popular post-American Civil War slogan

SEE ALSO: ARQUEBUS, GUN, GUNPOWDER, FLINTLOCK, GUN SILENCER, MUSKET, REPEATING RIFLE

◪ *The Colt Navy .36 of 1851 was one of the most popular Colt revolvers; it was produced for twenty-three years.*

Soda Syphon (1837)

Perpigna puts a new fizz into everyday drinks.

Carbonated drinks were known in France as early as 1790, but the soda syphon or seltzer bottle was developed by Deleuze and Dutillet in 1829. This featured a hollow corkscrew that allowed some of the contents to be dispensed, via a valve, while keeping the bottle under pressure and its contents fizzy.

The modern bottle, which remains essentially unchanged today, was patented by Antoine Perpigna in 1837 and was known as the "Vase Syphoide." In an improvement on the earlier design, the head featured a valve that was closed by a spring. The drink is carbonated by pressurized carbon dioxide. The excess pressure causes more of the gas than usual to dissolve in the liquid. When the liquid is returned to atmospheric pressure in the glass, this gas comes out of solution as bubbles.

Syphon bottles were made of glass initially and had to be refilled by hand using specialized pumps. The process was further complicated by the need for the valve to be depressed in order for the syphon to fill. Explosions were not uncommon with the pressurized bottles, which were generally filled to a pressure of 140–160 pounds per square inch (10–11 kg per sq cm). The syphon was improved by the use of gas canisters to carbonate the liquid, with several patents being filed on this development.

The soda syphon continued to grow in popularity until the 1920s and 1930s. However, in the 1940s the seltzer trade virtually collapsed with the major bottle-making centers in Czechoslovakia being devastated by World War II and American bottle production diverted to making war materials. **HP**

SEE ALSO: CORKSCREW, CROWN BOTTLE TOP, GLASS, SODA WATER

↗ *Many of the old styles of soda syphon were attractively made from different colors of glass.*

"A rather unusual advantage of the syphon bottle was its use as a fire extinguisher."

Bryan Grapentine, antique collector

Standard Diving Suit (1837)

Siebe's new suit revolutionizes diving technology.

The standard diving suit, or "hard-hat" diving suit, was a major advance in diving technology. Early diving suits were crude and inflexible and imposed major restrictions on divers' movements, such as an inability to invert. A brilliant German inventor by the name of Augustus Siebe (1788–1872) changed all that with his innovative "closed" helmet suit—a design that remained essentially unchanged until fiberglass SCUBA suits arrived in the 1960s.

After learning metal craft and working as a watchmaker, and following service as an artillery officer in the Prussian army at the Battle of Waterloo, Siebe moved to England in 1815. While living in London he stumbled across the solution to creating a more practical diving suit. Previous designs—so-called "open dress" suits—were simple diving bells that trapped air for breathing, but these took in water when the diver left the vertical position.

Siebe took a large metallic helmet and bolted it to a canvas diving suit to create a perfect water-sealed suit. A length of tubing supplied air to the diver from the surface. Siebe created many prototypes of his invention before settling on the eventual design. The final version of the suit was first used to investigate the wreck of HMS *Royal George*, which had sunk off the coast of Spithead, Isle of Wight, England. It was later suggested that the helmet be detachable from the suit, and so the standard diving suit was born.

Siebe's suit heralded a revolution in diving, enabling underwater exploration to take place with a much greater level of freedom than had previously been available. It also provided the basis upon which future suits were designed, up until the introduction of the SCUBA suit, which eventually succeeded Siebe's great invention. **SR**

> *"Siebe's design . . . remained in use essentially unchanged by the Royal Navy until 1989."*

English Heritage

⬆ *Siebe's diving helmet was watertight because it was sealed to the diving suit.*

➡ *This drawing from 1830 shows a demonstration of an earlier open style of diving helmet with air pump.*

SEE ALSO: CLOTHES, HELMET, SUBMARINE, REBREATHER, BATHYSPHERE

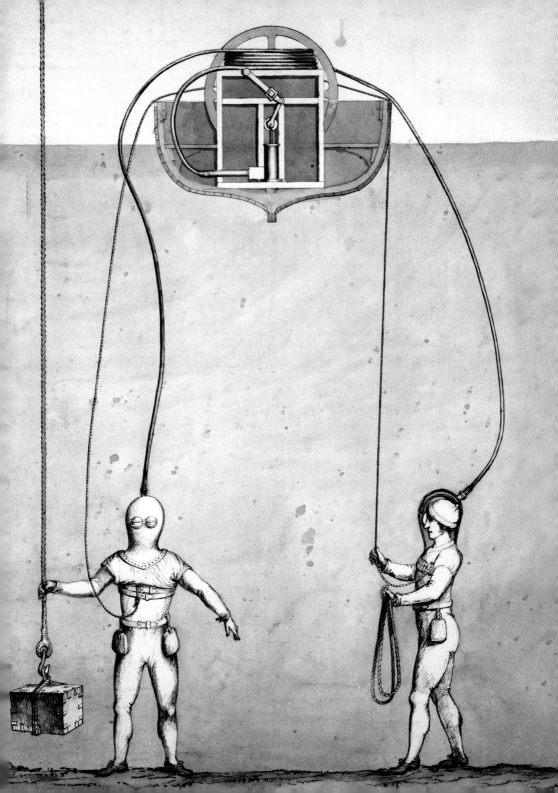

Steam Hammer (1837)

Nasmyth paves the way to large-scale iron production.

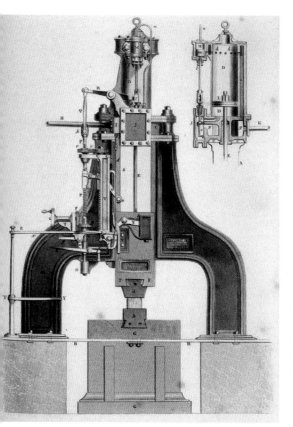

In the early days of the Industrial Revolution, large pieces of metal were made by forging multiple small segments then welding them together into a finished product. Metalworking hammers of the day were capable of forging small items efficiently, but when a large item was placed in them, they had little room to maneuver and therefore little force.

This fact became painfully obvious when the Great Western Steamship Company began to build the SS *Great Britain*. The engineer in charge found it impossible to locate a hammer capable of producing the mammoth paddle wheels that were to propel the ship. Edinburgh-born engineer James Nasmyth (1808–1890) heard about the problem and sketched out a design for a large steam-powered hammer that would be capable of producing just such a piece of equipment. Nasmyth's hammer consisted of a piston attached to a hammer head. By admitting steam to the piston, the hammer head was raised to a certain height and then released, falling with tremendous force before being rapidly raised again. Unfortunately for the steam hammer and the engineer, the design of the *Great Britain* was changed from a paddle-wheel design to that of screw drive, rendering redundant the production of the hammer.

Nasmyth did not attempt to manufacture or even patent his hammer until he visited the Le Creusot ironworks in France where, to his surprise, he found his invention in action. He quickly patented his design back home in England and then set about production. His hammer proved so well designed and easy to control that it reduced production costs of forged iron by as much as 50 percent and was also so adaptable that it could be used to forge very large items or something as small as a nail. **BMcC**

> "In 1854 . . . I took out a patent for puddling iron by means of steam."
>
> James Nasmyth

↑ *Engraving from 1851 of James Nasmyth's steam hammer, which he did not patent until 1854.*

→ *A steam hammer in use at a British factory in c. 1855. It enabled workers to forge larger pieces of metal.*

SEE ALSO: COKE-BASED IRON SMELTING, HAMMER, STEAMBOAT

Steel Plow (1837)

Deere improves plow durability.

When John Deere (1804–1886) developed and manufactured the first commercially successful steel plow in 1837, he greatly enhanced agricultural methods in regions encumbered by heavy soils. Since the earliest recorded usage of an effective plow in 5500 BCE by the Sumerians and Babylonians, the moldboard and cast-iron plow were the most notable innovations in design, but neither could effectively counteract the problems of sticky soil.

It was the threat of bankruptcy in his native Vermont that persuaded Deere to seek his fortune out west, journeying to Grand Detour, Illinois, where he opened a blacksmith's shop. He soon learned through the frequent repairs he had to undertake that the cast-iron plow that performed so well in the light, sandy New England soil was not suited to the heavy, sticky soils of the Midwest. Farmers complained that the soil had to be continually removed from the bottom of the plow by hand, making the work arduous and time-consuming. Deere began experimenting with a design that incorporated a highly polished surface that prevented soil from clinging to the bottom. Three newly fashioned plows were produced and sold in 1837, encouraging Deere to create further improvements. These proved so popular that by 1846 his annual output approached 1,000 plows.

In 1848 Deere moved his operation to Moline, Illinois, to take advantage of its superior transport links alongside the Mississippi River. It was here that he negotiated with Pittsburgh manufacturers to cut steel to his own specifications and began rolling out plows across the United States. By 1857 Deere's company was selling 10,000 plows annually and farmers in the Midwest were reaping their own rewards from the newly cultivated land. **SG**

SEE ALSO: CAST IRON PLOW, MOLDBOARD PLOW, SCRATCH PLOW, SEED DRILL, STEEL

Composting Toilet (1838)

Swinburne challenges the water closet.

Although the first patent application for a composting toilet was filed by Thomas Swinburne in 1838, the idea of recycling human waste was by no means new at this time. The Chinese had already been composting human and animal waste for use as fertilizer for thousands of years. However unattractive this process may seem, it is certainly more environmentally viable than a modern sewage system. Composting cuts down on water pollution, and human excreta contain valuable nutrients that can help farmers reduce the amount of chemical fertilizers they use on crops.

Swinburne's "earth closet" was the first device that can be described as a composting toilet, that is, one that deposits earth or peat onto feces to kickstart the composting process. But it was not until Henry Moule established the Moule Patent Earth-Closet Company in the 1860s that they became more commonly used. Some schools and military camps reportedly preferred composting toilets because they were cheaper to install than the water-closet alternative.

A variety of designs for composting toilets now exist, including the popular "Rota-Loo," which houses a rotating tank underneath the toilet. Modern composting toilets can even combine vaults for kitchen and toilet waste into a single storage space, which is emptied roughly once a year.

There are many areas of the world where communities would benefit from composting toilets. World Health Organization figures from 2004 estimate that as many as 2.6 billion people are without "improved" sanitation, although the definition of "improved" is still debated. Some communities in developing countries use simple "drop and store" facilities as latrines, but these are often unhygienic and can be a source of contamination and disease. **HB**

SEE ALSO: FLUSH TOILET, SEWAGE SYSTEM, S-TRAP FOR TOILET, TOILET PAPER

Daguerreotype Process (1839)

Daguerre enables the development of photography as an art form.

Dating back to 1839, the daguerreotype is one of the oldest known forms of photography. It was the first process that did not require excessively long exposure times, making it ideal for portrait photography.

Louis Daguerre (1787–1851) had been trying since 1829 to capture the images he viewed through his camera obscura, a wood box that produced an image on a sheet of frosted glass via a lens at one end. In 1839, after a decade of painstaking work, he presented his daguerreotypes to a joint session of the Académie des Sciences and the Académie des Beaux-Arts. The pictures included images of shells, fossils, and a dead spider, photographed through a microscope.

The process for developing the pictures was long and laborious. The plates had to be prepared from a sheet of copper coated with a thin layer of silver. The silver surface had to be polished until it was mirrorlike. It was then exposed to iodine vapors to form a surface of silver halide that would react with light to produce the image. The plate was then exposed in a large, boxlike camera, after which the image was developed using mercury fumes. The resulting picture was fixed using a water and sodium thiosulfate solution.

Finished daguerreotype images are laterally inverted from the original object and are reflective, like a mirror. They are mounted and placed behind glass sealed with paper tape to avoid tarnishing. Daguerreotypes are not reproducible—each image is a one-off; the only way to copy a daguerreotype is to take another daguerreotype of that image. **HP**

SEE ALSO: CALOTYPE PROCESS, PHOTOGRAPHIC FILM, PHOTOGRAPHY, SELF-DEVELOPING FILM CAMERA

↗ *This picture features all the equipment necessary for the daguerrotype process.*

⇥ *Example of a daguerrotype by an unknown photographer, from around 1840.*

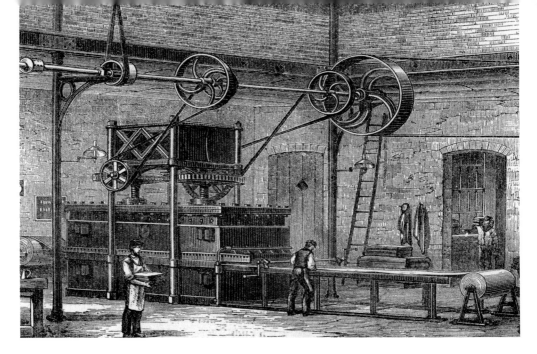

Vulcanization (1839)

Goodyear uses heat and sulfur to increase the usefulness of rubber.

The idea of curing rubber goes back to prehistoric times. The Aztecs, for example, processed rubber by mixing latex with vine juice. However, vulcanization, as we now think of it, was invented by Charles Goodyear (1800–1860) in 1839 (according to his own account). Although Goodyear patented his invention in 1844, he died with huge debts in 1860.

Named after Vulcan, the Roman god of fire, vulcanization is a process of curing rubber using high temperatures and the addition of sulfur. In its natural state, rubber is sticky, deforms when warm, and is brittle when cold. But by linking together the polymer molecules in rubber with bridges of sulfur atoms, it becomes much harder, less sticky, more resistant, and much more durable. As a result, vulcanized rubber has a whole range of useful applications, not least in sealing gaps between moving parts, thereby playing a crucial part in the development of more efficient industrial machines. The ability of vulcanized rubber to absorb stress then revert to its original shape has made it an ideal material for products such as tires, shoes, and rubber bands, while its water-resistant qualities make it perfect for boots, wet weather clothes, and waterproof linings and coatings.

While vulcanization revolutionized manufacturing processes in the nineteenth and twentieth centuries, environmental concerns now focus on how to recycle vulcanized rubber. Some one billion rubber tires, for example, are produced annually but only about 10 percent of waste rubber is reused. Efforts are now being made to find ways to devulcanize rubber without losing all those important properties. **RBd**

SEE ALSO: RUBBER BALL, PNEUMATIC TIRE, RADIAL TIRE, RUBBER BANDS, RUBBER WELLINGTON BOOTS, TIRE VALVE

⬆ *This woodcut from 1867 shows the process of vulcanization at a German rubber factory.*

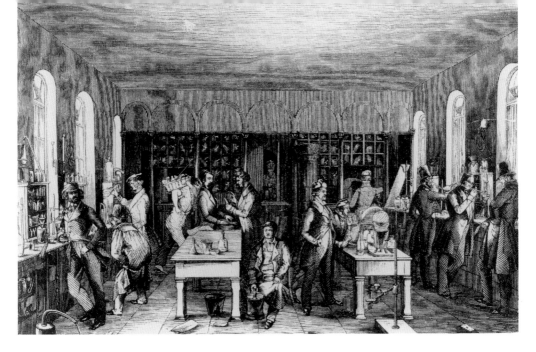

Artificial Fertilizer (*c.* 1839)

Von Liebig makes a scientific approach to maintaining soil fertility.

Until Justus von Liebig (1803–1873) investigated continuance of soil fertility, agricultural fertilizer came in the form of animal dung, cinder, or ironmaking slag. Von Liebig, a chemist by training, rejected the hypothesis that plants gained all their nutrition from partially or completely decayed animal and vegetable matter in the soil, suggesting instead that plants gained carbon and nitrogen from carbon dioxide and ammonia in the atmosphere. These gases would be in plentiful supply since the process of decay would return the gases to the atmosphere, so the only nutrients supplied by the soil were minerals. Since these would be the limiting factor in the growth of crops, all a farmer had to do was determine which mineral each plant needed, by analysis of its ashes, and provide the soil with the requisite chemical compounds.

Von Liebig's initial experiments in this area were a failure as he had taken pains to add alkaline compounds in an insoluble form, which—while avoiding their being washed away by rain—also precluded their being taken up by plant roots.

Von Liebig's work led directly to the development of artificial fertilizers and their widespread use in farming. Unfortunately the overuse of chemical fertilizers is now thought to have a widespread environmental impact, being held responsible for the accumulation of heavy metals in soil. The leaching of nitrates into waterways leads to the process of eutrophication, causing a sudden growth of algae that removes oxygen from the water, killing fish and plants.

Currently efforts are being made to reduce fertilizer use worldwide by applying it more efficiently. **HP**

SEE ALSO: PESTICIDE

⬆ *Drawing of the first practical teaching laboratory used by Von Liebig at the University of Giessen, Germany.*

Fuel Cell
(1839)

Grove generates energy in a cell.

Early telegraphs led to improved communication but were limited by a lack of readily available power. In 1839 Welsh scientist Sir William Grove (1811–1896) tried to tackle this problem by designing a device that could generate a strong flow of electricity. Grove's electrochemical device harnessed the energy released by a chemical reaction to generate electricity.

Grove's first attempt consisted of zinc in dilute sulfuric acid and platinum in concentrated nitric acid, separated by a porous pot. The "Grove Cell" was the favored power source in the mid-nineteenth century because it produced a strong current. However, as telegraph traffic increased, it soon became apparent that the cells were releasing poisonous, nitric oxide gas. Large telegraph offices were filled with smoke from rows of hissing Grove cells.

Grove's second electrochemical cell, the "Gas Voltaic Battery," provided the basis for the modern fuel cell. His idea was based on the fact that sending an electrical current through water splits the molecules into hydrogen and oxygen. Grove's theory was that if you could reverse the reaction, by combining hydrogen and oxygen using a catalyst like platinum, then you could potentially generate electricity and water. Grove immersed two platinum strips, each encased in a tube of either hydrogen or oxygen, in a tank of sulfuric acid. After completing the circuit, Grove showed that electricity was flowing by using it to convert water back into hydrogen and oxygen.

Grove had proved that his fuel cell worked, but he was not an entrepreneur and his idea lay dormant for another 130 years. The term "fuel cell" was eventually coined in 1889 by Ludwig Mond and Charles Langer, who developed the first practical device. **RB**

SEE ALSO: BATTERY, DRY-CELL BATTERY, PHOTOVOLTAIC (SOLAR) CELL, RECHARGEABLE STORAGE BATTERY

Compound Camera Lens (1840)

Petzval speeds up photograph-taking.

In 1839 portrait photographs took an age using simple meniscus lenses. All that changed when Hungarian mathematician Jozef Petzval (1807–1891) designed the first compound camera lens. The Petzval lens dramatically cut exposure times, boosted camera performance, and revolutionized photography.

The "daguerreotype system," developed by Frenchman Louis Daguerre, was the forerunner to the Petzval lens. Requiring around half an hour of exposure time, this was still an improvement over existing techniques that needed several hours for successful exposure. However, this was still too long for taking portrait shots, which inevitably blurred with

---

"[Petzval] took on shortening [the daguerreotype's] exposure time from minutes to seconds."

Slovakia Today

---

the slightest movement of the subject.

Working with Friedrich Voigtländer at the University of Vienna, Petzval performed calculations that led him to create an achromatic portrait lens with four lenses arranged in two groups, providing six times the luminosity and an undistorted image for the first time. Crucially, the lens allowed photographs to be taken at around twenty times the speed of the daguerreotype lens. Petzval won many accolades for his invention, but his success was clouded by a dispute with Voigtländer over the right to supply the new lens. Voigtländer went on to mass-produce the invention, but it is Petzval who pioneered the modern camera lens. **SR**

SEE ALSO: DIGITAL CAMERA LENS, PHOTOGRAPHY, DAGUERREOTYPE PROCESS, SINGLE-LENS REFLEX CAMERA

Postage Stamp

(1840)

Hill simplifies the postal system with the introduction of his iconic Penny Black.

On May 6, 1840, in the greatest single reform of the English postal system since its inception in 1510, the world's first prepaid adhesive postage stamp, the Penny Black, was issued. The stamp ushered in the era of prepaid post and was instrumental in lowering postal rates. By 1856 the number of letters posted annually in Britain had grown to almost 400 million, compared to a mere 76 million in 1839.

The Penny Black, graced by the profile of England's young Queen Victoria, was designed by postal service employee Rowland Hill (1795–1879). In 1837 Hill authored and circulated a pamphlet titled *Post Office Reform: Its Importance & Practicality*, in which he tackled the reasons for the complexity and often prohibitive costs involved in delivering Britain's mail. Prior to 1840, prepaid postage was purely voluntary. However, where the letter was not already paid for, the cost of delivering it had to be redeemed by the postal worker upon delivery, which sometimes resulted in several attempts before the addressee was found. In addition, cryptic messages were often written on the envelope's face, which the receiver could see, read, and decipher, and then promptly refuse delivery.

The Penny Black also eliminated the laborious criteria involved in determining the cost of a letter, which varied according to the number of sheets used and the distance the item traveled. Now one stamp, costing a single penny, could send a letter weighing less than half an ounce (14 g) anywhere in Britain. **BS**

SEE ALSO: E-MAIL, GLUE, POST-IT® NOTES

↗ *The Penny Black was only in use for a year because the red cancellation mark was hard to see on black.*

➜ *A Penny Black with a black cancellation mark; the stamp was replaced by the Penny Red in 1841.*

Standardized Screw System (1841)

Whitworth introduces standardized sizes for nuts and bolts.

Sir Joseph Whitworth (1803–1887) had a lifelong desire to improve on existing technology. By the time of his death, his interests and efforts had amassed him assets worth more than £150 million ($300 million) in today's money and a large manufacturing complex in Manchester, England.

En route, Whitworth had produced pretty much everything from road-sweeping machinery to firearms. Moreover, by absorbing the ideas and philosophies to which he was exposed while working at the renowned Henry Maudslay works at Lambeth Marsh, London, and rejecting the often shoddy standards of the day, this perfectionist had established himself as the father of precision engineering.

One Whitworth contribution to the modern world that underpinned so many others, and one that would keep his name on the lips of mechanics and engineers alike for many years, was the Whitworth Thread, which he proposed in 1841. Known subsequently as the British Standard Whitworth (BSW) Thread, it featured the first standardized thread pitch for all the various sizes of nuts and bolts in its range—in other words any BSW nut would fit precisely any BSW bolt of the same size. This sounds pretty obvious now, but prior to Whitworth, in an age when machines were still evolving, nuts and bolts of the "same" size were custom-engineered to fit each other, as a pair; they were not interchangeable, and a nut of a given size from one source would not necessarily fit a bolt of the same specified size from elsewhere. The BSW thread is now effectively redundant, having been superseded by "Unified" and "Metric" types. **MD**

SEE ALSO: METRIC SYSTEM, SCREW, PHILLIPS SCREW, SQUARE-HEADED SCREW

⬆ *Whitworth's company in England made standard gauges for checking and adjusting wire gauges.*

Facsimile Machine (1842)

Bain invents the recording telegraph, predecessor of the fax machine.

No office is complete without a fax machine, so you would probably think that it was a modern invention. But the earliest facsimile machine was actually invented thirty years before the telephone, in 1842.

Today faxes run a sheet of paper through rolls, using optical chips to record the image. These chips did not exist until the late 1960s, but machines using photoelectric cells, invented by Édouard Belin in 1907, sent the light and dark parts of a picture as electrical pulses. By 1902 Arthur Korn had already invented a similar machine. Even earlier, in 1898, Ernest A. Hummel invented the copying telegraph, which sent pictures between major newspapers in the United States.

But long before Hummel there was a way to transmit images. In 1855 Giovanni Caselli made the pantelegraph, synchronizing the sending and receiving machines with an electronic heartbeat and sending pulses between them. But even this was just an improvement on the original facsimile machine.

In 1842 the Scottish clockmaker Alexander Bain (1811–1877) invented the recording telegraph, the ancestor of the modern fax machine. Using a pendulum swung over metal type, the machine detected light and dark spots. It transmitted the spots as electrical pulses like Morse code and a receiver stained the image onto chemically treated paper. The machines were popular with newspapers, which needed images quickly, and with the Chinese and Japanese, who have pictograms rather than an alphabet and could not use Morse code. A synchronization of all fax machines in 1983 made them fast enough for the modern world. **DK**

SEE ALSO: TELEPHONE, COMPUTER SCANNER, E-MAIL, INTERNET, OPTICAL CHARACTER RECOGNITION

⬆ *This chemical telegraph, made by Bain in 1850, was used for recording Morse code signals.*

Grain Elevator (1842)

Dart speeds up grain deliveries.

The first grain elevator was built in Buffalo, New York, by Joseph Dart in 1842. Dart was a retail merchant who had seen Buffalo boom since the opening of the Erie Canal in 1825, linking the Midwest to New York. Initially grain was loaded and unloaded by hand, a back-breaking job that took several days. To overcome this problem Dart built the first wooden grain elevator.

The elevator consisted of a large wooden structure that served as a storage bin for the grain, to which a steam-driven belt with buckets was attached. The belt could be maneuvered into the hold of cargo vessels and activated, whereupon the buckets would scoop up the grain and deposit it into the storage bins.

> *"They serve as a monument to a bygone era ... waiting for ... freighters that no longer come."*

Grain Elevators, A History (website)

The elevator allowed ships to be unloaded at a rate of over 1,000 bushels (35,000 liters) per hour, leading to ships docking, unloading, and departing all within the same day. The elevator was also an ideal store for grain as it was cool, dry, and free of pests that could endanger the crop. The only risk was from fire in the wooden elevator since grain dust is highly flammable.

Joseph Dart claimed not to be the inventor of the elevator, saying instead that he had based his designs entirely on those of Oliver Evans. In the 1780s Evans had designed the principles for moving and storing grain using a conveyor fitted with buckets, although his invention was intended for use in flour mills. **HP**

SEE ALSO: PASSENGER ELEVATOR, CONVEYOR BELT

Electric Car (1842)

Davenport and Davidson try electricity.

Possibly in 1834, Robert Anderson of Scotland created the first electric carriage. The following year, a small electric car was built by the team of Professor Stratingh of Groningen, Netherlands, and his assistant, Christopher Becker. More practical electric vehicles were brought onto the road by American Thomas Davenport (1802–51) and Scotsman Robert Davidson (1804–1894) circa 1842. Both of these inventors introduced non-rechargeable electric cells in the electric car.

The Parisian engineer Charles Jentaud fitted a carriage with an electric motor in 1881. William Edward Ayrton and John Perry, professors at the London's City and Guilds Institute, began road trials with an electrical tricycle in 1882; three years later a battery-driven electric cab serviced Brighton. Around 1900, internal combustion engines were only one of three competing technologies for propelling cars. Steam engines were used, while electric vehicles were clean, quiet, and did not smell. In the United States, electric cabs dominated in major cities for several years.

Petrol-driven cars dominated in the twentieth century, but concerns about the sustainability of a petroleum-based economy led to a resurgence of interest in electric cars: in the late 1990s, Toyota and Honda introduced hybrid vehicles combining internal combustion engines and batteries, and models such as the Tesla Roadster (2008) and the Nissan Leaf (2009) have set a new standard. Barriers still remain to the rise of electric cars—not least their high cost and the scarcity of public charging points—but they are likely to outnumber petrol-driven cars by the 2050s. **TZ**

SEE ALSO: AIR CAR, BATTERY, CAR BATTERY, DRIVERLESS CAR, ELECTRIC MOTOR, MOTORCAR, ELECTRIC CAR BATTERY

▶ *A woman, circa 1910, using a mercury arc rectifier to provide direct current to charge her car's batteries.*

Ether Anesthetic (1842)

Long discovers a new medical painkiller.

The discovery of the properties of ether as an anesthetic was one of the major breakthroughs for the medical profession. Until then, patients undergoing surgery had to rely on hypnotism or alcohol.

American Crawford Long (1815–1878) is reputed to have first discovered the effects of ether when attending "laughing gas" parties and "ether frolics" during his years at medical school. There he noticed that those under the influence of nitrous oxide (laughing gas) or ether were unaware of pain through knocks and falls, until the effect had worn off.

Long established his rural practice in Jefferson, Georgia, and began to experiment with sulfuric ether as an anesthetic. The first procedure in which he used ether was an operation on March 30, 1842, to remove a tumor from a young man's neck; after the surgery the patient could not believe that it had been done. Long then began to use ether for women during childbirth, but he did not publish his findings.

In 1846 American dentist William Morton claimed to be the first to use ether as an anesthetic, which prompted Long to start recording his research. In 1849 he presented his findings and proof of his work to the Medical College of Georgia, and at the same time learned of two other doctors, Horace Wells and Charles Jackson, who were claiming to have discovered the use of ether as an anesthetic. Although Long's results were published in 1849, he did not receive official recognition during his lifetime. Finally, on June 18, 1879, just one year after his death, Long was credited as the father of anesthesia, and remains widely respected for his work in this field. **TP**

SEE ALSO: ANESTHESIA, CHLOROFORM ANESTHETIC, MODERN GENERAL ANESTHETIC, NITROUS OXIDE ANESTHETIC

⭱ *The "Letheon" inhaler of 1847 worked by drawing air from a jar of ether-soaked sponges to the patient.*

Aneroid Barometer (1843)

Vidie registers changes in barometric pressure.

The aneroid barometer is now a small, inexpensive, robust, accurate, lightweight, and portable instrument. It does away with the delicate glass tubes of weighty mercury and easily overturned reservoirs of the original instrument, relying instead on a sealed bellows-type flexible container. This container contracts or expands according to the pressure of the surrounding atmosphere. The pressure is displayed on the dial by a mechanically driven lever or pointer. This dial is usually fitted with a second pointer that can be set manually to indicate the current pressure. Thus the rate of pressure change can be assessed, as well as whether it is increasing or decreasing.

The concept of the aneroid barometer was suggested by Gottfried Wilhelm Leibniz in 1698 but the first working model was made and patented in 1843 by the French scientist and engineer Lucien Vidie (1805–1866). He developed the idea from his work on pressure-measuring manometers on steam-engine boilers. Vidie presented his instrument at the Great Exhibition in London in 1851, when he was awarded a Council Medal. The association between barometric pressure and the weather led to aneroid barometers being purchased by all meteorologists, and they also became part of the equipment of most sailors, explorers, and farmers. At sea the barometer was unaffected by the rocking of the boat.

A barograph is a type of aneroid barometer that records atmospheric pressure over time. It is fitted with inked recording needles and continuously moving recording paper. These usually record the atmospheric pressure for a whole week. **DH**

SEE ALSO: ANEMOMETER, BAROMETER, WEATHER RADAR

⬆ *On this aneroid barometer, high pressure generally indicates dry weather and low pressure, rainy weather.*

Variable Resistor (1843)

Wheatstone's invention regulates the volume of audio equipment.

A variable resistor—or rheostat—is a device that controls the flow of a current, *rheo* being Greek for "to flow." Rheostats are employed to adjust the current in electric machines, and to vary the resistance in electric circuits. Examples of their use are the dimming of lights and the controlling of a motor's speed. However, the way that most people encounter variable resistors is behind the knobs on radios or under the sliders on more complicated audio equipment. Turning the volume knob, you are moving the 'finger' of a variable resistor, changing the tapping point and therefore the resistance, supplying more or less power to the speakers, which makes the sound louder or quieter.

The variable resistor's design is based on the Wheatstone Bridge, which was invented by British mathematician Samuel Christie (1784–1865) but named after his compatriot Charles Wheatstone (1802–1875), the scientist who popularized it. It could measure electrical resistance by using four resistors, a battery, and a galvanometer (an instrument for detecting electric current and measuring it). From this starting point, Wheatstone devised the rheostat, which was able to measure (and control) an unknown resistance when being placed in series with a rheoscope (a device to measure electric current) and a rheomotor (a source of electric current). You had to record the reading of the rheoscope before inserting the rheostat into the circuit as a subsitute for the unknown resistance. The rheostat was then set to give the same current reading on the rheoscope as before.

Rheostats thus allow you to simulate the resistances of long cables without having to use them. **DK**

> *"The numbers [on the amplifier] all go to eleven. . . . Well, it's one louder, isn't it? It's not ten."*

This Is Spinal Tap (1984)

SEE ALSO: ELECTROMECHANICAL RELAY, ELECTRIC CHAIR, BLOW-DRY HAIR DRYER

◩ *Rheostats like this one typically consist of resistive wire wrapped to form a coil.*

Computer Program (1843)

Lovelace devises the first computer program.

Ada Byron (1815–1852) was the daughter of the English poet Lord Byron. Under her mother's guidance, Ada was tutored from an early age in mathematics and science; she later married the Earl of Lovelace.

In 1835 Ada was introduced to Charles Babbage and learned of his ideas for his "analytical engine." In 1842, Babbage was invited to give a seminar at the University of Turin in Italy. Interest in his work spread when an Italian engineer, Federico Luigi Menabrea, published an account of the lecture in a leading French scientific journal. Babbage asked Lovelace to translate Menabrea's work; the notes and comments she made were considerably more extensive than the

> "The Analytical Engine . . . can do whatever we know how to order it to perform."

Ada Lovelace

original paper and were published in their own right. In the last of her seven notes, Lovelace describes an algorithm that would enable the analytical engine to compute Bernoulli numbers. Given that this was designed specifically for use with a computing machine, it could be argued that she had created the first-ever computer program. But, the machine was never built, so the algorithm was never put to the test.

Not all historians credit Lovelace so highly. Some claim that she merely documented Babbage's work or, at best, that the algorithm was a collaboration. That said, Babbage believed that she had greater understanding than anyone of the potential for his work. **TB**

SEE ALSO: COMPUTER-ASSISTED INSTRUCTION, CAD, CAM, DIGITAL ELECTRONIC COMPUTER, MECHANICAL COMPUTER

Typewriter (1843)

Thurber patents a practical writing machine.

Quick and efficient written communication became essential in the mid-nineteenth century as the pace of business activity increased. Many attempts to mechanize writing are recorded in the patents of that period, although few went into production. Pinpointing one single inventor is, therefore, difficult but the "chirographer" or printing machine, patented in 1843 by Charles Thurber (1803–1886), was the first to be produced and sold commercially.

A wide variety of writing machines were designed in the early- to mid-nineteenth century, including Pellegrino Turri's 1808 machine to enable blind people to write, William Austin Burt's "typographer" of 1829, and the Hansen writing ball of 1864, which was probably the first typewriter that made it possible to write faster than by hand. The first commercially successful typewriter was developed by newspaper editor Christopher Scholes and others in Milwaukee, Wisconsin, in 1867. The patent was later taken up by Remingtons, a well-established sewing-machine company, which launched the Scholes and Glidden typewriter in 1873. This enabled operators to see the page as they worked. The keyboard incorporated the inventors' QWERTY layout for the keys, designed to prevent the letters from jamming, which became the universally accepted standard.

Electric typewriters were developed in the early twentieth century, but proved too expensive until International Business Machines (IBM) launched their first electric typewriter in 1935. Manufacturers progressively added further features, including proportional spacing, memory, and "automatic" correction until the typewriter became known as a "word processor," whose functions were taken over by the desktop computer during the 1980s. **EH**

SEE ALSO: CARBON PAPER, STENOTYPE MACHINE, DIGITAL ELECTRONIC COMPUTER, LAPTOP, PERSONAL COMPUTER

Ice-cream Maker (1843)

Johnson minimizes the preparation time of ice-cream.

In 1846 Williams & Company—a small Philadelphia-based wholesaler of kitchen appliances—purchased the patent for a new hand-cranked machine that could speed up the process of making ice-cream. Over the next thirty years there would be more than seventy improvements to the machine they had shrewdly procured for a mere $200; but what Williams & Company could not buy, however, was the credit for having invented the machine. That distinction belongs to a humble Philadelphia housewife named Nancy Johnson who lacked both the money and the business savvy to promote her invention and sold the patent to her machine, which went on to be marketed as "Johnson's Patent Ice-Cream Freezer." Nancy Johnson then proceeded to fade quietly back into the obscurity from whence she had come.

Johnson's innovative design involved placing the ice-cream's ingredients of ice, sugar, vanilla, eggs, and salt in a tin with a removable lid and a scraping device to prevent the mixture from sticking to the sides of the tin. Hand-cranking enabled the ice and salt to mix together and freeze more rapidly than previous methods of hand-stirring with a wooden spoon. Prior to Johnson's invention, making ice-cream was a laborious, labor-intensive process that saw the confection as a treat for the sole enjoyment of the upper classes. Once in the hands of Williams & Company, ice-cream makers were available for just $3 and were inexpensive enough finally to bring ice-cream to the masses. The first ice-cream factory opened in Baltimore in 1851, and smooth, textured ice-cream was developed in the United States in 1938. **BS**

SEE ALSO: POWDERED MILK, REFRIGERATOR

"Age does not diminish the . . . disappointment of having . . . ice-cream fall from the cone."

Jim Fiebig, columnist

◹ *This "Champion" ice-cream maker—patented in 1872—featured a flywheel for hand-cranking.*

Safety Match (1844)

Pasch reduces the risk of match accidents.

The development of the safety match in 1844 by the Swedish chemistry professor Gustaf Erik Pasch (1788–1862) followed the invention of the friction match. Pasch replaced the dangerous white phosphorus in the flammable mixture coating the match head with nontoxic red phosphorus, which was far less flammable. He also removed the phosphorus from the mixture at the head of the match and added it to a specially prepared striking surface. The striking surface was made from red phosphorus and powdered glass, leaving a composition of antimony(III) sulfide and potassium chlorate on the match head. Some of the red phosphorus was converted to white by friction heat as the match was struck. The small amount of white phosphorus then ignites, starting the combustion of the match. These matches were considered very safe, as they would ignite only when struck against the striking surface.

After obtaining a patent for the new safety match, Pasch manufactured them in a factory in Stockholm, but was eventually deterred by high costs. Another Swede, John Edvard Lundstrom, improved Pasch's safety match by placing the red phosphorus on sandpaper on the outer edge of the box, and in 1855 he obtained a patent for his new safety match. This model was manufactured on a large scale and Lundstrom, along with his bother Carl Frans, created a virtual global monopoly on safety matches. Although these matches were much safer than those used previously, they still contained poisonous material. The Diamond Match Company was the first to patent a nonpoisonous match in the United States in 1910. **CA**

SEE ALSO: CONTROLLED FIRE, FRICTION MATCH

⬈ From 1845, safety matches like these ones were ignited by striking a strip of red phosphorus.

"How is it that one careless match can start a forest fire, but it takes a whole box to start a campfire?"

Author unknown

Morse Code (1844)

Morse and Vail communicate electrically.

Named after Samuel Morse (1791–1872)—Morse code is a simple code in which letters are represented by dits and dahs. The code, which is a milestone in long-range communication, was designed so that telegraph operators could communicate via a series of electrical signals. In 1844, several years after Morse and his partner Alfred Vail had created the code, Morse demonstrated to the American Congress the power of the telegraph by using the code. To their awe, he sent the message "What hath God wrought" from Washington to Baltimore at a speed that no horse or car could hope to match.

Congress was skeptical of Morse for several years before they agreed to build the Washington-Baltimore line (at the enormous cost of $30,000). Morse code took off and was soon used internationally by militaries, railroads, and businesses. The first form of radio communication was "wireless telegraphy," which involved transmitting Morse code via radio.

Morse code assigns a single unit of time to a dit and three units of time to a dah. Between each letter are pauses of three time units and pauses of seven time units between words. The precise unit of time varies with the operator, as skilled Morse-code operators could regularly transmit and receive twenty to thirty words per minute.

Morse code has now been largely superseded but it is still used occasionally, especially among ships. The famous SOS signal (three dits, pause, three dahs, pause, three dits) is used as a distress signal. Amateur radio enthusiasts from all over the world communicate with the reliable dits and dahs of Morse code. **LW**

SEE ALSO: **COHERER, ELECTROMAGNETIC TELEGRAPH, SEMAPHORE, ARC TRANSMITTER**

← *The Communications Room onboard the ill-fated Titanic, which sent out the SOS signal before it sank.*

Portland Cement (1845)

Aspdin improves the constitution of cement.

Joseph Aspdin (1778–1855) invented the first Portland cement, named after the gray Portland stone it resembled, in 1824. However, his cement set too quickly and had poor early strength. Joseph's son William then noticed that when the constituent clay and limestone are burned together to combine the minerals, increased burning temperatures (exceeding 2,282°F/1,250°C) and greater proportions of limestone resulted in much stronger cements suitable for concrete, and hence construction.

The difference in strength was largely due to the calcium silicates present. The strength of Joseph's cement was derived from its belite (dicalcium silicate) content, which could take weeks to develop. For early

> *"Conditional co-operation is like adulterated cement which does not bind . . ."*
>
> Mahatma Gandhi

strength, cement has to contain alite (tricalcium silicate), which is precisely what William had produced by using higher burning temperatures. Despite having a different composition than his father's invention, William neither patented his idea nor changed its name. By not applying for a patent, claiming it was covered by his father's, he was able to keep his invention a secret from his competitors.

Due to the large amounts of carbon dioxide released during cement production, there are calls to find alternatives. However, our love affair with cement is unlikely to be extinguished any time soon as the alternatives are expensive and not as versatile. **RP**

SEE ALSO: **FIRED BRICKS, PLASTER, REINFORCED CONCRETE**

Electric Fire Alarm (1845)

Channing speeds response to fire warnings.

Fire alarms were originally raised by ringing church bells. Due to the nature of sound, bells had the huge disadvantage of being affected by environmental conditions, making finding the fire extremely difficult.

In 1845 American William Channing proposed using Samuel Morse's telegraph system to raise the alarm and coordinate a response. His system comprised signal boxes that would send automated messages of their location (and hence the fire) to a central office. The fire-alarm signal-box system is still used in the United States and is akin to the manual "break-glass" fire call point more commonly used in Europe. In the central office an operator would forward the message to all other signal boxes within the circuit. At the same time electrical impulses would be sent to automatic bell strikers to sound the alarm and alert the firefighters, who at that time were mostly volunteers. By going to their nearest signal station, the firefighters could communicate with the central office and act appropriately.

Channing collaborated with Moses Farmer, a pioneer of electrical engineering, to bring his vision to life. In 1852, after convincing the mayor of Boston to trial their invention, the world's first system was installed. After initial hiccups, their system was eventually put into service in 1854, and a patent was issued in 1857, by which time their system had been installed in more than forty towns and cities.

Although improvements were made, the system remained largely unchanged until the advent of digital systems in the late twentieth century, but the principles of their design remain valid today. **RP**

SEE ALSO: BELL, AUTOMATIC FIRE SPRINKLER, BURGLAR ALARM, FIRE EXTINGUISHER, MORSE CODE, SMOKE DETECTOR

◁ *The fire alarm-telegraph system enabled the public to summon the fire department quickly and efficiently.*

Rubber Band (1845)

Perry finds a clever new use for rubber.

Even though the Mesoamericans of Central America used rubber—made from the sap of the indiginous rubber tree—as early as 1600 BCE, it was not until the first half of the nineteenth century that American Charles Goodyear developed a chemical process to make commercially viable rubber. By adding sulfur to naturally milky latex and then heating the substance, he "vulcanized" the rubber, thereby making it harder and more durable. Six years later, in early 1845, Englishman Stephen Perry of Messers Perry and Co. used this process to produce the world's first vulcanized rubber band.

Previously, fellow Englishman Thomas Hancock had produced rubber bands by cutting Central

> *"Jules Verne's story of travel to the moon would be as much science fiction if they went by rubber band."*
>
> Philip K. Dick, science fiction writer

American rubber bottles into sheets and slicing these into thin strips. In order to dispose of the resulting waste he invented a machine—the masticator—to shred it. However, the end result after the waste had gone through the machine was a homogeneous mass of rubber. Hence the masticator can be considered a forerunner of the modern rubber milling machine used to make rubber bands.

Unlike most rubber products, rubber bands are still made from natural rather than synthetic latex due to the former's superior elasticity. Rubber bands are used throughout all societies and industries, the U.S. Post Office being the biggest consumer worldwide. **CB**

SEE ALSO: RUBBER BALL, CATAPULT, RUBBER (WELLINGTON) BOOTS, VULCANIZATION

Pneumatic Tire (1845)

Thomson's cushion of air guarantees a smooth ride.

Scottish-born engineer Robert William Thomson (c. 1822–1873) left school at age fourteen but within a couple of years he had managed to teach himself astronomy, chemistry, and the physics of electricity. By the time he was seventeen, he had his own workshop. Then at the tender age of twenty-three, he patented the "aerial wheel"—now known as the pneumatic tire. The tires consisted of a hollow belt of India rubber that could be inflated with "a cushion of air to the ground, rail or track on which they run." The idea was to provide people traveling over bumpy ground with a smoother ride.

Unfortunately for Thomson, in 1845 there were no cars to take advantage of the new tire; nor were there any bicycles. The only applications of his innovation were a few steam-powered carriages together with the traditional horsedrawn carriages. Thomson, however, arranged for some journalists to watch him demonstrate the advantages of his cushioned wheel at Regent's Park in London. He had two carriages, one with old wheels and one fitted with pneumatic tires. Many of the observers had presumed that the pneumatic tires would make it harder to pull a carriage because the tires were soft. However, Thomson proved them wrong and he also showed that his tires produced a tiny fraction of the noise made by the traditional solid wheels. A set of his tires ran for 1,200 miles (1,931 km) without noticeable deterioration.

Sadly, Thomson's invention also coincided with a shortage in the thin rubber necessary for the inner tires. In time, the frustrated young inventor, who never became rich from his idea, turned his attention to the then more popular solid rubber tires. It was another fifty years before John Boyd Dunlop rediscovered the idea and built a global brand from it. **DHk**

> *"[The wheels are like] a cushion of air to the ground, rail, or track on which they run."*

Robert William Thompson

⬆ *Robert Thomson's original design for his "aerial wheel"—or pneumatic tire.*

➡ *Dunlop produced pneumatic tires from 1890, just in time to coincide with the arrival of the automobile.*

SEE ALSO: WHEEL AND AXLE, MOTORCAR, RADIAL TIRE, TIRE VALVE, VULCANIZATION

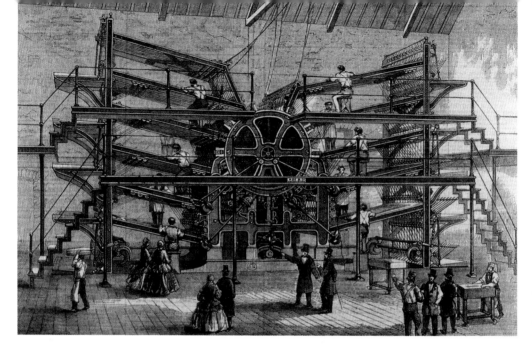

Rotary Printing Press (1846)

Hoe accelerates the printing process.

The rotary printing press developed by Richard March Hoe (1812–1886) was the key invention that led to the development of the mass media.

Hoe's father ran a factory for the production of printing presses in New York City, and Richard worked in the family business from a young age. Along with his father and brothers-in-law, he made various improvements to the traditional flat-bed press, which was based on the very earliest fifteenth-century presses from Europe.

Hoe's press was revolutionary because of the speed at which documents could be printed. Unlike the flat-bed press, which had to be reset for each new sheet of paper, the rotary or revolving printing press passed paper continuously through, using several cylinders to apply the type. As long as there was someone there to feed the paper, Hoe's original "lightning press" could produce up to 8,000 sheets per hour with a good quality of output.

The invention was timely. The newspaper industry in the United States was booming, and the new press allowed for enormous daily editions to be printed and circulated. The *Philadelphia Public Ledger* was the first newspaper to use the new press in 1847. Hoe continued to develop the rotary press until 1871 when he unveiled the "web perfecting" press, which printed both sides of the paper simultaneously using a continuous roll of paper, and was first used by the *New York Tribune*. This, combined with his innovations in cutting and folding the paper, established the modern printing press, which lasted until digital printing took over in the late twentieth century. **JG**

SEE ALSO: INK, PRINTING PRESS, WOODBLOCK PRINTING, COMPUTER LASER PRINTER, PRINTING ON DEMAND

⬆ *Visitors to the* Daily Telegraph *newspaper in London witness a Hoe rotary printing press in action in 1860.*

Hydraulic Crane (1846)

Armstrong lifts weights using water power.

Sir William George Armstrong (1810–1900), the first and last Baron Armstrong of Cragside, was an English industrialist who pioneered the use of hydraulic power to operate a wide variety of machinery, harnessing the power of water to feed the Industrial Age.

One of his first inventions using the resource of water was an improved rotary water motor, and soon after this innovation he designed a piston engine driven by water. He realized that his invention had the potential to be incorporated into a more efficient design of crane than those then in operation.

The first of Armstrong's hydraulic cranes was built on Newcastle docks in 1846 and was tremendously successful. It utilized the pressure from the town's mains water supply, acting on a piston inside a cylinder, this in turn moving gears that drove the crane. The design was so successful that the Newcastle Corporation ordered three more cranes soon afterward, and the success of the crane was secured.

Armstrong's original design, however, proved somewhat limiting as it relied on high water pressure, which had to be supplied either by mains or by the use of high water towers. To remedy this limitation, for situations where neither of these options was feasible, Armstrong designed the hydraulic accumulator, a device that used a weighted plunger on a large cylinder to generate the required pressure.

Armstrong's pioneering work in the field of hydraulics led to a greatly increased use and understanding of water as a power source. Cranes and other heavy machines utilizing hydraulic components continue to be used around the world to this day. **SB**

SEE ALSO: CRANE, HYDRAULIC BRAKE, HYDRAULIC JACK, HYDRAULIC PRESS

⬆ *Built in 1883, this fine example of Armstrong's crane engineering can be seen at the Arsenale in Venice, Italy.*

Printing Telegraph (Teleprinter) (1846)

House speeds written communication.

The electric type-printing telegraph of Royal Earl House (1814–1895) looked like the offspring of a record player and a piano. Several intricate devices sat on the telegraph's wooden base above a keyboard whose keys were labeled with the letter to which they corresponded. Despite its looks, the machine was not musical. It did, however, rely on the steady beat of its underlying clockwork-generated electricity to produce a message printed on a strip of paper.

House eventually shared his idea with Jacob Brett, a British electrical engineer, who built a working model. When a key on the machine's keyboard was pressed, an electric circuit would be temporarily broken at one of twenty-eight corresponding pins on an underlying rotating cylinder. The breaking of the circuit would stop the cylinder's motion, consequently stopping a synchronized electromagnet controlling the type-wheel. With the type-wheel halted on the proper letter, the connected apparatus then pressed a strip of paper to the type-wheel, depositing graphite ink onto the paper.

The European and American Electric Type-printing Telegraph Company and the Submarine Telegraph Company (S.T.C.) began using the device to send messages between France and England. The S.T.C. was particularly confident, proclaiming in 1850 that the device could send 100 messages, of fifteen words each, in 100 minutes.

House's electric type-printing telegraph was faster than the old needle telegrapher and was intricate and expensive by comparison. Not only was it a stylish commodity, but the convenience of its printed messages and speed earned its place as the new telegraphing standard. **LW**

SEE ALSO: DUPLEX TELEGRAPH, ELECTROMAGNETIC
TELEGRAPH, TELEPHONE

Chloroform Anesthetic (1847)

Simpson eases the pain of childbirth.

James Simpson (1811–1870), Professor of Obstetrics at the University of Edinburgh, Scotland, was the first doctor to use chloroform in childbirth. He had become dissatisfied with the use of ether and set about finding an alternative, using himself and friends as guinea pigs. On November 4, 1847, the group tried chloroform—tradition has it that Simpson's wife, on bringing in dinner, found them all asleep under the table. Simpson immediately recognized the advantages of chloroform over ether; it was inflammable, lighter, and easier to administer. Within a week he had administered the chemical to over thirty women in labor.

> *"Doctor Snow gave that blessed chloroform and the effect was soothing, quieting, and delightful."*
>
> Queen Victoria of the United Kingdom

Simpson's innovation brought down the wrath of the Church and the medical establishment. The Bible taught that women should bring forth in pain, and doctors claimed pain was a biological necessity. With such opposition, chloroform did not become widely established until 1853, when Queen Victoria used it for the delivery of her last child. The use of chloroform as an anesthetic was finally abandoned in the early twentieth century when it was shown to be responsible for a number of fatal heart attacks. **JF**

SEE ALSO: ANESTHESIA, ETHER ANESTHETIC, MODERN
GENERAL ANESTHETIC, NITROUS OXIDE ANESTHETIC

→ *Joseph Clover demonstrates his device of 1862, which administered chloroform in controlled doses.*

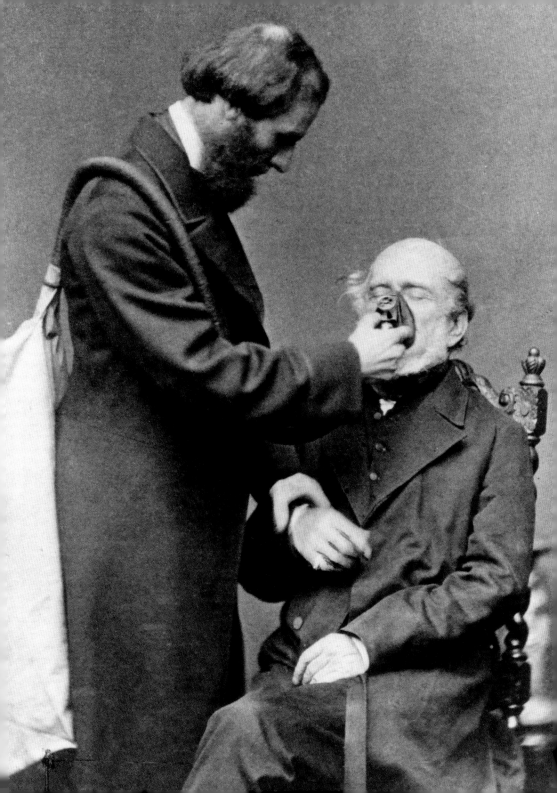

Chewing Gum (1848)

Curtis starts a masticatory fashion among North Americans.

"Chewing gum! A new and superior preparation of Spruce Gum."

Chicago Daily Democrat, October 25, 1850

Chewing gum is widely regarded as an American phenomenon, but the practice of chewing a form of gum actually dates back to prehistory and Europe. Thousands of years later the ancient Greeks were chewing mastiche, a resin from the mastic tree, and the ancient Mayans were chewing chicle, a rubbery sap from the sapodilla tree. Native Americans chewed a gum made from the resin of spruce trees, and it was from this that the American John Curtis invented his chewing gum. In 1848 he sold his "spruce gum" commercially, which started a fashion for the chewy substance. Gradually spruce gum was replaced by gum made from paraffin wax, which was then sweetened and sold by Curtis around 1848.

Modern chewing gum came into being rather by accident. The Mexican general Antonio Lopez de Santa Anna was searching for an alternative to rubber for manufacturing purposes and contacted the American inventor Thomas Adams to see if he could use chicle as a substitute. Adams experimented with the chicle to no avail, but subsequently he turned his attention to creating a sweetened and flavored chewing gum. Around the same time, William J. White experimented with adding corn syrup, sugar, and peppermint to gum, creating one of the most popular gum flavors. In 1891 William Wrigley Jr. founded his chewing gum enterprise, and today Wrigley's is the largest chewing-gum manufacturer in the United States.

Although chicle and other natural products are still sometimes used in the chewing-gum industry, man-made gum bases are more commonly used to keep up with the enormous demand. **TP**

SEE ALSO: RUBBER BALL, SACCHARIN

⬉ *Wrigley's popularized the use of chewing gum with brands such as classic spearmint, launched in 1893.*

Dry Cleaning (1849)

Jolly discovers cleaning with solvents.

Like so many inventions, the idea of dry cleaning was hit upon by accident. One morning, French dye-works owner Jean-Baptiste Jolly noticed a clean spot on a dirty tablecloth where his maid had knocked over a kerosene lamp. Since materials such as silk and wool blends shrink and lose their color when washed with water and soap, Jolly knew exactly how to capitalize on his discovery, and soon he had developed a way of cleaning clothes without using water. Jolly called the procedure "dry cleaning"—though this is something of a misnomer since the process does involve immersing dirty clothes in a liquid solvent.

Jolly started to clean delicate clothes with kerosene and gasoline. These early solvents were highly

---

"A hearty laugh gives one a dry cleaning, while a good cry is a wet wash."

Puzant Kevork Thomajan

---

flammable and the risk of fire drove the dry-cleaning industry to look for alternatives. Synthetic hydrocarbon solvents soon took over, although their health risks are arguably greater. When customers receive back their dry-cleaned goods, almost all of the solvent has been removed so risks are minimal, but some do complain of irritation to the eyes and nose. For dry-cleaning workers, the constant contact with the toxic solvents could pose much more serious health risks, including cancer and other effects on the nervous system and organs. Recently, in the United States, where dry cleaning is particularly popular, new laws will phase out dangerous chemicals completely. **LS**

SEE ALSO: CLOTHES, SOAP, WOVEN CLOTH, WASHING
MACHINE, SYNTHETIC DETERGENT

Francis Turbine (1849)

Francis improves the water-driven turbine.

The use of flowing water as a source of energy has been exploited for hundreds of years, with traditional water mills powering early industrial processes. Turbines are the modern equivalent of a water mill but run at much higher energy efficiency due to the efforts of engineer James Francis (1815–1892), who invented the Francis turbine, and whose calculations enabled modern, super-efficient turbines to be built, allowing clean, renewable energy production.

The first large-scale turbine for power generation was built by Benoît Fourneyron in 1827. His invention was an out-flow design, but it took another ten years for him to stabilize the turbine, at which point he managed 80 percent efficiency—that is, the turbine could usefully extract 80 percent of the kinetic energy theoretically contained within the running water.

Francis, an English engineer, moved to the United States to work for the proprietors of Locks and Canals Company in Lowell, Massachusetts. He began to study the hydraulics of turbines in detail and conducted many experiments into its properties. This extensive analysis led him to design a far more efficient inward-flow, reaction turbine than had been created previously—his invention reached upward of 90 percent efficiency. The turbine worked on the principle that, as the water passes through the turbine, it changes from high to low pressure, and the energy it releases in doing so is transferred to, or mopped up by, the blades of the turbine.

Francis gave his name to the new turbines and led the company for forty years. The new turbines helped transform Lowell into an industrial town. Today, Francis turbines are commonly employed as energy generators in hydroelectric dams and pumped-storage plants across the world. **SR**

SEE ALSO: CANAL, FREE-JET WATER TURBINE, GAS TURBINE,
WATER MILL, STEAM TURBINE, WIND-TURBINE GENERATOR

Safety Pin (1849)

Hunt devises a perfect temporary fastener.

Necessity is the mother of invention according to Plato, and this was certainly true for Walter Hunt (1796–1859) and his most famous invention—the humble safety pin. This useful object is found in households across the globe; it even gained status as a fashion accessory, with the Punk movement of the 1970s.

Walter Hunt was a New York mechanic who, in 1849, sat wondering how he could pay off a small debt. He spent around three hours twisting a length of wire in his fingers before he created the answer to his problems, the ubiquitous safety pin. Pins were by no means a new idea, having existed for centuries before Walter's twist on the design. However, his creation was unique as it provided a solution to the

---

"A man who could invent a safety pin ... was truly a mechanical genius ..."

New York Times

---

potential problem of pricking oneself with the old-style variety. His pin consisted of a length of wire coiled into a spring at the center, with a sharp point at one end and a safety clasp at the other.

Hunt's design was patented in April 1849, and he sold the rights to his creditor, clearing a $385 profit. Unfortunately Hunt had no idea how popular his invention was set to become. He also designed America's first sewing machine (with an eye-pointed needle) but, fearing the loss of jobs his creation may cause, he did not patent the idea. It was left to a fellow American, Elias Howe, to claim the credit for this invention some twenty years later. **JG**

SEE ALSO: BUTTON, BUCKLE, ZIP FASTENER

Oil Refinery (1850)

Young extracts crude oil from coal.

In 1848 Scottish chemist James Young (1811–1883) spotted the potential of a natural oil seepage at a Derbyshire colliery. By 1850 he had taken out a patent for a process of extracting crude oil from cannel coal. Young located a huge new source of coal, at Boghead Colliery in Bathgate, West Lothian, and in 1851 he built the world's first commercial oil refinery on the site.

Young began a major industry that was to continue in full production for another fifty years, until the arrival of crude oil from the United States and the Middle East. Young's Paraffin Light and Mineral Oil Company sold paraffin oil and lamps and also produced naphtha, gas, coke, and ammonium sulfate.

"Paraffin" Young, as he had become known, took

". . . with one distillation it gives a clear colorless liquid of brilliant illuminating power."

Lyon Playfair in a letter to James Young

---

out a U.S. patent for "parafinne oil" in 1852, which was to be a major blow to the business ambitions of the Canadian geologist Abraham Gesner, who had also learned to distill what he called kerosene. He and his associates began to market kerosene—made from albertite and cannel coal—in the United States in 1857, before having to admit defeat to both Young and the rapidly growing petroleum industry in Pennsylvania.

In the Austrian Empire, in the Carpathian region, Polish chemist Ignacy Lukasiewicz built an oil refinery in 1856, initially producing paraffin for local use, distilled from natural seep, or rock, oil. He would grow very wealthy on its and its successors' output. **AE-D**

SEE ALSO: THERMAL CRACKING OF OIL, GAS TURBINE, OIL LAMP, SULFUR LAMP

Trawler Fishing (1850)

English fishermen devise a catch-all method.

Trawling is a type of fishing in which one or more boats (trawlers) pull a fishing net through the water behind them. This can involve dragging the net along the sea floor (known as bottom trawling) or pulling it along higher in the water (pelagic trawling).

It is impossible to say when this method was first used in its simplest form, but there is evidence of concerns about its environmental impact as early as the fourteenth century, when fishermen protested against bottom trawling because of the indiscriminate way in which it caught all types and sizes of fish.

It was not until the late eighteenth century that modern beam trawling, a method of bottom trawling where the mouth of the net is held open by a solid metal beam, was widely used. The fishermen of both Barking and Brixham in England claimed credit for pioneering this technique, and it led to a much greater uptake of deep-sea trawling.

By the 1840s, trawling had become the principal method of fishing on a large scale. The arrival of the steam trawler in the 1880s gave a further boost to the trawling industry, giving fishermen the range to exploit more distant fishing grounds.

Environmental concerns have grown over the centuries, particularly to bottom trawling, because of its damage to the seabed and the fact that the nets catch the small fish needed to replenish stocks, and which are often discarded as uneconomic. Even the word "trawl" has itself come to mean a form of fast but indiscriminate collecting. Trawler fishing is now heavily regulated in most countries, but its impact remains the subject of protest and debate. **RBd**

SEE ALSO: DUGOUT CANOE, ROWING BOAT, RUDDER, SAILS, STEAMBOAT

↗ *This drawing illustrates the beam trawling process. The beam keeps the fishing net above the sea bed.*

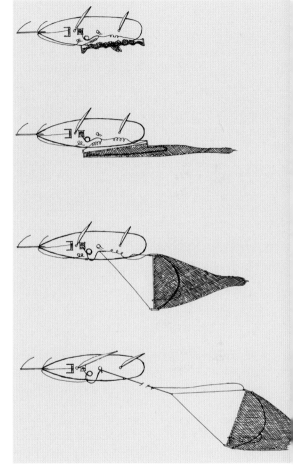

"Thou shalt have a fish . . . Thou shalt have a haddock when the boat comes in."

"Dance to Your Daddy," Northumbrian folksong

Hydraulic Jack (1851)

Dudgeon exploits water for heavy lifting.

Hydraulic means moved or operated by water or liquid and refers to a method of engineering that has been used since records began. The ancient Chinese and Egyptians used water as a method of transport and the Romans relied heavily on it to move weighty objects. The arrival of the hydraulic jack in 1851, therefore, was not especially novel, but it was extremely useful.

Essentially, a jack is a device that is capable of lifting heavy objects with relative ease. Hydraulic jacks make use of Pascal's Law that, in simple terms, says that if there is an increase in pressure at any point in a container of liquid, there is an equal increase in pressure at every other point within the container.

Richard Dudgeon, Inc., was founded as a machine shop in New York in the mid-1800s, and its owner, founder, and namesake Richard Dudgeon was given a patent in 1851 for his hydraulic jack—or, as he called it, a "portable hydraulic press." Dudgeon also takes credit for several other mechanical inventions, including roller boiler tube expanders, filter press jacks, pulling jacks, heavy plate hydraulic hole punches, and various other kinds of lifting jacks.

Dudgeon's hydraulic jack was vastly superior to the other jacks available at the time because of its seemingly infinite power. Compared with conventional screw jacks—which rely on a large hand-turned lead screw using compressive force to lift heavy weights— they are far easier to operate and offer a smoother and more powerful lifting mechanism. The hydraulic jack was capable of easily lifting cars and other heavy machinery, which made it incredibly useful—and the device remains widely in use today. **KB**

SEE ALSO: CRANE, HYDRAULIC CRANE, HYDRAULIC PRESS

⬆ *A series of hydraulic jacks and other equipment were used to move an Egyptian obelisk to New York in 1881.*

Airship (1852)

Henri Giffard invents the first navigable full-sized airship.

More than fifty years before the Wright brothers flew almost an entire minute in the world's first airplane, French engineer Henri Giffard (1825–1882) traveled 17 miles (27 kilometers) from Paris to Trappes in a lighter-than-air aircraft. Inspired by the streamlined model airship unveiled by his compatriot Pierre Jullien in 1850, Giffard built his 143 feet (44 meter) long, cigar-shaped dirigible and got it off the ground two years later. With its three-bladed propeller driven by a 3-horsepower (2.2 kilowatt) steam engine, it was the first passenger-carrying, powered, and steerable airship in history.

The world's most famous airship, the twentieth century zeppelin, was rigid with a shape determined by a skeletal structure. Giffard's design was non-rigid. Like a balloon, the envelope's shape depended on the pressure of the hydrogen inside that lifted the airship.

The dirigible's maiden flight took place on September 24, 1852, when Giffard—sitting in a gondola hanging from a net surrounding the balloon—used a rudimentary, sail-like rudder to steer the airship in the intended direction. He accelerated to a speed of 6 miles (10km) per hour. But unfortunately the wind was too strong to allow for an immediate return journey, as the combined weight of engine and boiler (350 pound/158 kg) only permitted travel in calm weather conditions.

In 1884, Giffard's fellow Frenchmen, Charles Renard and Arthur Krebs, made the first full round-trip flight in their airship *La France*. Airship travel reached its peak in the first half of the twentieth century until the disastrous last flight of the *Hindenburg* on May 3, 1937 shattered public confidence in airship travel. **DaH**

SEE ALSO: **HOT-AIR BALLOON, GLIDER, PARACHUTE, POWERED AIRPLANE, HELICOPTER**

⬆ *This dirigible was used in the first successful application of mechanical power to flight on September 24, 1852.*

Passenger Elevator
(1852)

Otis designs an elevator with a safety brake.

Who would now dare to travel in a high-rise elevator if their life literally hung by the elevator cable?

In 1852 Maize and Burns, a bedstead-making firm in Yonkers, New York, was faced with that problem—how to hoist its bedsteads up to the top floor of its premises without risk of them crashing down if the rope broke. Elisha Otis (1811–1861), the firm's master mechanic, had experience of designing safety brakes for railway wagons. He devised a system involving a platform that could move freely within an elevator shaft unless there was a cable failure, whereupon a tough steel-wagon spring would mesh with a ratchet in such a way as to catch and hold the heavy platform.

> *"If you die in an elevator, be sure to push the Up button."*
>
> Sam Levenson, American author and humorist

Otis left the company to market his invention. Interest was slow at first, but in 1854 he arranged a public display at New York's Crystal Palace. He had an elevator shaft built, open on one side for viewing. He was then hoisted to the height of a house before his assistant took an axe and cut the cable holding the elevator in place. The ratchets engaged, and the elevator and Otis, to great relief, remained suspended.

In 1903 the electric elevator ushered in the era of the skyscraper. Otis elevators remain essential to tall buildings, with 1.7 million in operation worldwide. **AE-D**

SEE ALSO: GRAIN ELEVATOR, STEEL, SKYSCRAPER

⬅ *Two intrepid women enter an Otis hydraulic lift in 1877, which an operator descends by hauling on a rope.*

Rubber (Wellington) Boots (1852)

Hutchinson produces waterproof boots.

The name for the ubiquitous rubber boot derives from the British Duke of Wellington, who had his bootmaker design a calfskin version of the hitherto popular Hessian boot. The "Wellington," as it became known, was both comfortable and hard-wearing.

However, the earliest recognizable "welly boots" were actually produced in France by a company set up by the American entrepreneur, Hiram Hutchinson (1808–1869). He had bought a patent from Charles Goodyear, who had invented the process of vulcanization in 1843; this enabled rubber to retain its elastic and waterproofing qualities even in extreme temperatures. Hutchinson had identified a huge

> *"Oh, wellies, they are wonderful, wellies they are swell ... they keep out the water, and keep in the smell."*
>
> Billy Connolly, British comedian

market in France where most of the population worked on the land with only wooden clogs for protection from the wet and the mud.

Henry Lee Norris, another U.S. entrepreneur, had also bought a patent from Goodyear and set up the North British Rubber Company in Edinburgh in 1856, bringing his skilled workers with him from the United States. This would later evolve into the current Hunter Boot company but originally produced mainly rubber goods other than boots. It took the appallingly muddy conditions of the World War I trenches to change that, when nearly two million pairs of boots were purchased by the British Army. **AE-D**

SEE ALSO: RUBBER BALL, SHOE, PNEUMATIC TIRE, RUBBER BANDS, VULCANIZATION

Hypodermic Syringe (1853)

Pravaz and Wood simultaneously develop a new means of administering medication.

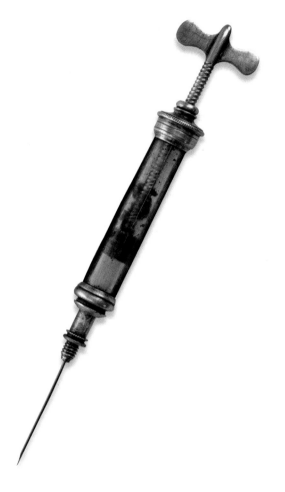

In 1853 the first practical hypodermic syringe, capable of penetrating the skin without the need for a prior incision, was developed simultaneously by the French surgeon Charles Gabriel Pravaz (1791–1853) working in Lyon, France, and Scottish physician Alexander Wood (1817–1884).

Pravaz's silver syringe included a piston with a screw adjustment to measure the administration of precise doses of blood-coagulating agents in treating aneurysms. Wood used a glass syringe that allowed him to monitor visually the injection of morphine in his treatment of patients with neuralgic disorders. Wood later added a graduated scale for more precise measurements. The syringe permitted for the first time the intravenous administration of anesthesia and helped eliminate many of the difficulties faced in the still experimental realm of blood transfusions. Neither versions, of course, would have been possible without the hollow-point needle, which was invented nine years earlier, in 1844, by Irish physician Francis Rynd.

Prior to the seventeenth century, urethral syringes made from pewter, bone, and silver were common, and by the mid-seventeenth century attempts were being made to deliver medication by intravenous means through animal skins. Sir Christopher Wren participated in experiments in which animals were injected via a hollow tube cut from a quill pen. In the early 1800s blisters were generated so the skin could then be peeled back and the drug administered. Post-1853 refinements included detachable needles and all-glass syringes, which greatly reduced the incidence of infections. **BS**

SEE ALSO: ANESTHESIA, ETHER ANESTHESIA, QUILL PEN, ANTISEPTIC SURGERY, TELESURGERY

"Sherlock Holmes took his ... hypodermic syringe from its neat morocco case."

Sir Arthur Conan Doyle, *The Sign of The Four* (1890)

◩ *Hypodermic syringes, like this glass and silver Pravaz-type, were used to inject fluids under a patient's skin.*

Burglar Alarm (1853)

Augustus Pope patents a thief deterrent.

On March 8, 1890, the *New York Times* published an article about two rival security companies. One, the Holmes Electric Protective Company, had been selling alarms for more than thirty years. The other had been doing so for just eighteen months. Now the presidents of the two companies were at war.

The president of the Electric Protective Company was Edwin Holmes, who regarded himself as having pioneered the electric burglar alarm. He had enjoyed a monopoly over the home security market for many years and viewed his opponents at the Metropolitan Burglar-Alarm Company as thieves who had stolen his invention. In fact, he had bought the patent for the burglar alarm from Augustus R. Pope in 1857. Pope's

> *"Absolute protection from the rascals with the dark lanterns and jimmies."*
>
> *New York Times*, March 8, 1890

original patent document of 1853 shows a simple system of magnetic contacts and switches attached to doors and windows. These were linked to a bell in the homeowner's bedroom that would ring if any of the switches were closed. Holmes took Pope's basic design, improved upon it, and then marketed it to New York's elite. His customers included banks, jewelers, and many expensive stores.

W. R. Alling's Metropolitan Company entered the marketplace in 1880, after an earlier skirmish with Holmes over patent infringement. The going rate for monthly security services swiftly plummeted from between $15 and $30 to less than $10. **HB**

SEE ALSO: ELECTRIC FIRE ALARM, CAR ALARM, YALE LOCK

Prism Binoculars (1854)

Porro makes distant things appear closer.

The first binoculars were really "opera glasses," and these small instruments consisted of two small Galilean telescopes side by side. The combination of a convex objective lens and a concave eyepiece produced a rather limited field of view and a magnification of about three.

By the 1790s the Venetian optician Lorenzo Selva had introduced a central adjustable hinge, enabling the binocular eyepieces to be moved apart or close together. The center-wheel focusing mechanism was introduced about 1830. A Keplerian telescope system of two convex lenses was used for astronomical observations but this had a huge disadvantage for terrestrial use as the image was inverted.

An Italian artillery officer, Paolo Ignazio Pietro Porro (1801–1875), overcame the problem of inversion by placing two prisms in a Z-shaped configuration between each objective lens and eyepiece. This widened the binoculars and separated the objective lenses. It also improved the user's stereoscopic vision and gave a better sensation of depth. The fact that the two optical paths were folded in on themselves also meant that the length of the instrument was less than the focal length of the objective, and this too made the binoculars more convenient. Porro patented his design in 1854, and it was subsequently refined by makers such as Carl Zeiss. The desire to avoid patent infringement eventually led manufacturers to introduce stereo prisms and roof prisms.

Widespread use of binoculars by the military in the two world wars pushed development of the technology further, with the introduction of antireflection lens coatings, wide-angle eyepieces, and slimline, lightweight versions with nitrogen-filled, rubber-coated, waterproof bodies. **DH**

SEE ALSO: GLASS, LENS, TELESCOPE

Bessemer Process (1855)

Bessemer provides cheap, plentiful steel for the Industrial Revolution.

English engineer Sir Henry Bessemer (1813–1898) was an inventor for all of his adult life. At seventeen he devised a counterfeit-proof embossed stamp for title deeds. He held more than 100 patents, including making lead for pencils, brass powder for use as "gold" paint, a hydraulic machine for extracting juice from sugar cane, and, most famously, the Bessemer process.

The Bessemer process was basically a considerably cheaper, faster, and more efficient way of making steel than the method then in use. Before Bessemer devised his process, steel was made by adding carbon to wrought iron. This process could take up to a week of continuous heat to produce and required immense amounts of fuel. The cost of steel for structural use in bridges or buildings, or on any mass-production scale, was therefore prohibitive.

Bessemer experimented with pig iron—a product of smelting iron ore that has a carbon content higher than that of steel—to see if he could remove the impurities and make steel. He finally realized that by blowing air through molten pig iron many of the impurities were removed by oxidation, leaving behind steel. He patented his idea in 1855 and it became known as the Bessemer process. Performed in large containers called Bessemer converters, the process made it possible to produce 30 tons of steel in about twenty minutes instead of a week.

The Bessemer process had a dramatic effect on the price and amount of steel, which had many applications in the buildings, weaponry, and machines of the Industrial Revolution. A similar process was discovered independently by U.S. metallurgist William Kelly. **JB**

SEE ALSO: WROUGHT IRON, STEEL, STAINLESS STEEL, ELECTRIC ARC FURNACE

> *"I had no fixed ideas ... and did not suffer from the general belief that whatever is, is right."*
>
> Sir Henry Bessemer

◹ *Bessemer's converter was created to make a strong cannon to fire new artillery shells in the Crimean War.*

Bunsen Burner (1855)

Bunsen improves a laboratory heat source.

Robert Wilhelm Bunsen (1811–1899) was appointed Professor of Chemistry and Medicine at the Ruprecht Karl University of Heidelberg in 1852. Before accepting the position, he negotiated the construction of a new laboratory building equipped with pipes for coal gas, which the city had begun to use to light the streets.

Bunsen was not happy with the equipment he had for heating samples in the laboratory. In 1827 Michael Faraday had written about a burner that used coal gas, but the flame produced too much soot as well as more light than heat. Bunsen's idea was to mix the coal gas with air before the flame rather than at the flame. Because the oxygen and gas would be well mixed at the point of combustion, the resultant flame would be hot rather than bright and almost free of soot.

Heidelberg University mechanic Peter Desaga developed the design for the burner based on Bunsen's idea. The burner is a vertical tube with a connection to a source of flammable gas—coal gas then, but these days usually methane, propane, or butane—with adjustable holes to admit air. The more gas used, the higher the flame will be, whereas the wider the air holes are opened, the more oxygen will be combined with the gas and the hotter the flame.

Together with physicist Gustav Kirchhoff, Bunsen used his burner to heat samples so they could study their emission spectra. They thus discovered the elements cesium (in 1860) and rubidium (in 1861).

Bunsen and Desaga's 1855 device is very similar to the burners used today. However, research laboratories rely increasingly on heating mantles or hot plates because they are safer and heat more evenly. **ES**

SEE ALSO: ARGAND LAMP, GAS MANTLE, GAS LIGHTING, SPECTROMETER

↗ *Bunsen burner with an adjustable air valve at the base to provide varying degrees of intensity.*

"The principle of this burner is simply that city gas is allowed to issue under such conditions."

Robert Bunsen

Rotary Clothes Line (1855)

Higgins revolutionizes clothes hanging.

The invention of the unassuming rotary clothes line —also known as the "Hills Hoist"—has a controversial and confusing history. It is named after Lance Hill, an Australian who developed the device in 1946. The Hills Hoist is considered an Australian icon and a symbol of Australian culture. With a winding mechanism that allowed the frame to be raised and lowered, it was extremely useful in the days of the baby boom when cloth diapers were abundant.

Many believe that this was the first rotary clothes line, but the invention was built upon earlier, less efficient, and more expensive models. Prior renditions of the clothes line include the James Hardie Company's 1925 clothes line called the "Drywell." In

---

> *"We must all hang together, or most assuredly we will all hang separately."*

Benjamin Franklin

1914 both a U.S. and an Australian company had come up with different versions of a rotary clothes line. Gilbert Toyne's 1912 rotary clothes line, built in his backyard in Geelong, Australia, is still in use today. A U.S. patent was issued in 1890 for a rotary clothes line developed in the United States. In circa 1870, *Cassells Household Guide* described a "drying machine" developed by a "Mr. Kent" of High Holborn.

However, the earliest version of the rotary clothes line, invented by James R. Higgins, appeared in an 1855 edition of *Scientific American*. Many are surprised at the early date of the invention, especially considering that many still credit Lance Hill with the creation. **RH**

SEE ALSO: CLOTHES, SOAP, WASHING MACHINE, SYNTHETIC DETERGENT, DRY CLEANING

Celluloid (1856)

Parkes fathers the plastics industry.

In the 1850s the search was on for synthetic alternatives to scarce, naturally occurring materials such as ivory. In Birmingham, England, metallurgist and inventor Alexander Parkes (1813–1890) was investigating ways to waterproof woven fabrics by using cellulose nitrate in combination with vegetable oils. This created a dough that dried into an ivorylike substance and that could be shaped under extreme temperatures; Parkes named it "Parkesine."

Parkes entered his discovery in the 1862 Industrial Exhibition in London. He won a bronze medal, but was nevertheless unable to market a product that was a long way from being refined, and was still inherently prone to warping and cracking.

> *". . . hard as horn, but flexible as leather, capable of being cast or stamped, painted, dyed or carved."*

Alexander Parkes praises Parkesine

The following year an English textile manufacturer, Daniel Spill, bought out Parkes, and renamed his invention "Xylonite." Then the U.S. brothers John Wesley Hyatt and Isaiah Hyatt together coined the word "celluloid" in 1872. They had long experimented with cellulose nitrate in their attempts to manufacture billiard balls, and had won a long-running legal dispute with Spill over patent infringement. Although it was the Hyatts who ultimately perfected Parkes's crude Parkesine, simplifying its manufacture and thus creating the first commercial plastic to be sold in the United States, it is Parkes who remains the undisputed inventor of celluloid. **BS**

SEE ALSO: CLOTHES, PHOTOGRAPHIC FILM, WATERPROOF RAINCOAT, PLASTIC BOTTLE (PET)

Mauveine (1856)

Perkin wins royal approval for his purple dye.

In 1856 eighteen-year-old William Henry Perkin (1838–1907) was assisting August Wilhelm von Hofmann at London's Royal College of Chemistry. Hofmann's particular interest lay in making synthetic versions of natural substances. Toiling away in a fume-filled, makeshift laboratory at the top of his London home, the young assistant was trying to synthesize the anti-malarial medicine quinine, which involved him working with coal tar. When a black sludge formed in the bottom of a flask he started to clean it out by adding alcohol and shaking up the mixture. A beautiful bluish color appeared. When tested on silk, Perkin found that it was a rich, purple, fast dye—the first synthetic dye after millennia of natural dyestuffs that often faded, ran, or were costly to produce.

By experimenting with a coal-tar-based substance, Perkin began a new industry that transformed the textile industry and helped advance other fields, from photography to medicine (including chemotherapy). He obtained a patent in 1856 and opened a factory. By 1860 he had found prosperity and fame, aided greatly by Queen Victoria and other royals, who developed a passion for the color. As purple became all the rage, Perkin's "mauveine" became "mauve"—the French word was perfect for this fashionable hue.

Perkin went on to create other synthetic dyes before returning to research in his mid-thirties. He was knighted in 1906. Meanwhile, the synthetic-dye trade gathered momentum everywhere, gradually replacing natural dyestuffs, though increasing environmental awareness in recent times has put natural options back in the spotlight. **AK**

SEE ALSO: CLOTHING, INK, SPINNING WHEEL, FLYING SHUTTLE

A bottle of the original mauve dye prepared in 1856. Perkin first made organic chemicals on a large scale.

"One would think that ... those purple ribbons were synonymous with 'Perkin for Ever.'"

All The Year Round, contemporary weekly journal

Military Camouflage (1857)

The British Army sees the benefit of khaki.

Although hunters and some elite military units had long made use of camouflage, most armies, until the end of the nineteenth century, wore brightly colored, distinctive uniforms to distinguish friend from foe on the battlefield among gunpowder smoke and to attract new recruits.

The British Army was the first to eschew bright uniforms in favor of those that would help to conceal their men. Heavy casualties during the Indian Revolt of 1857 forced them to dye their scarlet tunics a dull brown color, or "khaki." White summer uniforms were dyed using the simple and quintessentially British expedient of dipping them in tea. Khaki uniforms became standard in India in the 1880s and in the rest of the army during the Second Boer War in 1902.

At the beginning of World War I, the French Army wore blue shirts and red pants. This striking ensemble was meant to "dominate the morale" of the enemy as they advanced across the battlefield. The German machine guns were unimpressed. The French Section de Camouflage, established in 1915, was the first to use disruptive patterns to break up the outline of the human body and thus help soldiers to blend in with their environment. Artists were drafted in to hand-paint tents and coverings for artillery, as well as individual uniforms for elite troops. Today, most armies use patterns on their camouflage uniforms or equipment to set themselves apart from other armies.

A computer-generated version of camouflage was invented by the Canadian Army in the 1990s. It is said to be 40 percent more effective than traditional patterns at a distance of 656 feet (200 m). **BO**

SEE ALSO: NIGHT VISION GOGGLES, OPTICAL CAMOUFLAGE, COLOR NIGHT VISION

"Camouflage is a game we like to play, but our secrets are revealed by . . . what we want to conceal."

Russell Lynes, author and editor

↖ *"Why are you painting his uniform? So the police cannot see him!"* from La Baïomette Magazine, *1915.*

Foghorn (1859)

Foulis guides mariners into port using steam.

Before the invention of the foghorn by Scottish-born inventor, civil engineer, and artist Robert Foulis (1796–1866), harbor towns relied on cannons or bells to guide ships into port in foggy conditions. One such harbor town, Saint John in New Brunswick, Canada, welcomed a new civil engineer when Foulis settled there in 1825. It is likely that Foulis witnessed the 1832 installation of a "warning bell" at Partridge Island, just over half a mile (0.8 km) from Saint John.

Partridge Island's new bell, a behemoth at 0.5 tons, helped to bring ships into port, but the need for louder warnings, along with the limitations of bell construction, had mariners crying out for an alternative. Foulis's 1852 offering was a steam-

---

"On the whole coast of America there is not another alarm equal to the one spoken of … "

Captain Winchester, SS *Eastern City*, 1860

---

powered automated foghorn with a deep warning sound. Though louder than the big bells of the day and much smaller, it took years to convince the town of Saint John to trade in their bell for the horn.

Finally, in 1859, Foulis's horn began operating on Partridge Island and soon foghorns were being used worldwide. Foulis, however, was not getting credit for what has been called one of the greatest maritime inventions. He sought to set the record straight and the government of New Brunswick agreed—the foghorn was Foulis's invention. Unfortunately Foulis did not patent the horn and never saw a profit from his invention. **RBk**

SEE ALSO: FRESNEL LENS

Tachistoscope (1859)

Volkmann's machine influences subconscious.

In 1859 German doctor Alfred Wilhelm Volkmann (1800-1877) built the first tachistoscope, a device capable of influencing subliminal thought by flashing pictures for as little as ten milliseconds. Using mechanical shutters similar to those used in cameras, Volkmann exposed people to images of emotive words. The people would not notice them consciously, yet subconsciously they would project the corresponding emotions onto a subsequent image.

However, because of their mechanical nature the shutters can't be controlled with perfect precision, and hence flashtube tachistoscopes were developed. These use mirrors in a large box and a small lamp that goes from complete darkness to extreme brightness

---

"If the grace of God miraculously operates, it probably operates through the subliminal door."

William James, psychologist

---

and back in a few milliseconds. The person looking into the viewing hole gets the impression that every image appears in the same spot, which allows the person conducting the experiment to show not only the "subliminal" and target images, but also an unrelated, meaningless picture in between, which serves to disrupt the link between conscious and subconscious thought even further.

While tachistoscopes were used to help World War II fighter pilots distinguish the contours of enemy aircraft, and subliminal messages are employed in advertising, doubts remain as to whether this technique actually achieves any tangible success. **DaH**

SEE ALSO: XENON FLASH LAMP, POLYGRAPH (LIE DETECTOR)

Rechargeable Storage Battery (1859)

Planté invents the lead-acid ancestor of today's car battery.

Whether it is for a mobile phone or an iPod, the modern way of life relies upon rechargeable batteries. The first steps along this path were taken by the French physicist Gaston Planté (1834–1889).

Working in Paris in 1859, Planté invented the lead-acid cell. His device comprised a coil of lead for the negative plate and a coil of lead oxide for the positive plate; these were separated by rubber strips and bathed in a bath of dilute sulfuric acid. The sulfuric acid reacts with the lead, releasing electrons that pass to the positive plate, hence generating current. Existing cells ceased to produce current when the chemical reactants were spent. But, in the Planté arrangement, the reaction is reversed when current is added to the cell from an outside source, and in so doing the battery is recharged.

Within a year, Planté had fitted nine of these units into a protective box, producing a battery capable of generating a significant current. Being made of lead, these early rechargeable batteries were inconvenient for handheld applications, for which alternative cells were subsequently developed. Thomas Edison later developed an alkaline storage battery, using nickel-iron and nickel-cadmium cells. Iron-zinc batteries have since become popular, and are two to three times more efficient than the lead cell.

But Planté's cell found one very important use, where high currents are vital and weight issues are less pressing—the electric car. Lead-acid batteries remain the commonest type of vehicle battery to this day. The scions of Planté's invention are now linchpins of modern society. **MB**

SEE ALSO: BATTERY, CAR BATTERY, ELECTRIC CAR

⬆ *This Planté rechargeable battery from c. 1860 has twenty cells with a switch for connecting them.*

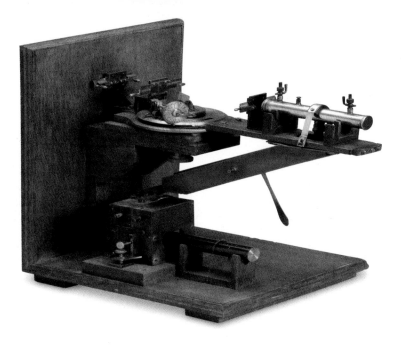

Spectrometer (1859)

Bunsen and Kirchhoff analyze light in terms of electromagnetic radiation.

English physicist and mathematician Sir Isaac Newton, working with an early spectrometer—a graduated glass prism—noted in 1666 that the seven rainbow colors, dispersed when white light was passed through the prism, could not be subdivided into more colors. A second breakthrough in light research occurred when German glassmaker Joseph von Fraunhofer found that the solar spectrum contained dark absorption lines of constant wavelength.

Robert Bunsen (1811–1899), Professor of Chemistry at the University of Heidelberg, working with Gustav Kirchhoff (1824–1887), used a prism spectrometer to reveal the spectral emission lines produced by elements when heated in a flame. In 1859 they became convinced that elements were uniquely characterized by their line spectra, and this led to the discovery of cesium and rubidium. The researchers also realized that the orange Fraunhofer D lines in the solar spectrum were at the same wavelength as the

lines emitted by laboratory sodium. So it was understood that spectrometers could be used to analyze the composition of the sun and the stars.

From the 1870s many European firms began to manufacture prism spectrometers. These were fitted with circular scales so that the positions (and thus wavelengths) of spectral lines could be measured. Resolution was improved by passing the light through sequences of prisms. At first, no one had any idea as to how spectra were produced. However, in 1913 the Danish physicist Niels Bohr introduced a model of the atom in which spectral lines resulted from electrons moving from one stable orbit to another. **DH**

SEE ALSO: BUNSEN BURNER, PHOTOGRAPHIC FILM, X-RAY PHOTOGRAPHY, ELECTRON SPECTROMETER

⬆ *This X-ray spectrometer was made in 1912 by William Bragg, a physicist at Cambridge University, England.*

The invention of the telephone and the motorcar in the 1870s and 80s was arguably the first major step toward the "global village" we live in today. The use of sound transmission in telecommunication also yielded a by-product that—along with its later incarnations—influenced many cultural movements of the following centuries: the phonograph. Meanwhile, the exodus to the cities and the invention of the elevator prompted the construction of the world's first skyscrapers.

The high-speed gas motor, invented by Gottlieb Daimler.

The AGE *of* EMPIRES

1860 to 1891

Linoleum (1860)

Walton makes a wipe-clean wonder flooring.

A few miles south of Manhattan, New York, is a town called Travis. Up until 1930, that town was known as Linoleumville—the home of America's first linoleum factory, owned by British inventor Frederick Walton.

Walton's love affair with linoleum began in the 1850s, when, as the story goes, he noticed a skin that had appeared around the top of an old paint can. This skin was the result of a simple reaction occurring between linseed oil in the paint and oxygen in the air. Most oil-based paints contain linseed oil; you've probably peeled off the rubbery solid that forms around the rim without even thinking about it. But Walton couldn't stop thinking about it. He embarked on a series of experiments that would eventually lead to a process for manufacturing floor tiles from linseed oil.

His road to success was a rocky one. In 1860, he filed a patent for a linoleum manufacturing method, but the method was by no means perfect. Walton had worked out that by adding other ingredients to the mix, such as lead acetate, he could get the linseed oil to react quicker. However, producing large amounts of the material was an arduous process and the big manufacturing companies were unwilling to invest. Even after setting up his own company, Walton found it hard to generate any interest.

Success came, finally, off the back of a mass marketing campaign, and in 1872, Walton's product went global when he opened his American factory. Today, lino is manufactured from a combination of resins, powdered limestone, cork, and of course, linseed oil. Sheets of the mixture are pressed onto a canvas backing material.

A waterproof, wipe-clean wonder, linoleum was the floor covering of choice until the 1960s, when it was ripped up and replaced by vinyl. **HB**

SEE ALSO: CARPET SWEEPER

Gas-fired Engine (1860)

Lenoir builds first internal combustion engine.

In 1860 Belgian engineer Étienne Lenoir (1822–1900) created the first gas-fired internal combustion engine, which was designed to provide an alternative to the rather cumbersome steam engines that were being used at that time. Since steam engines required furnaces, boilers, and large supplies of bulky fossil fuel to be of any use, Lenoir's idea of a more compact, workable gas engine was incredibly popular.

Lenoir earned the patent for his idea in 1860 and created his first actual engine the same year. Initial demand for the invention was high and production started in France, England, and the United States.

The gas engine worked by drawing a mixture of gas and uncompressed air into the cylinder during the

> *"The engineer is the key figure in the material progress of the world."*
>
> Sir Eric Ashby, British botanist and educator

first part of the piston's stroke, then igniting it with a jumping spark in the remaining part. The vapors were then admitted at the opposite end of the cylinder. This resulted in two explosions—one on either side of the piston—for each revolution of the crank.

Alas, the success of the Lenoir engine was limited since practical flaws soon became apparent. The gas that it ran on was, at that time, an expensive option. In addition to that, huge quantities of cooling water and oil were needed to keep the engine from overheating and seizing up. Fewer than 500 Lenoir engines were built before production was stopped, as other inventors were soon making more useful engines. **CB**

SEE ALSO: STEAM ENGINE, STIRLING ENGINE, FOUR-STROKE CYCLE, TWO-STROKE ENGINE, BOURKE ENGINE

Repeating Rifle (1860)

Henry accelerates the pace of war with a lever-action repeating rifle.

When repeater rifles came into their own in the mid-1800s, warfare changed forever. Single-shot rifles had long been predominant; the repeater, however, had a magazine holding several rounds of ammunition, so continual reloading was not necessary. Repeater models had been around since the seventeenth century, but they required separated-out ammunition elements rather than all-in-one metallic cartridges.

In 1860, New Hampshire-born gunsmith Benjamin Tyler Henry (1821–1898) patented a repeater that used a lever action to load its .44 caliber rimfire metallic cartridges—often seen as the first truly self-contained cartridges. It produced continuous fire that was much faster than that of the prevailing muzzle-loading musket, giving one man the firing power of several. The "Henry" had its shortcomings but was very popular with soldiers during the American Civil War, as was Christopher Spencer's slightly different repeater, which appeared at the same time as Henry's.

By the time Henry produced his rifle, he was working for shrewd American businessman Oliver F. Winchester, who built on Henry's model to develop the first classic Winchester repeater (1866). This eclipsed the Henry and incorporated a more sophisticated loading system, still used today. Henrys, and more especially Winchesters, were the guns of the often wild American West, while various repeaters were central to World War I. The Winchester remained a classic through many new incarnations, and Henry and Winchester repeaters are still made today. **AK**

SEE ALSO: MUSKET, BREECH-LOADING GUN, MACHINE GUN, PORTABLE AUTOMATIC MACHINE GUN

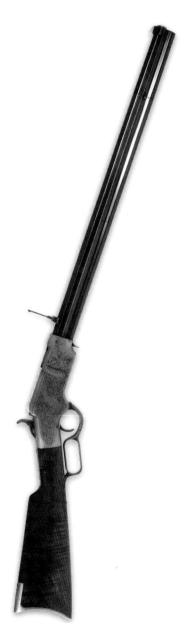

◤ *This reproduction Henry .44–.40 was owned and used in Western movies by American actor Tom Selleck.*

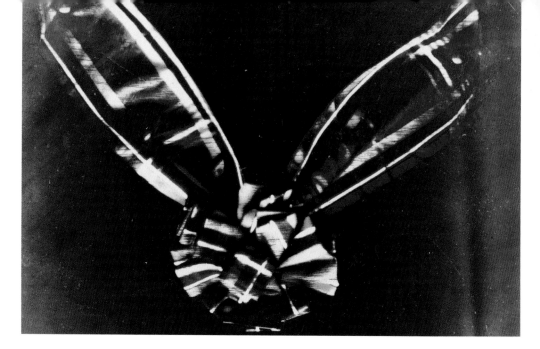

Color Photography (1861)

Maxwell develops the trichromatic process for producing color images.

British mathematician and physicist James Clerk Maxwell (1831–1879) was a giant of nineteenth-century science. Best known for his Maxwell equations, which were the best insight into electromagnetism of their day, his interests also included Saturn's rings and the human perception of color. It was this latter interest that led to the first color photograph in 1861.

In the manner of a true showman, Maxwell revealed his photograph of a tartan ribbon at the Royal Institution in London. His studies of human vision, including the condition of color blindness, had led him to conclude that color images were possible using a "trichromatic process." He had arranged for his tartan ribbon to be shot by professional photographer Thomas Sutton, the inventor of the single-lens reflex camera. The images were black and white, but, critically, Maxwell had three such images taken through red, green, and blue filters, respectively. Having turned the images into slides, he then projected them through the same filters in such a way that they were carefully superimposed on each other on the screen. The effect was a recognizable reproduction of the tartan in glorious color.

Maxwell was lucky. His demonstration should not really have worked at all because, unknown to him, his photographic emulsion was not sensitive to red light. Fortunately, the red in the tartan did reflect ultraviolet light, and this was picked up by the emulsion.

The trichromatic process became the foundation for all color photography, and Maxwell's three original slides of the tartan ribbon are now on display in a small museum in Edinburgh. **DHk**

SEE ALSO: PHOTOGRAPHY, CALOTYPE PROCESS, DAGUERREOTYPE PROCESS, INFRARED PHOTOGRAPHY

⬆ *Maxwell's composite image of three photographs shows tartan ribbon through filters of different colors.*

Single-lens Reflex Camera (1861)

Sutton ushers in modern photography with a new camera.

The modern era of photography began in 1861 with the invention and patenting of the world's first single-lens reflex (SLR) camera by photography expert Thomas Sutton (1819–1875). His prototype led to the creation of the first batch of SLR cameras in 1884, with a design that is still in use today. Sutton also assisted James Clerk Maxwell in his successful demonstration of color photography in 1861.

In non-SLR cameras, light enters the viewfinder at a slightly different angle to that at which it enters the lens, so the resulting photo can appear different to the intended composition. In SLR cameras, a mirror is positioned in front of the lens and directs light up into a pentaprism. The light bounces between its edges until it enters the viewfinder with correct orientation, as if the viewer is looking directly through the camera lens. When a photograph is taken, the mirror moves out of the way allowing light to reach the film or, with digital SLRs (DSLRs), the imaging sensor.

Now the most popular professional camera format, the SLR camera was the culmination of decades of photographic innovations that began with the production of Louis Daguerre's daguerrotype and Josef Maximilian Petzval's lens systems, which led to the first mass-produced cameras.

Although no record of the first production model exists, the camera was first commercially produced in the mid-1880s. By the 1930s it was extremely popular with photographers, allowing an undistorted view of the subject from the correct perspective. DSLRs have all but replaced the traditional SLR, but the principle that Sutton pioneered is still used today. **SR**

SEE ALSO: **PHOTOGRAPHY, COMPOUND CAMERA LENS, FILM CAMERA, PORTABLE "BROWNIE" CAMERA, 35-MM CAMERA**

⬆ *With decreasing costs, single-lens reflex cameras like this 1911 Adam's Minex became more popular.*

Twist Drill Bit (1861)

Morse modernizes the drilling process.

Stone Age man figured out that spinning a sharp rock on a wooden board would produce a circular hole—a useful discovery. The application of a bow to such a drill increased the rate of spinning and therefore the speed of boring the hole. By the nineteenth century, the bow had been replaced by geared machinery and the stone bit by metal, but otherwise the concept remained basically the same.

Prior to the early 1860s, the standard drill bit consisted simply of a flattened, sharpened piece of metal. These "spade bits" were notoriously imprecise and also prone to rapid dulling when used on hard surfaces. American Stephen Morse believed that he could improve the standard drill bit, and, in 1861, he

> *"I have little patience with scientists who . . . drill . . . holes where drilling is easy."*
>
> Albert Einstein, theoretical physicist

patented what has now become known as the twist drill. The device was the early version of the now-familiar helical drill. Consisting of a sharpened cutting edge with a spiral groove twisting upward along the shaft of the bit, it represented a great leap forward. The radial groove proved to serve as a guide, actually pulling the bit down through the substance being drilled and producing a round hole, but also routed the remnants of the bored substance up and out of the hole onto the surface by acting as a screw.

Though the bits that are around today are made of harder metal, the same basic concept predominates today, as does the company Morse founded. **BMcM**

SEE ALSO: STONE TOOLS, SHARP STONE BLADE, DRILL, METALWORKING, SEED DRILL, ELECTRIC DRILL

Solvay Process (1861)

Solvay synthesizes sodium carbonate.

Sodium carbonate is an alkaline powder familiar to many generations of laundry workers as washing soda. However, this is a versatile substance with myriad other uses. Most notably, in this age of the skyscraper, it forms glass when heated then rapidly cooled with sand and calcium carbonate. On a more grisly note, sodium carbonate is used in taxidermy to strip away flesh from bone. It is also a common food additive.

Traditionally, sodium carbonate was sourced from mining and from the ashes of plant matter (hence its other name, soda ash). Throughout the industrial era, efforts were made to find a synthetic process to produce the compound, which was in great demand for textiles and glass. The first attempt was made in 1790 by Frenchman Nicolas Leblanc (1742–1806), who found a way to convert salt into sodium carbonate using sulfuric acid, limestone, and coal. The method worked, and was used by industry for seventy years.

The Leblanc method had two drawbacks: expensive reactants and pollution. In 1861, Belgian chemist Ernest Solvay (1838–1922) worked out a more efficient process that bears his name and is still used today. Carbon dioxide is bubbled through a mixture of ammonia and salt water to form ammonium chloride and baking soda. The latter is then filtered out and heated to produce sodium carbonate.

Ernest and his brother Alfred founded their own company in 1863, and production of sodium carbonate began in 1865. It took until the end of that century for them to perfect it, but Solvay's process, which was more viable for large-scale commercial production, finally replaced the Leblanc process and came to dominate industry. Three-quarters of the world's sodium carbonate is now produced by Solvay's method, with the rest being mined. **MB**

SEE ALSO: DOW PROCESS (FOR EXTRACTING BROMINE), HABER PROCESS (MANUFACTURE OF AMMONIA)

Yale Lock (1861)

Yale improves on Egyptian pin-tumbler locks.

Securing one's valuables was big business in the nineteenth century. Locksmiths were determined to get one up on the opposition by devising a superior lock to their rivals' devices.

In 1847, Linus Yale Sr. (1797–1858) opened the Yale Lock Shop in Newport, New York. He was interested in bank safe locks and began looking into the pin tumbler lock that had been used by the Egyptians over 4,000 years ago. The Egyptians' locks had two major shortcomings—they were wooden and bulky, with some measuring 23.5 inches (60 cm) in length. His early attempts to improve the design focused on incorporating a pin tumbler into the case of the lock, which could then be opened with a round, fluted key. This was only a modest improvement, however, and it was Yale's son, Linus Yale Jr. (1821–1868), who eventually realized the full potential of this lock.

In 1861, Yale Jr. designed the cylinder pin tumbler lock, which was a lot less bulky than his father's design and could be opened with smaller keys. When a key is inserted into the cylinder lock, it pushes the bottom set of pins into alignment, allowing you to unlock it. Initially, the cylinder lock utilized flat keys, but these were soon replaced by ones with serrated edges like those in use today. As the keys were smaller and lighter in weight, it meant several could be carried at once. Cylinder locks were amenable to mass production, and this ensured that they were also affordably priced.

Not content with his achievements, Yale Jr. also devised the combination lock that obviated the use of keys. Both cylinder and combination locks are widely used today. **RB**

SEE ALSO: MECHANICAL LOCK, COMBINATION LOCK, PADLOCK, TUMBLER LOCK, SAFETY LOCK, TIMELOCK SAFE

↗ *The Yale lock is opened when its spring-loaded pin tumblers are raised by the key's serrated edge.*

"Everything we are today, we owe to the inspirational ingenuity of Linus Yale Sr. and Linus Yale Jr."

Yale website

Snellen Eye Test (1862)

Snellen standardizes tests of visual acuity.

Dutch ophthalmologist Hermann Snellen (1834–1908) first came up with the idea of a standardized test to measure how well a person can see and also allow comparisons of different people's visual capabilities.

The Snellen Chart, developed in 1862, consists of eleven rows of block letters. The first row consists of very large letters; subsequent rows decrease in size. A person taking the test covers one eye and reads aloud the letters of each row, beginning at the top. The smallest row that can be read accurately indicates the patient's visual acuity in that eye. The patient then reads the letters with the other eye, and then again with both eyes. A traditional Snellen Chart features only the letters C, D, E, F, L, N, O, P, T, and Z.

The relationship between the size of letters and the distance at which they are seen has become the standard method of recording visual acuity. Snellen determined that a letter that was approximately 0.3 inches (8.75 mm) high (size 20) could be identified by most people with "normal" eyesight at a distance of 20 feet (6 meters). Therefore, the Snellen fraction of 20/20 became the model for "normal" vision. The top number of the fraction is always 20 because it represents the testing distance, and the bottom number identifies the smallest letter the person taking the test can see at 20 feet. If the smallest letter that a person can identify is 0.6 inches (17.5 mm)—or twice the size of a size 20 letter—then the Snellen fraction for that person would be 20/40. The biggest letter on the chart represents an acuity of 20/200, the measurement that is considered "legally blind."

For children who cannot read, charts were made using pictures of common objects or broken circles with missing segments. The Snellen Chart remains the most popular chart design among eye doctors. **JF**

SEE ALSO: SPECTACLES, BIFOCALS, CONTACT LENSES, LASER EYE SURGERY, VISUAL PROSTHETIC (BIONIC EYE)

Machine Gun (1862)

Gatling revolutionizes warfare with rapid fire.

The invention of the machine gun by Richard J. Gatling (1818–1903) irreversibly changed the face of battle. Gatling took advantage of the newly invented brass cartridge (which, unlike the earlier paper cartridges, had its own percussion cap) to produce the first rapid-fire weapon in about 1862. The Gatling gun consisted of ten parallel barrels that could fire and reload brass cartridges at rapid speeds through the rotation of a hand-operated crank. With each rotation the firing and loading mechanism of each barrel came into contact with a series of cams. The first cam opened the bolt on the barrel, allowing the bullet to fall into a chamber, while another closed the bolt. A further cam released the firing pin with the final one opening the bolt and ejecting the spent case. The first successful model was deployed in a limited capacity during the American Civil War by Union troops.

The American inventor Hiram Stevens Maxim utilized the power of the cartridge explosion to design the first automatic machine gun in 1884. By using the recoil power of the fired bullet, the empty cartridge is expelled and the barrel reloaded. Both reliable and easily transportable, with a firing capacity of 600 rounds per minute, the Maxim was adopted by European armies.

Machine guns based on Maxim's design dominated fighting during World War I. In the static environment of trench warfare, the weapon caused heavy loss of life, earning the machine gun a fearsome reputation. It continued to produce deadly results in World War II, but found its popularity supplanted by the lighter and more mobile sub-machine gun. **SG**

SEE ALSO: ARQUEBUS, CANNON, GUN, MUSKET, PORTABLE AUTOMATIC MACHINE GUN

➡ *Gatling submitted this detailed drawing of his "battery gun" to the U.S. patent office in 1865.*

Received March 8 1865

Nº 47.631.

R.J. Gatling

Battery Gun.

1½ Inch to 1 Ft.   *Patented May. 9.1865.*

47 631

Fig.1

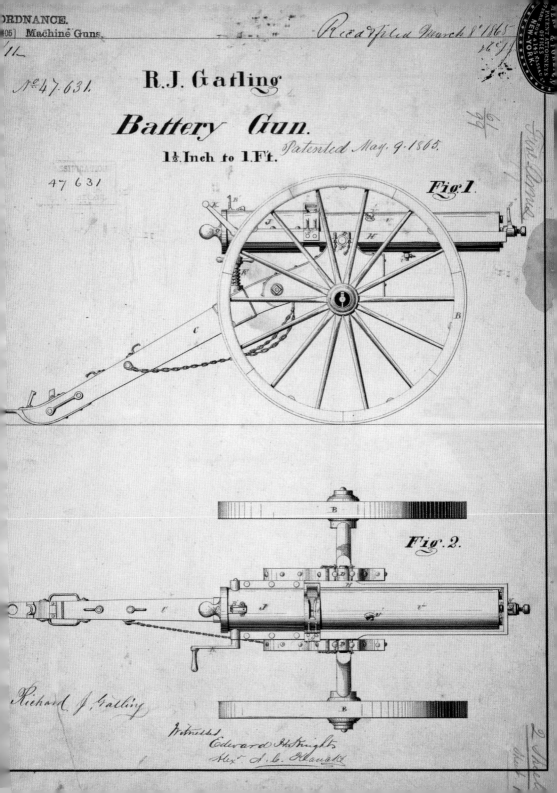

Fig.2

Richard J. Gatling

Witness
Edward H. Knight
Alexr. J. C. Klaudt

Pasteurization (1862)

Pasteur and Bernard destroy bacteria in milk.

At the beginning of the 1860s French chemist Louis Pasteur (1822–1895) observed that living cells, known commonly as yeast, were responsible for forming alcohol from sugar and, when contaminated, led to its souring. He was able to distinguish and separate these microorganisms, and, in 1862, he and French physiologist Claude Bernard (1813–1878) showed that most of the bacteria present in milk could be killed if the milk was heated to 145°F (63°C) for thirty minutes. The milk was then rapidly cooled to eliminate bacterial contamination. Pasteur applied the same principle to beer, the souring of which was a sore point in the French economy. This process of extended heating came to be known as "pasteurization."

"Did you ever observe to whom the accidents happen? Chance favors only the prepared mind."

Louis Pasteur, French chemist and microbiologist

Pasteur's revolutionary understanding of germ theory finally brought to an end the centuries-old notion of "spontaneous generation." Maggots did not "just appear" in decaying meat, according to Pasteur—they were the end products of microorganisms and bacterial infection. His proof that germs could cause fermentation and degeneration in liquids provided the framework for later research that led to cures for diseases in crops such as potato blight and various silkworm diseases, as well as anthrax and rabies. **BS**

SEE ALSO: POWDERED MILK, MILK/CREAM SEPARATOR, MILKING MACHINE

⬅ *Pasteur poses with his life-saving sterilizing apparatus at the Pasteur Institute in Paris, France.*

Modern Harpoon (1864)

Foyn develops modern whaling methods.

Norwegian Svend Foyn (1809–1894) is regarded as the father of modern whaling and is credited with the invention of the modern harpoon. Foyn's genius was to combine the use of fast steam-powered boats with deck-mounted cannons that could use both harpoons with strong lines attached to them and bomb lances to kill the target whale as quickly as possible. Foyn's purpose-built steam whale "catcher" *Spes et Fides* (*Hope and Faith*) first sailed in 1864 to northern Norway.

By 1868 Foyn had perfected the technique and "modern" whaling methods began to reap financial rewards. This success soon transformed Norway into the dominant force in whaling, an industry previously dominated by British, American, French, and German ships. During the nineteenth century, whale oil was much in demand for lighting and soap; whale meat, which has been eaten by some cultures since early times, became popular in the twentieth century.

Foyn's innovations were successful because they brought a whole new family of whales within the reach of the whalers—the Balaenopterinae, or rorquals. The rorquals, which include minke, sei, fin, and blue whales, were difficult to hunt because they are generally fast, large, and powerful, spend little time at the surface, and sink when harpooned. Another Foyn innovation was to inflate the carcass of the whale with air so that it floated and was easier to collect.

The success and efficiency of modern whaling was also in many ways its downfall. Populations of large mammals can be very fragile and whale populations soon crashed, eventually leading to a worldwide moratorium on whaling in 1982. Although Japan, Norway, and many other countries continue whaling on a small scale, there is strong international pressure to stop the practice altogether. **JB**

SEE ALSO: SPEAR, FISHHOOK, ATLATL

Torpedo (1864)

Giovanni Luppis creates the self-propelled underwater missile.

Despite its notoriety as a naval weapon, the first modern torpedo was developed in landlocked Austria, or rather by a retired army officer in what was then the Austrian Empire stretching down to the Adriatic Sea. In 1864 Giovanni Luppis (1813–1875) presented his idea of using small, unmanned boats carrying explosives against enemy ships to Robert Whitehead (1823–1905), an English engineer producing steam engines for the Austrian Navy. Similar devices (spar torpedoes) were also employed in the American Civil War taking place at the same time. However, those contraptions consisted of manually driven steam launches with explosives hanging from a long pole. In order to set them off the crew would ram the end of the spar into the target vessel and then back off again, thus pulling a mechanical trigger by disconnecting the cable linking the boat to the weapon.

Luppis's torpedo, on the other hand, was self-propelled and navigated from the land by ropes attached to it, thus greatly reducing the risk for those using the weapon. Whitehead was initially sceptical about the design's feasibility, fearing that a remotely controlled surface device powered by a clockwork motor would be too slow and clumsy to work efficiently. But he continued to mull over the concept, and two years later he created an improved automobile torpedo driven by compressed air and launched from an underwater tube that determined its trajectory.

Luppis and Whitehead thus had a major impact on the outcome of the two world wars, turning submarines into a terrifying prospect. Like an electric ray (or torpedo, the fish it has been named after), their explosive projectile weapon is capable of delivering a stunning shock to its target and is responsible for over 25 million tons of shipping rusting on the seabed. **DaH**

"Damn the torpedoes. . . . Captain Crayton, go ahead! Joucett, full speed!"

Admiral David Farragut, Battle of Mobile Bay, 1864

⬆ *Torpedoes are examined on the deck of a target ship after a test firing from HMS* Snapper *in 1940.*

➡ *The British 7,000-ton steamer* Beluchistan *sank after this torpedo strike by the German U-boat U-68 in 1942.*

SEE ALSO: GUNPOWDER, ROCKET, CANNON, LAND MINE, LASER-GUIDED BOMB

Barbed Wire (1865)

Jannin's wire fencing divides up the West.

Barbed wire is unusual among inventions because it appears to have been created on several separate occasions at roughly the same time in history. Frenchman Leonce Eugene Grassin-Baledans made the first attempt in 1860, with twisted strands of sheet metal. Another Frenchman, Louis Jannin, came up with what would become known as barbed wire in 1865. In 1867 several Americans followed in his footsteps with different yet fairly ineffective designs, using single-stranded wire. The final contribution in this flurry of inventiveness came from American Joseph Glidden (1813–1906), whose 1874 patented wire machine created practical, cheap, mass-produced barbed wire, leading to its widespread adoption.

> *"It was fearsome looking stuff, this barbed wire . . . the barbs looked . . . like miniature daggers . . . "*
>
> Cameron Judd, *Devil Wire*

The reason for this simultaneous outbreak of ingenuity was due in part to a need for cheap fencing after the 1862 Homestead Act allowed any resident outside of the thirteen original American colonies to lay claim to up to 160 acres (67.4 ha) of land.

Barbed wire changed lives across the world. In America it created land barriers that had previously not existed, opening up the Wild West to more settled farming communities and cattle ranches. Elsewhere, barbed wire was adapted for use in warfare—initially in the Boer War and World War I—and it has become a powerful symbol of oppression and conflict throughout the world. **JB**

SEE ALSO: **METALWORKING, WROUGHT IRON, STEEL**

Roller Coaster (1865)

Thompson thrills New Yorkers with a new ride.

The precursors of roller coasters first appeared in Russia in the fifteenth century, in the form of huge ice slides. The first modern coasters opened in Paris in 1817, but it took another sixty years and the invention of the inclined railway for the rides to really take off.

Regarded as the father of the roller coaster in the United States, La Marcus Adna Thompson filed his first patent for a roller coaster in 1865 and built the first such ride in 1884 at Coney Island, New York. Charging five cents a ride, he was soon grossing over $600 a day. Thompson's Switchback Railway consisted of two parallel tracks undulating over a wooden structure 600 foot (182.8 m) long. A small train with seats facing sideways for sightseeing started on a 50-foot (15 m) peak at one end of the beach, and rolled down the grade at around 6 miles (9.6 km) per hour until the momentum died at the other end.

Over the next few years there was a rapid development in roller coasters. Charles Alcoke created a continuous oval track, while Philip Hinkle added a hoist to pull cars to the top of a steep hill, and had the seats facing forward. Thompson, meanwhile, concentrated on building elaborate scenic railways with impressive indoor tableaux of biblical and ancient Egyptian scenes.

The 1920s were the golden age of the roller coaster, with nearly 2,000 being built, but depression and war led to a decline that continued until the opening of Racer in 1972 at King's Island in Cincinnati, Ohio. The revival continues to this day, with theme parks around the world striving to outdo each other in speed, complexity, and screams. **BO**

SEE ALSO: **FERRIS WHEEL**

➡ *Visitors enjoy the alarm of riders on the roller coaster at Lunar Park, Coney Island, New York, in 1910.*

Ice-making Machine
(1865)

Lowe revolutionizes food storage.

Thaddeus Lowe (1831–1913) did not only invent the humble ice-making machine; he also made waves in aeronautics, engineering, and chemistry. In the course of his work on the cooling properties of compressed gases, he became interested in carbon dioxide specifically and, putting his research into practice, developed the "Compression Ice Machine" in 1865.

After the American Civil War, Lowe began extensive research on the properties of gas. Refrigeration is essentially a process whereby heat is removed from an enclosed space and ejected somewhere else. Most

> " . . . it is with great pleasure and satisfaction that we welcome proof of [Lowe's] genius."

Professor Joseph Henry, Smithsonian Institution

systems work by using a chemical, usually gas, to remove the heat. As the gas expands, heat is turned into kinetic energy, cooling the air.

In 1869 Lowe and other investors purchased an old steamship equipped with refrigeration units and began shipping fresh fruit and meat from Texas to New York. The business failed, largely due to the group's lack of shipping knowledge and the public's skepticism about eating meat that had been so long out of the packing house. Despite this setback, refrigeration became massively popular and has revolutionized the way the world preserves food. **KB**

SEE ALSO: REFRIGERATOR, FAST-FROZEN FOOD, FREON

◄ *The manufacturing of ice was a major commercial activity until the advent of electric refrigerators.*

Dry-cell Battery
(1866)

Leclanché puts a seal on dry-cell power.

At the end of the eighteenth century, the Italian physicist Alessandro Volta invented the battery. By placing a pair of electrodes—one of zinc, the other of copper—in a solution that conducts electricity (sulfuric acid), he developed the principle that is still in use in today's dry-cell batteries.

More than half a century later there had been many new versions of the battery, but there were still problems with the original Voltaic design. The solution that conducts the electricity, usually some kind of acid, was too dangerous to touch and could spill out if the battery was tipped over. Nor were the electrodes very steady and were in danger of falling out if the battery was shaken too much. Furthermore, the batteries themselves were far too heavy to use around the house.

It was not until about 1866 that the French engineer Georges Leclanché (1839–1882) resolved these problems. Filling a porous pot with ammonium chloride, which is an alkaline rather than an acid, meant that a battery was less toxic. Replacing the lead electrodes used in earlier models with one of zinc and one of carbon-manganese dioxide created a much lighter battery. Finally, sealing the battery with a hard wax mixture, made it much more resilient against shocks and shakes and spillage.

Leclanché's design was tweaked again in 1887 by German scientist Carl Gassner who mixed the liquid ammonium chloride conductor with plaster of Paris to form a paste, which gave the world the first true dry-cell battery. Leclanché cells were extremely popular. They were used in doorbells, in cars, and, importantly, by Alexander Graham Bell in long-distance demonstrations of the telephone. **DK**

SEE ALSO: BATTERY, ELECTRIC CAR BATTERY, RECHARGEABLE STORAGE BATTERY

Dynamite

(1866)

Nobel introduces the era of high explosives.

Until the mid-nineteenth century, the most powerful known explosive was gunpowder. In 1846 Italian chemist Ascanio Sobrero discovered that by nitrating glycerin he could make a fearsomely explosive liquid. It was also frighteningly unstable. Alfred Nobel (1833–1896) undertook the dangerous task of turning nitroglycerin into a marketable product. Despite an explosion in 1864 that killed his brother and workers at the family's factory, he persisted with his experiments.

Nobel's first success came from combining nitroglycerin with a mercury fulminate detonator. He began producing this explosive in bulk, but the frequency of accidents soon saw it banned in many countries. In 1866 Nobel discovered a mixture of nitroglycerin with diatomaceous earth, a chalklike sedimentary rock. He christened this comparatively safe explosive *dynamite*, from the Greek word for "power." Packed into paper tubes, dynamite was soon selling in vast quantities and revolutionized activities such as tunnel-building and quarrying.

Dynamite, patented in 1867, started the era of "high explosives" as patents for similar, but even more powerful and reliable substances multiplied. These greatly increased the destructive power at the disposal of humankind—a power that was inevitably applied to warfare. The story goes that Nobel was shocked at reading his own premature obituary in 1888, in which he was described as a "merchant of death." This may have motivated him to leave his fortune to the funding of the annual Nobel prizes to those who have "conferred the greatest benefit on mankind." **RG**

SEE ALSO: GUNPOWDER, ROCKET, CANNON, LAND MINE

← *The number of paper-wrapped sticks of dynamite used would determine the force of the explosion.*

Folding Ironing Board

(1866)

Mort improves home laundering equipment.

Devices for getting wrinkles out of fabric have been around nearly as long as fabric itself. The Vikings used whalebone smoothing boards, the Chinese filled metal pots with hot coals to press cloth, and, in seventeenth-century England, the screw press was popular. By the nineteenth century, most metal smoothers (irons) had adopted their familiar shape, but ironing boards had not evolved at the same rate as fashion design and ironing was often carried out on tables or boards resting across two chairs.

Sleeves, pant legs, ruffles, pleats, pockets, buttons, curved seams—the more details were added to clothes, the more difficult it was to remove the

> *"In most Iowa homes this third day of the week [Tuesday] is reserved for ironing."*
>
> The Iowa Housewife, 1880

wrinkles after laundering. To help improve matters, inventors turned their attention from irons to ironing boards. The first U.S. patent for an "ironing table" was granted in 1858 to W. Vandenburg and J. Harvey. By the end of the century hundreds had been issued.

One inventor, Sarah Mort of Dayton, Ohio, got ironing off tables and out-of-sight by designing a folding ironing board. "When extended it can be placed anywhere, and when not in use can be folded to the capacity of an ordinary board, which renders it very convenient," explained Mort in her patent application. Although the new board was practical and easily hidden, ironing itself remained a chore. **RBk**

SEE ALSO: ELECTRIC IRON

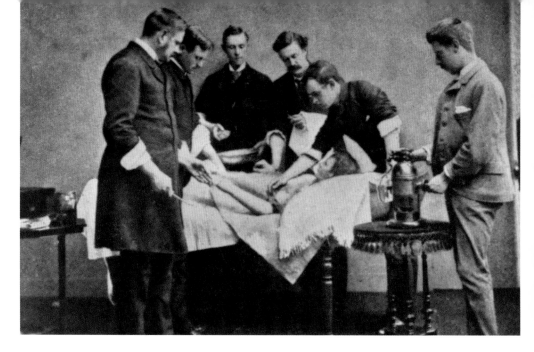

Antiseptic Surgery (1867)

Lister reduces wound infections with carbolic acid.

In the early 1860s, French microbiologist and chemist Louis Pasteur (1822–1895) showed what a few scientists had already begun to suspect—that the rotting of organic tissue (gangrene) was caused by bacteria rather than by chemicals in the air (or "miasma"), as had been previously thought.

After reading a paper by Pasteur, English surgeon Joseph Lister (1827–1912) set about conducting his own experiments. Reaching the same conclusion as Pasteur, Lister then sought a way to get rid of the microorganisms that caused gangrene by applying a chemical solution to wounds. Carbolic acid, or phenol, had been used for deodorizing sewage, so Lister applied a solution of carbolic acid to both surgical instruments and open wounds. He found that by doing so the incidence of gangrene in patients in his care was vastly reduced. He had created the first antiseptic to be used in medical practice.

In 1867 Lister outlined his findings in a series of articles for the medical journal *The Lancet*. He proposed that surgeons should wash their hands in a carbolic acid solution before carrying out surgical procedures, that they wear clean gloves, that operating rooms be sprayed with a carbolic acid solution, and that the handles of surgical instruments should not be made from porous materials such as wood and bone, as these could harbor infection-causing bacteria.

Lister's discovery marked an end to the days when a blood-soaked apron was a badge of medical expertise. Although he was laughed at by many of his contemporaries, his innovative ideas changed surgical methods for ever. **JL**

SEE ALSO: ANESTHESIA, VACCINATION, NITROUS OXIDE ANESTHESIA, ETHER ANESTHETIC

⬆ *During a surgical procedure, an attendant (right) uses Lister's antiseptic spray to reduce the risk of infection.*

Reinforced Concrete (1867)

Monier develops a new construction material.

Plain old concrete—undeniably useful and popular though it may be—is really not as spectacular a material as you might at first think. Yes, it is good for making good, hard pavements and keeping fence-posts firmly in place in the ground, but if you are hoping to build a multistory car park out of it, or an overpass that runs above a busy motorway, you will quickly discover that concrete itself is not enough.

In the 1860s, however, French gardener Joseph Monier (1823–1906) demonstrated the reinforced garden tubs he had made using ferroconcrete—a concrete and chicken-mesh combination that fellow Frenchman Joseph-Louis Lambot had pioneered. The garden tubs made their debut at the Paris Exposition of 1867, and Monier applied for a patent the same year. Monier was not the first to think of strengthening concrete with metal, but his patent design clearly established the principle of reinforced concrete for structural purposes.

Monier continued to apply for and secure patents throughout his career, all of which followed along strangely similar lines. They included the 1868 patent on iron-reinforced cement pipes; the 1869 patent on iron-reinforced cement panels for building facades; and the 1873 patent for the construction of bridges and footbridges made of iron-reinforced cement.

Reinforced concrete—still prevalent today, with 35 cubic feet (1 cubic m) of the stuff used for every person on the planet in 2006 alone—has undergone various improvements since its invention and is now used in all sorts of constructions—from street lights and swimming pools to dams and viaducts. **CL**

SEE ALSO: PLASTER, FIRED BRICKS, WATTLE AND DAUB, PORTLAND CEMENT

⊡ *In Paris, France, workers prepare metal armature for a reinforced concrete construction in 1913.*

Paper Clip (1867)

Fay's wire fasteners bring order to office work.

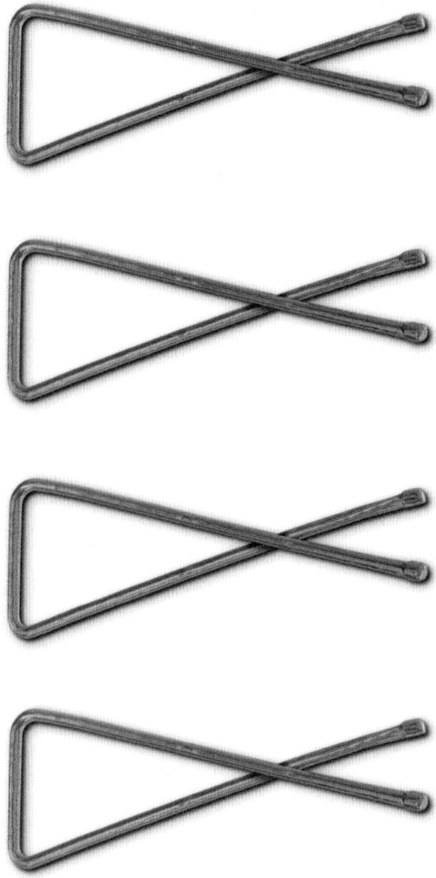

"The wire clip for holding office papers together has entirely superseded the use of the pin."

Business, March 1900

Steel wire was still a relatively new concept in the mid-nineteenth century when an American, Samuel Fay, patented a wire ticket fastener that he used to attach labels to garments in 1867. When Fay mentioned as an aside in his patent application that his fastener was also useful for holding together sheets of paper, his simple triangular-design wire fastener had unwittingly become the world's first bent-wire paper clip.

Prior to this, the straight pin was the preferred method of attaching labels to garments. The paper clip, a single length of wire bent at either end to create a simple cross, speeded up the fastening of labels and resulted in less damage to the item being tagged.

The destiny of the paper clip, however, was not to be found in the tagging of garments. Over the next thirty years, more than fifty paper clip designs were patented in the search for the best method of clipping together sheets of paper. Fay's original design proved awkward with its two exposed ends tending to damage and scratch documents. A plethora of designs soon appeared, including paper clips designed for specific purposes, such as Erlman Wright's newspaper clip from 1877. In the 1890s paper clips began to be widely used in offices. Apart from the paucity of steel wire, the other reason for the paper clip's relatively late emergence was that there'd been no machines available previously that could bend the wire into the right shape.

The first paper clips approaching the modern design were described in a patent issued to William Middlebrook of Connecticut in 1899, who also built and patented a machine to mass-produce them. **BS**

SEE ALSO: PAPER, STAPLER

⇱ *Paper clips only became common stationery items in the 1890s, shortly before mass production began.*

Flat-bottomed Paper Bag (1868)

Knight creates a carrier bag for bulky items.

Margaret Knight (1838–1914) was one of the first American women to be awarded a patent. She was a prolific inventor from the age of twelve, when an accident in a textile mill prompted her to design a safety feature to protect workers from the looms. The flat-bottomed paper bag, however, is her most widely remembered invention, as it endures to this day.

Knight was working in a paper-bag factory after the American Civil War when she saw the need for a different kind of bag. The factory produced flat bags, more like envelopes, which were unsuitable for bulky items. Square, flat-bottomed bags could be made, but only by hand. Although she had little education, Knight—after studying the factory machinery—built a working wooden prototype of a machine at home. Understanding that she would need an iron model for wider use and for a patent application, she sent her design to a machine shop. While it was there, Charles Annan, an employee of the shop, stole her design and filed for a patent. He was initially granted it, but the decision was overturned in court, and Knight was awarded her patent in 1873.

Initially Knight's invention was dismissed by her male colleagues because she was a woman, but then it quickly received attention from around the world and in 1871 she was decorated by Queen Victoria. She cofounded the Eastern Paper Bag Company in 1870, and the bags were used almost universally. Knight continued inventing, gaining twenty-seven patents in total, including a barbecue spit, window sash and frame, and a numbering device. **JG**

SEE ALSO: PAPER, CARBON PAPER, CARDBOARD BOX, TETRA PAK

⬈ *Knight's achievement was to produce a machine able to cut, fold, and paste the paper bag in one operation.*

> *"I was famous for my kites; and my sleds were the envy … of all the boys in town."*

Margaret Knight

Spring Tape Measure (1868)

Fellows invents a retractable, flexible rule.

Tools have been around for more than two million years, and over the centuries they have been designed to dig, lift, hammer, screw, propel, crush, launch, and measure. Measurement tools are of the utmost importance in modern construction.

Although standard weights and measures stretch back beyond the Babylonians—one of the earliest civilizations—the world's most popular measuring tool is a mere infant. The calibrated foot ruler was invented in 1675, before German Anton Ullrich (1826–1895) improved on it in 1851, when he invented the folding ruler. But an even more flexible and compact ruler was soon to arrive on the scene.

The spring tape measure was patented in 1868 by American Alvin J. Fellows. Although the tape was made of cloth rather than metal, Fellows' retractable tape measure is nearly identical to those found in today's toolboxes. Swapping fabric for steel, New Jersey engineer Hiram Farrand updated the tape

measure in 1919, when he filed his first rule patent. He received the patent in 1922. Farrand's tape measure was made famous by tool giant, Stanley®, which purchased Farrand's company in 1931 and in 1932 entered the tape-measure business.

Today's tape measures have become high-tech. The latest is the laser distance meter, complete with liquid crystal display and memory storage—effectively a computerized tape measure. Such gadgets, however, are not nearly as popular as their low-tech, low-cost counterpart, the retractable spring tape measure, which is cheap, compact, and flexible enough to be used to measure round curves or angles. **RBk**

SEE ALSO: STANDARD MEASURES, THEODOLITE, SLIDE RULE, VERNIER SCALE, METRIC SYSTEM, SI UNITS

⬆ *Easily fitted into a pocket, the retractable steel tape measure was highly convenient for many professions.*

Stapler (1868)

Gould develops a prototype paper binder.

The very first stapler is thought to have belonged to King Louis XV of France. The elaborate, handmade staples were imprinted with the royal insignia and were used to fasten together court documents.

In 1868 Charles Gould received a British patent for a wire stitcher that could be used to bind magazines. His invention used uncut wire that was then cut to length, the pointed ends forced through the paper, and the ends folded down. The device was a direct predecessor of the modern stapler.

In the United States, in 1868, Albert Kletzker patented a type of paper clip that used a single large staple to fasten together papers but did not crimp the ends, which had to be done by hand. The first machine that both inserted the staple and crimped it in one motion was patented by Henry R. Heil in 1877.

The first commercially successful stapler was produced by George W. McGill from 1879. Although the staples had to be loaded one at a time, the machine featured an anvil that bent the ends of the staple back on themselves as it was inserted. The fixed anvil can still be seen on modern staplers today.

Improvements to the basic designs were made to enable multiple staples to be loaded at once, either in the form of strips or singly. Eventually, in the 1930s, manufacturers produced strips of staples that were glued together in a "herringbone series," alongside staplers that swung open, allowing easy loading.

Staplers are ubiquitous in offices worldwide, a constant source of frustration to workers when they go missing from desks. Staples have also found a use in surgery as a replacement for sutures. **HP**

SEE ALSO: PAPER, PAPER CLIP

⬆ *The "Brinco" has a magazine of staples and is operated by pushing the black knob downward.*

Margarine (1868)

Mège-Mouriès develops a butter substitute.

In 1867 the French President, Napoleon III, offered a prize to the inventor of a butter substitute that would keep well, for use by the army, and be a cheap alternative for the poor.

French chemist, Hippolyte Mège-Mouriès won the prize with a substance he called oleomargarine, which he had developed from margaric acid, a mixture of palmitic and stearic acids. He established the first margarine factory in France and later expanded the business to the United States. But the business foundered and he died in obscurity in 1880.

Mège-Mouriès had, however, sold his product to the Dutch businessman Anton Jurgens who built a margarine business that merged with the Lever

---

"My mother would buy pale white margarine in a soft plastic pouch, with an orange dot in the middle."

Food Reference website

Brothers' business to form Unilever in 1929. Demand for margarine in Europe grew as World War II resulted in butter rationing.

The American margarine market also took off and by 1877 the dairy industry in the United States found itself engaged in a bitter struggle to protect its market position. By the mid-1880s the U.S. federal government had introduced an expensive license scheme for the manufacture and sale of margarine. A ban on the addition of yellow colorings to the naturally white margarine proved to be the most effective means of slowing the growth in sales and remained in force in some places for over a hundred years. **EH**

SEE ALSO: PASTEURIZATION, BREAKFAST CEREAL

Air Brake (1869)

Westinghouse revolutionizes rail safety.

Rail travel is one of the safest ways to get around in the modern age. The pioneer responsible for much of this safety record was the visionary inventor and industrialist George Westinghouse (1846–1914).

Before he invented his revolutionary air brake, slowing and stopping a train was an exercise fraught with risk. Each separate car of the train needed its own brakeman to manually operate brakes on its own set of wheels. Accidents caused by uncoordinated braking were frequent and Westinghouse realized that the poor safety of trains was holding up the whole industrialization of the United States.

He spent several years working on a replacement for brakemen's manual labor. Various models failed until, in 1868, he found a solution. He placed an air compressor inside the train driver's cabin and connected long air hoses to it. These hoses traveled the length of the train and were attached to brakes on each train carriage. This meant the driver could operate all the brakes on his own simply by allowing compressed air to pressurize the hoses that, in turn, provided the energy to activate the brake shoes. He obtained a patent for his air brakes in 1869 when he was just twenty-two years old, going on to found the Westinghouse Air Brake Company—the first of sixty Westinghouse companies.

The air brakes quickly became industry standard, and the vastly improved safety of trains meant they could travel much faster than before. By 1905 Westinghouse air brakes had been installed on more than two million train carriages as well as some 89,000 locomotive engines around the world. **DHk**

SEE ALSO: BAND BRAKE, DISC BRAKE, DRUM BRAKE, HYDRAULIC BRAKE, REGENERATIVE BRAKES

➦ *Should the Westinghouse brake system fail, reservoirs of compressed air automatically apply the brakes.*

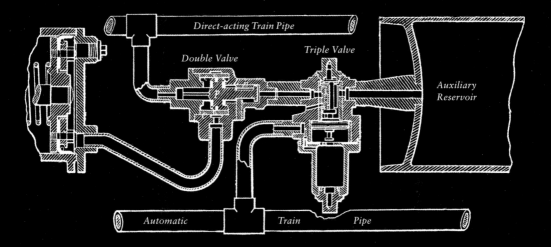

Direct-acting Train Pipe

Double Valve

Triple Valve

Auxiliary Reservoir

Automatic   Train   Pipe

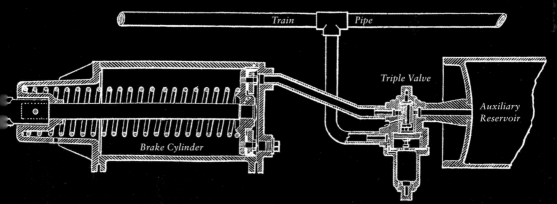

Train   Pipe

Triple Valve

Auxiliary Reservoir

Brake Cylinder

Electrical Generator (1869)

Gramme fulfills the dream of plentiful, cheaply produced electricity.

The dynamos produced by Michael Faraday and Joseph Henry in the 1830s were little more than laboratory curiosities. It was a Belgian industrialist and electrical engineer, Zénobe Théophile Gramme (1826–1901), who developed, in 1869, the first high-voltage, smooth, direct-current generator.

In 1871 Gramme and the French engineer Hippolyte Fontaine entered a manufacturing partnership. In 1873 the pair discovered that their dynamo machine was reversible and could thus be converted into an electrical motor. Their 1873 exhibit at the Vienna Exposition convinced the world of the ease of generating electricity and conversely that electricity could be reliably utilized to do heavy work.

By 1880 Sebastian Ziani de Ferranti had patented the Ferranti dynamo, a machine that he developed with the help of William Thomson (later Lord Kelvin). The London Electric Supply Corporation commissioned Ferranti to design the world's first modern power station, at Deptford, England. He designed the generating plant and its building and also the system for distributing the electricity it produced. Completed in 1891, this power station supplied high-voltage alternating current, the voltage being "stepped down" at street level for consumer use. The efficiency of the Ferranti system soon overcame the direct-current supply system favored by Thomas Edison and the Westinghouse Company.

The social and commercial implications of Faraday's revolutionary dynamo invention were huge—it was now possible to create energy in an efficient manner and distribute it on a large scale. **KB**

SEE ALSO: PUBLIC ELECTRICITY SUPPLY, ELECTRICITY METER

⬆ *In practical terms, Gramme's dynamo was the first to make an appreciable difference to human life.*

Cutting-wheel Can Opener (1870)

Lyman invents an efficient device for opening cans.

The invention of the tin can in 1810 was something of a revolution in the food industry, particularly for the armed forces and explorers. The only problem was how to open them. They were so thick and heavy that opening them was difficult, and was usually done with a hammer and a sharp instrument. In the 1850s cans were produced using thinner steel, and in 1858 the first can-opener patent was issued to Ezra Warner of Connecticut who devised an opener with a pointed blade and a guard to keep the blade from penetrating too far into the can. In 1868 J. Osterhoudt patented the keyed type of can opener used for sardine cans. It was not until 1870, however, and the efforts of William Lyman—also from Connecticut—that the "modern" version of the can opener came along.

Perhaps driven by an insatiable desire to taste the contents of a can without going through the cumbersome process of opening it, Lyman designed a user-friendly opener with a sharp rolling wheel that cut the rim of the can. This cutting wheel still forms the basis for can openers today. Lyman's invention was not without its problems however. The exact center of the can had to be pierced by one end of the opener, which then acted as the pivot for the cutting wheel—but this meant that it had to be adjusted for different sizes of can.

Lyman's design was improved and updated in 1925 when the Star Can Company of San Francisco produced a can opener that used two cutting wheels, the second one fitting below the rim of the can, which squeezed together with the top wheel to give a clean, steady, and smooth cut. **TP**

SEE ALSO: LEVER, CANNED FOOD, SELF-HEATING FOOD CAN, RING PULL

⬆ *With the point stuck in the can's center, the cutting wheel (under the wingnut) rotates around the edge.*

Sandblasting
(1870)

Tilghman improves the engraving process.

Legend has it that Benjamin Chew Tilghman (1821–1901), when based in the desert with the U.S. Army, noticed the effect that windblown sand had on the windows of the army buildings, and developed his ideas on sandblasting from there.

Tilghman proposed that grains of sand, or quartz, fired at high speed toward a hard surface would allow smoothing, shaping, cleaning, and engraving with greater accuracy and power than current methods.

Using compressed air, steam, or water to propel the grains, Tilghman was able to etch away at the surface of a substance harder than the grains themselves. Stone engraving was not new but

> "The sand had etched the glass ... and revealed the contrast against parts ... covered by steel mesh."
>
> H. J. Plaster, *A Tribute to Benjamin Chew Tilghman*

Tilghman's system replaced the expensive and time-consuming process of hand-chiseling, allowing the user to create effects on any hard surface. From then on, sandblasting has been used to clean buildings, carve glass, and engrave stone.

Since 1870 one minor adjustment has been made to Tilghman's original design. The silica dust produced during the sandblasting process was often inhaled and caused silicosis—a potentially fatal lung disease. Now, other materials tend to be used instead of sand. These include steel shots, glass beads, dry ice, baking soda, and walnut shells. Safety measures such as proper ventilation have also been adopted. **FS**

SEE ALSO: ICE RINK CLEANING MACHINE

Compressed Air Rock Drill (1871)

Ingersoll's drill revolutionizes mining.

The first step along the path to modern mining began about two and a half million years ago when humans first started using stone tools. Although stones would have been collected from the Earth's surface initially, there is evidence for some form of flint mining as early as one million years ago.

Early mining was extremely slow and labor-intensive. Although the discovery of metal and, much later, the introduction of explosives eased the process, the holes for the explosives still had to be made using hammers powered solely by brute force in order to drive a drill into the rock.

By the mid-1800s, efforts to develop a mechanical rock drill had begun in earnest, not only to increase the efficiency of the mining industry but also to help build tunnels for the first railways. In 1871 Simon Ingersoll (1818–1894) received a patent for his rock drill and, although not the first of its kind, it is considered to have revolutionized mining. Previous offerings were at best moderately efficient, cumbersome, and invariably unreliable. Ingersoll's drill surpassed the productivity of other drills due to its innovative lightweight design and was the first to be mounted on a tripod to resist the forces as the head recoiled. Initially steam-driven, a compressed air model was soon developed by the Ingersoll Rock Drill Company.

The increased productivity of mechanical rock drilling enabled the seemingly insatiable appetite of society for new technology, such as the construction of suspension bridges, ocean-going ships, roads, and railways, to be met. **RP**

SEE ALSO: DRILL, SEED DRILL, TWIST DRILL BIT, ELECTRIC DRILL

➔ *A team on the early 1900s' Minidoka Project in Idaho set up an Ingersoll drill to create an irrigation channel.*

Stock Ticker (1871)

Edison speeds transmission of stock prices.

"Good fortune is what happens when opportunity meets with planning."

Thomas Alva Edison

⬆ *This example of a stock ticker originally provided data for dealers in the New York Stock Exchange.*

➡ *A woman consults a pillar-mounted stock ticker in a photograph taken for an advertisement, circa 1890.*

The idea behind the stock ticker, which derived originally from telegraph technology, was to provide stock prices via a telegraph machine. It earned the name "ticker" because of the noise it made as prices came through, which also explains the name "tick" for the up or down movement in the price of a security.

E. A. Calahan of the American Telegraph Company invented the first stock ticker in 1867. Others came on the market shortly afterward, but it was not until Thomas Alva Edison (1847–1931) created the Universal Stock Ticker for Gold and Stock in 1871 that the machine's efficiency was greatly improved.

Edison created the "screw-thread unison" device that enabled stock tickers to be synchronized, allowing them to transmit the same information at the same time. This was a huge improvement on previous models, which had to be reset by hand when they fell behind the transmission they were receiving. This saved companies a huge amount of time and effort, because they did not have to send their employees running to the ticker every time they needed to be reset.

Edison improved the design further by changing the typewheel and paper feed, as well as devising a transmitter that worked like a typewriter keyboard. These improvements required much less battery power than previous tickers. Edison's ticker continued to be used until the 1930s, and the money he earned from this invention allowed him to set up his first laboratory and manufacturing plant in Newark, New Jersey. New tickers introduced in 1930 and 1964 were twice as fast as earlier models, but a fifteen to twenty minute delay between a transaction and the time of its announcement existed until real-time electronic tickers were launched in 1996. **RH**

SEE ALSO: ELECTROMAGNETIC TELEGRAPH, MORSE CODE, PRINTING TELEGRAPH (TELEPRINTER), DUPLEX TELEGRAPH

Automatic Lubrication
(1872)

McCoy automates locomotive oiling.

Elijah McCoy (1843–1929) was the son of former African slaves who had escaped to slavefree Canada. At around the age of sixteen he traveled to Edinburgh, Scotland, where he studied engineering. Upon his return to the United States he worked as a fireman and oilman on the Michigan Central Railroad. It was his job to ensure that the moving parts of the train's engine, as well as axles and bearings, were well lubricated.

Locomotive trains suffered considerable wear and tear on all their moving parts. Engineers had already devised a way to keep the axles lubricated by encasing them within oil-filled chambers. But since many parts of the engine ran under the immense pressure of

> *"We were satisfied that, with proper lubrication . . . , a little more power could be expected."*
>
> Orville Wright, aviation pioneer

steam, oil would tend to get propelled away from the moving parts. It was to this problem that McCoy turned his attention. His solution was a simple "lubricating cup"—a vessel containing oil that automatically dripped lubricant onto the moving parts at a rate that kept the engine oiled and prevented it from overheating.

McCoy produced more patents than any other black inventor before him. Attempts at selling imitations of his inventions almost certainly spawned a well-known catchphrase. Well-informed engineers would always seek out the best-designed model for their machinery and so they asked for "the real McCoy." **DHk**

SEE ALSO: LUBRICATING GREASE, OIL REFINERY

Vaseline®
(1872)

Chesebrough promotes a healing product.

Patented in 1872, Vaseline®—the trade name for soft paraffin, petroleum jelly, and petrolatum—was invented by the British-born chemist Robert Chesebrough (1837–1933). A semi-solid mixture of hydrocarbons, Vaseline was initially derived from the rod wax that came from the drilling of petroleum. Chesebrough created it by vacuum distillation of the crude rod wax, then filtered the residue from the still through bone char. The name is thought to come from the German for water (pronounced "vasser") and the Greek for oil (*elaion*).

The first Vaseline® factory opened in 1870 in Brooklyn, New York, and in 1911 the company built its first operation plants in Europe. Initially Chesebrough traveled around the United States selling his product by demonstrating its medicinal properties. He did this by burning his skin either with acid or with a naked flame, then rubbing the Vaseline over the wounds, and showing the healed areas of previous burns. It was primarily promoted as a treatment for grazes, burns, and cuts, but Vaseline® has no confirmed medicinal effects. Yet, it does help speed the healing process by sealing the wound from infection and moisture.

In the early twentieth century, it was mixed with additives such as beeswax, to make effective mustache waxes. Now used in many skin lotions and cosmetics, Vaseline® is a popular handbag essential, with uses ranging from lip balm to blister prevention.

Chesebrough lived to the age of ninety-six and claimed to have eaten a spoonful of Vaseline® a day. It is also said that when he was suffering from pleurisy he covered his body entirely in Vaseline®—and quickly recovered. **LC**

SEE ALSO: LUBRICATING GREASE, CHITOSAN BANDAGE

Cable Car

(1873)

Hallidie replaces horsepower with cables.

The first cable-operated railway was the London and Blackwall Railway, which opened in 1840. It consisted of a line 3.5 miles (5.6 km) long with hemp rope hauling the cars, but because the ropes wore out too quickly, it switched to steam locomotives in 1848.

In 1870 San Francisco attorney Benjamin Brooks proposed using cable cars to provide fast, inexpensive, and convenient access to the desirable heights of that hilly city. Horsedrawn cars worked well on level ground but had great difficulty with San Francisco's steep gradients. Brooks obtained a cable-line franchise from the city but was unable to obtain financing and sold it to Andrew Smith Hallidie (1836–1900).

Hallidie had been the first person in California to manufacture wire rope, and by 1871 he had two cable-car patents to his name. He hired engineer William Eppelsheimer, and the two designed the world's first practical cable car system. Their key innovation was a grip that allowed the car to stop by releasing the continuously moving wire cable and to start moving again by grabbing onto the cable. The San Francisco cable-car system began operating in 1873, proving both mechanically sound and financially successful.

For the next two decades, many cities around the world replaced their horse-driven cars with Hallidie cable-car systems. However the electric-powered trolleys available by the end of the nineteenth century proved to be less expensive to build and operate, and by 1957 only San Francisco retained an operating Hallidie system. It is still working today, but these days the tourists greatly outnumber the commuters. **ES**

SEE ALSO: LOCOMOTIVE, MONORAIL, ELECTRIC TRAM, MAGLEV TRAIN

↗ *Hallidie (standing at rear) controls San Francisco's first cable car on its inaugural journey in 1873.*

"I . . . have [decided to call my invention] the Improved Endless-Wire-Rope Way."

Andrew S. Hallidie

Modern DC Motor (1873)

Gramme produces movement from electricity.

Most kinds of mechanical movement in appliances are driven by electric motors—fans, fridges, and even computers are all powered in this way. In 1873 Frenchman Théophile-Zénobe Gramme (1826–1901) was the first to show that electricity could be used to move things efficiently. Semiliterate, and with only a grasp of simple arithmetic, he was not a typical inventor. However, his manual skill and logical thinking led to one of the most important applications of electricity.

A carpenter by trade, Gramme was appalled by the dirt produced by newly invented electric batteries and decided to concentrate his efforts on improving their design. It had not been long since Michael Faraday in Britain and Joseph Henry in the United States had created dynamos, which converted energy from movement into electricity. These are the devices that convert leg power to light in bicycle lamps and wind power into electricity. Gramme worked hard and greatly improved the original dynamo design. He opened his own dynamo-producing factory and buyers initially used his devices for electroplating and lighting.

Motors work in the opposite way to dynamos; they take electrical energy and turn it into something that can make things move. When showing off his improved dynamo design, Gramme connected a dynamo giving out electricity to a reversed device, which then turned the energy back into mechanical energy—demonstrating that it was possible to convert electricity into movement. This was the basis of the DC (direct current) motor. As electricity production became more efficient, the electric motor became an essential component in the development of most household appliances with moving parts—from washing machines to fans and blenders. **LS**

SEE ALSO: ELECTRIC MOTOR, DYNAMO, INDUCTION MOTOR, AC ELECTRIC POWER SYSTEM

Blue Jeans (1873)

Strauss and Davis combine denim and rivets.

The history of blue jeans can be traced to two men—Levi Strauss (1829–1902), a German who emigrated to the United States as a young boy, and the lesser known Latvian Jacob Davis (1834–1908), who moved to the United States in 1854.

In 1853 Levi Strauss moved to San Francisco where he set up a company, Levi Strauss & Co., selling buttons, scissors, bolts of cloth, and canvas. He also designed heavy-duty canvas work overalls for local miners. When his canvas supplies ran out, he began using heavyweight cotton twill, later known as denim. One of Strauss' customers was a tailor, Jacob Davis, who also made work trousers. His clients were complaining that the pockets kept ripping out, so Davis devised a method of strengthening the pocket corners and fly fastenings with metal rivets. This was an immediate success, but Davis did not have the money to obtain the patent, so he approached Strauss. An astute businessman, Strauss paid for the paperwork and the two men filed a joint patent for the new rivet-strengthened work trousers on May 20, 1873.

Davis went to work for Levi Strauss & Co., overseeing production of the new work trousers, which were not called jeans until the 1960s. They quickly became popular and their fame as the best work trousers spread throughout the United States. In around 1890 the patent ran out, allowing any company to manufacture riveted jeans. At the same time Levi Strauss & Co. assigned their jeans the number 501. The term *Levi's*, however, was not coined by the company but by the public, and the company trademarked the name. **TP**

SEE ALSO: CLOTHING, WOVEN CLOTH, BUTTON, BUCKLE, SEWING MACHINE, ZIP FASTENER

▣ *Jean-wearing miners pose by a wagon in the 1890s, a portrait tailor made for the rugged Levi's image.*

BLUE EYES
MINE

Automatic Fire Sprinkler (1874)

Parmelee begins a new era of fire protection with a permanently installed dousing system.

The automatic fire sprinkler can trace its earliest origins back to 1806 and an Englishman called John Carey. However, it was almost seventy years later that the first commercially viable fire sprinkler was invented.

In 1874 Henry S. Parmelee invented a sprinkler head for use in his piano factory. It had a single valve, plugged with a solder that would melt in a fire. Once melted, the valve was opened, releasing water through a perforated chamber. The design was such that only areas affected by fire would be doused.

A few modifications later and the Parmelee Sprinkler Company began to market the invention. After installation in a number of factories, where they were tested by real fires, the sprinklers soon spoke for themselves. Once insurance companies were on board, offering reduced premiums for factories with sprinklers installed, success was ensured. Parmelee took his sprinkler to the United Kingdom in 1881, establishing an industry there, too. From the time of its invention, Parmelee's fire sprinkler evolved constantly to achieve improved performance and reduced cost, resulting in ever-increasing sales.

From its invention, Parmelee's sprinkler evolved to achieve improved performance, reduced cost, and consequently greater sales. Today it is estimated that forty million sprinklers are installed around the world each year, massively reducing the financial, cultural, and personal losses caused by fires. However, many argue that their use is not widespread enough. Sprinkler installation remains a hot topic, with groups around the world campaigning for their compulsory fitting in both public and private buildings. **RP**

"Sprinklers typically reduce the chances of dying in a home fire by one half to two thirds."

National Fire Protection Association

SEE ALSO: CONTROLLED FIRE, FIRE EXTINGUISHER, SMOKE DETECTOR

◹ *While Parmelee's original sprinkler head resembled a salt shaker (top), later models were more streamlined.*

Electric Tram (1874)

Field opens up the American suburbs with fast commuting to the center of town.

To this day the tram remains the least glamorous of all methods of public transport, but at least American electrical engineer Stephen Dudley Field (1846–1913) tried to give the humble vehicle a little more pizzazz.

Field was not issued with a patent for his system of propelling railway cars by electromagnetism until 1880, even though it had been installed in New York City in 1874 by the twenty-eight-year-old inventor—a man often called the "father of the trolley car." The innovative method of providing electricity to the onboard motor worked by having a dynamo generate a current, which was conveyed via two metal wheels that linked the motor with either one of the rails. The system was pretty ineffective and actually highly dangerous, but it marked an important change in the way that people viewed public transport.

In the years between the implementation of the electric streetcar in many U.S. cities in the late 1880s and World War I (1914–1918), it became a very popular way of getting around. Because the electric streetcar traveled significantly faster than any of the other forms of transportation, the notion of commuting became a much more palatable option to city residents. Rather than having to inhabit houses right in the center of the cities, people could now move out along the lines of streetcar routes into the suburbs without causing too much disruption to their daily routine.

Although Dudley Field enjoyed great success in his lifetime, collecting more than 200 patents for various inventions, few can appreciate his work since none of them actually bear his name. Imagine—a streetcar named Dudley. **CL**

SEE ALSO: LOCOMOTIVE, MONORAIL, CABLE CAR, MAGLEV TRAIN

↗ The electric tram required the erection of a network of overhead cabling wherever it was introduced.

"You're talking about desire. The name of that . . . streetcar that bangs through the Quarter."

Tennessee Williams, *A Streetcar Named Desire* (1951)

Electric Dentist's Drill
(1875)

Green reduces the pain of dental treatment.

As long as 5,000 years ago, people were using bow drills to bore into teeth. Later, the Greeks and Romans drilled teeth for rudimentary purposes, but the art was lost in the Dark Ages. Precision drilling was reinvented by French physician Pierre Fauchard in 1728, but many crucial subsequent developments came from the United States. George Washington's dentist, John Greenwood, developed the first powered drill, enabling more accurate and faster drilling through the use of a foot treadle (taken from a spinning wheel).

Englishman George Harrington refined this with a motor in 1864, but the most important developments of the era were by an American engineer and inventor,

> *"The faster the drill rotated, the less discomfort the patient experienced."*
>
> Malvin E. Ring, dentistry historian

George F. Green from Kalamazoo, Michigan. In 1868 Green redesigned Harrington's drill using compressed air. He had worked on several prototypes while employed by the S. S. White Company of Philadelphia, the largest dental supplies manufacturer of the time. His pneumatic device was heavy and slow, but it improved upon the speed, 100 revolutions per minute (rpm), of hand drills at least fivefold.

By 1875, Green—whose other inventions included an electric railway and a grain binder—had perfected and patented an electric drill, which revolutionized dentistry. Forty years later speeds were up to 3,000 rpm, and today they can exceed 400,000 rpm. **AC**

SEE ALSO: TOOTHPASTE, TOOTHBRUSH, DENTAL FLOSS, DENTURES

Firearm Magazine
(1875)

Hotchkiss simplifies rifle loading.

The development of the firearm magazine was one of many improvements to weapons in the nineteenth century, and its creator's name is still recognized today by gun enthusiasts all over the world.

Benjamin Hotchkiss (1826–1885) worked as a gunmaker in Hartford, Connecticut, in the 1850s and 1860s. After the American Civil War the U.S. government had little interest in firearms, and like other famous firearm designers such as John Browning and Hiram Maxim, Hotchkiss moved to Europe to market his designs. He ended up in France in 1867 and set up a factory in St. Denis in 1875—the same year that he designed the bolt-action magazine rifle.

The story goes that Hotchkiss was on a train from Vienna to Bucharest when he became engaged in a conversation with a Romanian army officer who suggested the idea that bolt-action rifles needed to be developed further for military use. Hotchkiss took this idea and designed a novel loading system for the bolt-action rifle, building upon earlier magazine concepts from inventors such as Valentine Fogarty and Christopher Spencer. The new magazine consisted of a spring-loaded sleeve inside a cylinder that was attached to the buttstock of the gun. Hotchkiss designed the gun so that when the trigger was pulled, the firing pin was released and the new cartridge could move into the receiver.

Hotchkiss received a patent for his new model in 1876. Shortly afterward he sold these patent rights to the Winchester gun company, which began manufacturing the bolt-action magazine rifle in 1879. This gun, and further modifications of it, has been used extensively in the military and is also popular among hunters. **RH**

SEE ALSO: MUSKET, BREECH-LOADING GUN, REVOLVER, PORTABLE AUTOMATIC MACHINE GUN, GUN SILENCER

Safety Razor
(1875)

The Kampfe brothers take much of the risk out of shaving.

Before the safety razor was invented, the dangers of shaving were evident in the description of the traditional instrument as a "cut-throat" razor. Shaving was a tricky operation, typically carried out by barbers or trusted family members rather than individuals.

To protect the skin while shaving, a Frenchman, Jean-Jacques Perret, introduced one element of the safety razor in the late eighteenth century, namely a guard. Inspired by the design of a carpenter's plane, he used a wooden sleeve around the blade so that only the leading edge protruded. However, the first true safety razor, combining both a guard and a removable blade, was introduced in the United States in 1875 by the German Kampfe brothers (Frederick, Richard, and Otto). The Star Razor featured a hoe-type razor with a wedge-shaped blade with only one sharp edge.

In 1901 the American inventor King Gillette (1855–1932) and his colleague William Nickerson introduced the next innovation, the disposable blade. Defying skeptics who believed it impossible to create such blades, Gillette's ultra-thin, carbon steel, double-edged blades were a great success. This was reinforced when Gillette struck a deal to provide safety razors and blades to every member of the U.S. Army during World War I.

Later innovations included the introduction of longer-lasting stainless steel blades by the British company Wilkinson Sword (1965). This was followed by the even safer replaceable blade cartridge (1971), and the entirely disposable razor (1974). Environmental concerns have now prompted experiments with recyclable and biodegradable razors. **RBd**

SEE ALSO: DISPOSABLE RAZOR BLADE, ELECTRIC RAZOR

⊿ *Many patents for safety razors were awarded, including one for this Gillette model in 1931.*

"If I had been technically trained, I would have quit, or probably would have never begun."

King Gillette, American businessman

Four-stroke Cycle (1876)

Otto revisits the internal combustion engine.

German engineer Nikolaus Otto (1832–1891) was responsible for one of the great developments in motorized vehicles with the invention of his four-stroke cycle internal combustion engine.

After developing an interest in technology, he began designs for a four-stroke engine based on Lenoir's earlier design for a two-stroke cycle. In 1864 he set up N. A. Otto and Cie alongside Eugene Langer, creating the world's first engine manufacturers. In 1872, he employed Gottlieb Daimler and Wilhelm Maybach as technical director and chief designer, respectively.

In 1876 the first practical four-stroke engine was constructed. The four strokes are an intake stroke, where the piston moves down to allow a fuel-air mixture into the combustion chamber; a compression stroke, where the piston moves back up to compress the gases; a combustion or power stroke, where a spark ignites the fuel and the piston is forced down again; and a final exhaust stroke, where the piston

moves up to expel spent fuel via the exhaust valve. Although Otto patented his design in 1877 the patent was overturned in 1886 and instead granted to Frenchman Alphonse Beau de Rochas, who never built a working engine.

Initially combustion engines were stationary as they could not be adapted to run on liquid fuel and so required a pilot light. Otto solved this problem in 1884 with the invention of a magneto ignition system that created the spark needed for the power stroke. This increased the practicality of the four-stroke engine and allowed it to be used by Daimler and Maybach in the first motorcycles and automobiles. **HP**

SEE ALSO: INTERNAL COMBUSTION ENGINE, STIRLING ENGINE, GAS-FIRED ENGINE, TWO-STROKE ENGINE

↑ *The cycle of Otto's four-stroke engine is commonly simplified as "suck, squeeze, bang, and blow."*

Loudspeaker (1876)

Bell converts electrical signals to sound.

As a conduit for the output of all electronically created sound, the loudspeaker is one of the most significant inventions of the past 150 years. Indeed, in one form or another loudspeakers have been at the heart of much of the technology that has since emerged—from telephone, radio, and television to hi-fi music systems.

It was Alexander Graham Bell (1847–1922) who patented the first electrical loudspeaker in 1876, as part of his telephone system. In conjunction with his assistant, Thomas A. Watson, Bell created a simple design. A drum was covered with a tightly stretched goldbeater's skin (diaphragm) and a magnetized free-floating armature was placed at its center. The armature was able to vibrate against the skin and responded to changes in a magnetic field. This device was connected to Bell's "liquid" transmitter into which he uttered words that were heard clearly by his assistant in the next room. It was an extremely rudimentary design, but in principle the device

performed the same function as most contemporary audio systems: It turned electrical signals—those that had been derived and converted from an original source—into audible sound.

Bell himself was not specifically interested in the loudspeaker function, and he left it to others to make improvements to the device. A year later, German inventor and industrialist, Werner von Siemens (1816–1892), patented a greatly more sophisticated idea—a loudspeaker cone with a diaphragm controlled by an electromechanical transducer. This would eventually evolve into the moving coil principle on which most loudspeakers have since been designed. **TB**

SEE ALSO: FILM SOUND, STEREO SOUND, SURROUND SOUND, DOLBY NOISE REDUCTION

⬆ *In the 1920s much effort went into designing a horn that perfectly expressed sounds from the loudspeaker.*

Telephone (1876)

Bell propels communications into the modern era.

*"An amazing invention—
but who would ever
want to use one?"*

Rutherford B. Hayes, U.S. president (1877–1881)

↑ *Sixteen years after inventing the telephone, Bell
makes the first call from New York to Chicago in 1892.*

→ *Technical illustrations of telephone components
designed by both Bell and Thomas Edison.*

In the 1870s Edinburgh-born Alexander Graham Bell (1847–1922) was working on a way to improve the telegraph. Although this was well established as a means of long-distance communication, the fact that only one message could be sent at any one time made it extremely limited. Bell's original idea was to develop a "harmonic telegram," using multiple pitches to transmit more than one message at the same time. While working on this, an idea came to him for a more elaborate system—one that could transmit not only the dots and dashes of Morse code, but actual speech.

Several other teams were also pushing to transmit sounds via electricity, and there remains some controversy as to whether ideas were "borrowed" from other inventors, but it is undisputed that it was Bell who built the first working model—and all before his thirtieth birthday. Along with his assistant Thomas Watson, Bell honed his ideas and on March 10, 1876, he made the first-ever telephone call—"Mr. Watson, come here, I want to see you." The call was to his assistant in the next room, and according to Bell's own accounts, he had to shout into the apparatus to get it to work, but the technology had been proven. The Bell Telephone Company was founded the following year and within ten years some 150,000 households in the United States owned telephones.

Early telephones were far from practical—the first systems requiring battery acid—and they were used merely as a curiosity, with Watson and Bell putting on displays of the novelty aspect of their invention. Soon, however, they developed their invention into one of the most important modes of communication in the modern world, and the technology for sending information via electrical signal forms the basis for modern communications systems. **SB**

SEE ALSO: SEMAPHORE, ELECTROMAGNETIC TELEGRAPH,
CELLULAR MOBILE PHONE, SMARTPHONE

TELEPHONE

Graham Bell's first telephone

Transmitter   Receiver

Instruments Exhibited at Philadelphia in 1876

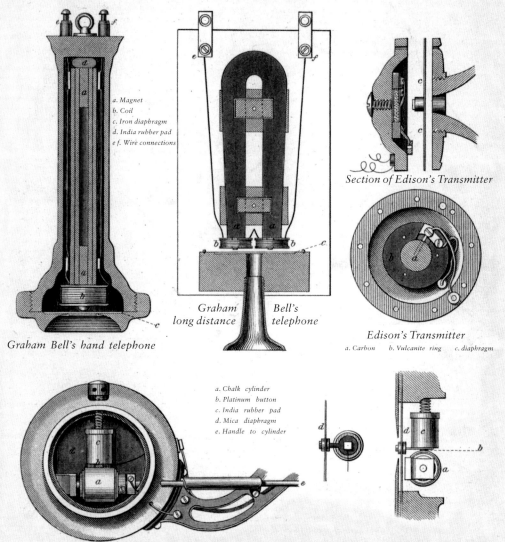

a. Magnet
b. Coil
c. Iron diaphragm
d. India rubber pad
e f. Wire connections

Graham Bell's hand telephone

Graham long distance Bell's telephone

Section of Edison's Transmitter

Edison's Transmitter

a. Carbon   b. Vulcanite ring   c. diaphragm

a. Chalk cylinder
b. Platinum button
c. India rubber pad
d. Mica diaphragm
e. Handle to cylinder

Receiver of Edison's Loud speaking Telephone

Carpet Sweeper (1876)

Bissell develops a better cleaning machine.

In the nineteenth century, the large Turkish or Axminster carpets that covered floors were not easily cleaned. They had to be taken outside and beaten vigorously with a rug beater to remove the dust.

American Melville Bissell (1843–1889) and his wife Anna noticed that the dust in their carpets irritated her and adversely affected his health. To alleviate the dust problem, Melville designed a long-handled carpet sweeper with rotating bristles that bent as they scooped up dirt from a carpet, then flicked it into a compartment inside the device. The rotation was powered by the sweeper's movement across the floor, and the head could be adjusted to clean bare floors, carpets, rugs, and uneven surfaces.

> *"The object of my invention is to adapt it to sweeping carpets and floors with uneven surfaces ... "*
>
> Melville Bissell

News of the Bissell sweeper spread. Local women began to manufacture the bristles at home and the sweepers were assembled in the shop before being sold door-to-door. He patented his design in 1876 and opened his first manufacturing plant in Grand Rapids in 1883 The company expanded internationally and had its big break in Britain after Queen Victoria allowed her palace carpets to be "Bisselled."

The advent of the home electrical vacuum cleaner saw sales of carpet sweepers gradually decline and the Bissell sweeper, although still available today as a rechargeable, motorized variant, has been relegated to something of a novelty item. **HP**

SEE ALSO: ELECTRIC VACUUM CLEANER, CYCLONE VACUUM CLEANER, SELF-CLEANING WINDOWS

Microphone (1876)

Berliner improves Bell's telephone system.

In 1870 nineteen-year-old Emile Berliner (1851–1929) left his native Germany and emigrated to the United States where he worked in a livery stable. There was nothing in his background or education—he had only the most rudimentary knowledge of electricity and physics—to suggest that he might have any impact on the emerging technology of the day. However, in 1876 he was so inspired by a demonstration of Alexander Graham Bell's telephone during the U.S. centennial celebrations that he decided to study the instrument. He discerned that its main weakness was the sound detector—the mouthpiece. The following year, working alone in his boarding house, Berliner created a new "loose contact" detector. This was arguably the earliest microphone because it increased the volume of the transmitted voice.

At this time, Alexander Graham Bell, who had recently founded the Bell Telephone Company, became aware that a young unknown inventor had submitted a patent covering a new transmitter for his telephone system, and he dispatched his assistant, Thomas Watson, to Washington to investigate. So impressed was he that Bell bought the rights to the invention for $50,000 and hired him as a researcher.

After seven years working with Bell, Berliner left to set up independently. After selling a number of his ideas to his former employer he began working in the field for which he is now best remembered—the early development of the gramophone record. In 1887 he worked out a way of recording onto a flat phonograph disc, over which a stylus moved horizontally, rather than vertically as in a cylinder. **TB**

SEE ALSO: LOUDSPEAKER, TELEPHONE, PHONOGRAPH, PHONOGRAPH RECORD

➡ *Using a microphone supported by a cord, Australian soprano Nellie Melba makes a recording in 1920.*

Phonograph (1877)

Edison unveils a device for recording and replaying sound.

"If we did all the things we are capable of, we would literally astound ourselves."

Thomas Alva Edison

⬆ *Edison's phonograph incorporated two heads, one for recording and one for reproducing sounds.*

➡ *Edison demonstrates his machine, which transmitted sounds via stethoscope-like ear tubes.*

On November 21, 1877, Thomas Alva Edison (1847–1931) announced the invention of the first device for recording and replaying sound—the "phonograph." Like the development of photography, it was a landmark invention that allowed for moments or periods in time to be captured in perpetuity. This worked by engraving a visual representation of a sound wave on a sheet of tinfoil wrapped around a grooved cylinder; the sound was captured as a series of indentations in the foil using a cutting stylus that responded to the vibrations of the sound being recorded. When a playback stylus passed over the cylinder a crude representation of the original recording could be heard.

As with so many of his inventions, Edison was spurred on in his efforts by his own hearing difficulties. The inventor's first recorded words were the nursery rhyme "Mary Had a Little Lamb." Edison considered cutting the indentations into a spiral groove on a flat disk, but he instead chose the outside of a rotating cylinder as it provided the stylus with a constant speed in the groove. Nevertheless, the principles that governed his invention would later evolve into the mass-produced gramophone record.

While Edison was working in New Jersey, French scientist, Charles Cros, was working on much the same concept. Cros had published his theories in April 1877, but he did not build such a machine. By the time he submitted his paper to the French Academy of Sciences, however, Edison had already given a practical demonstration of his invention, guaranteeing his place in the history books. The phonograph was the invention which made Edison famous, and he received a patent for it in February 1878, leaving Cros little more than a minor footnote. **TB**

SEE ALSO: PHONOGRAPH RECORD, MAGNETIC RECORDING (TELEGRAPHONE), VINYL PHONOGRAPH RECORD

Milk/Cream Separator

(1878)

Laval speeds up cream extraction.

Before 1878, the separation of cream from milk occurred through nothing more than the force of gravity. This process took time and was also inefficient since it limited the amount of cream that could be processed from the milk. Cream is formed when the lighter fat molecules of cream rise to the top of the milk through its heavier water-based fraction. This process happens naturally in raw milk when it is left to settle for a period of at least twenty-four hours. Cream can then be skimmed off or the milk can be drained from underneath, leaving only the cream behind.

Fortunately, help was at hand for the dairy industry. Stockholm Institute of Technology graduate Carl Gustaf de Laval (1845–1913) had been experimenting with centrifugal force as a way of separating fluids. In 1878 he invented a steam-driven device that spun a raw-milk sample at 4,000 revolutions per minute. Under the force exerted through the centrifugal effect of the device, heavier fraction is spun toward the outside of the rotating vessel while the lighter cream is kept in the center. Not only did Gustaf's machine extract cream faster than the conventional method, it was also more efficient, leaving less than 0.1 percent of cream in the milk.

Gustaf was a prolific inventor, perhaps owing to his background in both mechanical engineering and chemistry, and he went on to create many useful devices, including a miniature steam turbine and milking machines (first patented in 1894), with which his name is still associated. Laval's company, Alfa Laval, marketed the first commercially viable milking machine in 1918, five years after the inventor's death. But most importantly Laval invented a device that made making ice cream easier. **BG**

SEE ALSO: POWDERED MILK, PASTEURIZATION,
MILKING MACHINE

Time Zones

(1878)

Fleming coordinates timekeeping worldwide.

A time zone is a longitude band around the globe in which everyone sets their clocks to the same time—regulated by the movement of the sun. Before time zones were introduced, every town kept its own local time. But with the advent of the railways this system became very inconvenient, as the time at the starting point of a journey might well differ from the one at the terminus. In 1847 the railway companies in Great Britain recommended that all clocks should be set using the same time marker. Noon on the Greenwich Meridian (0 degrees longitude) was chosen, which is known as Greenwich Mean Time (G.M.T.).

Because Earth spins every twenty-four hours, local

"What then is time? If no one asks me, I know what it is. If I wish to explain it, ... I do not know."

Saint Augustine, theologian

---

time varies by one hour for every 15-degree change in longitude. Scottish-born engineer and inventor Sir Sandford Fleming (1827–1915) recognized this and suggested that not only should continental America have four meridians, but that similar 15-degree time zones should be established all over the globe.

In 1884 the Fleming plan was accepted for the whole world at the International Meridian Conference in Washington, D.C., and the Greenwich Meridian was chosen as the prime meridian. **DH**

SEE ALSO: LUNAR CALENDAR, GREGORIAN CALENDAR

➔ *This clock at the Royal Observatory at Greenwich, London, marks time at the Greenwich Meridian.*

Cathode Ray Tube (1878)

Crookes discovers that streams of light become streams of matter.

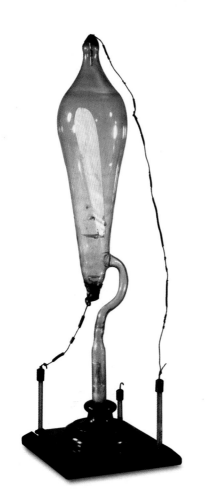

The mid- to late-1800s was a period of scientific revolution with physical processes such as electricity beginning to reveal their secrets. Early investigations into electricity led to the development of the cathode ray tube, which would eventually lead to the discovery of the electron, as well as to the invention of television.

Michael Faraday (1791–1867) noticed that after removing most of the air in a glass tube containing a cathode and anode, a faint glow could be seen between the positive and negative electrodes. Faraday's work was limited initially by the inability to create more than a partial vacuum.

Around 1855 German scientists Heinrich Geissler and Julius Plucker improved vacuum technology and were able to remove more of the air from inside such a tube. With the improved vacuum, Plucker was able to produce a much brighter glow between the electrodes. He was also able to demonstrate that the glow responded to the effects of a magnetic field.

English chemist and physicist William Crookes (1832–1919) had already made significant discoveries before he turned his attention to vacuum tubes. Using his Crookes tubes, which created an even better vacuum than those of Geissler and Plucker, he showed that the "cathode rays" that cause the glow traveled in straight lines, causing phosphorescence upon striking certain materials. Crookes also installed tiny vanes that would turn in the tubes as the current was applied. He thought he had discovered a fourth state of matter— "radiant matter"—but the true nature of this phenomenon had to wait until J. J. Thompson's studies showed these particles to be subatomic. **DHk**

SEE ALSO: TELEVISION, COLOR TELEVISION

"It [radiant matter] is projected with great velocity from the negative pole."

William Crookes, English chemist and physicist

⬉ *Crookes's tube revealed properties of the rays that pass from a cathode to an anode in a vacuum.*

Rebreather (1878)

Fleuss makes underwater breathing possible.

Scuba diving usually involves filling a tank with air, strapping it to your back, and breathing from it underwater. This simple system is called open circuit scuba (or self-contained underwater-breathing apparatus). Before this method of scuba diving caught on, however, people were using rebreathers. In 1878, Henry Fleuss built a diving system that allowed the user to breathe the same air over and over again. Using a rubber mask, a breathing bag, a copper tank, and a bit of string, he constructed the first scuba rebreather.

A rebreather works by removing carbon dioxide from the diver's exhaled gas and recycling its usable components. The contraption uses an expandable breathing bag to hold the exhaled gas and a system of valves to keep the gas flowing in only one direction. A carbon dioxide scrubber then filters the exhaled gas into a breathable form.

In the carbon dioxide scrubber, the exhaled air moves past an absorbent mixture often containing soda lime. The carbon dioxide reacts with the soda lime, and the other components of air pass through to be inhaled again. Air lasts longer with a rebreather, so users can use smaller tanks.

The rebreather is more efficient than normal scuba systems, whose users lose 75 percent of the available oxygen when exhaling. Rebreathers also produce dramatically fewer bubbles than their counterparts, thus helping naval divers to remain inconspicuous. Similar systems are used in space suits and by firefighters. Rebreathers are also now affordable enough for recreational divers to experiment without paying exorbitant costs. **LW**

SEE ALSO: STANDARD DIVING SUIT, AQUALUNG

↗ The somewhat cumbersome rebreathing apparatus; diving wear would replace the smart suit in actual use.

"When we exhale, a large portion of the oxygen we inhaled, around 80 percent, is exhaled . . . "

Adam Altman, Long Island Divers Association

Cash Register (1879)

Ritty puts a stop to petty pilfering.

Before the cash register came into being, short of sitting and watching every transaction taking place, there was no way for the boss or manager of a shop or other establishment to check exactly how much money was being taken over the counter each day.

Sick of his light-fingered staff pilfering the takings from his saloon in Dayton, Ohio, proprietor James Ritty (1836–1918) decided that he would design a system to put a stop to employee embezzlement.

The inspiration for the machine came to Ritty after he traveled by ship to Europe. While on board he became fascinated by a contraption that kept a record of how many rotations the ship's propeller had made. With this as inspiration, Ritty set to work on the prototype for the first cash register.

Ritty was trained as a mechanic, but he had given up manual labor in favor of running his own business. It therefore took him several attempts to create a working model but, with the help of his mechanic brother, the first cash register—"Ritty's Incorruptible Cashier"—was patented in 1879.

Ritty's cash register differed from today's versions significantly. For example, it had no cash drawer. The fledgling machines were simply devices to record that a transaction had taken place and counted the grand total on a dial like a clock face. Ritty continued to develop the machine, later adding paper rolls and pins to make a physical record of the transaction, and he began to sell them to the public.

Unable to run both the saloon and his cash register business, Ritty sold the company to Jacob H. Eckhart for the sum of $1,000. Eckhart in turn, sold it on to John H. Patterson. In 1884, the company was renamed the National Cash Register Company, which is still in operation today as NCR Corporation. **CL**

> *"Life is like a cash register, in that every thought, every deed, like every sale, is . . . recorded."*

Fulton J. Sheen, archbishop

⬆ *A 1910 product of the U.S. National Cash Register Company, set to sterling for the British market.*

➡ *An advertisement for the U.S.-made Union Cashier mentions the mental laxity allowed by the machine.*

SEE ALSO: TALLY STICK, BANKNOTE, CREDIT CARD, CASH MACHINE (ATM)

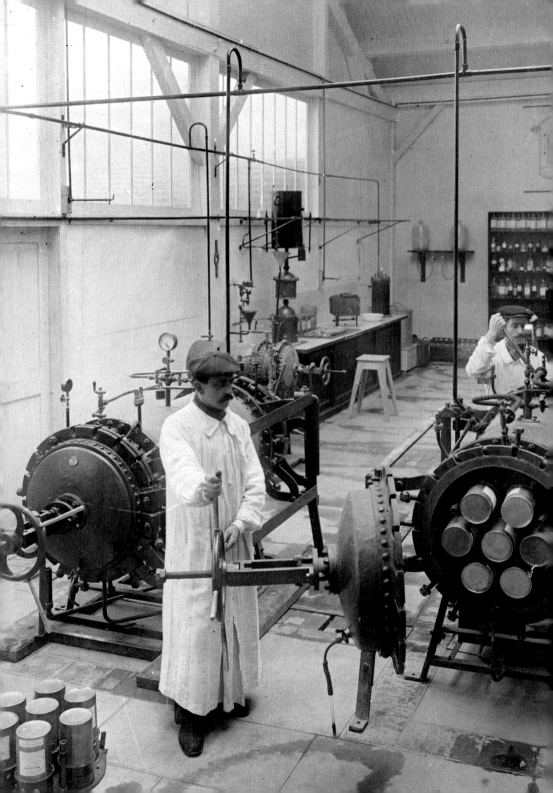

Medical Autoclave (1879)

Chamberland sterilizes surgical instruments.

In the 1870s, French microbiologist Louis Pasteur (1822–95) was still trying to disprove criticism of his germ theory. He had reported that boiled fluids, such as broth and urine, do not support bacterial growth if kept free of contamination. The British physician Harry Bastian, an outspoken critic, countered that boiled urine could indeed grow bacteria. Pasteur realized that to prove his germ theory, he needed to achieve temperatures greater than 212°F (100°C) and charged French microbiologist Charles Chamberland (1851–1908) with creating such a device.

Chamberland knew that if water were boiled under pressure, it could reach 250°F (121°C). Fifteen minutes at this temperature killed all known bacteria. He

> *"Will you have some microbe? … The Microbe alone is true, and Pasteur is its prophet."*
>
> French journalist mocking Pasteur in 1881

devised an autoclave, or "self lock," in 1879, for sterilizing surgical instruments. Chamberland's device was based on the 1679 "steam digester" of Denis Papin—the first pressure cooker.

Both the autoclave and the pressure cooker remain in use today. Sadly, the autoclave does not protect against the food-borne pathogen leading to "mad cow disease" or the variant Creutzfeldt-Jakob disease because prions, the causative agent, can survive ordinary autoclaving temperatures. **SS**

SEE ALSO: PRESSURE COOKER, ANTISEPTIC SURGERY,

CHITOSAN BANDAGES

◄ *French workers in 1905 place tubes of bandages inside an autoclave for sterilization at 250°F (121°C).*

Saccharin (1879)

Remsen and Fahlberg make a sugar substitute.

On February 27, 1879, Ira Remsen (1846–1927) and Constantin Fahlberg (1850–1910), two chemists from Johns Hopkins University in Baltimore, Maryland, were working on the oxidation of o-toluenesulfonamide, a coal tar derivative. Legend has it that both scientists went home for dinner and tasted a sweet residue on their foods, which originated from their unwashed hands. The next day they compared notes on this mysterious sweet chemical and checked their unwashed equipment. The result was a calorie-free, artificial sweetener that they later named saccharin. The scientists published their findings in 1880, though Fahlberg alone pursued a patent.

Due to its lack of calories and glucose, saccharin

> *"[I]njurious to health? … Anybody who says saccharin is injurious to health is an idiot!"*
>
> President Theodore Roosevelt

proved very successful with consumers. However, its safety has always been controversial. In 1907, food safety officials tried to ban its use, only to be thwarted by President Theodore Roosevelt, who was a fan. Its use was limited in 1911, but the restriction was lifted during World Wars I and II due to sugar shortages.

In the 1970s, rats fed large amounts of saccharin appeared to be at risk of developing bladder cancer. However, some scientists argued that this was due to the impurities in saccharin, not the saccharin itself. In 2000, President Bill Clinton signed a bill to remove warning labels from saccharin products. Saccharin's safety, however, continues to be debated. **RH**

SEE ALSO: EDIBLE MYCOPROTEIN, BREAKFAST CEREAL

Seismograph (1880)

Milne creates a sensitive device to record earth movements.

During an earthquake the ground moves up and down and from side to side as a result of a release of energy from the Earth's crust. The seismograph is an instrument that continuously records this movement (seismic waves) as a function of time. Crude seismoscopes were invented by the Chinese around 132 CE but these merely indicated the direction of the earthquake's epicenter. Later Iranian and Italian instruments containing mercury baths that spilled in measured ways when the earthquake occurred indicated both direction of source and magnitude of movement, but not time of occurrence.

British scientists Sir James Alfred Ewing, Thomas Gray, and John Milne (1850–1913) studied earthquakes and devices to record them whilst working in Japan. This resulted in the invention of Milne's horizontal pendulum seismograph in 1880. The idea of having networks of standard seismographs all over the world was promoted by Milne in the late 1800s.

Seismographs are designed to measure the movement of the Earth's crust during a quake. The main component is a large inertial mass suspended by a spring in a frame attached to the bedrock. When the ground shakes, the frame moves and the inertial mass does not. The varying distance between the mass and the frame is recorded, and this provides details of the quake. Initially the recording was done by pen on a moving paper chart, but today digital records are produced. Seismograph networks led to the discovery of the Earth's liquid core in 1905, the discontinuity between the crust and mantle in 1909, and the small solid inner core in 1936. **DH**

SEE ALSO: POLYGRAPH (LIE DETECTOR)

"Why have I stopped writing? I'd rather be a lightning rod than a seismograph . . ."

Ken Kesey, writer

Made by James White of Glasgow in 1885, this model records traces with three pens on one roll of paper.

Spring-loaded Mousetrap (1880)

Maxim develops a simple but effective form of rodent control.

Hiram Stevens Maxim (1840–1916) was an eccentric inventor who produced numerous creations throughout his life. He filed 271 patents in total, and his work included such diverse objects as a gun, a flying machine, curling irons, and a coffee substitute. Arguably one of his most famous inventions was also one of the simplest—the spring-loaded mousetrap.

Maxim is said to have built his automatic mousetrap at the tender age of fourteen, while he was working as an apprentice to a local carriage builder. His spring-loaded trap was tested at a local grist mill, which was subsequently rodent-free. He created a trap design that is still familiar today. It has a baited trip that releases a heavy spring-loaded bar trapping the mouse in its tracks. The device thus damages the mouse's spinal cord, skull, or ribs. However, contrary to popular belief the traditional Swiss cheese bait was found to be less effective than morsels such as peanut butter, chocolate, or meat.

Various improvements have been made to the mousetrap since the 1900s, with features such as the terrifying sounding mouth, as well as electric and bucket mousetraps. Humane traps are becoming more popular, allowing mice to be caught and returned to the wild without harm. A modern take involves an inert gas mousetrap, the RADAR device developed by Rentokil, which uses carbon dioxide gas to rapidly and painlessly kill the mouse. The users then get an e-mail alerting them to empty and reset the device.

Despite the plethora of designs on the market today, Maxim's classic spring-loaded trap remains instantly recognizable and is still widely used. **JG**

SEE ALSO: FISHHOOK

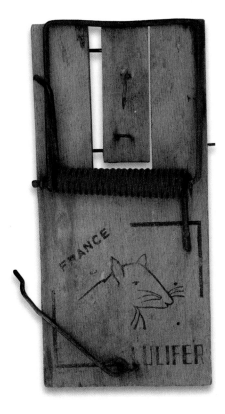

↗ Maxim's mousetrap design was simple to manufacture, as this early French example testifies.

"If a man makes a better mousetrap, . . . the world will [beat a] path to his door."

Ralph Waldo Emerson, writer

Cholera Vaccine (1880)

Pasteur introduces a new vaccine.

French chemist Louis Pasteur (1822–1895) was one of the truly brilliant minds of the nineteenth century, whose groundbreaking work in the field of microbiology and chemistry led to a long list of scientific discoveries. He is perhaps best remembered today for his development of "pasteurization" in milk, and the invention of a number of vaccines, including rabies, and anthrax. His discovery of a vaccination for cholera, however, was something of an accident, although he was aware of the work of Edward Jenner (1749–1823), who pioneered the smallpox vaccine.

In the summer of 1880, Pasteur was conducting experiments on chickens with cholera, and had instructed his assistant Charles Chamberland to inoculate the birds with a culture of cholera bacteria. Chamberland failed to do so, and a month later the culture, which had now spoiled, was used on the birds. They became ill but did not die, so Pasteur introduced a new group of chickens and inoculated them all with a fresh culture of the bacteria. Those that had received the old culture survived, but the new group of chickens all died, which led Pasteur, like Jenner before him, to realize that the weakened bacteria produced immunity in the subject. The difference between Jenner and Pasteur's work, however, was that Pasteur was using an artificially generated form of the bacteria. This allowed the vaccine to be created in large quantities and revolutionized the prevention of disease. Following his successes with cholera, Pasteur went on to create the vaccination for anthrax (1881) and then rabies (1895).

Despite Pasteur's successes and fame, there were still sceptics within the medical world during his day who failed to appreciate his advancements in the treatment of infectious diseases. **TP**

SEE ALSO: VACCINATION, RABIES VACCINE, ANTHRAX VACCINE, POLIO VACCINE, RUBELLA VACCINE, HEPATITIS-B VACCINE

Incubator (1880)

Tarnier and Martin help save premature babies.

Inspired by chicken incubators, which had been based on those depicted in Egyptian hieroglyphs, a French obstetrician by the name of Étienne Stéphane Tarnier enlisted the help of a poultry raiser, Odile Martin, to construct incubators suitable for human infants. This 1880 adaptation of an ancient design has gone on to save millions of lives.

The design was very simple: two chambers, one on top of the other with space for a baby in the upper chamber, and water heated by an oil lamp in the lower chamber. The lower chamber gently warmed the upper chamber whereas an opening in the uppermost compartment ensured that the infant could breathe. Since 1880 incubators have changed hugely with the

---

"The person who has health has hope; and the person who has hope has everything."

Arabic proverb

---

modern version housing some of the most sophisticated equipment humans have devised.

With around fourteen million babies born prematurely worldwide each year the need for a means to support fragile infants is clear. The exact number of lives this invention has saved is harder to determine. It is hard to imagine that just over a century ago premature babies were being placed in jars filled with feathers to help them through their perilous inaugurations into this world. **CB**

SEE ALSO: IRON LUNG, KIDNEY DIALYSIS, HEART-LUNG MACHINE, VENTOUSE

➡ *Dr. How's Electric Infant Incubator, shown here in 1903, progressed the ideas of Tanier and Martin.*

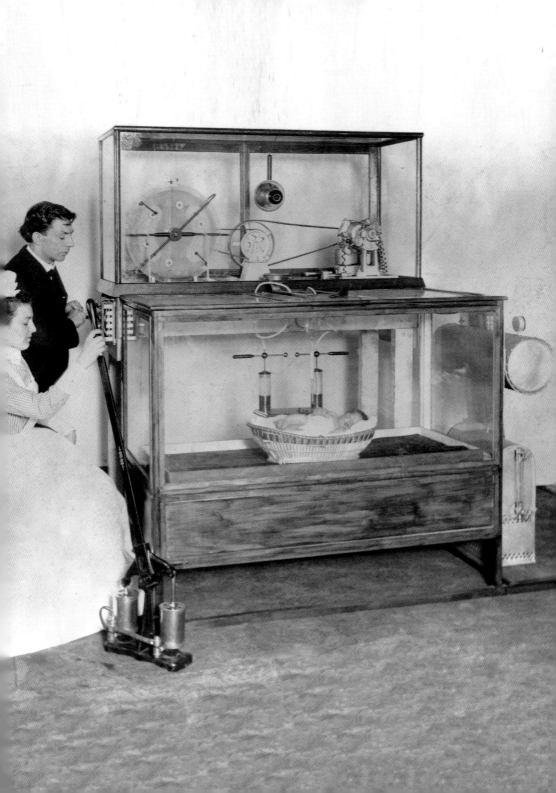

Ballcock

(1880)

Crapper devises a system to govern the flow of water in the toilet cistern.

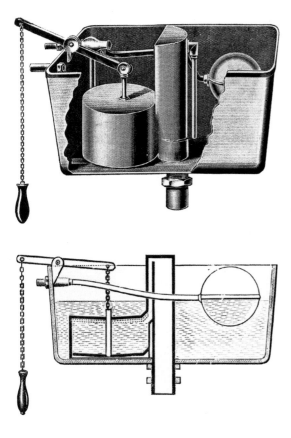

With his name as manufacturer proudly cast in the iron of countless toilet cisterns the world over, Thomas Crapper (1836–1910) is unlikely ever to be dissociated with visits to the lavatory. But one of the most disappointing of facts relating to inventions is that he did not himself invent the flush toilet. He did, however, popularize and endorse the flush toilet, and he was indeed a plumber. As a young child in the mid-nineteenth century he was apprenticed to a master plumber, gaining the same title himself at the age of twenty. His plumbing was exceptional and he even did work for members of the Royal Family.

Crapper's invention of the ballcock was one of nine patents he received, and one of three that were related to improving toilet design. The ballcock was a simple, air-filled, and watertight float that was attached to a valve inside a toilet cistern by means of an armature. As the cistern gradually filled up with water from the mains, the float would be lifted by the rising water level; at a preset point it caused the valve to close and stop the cistern from overflowing.

Crapper was a shrewd businessman and formed his own plumbing and bathroom fixtures company, one of the first in the world to feature a public showroom. Thomas Crapper and Co. Ltd. continues to make high-quality bathroom fittings, including faithful replicas of the originals designed by the man himself.

Today in Westminster Abbey, London, there are manhole covers bearing Crapper's company name. Fascinating to some, they have become a tourist attraction in their own right. **DHk**

"Raise your glasses, please, to the Crapper who installed the royal flush!"

Adam Hart-Davis, English scientist and broadcaster

SEE ALSO: FLUSH TOILET, SEWAGE SYSTEM, TOILET PAPER, S-TRAP FOR TOILET, COMPOSTING TOILET

▣ *Crapper's "water waste preventer" depended on a float (the ball on right) to control the inflow of water.*

Free-jet Water Turbine
(1880)

Pelton improves waterwheel efficiency.

During the American gold rush (1848–1855), people flocked from all over the world to California with dreams of finding great wealth. Lester Pelton (1829–1908) of Ohio was one of these migrants, but it was not in gold that he found success. It was with his free-jet water turbine, which he first patented in 1880, that he found fame and fortune.

Gold mining was becoming a large-scale industry and required ever-increasing amounts of power. As firewood supplies dwindled, steam power became very expensive and mining companies looked for an alternative energy source in the creeks and waterfalls surrounding the mines.

In 1866 Samuel Knight invented a water turbine that replaced the paddles of waterwheels—like those used in rivers to power flour mills—with cups to catch jets of water directed from above. While watching a misaligned turbine, Pelton noticed that the water ran down the edge of the cup rather than hitting the middle, making the wheel turn faster. This is because the amount of force generated by the water jet increases with the distance over which it is applied. To take advantage of this, Pelton split the cups in two with a metal wedge so that the water would hit the wedge and move down on either side and thus travel further. Pelton's design—the Pelton wheel—was more than 90 percent efficient, an improvement of about 14 percent on its closest rival.

The Pelton wheel remained the standard for decades and was the basis for later water turbines. These are still manufactured today, and many Pelton wheels still function around the world. Maybe the search for renewable energy sources will spark a revival in this method of water power. **RP**

SEE ALSO: WATERMILL, BALL BEARING, TIDAL MILL, FRANCIS TURBINE, STEAM TURBINE

Metal Detector
(1881)

Bell uses magnetism to find metal.

When U.S. President James Garfield was shot in an assassination attempt in 1881 the doctors called on Alexander Graham Bell (1847–1922) to find the bullet in his dying body. For this purpose Bell quickly threw together a makeshift instrument that could detect metal, but unfortunately the President was lying in a bed with a metal frame, unusual at the time, and this interfered with the instrument. Bell eventually realized why his detector was not functioning properly, but it was too late and President Garfield died shortly after.

Metal detectors rely on the relationship between electricity and magnetism. A current is run through a coil of wire, and this causes a magnetic field. If a metal

---

*"Don't ask me.
Ask the metal detector,
it's supposed to work…"*

Manny, from American crime drama series *CSI*

---

object passes through the field, then an electric current will be generated on the object. This, in turn, causes an opposite signal in the coil and this change alerts the user to the presence of metal.

Bell's crude metal detector was eventually improved upon, and the modern detector dates from the 1930s. Metal detectors are commonly seen today in airports, where they are used to detect concealed weapons. They are also used to detect land mines on the battlefield and to check food for shards of metal in factories. Simple metal detectors are cheap to buy and easy to make, and some take up metal detecting as a hobby to look for coins and other metal items. **LS**

SEE ALSO: METALWORKING, WROUGHT IRON, STEEL, POLYGRAPH (LIE DETECTOR)

Septic Tank

(1881)

Mouras's tank deals with human waste.

The treatment of human waste has varied throughout history from simply throwing it out on the street to complex modern sewage systems. Frenchman Jean-Louis Mouras invented a new type of waste system in the late nineteenth century, which is still used today by communities not connected to main sewer lines.

During the 1860s Mouras built a masonry tank attached to his house to collect human waste, which then overflowed into a regular cesspool. After about twelve years the tank was opened and Mouras discovered to his surprise that there were almost no solids in the tank. Mouras, along with scientist and priest Abbe Moigno, patented the tank in 1881.

"A mysterious contrivance consisting of a vault hermetically closed by a hydraulic seal . . ."

Jean-Louis Mouras

The working of the septic tank is simple in concept. Waste enters the tank at one end, is allowed to sit for a period, and is then discharged at the other end. The tank operates when full, so that the inlet liquid pushes out the same amount of outlet liquid. The liquid inside the tank consists of a scum layer on top, a sludge layer that settles to the bottom, and a fairly clear liquid layer in the middle. Decomposition happens via bacteria but solids eventually have to be removed. The bacteria in the tank and the retention of solids are both types of treatment of the waste. The partially treated waste is then released into a carefully situated drainfield, so as to not contaminate groundwater. **RH**

SEE ALSO: FLUSH TOILET, SEWAGE SYSTEM, COMPOSTING TOILET

Halftone Engraving

(1881)

Ives shows the way to print photographs.

The popularity of photography soared during the 1800s and with that grew a desire to print photographs in books and newspapers. However, the printing press was not capable of producing the continuous tone images, with infinite shades of gray, of photographs. It was not until 1881 that American Frederic Ives (1856–1937) developed the first successful halftone process.

Halftone printing involves converting continuous tone images to images made of dots of various sizes. The key to this system is that it exploits the limitations of the human eye. With the right resolution, the individual dots cannot be seen, resulting in the illusion of shades of gray, with larger dots appearing as darker shades.

"The truth is, a halftone is nothing more than a kind of magic trick."

Bill Stephens, printer

The first step is to produce a negative using a process camera. The process camera has a screen between the lens and the film that contains a grid. The grid divides the image into small squares through which light passes, resulting in discrete spots. The different size spots are made as a result of different amounts of light landing on the film; more light results in bigger spots. The negative is then used to produce an engraving, often by acid etching of a metal plate, with image areas left standing proud of the surface. The resulting plates are then coated with ink and the image printed. This cheap and effective halftone process is still used today, mostly in newspapers. **RP**

SEE ALSO: PRINTING PRESS, PHOTOGRAPHIC FILM, FILM CAMERA/PROJECTOR, HOLOGRAPHY, DIGITAL CAMERA

Revolving Door

(1881)

Bockhacker's idea stops people walking into each other going in and out of buildings.

The concept of a revolving door is not, for want of a better word, revolutionary. It is simply a rotating door made from several "wings" as opposed to one flat panel. But the architectural, social, and environmental implications of revolving doors are rather intriguing.

The first patent for a revolving door was awarded to H. Bockhacker in 1881. Although they did not become commonplace until much later, several patent applications were filed for revolving doors before the close of the Victorian age. They included one filed by Theophilus Van Kannel in August 1888 for a three-winged "storm door structure" to guard against the elements. Van Kannel also highlighted the fact that his door only rotated in one direction and could therefore control the flow of people traffic and minimize the risk of a collision.

A well-designed door can do more than stop people walking into each other. The most frequently made claim about revolving doors is that they reduce energy loss by keeping warm air from escaping—four rotating wings mean that there is never a direct passage between inside and out. Research carried out at the Massachusetts Institute of Technology (MIT) showed that revolving doors can actually save significant amounts of energy.

In 2008, designers at Fluxxlab in New York unveiled plans for a revolving door that not only saves energy but creates it. Their "Revolution Door" technology, which works in a similar way to a wind turbine, captures the kinetic energy generated by the movement of a revolving door and turns it into electricity. **HB**

SEE ALSO: STEEL-GIRDER SKYSCRAPER, ESCALATOR, AUTOMATIC DOOR

⬈ *Theophilus Van Kannel's storm door structure was one of several that exploited Bockhacker's concept.*

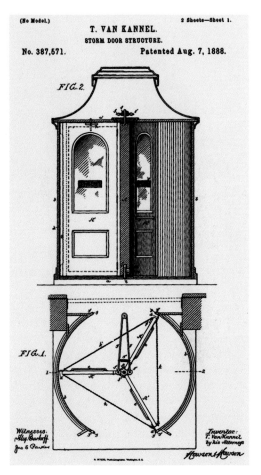

(No Model.)   2 Sheets—Sheet 1.

T. VAN KANNEL.

STORM DOOR STRUCTURE.

No. 387,571.   Patented Aug. 7, 1888.

FIG. 2.

FIG. 1.

Witnesses:
Inventor:
T. Van Kannel
by his attorneys

"Great works are often born on a street corner or in a restaurant's revolving door."

Albert Camus, writer

Blowtorch (1881)

Nyberg invents a compact and highly versatile tool for directing heat.

From the late 1890s Swedish inventor Carl Nyberg (1858–1939) was interested in solving the problem of manned flight. His numerous experiments with his Flugan (The Fly) flying machine generally proved unsuccessful, much to the amusement of local onlookers. Nyberg may not have suceeded in achieving his ambition of flight, but his flying machine was powered by a steam engine heated by four blowtorches, and it was the latter—a handy tool still widely used today—that he gave to the world.

Nyberg, a prolific inventor who also worked on cookers, steam engines, and boat propellers, invented the blowtorch in 1881, although the actual patent application was made by business entrepreneur Max Sievert who showed an interest in Nyberg's invention and began selling it from about 1886.

The blowtorch consists of a cylinder filled with fuel, usually propane, butane, or liquid petrol gas. This is vaporized and then mixed with oxygen (from the air) in a combustion chamber before being ignited to form a flame. The pressurized fuel, which issues from a small nozzle, gives the flame its direction and "strength," allowing the intense heat from the blowtorch to be applied to relatively small areas.

Blowtorches perform a wide variety of functions. Because the flame is substantially cooler than that of an oxyacetylene torch, the blowtorch cannot be used for welding or cutting, but almost anything else is possible. Blowtorches of various sizes can be found everywhere from the kitchen to the toolbox of the professional mechanic, performing tasks as diverse as crisping up a crème brûlée and soldering metal. **BG**

SEE ALSO: CONTROLLED FIRE, LIQUEFACTION OF AIR, MODERN FLAME THROWER, WELDING

⬆ *The design of the blowtorch has changed little since its introduction by Nyberg in 1881.*

Electric Iron (1882)

Seely makes light of ironing with his labor-saving device.

Before the arrival of the electric iron, various methods were used to smooth out washed and wrinkled clothes. Charcoal-filled pans dating back to ancient China lasted through to the seventeenth century when they were replaced by cast iron flat irons, which were heated up in open fires. By the late nineteenth century, flat irons were being heated by a range of fuels including kerosene and animal oils.

But ironing was a sweaty, tiring, and dirty task typically involving a hot coal stove and numerous flat irons, which required continuous heating. With the coming of electricity it was inevitable that someone would spot an opportunity. Henry W. Seely, an American inventor based in New York, was the first to develop and patent an electric iron in 1882. The electric iron uses resistive heating—heat produced by resistance to an electric current. This is used to warm a metal hot plate, today made of aluminum or stainless steel. The plate is attached to a handle to make it portable. It was difficult to control the temperature of the plate of early electric irons and they were dangerous to handle. Seely and his partner Richard Dyer (Thomas Edison's patent lawyer) responded by developing a type of cordless iron that sat, when not in use, on a stand heated by electricity.

By the 1920s most homes in developed countries had electricity, and electric irons were very popular, being fast, efficient, and clean. A thermostat that controlled the temperature by switching the current on and off was introduced in the 1930s, followed by the modern steam iron. By 1941 more than three-quarters of all U.S. homes had an electric iron. **RBd**

SEE ALSO: FOLDING IRONING BOARD

⤒ *This early electric iron, heated by carbon arc electrodes, was produced in France circa 1905.*

Cantilever Bridge (1882)

Baker and Fowler accomplish a daring feat of structural engineering.

Between 1882 and 1890, construction of one of the most ambitious engineering projects of the time took place near Edinburgh, Scotland. The project was to create a railway bridge that would span the Firth of Forth, one of Scotland's major tributaries, and connect the northeast and southeast of the country. The men who stepped forward to take up this challenge were Benjamin Baker (1840–1907) and John Fowler (1817–1898). Artist William Morris described it as "the supremest specimen of all ugliness," but their design became a national icon and set a new standard in engineering.

Baker and Fowler were chosen in 1882 to replace the previous designer of the Forth Rail Bridge, Sir Thomas Bouch, when one of his projects, the Tay Bridge, collapsed in 1879 killing seventy-five people. Baker and Fowler had an established pedigree of engineering in Victorian Britain, their achievements including the construction of the Metropolitan Line, the first underground line in London, as well as many other railway bridges. They opted to design a cantilever bridge to span the Firth of Forth, using 64,000 tons of steel as their building material. This was the first bridge to be built from this material.

The principle behind a cantilever bridge is one of balance. The bridge is projected out over the gap that needs to be spanned and counterbalanced at the shore end. Often two bridges are used, one from each side, and where the ends meet there is a third section, a simple beam bridge, to cover the gap. The idea of using cantilevers in bridges was not an entirely new one. But what made Baker and Fowler's construction unique was the sheer scope of the project. **BG**

SEE ALSO: ARCHED BRIDGE, PONTOON BRIDGE, SUSPENSION BRIDGE, TRUSS BRIDGE

⊞ *At more than 3,280 feet (1,000 m), the Forth Rail Bridge is the second longest cantilever bridge in the world.*

Public Electricity Supply (1882)

Edison switches on electricity for the masses.

In the fall of 1882 part of New York's lower Manhattan flipped the switch on what seemed to many to be an unholy miracle—a centralized, commercial electrical system providing both power and light. The power station at its hub stood on Pearl Street, in the capital's financial district. This was the first permanent system of its kind. It used direct (as opposed to alternating) current and 3,000 electric lamps. The man behind it was the irrepressible multiple inventor and "wizard of Menlo Park" Thomas A. Edison (1847–1931).

In the late 1870s one of the greatest quests of practical science had been to replace large, powerful electric arc lamps, which overheated easily, with smaller, safer lights. Edison's Electric Light Company, backed by a brace of prominent financiers including J. P. Morgan and the Vanderbilts, set about creating a parallel circuit where the current was divided between a string of small lamps (as opposed to the series circuit of the arc lamp). The aim was to prevent the entire

circuit from blowing if one lamp failed. Using the type of carbon technology he had pursued for his carbon-button transmitter and phonograph of 1877, Edison produced carbonized bamboo bulb filaments. He also devised generators, junction boxes, safety fuses, sockets, and other related equipment to create a whole system to be used at Pearl Street.

Extensive networks with large central power stations, as opposed to those in individual buildings, took time to take off, and gas lights remained common for some time. However, a revolution had been set in motion. Tungsten bulbs appeared around 1915 and produced a much whiter light. **AK**

SEE ALSO: **ARC LAMP, INCANDESCENT LIGHT BULB, GAS MANTLE, TUNGSTEN FILAMENT, ANGLEPOISE® LAMP, HALOGEN LAMP**

⊡ *An 1882 illustration of the inside of Edison's Dynamo Room at the Pearl Street Station in New York City.*

Electric Fan

(1882)

Wheeler turns down the heat with his electric desktop fan.

Being too hot must have been a major problem for people before the late 1800s. As soon as electrical power was introduced, inventors started to work on ideas for the electric fan.

Dr. Schuyler Skaats Wheeler (1860–1923) was the American engineer responsible for creating the personal two-blade desk fan—an invention beloved of anyone who has ever held down an indoor job in the summer months. Invented by Wheeler at the tender age of twenty-two, the fan was made of brass, with no protective caging surrounding the rotating blades, resulting in a product that was both stylish and dangerous in equal measure.

However, like most inventions of that time that used electricity, when they were first introduced these fans were the reserve of the rich and the powerful. It was not until the 1920s, when industrial advances meant that fan blades could be mass-produced from steel, that prices started to drop and the ordinary homeowner could afford one.

Aside from his fan, Wheeler also became known for employing a large workforce of sightless people. He noticed that his sighted employees who were skilled at winding coil did so without ever looking at their hands. He blindfolded himself to see if he could wind coil without looking and found that, with a little practice, he could. The number of blind individuals in the population had increased as a result of World War I. Wheeler set up a department at his factory that employed only sightless men and women, putting them on a par with their sighted contemporaries. **CL**

"What is my loftiest ambition? I've always wanted to throw an egg at an electric fan."

Oliver Herford, writer, artist, and illustrator

SEE ALSO: KITCHEN EXTRACTOR, JET ENGINE, HOVERCRAFT

◤ *This fan, patented in Italy in 1927, was called the Nuvol Acchiappamosche (New Fly Catcher).*

Induction Motor
(1883)

Tesla's motor uses alternating current

One of the most important inventors in history, Nikola Tesla, was born in 1856 in Smiljan, Croatia. His inventions would revolutionize our world. Among his almost 300 patents were wireless communication, the alternating current, and the induction motor.

Tesla built the first working induction motor in 1883. Michael Faraday had demonstrated an electric motor in 1821 and Zénobe Gramme went on to invent the modern direct current motor in 1873, but it is Tesla's motor that most of our household appliances rely on.

The induction motor works using alternating current rather than direct current. It has a simple design and is significantly less expensive to

"The practical success of an idea . . .is dependent on the attitude of its contemporaries."

Nikola Tesla

manufacture than the direct current motors. It also has fewer parts to wear out and is thus more reliable.

The induction motor does more than run your vacuum cleaner. It is also used extensively in industrial settings to power machine tools, conveyer belts, and a variety of other applications. However, induction motors are not well suited to applications that require precise speed control or low-speed operations. For that reason, computer disk drives, laser printers, and photocopiers typically use direct current motors.

Because of Tesla's inventions, we have electricity in our homes and the means to convert this electricity into useful work. **ES**

SEE ALSO: ELECTRIC MOTOR, PUBLIC ELECTRICITY SUPPLY,
AC ELECTRIC POWER SYSTEM, ELECTRIC VACUUM CLEANER

Image Rasterization
(1884)

Nipkow's disk paves the way for television.

Selenium has a lower electrical resistance when it is soaked in bright light than it does when it is in darkness. This means that by altering the light shone onto a selenium cell, the amount of electricity that can pass through it can be changed. This, combined with the knowledge that all we see is made up of different shades of light and dark, allowed German engineer Paul Nipkow (1860–1940) to come up with a way to convert pictures into electrical signals.

Nipkow's system worked by using a rotating disk that has a spiraled series of holes punched in it. When the disk turns, the moving holes break up the image into a series of varying light signals. When light passes

"My father hated radio; he couldn't wait for television to be invented so he could hate that too."

Peter De Vries, writer

through the holes onto the selenium cells, the selenium reacts. The electrical signal created by passing a current through these selenium cells can then travel by wire to power a lamp set up elsewhere. If a second disk rotating in sync with the first is placed in front of this lamp, the light signals will match up with the holes in the second disk, creating a replication of the original image. If this is done at high speed, the eye will no longer be able to recognize that the light signals are being emitted in quick succession and will instead see one whole image. This process of breaking up an image into small dots, or pixels, is the basic principle on which television is based. **CL**

SEE ALSO: KINETOSCOPE, ICONOSCOPE

Punchcard Accounting

(1884)

Hollerith automates information processing.

The first job German-American statistician Herman Hollerith (1869–1929) had was with the U.S. Census Bureau. His task was to collate information gained from the 1880 census by hand and he quickly realized that the process would be much faster and less prone to error if it was automated.

Trying to work out a solution to the problem, he arrived at the idea for a system based on punchcards while observing a bus conductor punch holes in tickets. He filed his first patent in 1884.

The punchcards were not entirely novel—the French weaver Joseph-Marie Jacquard had invented a way of controlling the warp and weft on his loom by patterns of holes in cards—but Hollerith's design for a tabulator and sorter were original.

The information was read from each card using an array of spring-mounted brass pins that formed an electrical connection through any holes in the card. The tabulator then had many output dials that could display the information contained on each card. In addition to reading information, the sorter part of the machine enabled the operator to select for certain characteristics such as sex, marital status, or profession using an array of switches. Cards that matched the specified criteria would be automatically gathered in a special container, allowing statisticians to gather data to their hearts' content for the first time.

Hollerith's machine was used in the 1890 census and reduced the time spent collating the data by over half, saving the country $5 million. Hollerith formed the Tabulating Machine Company in 1896 to market his invention, which in 1924 changed its name to International Business Machines (IBM). His system was used in computers until the late 1970s. **HP**

SEE ALSO: PUNCHED CARD, POWER LOOM, MECHANICAL COMPUTER, COMPUTER PROGRAM

Linotype Machine

(1884)

Mergenthaler prints the news for the masses.

The basic process of typesetting using movable type—handpicking metal letters to mount them in a rack or plate for printing—developed little in the 400 years following Gutenberg's printing press of 1436. In 1822, William Church of Boston, patented a machine that chose brass-reversed letters from a bank to create a continuous line of text, which had to be finished by hand with spacing, line breaks, and justification. However, it was not commercially successful.

In Baltimore in 1884 German-born Ottmar Mergenthaler (1854–1899) patented designs for the Linotype machine, which used a ninety-character keyboard to select brass molds (called "matrices") for letters and other characters from a font magazine and mounted these into an assembler, creating one short line of text (hence "lin'o'type"). Tapered spacebands, larger than the matrices, were added between words and used to justify the line by wedging words apart. Assembled lines were then used to cast thin slugs of molten lead alloy that was cooled quickly in water, creating lines of text that were mounted into a plate and used for printing. The matrices were notched to identify their characters, so that they could be mechanically sorted and returned to the correct compartment in the magazine. After printing, the alloy slugs were melted down for reuse.

After being installed at the *New York Tribune* in 1886, the Linotype machine was rapidly adopted by the newspaper industry throughout the world. It speeded up the composition process and reduced the amount of skilled labor required, bringing down costs and accelerating an expansion in newspaper and magazine publishing that continued well into the twentieth century. **EH**

SEE ALSO: ROTARY PRINTING PRESS, PRINTING PRESS, HALFTONE ENGRAVING, MONOTYPE MACHINE

Portable Automatic Machine Gun

(1884)

Maxim's gun changes warfare forever.

American inventor Hiram Maxim (1840–1916) would have been much more famous had he won the battle with Thomas Edison over the credit for the invention of the electric light. But his other inventions proved to be just as important, at least in the development of firearms in the late nineteenth century.

Maxim first introduced the principle of his portable, automatic machine gun in an 1883 patent. Several features of Maxim's gun made it innovative. The action was completely automatic and the user needed only to keep his finger pressed on the trigger to fire the gun. The recoil energy of the shot was used to extract the old shell, reload a new one, and automatically fire. The gun had a single barrel that was surrounded by a jacket filled with water to cool the barrel after it was heated by firing. The gun was lighter and more portable than previous weapons.

The U.S. government wasn't interested in buying Maxim's invention, so he took it to Europe where he was able to market it to both the English and the Germans. Maxim became a British citizen in 1901 and was knighted by Queen Victoria. His company, Vickers, Son, & Maxim, was very successful at marketing the deadly device throughout Europe, especially Germany.

The guns were licenced and manufactured by companies in Britain and Germany and the first of the so-called Maxims was delivered to the German navy in 1894. Both sides used Maxim machine guns during World War I. The Maxim guns evolved over the years to become even more effective, and served as the prototypes for today's modern machine guns. **RH**

SEE ALSO: GUN, MUSKET, CANNON, ARQUEBUS, FLINTLOCK MECHANISM, BREECH-LOADING GUN

↗ *Sir Hiram Maxim with one of his Maxim machine guns, which could fire 500 rounds per minute.*

"[It will] make it easier and quicker for these Europeans to cut each others' throats."

Anonymous acquaintance of Hiram Maxim

Kraft Process (1884)

Dahl's process leads to stronger, better paper.

The first modern paper was invented in 105 CE by a Chinese court official called Ts'ai Lun. He made sheets out of mulberry bark, rags, and hemp waste mixed with water. Paper continued to be made primarily of rags until the early nineteenth century, when mechanized papermaking took off. Soon, demand for paper far outweighed the supply of rags, and wood was explored as a substitute.

In order to make paper from timber, the plant fibers must be turned into a pulp, which is then spread onto a flat screen. When the fibers dry, they stick together forming a sheet of paper.

In 1866, Benjamin C. Tilghman invented the sulphite process, where wood is heated in a liquor

> *"My head is full of [ideas] . . . but they serve no purpose there. They must be put down on paper."*
>
> Camilo José Cela, writer

containing an excess of sulfur dioxide to create pulp. Then in 1884, German inventor Carl F. Dahl (c. 1813–1865) found that using caustic soda and sodium sulphate in a "white liquor" resulted in a much stronger pulp. This pulp produced paper that was more resistant to tearing and so the process was named kraft after the German word meaning "strength".

The kraft process had the advantage of being capable of pulping pine trees. It also had higher recoverability than the sulphite process so the chemicals could be recovered for future use, making it more efficient. It overtook the sulphite process as the dominant form of pulping and is still used today. **HI**

SEE ALSO: PARCHMENT, PAPER, FOURDRINIER MACHINE

Dissolvable Pill (1884)

Upjohn's pill eases taking the medicine.

In 1884 William Upjohn (1853–1932), a U.S. doctor from Michigan, invented the first pill that could dissolve in the stomach. Upjohn had the innovative idea of introducing a starter particle into a revolving pan. As the pan turned, the starter was sprayed with powdered medicine, building up the new pill layer by layer. The number of layers controlled the strength of the pill. The resulting "friable" (easily crushable) pill dissolved when ingested. Prior to Upjohn's development, patients had to ingest drugs in liquid form or as pills with hard coatings. The problems were that the dosage of liquids was inconsistent and the pill coatings were so hard that they did not always dissolve, meaning that patients received no benefit.

> *"Never under any circumstances take a sleeping pill and a laxative on the same night."*
>
> Dave Barry, humorist

After patenting his invention in 1886, Upjohn developed a machine to mass-produce his pills. With his brother he set up the Upjohn Pill and Granule Company. The reputation of the friable pill quickly spread, thanks to Upjohn's clever marketing strategy. He sent small pine boards to thousands of doctors, together with samples of his friable pills and his rivals' hard pills, inviting the doctors to crush the pills into the boards to see which one was the most digestible. Early Upjohn products included quinine pills and the first candy laxatives. Over the next century the company manufactured 186 different medications in pill form. The dissolvable pill is still in use today. **JF**

SEE ALSO: ASPIRIN, BIRTH CONTROL PILL

Franking Machine (1884)

Bushe's machine puts a stamp on the business of mail.

Sticking a stamp to a letter is a fairly trivial matter if you only have one letter. But if you have hundreds of letters then the process of licking and sticking each individual stamp becomes very time-consuming.

In the late nineteenth century, Frenchman Carle Bushe first conceived and patented a machine that would print a stamp on an envelope and record the amount of postage payable on a meter. Today the post office supplies users with franking machines, set up with a pre-paid credit limit, which stamps envelopes and registers the cost of the postage used. The credit is then topped up when necessary.

Franking originally dates back to the seventeenth century when Members of Parliament (MPs) regularly sent hundreds of official letters and were given the privilege of free postage. In those days they inscribed their signature on the top of the envelope to circumvent the stamp and to get free postage, but this was open to abuse, and there were reports of some providing their signatures to friends and family and of others writing to their constituents as part of their re-election campaigns.

Bushe designed his machine in 1884 but sadly his invention remained on paper, and no working model of the machine is known to have been built. In his patent application he clearly recognized the time savings of such a machine. It was left to Chicago inventor, Arthur Hill Pitney, to successfully gain his patent for a mailing system in 1902. It consisted of a manual crank, chain action, printing die, counter, and lockout device. He formed a company, which became the American Postage Meter Company in 1912. **RB**

SEE ALSO: POSTAGE STAMP

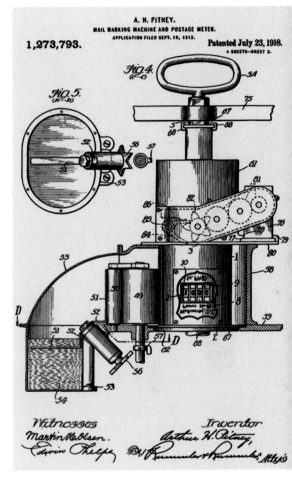

⬈ *This franking machine and postage meter improved upon Bushe's design and was patented in 1918.*

"Adhesive stamps . . . [entail] a serious loss of time when hundreds of letters have to be despatched."

Carl Bushe

Steel-girder Skyscraper (1884)

Jenney devises a way of building in the sky.

"The skyscraper establishes the block, the block creates the street, the street offers itself to man."

Roland Barthes, literary and social theorist

⤒ *A 1952 advertisement for the Otis Elevator Company glorifying the future possibilities of skyscrapers.*

⇥ *The Home Insurance Company Building was the world's first skyscraper; it was demolished in 1931.*

Before the advent of the skyscraper, tall buildings were built to showcase great wealth, power, or religious beliefs. For the architect and civil engineer William Le Baron Jenney (1832–1907), the urge to build great edifices was born from a necessity to solve commercial (and later residential) needs in his native Chicago, where ground space was at a premium.

Two obstacles to the construction of highrise buildings were overcome in the mid-nineteenth century, paving the way for the skyscraper. In 1853 Elisha Graves Otis devised a mechanism to prevent elevators from falling if their cable broke, enabling passengers to be transported upward safely. The second breakthrough came with a steel-framed structure that could support the entire weight of its walls, instead of the traditional load-bearing walls that carry the building's weight.

Jenney's ten-story Home Insurance Company Building, built in Chicago in 1884 and 1885, was the first to use an internal framework, or skeleton, made from steel columns and girders as well as a curtain wall that was fixed to the steel structure. Architects were soon racing to design bigger skyscrapers, especially in New York where there were no laws restricting height.

Since Jenney's design, skyscrapers built with glass have been able to withstand severe weather, including earthquakes. Buildings have incorporated plazas and parks alongside numerous entertainment and consumer venues at street level. Energy conservation is paramount to all future design in the twenty-first century. Today the skyscraper is an increasingly familiar sight, springing forth in growing numbers in cities all across the world—especially Hong Kong, Shanghai, and Dubai, as well as Chicago once again—shaping the way we live within urban centers. **SG**

SEE ALSO: PASSENGER ELEVATOR

Steam Turbine (1884)

Parsons refines the use of steam as a power source.

Steam engines have been around since the mid-seventeenth century. Noisy monsters, they used steam pressure to push pistons and turn engines, but they were massively inefficient and expensive to run. Engineers and inventors were aware that their machines made inefficient use of steam, and were looking for a better system to harness all that energy. In 1884 British engineer Charles Parsons (1854–1931), head of the electrical section of a ship manufacturer, patented the first steam turbine, which he used as the power source for an electrical generator that he had also built.

Using incredibly fast jets of steam had been causing inventors problems because they were just too rapid to use. At low pressure, steam jets were still reaching speeds of more than 1,000 miles per hour (1,610 kph), and at high pressure as much as twice that.

Spinning a turbine blade at such a speed would rip it apart. However, Parsons managed to slow down slightly the movement of steam from areas of high pressure to areas of low pressure. By setting up the turbine in a series of stages, he made the differences in pressure small enough to move the steam fast, but not too fast. The resulting engine turned over fifty times as quickly as the best of the old designs, which marked an enormous leap forward.

Parsons's steam turbine was fitted for use in electrical generating stations in 1891. Its application to marine propulsion meant that steamers and warships could travel much faster than before—in 1897 the ship *Turbinia* achieved the speed of 34.5 knots, making it the fastest ship in the world at that time. Steam turbines are still used in power stations today. **DK**

SEE ALSO: STEAM ENGINE, STEAMBOAT, GAS TURBINE, INTERNAL COMBUSTION ENGINE, FRANCIS TURBINE

⬆ *Parsons's 1884 prototype steam turbine is the forerunner of today's electricity turbo-generators.*

Photovoltaic (Solar) Cell (1884)

Fritts harnesses the power of light.

The image of a solar cell glistening in the sun illustrates many a magazine article on modern technology. But the means of converting light energy into electrical energy is nothing new. A "photovoltaic effect," whereby light hitting an electrode immersed in an electrolyte produces a current, was first observed in 1839 by A. E. Becquerel. This phenomenon was harnessed in 1884 when the first solar cell was built by American scientist Charles Fritts.

Fritts used the less than economic design of semiconductive selenium coated in a thin layer of gold to achieve the conversion, at an efficiency of just 1 percent. The cell works by absorbing energy in the form of photons of light, which then displace electrons in the semiconductor, generating a current.

Despite Fritts's optimism that the technology might replace centralized power plants, no one was ever going to power their home with only this kind of efficiency. But Fritts's ideas did find applications in

photography. Selenium, and later copper, cells made convenient sensors for measuring lighting levels in twentieth-century cameras.

Solar technology benefited from the introduction of silicon semiconductors in 1941. Developments in the 1950s and 1960s increased efficiency to levels where domestic applications became feasible, and solar panels became a lightweight power source for spacecraft and satellites. The technology is still inefficient today, with commercial cells performing at only around 13 percent, and the world record standing at 42.8 percent. Still, this is enough to power homes, outdoor gadgets, and spacecraft. **MB**

SEE ALSO: BATTERY, FUEL CELL

⬆ *A modern-day selenium photocell that operates on the same principles as photovoltaic cells.*

Gas Mantle

(1885)

Welsbach paves the way for the light bulb.

In 1807 Pall Mall in London became the first public street to be lit with coal gas, and other countries soon followed. Although gaslight was much cheaper than oil lamps or candles, it also produced smoke, bad smells, and lots of heat. These problems were resolved by an Austrian chemist.

Carl Auer von Welsbach (1858–1929) had studied chemistry at the University of Heidelberg under Robert Bunsen (co-inventor of the Bunsen burner). In 1885, Welsbach discovered the rare-earth elements neodymium and praseodymium. In the course of his research he found that some rare-earths produced a bright light when heated in a Bunsen burner.

> "It was evident that it was much more economical to renounce the lighting power of the open flame."
>
> Carl Auer von Welsbach

In 1885, he patented a mantle, a sheath of metallic threads, made of 60 percent magnesium oxide, 20 percent yttrium oxide, and 20 percent lanthanum oxide. It was not a commercial success. Five years later, he developed a mantle made of 99 percent thorium dioxide and 1 percent cerium dioxide. These mantles lasted longer, produced a brighter white light, and were soon used to light streets, factories, and homes.

Welsbach mantles still produced soot and heat. The solution to this was found in incandescent electric lamps, and most gaslights were replaced by electric lights in the early twentieth century. Welsbach mantles are still used today for camping. **ES**

SEE ALSO: BUNSEN BURNER, INCANDESCENT LIGHT BULB

Electric Arc Welding

(1885)

Two Eastern Europeans improve welding.

Welding is one of the processes whereby two metals are joined together. The two metals to be joined are melted (sometimes in the presence of a molten filler metal) and made to intermingle by applying intense heat. The bond formed between the two metals, being made of a mix of both the metals, is incredibly strong. This process is different from soldering and brazing where the joining metal is a different metal with different properties.

Electric arc welding utilizes the incredible heat generated by an electric arc as the means of melting the metals. A power source is linked to the metal to be worked on and, at the other end of the circuit, to an electrode of some kind. It is between this electrode and the surface of the work metal that the electric arc forms.

Like so many inventions, its creation cannot strictly be allotted to one person or group of people; the arc welder as it exists today is the product of many individually patented innovations, all implemented to improve on the arc welder's function. However, the first patent relating to the application of an electric arc in the welding process is best attributed to Russian Nikolai Benardos and Pole Stanislaus Olszewski in 1885.

Auguste De Meritens had been using electric arcs to join lead plates together. Benardos, his student at the Cabot Laboratory, took this to the next level, creating and patenting, with Olszewski, an electrode holder—named an Electrogefest—that formed the basis for carbon arc welding. **BG**

SEE ALSO: WELDING

➡ *In 1942 a British arc welder welds a sprocket wheel of a Valentine tank for service with the 8th Army.*

Modern Safety Bicycle (1885)

Starley builds a prototype of the modern bicycle.

Debate rages over the starting point of the bicycle, but most experts agree that the nineteenth century was the great era for the development of the bike. British cycle-maker John Kemp Starley (1855–1901) claimed the safety bicycle as his invention, first demonstrating it in 1885. Questions have been raised over Starley's right to this invention as other similar models appeared around the same time, but his version was undoubtedly the best. Starley's Rover Safety model featured spoked wheels of almost equal size, a diamond-type frame, J. H. Lawson's recent invention of the chain-drive (which powered the rear wheel), and an easily adjusted seat and handlebar.

The word "safety" was used here for a reason. Previous bicycles were perilous contraptions, especially the penny-farthing (developed by Starley's uncle, James Starley) that immediately preceded the Rover. The penny-farthing's giant front wheel and tiny rear one made it a strange-looking device, with riders perched precariously a long way off the ground. The safety bike was a more sensible height and its design and weight distribution were far better balanced.

The new bike featured tangentially spoked wheels—one of James Starley's innovations—but hard rubber tires still made for an uncomfortable ride. When John Dunlop's pneumatic tires were incorporated, however, the safety bicycle became a massive hit. It was also less costly than its predecessors, allowing cycling to become a more widespread activity. Many refinements in the bicycle have since taken place, significantly with the advent of the mountain bike in California during the 1970s. **AK**

SEE ALSO: DRAISINE, PNEUMATIC TIRE, MOTORCYCLE

↑ *Even without brakes, the safety bicycle was much less hazardous to ride than the penny-farthing.*

Motorcycle (1885)

Daimler and Maybach add the power of gas combustion to the safety bicycle.

When the German inventor Nikolaus Otto produced the first four-stroke internal combustion engine in the late nineteenth century, he inspired Gottlieb Daimler (1834–1900) and Willhelm Maybach (1846–1929) to produce an exciting form of transport. The concept was every parent's worst nightmare: to combine a strengthened but still dangerously unstable "bone-crusher" bicycle with a gas engine.

Two-wheeled, powered transport was not itself a novelty. Steam-powered bicycles had been around since 1867, and the Michaux-Perreaux steam bicycle—with its front wheel larger than its back wheel and the steam engine mounted under the saddle—went into production in 1868. But when Daimler and Maybach produced their gas-based version in 1885 they unveiled what was to be recorded by historians as the world's first motorcycle. Maybach drove the prototype from Cannstatt to Untertürkheim, a distance of nearly 2 miles (3 km), reaching a speed of 7.5 miles per hour (12 kph). While modern motorbikes resemble powerful jet engines on wheels, with leather jacket-clad riders confidently astride them, the original two-wheeled motorcycle actually had additional safety wheels protruding from each side of the chassis to provide a stable riding platform. Only in later models, as riders mastered the techniques of controlling fast movement on two wheels, were the stabilizers dispensed with.

Despite its cautious beginnings, the motorcycle has evolved into a highly adaptable road vehicle. In the twenty-first century, the need to minimize carbon footprints will only further increase the popularity of this environmentally efficient invention. **CB**

SEE ALSO: FOUR-STROKE CYCLE, INTERNAL COMBUSTION ENGINE, TWO-STROKE ENGINE

⬆ *With its sturdy stabilizers, the 1885 Daimler-Maybach motorcycle was effectively a four-wheeled machine.*

Rabies Vaccine

(1885)

Pasteur finds an antidote to a deadly disease.

Since antiquity, rabies had been feared as a death sentence. In 1884 Louis Pasteur (1822–1895) injected material from rabid dogs into rabbits, removing their spinal cords after they had died of the disease. When the cords were suspended over a vapor of potassium hydroxide, Pasteur found the more the cords dried, the fewer infectious agents survived.

He made a series of graduated vaccines, the strongest comprising spinal cord dried for just one day, and the weakest, cord dried for fourteen days. The vaccine was tested in forty-two dogs, twenty-three of whom received fourteen injections (starting with the weakest vaccine, and ending with the strongest),

---

"I got rabies shots for biting the head off a bat but that's okay— the bat had to get Ozzy shots."

Ozzy Osbourne, rock vocalist

---

while nineteen received no treatment. At the end of the experiment, all the dogs were exposed to rabies; none of the immunized dogs got the disease, while thirteen of the control group did. Pasteur's vaccine was tested in 1885 when a woman from Alsace turned up at his laboratory with her nine-year-old son who had been bitten by a rabid dog two days earlier. Pasteur ordered a fourteen-day course of increasingly virulent injections, and the boy stayed well.

In 1915 a ten-year study confirmed that, of 6,000 people bitten by a confirmed rabid animal, only 0.6 percent of those who had received the vaccine died, compared to 16 percent of those who had not. **JF**

SEE ALSO: INOCULATION, VACCINATION, CHOLERA VACCINE, ANTHRAX VACCINE, POLIO VACCINE, RUBELLA VACCINE

Transformer

(1885)

Stanley introduces the modern transformer.

Transformers convert alternating current (AC) from one voltage to another without changing the frequency. When American William Stanley, Jr. (1858–1916) invented this master of conversion (based on an idea of Lucien Gaulard and John Dixon Gibbs) in 1885, he paved the way for televisions, computers, battery chargers, and lamps. As a result, Stanley was invited to go and work for the entrepreneur George Westinghouse.

Transformers take advantage of Michael Faraday's principle of mutual inductance, which enables one coil to induce a current in another coil. The ratio between the input and output currents is determined by the number of loops in the two respective coils. Thus a current can be raised from low voltage to high voltage with relative ease, the significance of which is driven by the fact that a low voltage transmitted over a large distance will dissipate much of its energy, whereas high voltages retain most of their energy.

As wonderful as it would be to have hundreds or even thousands of volts of current streaming through our walls, it would be very dangerous, and for this reason it was recommended that Stanley's invention be used to return the current to appropriate voltages. Had the current been transmitted at these lower voltages, it would have been leaking energy like a hosepipe made out of teabags.

Stanley's transformers were first put to use in March 1886 when they powered businesses along the main street in Great Barrington, Massachusetts. They were a huge success, and the basic design is still in use well over a hundred years since they first appeared, even though the appliances they power have themselves been transformed again and again. **CB**

SEE ALSO: TELEVISION, DISTRIBUTOR

Supercharger
(1885)

Daimler increases the power of the internal combustion engine.

Superchargers, also called blowers, are used in cars to increase the power of internal combustion engines. German automobile-maker Gottlieb Daimler (1834–1900) first came up with the idea of pumping extra air into the engine to increase the horsepower. This effectively makes an engine larger for less weight, which is ideal for racing cars and aircraft. Initially engines relied on atmospheric pressure to keep air inside the engine. By pumping extra air into the engine, the amount of oxygen was increased, and this burned more fuel, giving the vehicle a power boost.

Daimler's design was based on twin-rotor air pumps that forced extra air into the system. His design went into production in Mercedes and Bentley cars in the 1920s and was essential to World War II aircraft. Today, superchargers are defined as any gas compressor driven directly by the engine. The supercharger's more powerful rival, the turbocharger, uses the exhaust gases to drive a turbine to blow in clean air. Both designs are essential in airplanes, since they experience a loss in atmospheric pressure as they climb to higher altitudes. At 18,000 feet (5,500 m) the air is at half the pressure of sea level, and hence twice as much air has to be blown in to keep a constant power.

The complexity and price of superchargers have restricted them to the more expensive cars on the market. Adding one to your current set of wheels will provide more horsepower per pound than any other component, perhaps bar the turbocharger. **LS**

SEE ALSO: INTERNAL COMBUSTION ENGINE, MOTORCAR, TURBOCHARGER

⬀ *This 6/25/40 hp Mercedes engine of 1921 (top) was part of the first supercharged car.*

⬀ *Mechanical supercharging was the hallmark of this engine, the 6/40/65 hp Mercedes of 1924 (bottom).*

> *"No other device since the shields and lances of the ancient knights fulfills a man's ego like [a car]."*

Sir William Rootes, automobile manufacturer

Hall-Héroult Process
(1886)

Hall and Héroult boost aluminum production.

Aluminum has not always been the light, cheap metal it is now. Chemists once painstakingly toiled to produce even small amounts, largely because it quickly burned when heated to high temperatures. The Washington Monument was topped with aluminum at the end of its construction in 1884. The 6.1-pound (2.8 kg) pyramid was one of the biggest pieces created.

In 1886, aluminum alchemists Charles Martin Hall and Paul Héroult discovered, independently, a process for cheap aluminum production. The twenty-two-year-olds, from the United States and France respectively, found that molten cryolite was the optimal environment for a chemical reaction to create large amounts of aluminum.

Before being put through the Hall-Héroult process, bauxite ore must first be changed into aluminum oxide. In the process, powdered aluminum oxide is dissolved in molten cryolite, a substance made up of sodium, aluminum, and fluoride. In the cryolite, aluminum oxide separates into highly reactive ions (charged atoms). The oxygen ions contain too many electrons, giving them a negative charge that pulls them toward positively charged carbon rods. The oxygen ions combine with the positively charged carbon to form carbon dioxide. Simultaneously, the aluminum ions are drawn toward negatively charged carbon lining the reaction container. When the aluminum contacts the carbon, it steals the carbon's excess electrons and thus becomes stable aluminum.

Aluminum is used to make food packaging and kitchen utensils, as well as having many specialized uses as a sturdy, lightweight material. **LW**

SEE ALSO: ALUMINUM FOIL, ANODIZED ALUMINUM

Dishwasher
(1886)

Cochrane cuts out washing the dishes.

The dishwasher was invented in the nineteenth century, not by a busy housewife or a restaurant owner looking to speed up the kitchen dishwashing, but by a well-to-do American socialite who was tired of her servants chipping her plates.

The first patent for a dishwasher was granted in 1850 to John Houghton, but his design proved to be impractical. Josephine Cochrane (1839–1913), the daughter of a civil engineer, came up with a machine, patented in 1886, that was not dissimilar to the dishwasher we use today. Plates and cups were supported in a wire rack lying flat in a copper boiler and blasted with pressurized water until clean. The

---

"1. Find out what the typical housewife really wants. 2. Produce it!"

Kenwood dishwasher ad, 1965

---

machine created interest among Josephine's friends who began putting in their orders, and, following a showing at the Chicago World's Fair in 1893, Josephine was soon supplying machines to restaurants and hotels in Illinois, trading under the name of Cochrane's Crescent Washing Machine Company.

It was not until the 1950s that dishwashers became more popular, with the advent of permanent plumbing and electronic motors. Today a dishwasher is a common feature in Western kitchens. **BG**

SEE ALSO: SYNTHETIC DETERGENT, WASHING MACHINE

▣ *A 1929 dishwashing machine advertisement from the French magazine "L'Art Manager."*

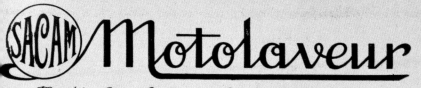

SACAM Motolaveur

Machines perfectionnées à laver la vaisselle

Brevetées en France & à l'Étranger

ur laver votre vaisselle
IL SUFFIT :

1°. — **De disposer la vaisselle à laver dans cette cuve.**

2°. — **De fermer cet interrupteur et de laisser le lavage s'effectuer seul pendant une minute.**

3°. — **D'ouvrir ce robinet pendant 10 secondes pour rincer la vaisselle avec de l'eau bouillante strictement pure.**

C'EST FINI

Une minute après vous n'avez plus qu'à retirer la vaisselle LAVÉE, RINCÉE, STÉRILISÉE et SÉCHÉE.

A tout moment vous pouvez de plus puiser à ce robinet de l'eau chaude accumulée

dans ce chauffe-eau à gaz à chauffage lent et à accumulation.

SOCIÉTÉ ANONYME DE CONSTRUCTION DES APPAREILS MÉNAGERS

AU CAPITAL DE 4.000.000 DE FRANCS

89, Rue de Sèvres - PARIS

R. C Seine 214.268 B

Tél. : LITTRÉ 91-10

Motorcar (1886)

Daimler, Maybach, and Benz introduce the first practical gasoline-engine automobile.

KAISERLICHES **[eagle emblem]** PATENTAMT.

PATENTSCHRIFT
— № 37435 —

KLASSE 46: Luft- und Gaskraftmaschinen.

BENZ & CO in MANNHEIM
Fahrzeug mit Gasmotorenbetrieb.

Patentirt im Deutschen Reiche vom 29. Januar 1886 ab.

"No invention of such far-reaching importance ... so quickly exerted influences [on] the national culture."

U.S. government commission, 1931

⬆ *The patent document awarded to Karl Benz in 1886 featured an illustration of his three-wheeled vehicle.*

➡ *With his wife Bertha and others, Karl Benz controls the progress of a four-wheeled Benz Viktoria in 1893.*

It is difficult, if not impossible, to imagine a world without the motorcar. When German engineer Karl Benz (1844–1929) drove a motorcar tricycle in 1885 and fellow Germans Gottlieb Daimler (1834–1900) and Wilhelm Maybach (1846–1929) converted a horse-drawn carriage into a four-wheeled motorcar in August 1886, none of them could have foreseen the effects of their new invention.

Benz recognized the great potential of petrol as a fuel. His three-wheeled car had a top speed of just ten miles (16 km) per hour with its four-stroke, one-cylinder engine. After receiving his patent in January 1886, he began selling the Benz Velo, but the public doubted its reliability. Benz's wife Bertha had a brilliant idea to advertise the new car. In 1888 she took it on a 60-mile (100 km) trip from Mannheim to near Stuttgart. Despite having to push the car up hills, the success of the journey proved to a skeptical public that this was a reliable mode of transport.

Daimler and Maybach did not produce commercially feasible cars until 1889. Initially the German inventions did not meet with much demand, and it was French companies like Panhard et Levassor that redesigned and popularized the automobile. In 1926 Benz's company merged to form the Daimler-Benz company. Benz had left his company in 1906 and, remarkably, he and Daimler never met. Due to higher incomes and cheaper, mass-produced cars, the United States led in terms of motorization for much of the twentieth century.

This kind of movement has, however, come at a cost. Some 25 million people are estimated to have died in car accidents worldwide during the twentieth century. Climate-changing exhaust gases and suburban sprawl are but two more of the consequences of a heavy reliance on the automobile. **TZ**

SEE ALSO: CLUTCH, MOTORCAR GEARS, ACCELERATOR, STEERING WHEEL, ASSEMBLY LINE, AUTOMATIC TRANSMISSION

Contact Lenses

(1887)

Fick, Kalt, and Müller aid vision correction.

Bad eyesight has likely plagued humans since they first stood upright. Real solutions were not available until the thirteenth century when eye glasses were invented.

Late in the 1880s, two eye doctors and a medical student independently invented contact lenses. Doctors Adolf E. Fick and Eugène Kalt set out to help their patients whereas medical student August Müller wanted to correct his own near-sightedness.

Early lenses were literally a glass lens in direct contact with the eye. For their comfort and health, users could only wear them for brief periods as the lenses caused pain, swelling, and cornea hypoxia. Despite these drawbacks, more than 10,000 pairs sold

---

"A pair of powerful spectacles has sometimes sufficed to cure a person in love."

Friedrich Nietzsche, philosopher

---

in the United States from 1935 to 1939. By 1949, sales had reached 200,000 thanks to the plastic polymethylmethylpropenoate (PMMA) and Kevin Tuohy's 1948 invention of plastic contact lenses. PMMA still caused cornea hypoxia and was replaced in the 1950s by hydroxyethylmethacrylate (HEMA), but these lenses required polishing. Bausch & Lomb introduced their revolutionary SofLens® in 1971. By the twenty-first century, the number of people who wear contact lenses surpassed 100 million. **RBk**

SEE ALSO: LENS, SPECTACLES, BIFOCALS, INTRAOCULAR LENS, LASER EYE SURGERY

◨ *A blown glass contact lens made by Muller Sohne of Wiesbaden, Germany, dating from circa 1930.*

Monotype Machine

(1887)

Lanston's machine enhances typesetting.

At the same time that the Linotype machine was being developed, Tolbert Lanston (1844–1914), a government clerk in the United States, was inventing another composition system that he called Monotype. Lanston's initial patent was awarded in 1885, but he was not successful until he founded the Lanston Monotype Machine Company in Washington in 1887.

In Lanston's system, letters, spaces, and other characters were selected mechanically from instructions contained on a paper tape into which patterns of holes, each representing a different character or space, had been punched using a keyboard. Although typesetting using Monotype was not as quick as Linotype, where complete lines of text were cast, Monotype text could be corrected more easily, spacing could be more finely controlled, and its versatility made complex setting possible.

Lanston used cold metal strips into which letters were punched to produce raised reverse type for printing, but realized that much finer definition could be achieved by casting the letters in hot metal. In 1896 he patented a hot-metal machine using copper molds for each character.

Now it was possible to hand-cut the molds for the characters and typography began to become an art form. The first Monotype type face was Modern Condensed. In 1920, Frederic Goudy became Art Director at Monotype and he designed over a hundred typefaces. In Britain, Stanley Morison (1889–1967) also came up with outstanding new typefaces. Although hot-metal typesetting was superseded by phototypesetting in the 1960s, and rendered obsolete by digital typesetting, Monotype typography continues to influence the way we read. **EH**

SEE ALSO: WOODBLOCK PRINTING, MOVABLE TYPE, PRINTING PRESS

Motorboat
(1887)

Duo's early car engine powers boats.

Paddle steamboats had been around for a century and propellers for half a century before the first proper motorboat took to the River Neckar near Stuttgart, Germany, in 1887. It had a petrol-driven internal combustion engine and was built by Gottlieb Daimler (1834–1900) and Wilhelm Maybach (1846–1929).

Rumor had it that the two inventors were looking for a less risky vehicle for their new engine than the old stagecoach they had recently motorized. The 14.7 foot (4.5 m) long boat traveled at a maximum speed of 6 knots. Daimler practiced a mild deceit on his nervous first customers, by concealing the engine with a ceramic cover and informing them that it was "oil-electrical," which sounded a great deal safer than the potentially explosive petrol. The deceit clearly worked because the Neckar, as it was called, sold well. It was produced by the recently formed Daimler-Motoren-Gesellschaft (DMG), and sales were undoubtedly helped by the poor state of German roads at the time.

Yet another German inventor, Rudolph Diesel, followed close behind Daimler and Maybach, introducing a diesel engine into boats in 1903 with instant commercial success. Indeed, for marine use diesel engines were to have a greater long-term future than petrol, proving to be more reliable and safer as they used a less flammable fuel.

The inventor of the first British motorboat was Frederick W. Lanchester, who had built the first gasoline-driven car in England in 1896. His design—launched in Oxford in 1904 and built in his back garden—had a stern paddle wheel and was powered by an engine with a wick-based carburetor that, despite being described as looking like a bed of celery when opened up, worked very efficiently. **AE-D**

SEE ALSO: DUGOUT CANOE, ROWING BOAT, SAILS, RUDDER, STEAMBOAT, JET BOAT

Phonograph Record
(1887)

Berliner invents a sound recording medium.

The phonograph (gramophone) record was invented in 1887 by German-born, American inventor Emile Berliner (1851–1929). His flat, rotating disc involved the stylus moving horizontally across the record rather than vertically, as with the cylinders used previously. Berliner's recording stylus cut down on sound distortion and was easier to manufacture than the cumbersome wax cylinders.

Early phonograph records were made from a mixture of hard rubber, cotton, and powdered slate, although shellac (a form of commercial resin) was later used after its introduction in 1896. Phonograph records were initially single sided but the double-sided disc became common after about 1923. They usually came in three standard sizes, 7-inch, 10-inch, and 12-inch, and originally they revolved at between seventy-five and eighty revolutions per minute (rpm). Most producers eventually settled on 78 rpm. A 10-inch disc would record about three minutes of music on each side. The discs used a spiral groove to record the sound in an analog fashion. This groove began at the outer edge and slowly worked its way in toward the center. The sound was transferred from the disc using a sharp needle that was forced to oscillate at audio frequencies from side to side by the groove.

The much loved long-play, 12-inch vinyl record was introduced in 1932. This revolved at 33 rpm, and twenty-five minutes of sound could be recorded on each side. Vinyl records were sturdier than their brittle shellac predecessors. Vinyl 7-inch, 45-rpm records took over from the shellac 78s with the advent of rock 'n' roll. **DH**

SEE ALSO: PHONOGRAPH, VINYL PHONOGRAPH RECORD, FLOPPY DISC, DIGITAL AUDIOTAPE (DAT)

➡ *Columbia Records, founded in 1888, switched from phonograph cylinders to records in the 1890s.*

Ballpoint Pen (1888)

Loud invents a practical writing instrument.

John J. Loud's invention, the ballpoint pen, was far from perfect. It leaked and smudged documents and was too crude for standard letter-writing. Nevertheless, Loud—a leather tanner from Massachusetts who wanted something that could write on leather and wood—patented his new writing instrument on October 30, 1888. With a tiny steel ball bearing in the nib and three smaller balls aligned above it to try to regulate ink flow, all refreshed with ink from a reservoir above, Loud had invented the world's first pen that did not constantly require dipping or refilling.

However, Loud was unable to control the flow of ink, which contributed to the pen never being sold commercially. If the ink were too thin the pen would

> ## "There is no lighter burden, nor more agreeable, than a pen."
>
> Petrarch, poet

leak; if it were too thick, it would clog. Depending on the temperature, the pen would sometimes do both. Loud's pen also relied solely upon gravity to deliver ink to the nib and had to be held in an almost vertical position for it to write. Eventually Loud's patent expired, making sure that the fountain pen prevailed.

It was only in 1938 that Hungarian newspaper editor László Bíró pressurized the ink reservoir. His pen used capillary action for ink delivery, solving the flow problems. In 1943 Bíró added a gravity-fed feeder tube. His "biro" amounted to a reinvention of the ballpoint pen, sold with a promise that it could write for a year without the need for refilling. **BS**

SEE ALSO: INK, PARCHMENT, PAPER, QUILL PEN, FOUNTAIN PEN, PENCIL

Deodorant (1888)

Philadelphia man stops an age-old problem.

Every civilization throughout history has searched for a remedy to the perennial problem of body odor. The ancient Egyptians and Greeks applied mixtures of carob, cinnamon, incense, and various citrus juices, though such attempts did little more than temporarily mask body odor. And the Roman historian Plinus wrote of a deodorizing salt made from a mix of potassium and aluminum. But, it was not until the nineteenth century that the eccrine gland was found to be responsible for the production of body odor.

An unknown Philadelphia man created the world's first trademarked deodorant in 1888. Named "Mum," it was sold as a rather difficult-to-apply, waxlike cream in a glass jar, and had zinc chloride as its principal drying-

> ## "Women who realize the importance of daintiness are grateful to Mum."
>
> Mum advertisement, 1926

agent. It worked to inhibit the growth of bacteria present in moist, warm areas of the body, such as the armpits, that were conducive to body odor.

In the early twentieth century powdered sodium bicarbonate diluted with talc became a common deodorant. In the late 1940s Helen Barnett Diserens, a researcher for Bristol-Myers, developed a solid aluminum chloride-based deodorant stick that, while easy-to-use, resulted in complaints of hairs being pulled during application. Diserens solved this problem in 1952 with the first roll-on that delivered deodorant to the skin via a plastic roller, using the same principle as the recently invented ballpoint pen. **BS**

SEE ALSO: ROLL-ON DEODORANT

Vending Machine (1888)

Adams popularizes a convenient way of selling products.

The concept for a vending machine actually dates back to ancient Greece, when the great experimenter, Hero of Alexandria, had the simple idea for a device that accepted a coin and dispensed holy water. The weight of the coin tipped a balance that opened a valve, letting water out until the coin slipped from the balance pan and a counterweight returned the mechanism and closed the valve. This neat idea formed the basis for the modern vending machine.

Various contraptions for vending began to be produced in the 1800s, including some designed to get around the legal ramifications of selling illicit merchandise, such as the machines designed by Robert Carlile for selling banned books. Another example is Simeon Denham's design for a machine to dispense postage stamps for a penny. But it is unclear whether this machine ever got beyond the idea stage.

It was not until the 1880s that vending machines became more of a commercial reality. Thomas Adams needed a quick and convenient method for dispensing his new chewing gum, and in 1888 he invented a machine to do just that. His machines were soon found on train platforms across the New York rail network. The design was soon improved (early models could be easily picked with a bent hairpin) and spread to the distribution of other items such as chocolate bars and peanuts. The idea of vending machines in turn inspired slot and pinball machines. Today, vending machines serve anything from snacks and hot coffee to umbrellas and toilet paper. In short, almost anything that comes in a small package can now be conveniently purchased from a machine. **SB**

SEE ALSO: CHEWING GUM, JUKEBOX, PAYPHONE

⬈ *This Zeno chewing gum slot machine dates from the 1900s and has a clockwork mechanism.*

"Change is inevitable— except from a vending machine."

Robert C. Gallagher

Dictating Machine (1888)

Edison creates a portable recording device.

By the age of 40, inventor Thomas Edison (1847–1931) already had around 400 patents under his belt. But he was never a man to rest on his laurels. His next project, which was in fact a return to a previous project, was to try to make a success of his dictation machine.

The device was an improved version of his earlier phonograph—a voice recording machine that turned speech vibrations into indents on a tin foil recording tape. In 1878, in an article in the *North American Review*, Edison had suggested that possible uses of his machine might include letter-writing, the teaching of spelling, and recording "the last words of dying persons." But he abandoned these ideas when the machine failed to take off.

> "I readily absorb ideas from every source, frequently starting where the last person left off."

Thomas Edison, inventor

After a rival group improved on his design with their "graphophone" in the 1880s, Edison saw potential in his phonograph once more. Picking up where his competitors had left off, he worked hard to create a more practical dictation machine. Chichester Bell and Charles Tainter had already developed a method for recording using a floating stylus and a wax cylinder, so Edison simply improved on their design, marketing his "Improved Phonograph" in 1888. Initially his machine was not a success and he faced great opposition from stenographers (shorthand typists). Sales of his later "Ediphone" grew after World War I, with the help of an advertising film called "The Stenographer's Friend." **HB**

SEE ALSO: PHONOGRAPH, PHONOGRAPH RECORD, VINYL PHONOGRAPH RECORD, MINIDISC

AC Electric Power (1888)

Tesla delivers cheaper, more reliable electricity.

When Serbian immigrant Nikola Tesla (1856–1943) began work at Edison's DC (direct current) power plant in the United States, his new employer was not interested in his ideas for a new type of power—AC (alternating current). At the time DC was the only electrical supply, but it could only be transmitted across short distances before it lost power. To Edison, AC sounded like competition and he persuaded Tesla to work on improving his DC system by offering him a huge sum of money. But when Tesla had done what he had been asked, Edison reneged on his promise. Tesla resigned and returned to his AC power concepts.

DC power is constant and moves in one direction and the resistance in wires causes it to lose power over

> "Alternating current is like a torrent rushing violently over a precipice."

Thomas Edison, inventor

distance. AC power does not have this problem as it varies in current so the resistance is less, and yet it delivers the same amount of power. This made AC power more cost-effective, as fewer power plants were needed. Entrepreneur George Westinghouse saw the potential of Tesla's AC power and bought his patents for AC motors. Edison began a propaganda war in an attempt to keep DC power on top, but it was inevitable that AC power would win. Almost all electricity around the world is now delivered as Tesla's AC power. **LS**

SEE ALSO: LEYDEN JAR, ELECTRIC MOTOR, ELECTRICAL GENERATOR, ELECTRICITY METER, PUBLIC ELECTRICITY SUPPLY

▣ *The Serbian inventor and engineer, Nikola Tesla, who built the first AC power system.*

Electricity Meter (1888)

Shallenberger's meter supplies AC power.

In the late nineteenth century, George Westinghouse was a happy man. He had just demonstrated that an alternating current (AC) generator could be used to power lights a mile away. It now dawned on him that he could make a fortune charging for AC electricity. All he needed was an electricity meter based on an alternating current. Thomas Edison was also fairly content because his company, General Electric, was generating the more popular direct current (DC) electricity and he was charging by the lamp.

Oliver Shallenberger (1860–1898), a graduate of the US Naval Academy, had been watching the developments in AC generation closely and had been working on an AC electrical meter. He had taken his early ideas to Thomas Edison, but as the meter was not DC-related, Edison was not interested. Shallenberger then visited Westinghouse to present his ideas.

Westinghouse was baffled by Shallenberger's drawings, but after half an hour gave him the job of chief electrician at his company. So it was that Shallenberger left the Navy in 1884 and joined the Westinghouse company.

In 1888, Shallenberger was working on a new AC lamp when a spring fell out and landed on the inside ledge of the lamp. He noticed that the spring was moving under the force of the nearby electric fields. He took this idea for his electricity meter, which became the industrial standard. The design was very similar to that of a gas meter. The same basic meter techonology is still used today. Nikola Tesla later showed that Shallenberger's meter was actually a type of AC electrical motor. **RB**

"Faith is like electricity. You can't see it, but you can see the light."

Unknown author, Poor Man's College

SEE ALSO: LEYDEN JAR, ELECTRIC MOTOR, ELECTRICAL GENERATOR, PUBLIC ELECTRICITY SUPPLY, AC ELECTRIC POWER

This Edison electrolyte meter from circa 1881 uses electrolysis to measure the consumption of electricity.

Kinematograph (1888)

Le Prince's groundbreaking work leads the way to motion pictures.

French-born Louis Augustin Le Prince (1842–1890) is fittingly remembered as the star of a mystery akin to a tragic silent movie. A photographer trained by Daguerre himself, as well as a chemist and artist, he worked secretively on pioneering moving-image experiments before disappearing just before revealing his findings. Many feel that Edison and the Lumière brothers wrongly displaced him as the inventors of motion pictures. Some even suggest that he was killed by rivals in this fiercely competitive race.

In the 1880s, Le Prince was one of several people working on Kinematographs—early machines for capturing and showing moving images. In 1888, he patented a sixteen-lens Kinematograph that probably never worked properly. He also produced a single-lens device incorporating Eastman paper film and, in the fall of 1888, used it to film horses on Leeds Bridge, England.

After probably improving his projection by replacing glass slides with the amazing new celluloid film, Le Prince felt ready to tell the world. His wife rented a mansion in New York for the debut. Her husband boarded a train at Dijon on September 16, 1890, bound for a French vacation before the unveiling, and no trace of him has been found since.

Edison's early cameras and projectors were soon making a splash and the science of movie equipment gathered momentum. Le Prince's family became embroiled in a court case that finally ruled that Edison was not the single inventor of moving images. One leading scholar concludes that Le Prince is well placed to take such credit but wonders whether his fatal error was being too slow to reveal all. **AK**

SEE ALSO: CELLULOID, PHOTOGRAPHIC FILM, KINETOSCOPE, FILM CAMERA/PROJECTOR, FILM SOUND

🡢 *Robert Paul's Kinematograph from 1896, which was used to film Queen Victoria's Jubilee in 1897.*

"The Method of and Apparatus for Producing Animated Pictures of Natural Scenery and Life."

Le Prince's 1888 patent

Electric Speedometer
(1888)

Belusic's device measures driving speed.

Useful as it might have been, an indication to drivers of how fast their automobiles were actually moving was not an option for pioneer motorists. The first production cars sported no such frivolous extras, but the race to develop suitable technology had begun. Extrapolating speed from the time taken to travel a given distance had been used for centuries, but it was a system for indicating an automobile's speed in real time—the speedometer proper—that was needed.

The electric speedometer, in a form that would be recognized today, appeared between the late nineteenth and early twentieth centuries. Josip Belusic, a Croatian professor from the (then) Austro-Hungarian region of Labin, was granted a patent for his electric "velocimeter" as early as 1888. Other inventors were to produce various speedometers over the years that followed, and although most would achieve the required objective—to measure the rotational speed of the wheels or some rotating drive component, and relay that information to a calibrated gauge—it was the electric (electromagnetic) speedometer that claimed pole position. Otto Schulze's patent for such a device was issued on October 7, 1902, in Berlin.

In a typical assembly, one end of a flexible, rotating shaft is coupled to a set of gears driven by the vehicle's transmission. The other end connects to a permanent magnet housed in a small metal "speedcup." The speedometer needle is fixed to the other side of this cup. As the cable rotates, so the magnet rotates proportionately, generating eddy currents and thereby dragging the cup and its attached needle with it. A spring counteracts the turning force and the needle remains steady as speed varies. **MD**

SEE ALSO: ELECTROMAGNET, MOTORCYCLE, MOTORCAR, ACCELERATOR, SPARK IGNITION

Wind Turbine Generator (1888)

Brush harnesses a natural power source.

Windmills had been in use for some 2,000 years before, in 1888, Charles F. Brush (1849–1929) linked a wind-powered turbine and a generator to provide power for the lights on his estate in Ohio. The 144-bladed rotor was based on wind-driven water pumps, and bore little resemblance to today's turbines.

Perhaps more familiar would be the turbines developed by Paul LaCour who, from 1891, developed wind power as a means of supplying farms and villages in his native Denmark. His four-bladed turbines could provide up to 25 kilowatts of power and were generating in their hundreds by 1910. In the early twentieth century, propeller-like turbines with only

> *"Industrial wind energy is a symptom of, not a solution to, our energy problems."*
>
> Eric Rosenbloom, National Wind Watch

two or three blades appeared. In 1931 the first large-scale turbine, a 100-kilowatt Russian device, was connected directly into an existing power supply network. Despite its consistent performance, the Russians abandoned the experiment at the start of World War II.

As fossil fuel supplies dwindle in the twenty-first century, huge wind farms have sprung up on land and offshore around the world. However, there are fierce debates about the commercial viability of such wind farms and whether their energy savings actually outweigh the ecological damage they have done to the rural landscape and wildlife. **FW**

SEE ALSO: WINDMILL, ELECTRICAL GENERATOR, PUBLIC ELECTRICITY SUPPLY

Brassiere

(1889)

Cadolle introduces a revolutionary new undergarment for women.

Before bras came along, corsets were the garment of choice for women to provide support and create a shapely silhouette of their figures. However, the whalebone reinforcement in corsets made them an uncomfortable and restricting item to wear. When Frenchwoman Herminie Cadolle (1845–1926) cut the corset in half in 1889, she created the very first bra.

In the late nineteenth century, Cadolle moved to Buenos Aires in Argentina, where she opened a lingerie shop. There she had the idea of separating the corset into two parts. She unveiled her new design, which used shoulder straps for support and was called the *corselet gorge*, at the Great Universal Exhibition of 1889 in Paris. (Another feat of engineering—the Eiffel Tower—was also constructed for this event.)

Cadolle's designs were among the first to use rubber thread or elastic. She used the new material to replace the old whalebones and lacing of corsets. Cadolle's prototype bra received an update in 1910, when New York socialite Mary Phelps Jacob fashioned her own version that could be worn under a sheer evening dress. It consisted of two silk handkerchiefs tied to a length of ribbon and Phelps Jacob eventually patented it in 1914 under the name "brassiere."

The bra did not become popular until World War I, when men left to fight and women took jobs in factories and required a more practical undergarment. Women were asked to stop wearing corsets because industry needed the metal used in them for the war effort. Cup sizes were not invented until the 1920s, by Russian immigrants Ida and William Rosenthal. **HI**

SEE ALSO: CLOTHING, SEWING MACHINE, BREAST IMPLANT, SYNTHETIC RUBBER

↗ *New Yorker Henry Lesher invented a rubber and cloth bra-style garment in 1859. The design never caught on.*

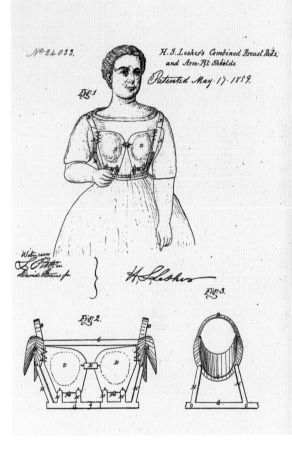

"Herminie had a simply ingenious idea. For women's comfort she had cut in two the traditional corset."

Cadolle website

Electric Drill (1889)

Arnot kickstarts the home improvement craze.

The electric drill embodies do-it-yourself (DIY). Factor in its significance in the construction and manufacturing industries, and the invention of the electric drill is highly important. For this we can thank electrical engineer Arthur Arnot (1865–1946), who built the first electric drill for use in the mining industry.

The concept of the drill is thousands of years old—bow drills were used by the Egyptians to make fire through friction as well as boring holes in wood. The action of a rotating head, moved by a "bow" wrapped around the shaft of the drill, was sufficient to make a hole into the surface. In essence, modern electric drills operate on a similar principle, except the power is supplied by electricity rather than by hand.

Arnot was born in Scotland and studied as an electrical engineer before going to work at the Grosvenor Gallery Power Station in 1885. Four years later, Arnot traveled to Melborne, Australia, to build an electricity plant for the Union Electric Company. While completing this contract and also providing Melborne with electric street lights, Arnot conceived his electric drill, which he patented in August 1889. His intention was for the drill to be used in the mining industry, where it would allow for more ambitious drilling projects. However, his invention was soon followed by the creation of the first portable electric drill by German Wilhelm Fein in 1895.

By 1917, in the United States, the newly formed company Black & Decker patented and sold the first electric hand drill, along with the pistol and trigger arrangement, now commonplace in homes and toolboxes everywhere. **SR**

SEE ALSO: DRILL, SEED DRILL, TWIST DRILL BIT

⬆ *Mining for coal in the 1920s using a high powered electric drill.*

Jukebox (1889)

Glass introduces music at the drop of a coin.

Forty-five years before the word "jukebox" was first coined, the world's first "nickel-in-the-slot phonograph" (later abbreviated to nickelodeon) was demonstrated to a small group of patrons at the Palais Royale restaurant on San Francisco's Sutter Street by its creator Louis Glass (1845–1924), General Manager of the Pacific Phonograph Company.

Glass attached a coin mechanism designed to accept nickels to an Edison Class M electric cylinder phonograph housed in an oak cabinet. With amplification still waiting to be discovered, prerecorded music from a single tin foil and wax cylinder was transmitted to a group of four people via listening tubes resembling stethoscopes held to their ears. Towels were supplied to wipe patrons' sweat from their earpiece.

Only a single cylinder could be played at any one time. This limitation nevertheless failed to dent the new machine's popularity, which went on to earn more than $1,000 in its first six months of service. Glass's nickelodeons were soon installed in public saloons and in the waiting rooms of several San Francisco-Oakland Bay ferry services. The advent of the "nickel-in-the-slot phonograph" also brought to an end the dominance of the player piano, which until 1889 had been the most popular method of playing music to large groups of people.

By the 1910s, the cylinder was gradually superseded by the gramophone record. The shellac 78-rpm record dominated jukeboxes until the introduction of 45-rpm vinyl record jukeboxes in the 1950s. The 45-rpm jukebox became the standard for years to come. **BS**

SEE ALSO: **PHONOGRAPH, VENDING MACHINE, PHONOGRAPH RECORD, VINYL PHONOGRAPH RECORD**

⬆ *For a nickel each, an excited group listen to a jukebox in Salina, Kansas, in the 1890s.*

Clutch (1889)

Daimler and Maybach help cars run smoothly.

Like most "firsts" in automotive history, just who devised the first clutch is debatable. Almost all historians agree that the clutch was developed in Germany in the 1880s and some of them credit Gottlieb Daimler (1834–1900) and Wilhelm Maybach (1846–1929) with its invention.

Daimler met Maybach while they were working for Nikolaus Otto, the inventor of the internal combustion engine. In 1882 the two set up their own company, and from 1885 to 1886 they built a four-wheeled vehicle with a petrol engine and multiple gears. The gears were external, however, and engaged by winding belts over pulleys to drive each selected gear. In 1889, they developed a closed four-speed gearbox and a

> *"We shall never run the risk of being confined in a coupé with insufferable people."*
>
> Otto Julius on automobile travel versus train

friction clutch to power the gears. This car was the first to be marketed by the Daimler Motor Company in 1890.

Without a clutch, if the car engine is running the wheels keep turning. For the car to stop without stalling, the wheels and engine must be separated by a clutch. The clutch also allows power to gradually be applied to the wheels, allowing for smoother starts and helping to avoid grinding gears when shifting.

A friction clutch consists of a flywheel mounted to the engine side. The clutch originates from the drive shaft and is a large metal plate covered with a frictional material. When the flywheel and clutch make contact, power is then transmitted to the wheels. **SS**

SEE ALSO: AUTOMATIC TRANSMISSION, CENTRIFUGAL CLUTCH

Motorcar Gears (1889)

Benz increases the car's speed and efficiency.

The invention of the gear system used in motorcars followed hot on the heels of the construction of the first automobiles. Karl Benz was the first to add a second gear to his machine and also invented the gear shift to transfer between the two. The suggestion for this additional gear came from Benz's wife, Bertha, who drove the three-wheeled Motorwagen 65 miles from Mannheim to Pforzheim—the first long-distance automobile trip.

The need for gears arises from the specifics of an internal combustion engine. The engine has a narrow range of revolutions per minute (rpm), where the horsepower and torque produced by the engine is at a maximum. The gears allow the engine to be

> *"One thing I feel most passionately about: love of invention will never die."*
>
> Karl Benz

maintained at its most efficient rpm while altering the relative speed of the drive shaft to the wheels. This allows the car to speed up or slow down while operating the engine in its optimal range.

Gears originally required "double clutching," where the clutch had to be depressed to disengage the first gear from the drive shaft, then released to allow the correct rpm for the new gear to be selected. The clutch was then pressed again to engage the drive shaft with the new gear. Modern cars use synchronizers which use friction to match the speeds of the new gear and drive shaft before the teeth of the gears engage, meaning that the clutch only needs to be pressed once. **HP**

SEE ALSO: CLUTCH, AUTOMATIC TRANSMISSION, SYNCHROMESH GEARS, CENTRIFUGAL CLUTCH

Two-Stroke Engine (1889)

Day invents a useful engine for small motorized devices from garden trimmers to chain saws.

To some it is the one of the most annoying sounds made by machines, while to others it has an almost musical quality. We are talking, of course, about the buzzing noise of the two-stroke engine. From motorbikes to lawn mowers this invention has been the model of simplicity in the world of internal combustion, and its modern design originated with British engineer Joseph Day (1855–1946).

An earlier two-stroke engine had been devised by Dugald Clark in 1880, but it lacked the simplicity of Day's version. Using just three moving parts, Day's engine used the pressure below the piston to force the fuel and air into the combustion chamber while simultaneously pushing the exhaust gases out. This meant that a pulse of power could be sent along the drive shaft with every revolution—an efficiency improvement over the four-stroke, which sends a pulse of power every two revolutions.

The simple design of the two-stroke has led to its wide popularity. Most high-performance motorbikes use them, as do outboard motors for boats. But despite the advantages of its high power-to-weight ratio, the two-stroke is actually more polluting than the four-stroke. Each time a fresh charge of fuel/air enters the combustion chamber, a small amount leaks out with the exhaust, which can be seen as the oily sheen on the water around an outboard motor and smelled as a distinctly oily aroma at go-kart tracks. Current environmental concerns mean that, unless significant improvements are made to two-stroke technology, we may not see them around for much longer. **DHk**

SEE ALSO: INTERNAL COMBUSTION ENGINE, STIRLING ENGINE, GAS-FIRED ENGINE, FOUR-STROKE CYCLE, BOURKE ENGINE

↗ *This internal combustion two-stroke valveless motor was made by the Day Motor Company in 1891.*

"Spring hasn't really arrived until you are awakened by the first lawn mower."

Source unknown

Photographic Film (1889)

Eastman's innovation helps spark amateur photography.

American entrepreneur and inventor George Eastman (1854–1932) became fascinated by photography in 1874. He pioneered a process making dry plates for photography and formed the Eastman Dry Plate and Film Company to produce them.

In 1889 he patented a system where photographic emulsion was coated onto a roll of flexible paper. The basis of this process was devised by a chemistry student, Henry Reichenbach, at Eastman's company. This convenient roll film on a transparent base replaced the fragile, unwieldy glass plate. Paper film created an unsatisfactory grainy image and was later replaced by a cellulose nitrate film that was unbreakable.

Eastman came up with the trademark "Kodak." His early Kodak camera was a wooden, handheld box with a simple lens, and a fraction-of-a-second shutter. This box camera was preloaded with enough film for 100 exposures. After all the negatives had been exposed, the whole camera was returned to the factory and the film developed. Prints were made and returned to the owner together with the negatives and the reloaded camera. By 1901 Eastman had introduced the "Brownie" roll-film camera. He had successfully wrested photography from the elite: everybody could now become a photographer.

Eastman's flexible transparent film was the basis of the nascent motion picture industry, used in Thomas Edison's kinetoscope from 1891. The introduction of Kodachrome color film with three dye layers in 1935 popularized color photography. **DH**

SEE ALSO: PHOTOGRAPHY, SINGLE-LENS REFLEX CAMERA, PORTABLE "BROWNIE" CAMERA, 35-MM CAMERA

◩ *A roll of transparent film for use in the No. 1 Kodak camera invented by George Eastman in 1888.*

◪ *The "Kodak Girl" demonstrating the Kodak Folding Pocket No. 3 camera in 1909.*

Payphone (1889)

Gray's idea creates a demand for telephones.

Superheroes may use phone booths to change into spandex, prostitutes may use them to place calling cards, and hooligans occasionally abuse them as urinals, but the impetus for the first payphone initially sprang out of an individual's desperate need for the use of a telephone and not being able to find one.

Conscious of this need, U.S. inventor William Gray devised the first coin-operated pay phone in 1889. It was a post-pay machine so the money was paid after the call to an attendant. Demand for the device was not immediate, but he managed to interest telephone companies, hotels, and shops. As people traveled more, the need for public payphones grew, and Gray's idea fueled the demand for household telephones.

> *"The telephone is a good way to talk to people without having to offer them a drink."*
>
> Fran Lebowitz, writer

Phone booths have undergone more design changes over the years than the actual phones themselves. Early indoor booths were often luxurious, crafted from wood like mahogany and carpeted inside. In London, England, the red public phone kiosk, designed by Sir Giles Gilbert Scott in 1926, became an iconic feature of the city. Across the Atlantic in New York's Chinatown one booth was even designed in the form of a pagoda.

By 1998 Gray's invention had reached its peak with 2.6 million payphones in the United States alone. This figure is now in rapid decline because of the growing popularity of mobile telephones. **CB**

SEE ALSO: TELEPHONE, VENDING MACHINE, DIGITAL
CELL PHONE, SMARTPHONE, IPHONE

Accelerator/Throttle (1890)

Benz's device enables drivers to control speed.

From time to time a talented inventor comes along who is so prolific that they almost redefine an industry single-handedly. German engineer Karl Benz (1844–1929) was one such inventor.

During the 1870s and 1880s he secured many patents—including the speed regulation system known as the accelerator or throttle in 1890—that represented significant developments in the technology of the automobile. Being one of the first to patent on many aspects of the design of the internal combustion engine eventually led Benz to become a leader in the field of automotive design.

The throttle performs a simple function in the internal combustion engine. The fuel—usually

> *"Speed has never killed anyone, suddenly becoming stationary … That's what gets you."*
>
> Jeremy Clarkson, motoring broadcaster

gasoline—is mixed with air before being ignited in the cylinders to produce the small explosion that fires the piston, which in turn rotates the drive shaft turning the wheels. The accelerator controls the fuel/air ratio, thereby determining the power output of the engine and as a consequence the speed of the car.

In the United Kingdom, automotive engineer Frederick Lanchester (1868–1946) introduced the accelerator or gas pedal between 1900 and 1904.

There have been some modifications to the basic design of the accelerator over the years, but the fundamental design has not changed and remains one of the pivotal elements of powered vehicles. **BG**

SEE ALSO: MOTORCYCLE, MOTORCAR, MOTORCAR GEARS,
SPARK IGNITION

Coherer
(1890)

Branly initiates wireless telegraphy.

The science behind the coherer had long been observed by various people before anybody actually managed to put it to any practical use. The effects of electrical charge on small specks of certain matter, such as dust, was noted in around 1850 by a man named Guitard. He spotted that when dusty air was electrified, the particles of dust would gather together to form a sort of stringed formation. It was not until 1890 that Edouard Branly (1844–1940) found a way for this unusual phenomenon to be exploited.

Experimenting with thin pieces of platinum film on glass, Branly discovered that there was a massive variation in the film's electrical resistance when it was

> *"Applications of science to warfare and materialistic enjoyment will be the downfall of mankind."*
>
> Edouard Branly

subjected to electromagnetic waves (known at the time as Hertzian waves, known today more commonly as radio waves). This discovery led Branly to create the coherer—a piece of equipment that consists of little more than a few metal filings lying loosely between two metallic electrodes in a glass tube. In its normal state, the coherer has a huge resistance. In the presence of Hertzian waves, the filings allow current to pass through them. Connected to a circuit, the coherer can then be used to detect electromagnetic signals.

This device became the basis of radiotelegraphy and was used in the development of wireless telegraphy, most notably by Marconi. **CL**

SEE ALSO: WIRELESS COMMUNICATION, IMPROVED RADIO
TRANSMITTER, CRYSTAL SET RADIO

Automatic Telephone Exchange (1890)

Strowger ensures privacy for phone calls.

The invention of the telephone was a gigantic leap forward. However, early models were connected by a wire in pairs, meaning that you could only connect to one other telephone. This was not useful for keeping in contact with lots of different people. Early phones did not even use a ringing bell—you had to whistle into the speaker to get the other person's attention.

The telephone had a long way to go, and it was Almon Strowger (1839–1902), a Missouri undertaker in the United States who devised a big improvement—the automatic telephone exchange. His local exchange operator was the wife of a rival undertaker and would divert calls away from Strowger's business.

> *"One day there will be a telephone in every major town in America."*
>
> Alexander Graham Bell, inventor

This motivated Strowger to invent an automated exchange that recognized the number dialed in by the caller. The earliest model was a hollow cylinder with a shaft in the middle capable of moving up and down and also rotating. A connector in the shaft could make contact with many different contacts on the inside of the cylinder and so send this down the line as a pulse, hence pulse dialing and rotary dial telephones. This system ended the need for manually connected calls until it too was superseded by digital systems. **JB**

SEE ALSO: ELECTROMAGNETIC TELEGRAPH, TELEPHONE,
DIGITAL CELL PHONE

☑ *An engineer testing London's revolutionary new automatic telephone exchange in 1929.*

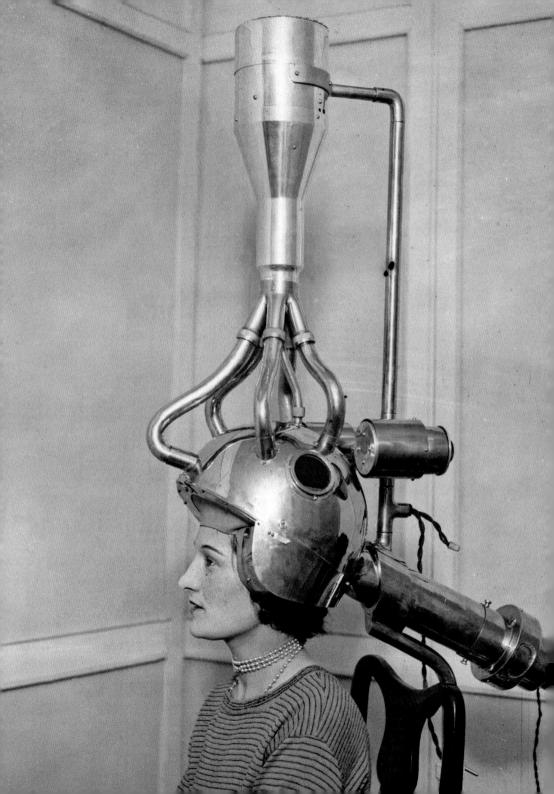

Blow-dry Hair Dryer
(1890)

Godefroy gives hairdressing a styling tool.

Hair salons were exclusively for rich women of the 1890s, and it was in that decade that French hairdresser Alexandre Godefroy introduced the first blow-drying hair dryers to put in his salon. His makeshift dryer was simply a bonnet attached to a flexible chimney stuck onto a gas stove.

It is hydrogen bonds that determine whether your hair looks like you have been dragged through a hedge backward or just walked out of a shampoo ad. Blow-dried hair generally looks better because it accelerates the temporary hydrogen bonds that reside in each strand of hair, allowing better control of shape and style. Curlers, straighteners, and all other hair

> "Hair brings one's self-image into focus; it is vanity's proving ground."
>
> Shana Alexander, journalist

appliances that work by heating the hair to change its shape are simply controlling the hydrogen bonding. Strong, hydrogen bonds are susceptible to humidity and completely disappear when the hair becomes damp or wet.

Hair dryers quickly evolved from Godefroy's design, and instead of having to sit under the hood of a heating device, today's hair dryers are handheld and boast a dizzying array of special features, the latest fad being "deionizing" hair dryers that reduce frizziness. **LS**

SEE ALSO: VARIABLE RESISTOR, HAIR SPRAY

⬅ *A practical hair drying demonstration at the annual White City Fair in London, England, circa 1930.*

Lawn Sprinkler
(1890)

Oswald whets the appetite for gardening.

During the nineteenth century, the United States saw many people moving out of the city centers into the burgeoning suburbs. With more space, people began to develop an interest in cultivating gardens. The widespread introduction of city water systems also brought water to more homes.

There are thousands of sprinkler patents, and it is difficult to credit the invention to one person. American Joseph Oswald's sprinkler was not the first to be patented, and his application described it as making "improvements in lawn sprinklers." His design improved the way in which the head of the sprinkler moved and also dealt with the issues of wear and tear

> "A garden sprinkler [in one hour] uses the same amount of water as a family of four in one day."
>
> Caroline Roux, journalist

that occurred because of the constant motion of the sprinkler. Earlier sprinklers had often been static, but Oswald's was one of the first to have a rotating mechanism propelled by the water it distributed.

The classic design of the domestic lawn sprinkler has a hosepipe supplying water to a metal arm that spins on a base, spreading water thinly over a large area. Within the mechanism of the sprinkler, there is a system of gears controlling the speed of the arm to keep it from moving too quickly.

Lawn sprinklers are popular throughout the world. However, in countries with water shortages, their use has been criticized as wasting a valuable resource. **JG**

SEE ALSO: IRRIGATION, LAWN MOWER

Spark Ignition (1890)

Benz starts cars with the press of a button.

Spark ignition may be regarded as the process by which a farmer uses a cattle prod to put his herd in motion. It is also the process that enables an internal combustion engine to run on gasoline.

Spark ignition works by passing an electric current through a carefully wired system and into a spark plug, which does what the first part of its name suggests by igniting the mixture of air and fuel in the chamber of the engine. In 1890 Karl Benz (1844–1929) sparked a new age of civilized society with the invention of spark ignition. This development in the automobile world—a world in which Benz was already famous for having invented the motorcar—made the gasoline-fueled internal combustion engine possible. Credit is also

> *"I would rather that my spark should burn out in a brilliant blaze than it be stifled by dry-rot."*
>
> Attributed to Jack London

given to Oliver Joseph Lodge, who made a major contribution to electric spark ignition for the internal combustion engine and whose sons, early in the twentieth century developed his ideas and formed the Lodge Plug Company, which sold spark plugs. However, it was only Gottlob Honold's later 1902 invention of the first commercially viable high-voltage spark plug as part of a magneto-based ignition system that made possible the development of the internal combustion engine.

The ability to improve the reliability of engines has made possible much of the mechanical technology that we associate with modern society. **CB**

SEE ALSO: INTERNAL COMBUSTION ENGINE, MOTORCYCLE,
MOTORCAR, MOTORCAR GEARS, DIESEL ENGINE

Electric Chair (1890)

Brown facilitates execution by electricity.

The word *electrocution* is actually a combination of the words *electricity* and *execution*, and was initially only applied to executions performed with electricity. It was only later that this word was adopted to describe any death caused by electricity.

The original idea for the electric chair can be credited to Alfred P. Southwick, a dentist who observed the death of a man who touched a live electrical terminal while sitting in his dentist's chair. At the same time, a fierce rivalry was developing between Edison and Westinghouse, both of them trying to market their own forms of electricity to the public. Edison (who favored direct current) wanted to show that alternating current was unsafe by having it

> *"… alternating current will undoubtedly drive the hangmen out of business in this state."*
>
> *New York Times*, 1888, on a prototype

used for executions. In the late 1880s, he and Harold Pitney Brown, holder of the electric chair patent, "researched" the viability of the invention by killing many different animals with electricity. From this they concluded that alternating current electricity worked best, and so the electric chair was made to run on AC.

The first electrocution, of William Kemler, took place in 1890 at Auburn prison. "The Chair" remained the most popular form of execution in the eastern United States and the Philippines for many years. **BG**

SEE ALSO: ELECTRICAL MOTOR, VARIABLE RESISTOR,
AC ELECTRIC POWER, PUBLIC ELECTRICITY SUPPLY

→ *A man strapped into an electric chair awaits his execution by electrocution in the United States in 1908.*

Pneumatic Hammer (1890)

King supplements human muscle power with air power.

The use of hand tools by humans has undeniably been a major factor in our evolution from hunter-gatherers to builders of civilizations. On many occasions a new design of tool has come at just the right time to add a boost of acceleration to the history of our technology.

The invention of the pneumatic hammer was one of those very events. Charles Brady King (1869–1957) had an eclectic range of interests. He considered himself a mystic, as well as a musician, artist, poet, and architect. He also had a degree in mechanical engineering from Cornell University, New York. His invention of the pneumatic hammer was perfectly timed. The 1890s were a time of rapid growth in construction, shipbuilding, mining, and the new automotive industry. Any invention that increased the productivity of workers was bound to be a success.

King's hammer used the power of compressed air and was extremely simple in concept. The input of the compressed air would drive a piston forward with great power, and this could be attached to any tool that required a to-and-fro percussive movement, such as a hammer. It was perfect for adding large numbers of rivets in construction, as well as for caulking.

King demonstrated his invention at the Chicago World's Fair in 1893 and went on to focus his energies on the car industry. His pneumatic hammer was further developed by what was to become the Chicago Pneumatic Tool Company. A famous iconic painting showing a female worker with a pneumatic riveter became a symbol of courage during World War II. The painting *Rosie the Riveter* by Norman Rockwell (1943) was sold in 2002 for $5 million. **DHk**

> *"If the only tool you have is a hammer, you tend to see every problem as a nail."*
>
> Abraham Maslow, psychologist

SEE ALSO: STONE TOOLS, METALWORKING, CLAW HAMMER, STEAM HAMMER, MOTORCAR

◩ *Pneumatic hammers being used at the Midvale Steel and Ordnance Company, Pennsylvania, in 1918.*

Frasch Process (1891)

Frasch fuels the U.S. chemical industry.

Sulfur is an important precursor to many industrial processes. Much of it goes into making sulfuric acid, a common reactant and a component of fertilizer. In the nineteenth century, Sicily dominated the production of this element. Deposits were found in the United States, too—notably in Louisiana and Texas—but they were much deeper and more difficult to mine. It may not have been gold, but unlocking this yellow substance could be a deeply profitable business.

One man had the answer, and from it derived the process that now bears his name. German-born Herman Frasch (1851–1914) settled in the United States and made a name for himself removing unwanted sulfur from petroleum. In the 1890s sulfur itself became the focus of his attention, and he sought a way to mine the deep-set mineral.

His solution was to bore a drill hole down to the layer containing sulfur. An arrangement of three concentric pipes is fed into the hole. Superheated water at 338°F (170°C) is pumped through the central pipe, down to the rock layer. Sulfur melts at around 239°F (115°C), so it becomes liquid when in contact with the superheated water. Forcing compressed air down another tube causes the molten sulfur to ascend as a slurry to the surface, where it solidifies into a highly pure form readily usable by industry.

Although Frasch was unable to make a sustainable sulfur-mining business for himself, the process he developed eventually enabled the United States to become self-sufficient for sulfur production, fueling a boom in its chemical industry sector. The Frasch process remained the world's dominant form of sulfur production until the 1970s, when the element was more economically obtained as a by-product of natural-gas and oil refining. **MB**

SEE ALSO: GUNPOWDER, FIREWORKS, VULCANIZATION

Carborundum (1891)

Acheson creates an extremely hard abrasive.

In 1880 Edward Goodrich Acheson (1856–1931) developed an electrical battery that he tried to sell to the inventor and entrepreneur Thomas Edison. Rather than buying his battery, Edison gave Acheson a job.

Despite rapid promotion, Acheson left to become an independent inventor. As the superintendent of a factory that made lamps, he could conduct his own experiments. He wanted to recreate diamonds in the laboratory, but his processes for heating carbon failed. Next he tried mixing clay and carbon together by electrically fusing the two. The fused mass he created had dark shiny specks in it, and examining these further he found they were extremely hard.

Acheson had created silicon carbide, which he

> *"Illegitemi non carborundum. Mock Latin slogan: Don't let the bastards grind you down."*
>
> Source unknown

named carborundum, wrongly thinking it to be a compound of carbon and aluminum. In 1893 he received a patent for his discovery. It was the hardest substance made by humankind, second only to diamond, and many realized its potential for the manufacturing industry. Its extremely high abrasiveness made it perfect for making precision-ground machine tools, and Acheson's initial production facility was soon swamped with orders.

In the mid-1890s Acheson discovered that by overheating carborundum he could make graphite. Today carborundum is used in sandpaper, disk brakes, ceramic membranes, and bulletproof vests. **DHk**

SEE ALSO: SYNTHETIC DIAMOND

Electric Car Battery
(1891)

Morrison powers the first electric car.

Although the earliest efforts to develop electric vehicles were made in France and Britain, the honor of the first truly successful electric car goes to Scottish-born William Morrison in the United States.

Morrison's passion was in fact for storage batteries—he only built the car, a surrey-type high-wheel carriage, to show off his latest battery. The battery holds the key to the success of any electric car, determining the speed and range of the vehicle. Morrison's car battery comprised twenty-four cells and contributed more than half of the total weight of the vehicle. It was claimed that the battery was capable of powering the vehicle for thirteen hours at

---

> *"It may take years, but the battery-electric car will eventually be back stronger than ever…"*

Chris Payne, dir. *Who Killed the Electric Car?* (2006)

---

speeds of up to 14 miles (22.5 km) per hour on just one overnight charge. Though far superior to the other car batteries available, these statistics are disputed.

Morrison's car was first shown to the world in 1890 at the Seni Om Sed parade in Des Moines, Iowa, and in 1891 he received a patent for his battery. Morrison gave his designs to the American Battery Company, who used the car to bolster their battery sales, and showed the car at the Chicago World's Fair in 1893.

By the mid-1930s electric cars had practically disappeared, having been replaced by the gas-powered vehicles that still predominate today. However, the car battery is still required for starting the engine. **RP**

SEE ALSO: BATTERY, ELECTRIC MOTOR, MOTORCAR, MOTORCAR GEARS, DRIVERLESS CAR, AIR CAR

Dow Process
(1891)

Dow extracts bromine from brine.

While studying for his degree, Herbert Henry Dow (1866–1930) became interested in developing more economical ways of extracting bromine from the underground brine reservoirs of Michigan in the United States. Dow, a man with a business acumen that matched his intellect, recognized that the use of bromine in medicine and the photographic industry meant there was a potential for massive profits. By 1889 he received his first patent for a new extraction process and immediately set up his own company, which went bankrupt within the year. Undeterred, Dow continued his investigations and by 1891 had patented the Dow Process. If a current is run through the brine, a process known as electrolysis, the negative bromide ions collect at the positive electrode, from where they can be collected.

Dow knew that he had to compete in the European market, where prices were controlled by a cartel of German companies called the Bromkonvention. Determined to make his mark, Dow ignored their rules and undercut them. The Bromkonvention responded by flooding the U.S. market with below-production-cost bromine in an attempt to put him out of business. However, they did not account for Dow's savvy nature; he hired English and German agents to buy the cut-price bromine and continued selling to the Europeans, undercutting their fixed prices. By the time Dow was found out, around four years later, there was little the Bromkonvention could do but invite him in.

Today the Dow chemical company is one of the largest in the world, and bromine production continues to grow, as do its uses, which include brominated flame retardants, dyes, agrichemicals, and pharmaceuticals, to name but a few. **RP**

SEE ALSO: PHOTOGRAPHY, DDT, METERED DOSE INHALER

Kinetoscope
(1891)

Edison and Dickson create the forerunner to the motion-picture projector.

Motion pictures and even television may never have become a reality if it were not for the creation of the Kinetoscope by Thomas Alva Edison (1847–1931) and his deputy William Dickson (1860–1935) in 1891. The idea of moving still frames in quick succession to give the illusion of a moving image had already been demonstrated. Edison and Dickson took the principle and built a machine that could show long rolls of film.

Edison was apparently inspired by a demonstration of the Zoopraxiscope, which used a fast-spinning disc with images around the outside to give the illusion of movement. Edison took it upon himself to develop a system that brought together moving images and sound; he called it the Kinetoscope. Edison set Dickson and his team to work on the project.

Edison and the team devised a system of printing images onto thin celluloid sheets cut into narrow rolls with perforated edges. Sprockets that engaged with the perforations enabled the rolls to be fed at a uniform speed in front of a lit bulb. A shutter would flash light through the "film" at just the right time so that each image was exposed for an instant. When all this was carried out at speed, the successive images gave the illusion of movement.

The final design of the Kinetoscope was unveiled in 1894 to great applause; the public was captivated by the invention. The initial peep-hole design of viewing led to the development of the projecting Kinetoscope. The work of Edison and his team led to the start of the motion picture industry, and entertainment was changed forever. **SR**

SEE ALSO: CELLULOID, KINEMATOGRAPH, PHOTOGRAPHIC FILM, FILM CAMERA/PROJECTOR, FILM SOUND

↗ *William Dickson's Kinetoscope of 1894 was the first device to show motion pictures.*

"We may see and hear a whole Opera as perfectly as if actually present."

U.S. Patent Office description for the Kinetoscope

Wireless Communication (1891)

Tesla patents practical uses for radio.

In 1887 David Hughes transmitted Morse code over a short distance. A year later, Heinrich Hertz produced and detected his own radio waves, but he did not realize their practical use. It was only in 1891, when Nikola Tesla (1856–1943) began his researches, that radio technology started to come into its own.

Serbian-born Tesla, by then living in New York, was a practical inventor. He quickly saw the potential in the strange resonances and interactions caused by the alternating currents of his electrical experiments. He worked on radio for several years, filed several patents, presented his ideas in London, and created a working long-distance radio system in New York.

> *"The scientific man . . . does not expect that his advanced ideas will be readily taken up."*
>
> Nikola Tesla

Tesla was distracted by the long, bitter war between his alternating-current electricity system and Thomas Edison's rival direct-current system, which left the way open to Guglielmo Marconi to take the public glory—and the money—from radio. Although the U.S. Patent Office did finally recognize Tesla's invention of radio, it was not until 1943, after Tesla's death, that they overturned Marconi's 1900 patent in favor of Tesla's earlier claims. One of the most world-changing inventions of all time, radio has given us television, the mobile phone, RADAR, wireless internet access, satellite navigation, radio telescopes, and even the microwave oven. **MG**

SEE ALSO: COHERER, IMPROVED RADIO TRANSMITTER, CRYSTAL SET RADIO

Escalator (1891)

Reno patents the moving stairway.

American Jesse W. Reno (1861–1947) came up with the notion of an "inclined elevator" at the age of sixteen and patented the idea in 1891. It was not the first idea of its type because a patent had been granted earlier for a steam-driven design, but this was never built. In 1895 Reno's moving stairway was built as an attraction at New York's Coney Island amusement park.

The term "escalator" was not attached to the invention until 1897, when Charles Seeberger combined *scala* (Latin for "stairs") and elevator, the name of a device invented some years previously. Seeberger redesigned the escalator, and it was built in the Otis factory, New York. This became the first

> *"I like an escalator because an escalator can never break, it can only become stairs."*
>
> Mitch Hedberg, comedian

commercial escalator, winning first prize at the Paris Exposition in 1900. The Otis Elevator Company bought the rights to both Reno's and Seeberger's designs and became the world's leading escalator manufacturers.

Reno continued to invent, and after a rejected bid to redesign New York's subway system, he moved to London, and with his new company, began manufacturing a spiral escalator for the London Underground train network. This was less successful and never used by the general public. **SB**

SEE ALSO: REVOLVING DOOR, STEEL-GIRDER SKYSCRAPER, PASSENGER ELEVATOR

→ *An early escalator by the Otis Elevator Company is shown here in this 1900 depiction.*

Swiss Army Knife (1891)

Elsener creates the multifunctional pocket knife.

When the Swiss Karl Elsener (1860–1918), owner of a company that made surgical equipment, found that the pocket knives supplied to the Swiss army were, in fact, made in Germany, Elsener decided to make them in Switzerland. With help from engineer Jeannine Keller, he launched a multifunctional pocket knife in 1891. Called the "Soldier's Knife," it had a wooden handle and incorporated a cutting blade, screwdriver, can opener, and punch. The knife was adopted by the Swiss army, but Elsener continued to work on the design.

In 1896 he developed a knife with blades at either end of the wooden handle and a spring to hold them in place. This innovation enabled Elsener to increase the number of useful tools in the handle for the Swiss army, so he added another blade and a corkscrew. The knives proved very popular, and when Elsener found himself facing rival manufacturers he marked all his knives with a symbol—the Swiss flag's cross in a shield—so that the original Swiss army knives could be clearly identified as such.

In 1921 the company adopted newly invented stainless steel to make the knives stronger and easier to clean. Stainless steel was also known as "inox," short for the French name for the metal, *acier inoxydable*. In memory of his mother, who had died in 1909, Elsener named his company "Victorianox."

The essential design of Swiss army knives has changed little since the original versions, though plastic is now used for the handle, and the range has been developed. In 2006 a special 100th-anniversary commemorative knife with eighty-five different tools was offered as a collector's piece. **SC**

> *"Fill your bowl to the brim and it will spill. Keep sharpening your knife and it will blunt."*
>
> Lao Tzu, Taoist philosopher

SEE ALSO: STONE TOOLS, SHARP STONE BLADE, METALWORKING, CHISEL

◥ *The original Swiss Army Knife (1891) had a blade, screwdriver, can opener, and hole punch.*

Adjustable Wrench (1891)

Johansson devises a tool for every nut.

Johann Petter Johannson (1853–1943) was one of Sweden's most prolific inventors. He amassed an impressive 118 patents over his lifetime, including one for tongs to enable people to put sugar cleanly into their tea. However, undoubtedly Johannson's greatest contribution to manufacturing and engineering was his adjustable wrench.

When Johannson opened his first workshop in Enkoping, Sweden, in 1886, there were no standard sizes or gauges for nuts, bolts, and screws, nor for the tools that manipulated them. Johannson became literally overloaded when he went about his business because his handcart had to accommodate ever-increasing numbers of wrenches to fit his various jobs. He practically had to make a new size of wrench for each task as machines and their components were built to their own bespoke specifications.

Johannson decided to make a single tool that would have some of the dexterity of a human hand. "Iron Hand," his tool of 1888, became what is now known as the pipe wrench. While eventually becoming indispensable, the pipe wrench would often ruin rusty or stuck bolts, so in 1891 he refined the design to produce the adjustable wrench.

Basically the same shape as a normal wrench, Johannson's device had an integral screw thread that, when turned, would adjust the distance between the opposing "jaws" of the wrench's working end. This meant that the tool could be offered up to the head of any nut and then adjusted to get a perfect fit—eliminating the inventor's need to carry a whole bunch of different wrenches around with him. **DHk**

SEE ALSO: METALWORKING, SCREW, ARCHIMEDES SCREW, ALLEN KEY

↗ *An example of one of the earliest adjustable wrenches, from 1892.*

"Man is a tool-using animal. . . . Without tools he is nothing, with tools he is all."

Thomas Carlyle, historian

Thermal Cracking of Oil (1891)

Shukhov converts crude oil into usable fuels.

It is thanks to Russian Vladimir Shukhov (1853–1939) that we can meet the fuel demand for modern engines. In 1891 he designed a refinery to convert crude oil into more useful things like gasoline and kerosene. Crude oil is a naturally occurring fluid that consists of a mix of hydrocarbons of various molecular lengths. Crude oil straight from the ground does not burn well, although its smaller molecules, which burn more easily, can be extracted by using fractional distillation. However, a large portion always remains as larger molecules. To make these long-chain molecules burn more easily, they can be broken into shorter chains by a process called thermal cracking.

> "... we [must] diversify our energy sources and reduce our dependency on foreign oil."
>
> Mary Bono Mack, U.S. politician

Shukhov patented a method of heating and pressurizing the oil to the extent that it would start to break up into smaller molecules. This thermal cracking was a precursor to the Burton Process, developed by William Merriam Burton in 1913, which doubled the production of gasoline. In 1937 both of these methods were superseded by catalytic cracking, which works in a similar way but with a catalyst to help the reaction. Catalytic cracking is more economical because it can take place at lower temperatures and pressures.

The oil industry is looking for even more efficient ways to extract and process crude oil before economic sources run out and oil refineries become obsolete. **LS**

SEE ALSO: BIODIESEL, BIOETHANOL

Tire Valve (1891)

Schrader creates a perennial survivor.

There is a part on all cars and bicycles whose design has not changed in more than a century: the tire valve. It was created by August Schrader (1807–94), a German immigrant who owned rubber depots and warehouses in Manhattan. He became intimately linked with the rubber industry, making molds and brass fittings for Goodyear and the Union India Rubber Company. In 1890 Schrader received a request to research and develop an airtight seal for the newly developed pneumatic tire. Two years later, August and his son George applied for a patent for the Schrader tire valve.

The valve consists of a small brass air tube with its outside surface threaded like a screw. The hollow tube

> "... the only standard in worldwide use in the automotive industry."
>
> David Beecroft, Society of Automotive Engineers

contains a metal pin in its center, running parallel to the tube. Screwing a pump onto the valve creates an airtight seal. Pumping air into the tire depresses the metal pin and opens the seal between the pump and the pressurized interior of the tire. Once the desired pressure in the tire is achieved, the pump is removed and the internal pressure of the tire, aided by a spring, seals the air tube from the inside. If the tire is inflated too hard, depressing the metal pin without a pump opens the valve and allows air to escape.

August Schrader's company is still trading under the name Schrader-Bridgeport and today is still making its universally employed tire valves. **DHk**

SEE ALSO: VULCANIZATION, PNEUMATIC TIRE, RADIAL TIRE

Crown Bottle Cap

(1891)

Painter invents an effective bottle-sealer.

When William Painter (1838–1906) left his home in Ireland in search of better opportunities in the United States, he could not have imagined that it would be a small, simple, and seemingly insignificant invention that would make him a very wealthy man and revolutionize the bottling industry.

With hundreds upon hundreds of patents received for bottle-sealing devices, the trade press wrote that it would be difficult to come up with a novel idea. Painter took up the challenge, believing that "the only way to do a thing is to do it," and do it he did, filing a patent for the "Crown Cork" in 1891.

The cap was metal with twenty-four teeth that gripped a flange around the neck of the bottle. It had a cork stopper on the inside to prevent the drink from leaking, going flat, and coming into contact with the toxic metal. The cap could be removed with a conventional corkscrew or prized open by any flat object, such as a pocket knife. Painter shrewdly mentioned in his patent application and marketing material that an opening device would be the most efficient method, and accordingly, in 1894, he received a patent for the tool he had designed.

To convince brewers and bottlers to change to his system, which required new bottles and machinery, Painter shipped some beer sealed with the Crown Cork from North America to South America. The results spoke for themselves. By 1906 the Crown Cork and Seal Company was opening factories around the world, and by the 1930s it was supplying almost half of the world's bottle caps. **RP**

SEE ALSO: GLASS, SODA WATER, CORKSCREW, SODA SYPHON, VACUUM SEALED JARS

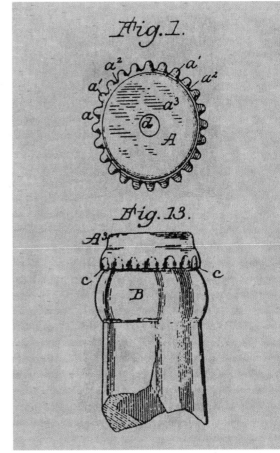

↗ *The patent application for the twenty-four-tooth "Crown Cork" from 1891.*

"Beer is proof that God loves us and wants us to be happy."

Attributed to Benjamin Franklin

In the 1890s and early 1900s, the car industry was still in its infancy, and several related products—such as the steering wheel—were developed, while the Wright brothers' airplane truly got the twentieth century underway. Wilhelm Röntgen's discovery of X-rays changed the way medical diagnoses are made, and radiation was used for the first time to treat diseases. Aspirin made its first appearance in pharmacies, but heroin exemplified that not all pharmaceutical "improvements" stand the test of time.

Birth of the
MODERN AGE

← The propeller of a World War I Sopwith Camel fighter biplane.

1892 to 1924

Fingerprinting (1892)

Vucetich adds a key component to forensics.

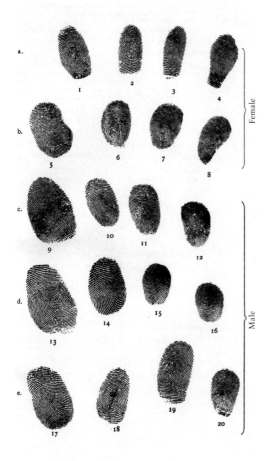

When two boys were brutally murdered near Buenos Aires in 1892, the police quickly named their mother's suitor, a man called Velasquez, as the only suspect. However, only a few days later police officer Juan Vucetich (1858–1925) proved beyond doubt that the murderer was, in fact, Francisca Rojas, the mother.

Working at the La Plata Police Office of Identification and Statistics, Vucetich's task was to identify criminals using anthropometry. Less than a decade previously, Frenchman Alphonse Bertillon had established that the measurements of certain parts of the human body never alter, therefore giving each individual—in addition to their personality traits and peculiar markings such as tattoos and scars—a distinctive anthropometric identity. His approach, named "Bertillonage," was widely adopted by police forces as a more reliable system of identification than mere eyewitness accounts and photos.

Being familiar with English scientist Sir Francis Galton's highly regarded research into fingerprinting (a technique first used in ninth-century China to authenticate records of debt), Vucetich became convinced that this was an equally foolproof yet less cumbersome way of identifying criminals. After contacting Galton to enquire whether fingerprinting would have any usefulness in forensics, he started to collect fingerprints from arrested men and classified them, calling his system "dactyloscopy."

Vucetich's easily executed system gradually soon replaced Bertillonage, while *Dactiloscopía Comparada* ("Comparative Dactyloscopy"), his acclaimed 1904 work, won him several awards. **DaH**

SEE ALSO: DNA FINGERPRINTING

> *"I've come to believe that each of us has a personal calling that is as unique as a fingerprint."*
>
> Oprah Winfrey, *O Magazine* (September 2002)

◥ *Five sets of fingerprints from Dr. Henry Faulds's Guide to Finger-print Identification, 1905.*

Vacuum Flask (1892)

Dewar invents the forerunner to Thermos®.

Needing a container capable of storing liquid forms of chemicals, Scottish physicist and chemist James Dewar (1842–1923) designed the vacuum flask that came to bear his name.

In 1892, Dewar put one flask inside another and then sucked out the air between the flasks creating a vacuum. The double-walled vessel proved a superior insulator and perfect for a focus of Dewar's low temperature phenomena, and subsequently led to the invention of the Thermos® flask.

For a few years, Dewar's flasks were on laboratory shelves not store shelves. The flasks' commercial potential was recognized by a glass blower employed by Dewar to fabricate the flasks, Reinhold Burger. He realized that Dewar's flasks could also prove useful for keeping food and drink either hot or cold. After adapting Dewar's flasks for household use, Burger obtained a German patent, founded Thermos GmbH, and began selling "thermoses" in 1904. The name Thermos® (coming from the Greek *therme* meaning hot) was suggested by a Munich resident in a competition. Burger received a U.S. patent in 1907, which described the flask as "a double walled vessel with a space for a vacuum between the walls."

Dewar never sought a patent for his flask and he lost a court case attempting to prevent Thermos from using his design. However, he gained many awards for his contributions to science. He was the first to produce hydrogen in liquid form and to solidify it, and he coinvented the smokeless gunpowder, cordite, with Sir Frederick Abel. Dewar was also knighted the same year the Thermos® flask first went on sale. **RBk**

SEE ALSO: VACUUM PUMP, VACUUM-SEALED JARS, ELECTRIC VACUUM CLEANER, SELF-HEATING FOOD CAN

↗ *A reproduction of one of Dewar's experimental flasks, cut away to show the space between the two walls.*

"The mirrored dome of an upturned Dewar flask is a thing of beauty, which every chemist should own."

Andrea Sella, Royal Society of Chemistry website

Duplex Telegraph
(1892)

Edison speeds up telegraph communication.

American inventor Thomas Edison (1847–1931) filed more than 1,000 patents during his lifetime. He fell into the world of telegraphy after he saved a three-year-old boy from being struck by a train. To repay his good deed the child's father agreed to teach him railroad telegraphy. Edison quickly picked up the skill and went to work as a telegraph operator at Western Union. Although he was eventually fired from his job, Edison's interest in telegraphy was born.

At the time, one wire could only send one telegram in one direction at a time; consequently sending and receiving messages was a slow process. Edison perfected the existing system to one signal in one

> *"Results! Why man, I have gotten a lot of results. I know several thousand things that won't work."*
>
> Thomas Edison

direction and one signal in the other, and he called this duplex telegraphy. He later made it possible to have two signals going down one wire in both directions, quadruplexing, and this was a fundamental concept for the development of the telephone in 1876.

With the help of mentor and fellow telegrapher, Franklin Leonard Pope, Edison devoted himself to inventing. He sold his patent rights for the quadruplex to a rival company, sparking a series of court battles that Western Union eventually won. However, the sale enabled Edison to set up his research laboratory at Menlo Park and his telegraphy work proved an essential step in the communications revolution. **LS**

SEE ALSO: ELECTROMAGNETIC TELEGRAPH, MORSE CODE, FACSIMILE MACHINE, PRINTING TELEGRAPH (TELEPRINTER)

Electric Arc Furnace
(1892)

Moissan aids the production of chemicals.

In 1886, Ferdinand Frédéric Henri Moissan (1852–1907) became the first person to isolate fluorine gas, for which he won the Nobel Prize in Chemistry. Six years later, he designed an electric arc furnace with the intention of turning iron and sugar into diamonds by heating them to temperatures of 3500°C. It is doubtful that he ever succeeded in this endeavor; however, he did discover other high-temperature chemical reactions, including a practical method of producing acetylene.

Moissan constructed his furnace using two blocks of limestone with a hollow cavity between them, into which he inserted two carbon rods. The sample to be heated was placed in the cavity and then an electric current of hundreds of amperes was put on the rods, creating an energetic stream—or arc—of vaporized carbon between them that produced temperatures of thousands of degrees. To make acetylene, Moissan mixed limestone and coal at high temperature to create calcium carbide, which he then combined with water. This is still the basic process of manufacturing acetylene, which both fuels high-temperature oxyacetylene welding and is used in industry to produce numerous chemicals.

Although the Moissan furnace was important, his contribution was only part of the story of its development. In 1800, Sir Humphry Davy had used one of Volta's new batteries to produce an arc between two charged carbon rods. In 1878, Sir William Siemens was the first person to develop a practical electric arc furnace, in which he was able to melt steel and platinum. In 1900, Paul Louis-Toussaint Héroult invented an electric arc furnace to smelt high-quality steel out of scrap steel and pig iron. **ES**

SEE ALSO: CONTROLLED FIRE, OVEN, BLAST FURNACE

Ferris Wheel
(1893)

Ferris creates an engineering wonder to put Chicago on the entertainment map.

How do you outdo the Eiffel Tower? This was the problem facing the organizers of the Chicago World's Fair of 1893. The solution was presented by George Ferris (1859–1896), who submitted details right down to ticket prices. Ferris was a civil engineer with an interest in railroads and bridge building. His design was for a wheel, 262 feet (80 m) tall, capable of carrying 2,000 people up for a view of the whole fair.

Smaller, wooden "pleasure wheels" had already appeared around the world, but this design was on a much grander scale. However, the fair organizers thought it could not be done and rejected Ferris. Not a man to be easily deterred, he returned after convincing fellow engineers to endorse the design and local investors to cover the $400,000 cost. This time he won approval and the Observation Wheel opened on June 21, 1893. There were thirty-six cars, one of which contained a band that played when the wheel moved, and each containing a conductor to make sure that passengers did not become hysterical and jump out of the windows. Visitors paid fifty cents for a twenty-minute ride and, by the time it was dismantled, the wheel had carried more than one and a half million people. The dismantled pieces were later used to build local bridges.

Since the 1890s a number of Ferris wheels have been built. Smaller wheels are common as rides in theme parks and funfairs, whereas larger wheels act as observation platforms in big cities around the world. Among the largest are the London Eye, the Star of Nanchang, and the Singapore Flyer. **DK**

SEE ALSO: CABLE-CAR (RAILWAY), ROLLER COASTER

⊿ The Ferris wheel was a major draw at the Chicago World's Fair, contributing to its economic success.

"I see nothing in space as promising as the view from a Ferris wheel."

E. B. White, writer

Carburetor (1893)

Bánki and Csonka fuel the newly invented motor car.

The development of the internal-combustion engine at the end of the nineteenth century was a long process, and hundreds of problems had to be overcome before a working engine actually appeared.

One of these problems was getting the fuel into the engine. Fuel, oxygen, and heat are needed for fire to happen, but producing the right mix inside the engine was difficult. The gas pump in the first engines vaporized the gas and mixed it with air, but the proportion of air to gas was not controlled, so the combustion was sometimes big, sometimes small, and this made the engine very unstable.

The solution to this problem was found by two Hungarian engine manufacturers, Donát Bánki (1859–1922) and János Csonka (1852–1939). Bánki was responsible for a number of advancements in automobiles, and together with Csonka he eliminated the problem of mixing gas and air. Inspiration came, as inspiration often does, from an unlikely place. While passing a florist, the two men noticed the flower girl spraying water on plants. Seeing her force water through a glass blowpipe to produce a mist to freshen the flowers, they reasoned that a similar technology could do the same work in an engine.

So, in 1893, the carburetor was born. This vital engine part sprays a small but regulated amount of fuel into the air and this is sucked into the engine, where it is combusted. The spray keeps the mix of air and fuel constant, fixing the problem. Carburetors based on this theory ended up in almost everything with an engine—cars, airplanes, boats—and the breakthrough was all thanks to Bánki and Csonka. **DK**

SEE ALSO: TWO-STROKE ENGINE, SPARK IGNITION, DIESEL ENGINE, TURBOCHARGER, FUEL INJECTION, STARTER MOTOR

⊞ *In the twin-choke carburetor, gas and air are mixed in two separate chambers with their own fuel jets.*

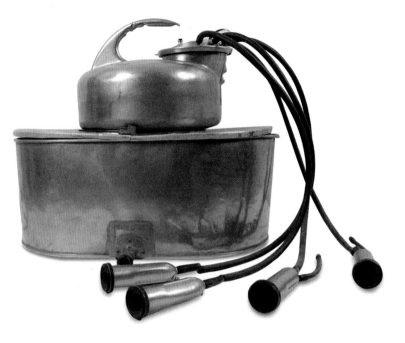

Milking Machine (1895)

The Thistle company devises a high-yielding, safer machine.

During the second half of the nineteenth century, more than a hundred patents for milking machines were applied for in the United States alone. The proliferation of people trying to develop this device, as with any that sees development across countries, makes it difficult to attribute the modern product to one person or organization.

The problem with all kinds of automatic milking processes is that there is potential for damage to the cow's teats. A viable milking machine needs to strike a balance between producing a good yield of milk and safeguarding the cow, either from injury or infection caused by the equipment.

Many inventors made contributions to the field. Foremost is the DeLaval Company of Gustaf de Laval, famous for inventing a device for separating milk and cream through centrifugal force, which researched almost every type of automatic milker in existence, including ones that simulated the action of the human hand. Other key inventors included L. O. Colvin, who created one of the first viable vacuum milkers, and William Mehring, whose foot-powered vacuum milking machine also received notable popularity.

The closest ancestor of the modern milking machine is probably the "Thistle" machine of 1895, made by the Scottish Thistle company; its steam-driven pump resulted in a distinctive broken flow, called the pulsator. The regulation of the vacuum being applied to the cow's teat is essential to stop blood pooling in the teat, which can lead to injury; this key feature of modern machines is what links them back to the "Thistle." **BG**

SEE ALSO: POWDERED MILK, PASTEURIZATION, MILK/CREAM SEPARATOR

⊡ *The Thistle was replaced by the 1923 Surge Milker, which proved more efficient and easier to clean.*

Breakfast Cereal (1895)

The Kelloggs invent a popular breakfast food.

Cornflakes were invented by accident in 1894 by John Harvey Kellogg (1852–1943) and his brother Will Keith Kellogg (1860–1951). A group of Seventh Day Adventists, including the Kellogg brothers, were trying to develop new foods to conform to a strict vegan diet, in the belief that this was beneficial to health. As the superintendent of the Battle Creek Sanatorium, a hospital and health spa for wealthy customers, John Kellogg tested out the foods on his guests.

Grains of all kinds were known to be nutritious. On one occasion, the Kellogg brothers left some cooked wheat to attend to something else, and when they returned the wheat had dried out. Not wanting to

"At the time I little realized the extent to which the food business might develop in Battle Creek."

Will Keith Kellogg

waste it, they pressed the wheat with rollers to try and make flat dough. But the grains turned into flakes, which the brothers then toasted. The flakes, served with milk and marshmallows, proved popular with the sanatorium guests. Patenting their invention as "Granose," the brothers founded the Battle Creek Toasted Corn Flake Company, led by Will Kellogg.

Will developed similar recipes using other types of grain, including corn. He started to manufacture Granose in 1906 and—to his brother John's horror—added sugar to the flakes to make them more palatable. John Kellogg viewed the cereals business as a sideline and sold his shares, which his brother acquired covertly until he had a majority stake. **EH**

SEE ALSO: SANDWICH, POWDERED MILK, CANNED FOOD, ICE-MAKING MACHINE, SACCHARIN, FROZEN FOOD

Breast Implant (1895)

Czerny introduces plastic surgery.

The earliest known breast implant was undertaken in 1895 by Austrian surgeon Vincenz Czerny (1842–1916). Czerny transplanted a large lipoma (a benign tumor composed of fatty tissue) from a patient's flank to create new breasts in a woman who had undergone a mastectomy. Details of the outcome went unrecorded.

In the early twentieth century paraffin-wax injections were used to augment breast size, but discontinued due to disastrous complications such as "wax cancer." Other substances tried included ivory, glass balls, ground rubber, and ox cartilage.

In the 1920s transplants of fatty tissue were attempted, where fat was surgically removed from the abdomen and buttock area and transferred to the breasts. The procedure was unsuccessful since the body quickly absorbed the fat, leaving the breasts in a lumpy, asymmetrical condition. However, modern transplants using the patient's own fat or muscle tissue are now much more likely to produce good results.

Otherwise, there are two main types of breast implant: saline-filled and silicone-gel-filled implants. Saline implants, first manufactured in France in 1964, use a silicone elastomer shell filled with salt water. These implants are empty when they are inserted, so the scar is smaller than that for silicone implants.

Silicone implants, first developed by plastic surgeons Thomas Cronin and Frank Gerow in 1962, have a silicone shell that is filled with silicone prior to the surgery being performed. After sporadic reports of silicone gel causing connective tissue disorders, implants have been made of more stable substances to eliminate the possibility of silicone migration. **JF**

SEE ALSO: SILICONE RUBBER, ARTIFICIAL HEART, POWERED PROSTHESIS, ARTIFICIAL SKIN, ARTIFICIAL LIVER, LIPOSUCTION

▣ *Czerny (center, in white gown) was an eminent surgeon who founded a cancer research institute.*

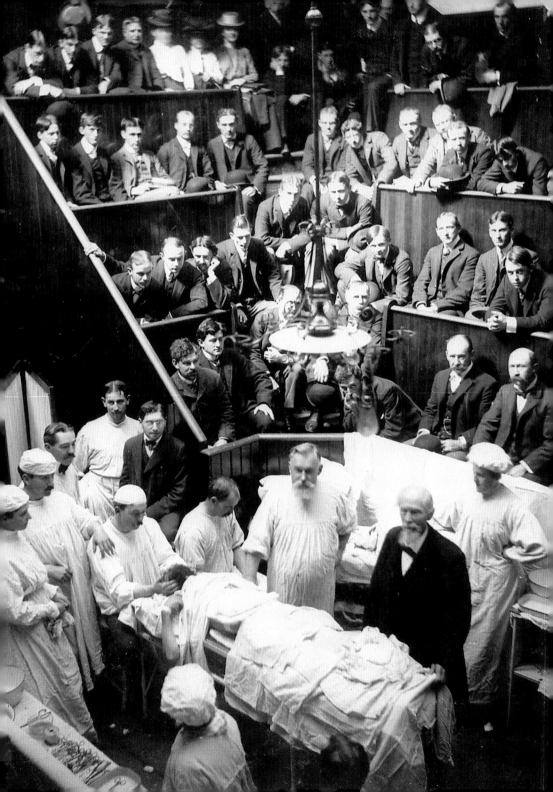

Diesel Engine (1895)

Diesel patents a fuel-economic engine.

"The fact that fat oils from vegetable sources can be used may seem insignificant today . . ."

Rudolf Diesel, 1912

While it can be safely asserted that the Diesel engine was, indeed, invented by Paris-born inventor Rudolf Diesel (1858–1913), it is not the case that we can attribute to him the very first "diesel" engine.

"Diesel engine" has for some time been the generic term used to describe any compression ignition (CI), internal-combustion engine, that is an engine that has no carburetor or spark plugs but instead injects a fuel oil directly into the cylinder. Because the piston has compressed the air therein so tightly, it is hot enough to ignite the fuel with no spark. As a cold engine cannot ignite the diesel fuel, glow plugs are sometimes used to preheat the cylinder/mixture.

Diesel researched CI engine technology for many years, testing a variety of fuels ranging from coal dust to thick tar-type oil. His engine design was patented in 1892—some two years after a patent for a CI engine had already been issued to one Herbert Akroyd Stuart (1864–1927), of Buckinghamshire, England. Stuart's patent is but one of several related patents jointly filed by him and Charles Richard Binney.

It was following demonstrations of CI engines at the Munich Exhibition of 1898, and the Paris Exhibition of 1900, that Diesel's name became synonymous with both the style of engine and the kerosene-type fuel ultimately selected to power it. Excellent fuel economy and the non-explosive nature of the fuel itself ensured the engine's widespread success.

Diesel actually foresaw the potential of (and ultimate need for) organically derived fuel sources. He died at sea after falling from the steamer, "Dresden," in September, 1913. **MD**

SEE ALSO: INTERNAL-COMBUSTION ENGINE, TWO-STROKE ENGINE, TURBOCHARGER, BIODIESEL, BIOETHANOL

◤ *The first diesel motor built by Krupp Industries in 1897 and successfully used for more than forty years.*

Film Camera/Projector (1895)

The Lumière brothers popularize movies.

In just over 100 years, "moving pictures" have evolved from peep-show parlors into a vast, multibillion dollar industry that spans the globe.

Today's movie industry has its roots in numerous nineteenth-century innovations. Thomas Edison sought to develop a device ". . . which does for the eye what the phonograph does for the ear." Attempts to duplicate the cylinder format of Edison's phonograph proved a dead-end, but Edison's "assistant," W. K. L. Dickson, eventually developed the Kinetoscope for viewing pictures in a "peep-show" format, and the Kinetograph camera with which to create footage. The viewing method of this equipment, one person at a time, had obvious limitations.

It was the Lumière brothers—Auguste (1862–1954) and Louis (1864–1948)—who created the first practical film camera/projector, the Cinematograph. Despite the fact that it incorporated both projection and filming functions in one (and was a printer too), it was much smaller than Edison's large, bulky Kinetograph. The machine also had the massive advantage of projecting pictures that many people could view at once. It is this similarity to modern projectors that establishes it as the first proper film camera/projector.

The first public showing, on December 28, 1895, featured ten short films including the Lumière's first film *Workers Leaving the Lumière Factory*. Each film was hand-cranked through a projector and lasted for approximately forty-six seconds. Film historians generally consider this historic screening at the Grand Café in Paris to mark the birth of cinema as a commercial medium. **BG**

SEE ALSO: ANIMATION, KINEMATOGRAPH, KINETOSCOPE, TECHNICOLOR, FILM SOUND

↗ *The Lumiere's Cinematograph, a combined camera, projector, and printer, is set up here for projection.*

> *"If my films make even one more person feel miserable, I'll feel I've done my job."*
>
> Woody Allen, movie director

X-ray Photography (1895)

Röntgen discovers how to photograph inside the bodies of living things.

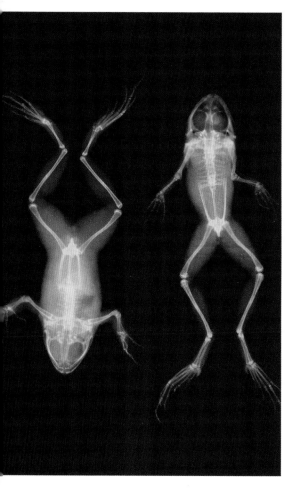

An X-ray is a form of electromagnetic radiation with a very short wavelength, in the range 10 to 0.01 nanometers. German physicist Wilhelm Röntgen (1845–1923) was experimenting with cathode rays in 1895 when he realized that these produced another form of radiation when they hit the glass of the cathode ray tube. He called them X-rays, as "X" stands for the unknown in mathematics.

Röntgen discovered that X-rays passed through soft materials, such as paper, card, and fabric, and produced fluorescence and can be used to form images on a barium-coated photographic plate. His next experiments involved human tissue. He asked his wife to put her hand on the photographic plate and discovered that the X-rays passed through the flesh, but not through her bones, or her ring. Röntgen was awarded the first Nobel Prize for Physics in 1901.

X-rays are a form of ionizing radiation, causing ions to be produced from the atoms that they hit, which is why they produce fluorescence. It was not known until the 1950s that X-rays can damage living cells, causing cancer. Some, but not all, who worked with X-rays before this, including Röntgen, protected themselves from the radiation by using lead shields.

We now take for granted the crucial role X-rays perform in medical diagnosis, and this application developed rapidly after Röntgen's initial discovery. X-rays also proved invaluable to crystallographers who used the scatter-patterns to investigate the structure of materials. Soon after their discovery, the harmful effects of X-rays were also harnessed to develop radiographic treatments for cancer. **EH**

"In recognition of the extraordinary services he has rendered by the discovery of the remarkable rays."

The Royal Swedish Academy of Sciences, 1901

SEE ALSO: X-RAY TUBE, INFRARED PHOTOGRAPHY, MAGNETIC RESONANCE IMAGING, X-RAY TELESCOPE

▣ *Photogravure from the original X-ray by Josef M. Eder and E. Valenta.*

Liquefaction of Air (1895)

Von Linde founds modern refrigeration.

German scientist Carl von Linde (1842–1934) first invented a way to turn gases into liquid. His discovery would form the basis of modern refrigeration and liquid air production.

From the 1600s scientists had known that temperature affects whether a substance is in a gaseous or liquid state. By cooling and pressurizing a gas, it is possible to slow the molecules down and compress them until they collapse and form a liquid. French mathematician Gaspard Monge first produced liquid sulfur dioxide in 1784, and by the late 1800s most gases had been successfully liquefied. However, it was still not possible to produce large quantities suitable for commercial use.

Von Linde had the bright idea of using air itself as a coolant. In 1894, in response to a request from the Guinness brewery in Dublin, Ireland, to create a new refrigeration system, von Linde succeeded in liquefying air by first compressing it and then letting it expand quickly, causing it to cool. The cold air was constantly recycled to cool more incoming air until eventually the gas liquefied.

The "Linde technique" allowed the continuous production of large quantities of liquid gases such as oxygen and nitrogen, and became an immediate commercial success. The biggest use of liquid oxygen was initially for the oxyacetylene torch, invented in France in 1904 and used for cutting and welding metal. This revolutionized the construction of ships, skyscrapers, and other ironworks during the early twentieth century.

Later applications of liquefaction of air included liquid oxygen for distribution to hospitals and industries, and liquid nitrogen for cooling blood and tissues in biological applications. **HI**

SEE ALSO: THERMAL CRACKING OF OIL, LIQUID FUEL ROCKET, REFRIGERATOR, FLUIDIZED BED REACTOR, AEROSOL

Electric Stove (1896)

Hadaway patents a new cooking device.

From the eighteenth century, people began to cook with stoves rather than using open fire, which was dirty, dangerous, and inefficient. The first electric stove was a little longer in coming, although exactly who invented it and when is a matter of some debate.

Although the Carpenter Electric Heating Manufacturing Company produced an electric stove in 1891, the first patent for an electric stove was granted to William Hadaway in 1896. Another claim to the invention comes from Thomas Ahearn (1855–1938), a Canadian businessman and inventor, who set up the Ottawa Electric Company in 1882. Ahearn was reputedly the first person to cook a meal using electricity in 1892.

> *"The maximum warmth produced by the two heaters [was] literally sufficient to roast an ox … "*
>
> The Evening Journal (August 29, 1892)

The early electric stove worked by running electricity through a resistance coil; the coil heats up, which in turn heats up an iron plate upon which a cooking vessel, with food inside it, sits. Today electric hobs use glass-ceramic tops instead of iron and the heat often comes from a halogen bulb instead of a resistance ring, but the principle is basically the same.

The electric stove was slow to take off because few people had access to electricity at the time. A fully electric kitchen was unveiled at the Chicago World's Fair in 1893, and featured an electric stove. But, it was not until the 1920s that the electric stove began to be a serious competitor to the gas stove. **BG**

SEE ALSO: OVEN, PRESSURE COOKER, GAS STOVE, SELF-HEATING FOOD CAN, MICROWAVE OVEN

Magnetic Recording (1898)

Poulsen's invention achieves the first magnetic audio recording.

Danish engineer Valdemar Poulsen (1869–1942) developed the first magnetic recording system. It was the direct antecedent of analog tape recording, the medium for which most audio was captured throughout the twentieth century.

The operating principles differed little from later analog recording systems. A motorized assembly pulled a spool of hair-thin wire at a constant speed across a magnetic recording head. The sound intended for capture was converted to electrical pulses that were then fed through a record head that imposed a pattern of magnetization onto the wire analogous to the original signal. The spool was then rewound, and the wire passed across a playback head that detected changes in the magnetic field stored on the wire; these were converted back to a continuous electrical signal that could then be heard.

Poulsen patented his Telegraphone in 1898. Two years later he took it to the World Exposition in Paris, where he recorded the voice of Austrian emperor Franz Joseph, which remains the oldest surviving magnetic audio recording. Poulsen was awarded the Grand Prix for scientific invention and found licensees for production in Europe and the United States.

In the first half of the twentieth century, the wire recorder was used as an office dictation machine. By the 1940s, the audio quality had improved and wire recorders were used in radio broadcasting. After World War II, manufacturers tried to introduce domestic wire recorders for home entertainment use, but the development of the magnetic tape recorder in the early 1950s all but rendered the medium obsolete. **TB**

SEE ALSO: ANSWERING MACHINE, ARC TRANSMITTER, WIRELESS COMMUNICATION, AUDIOTAPE RECORDING

⊡ *Poulsen's clockwork driven Telegraphone was a forerunner to the modern dictaphone.*

Powered Hearing Aid (1898)

Hutchinson's invention lends power to the hard of hearing.

Although hearing aids have probably been around for centuries, they are first mentioned in Giambattista della Porta's *Magia Naturalis* (1598). These early devices were made from wood and were carved to resemble the ears of animals known to have acute hearing.

By the late 1700s, ear trumpets were widely available in an array of different shapes, sizes, and materials. These devices all served to passively gather sound waves and direct them to the ear canal. However, in 1819 F. C. Rein made an acoustic throne for King Goa of Portugal. It had carved lions' heads for arm rests and concealed in the heads were resonating chambers that led to a hearing tube by the king's head.

In the 1890s there were numerous attempts to develop a powered hearing aid, using the recently invented storage battery. The first commercially successful powered hearing aid was the "Akoullallion" developed in 1898 by Dr. Miller Reese Hutchison (1876–1944), and patented in 1899. This early model was bulky, being designed to sit on a tabletop, and very expensive. Battery life was very limited, as was the range of frequencies. It produced only modest amplification, restricting it to those with only mild to moderate hearing loss. In 1902, he traveled to London and presented a model to Queen Consort Alexandra, who was becoming hard of hearing. By 1903, he had developed the portable "Acousticon."

Hutchison is also known for developing the electric klaxon horn, which saw widespread use in early automobiles. This has led some to quip that he invented the horn to deafen people so they would have to buy more Acousticons. **SS**

SEE ALSO: COCHLEAR IMPLANT

⬆ *This battery-powered hearing aid, made in c. 1929, has an earplug and hook for hanging over the ear.*

Heroin

(1898)

Hoffmann develops a drug initially thought suitable to replace morphine.

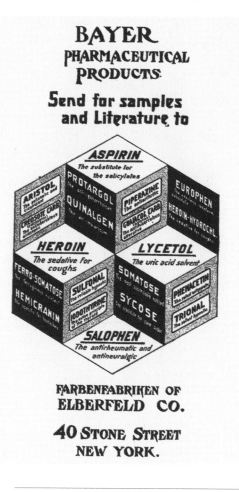

BAYER
PHARMACEUTICAL
PRODUCTS

Send for samples
and Literature to

FARBENFABRIKEN OF
ELBERFELD CO.

40 STONE STREET
NEW YORK.

*"[Heroin is] not hypnotic,
and there is no danger
of acquiring a habit."*

Boston Medical and Surgical Journal (1900)

Throughout the nineteenth century, scientists sought a non-addictive substitute for the painkiller morphine. Morphine itself was first derived from the seeds of the white Indian poppy (*Papaver somniferum*) in 1803 by German pharmacist Friedrich Sertürner. Heroin was first processed in 1874 by C. R. Alder Wright, a chemist working at St. Mary's Hospital Medical School, London. He boiled anhydrous morphine alkaloid with acetic anhydride over a stove for several hours and produced a more potent form of morphine, diacetylmorphine.

But heroin only became popular after it was independently re-synthesized twenty-three years later by Felix Hoffmann (1868–1946), a German chemist working at the Bayer pharmaceutical company. When the drug was tested on Bayer workers, they said it made them feel "heroic," leading to the name heroin. From 1898 to 1910 Bayer marketed heroin as a cure for morphine addiction, and as a component of cough pastilles and elixir because it decreased respiration. By 1899, Bayer was producing around a ton of heroin a year, and exporting it to twenty-three countries.

Then, somewhat to Bayer's embarrassment, it was found that heroin is converted to morphine when metabolized in the liver, and is basically only a quicker acting, more potent form of the drug. In 1914 use of heroin without prescription was outlawed in the United States, with a court ruling in 1919 determining that it was illegal for doctors to prescribe it to addicts.

Today heroin abuse is a serious problem for many countries in the world. Its use in medicine is also strictly controlled for instances of severe pain. **JF**

SEE ALSO: ASPIRIN

◩ *Bayer did not advertise heroin to the public but sent flyers to physicians; the company stopped selling it in 1913.*

Portable Electric Flashlight (1898)

Misell and Hubert provide handheld light.

Before the invention of the torch or flashlight, it was impossible for children to experience the guilty pleasure of staying up late at night reading a book under the covers. It began life as the idea of Joshua Lionel Cowen, who wanted to make flowerpots light up as a decorative gimmick. He sold his company (and his idea) to Conrad Hubert (1856–1928), a manufacturer of Christmas lights and other electric novelties.

Hubert decided to reinvent Cowen's light without the flowerpot. One of his shop workers, British inventor David Misell came up with a basic tube made from paper and other fibers, complete with a bulb and reflector in 1898. In an astute marketing move, New

---

"Knowledge belongs to humanity, and is the torch which illuminates the world."

Louis Pasteur, scientist

York's policemen were given the flashlights to try out, and soon everyone wanted one. Hubert's company became Eveready and its catalog pictured the new flashlight alongside the words "Let there be light."

Despite weak early batteries and inefficient carbon filament bulbs, the new flashlights were popular. In 1910 tungsten filament bulbs vastly improved the brightness and power usage of flashlights, and this technology dominated the market for more than fifty years. Fluorescent bulbs were introduced in 1968, followed by halogen bulbs in 1984. Today the best bulbs are highly efficient white LEDs, which can shine for up to thirty-five hours on one set of batteries. **DHk**

SEE ALSO: BATTERY, NIGHT VISION GOGGLES

Steering Wheel (1899)

Winton's system gives drivers more control.

It is estimated that over 100,000 patents went into the creation of the first practical automobile. At the end of the nineteenth century, the steam car had evolved considerably and was being sold commercially in the United States and the United Kingdom. But manufacturers were still using a lever device, called the steering tiller, to direct the vehicle, which made steering motor cars difficult and strenuous.

Alexander Winton (1860–1932), a keen cyclist and owner of Winton Automobiles, had been trying to replace the tiller on his car with a system modeled on a bicycle's steering. He came up with a circular wheel with a tube running down to a steering box linked to all four wheels. The mechanism in the steering box translated rotation of the wheel into linear action, and gave drivers increased control of their vehicles.

Even Henry Ford, inventor and founder of the Ford Motor Company, was converted to the new steering system. Prior to the Grosse Pointe Race in 1901, he had been given one of Winton's steering mechanisms, complete with steering wheel assembly, because Winton believed that Ford's device was dangerous. Ford went on to beat Winton, who had been tipped to win the race, but using Winton's steering system.

Unfortunately, Winton's local competitors were simultaneously working on a very similar system and beat him to the patent. The Ohio Automobile Company, later renamed the Packard Motor Car Company, added their version of the steering wheel, based on Winton's early developments, to the second car they launched in 1899. It was immediately successful, and Winton, whose company custom-made every vehicle, found the competition difficult and was forced to stop production in 1924. **SD**

SEE ALSO: POWER STEERING

Self-heating Food Can (*c.* 1899)

Unknown inventor makes hot food on the go possible.

Thanks to some unknown inventor, self-heating cans first appeared around 1900 for use by mountaineers and explorers. The most common versions involve a can with two chambers—one for the food and one for the heating unit. The heat is generated by an exothermic (heat-producing) chemical reaction between calcium oxide (quicklime) and water.

The heating unit is contained in either an outer chamber surrounding the food, or an inner compartment immersed in the food or drink. It is activated by pressing a button or poking holes to break the seal between the water and quicklime. The reaction heats up within a few seconds, and heats the food inside the can in minutes.

During World War II, Heinz manufactured self-heating beverages with a cordite stick down the center that heated the contents when lit, but they were not always reliable. Legend has it that the first recorded casualty of the D-day invasion of Normandy was a British soldier whose self-heating meal exploded, covering him in tomato soup when he tried to light the cordite stick.

More recently, some companies have revived the concept. In 2002, Nescafé tested a self-heating coffee can in the United Kingdom, but found that the can did not heat the liquid to a consistent temperature. In the United States, celebrity chef Wolfgang Puck began selling self-heating coffees and lattes in 2005, but had to recall them the next year after they failed to heat evenly, or in some disturbing cases, exploded or melted. It seems that the world may have to wait a few more years for this food of the future. **BO**

SEE ALSO: VACUUM FLASK, CANNED FOOD, FROZEN FOOD, TETRA PAK

⊡ *Self-heating cans like these 1940s models won rave reviews in the* New York Times *in 1941.*

Aspirin (1899)

Hoffman invents one of the most popular painkillers of all time.

The effects of aspirinlike substances have been known since ancient times. Romans recorded the use of willow bark as a means of fighting fever. In the early nineteenth century it was discovered that the leaves and bark of the willow tree contain a substance called salicylic acid. Although salicylic acid reduced pain and fever, it also produced severe stomach upsets.

In 1832 French chemist Charles Frédéric Gerhardt tried to eliminate these side effects by combining salicylic acid with acetyl chloride. But he found the process too time-consuming and gave up. In 1899 German chemist Felix Hoffmann (1868–1946), who worked for the Bayer pharmaceutical company, became aware of Gerhardt's work and sought the drug to relieve his father's arthritis symptoms. Hoffmann simplified the method and came up with acetylsalicylic acid. Hoffmann took a small phial home for his father, who had his first pain-free night for years. Hoffmann named the drug aspirin (taking the "a" from acetyl chloride and "spir" from *Spiraea ulmaria*, the plant from which salicylic acid is extracted).

Bayer began marketing aspirin in July 1899 with instant success. It was originally sold as a powder, but in 1914 the company introduced aspirin tablets. It was not until 1971 that British scientist John Vane identified how aspirin works; it reduces the production of certain prostaglandins (hormonelike chemicals) that are responsible for inflammation, pain, fever, and the clumping of blood platelets.

As well as its effects on pain and fever, aspirin is now known to reduce the risk of heart disease, pre-eclampsia during pregnancy, and even colon cancer. **JF**

SEE ALSO: HEROIN, DISSOLVABLE PILL, ACETAMINOPHEN

⬆ *Soluble aspirin powder from c. 1900. Aspirin remains of the most widely used drugs for minor pain relief.*

Band Brake (1899)

Daimler makes driving more secure.

When pioneer automobile makers, searching for an effective braking system, looked back to the horse-powered carriages on which their vehicles were based, they realized, unfortunately, that no one had ever actually managed to design one. So for several years, brake technology dragged along behind that of engines and transmissions.

Carriage brake evolution had peaked in 1838 with the spoon brake—in essence, a crude lever and shoe system that forced a block of wood directly against the tire. Henry Ford avoided the dilemma of which braking system to fit to his Quadricycle in 1896 by not installing any. Instead, a lever released a drive belt and the vehicle (eventually) coasted to a stop. Drivers could assist the process by pressing their feet against the front wheels.

The introduction of pneumatic tires rendered a block-on-tire system impracticable. A leap forward came with the externally contracting band brake. Here, a flexible metal band or cable, covered with friction material, is forced via a lever or pedal, to contract tightly around a rotating drum or wheel hub, thereby slowing the rotation. While a great improvement, it wore rapidly and could malfunction due to dirt and bad weather. It was also far from effective at holding a heavy vehicle on an incline.

In 1899, Gottlieb Daimler (1834–1900), en route to deploying an internally expanding drum brake a year or so later, used a band brake incorporating a chassis-anchored cable, wound around a drum. Providing the vehicle was moving forward, the stopping power was increased, without any extra effort from the driver, as the rotation of the drum tried to drag the cable around with it, increasing the force holding them together—in essence, a mechanical servo. **MD**

SEE ALSO: DISC BRAKE, DRUM BRAKE, HYDRAULIC BRAKE, ANTI-LOCK BRAKING SYSTEM, REGENERATIVE BRAKES

Assembly Line (1901)

Olds invents a revolutionary system for industry.

Just as motor cars were appearing on the market, Ransom Eli Olds (1864–1950) had an idea that was to revolutionize industry—the assembly line. After building his first gasoline-powered car in 1896, Olds set out to mass-produce successors to his beloved "Oldsmobile." Spreading himself thinly, Olds tried to produce a large range of models. Then, in March 1901, his company burned to the ground. The fire destroyed all but one of his models, the "Curved Dash" Oldsmobile. Olds focused on producing this model exclusively and made a phoenixlike comeback. He soon had more orders than he could actually meet.

Recalling how he had watched workers at a musket rifle factory assemble guns in assigned stations,

> *"Ransom Olds used [the assembly line] to jump his production from 425 cars in 1901 to 2,500 in 1902."*

Curtis Redgap, automobile enthusiast

Olds came up with an ingenious scheme for a car assembly line. He spent the rest of 1901 working to implement the idea.

The new technique proved to be an effective one, increasing his car output dramatically. Several years later, car manufacturer Henry Ford adopted and reworked the concept for even more efficiency. Machines have now largely replaced human laborers, but virtually all mass-produced products rely on some form of assembly line. **LW**

SEE ALSO: PRODUCTION LINE, AUTOMATON, AUTOMATIC FLOUR MILL, INDUSTRIAL ROBOT

⇨ *Photograph of early automobiles being produced on a factory assembly line in the United States.*

Portable "Brownie" Camera (1901)

Brownell and Kodak develop a camera for use by everybody.

The Kodak Brownie was the first handheld camera suitable for use by everyone, including children. It cost just one dollar and it was designed by camera-maker Frank Brownell who had been asked to invent the cheapest camera possible, without compromising its reliability and quality by George Eastman, the founder of Kodak. However, it was not Brownell after whom the camera was named.

During the 1890s, children's author and illustrator Palmer Cox was the Walt Disney of his day. His Brownie characters were so popular that they were used to advertise everything from sweets and dolls, to trading cards and cigars. Eastman thought that branding the new camera with the Brownie name would ensure its success. And he was right because the Brownie name became synonymous with popular photography for the next eighty years.

It was a simple device consisting of a cardboard box and a meniscus lens, which was curved on both sides. It took square images that were 2¼ inches (5.7 cm) across. It was only in production for about four months, but some 150,000 were sold. In 1901, the larger Brownie No. 2 model was released, still costing $1.

In 1930 Kodak produced a new model to celebrate the company's fiftieth anniversary. It gave away half a million Brownies, along with a roll of film, to American children who turned twelve years old that year. In total nearly 100 different models of the popular Brownie had been released by the time Kodak finally ceased production in 1980. **DHk**

SEE ALSO: SINGLE-LENS REFLEX CAMERA, PHOTOGRAPHIC FILM, 35-MM CAMERA, SELF-DEVELOPING FILM CAMERA

↗ *An original Brownie camera in its cardboard box featuring the familiar pixielike character.*
⮕ *Historians have found no evidence that Eastman sought Palmer Cox's permission to use his Brownie.*

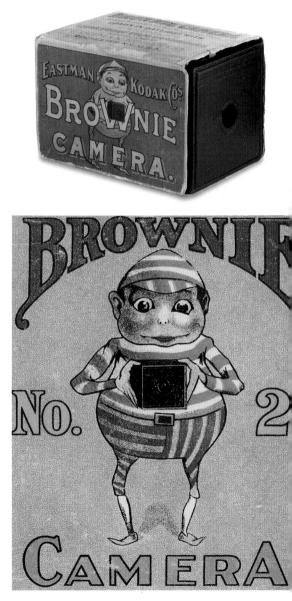

Instant Coffee (1901)

Kato invents the quick caffeine fix.

The cultivation of the coffee bean can be traced back to tenth-century Ethiopia. It was introduced to Europe in the sixteenth century and the Americas in the mid-1600s, after which it proved an extremely popular beverage. But preparing it correctly, brewed from ground coffee beans, could be time-consuming.

It was not until 1901 that a Japanese-American chemist named Satori Kato, using an earlier process he pioneered for making instant tea, created the world's first soluble instant coffee. Kato's coffee, which he called "Sanka," though initially bitter and pungent, was a concentrated solution made from coffee beans and water that was dehydrated leaving a powdery residue, which dissolves easily in hot water.

> *"A mathematician is a device for turning coffee into theorems."*
>
> Albert Einstein, scientist

While living in Guatemala in 1909, Belgian-born chemist George C. Washington was the first to market mass-produced coffee with his "Red E Coffee" brand, after observing dried coffee forming on the spouts of coffee pots. It dominated the instant coffee market in the United States for the next thirty years.

Although coffee sales boomed in the United States in the 1920s due to prohibition, it was not until 1938 that the Brazilian government approached the Nestlé Company for assistance in reducing its massive coffee surplus. Nestlé developed a spray-dried process that resulted in a more palatable, dehydrated coffee using soluble carbohydrates. **BS**

SEE ALSO: VENDING MACHINE, TEA BAG, POWDERED MILK, COFFEE FILTER

Flame Thrower (1901)

Fiedler conceives a devastating weapon.

The modern flame thrower was not particularly innovative—it simply launched burning fuel to spread fire. However, when used on the battlefield, its effect was devastating, and it is remembered as the most demoralizing infantry weapon ever used.

At the turn of the twentieth century, German inventor Richard Fiedler experimented with two types of flame thrower. The *Flammenwerfer* was a smaller, handheld weapon that used pressurized gas to push out streams of burning oil. The larger model was not as portable, but had a range of 118 feet (36 m) and could produce a continuous stream for forty seconds. When used in warfare, the weapons were highly dangerous, both for the enemy and the users, as the pressurized gas cylinders were prone to explosion.

Modern flame throwers consist of a backpack containing a tank with a flammable liquid, often napalm, and a tank with compressed gas. When fired the pressurized gas forces the fuel through the handheld gun where it is ignited at the tip and pushed out in a fiery display. The Germans adapted the flame throwers for use in World War I. More sophisticated and effective flame throwers caused massive destruction during the Vietnam War, when they were used to flush out enemy soldiers hiding in tunnels. However, countless civilians also died in the process as the fires often got out of control.

Flame throwers are rarely used by military forces today, although their use in warfare has not been banned. However, the modern flame thrower has found a more benign use by foresters and farmers who use them to clear land. **LS**

SEE ALSO: CONTROLLED FIRE, BLOW TORCH

➡ *French troops use a flame thrower to project liquid fuel across a railway line in August 1915.*

Toy Electric Train Set (1901)

Cowen invents a timeless toy for young and old.

As long as railways have existed, there have been model versions. Initially, the trains were just pull-along, but they gradually developed to use miniature steam or clockwork engines. By 1891, the German company Märklin was selling a wind-up locomotive together with an expandable track system. Then, in 1896, Carlisle and Finch produced the first electric toy train powered by batteries.

American entrepreneur Joshua Lionel Cowen (1877–1965) had a long-term interest in trains. At the age of seven he had whittled a locomotive out of wood, which exploded when he tried to fit it with a steam engine. Years later when he was researching products for his manufacturing company in New York to sell, he spotted a push train in a shop window and had the idea for a toy train that could run without supervision on a track. He initially envisaged it as an eye-catching window display for a toy shop. In 1901 he fitted a small motor under a model of a train and the Lionel "Electric Express" was born. The first versions had a battery like the Carlisle and Finch engines, but they were soon replaced by trains that ran on electricity. Over the next decade, Lionel made more extravagant train sets incorporating different engines, passenger cars, stations, bridges, and tunnels.

By 1909, many different sizes of toy electric train existed. In a smart marketing move, Cowen's company began to sell their trains as standard gauge, forcing the other American companies to adopt their scales. In England, although Frank Hornby's Meccano® set came out the same year as the Lionel train, Hornby did not produce his electric train set until 1925. **DK**

SEE ALSO: LOCOMOTIVE, MECCANO®, LEGO®

⬆ *Scale model of a railway station at the Model Railways Exhibition in London in 1948.*

Improved Radio Transmitter (1901)

Fessenden sends the first audio radio transmission.

Canadian-born Reginald Fessenden (1866–1932) caused a landmark in the development of radio when he transmitted his own voice over radio waves late in 1900, a feat not even Marconi had achieved. At the time, Fessenden was working for the United States Weather Bureau to develop wireless technology for weather forecasting. On December 23, at his station on Cobb Island, Maryland, Fessenden transmitted what is considered the first wireless transmission carrying audio sound. "Hello, one, two, three, four. Is it snowing where you are Mr. Thiessen? If it is, telegraph back and let me know," he shouted into the microphone. Thiessen excitedly telegraphed back that it was.

The transmitter that Fessenden used was a spark transmitter, a device developed in the late nineteenth century by radio pioneers Hertz, Marconi, and Braun to generate radio frequency electromagnetic waves. Fessenden had modified it so that the sparks produced more continuous waves rather than ones that died away quickly. He also placed a carbon microphone directly in series with the antenna lead. He applied for a U.S. patent for the transmitter in 1901, and it was the first to use the same principles that AM (mediumwave) radio stations use today .

This milestone was only the start of the influence Fessenden was to have on radio history. After further improvements to his transmitter, he made the first transatlantic voice transmission from Massachusetts to Scotland, as well as the world's first radio broadcast on Christmas Eve, 1906. The broadcast consisted of music and readings, and was heard mostly by ship operators, some as far away as the West Indies. **RH**

SEE ALSO: WIRELESS COMMUNICATION, ARC TRANSMITTER, AMPLITUDE MODULATION

⬆ *Fessenden paves the way for radio broadcasting when he transmits sound over the radio in 1906.*

Mercury Vapor Lamp (1901)

Hewitt pioneers fluorescent lighting.

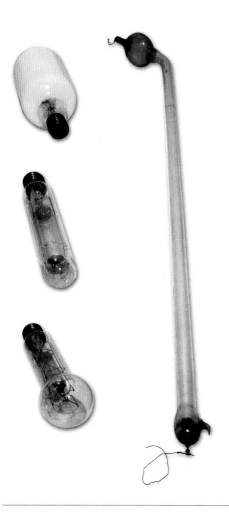

During the latter half of the nineteenth century, there was considerable interest in the effects of electric currents being passed through gases at low pressures confined in glass tubes. Germans Julius Plücker and Heinrich Geissler noticed that if vaporized mercury was used as the gas, a bluish glow was emitted.

An electrical engineer from the United States, Peter Cooper Hewitt (1861–1921), began to experiment with these mercury tubes and produced ones that emitted a great deal of an unflattering bluish-green light. Some of this light was in the ultraviolet and thus could not be seen. The fact that there is no emission in the red meant that the light made people appear like "bloodless corpses." The dangerous effects of the ultraviolet emission could be alleviated by coating the tube with a fluorescent chemical that absorbed the ultraviolet and then emitted the energy at higher wavelengths.

Hewitt patented his mercury vapor lamp in 1901 and formed a company with the American entrepreneur George Westinghouse to produce the lamps. Although, their primary use at the time was industrial and in photographic studios, Hewitt's mercury lamp was actually the precursor to modern fluorescent lighting.

Mercury lamps are cheap to make and have a long life. It takes quite a while for the mercury to warm up and auxiliary electrodes are needed to start the excitation. Mercury lamps have a rather low luminous efficiency when compared to other discharge lamps. They also have the disadvantage in that they cannot be switched on and off quickly. **DH**

SEE ALSO: INCANDESCENT LIGHT, NEON LAMP, FLUORESCENT LAMP, ANGLEPOISE® LAMP, HALOGEN LAMP, LAVA LAMP

"Experience is a dim lamp, which only lights the one who bears it. "

Louis-Ferdinand Celine, writer and physician

◤ *A low pressure mercury vapor lamp (right) alongside three improved high pressure lamps from the 1930s.*

Electric Vacuum Cleaner (1901)

Booth's machine makes light of housework.

In 1901 mechanical engineer Hubert Cecil Booth (1871–1955) watched a railway carriage being cleaned at St. Pancras Station in London by a series of high-pressure hoses using compressed air to blow away debris. Dining with friends afterward, he impulsively covered his mouth with a moistened handkerchief, placed his mouth against the cover of his cloth chair and inhaled, trapping dust against the outer lining of the handkerchief. Convincing himself that a device using reverse pressure and equipped with a filter would effectively capture and store dust, he set about creating a cleaning machine using suction rather than simply blowing particles into the air.

Together with his friend F. R. Simms, who designed a water-cooled, six-horsepower piston engine driven by an electric motor, to which Booth attached a simple vacuum pump, he produced the world's first electric vacuum cleaner. Booth set up the Booth Vacuum Company and began a cleaning service. However, with few Victorian homes and businesses having the luxury of electricity, his machine needed to be mobile. He parked his carriage, which he named Puffing Billy, on the street with hoses passed in through the windows of his clients' premises. He used transparent tubes so that skeptical onlookers could witness the dust particles being drawn into his machine.

His invention proved such a success that he gained a commission to clean the ceremonial carpets in London's Westminster Abbey prior to the coronation of King Edward VII in 1902. The king was so impressed he gave Booth a royal warrant to provide vacuum cleaners to Windsor Castle and Buckingham Palace. **BS**

SEE ALSO: CARPET SWEEPER

◪ Booth's Puffing Billy machine of 1901 was successful because it had an effective source of power.

"Sucking out dust is impossible. It has been tried over and over without success . . ."

John S. Thurman, inventor

No. 696,788.

C. C. ALLEN.
OPTICAL OBJECTIVE.
(Application filed Feb. 25, 1901.)

Patented Apr. 1, 1902.

(No Model.)

2 Sheets—Sheet 1.

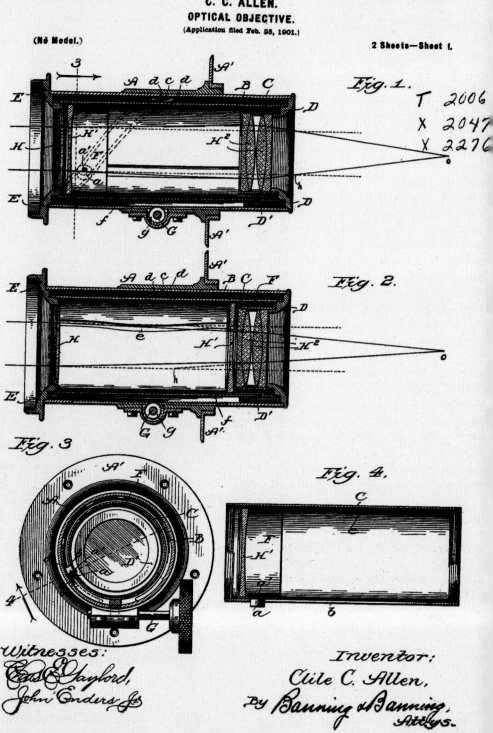

Fig. 1.

Fig. 2.

Fig. 3.

Fig. 4.

Witnesses:

Inventor:
Clile C. Allen,
By Banning & Banning,
Attys.

Zoom Lens (1901)

Allen brings a sharper focus to photography.

Although the first zoom lens was invented in 1901 by Clile C. Allen, its popularity in sales terms only came in the late 1980s. Prior to this the fixed-focus lens was the standard; however, this was pushed aside because of the physical effort required to photograph different-sized subjects. With a fixed-focus lens, the only way to adjust the amount captured in a photograph is to physically move backward or forward with the camera.

A zoom lens works by allowing the photographer to widen or shorten the focal length, which increases or decreases the magnification of the subject. The focal length is the distance between the lens and the point in the camera at which the light rays converge. Changing this length allows you to zoom in and out when taking a photograph.

There are many different designs of zoom lens. Some may be composed of up to thirty individual lenses among other moving parts. However, all of these lenses feature a final focusing lens that ensures the image remains crisp and sharp as the focal length changes. This was a major improvement on the nineteenth-century varifocal lenses, which did not remain focused after changes to the focal length and had to be refocused each time. Allen's zoom lens was therefore instrumental in making photography both more efficient and relatively effortless.

Early zoom lenses tended to produce lower quality images, compared with those obtained using a fixed-focus lens. However, with the advent of digital photography and computer optimization, modern zoom lens cameras have quickly caught up with their fixed-focus lens counterparts. **RB**

SEE ALSO: PHOTOGRAPHY, COMPOUND CAMERA LENS, SINGLE-LENS REFLEX CAMERA

⬅ *Allen's patent application for an "optical objective," approved April 1, 1902.*

Laparoscope (1901)

Kelling revolutionizes abdominal surgery.

The laparoscope is the James Bond-like gadget of the surgeon's repertoire of instruments. Only a small incision through the patient's abdominal wall is made into which the surgeon puffs carbon dioxide to open up the passage. Using a laparoscope, a visual assessment and diagnosis, and even surgery can then be performed using tiny tools. This surgery causes less physiological damage, reduces patients' pain and speeds their recovery leading to shorter hospital stays.

In the early 1900s, Germany's Georg Kelling (1866–1945) developed a surgical technique in which he injected air into the abdominal cavity and inserted a cytoscope—a tubelike viewing scope—to assess the patient's innards. In late 1901, he began experimenting

> *"After a complex laparoscopic operation, the 65-year-old patient was home in time for dinner."*
>
> Elisa Birnbaum, surgeon

and successfully peered into a dog's abdominal cavity using the technique.

Without cameras, laparoscopy's use was limited to diagnostic procedures carried out by gynecologists and gastroenterologists. By the 1980s, improvements in miniature video devices and fiber optics inspired surgeons to embrace minimally invasive surgery. In 1996, the first live Internet broadcast of a laparoscopy took place. A year later, Dr. J. Himpens used a computer-controlled robotic system to aid in laparoscopy. This type of surgery is now used for gall bladder removal as well as for the diagnoses and surgeries of fertility disorders, cancer, and hernias. **LW**

SEE ALSO: ENDOSCOPE, SURGICAL ROBOT

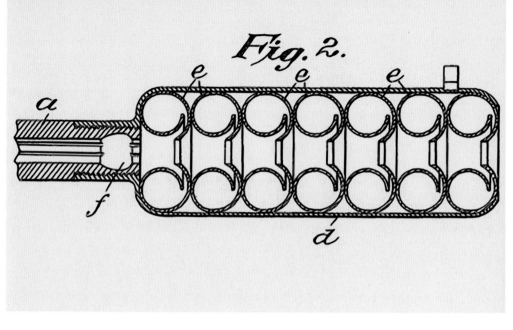

Gun Silencer (1902)

Maxim silences his father's invention, the Maxim gun.

Few other names are as synonymous with invention and innovation in the field of munitions and weapons as that of the Maxim family. This remarkably inventive family's output was not exclusively weapon-related. Indeed, Hiram Percy Maxim's (1869–1936) creation—which enabled a firearm to be discharged without the traditional loud bang—initially stemmed from his interest in automobile design, specifically exhaust silencers or "mufflers."

The bulk of noise produced by a firearm discharging subsonic ammunition is caused by a massive and rapid expansion of propellant gases leaving the muzzle, a bit like uncorking a bottle of fizzy champagne. Maxim's silencer—which, like all such devices, suppresses rather than truly silences—attaches to the barrel of a firearm. Essentially a cylindrical casing of much larger capacity than the barrel, and containing a series of baffles, it allows the explosive gases from the cartridge to expand and

slow prior to hitting the surrounding air, producing far less noise. Muzzle flash is also reduced.

First designed around 1902 and patented in 1909, a unique feature of the Maxim unit was that it was not concentric to the bore, as are the majority of today's silencers, but was offset to allow the use of the original weapon sights without modification. Maxim's silencer was marketed quite creatively as a gentlemanly way of target shooting. Although it was later adopted by police and military forces across the world for use on a variety of weapons, fears were expressed at the time that the availability of a "noiseless weapon" could prove a major boon to criminals and gangsters. **MD**

SEE ALSO: GUN, MUSKET, BREECH-LOADING GUN, MACHINE GUN, PORTABLE AUTOMATIC MACHINE GUN

⬆ *This illustration was part of the patent application Maxim submitted in 1908.*

Air Conditioning (1902)

Carrier makes the world a much cooler place to live.

The ancient Romans tried to keep their buildings cool during hot weather by pumping water from aqueducts through the walls of their houses, whereas in Southeast Asia people hung wet grass mats over the windows to lower the temperature of air inside. Modern air conditioning, which arrived in 1902, is the continuation of this rudimentary principle.

Willis Carrier (1876–1950) of Buffalo, New York, developed the fundamental scientific theories of air conditioning. His first system was designed for use in a printing plant. Changes in the temperature and humidity of the plant were causing the ink nozzles to be out of line, which made color printing problematic. Carrier was assigned with the task of fixing this problem. His early system, which made use of spraying nozzles to cool and dehumidify the air, was large, extremely expensive, and rather dangerous because it relied on the use of ammonia as a coolant. For years, his machines were just used to cool machines, but when the potential to cool people was spotted, they began to be installed in other commercial buildings, such as offices, hotels, and hospitals. He even went on to place units in the United States Senate and the White House. Air conditioning has made life more comfortable and led to greater economic activity in the summer months. It can also reduce death rates from heat-related illnesses by up to 40 percent.

Carrier refined his designs and in 1922 installed a system at a Los Angeles theater where conditioned air was fed in from the ceiling and exhausted at floor level. The Carrier Corporation continues to make and install air conditioning systems all over the world. **SB**

SEE ALSO: REFRIGERATOR

⊤ *Absorption chillers like Carrier's model above use the waste heat from gas turbines to cool buildings.*

Drum Brake (1902)

Renault uses friction to slow down motor cars.

Louis Renault (1877–1944) is a well-known name in motoring, and in 1902 the French engineer invented the drum brake for cars.

Drum brakes work in a similar way to bicycle brakes, using friction to slow and eventually stop a fast moving wheel. In cars, the friction is caused by pads that press against the inside surface of the rotating drum that is connected to the car wheel. The pads provide a wedging action that stops the wheel spinning and thereby stops the car moving.

To function correctly, the brake pads need to linger close to the drum without scraping it during normal driving. If the pads are too far away—for example, when the pads wear down from use—then the brake pedal has to be pressed deeper to get a braking action. To avoid this, most drum brakes have an automatic adjuster that regulates the distance between the pads and drum. Even though drum brakes were very successful in cars for many years, in the 1960s they began to be replaced in some car models by disk brakes.

The first drum brakes were mechanically operated but later used oil pressure pistons. Although drum brakes are still used today they have the major disadvantage of needing frequent adjusting and replacing. The force of friction on the pads wears them away fairly quickly. If a driver is heavy on the brakes then inevitably the pads will need replacing more often, which adds to the cost of upkeep. However, because drum brakes are often less costly and easier to produce than alternative brake designs, they are favored by some car manufacturers. **LS**

SEE ALSO: MOTORCAR, ACCELERATOR, BAND BRAKE, DISK BRAKE, HYDRAULIC BRAKE

⊞ *The drum brake, favored for many years, presses the brake pads against the drum rather than the rotor.*

Disk Brake (1902)

Lanchester introduces a modern car braking system ahead of its time.

The modern system of automobile brakes was patented in 1902 by British car manufacturer Frederick William Lanchester (1868–1946). He took the disk brakes then available and radically improved their design. But, competition with the new drum brakes meant that disk brakes took almost fifty years to become a reality on a mass-produced car, and another five years before the concept became widespread.

Lanchester and his three brothers formed the Lanchester Engine Company in 1899. He had been designing cars and engines since 1889, and had built the United Kingdom's first four-wheel drive car in 1895. In 1902 their latest prototype, which contained a ten horse-powered, twin cylinder engine, was fitted with the new system of disk brakes.

Disk brakes slow a car by removing energy from the rotating wheel; brake pads squeeze both sides of the rotor connected to the wheel. The friction of this generates heat, which escapes via air vents, and the car slows down. The brake pads are activated by hydraulic pressure through a piston that presses the brake pad onto the rotor. The drum brake, in contrast, uses pistons and brake pads but applies its force via brake "shoes" against the inside of the wheel drum.

Within two years of patenting his idea, Lanchester's company collapsed due to financial difficulties. Car manufacturers favored the lighter, cheaper drum brake created in the same year, despite the disk brake being more responsive. It took until the 1950s for the disk brake to be rediscovered in the United Kingdom, and since then it has been the standard brake design on cars in Europe and the United States. **SR**

SEE ALSO: MOTORCAR, ACCELERATOR, BAND BRAKE, DRUM BRAKE, HYDRAULIC BRAKE

⬆ *Lanchester's disk brake is still widely used in modern vehicles; this one is fitted to a 2002 Ford.*

Ostwald Process (1902)

Ostwald manufactures nitric acid.

The process for fixing nitrogen from ammonia into nitric acid was a key development in the industrial production of fertilizers and explosives. It was patented in 1902 by Russian–German chemist Wilhelm Ostwald (1853–1932). One of the founders of the field of physical chemistry, Ostwald received the 1909 Nobel Prize in Chemistry for his work on catalysis, chemical equilibria, and reaction velocities. His process remains fundamental to the modern chemical industry.

During the Ostwald Process, ammonia is heated in the presence of a platinum-rhodium catalyst to form nitric oxide, which is then oxidized to yield nitrogen dioxide, which in turn reacts with water to produce nitric acid and nitric oxide.

Ostwald's major breakthrough was his discovery that the length of time the reactants are in contact with the catalyst affects the yield of the reaction. Leave them there too long, and the nitric acid degrades back into nitrogen. Ostwald passed the ammonia and oxygen gases over the catalyst at a speed slow enough to allow the reaction to take place, but fast enough to protect the acid from degrading.

An earlier patent—by Kuhlmann in 1838—had described the basic chemistry of the process, but was of purely academic interest because of the scarcity of ammonia from animal sources at the time. Once the Haber Process for producing ammonia was commercialized in the early twentieth century, the large-scale industrial manufacture of nitric acid could commence.

A negative effect of the development of the Ostwald and Haber processes is that it almost certainly prolonged World War I, allowing Germany to continue making explosives when its supplies of sodium nitrate from Chile were cut off. **BO**

SEE ALSO: HABER PROCESS

Teddy Bear (1902)

Iconic children's toy is named after Roosevelt.

In 1902 American president Theodore Roosevelt went on a bear hunting expedition in Mississippi that led to the invention of perhaps the most iconic children's toy in history. Holt Collier, a former slave and prodigious huntsman, was charged with organizing the chase. In order to provide Roosevelt with a clear shot, Collier and his hounds tracked down a bear and drove it to the stand where the president was waiting. However, when Collier arrived with the bear, Roosevelt had left for lunch. In the ensuing confusion the bear attacked one of the hunting dogs. But unwilling to kill the beast he'd promised to the president, Collier simply knocked it out with his rifle and tied it to a tree. When Roosevelt returned a short while later he was impressed by Collier's feat, but refused to kill the defenseless bear.

The episode gained widespread media attention, and in November 1902 the *Washington Post* ran a series of Clifford Berryman cartoons, initially featuring Roosevelt and an adult bear tied to a tree. In subsequent cartoons the bear was depicted as an endearing little cub, and inspired by the drawings, Morris and Rose Michtom—two shopkeepers from Brooklyn, New York—created a toy bear. Named after "Teddy" Roosevelt, theirs was not a real-life representation, and its sweet and innocent looks became a huge hit with society ladies and children. The Michtoms subsequently founded the Ideal Novelty and Toy Company, which still exists today.

Coincidentally, German seamstress Margarete Steiff also began producing toy bears at around the same time. Initially shunned in Europe, they flew off the shelves in the United States. Steiff bears—with their trademark button in the left ear—were always the most expensive ones, and hence their antique models still command the highest prices among collectors. **DaH**

SEE ALSO: TOY BALLOON, MECCANO®, TOY ELECTRIC TRAIN SET, CRAYONS, LEGO®

Spark Plug (1902)

Honold paves the way for engines with higher operating speeds.

Gottlob Honold's (1876–1923) career as an engineer and inventor started with a lucky break. His father was friends with the father of Robert Bosch, founder of the great German technology company. At the age of fourteen, Honold was given his first job in Bosch's workshop in Stuttgart, where he began to hone his technical prowess. He later left to study engineering at Stuttgart University, but returned to Bosch in 1901 as their technical manager. It was then that he made important changes to the concept of the spark plug.

Ignition systems for cars had been around for some time, but none of them were reliable. Some systems rapidly drained the car's battery, whereas the Daimler glow tube ignition system sometimes even set fire to engines. The issue of inventing a reliable ignition system for cars was described as the "problem of all problems" by automotive pioneer Karl Benz.

Honold developed a high-voltage magneto ignition unit complete with a spark plug. Previously, magneto ignition units had only been used on stationary engines but when Bosch fitted one to a motor tricycle in 1897, it was able to reach speeds of up to 50 mph (80 kph)—extremely high for the time.

Honold's souped-up version, in which the electric charge for the spark plug is generated by the movement of magnets within the engine itself, allowed the development of engines with higher operating speeds of around 1,000 rpm. It coincided with the increase in demand for cars and Honold's design soon became standard. In 1902 Bosch made some 300 spark plugs; a century later worldwide production had reached more than 350 million. **DHk**

SEE ALSO: MOTORCYCLE, MOTORCAR, SPARK IGNITION, CAR BATTERY, CARBURETOR, DIESEL ENGINE, TURBOCHARGER

↗ *Spark plugs were made earlier, but only Honold's magneto-based ignition system was practical.*

"The high-voltage magneto and spark plug ... became available just as serial car production [took off]."

Vaclav Smil, *Creating the Twentieth Century*

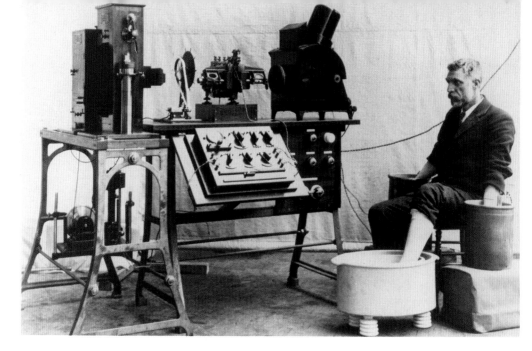

Electrocardiograph (1903)

Einthoven's machine provides a major boost to diagnosing heart problems.

The electrocardiograph—also known as EKG or ECG—is an instrument designed to record minute electric currents generated within the heart, which are used to diagnose different types of heart disease. At the end of the nineteenth century, physiologists understood that beating hearts produced electrical currents, but they could only measure them by placing electrodes directly on heart muscle.

Dutch physician and physiologist Willem Einthoven (1860–1927) adapted the string galvanometer for use in cardiology. String galvanometers had first been used to amplify electrical signals transmitted along undersea cables. Einthoven's galvanometer, which he produced in 1903, consisted of a microscopic thread of quartz known as a "string" that was vertically suspended in a strong magnetic field. When minute currents passed through the string it deflected and obstructed a beam of light, allowing the shadow to be recorded on photographic paper.

Early prototypes were unwieldy—weighing 600 pounds (272 kg), they needed to be operated by five technicians and required patients to place both hands and feet in buckets of cold water. The instrument was, however, sufficiently sensitive to detect electrical impulses as the heart contracted and relaxed. Einthoven studied both normal and abnormal electrocardiograms to give doctors reference points for interpreting the results. By the 1920s heart attacks could be diagnosed from characteristic abnormal patterns. Portable, lighter EKGs eventually came along and after World War II the string galvanometer was superseded by direct writing equipment. **JF**

SEE ALSO: X-RAY PHOTOGRAPHY, ULTRASOUND, MAGNETIC RESONANCE IMAGING, COMPUTED TOMOGRAPHY (CAT SCANS)

⬆ *The table-model of Einthoven's electrocardiograph was made by Cambridge Scientific Instruments in 1911.*

Blink Comparator (1903)

Pulfrich's apparatus improves observation of the skies.

A blink comparator enables astronomers to look at two different photographic plates taken of the same region of the sky on different nights, using the same telescope and plate exposure. If something "blinks" as the view rapidly switches from one illuminated plate to the other, the object has either changed brightness or moved.

This apparatus and technique has been used to detect asteroids, comets, and variable stars. Plates taken a few years apart have been used to detect nearby fast-moving stars or to distinguish between binary stars that orbit a common center of mass, and two stars that happen to be close to the same line of sight, or an optical double.

The German physicist Carl Pulfrich (1858–1927) developed the device while working for the Carl Zeiss Optical Workshop. Blink comparators were soon being used by observatories around the world and led to the discovery of hundreds of variable stars. The most

important discovery made with a blink comparator was the existence of Pluto, by Clyde W. Tombaugh in March 1930. At the time, "planet X" was believed to be the gravitational perturber of Uranus and Neptune, orbiting near the outer edge of the solar system. More recently, Pluto has been found to be a low-mass object and has been demoted from planetary status.

Blink comparators are still used today, although the photographic plates have been replaced by digital images that can be stored on a computer. A new application has been proposed to allow radiologists to more accurately compare new and old X-rays, and CAT and MRI scans. **DH**

SEE ALSO: TELESCOPE, RADIO TELESCOPE, SPACE OBSERVATORY, X-RAY TELESCOPE, HUBBLE SPACE TELESCOPE

⬆ *Clyde Tombaugh with the blink comparator he used to discover Pluto, as well as some 3,000 asteroids.*

Arc Transmitter (1903)

Poulsen creates a portable radio system.

In 1906 the Amalgamated Radio Telegraph Company was founded as a merger between the UK De Forest Wireless Telegraph Syndicate and the fledgling operation run by Danish inventor Valdemar Poulsen (1869-1942). Before too long they had successfully established an experimental wireless telegraphy link between Newcastle, England, and Denmark. Unfortunately, the Amalgamated Radio Telegraph Company went bankrupt in 1907 before any commercial operation could be set up.

The historic event had been made possible by Poulsen's invention of the arc transmitter in 1903. Pouslen was an electrical engineer and prolific inventor who, by 1898, had invented the first device to use magnetic sound recording—the "Telegraphone." Poulsen did not stop there. He became interested in the work of British inventor William Duddell who used a carbon arc lamp to make a resonant circuit that could "sing." Duddell's musical arc resonated at audible frequencies and he adapted this into a crude electronic musical instrument in 1899. The problem was that, when he tried to increase the operating frequency, its efficiency plummeted, and it seemed destined to remain as a gimmick.

Poulsen attacked the problem and, by making the arc happen in hydrogen gas, and using a water-cooled copper anode, he was able to make the arc "sing" at radio frequencies. Unlike all previous radio transmitters, Poulsen's arc transmitter generated continuous waves. Despite the failure of Poulsen's Amalgamated Radio Telegraph Company, his invention was taken up by the U.S. Navy for their communications. Poulsen's arc transmitter was the best portable radio system for a decade before the introduction of vacuum tube systems. **DHk**

SEE ALSO: ELECTROMAGNETIC TELEGRAPH, COHERER, WIRELESS COMMUNICATION

Crayons (1903)

Binney and Smith bring out the artist in kids.

They have brought color and creativity to generations of children across the world, but the crayon family began with just one color, black. Originally based on a mixture of charcoal and oil, the oil was soon replaced with wax to make the crayon stronger and easier to hold.

American cousins Edwin Binney (1866–1934) and Harold Smith (1860–1931) had a company that sold paint pigments—red for painting barns and black for car tires. In 1900 they began to make pencils and came up with the useful idea of dustless chalk for teachers. While touring schools with these products, the cousins spotted a gap in the market for a new drawing and writing implement. They set about making a safe, nontoxic toy for children and, in 1903, the first box of

---

"Sunglow, unmellow yellow, atomic tangerine, purple pizzazz, razzle dazzle rose, neon carrot…"

Names of crayon colors added to the 1990 box

---

modern crayons was born. Containing eight colors—black, brown, blue, red, green, orange, yellow, and violet—the box cost a nickel. This was the start of Crayola crayons, a brand name Binney's wife coined by combining the French words for chalk and oily. Over the years, more exotic-sounding colors appeared, such as thistle, melon, and burnt sienna. Fluorescents were added in the 1970s. Today, crayons can glow in the dark or smell like flowers, but they are still pretty good for simply drawing. **DK**

SEE ALSO: INK, QUILL PEN, GRAPHITE PENCIL, FOUNTAIN PEN, PENCIL SHARPENER, BALLPOINT PEN

➲ *Binney and Smith reissued the original eight colors of Crayola crayons in 1991.*

Wire Coat Hanger (1903)

Parkhouse's idea keeps clothes crease-free.

Patented more than 200 times in the United States alone, the humble coat hanger has undergone many transformations to reach its modern incarnation.

Various methods of hanging clothes had probably existed before Britain's Queen Victoria was gifted a set of wooden coat hangers for her wedding in 1840, however, the mass-market wire hanger was not invented until 1903. The story goes that Albert Parkhouse, an employee of the Timberlake Wire and Novelty company, a Michigan-based firm that specialized in wire lampshade frames, was irritated by arriving at work one day to find that all the coat hooks were in use. Seizing a piece of wire, he bent it into two large oblong hoops and then twisted both ends at the

---

"Left all alone in some punkerish place, like a rusty tin coat hanger hanging in space . . ."

Dr. Seuss, *Did I Ever Tell You How Lucky You Are?* (1973)

---

center into a hook. As was common practice among nineteenth-century companies, Timberlake patented the wire coat hanger in January 1904, and made a fortune whereas Parkhouse never got a penny.

The Timberlake wire coat hanger soon proved very popular. By 1906 it was being used by Meyer May, a men's clothier, to display clothes in store. In 1932, Schuyler C. Hulett added a cardboard tube mounted on the upper and lower parts of the wire to prevent the wire from marking the clothes. In 1935, Elmer D. Rogers brought the coat hanger closer to what we know by adding a tube on the lower bar. Today, hangers are made from wood, metal, and plastic. **SD**

SEE ALSO: CLOTHING, SEWING, FOLDING IRONING BOARD, BLUE JEANS, ELECTRIC IRON

Windshield Wiper (1903)

Anderson gives drivers a clear view.

At the beginning of the twentieth century, New York City may have been very picturesque in the snow and ice, however, it was not much fun if you were trying to drive a trolley car at the time. Then the only solution was for the driver to keep the windshield up and constantly get out every few minutes to clear the slush that built up on the windshield.

Mary Anderson (1866–1953) was visiting New York when she noticed this predicament, and when she returned home she jotted down a solution. Her device was a swinging arm with a rubber blade that is moved by a lever inside the car, keeping the driver warm and the window clear. In 1903 she received a patent for the novel device, but when she tried to sell it in 1905 nobody was interested.

It was still three years before the Ford Model T and other automobiles for the masses were introduced. However, by 1916 windshield wipers would become regular equipment on all cars. In England a patent for wipers was registered in 1911 by Gladstone Adams although Mills Munitions of Birmingham also claim to have patented the invention first. Automatic electric wipers eventually followed, using rollers rather than blades. These were invented by Charlotte Bridgwood in 1917, and these had become standard in cars by the early 1920s. Windshield wipers usually work in conjunction with a windshield washer that operates using a pump dispensing liquid detergent and water.

Robert Kearns invented and patented intermittent windshield wipers in 1969. He took his idea to the car companies, who showed no interest but then installed similar wipers in their vehicles. Kearns took Ford and Chrysler to court and won. Rain-sensing wipers were developed in the late twentieth century, and hence we don't even have to switch the wipers on these days. **DK**

SEE ALSO: INDICATORS, CATSEYES

Glass Bottle-making Machine (1903)

Owens invention does more to eliminate child labor than previous legislative efforts.

Michael Owens's (1859–1923) automatic glass bottle-making machine not only revolutionized the glass industry by speeding up the process of bottle-making and reducing its cost, it also helped the growth of several related sectors and eradicated child labor in the industry. At the time glassblowing was one of the most highly paid crafts, and children were often employed as cheap labor. In fact, Owens—who never received any formal education—started working at a West Virginia glass factory at the age of ten to support his family. He subsequently moved to Toledo, Ohio, to work for entrepreneur Edward Libbey, who gave him the opportunity to realize his inventive potential.

Building on existing concepts of similar semi-automatic machines (operated by five people), he conceived a fully automatic device in 1903. The suction of a vacuum—created by withdrawing the piston rod on the machine's hand pump—automatically sucked the required amount of glass into a mold. The resulting neck of the bottle was put in a body mold where the glass was automatically blown into the right shape. A conveyor belt then passed the bottles through a tempering oven to slowly cool them.

Owens's first device—which only required two workmen to operate it—had five pumps on a circular rotating frame and could produce about 17,000 bottles a day (six times more than the semi-automatic ones). At a time when glass was still a luxury item, Owens made it possible to produce bottles of identical sizes. With Libbey's help, he founded his own bottle-making company, which still exists today. Modern machines are able to produce one million bottles a day. **DaH**

SEE ALSO: GLASS, BLOWN GLASS, VACUUM-SEALED JAR, FIBERGLASS, PLASTIC BOTTLE (PET)

⤢ *Owens received the patent for his machine in 1904. He eventually held forty-five U.S. patents.*

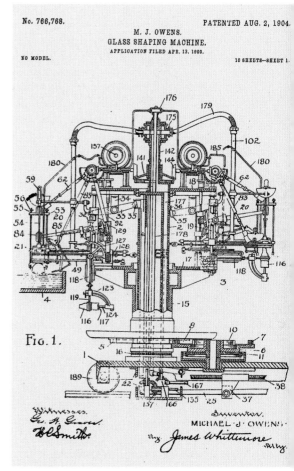

"Owens was an inventor. He was no designer, but he could direct engineers."

Richard LaFrance, Owens's chief of engineering

Powered Airplane (1903)

The Wright brothers build the first winged aircraft capable of sustained flight.

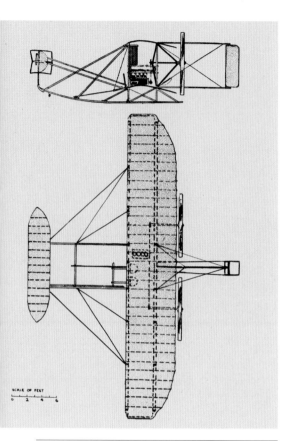

SCALE OF FEET
0 2 4 6

"For some years I have been afflicted with the belief that flight is possible to man."

Wilbur Wright in 1900

⬆ *Between 1907 and 1909, this was the general arrangement of Wright airplanes.*

➡ *In 1909 the Wright brothers demonstrated their airplane to the U.S. Army at Fort Myer, Virginia.*

On the morning of December 17, 1903, amid the sand dunes of Kill Devil Hills, North Carolina, Orville Wright (1871–1948) took off into a gale-force wind aboard the gasoline-powered biplane, *Wright Flyer*. Orville flew for only twelve seconds, but later, in the fourth flight attempt of the day, his brother Wilbur (1867–1912) stayed aloft for fifty-nine seconds, traveling a distance of 852 feet (260 m). It was enough to constitute the holy grail of flight experimenters—sustained, controlled, powered, heavier-than-air flight.

Bicycle manufacturers from Dayton, Ohio, the Wright brothers had approached the problem of flight with a combination of practical hands-on experimentation and scientific rigor. After absorbing all published information on previous flight experiments, in 1900 they built the first of a series of gliders that they tested each summer in North Carolina. Each winter, back in Dayton, they refined their design, building their own crude wind tunnel to test different wing shapes and angles. They created an ingenious control system with a front elevator for pitch, a rudder for horizontal yaw, and wing-warping—the bending of the wing-tips—to control roll. Through their glider experiments they taught themselves to fly, lying prone to reduce drag. When they felt ready for powered flight, they designed and built their own engine and their own propellers.

The *Wright Flyer* was underpowered. For further flight experiments in Dayton in 1904 and 1905, without the aid of a strong headwind, the Wrights built a catapult to help the aircraft off the ground. With this assistance, they made flights of up to thirty-eight minutes at a time when no other flight experimenter even claimed to have achieved more than a brief hop in a straight line. **RG**

SEE ALSO: GLIDER, AIRSHIP, JET ENGINE, TURBOPROP ENGINE, HELICOPTER, SUPERSONIC AIRPLANE, SCRAMJET

79

117484

117480

right aeroplane
ng the starting
k (Orville Wright
chine) Ft. Myer, Va.

The Wright aeroplane on
the starting derrick (Orville
Wright in machine - Wilbur
Wright in white shirt) Ft. Myer
Va.

12577 Oct. 449 R

117478

117483

Sept 1909 R
ght turning flight -
ght aeroplane going
Myer Va.

1909 R
The Wright aeroplane leaving the starting ra
Ft. Myer, Va.

117727

117726

Tea Bag (1903)

Sullivan accidentally discovers a way to brew an instant cup of tea.

Eager to boost orders for his teas, New York tea merchant Thomas Sullivan devised a new method of distributing samples of tea to his customers. He stitched them into small silk muslin bags, making them easy to ship and less messy for the recipient to unpack. Some customers did not bother opening the cloth bag and simply poured boiling water over them. Sullivan was inundated with orders for more tea packaged in this way. Responding to suggestions that the silk mesh was too fine, Sullivan used cotton gauze instead and began to sell the bags commercially.

The patent for the tea bag was registered by Sullivan in 1903. By 1920, tea bags were in wide use by the catering trade in the United States. Later, paper was used instead of cotton and a fine string and decorated tag were sometimes added, making them more convenient for drinkers making a single cup.

The American market for tea bags was well developed by the time they were introduced into Britain several decades later by Joseph Tetley and Company in 1953. It is not clear whether this delay was a result of war-time shortages of materials, or of initial resistance to change on the part of the British tea-drinker. However, once they were launched, tea bags took off rapidly in Britain, as one of the many labor-saving products that emerged in the postwar period, offering convenience and saving time.

By the 1970s, tea bags were also being used for herbal teas. Tea bag technology continues to develop, with new tea bag shapes to increase the effectiveness of the infusion process, such as the pyramid, and to attract the interest of the tea-drinker. **EH**

SEE ALSO: **SODA WATER, POWDERED MILK, SACCHARIN, INSTANT COFFEE, COFFEE FILTER**

⊡ *A breakthrough in tea bag history came in 1930 when the heat-sealed paper fiber tea bag was produced.*

Automatic Transmission (1904)

Sturtevant renders gear sticks superfluous.

The Sturtevant family business was founded in 1883 by Thomas L. Sturtevant, with the aim of satisfying the increasing need for mechanization in the fertilizer industry. Thomas's son Lawrence, and his nephew Thomas J. Sturtevant, came to work for him—Thomas bringing with him a degree from the Massachusetts Institute of Technology. Sturtevant soon branched out into the automotive field and designed various improvements including vacuum brakes and automatic engine lubrication. However, it was his bold invention of the automatic transmission that paved the way for today's automatic cars.

At the time, one of the biggest headaches for designers in the fast evolving car industry was simply getting the power from the engine to the wheels. Sturtevant wanted a way to change gear without having to depress a clutch and temporarily disengage the engine from the wheels. His solution was innovative, but initially a failure. His first automatic car

of 1904 used the centrifugal force of spinning weights to change gear. As the speed of the car increased, the spinning action caused these weights to swing outward, where they would eventually engage with a band that shifted the car from low to high gear. The design was flawed and often the weights would fly apart under the stress. But the concept of a car automatically changing gear was proven.

Despite various experimental attempts to come up with a practical automatic car, the initial costs, reliability, and lack of demand meant that it was several decades before the automatic became a common site on the road. **DHk**

SEE ALSO: MOTORCAR GEARS, CLUTCH, SYNCHROMESH GEARS, CENTRIFUGAL CLUTCH

⬆ *A student at a technical college in the United States studies an automatic transmission in 1958.*

Answering Machine (1904)

Poulsen develops a solution for missed phone calls.

Dane Valdemar Poulsen (1869–1942) shaped a surprising amount of the modern world with the invention of magnetically recorded sound in 1898. It was an incredibly useful innovation that has been used in tape recordings, hard disks, floppy disks, and credit cards. It also led him to create, in 1904, the world's first "telephone answering device."

Modern society relies on communication tools such as the telephone to function, and today it is very unusual to encounter a telephone that does not have some form of answering phone or voicemail. After its invention in 1876, the telephone became a world-changing tool, allowing anyone in the world to have a conversation with anyone else, immediately. It was only a matter of time before somebody had the idea for an answering machine.

Poulsen's magnetic wire recording device was initially used in the answering machine but later versions used magnetic tape to record the telephone message. Today, equipment tends to use solid-state memory storage. The first digital answer-phone was invented in the United States by Kazuo Hashimoto in 1983.

Telephones are, of course, intrinsically rude devices that pay no respect to normal methods of adult communication, and they interrupt whatever activity is going on. The answering machine—along with its related modern counterparts of voicemail, call registers, and text messages—gives people some relief from this impolite badgering, by allowing them to know who has been trying to talk to them and deciding whether they want to talk back. **JB**

SEE ALSO: TELEPHONE, MAGNETIC RECORDING, AUDIO TAPE RECORDING, CREDIT CARD, FLOPPY DISK

⬆ *This version of Poulsen's Telegraphone was used in one of the Royal Dockyards, England, in the early 1900s.*

Valve Diode (1904)

Fleming invents the first true electronic device.

When British electrical engineer John Ambrose Fleming (1849–1945) invented the first thermionic diode, he had no idea of the impact it would have on the technology of the twentieth century.

The diode is used in many electrical and electronic systems. In its early valve form, the diode used thermionic emission—the flow of electrons within a sealed vacuum—to create a one-way valve for electrical current. This is critical in regulating voltages, processing high-frequency signals, and in converting AC (alternating current) to DC (direct current).

Predictably, Thomas Edison, was one of the first to discover the underlying principles. In 1883 he created a kind of diode by modifying an electric light bulb. He patented what he called his "Edison Effect," but ultimately saw little potential in its development. In 1904, having acquired a number of Edison Effect bulbs from the United States, Fleming developed an "oscillation valve," which he used to convert radio waves into a one-way flow of current that could be measured on a galvanometer. His invention, known as the "Fleming Valve"—the term *diode* only came to be used much later—would make the early radio and phonographic electronics industries a commercial reality.

For much of the twentieth century, thermionic valve diodes were used in analog signal applications and as rectifiers in power supplies. The development of the smaller, more predictable, transistor in 1947 rendered the valve obsolete for most of its existing uses. Today, valve diodes are only used in niche applications, such as guitar amplifiers. **TB**

SEE ALSO: INCANDESCENT LIGHT BULB, TRIODE, FLIP-FLOP CIRCUIT, SEMICONDUCTOR DIODE, TRANSISTOR, LED

⬆ *Fleming's valve has a glass bulb with a carbon filament inside, surrounded by an open-ended aluminum tube.*

Tractor (1904)

Holt invents caterpillar tracks.

When engines were invented in the early nineteenth century, they were quickly adapted for use in farming—at first just to drive farm machinery, using the engine to move other equipment, but not itself.

When steam-traction engines were introduced in 1868, they were used only on the roads to haul timber and other heavy loads around. Gradually, however, they came to be used in the fields, dragging plows behind them. One of the biggest obstacles facing the traction engines was their wheels. On soft soil, thin wheels just sank, so the wheels were fitted with wide metal tires to spread out the weight. These wheels lacked grip and got people looking for other ways to spread the weight. In 1904, Benjamin Holt (1849–1920) tested his first tractor with tracks instead of wheels and went on to form a company that became Caterpillar. In 1932 the metal tires were replaced with rubber ones, increasing grip and decreasing weight.

Engine-wise, the Charter Gasoline Engine Company created a gasoline-fueled engine in 1887 that they adapted to drive a traction engine. In 1892, American inventor John Froelich built his own version in Iowa and his design became the first successful gasoline tractor—and father to many others. Froelich's success made other companies follow suit. Hart-Parr, founded at Madison, Wisconsin, in 1897 before moving to Iowa in 1905, became known as the "Founders of the Tractor Industry" because their factory was the first to be used continually and exclusively to make tractors. In 1906, their sales manager decided "traction engine" was too long and vague a description and shortened it to "tractor." **DK**

SEE ALSO: COMBINE HARVESTER, SCRATCH PLOW, REAPER, SEED DRILL, THRESHING MACHINE (THRESHER), STEEL PLOW

◁ *The Ivel Agricultural Tractor was the first successful agricultural tractor with internal-combustion engine.*

Derailleur Gears (1905)

De Vivie makes cycling easier.

Paul de Vivie (1853–1930) did not buy his first bicycle until he was twenty-eight, but his passion for cycling would eventually take over his life, and led to the invention of a new system of variable speed cycles.

De Vivie's first bike was an "ordinary" high-wheel, or penny-farthing as it's more commonly known. The pedals of this bike were attached directly to the wheels so that one turn of the pedals equaled one turn of the wheel. De Vivie sought a way to improve this ratio to make cycling more energy-efficient. In 1887, he set up a cycle shop in the mountainous region of Saint-Etienne, France, and launched a magazine, *Le Cycliste*, in which he wrote passionately about cycling, under the pen-name "Velocio."

> *"He [Paul de Vivie] was a man who devoted a lifetime to the perfection of the bicycle … "*
> Clifford L. Graves, writer

De Vivie's first attempt at creating gears for a bicycle involved two concentric chain wheels with a chain that had to be lifted manually from one to the other. In 1905, he tested a two-speed derailleur gear, but the cycling world was reluctant to buy into the idea, dismissing it as an easy way out. To win over the skeptics, de Vivie organized a mountain race between a male cyclist on a single-speed bike and a female cyclist on a three-speed derailleur bike. Much to his delight, the woman won.

Today, modern bicycles frequently may have more than twenty different derailleur gears, making riding in even the most varied terrain a breeze. **HI**

SEE ALSO: SAFETY BICYCLE (MODERN BICYCLE)

Turbocharger

(1905)

Büchi reclaims lost energy to make engines more efficient.

The turbocharger is similar to the supercharger but, instead of mechanically forcing extra air into the engine, it uses the exhaust to drive a turbine that boosts air in. Swiss engineer Alfred Büchi (1879–1959) realized that using a turbine to make use of engine exhaust would actually recover otherwise lost energy to make the combustion cycle much more efficient.

Turbochargers and diesel engines fit together perfectly since diesel engines have no throttle to stall the air flow to the turbine. They were first implemented commercially in two German passenger ships, in which the addition of the turbocharger increased the horsepower from 1,750 to 2,500.

Putting turbochargers into cars, however, proved a little harder. As it is the hot air from the combustion that drives the turbine, the materials that make it must be able to withstand high temperatures up to 1,832°F (1,000°C). This has been the issue that held back turbochargers from the mass market, but recently manufacturers have been taking technology straight from space rockets to try and improve the material design. As rockets need to withstand much higher temperatures than 1,832°F (1,000°C), the hope is that their designs can be scaled down enough to make this a viable option for cars.

The most talked about car designs for the future, such as hybrids and fuel-cell-based vehicles, would benefit from the incorporation of a turbocharger into the plan. With their green design, which can save a third of engine energy that would normally be wasted, it looks as if the turbocharger is here to stay. **LS**

"From backyard tuners to luxury limousine manufacturers, they've all relied on Büchi's turbocharger."

Don Sherman, *Automobile* magazine

SEE ALSO: INTERNAL COMBUSTION ENGINE, SUPERCHARGER, TWO-STROKE ENGINE

◥ *The purpose of a turbocharger is to increase the mass of air entering an engine to create more power.*

Die-casting Machine
(1905)

Doehler mass-produces metal parts.

Die-casting is the name given to a process of producing identical and often complicated metal parts by forcing molten metal under high pressure into a reusable type of mold, which is known as a die. It is a four-step process. First the inside of the die is coated with a lubricant—used partly to help control the temperature of the mold, and partly to aid the removal of the cast when complete. Molten metal is then injected into the die under high pressure. Generally non ferrous metals are used in die-casting. Zinc is popular as it is easy to cast and has a low melting point, which increases the working lifespan of

---

"Engineering is the . . . art of applying science to the optimum conversion of natural resources."

Professor Ralph J. Smith, Stanford University

---

the die. Aluminum is also used as it is very lightweight. The metal is kept under pressure until the casting has solidified. The die is then opened, the molding scrap is removed (which will then be remelted and recycled), and the piece is generally complete, although various finishes may be applied.

German-born mechanic Herman Doehler (1872–1964), founder of the Doehler Die Casting Company, created the first die-casting machine in 1905; he was issued a patent for this process in 1906 and set up his company in Brooklyn in 1907, moving to Batavia, New York, in 1921. Among other achievements in the field of die-casting, he went on to become the largest producer of hood ornaments for American cars. **CL**

SEE ALSO: METALWORKING

Amplitude Modulation
(1906)

Fessenden pioneers radio broadcasting.

It is a common misunderstanding that radio waves are sound waves. In fact, they are part of the electromagnetic spectrum—just another type of light, with a long wavelength. The clever bit is that if the waves are varied in height or length, they can carry information. Of course the most common example of this is to carry audio information, which is then deciphered by your radio and turned back into sound.

There are two ways of varying radio waves and these are called AM (amplitude modulation) and FM (frequency modulation), which are the two most common modes on radio. The first changes the intensity of the transmission—or if you think in terms of waves, it changes their height—and it was this method that Canadian inventor Reginald Fessenden (1866–1932) came up with in 1906.

Fessenden's set-up used a microphone to convert sound into an electric signal that was then combined with continuous radio waves. These modulated waves were transmitted through an antenna and received by a second antenna some distance away. To hear the original sound, the waves were turned back into electric signals. This process is called demodulation. Finally the signal was sent to a loudspeaker, which converted it back to sound for ears to hear.

It was on Christmas Eve of 1906 that Fessenden made the world's first radio broadcast using his amplitude modulation technique. He transmitted his voice and recorded music over several hundred miles from Brant Rock, Plymouth County, Massachusetts, to ships in the Atlantic Ocean. This was the start of broadcasting, and it was not long before every self-respecting household was gathering around the radio in the evenings to listen to broadcasts. **LS**

SEE ALSO: COHERER, FM RADIO, IMPROVED RADIO
TRANSMITTER, PODCAST, SATELLITE RADIO BROADCASTING

Crystal Set Radio (1906)

Pickard brings broadcasts to the masses with a homemade receiver.

Greenleaf Pickard (1877–1956) was a pioneer in the early days of wireless. He experimented with methods of receiving radio signals, using mineral crystals to filter out noise in the signal. After testing more than 30,000 combinations of materials, he finally patented his "crystal detector" in 1906. This featured a crystal of silicon that was later connected to the radio circuit via a fine, sharp piece of wire known as a cat's whisker.

A crystal set radio is the simplest form of radio as it needs no power source to receive signals. A typical set consists of an antenna, a detector, a tuner, and an audio output—normally an earpiece. The antenna can be any piece of wire or metal.

As radio waves pass through the antenna, tiny voltages are created from all radio stations broadcasting. The antenna is connected to a tuning circuit comprising a coil and a capacitor. The capacitance can be varied and, in conjunction with the coil, acts as a filter to allow only the voltages from certain stations to pass. The filtered signal is then passed on to the detector. The detector performs the demodulation of the signal, letting only the radio program pass through to the earpiece, where it can be enjoyed. The cat's whisker needed to be moved across the surface of the crystal to form the best contact and so allow the best clarity of sound.

Crystal sets were particularly popular in the early 1920s when radio broadcasting was beginning to take off but factory-made radios were still prohibitively expensive. At the time, instructions on how to make crystal sets out of household items such as cereal boxes and baseball bats were published. **HP**

" … with enough ingenuity, one could tickle the crystal with a cat's whisker and pick up anything."

Theodore H. White, journalist, historian, and author

SEE ALSO: AMPLITUDE MODULATION, COHERER, FM RADIO, PODCAST, SATELLITE RADIO BROADCASTING

⬚ *The crystal set radio needs no power source except the power from radio waves. This set is from 1948.*

Triode (1906)

De Forest creates the first valve amplifier.

In 1904, British physicist John Ambrose Fleming (1849–1945) developed what he called an "oscillation valve." Later rechristened the "diode," it is universally recognized as the first true electronic device. Two years later, building on Fleming's work, American electrical engineer Lee De Forest (1873–1961) created the Audion vacuum tube—the first valve amplifier.

De Forest's major innovation was in creating a valve that would not only rectify the AC current, but boost it. The Audion contained the same filament/cathode and plate/anode design of the Fleming valve, but placed between them was a zigzag of wire called a grid. A small electric current applied to the grid would result in much current shifting from the filament to the plate. Thus was born the first electrical amplifier. With its three active electrodes, De Forest's valve amplifier evolved into what would later become known as the triode, and would be central to the creation of most of the important developments in electronics over the next half century.

De Forest frequently ran into patent disputes, and evidently spent much of the fortune he amassed in assorted lawsuits. There was a clear similarity between the Fleming tube and the first Audion valves, and decades of costly and disruptive litigation followed. It was not until 1943 that, having first sided with Fleming, the U.S. Supreme Court ruled that, on a legal technicality, Fleming's patent was invalid. In 1946, the importance of De Forest's invention was acknowledged when his work on the triode was awarded the prestigious Edison Medal of the American Institute of Electrical Engineers. **TB**

SEE ALSO: INCANDESCENT LIGHT, LED (LIGHT-EMITTING DIODE), SEMICONDUCTOR DIODE, TRANSISTOR, VALVE DIODE

↗ *The triode was able to amplify weak electrical signals, making it crucial in the development of radio.*

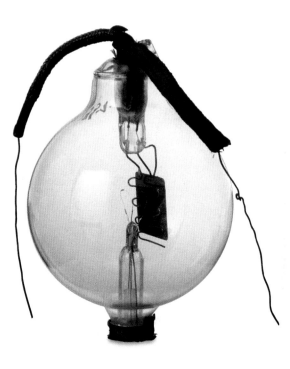

"What have you done with my child [the radio broadcast]? You have debased this child …"

Lee De Forest, *Chicago Tribune* (1946)

Sonar

(1906)

Nixon uses echoes to pinpoint objects at sea.

Sonar (which started as an acronym for sound navigation and ranging) is a technique widely used in shipping to detect nearby vessels and underwater obstructions. The word was coined by the Americans during World War II. The British also call Sonar, ASDIC, which has been claimed to stand for Anti-Submarine Detection Investigation Committee.

There are two major kinds of sonar, active and passive. Active sonar produces a sound "ping"' and then measures how long it takes a reflected pulse to return to the ship. The sound source and receiver are continually rotated, so that the direction of the echoing body can be found. Lower frequencies are used if the reflector is a long way off. Passive sonars listen without transmitting. The first Sonar devices were passive listening devices—no signals were sent out. They are usually used by the military (although some are scientific).

Lewis Nixon (1861–1940), a U.S. shipbuilding executive, introduced a passive sonar system. There were two driving forces: One was the hope of hearing the grinding sounds of nearby icebergs; the other was hearing submarines. Active sonar development accelerated due to the sinking of the *Titanic* in April 1912, and naval activities during World War I. By the time of World War II all ships were fitted with sonar.

The system is complicated by the fact that the speed of sound depends on the temperature and salinity of the water, and thermal gradients can bend the sound waves in peculiar ways. Small sonar systems are used on fishing boats to detect shoals of fish. **DH**

SEE ALSO: RADAR, SUBMARINES

"Publishing [verse] is like dropping a rose petal down the Grand Canyon and waiting for the echo."

Don Marquis, U.S. humorist, journalist, and author

⬉ *A sonar operator on board the U.S. submarine chaser PC-488 during World War II.*

Indicators / Electric Turn Signals (1907)

Douglas-Hamilton develops automatic signals.

Electric turn signals, or indicators, are now a standard feature of virtually all motor vehicles. They are essential for warning fellow road users of intended movements to avoid collisions. Before their invention, drivers had to rely on good old-fashioned hand signals to let others know when they intended to turn.

It is thought that the first automatic turn signals were patented in 1907 by an Englishman called Percy Seymour Douglas-Hamilton. His so-called "devices to indicate the intended movement of vehicles" were in the shape of hands to mimic the manual signals in use at the time. Some years later, in 1925, Edgar A. Walz Jr. obtained a patent for a modern turn signal, but car manufacturers at that time were not interested.

The first company to incorporate electric turn signals into a commercial car was Buick. They introduced the device in 1938 as a safety feature and advertised it as the "flash-way directional signal." The driver would simply flick a switch on the steering column, which would illuminate a flashing directional arrow on the rear of the vehicle. In 1940, Buick improved upon their design by introducing self-canceling turn signals.

Since then, the design and function of the electric turn signal has remained virtually unchanged. Most modern cars have indicator lights mounted at the front and rear corners, and occasionally at the sides. Another improvement is the introduction of amber lights, which allow the signals to be visible in bright sunlight. The signal is usually operated by a lever mounted on the steering column, with upward or downward movements activating right- or left-hand turn signals. The signals flash at a steady rate of between sixty and 120 per minute. **HI**

SEE ALSO: ACCELERATOR, CLUTCH, MOTORCAR GEARS, STEERING WHEEL, WINDSHIELD WIPER

Synthetic Detergent (1907)

Henkel introduces washing powder.

In 1907 the technology company Henkel launched Persil—the world's first self-acting detergent. This new type of washing powder allowed stubborn stains to be removed from clothes, without the need for rubbing or bleaching.

Henkel & Cie was established in Aachen, Germany, in 1876 by Fritz Henkel (1848–1930) and his business partners and the first product they marketed was a universal detergent based on silicate. In 1905, Henkel's youngest son joined the company and began to study the chemistry of washing, bleaching, and peroxide.

Persil was one of the first-ever branded products. The name was derived from two of the detergent's

> *"When I'm watchin' my TV / And that man comes on to tell me how white my shirts can be …"*
>
> The Rolling Stones, "Satisfaction"

ingredients: perborate and silicate. Though Persil proved to be hugely successful commercially, synthetic detergents did not develop until World War I, when problems securing a supply of fats and oils led scientists to seek alternative ingredients.

In 1946, a breakthrough launch of "built" detergents gave rise to products that performed better at cleaning heavily soiled garments. Detergents for automatic washing machines were introduced in the 1950s, and the first washing powders to use enzymes were launched in the following decade. The emergence of the dishwasher in the 1980s initiated the birth of a whole new class of detergents. **HI**

SEE ALSO: DRY CLEANING, LIQUID SOAP, SOAP

Distributor (1907)

Kent develops an ignition system.

American inventor Arthur Atwater Kent (1873–1949) had a fascination with electrical and mechanical gadgets. At the age of thirty, he set himself up as a battery maker in a run-down factory where, it was said, the cracks in the floor were so big that he never needed a dustpan.

With his first earnings, Kent bought a one-cylinder automobile, and tinkering with it led him to develop his own ignition system. The ignition system is an essential part of any engine. An ignition coil turns the 12 volts in the car battery into the thousands of volts needed for the spark plugs to produce a spark to ignite the fuel in the tank. It works like a high voltage transformer, consisting of a primary and secondary winding wrapped around a metal core. The primary coil has a lower number of turns than the secondary coil and when the circuit is broken the magnetic field of the primary coil disappears. This causes the secondary coil to be overwhelmed by a large changing magnetic field that induces a current with enough power to create a spark to ignite the fuel. Before the high voltage from the coil goes to the spark plugs it needs to be distributed to the right cylinder. This is done by a rotating mechanism that touches one contact per cylinder, distributing the electricity. It was this mechanism that Kent came up with in 1907, and it is an essential part of the car ignition system.

Although Kent's ignition system was ingenious, he actually made his fortune in radio manufacturing. Visitors to his factory could watch through a special window to see gold bars dissolved in acid to make the plating for the Atwater Kent trademark. **LS**

SEE ALSO: FOUR-STROKE CYCLE, SPARK IGNITION, SPARK PLUG, TWO-STROKE ENGINE

⭱ *The distributor from the ignition system of an early automobile engine.*

Self-aligning Ball Bearing (1907)

Wingquist smooths mechanical movement.

The principle of bearings has been around for as long as the Pyramids of Giza. Bearings enable easy smooth movement between two objects, and the builders of the pyramids applied this principle by using rolling tree trunks laid under planks to move heavy loads in the construction of these wonders of the ancient world.

Early bearings were linear (allowing movement in a straight line, like the opening and closing of a drawer) and were made of wood, stone, sapphire, and glass. However, advances in technology demanded new and improved bearings to allow smoother mechanical movement.

In 1907 Swede Sven Wingquist (1876–1953) patented a design for a multi-row self-aligning ball bearing. Made of steel to lower friction, the bearing comprised two rows of balls in a concave raceway. Wingquist's design was structurally superior to earlier bearing designs and was rotary—allowing motion around a center, such as in wheel axles. Self-alignment

meant the bearings had the lowest friction of all rolling bearings allowing them to run at high speeds and still keep cool. The self-aligning quality of the ball bearing also meant it could absorb some shaft misalignment without lowering its endurance. The ball bearing has come to be thought of as emblematic of the Machine Age of the 1920s and 1930s.

Wingquist's company SKF (Svenska Kullager-fabriken AB) was founded on the patent, and is now the world's largest bearing manufacturer. SKF are looking to produce energy-efficient bearings that will reduce the energy consumption of machines by up to 30 percent. **FS**

SEE ALSO: BALL BEARING, ACCELERATOR, CARBURETOR, CLUTCH, MOTORCAR, MOTORCAR GEARS

⊥ *A Sven Wingquist-designed self-aligning ball bearing from 1907 in chrome-plated steel.*

Outboard Motor (1907)

Evinrude speeds up pleasure boating.

Ole Evinrude (1877–1934) did not invent the first outboard motor. However, his was the first to achieve great commercial success. He built his first motor in 1907, received a patent for the device in 1911, and by 1913 the Evinrude Detachable Row Boat Company was selling almost 10,000 outboard motors a year.

Some consider the inventor of the outboard motor to be Cameron B. Waterman, who received a patent for his version in 1907. In 1905, he had attached a propeller to a motorcycle engine and hung it off the back of a rowboat. But Waterman was not the first inventor either—the American Motors Company had built a "portable boat motor with reversible propeller" as early as 1896.

"The . . . Evinrude rowboat motor . . . had a wooden 'knuckle buster' starting knob on the flywheel."

The Practical Encyclopedia of Boating

Whoever the original inventor, the outboard motor revolutionized pleasure boating. Typical outboard motors propel boats at 9 miles per hour (15 kph)—about three times the speed of rowing. The outboard motor also acts as a rudder to steer the boat, and there is plenty of water for cooling, simplifying the design of the engine and lowering the cost. Outboard motors are much cheaper than built-in motors and are light enough to be carried from one boat to another.

More than fifty companies began making outboard motors between 1910 and 1930, and by the 1950s, more than 500,000 outboard motors were being sold each year. **ES**

SEE ALSO: DUGOUT CANOE, JET BOAT, MOTORBOAT, ROWBOAT, RUDDER, SAIL, STEAMBOAT, STEAM TURBINE

Cellophane (1908)

Brandenberger revolutionizes food packaging.

Both waterproof and airtight, cellophane is now used for everything from food packaging to sticky tape. The man who invented it—Swiss textiles engineer Jacques E. Brandenberger (1872–1954)—initially wanted to develop a clear coating for cloth to make it waterproof after witnessing a wine spill on a restaurant tablecloth. He tried coating cloth with a thin sheet of viscose, but viscose made the cloth too stiff. The transparent sheet of film separated easily from the cloth and Brandenberger soon realized that the film itself had more potential than the waterproofed cloth.

To create cellophane, Brandenberger dissolved cellulose fibers from materials such as celery, wood, cotton, or hemp in alkali and carbon disulfide to make viscose, which is then extruded through a slit into an acid bath to reconvert the viscose back into cellulose. The acid regenerates the cellulose, which forms a film, and further treatment—for example washing and bleaching—produces cellophane. (Rayon is made by a similar process, but the viscose is extruded through a hole rather than a slit.)

Brandenberger named the substance cellophane after "cello," from cellulose, and "phane" from the French word *diaphane*, meaning transparent. It took the scientist almost a decade to perfect the substance and produce it commercially, and, in 1927, the invention of a waterproof lacquer created moistureproof cellophane, which meant that it could be used to package food.

The use of cellophane for packaging has decreased since the 1960s, and DuPont, the company that introduced it to the United States, even discontinued the product in 1986. Yet, despite this, cellophane's 100 percent biodegradability means that it could be due for a comeback. **SD**

SEE ALSO: CELLULOID, RAYON CELLULOSE ESTER, TAPE

Food Mixer (1908)

Beach speeds up kitchen tasks with a lightweight motor.

Chester Beach grew up on a Wisconsin farm, and it was there that his natural aptitude for repairing and fixing machinery was nurtured. He met his future business partner L. H. Hamilton, when they were both working at an electrical motor company in Wisconsin in the early 1900s. Realizing their potential for a profitable future, the pair formed the Hamilton Beach Manufacturing Company.

What made the company so successful was essentially Beach's invention of a high-speed, lightweight universal motor. The motor was able to safely and consistently achieve up to 7,200 revolutions per minute and its ability to run on both AC (alternating current) and DC (direct current) meant that it was extremely adaptable. The motor was used in their food mixer—which was a huge success for the company—and later in their equally profitable drink mixer, as well as many other appliances. The introduction of the food mixer in 1908 revolutionized home cooking, allowing for speedier preparation and larger quantities of food to be made in the home.

The company was fortunate in the timing of the introduction of their drink mixer in 1911. Doctors in the United States had begun to publicize the health benefits of malted milkshakes, and the appliance soon became a must-have for every self-respecting suburban American household. Beach and Hamilton also came up with a range of other electric-powered domestic products, including hair dryers, as well as other kitchen utensils such as a batter mixer. The company still trades today under the Hamilton Beach name, producing a wide range of products. **KB**

SEE ALSO: OVEN, PRESSURE COOKER, ELECTRIC STOVE, FOOD PROCESSOR, MICROWAVE OVEN

↗ *This food mixer from 1918 was made by Landers, Frary and Clark, in Connecticut, United States.*

"[The Hamilton Beach drink mixer] became as much a tradition as hot dogs, apple pie, or baseball."

Goodman's online shopping catalog

Geiger Counter (1908)

Geiger and Rutherford invent a device for detecting radiation.

In 1908 physicists Hans Geiger (1882–1945) and Ernest Rutherford (1871–1937) were observing ionized helium atoms at Manchester University and wanted to confirm the data from their scintillation crystal counters. Their new "Geiger" counter consisted of a high-voltage wire running along the central axis of a sealed brass cylindrical tube containing low-pressure carbon dioxide. When charged particles entered through a window in the chamber they collided with and ionized CO_2 molecules leading to a voltage change on the central wire that was registered on a galvanometer. A count rate of five to ten per minute could be registered. Soon more sophisticated detectors using helium gas and a photographic voltage registration were introduced, increasing the count rate to about a thousand per minute. In 1928 improvements made in the electrical characteristics by Walther Müller (one of Geiger's PhD students) enabled the instrument to detect electrons.

The main problem was the recovery time—the "dead time" between recording a particle and being ready to record the next one. This was considerably reduced in the 1930s by adding ethyl alcohol to the CO_2 as a quenching agent thus quickly gathering up the ionization after a particle had passed through.

Many Geiger counters were used in the nuclear weapons industry during World War II, and the nuclear power industry afterward. They had been developed into cheap and robust instruments, and their urgent clicking sound, indicating the presence and intensity of ionizing particle radiation, has become a common feature in cold war and science-fiction films. **DH**

SEE ALSO: ELECTRON SPECTROMETER, RADIOTHERAPY, SPECTROPHOTOMETER

⊡ *This Geiger counter, made by Hans Geiger in 1932, was used by James Chadwick, the discoverer of the neutron.*

Gyrocompass (1908)

Anschütz-Kaempfe improves navigation at sea and in the air.

Gyrocompasses have two great advantages over magnetic needle compasses: They point to the spin pole of Earth as opposed to the magnetic pole; and they are completely unaffected by the ferrous metal of a ship's hull or the magnetic fields produced by electrical currents running through nearby wires. Their main component is a motorized, fast-spinning, damped, gimballed wheel. When this wheel is not spinning exactly in the plane containing Earth's spin axis, an interaction between the angular momentum of the wheel and the angular momentum of Earth produces a restoring torque that pushes the wheel back into the true north-south orientation.

A ship's gyrocompass is mounted in a complex set of gimbals that isolate the instrument from the ever-present pitching, yawing, and rolling. Aircraft gyrocompasses are even more complicated due to the higher velocity of the plane and the speedy changes in altitude during takeoff and landing.

German scientist Hermann Anschütz-Kaempfe (1872–1931) started working on the device because he needed a compass to guide a submarine on a planned expedition underneath the north polar ice. In partnership with his cousin, Max Schuler, Anschütz-Kaempfe sold a prototype to the German Navy in 1908. Schuler found that if the pendulous suspension of the gyroscope had a period of eighty-four minutes (the period of a pendulum of length equal to Earth's radius), disturbances due to the ship's acceleration were canceled out. Elmer Sperry, the American inventor, also produced gyrocompasses, and patent disputes ensued. **DH**

SEE ALSO: ASTROLABE, GLOBAL POSITIONING SYSTEM (GPS), MAGNETIC COMPASS, MAGNETOMETER

⬆ *This type of gyrocompass from 1940 was installed to control the flight paths of German V2 rockets.*

Coffee Filter
(1908)

Bentz improves the coffee-making process.

Melitta Bentz (1873–1950) invented the coffee filter to solve a simple household need, and it resulted in a hugely successful company and her filter being used throughout the world.

Bentz wanted to find a way to produce coffee without the grounds in it and began to experiment with passing the coffee through various types of filters, eventually trying the blotting paper that her children used when doing homework. By putting a circle of the blotting paper into a metal cup with holes in it, Bentz could pass hot water over the coffee, then let it drain into another cup, with the grounds left behind in the filter. In 1908, Bentz filed a patent for her

> *"I never drink coffee at lunch. I find it keeps me awake for the afternoon."*
>
> Ronald Reagan, U.S. President 1981–1989

invention and, with her husband, formed the Melitta Bentz Company to promote it. The invention took off at the Leipzig Trade Fair the following year, with over 1,000 coffee-makers sold. The company later began to produce its own filters, rather than using blotting paper. Like those used today, these had a great "wet strength" and would not fall apart when wet.

The Melitta Bentz Company later developed filter bags, as used in modern filtered coffee. Although many rival methods of coffee-making have been developed since 1908, Bentz's filtering system still holds its own, and continues to be widely used in homes as well as in commercial catering. **JG**

SEE ALSO: INSTANT COFFEE, TEABAG

Square-headed Screw
(1908)

Robertson invents a safer and faster screw.

Home improvement work is risky business. Children's cartoons serve as continual public service announcements of the hazards of "do-it-yourself"—stepping on garden rakes and putting hammers through thumbs tend to top the charts in terms of accidents. For traveling salesman Peter L. Robertson (1879–1951), however, a real accident in 1906, when he put a sharp screwdriver through his hand, prompted him to find a way of avoiding such accidents. By 1908 Robertson was manufacturing a screw that was to revolutionize the industry.

In Henry Petroski's book on the evolution of useful things he talks about the square-headed screw being invented to improve on existing designs—specifically the elimination or reduction of the risk of a slip. However, another improvement that Robertson made on the traditional screw was that his version could be fastened tighter than its rivals and operated with one hand—a very useful advantage for mechanics and other craftsmen alike.

The square-headed screw was simply a screw with a square cavity in its head and a pointed indentation at the deepest point of the hole. This guided the screwdriver into the cavity, juxtaposing it with the screw. This tight fit also provided sufficient torque to safely secure the screw into place more quickly than had previously been possible. It did not take long before the commercial benefits of this were recognized and, as a result, there were over 700 Robertson screws in each Model T Ford motor car that came off the production line. What had been a lack of sleight on Robertson's part in 1906 turned out to be a twist of fate that would earn this Canadian entrepreneur his fame. **CB**

SEE ALSO: CLAW HAMMER, NAIL, NAIL-MAKING MACHINE, SCREW-CUTTING LATHE, UNIVERSAL JOINT

Washing Machine

(1908)

Fisher minimizes the physical work of the household washday.

Washing clothes used to be work of drudgery, dubbed "the American housekeeper's hardest problem," but this problem has been largely solved by modern technology—in the shape of the washing machine.

Throughout history, various devices—from the washboard to the mangle—have been invented to wash clothes more effectively and with less effort, but these still required manual labor by the user.

In 1858, Hamilton Smith patented a rotary washing machine. Cylindrical, with "agitated water" and revolving paddles, it was a first step toward modern machines. Others followed from other manufacturers, some combining the machine with the mangle. These early machines were still hand-cranked, but it was not long before motorized versions became available, either with fuel-burning or electric motors.

It was not until the early 1900s, however, that the Hurley Company first produced the "Thor." Designed by Alva Fisher (1862–1947), it was the first electric-driven washing machine to be mass-marketed, and contained a novel self-reversing gearbox that stopped the clothes from becoming compacted during the washing process. It was slow to catch on, partly because the electric motor was unprotected and spilled water could cause it to fail, and partly because any household rich enough to have electricity at the time was also likely to have servants, therefore negating the need for such a machine. As electricity became more widespread, however, the washing machine followed suit and is now an indispensable part of the household arsenal. **SB**

SEE ALSO: DISHWASHER, DRY CLEANING, ELECTRIC MOTOR, SYNTHETIC DETERGENT

↗ *An electrically-driven domestic washing machine, made circa 1920 by Beatty Bros. of Canada.*

"Technology has taken us from the rock, used to pound the clothes, to the modern rectangular washer."

Lee Maxwell, Washing Machine Museum owner

Haber Process (1908)

Haber makes ammonia readily available.

The Haber process (sometimes called the Haber-Bosch process)—invented by the German chemist Fritz Haber (1868–1934) in 1908—may be the most important technological advance of the twentieth century. At that time, the main way of obtaining large quantities of ammonia was from naturally occurring saltpeter. Ammonia was an incredibly useful substance, with uses ranging from cleaning to fertilizer and explosives. But saltpeter could be difficult to harvest, with deposits occurring on the walls of caves, and making it required the large-scale decomposition of piles of animal dung.

In the first decade of the twentieth century, increasing global agriculture was putting a large strain on the supplies of ammonia, and there were fears that the supply would not be able to keep up with the demand. What Haber created was a means of making ammonia that would make it a plentiful resource. He extracted hydrogen gas from methane and made it chemically react with nitrogen from the atmosphere. To do this he needed a catalyst—a substance that promotes certain chemical reactions. During his experiments he found that iron was the perfect catalyst and, by mixing the nitrogen and hydrogen under high pressure, in the presence of iron, he could make NH_3 (ammonia) in large quantities.

Just two years after Haber's breakthrough, German chemist Carl Bosch (1874–1940) was able to commercialize the process in 1910, while working for the chemical company BASF—and suddenly German industry had plentiful supplies of ammonia.

Fritz Haber won the Nobel Prize in Chemistry in 1918 for his work. These days more than 500 million tons of artificial fertilizer are produced worldwide using the Haber process. **DHk**

SEE ALSO: DOW PROCESS, FRASCH PROCESS, OSTWALD PROCESS, REFRIGERATOR, SOLVAY PROCESS

Bakelite (1909)

Baekeland creates synthetic shellac.

A few lucky individuals change history and, at the same time, make a lot of money. Such was the case with Belgian-born Leo Hendrik Baekeland (1863–1944), who proved to have a keen eye for opportunity and the drive and know-how to see a project to fruition.

Baekeland's first major invention was a dramatic improvement in photographic paper. Eastman Kodak purchased his invention (dubbed "Velox") for upwards of three-quarters of a million dollars in 1898. Freed from financial worry, Baekeland then had his days free to experiment in his laboratory. As the electronics industry grew in the late nineteenth and early twentieth centuries, so did the demand for insulators. Greatly in demand at this time was shellac, a natural

> *"Well, it was kind of an accident, because plastic is not what I meant to invent."*
>
> Leo Baekeland, Belgian-American chemist

product made from a beetle indigenous to Asia. It also turned out to be a great insulator. Baekeland realized that given its high demand and limited supply, synthetic shellac would be a goldmine.

By mixing carbolic acid and formaldehyde in a controlled environment, Baekeland soon created a product that could be painted onto a surface, just like shellac, or be molded into almost any shape or form. The product, named Bakelite, was the first totally synthetic plastic and the forefather of the same product that is used today in bottles, computer keyboards, light switches, and countless other everyday products. **BMcC**

SEE ALSO: CELLULOID, POLYSTYRENE, PERSPEX

Synthetic Rubber (1910)

Lebedev cooks up polybutadiene.

Hevea brasiliensis—otherwise known as the "rubber tree"—had been tapped for thousands of years to harvest its saplike extract (latex, the source of natural rubber) long before anyone knew what rubber was. Rubber's chemical composition was finally cracked by British chemist and physicist Michael Faraday (1791–1867) in 1829. Faraday determined natural rubber's basic chemical formula—C_5H_8—a molecule that was later named "isoprene." Rubber is composed of many isoprene units strung together, forming a polymer—polyisoprene.

Attempting to mimic the natural product of *Hevea brasiliensis*, scientists set out to reproduce polyisoprene in the laboratory—and met mostly with failure. Fed-up with isoprene, scientists ditched it for butadiene (C_4H_6), hoping the change would be more successful in yielding a synthetic polymer similar to rubber. Working in St. Petersburg, and using ethanol to make butadiene and sodium to kick-start the polymerization, Russian-Soviet chemist Sergey Lebedev (1874–1934) cooked up polybutadiene, producing the first fake rubber in 1910. By the 1940s, the Soviet Union had the largest synthetic rubber industry in the world.

The discovery of polybutadiene, which is known today as butadiene rubber (BR), sparked the synthetic rubber industry. Other synthetic rubbers followed BR; styrene-butadiene rubber (SBR) in the 1930s and silicone rubber (SR) in the 1940s. By the end of the 1960s, synthetic rubber had natural rubber on the ropes, and today synthetic rubber beats natural rubber in both production and usage. All thanks to Levedev who showed us how to fake it. **RBk**

SEE ALSO: NEOPRENE, RUBBER BAND, RUBBER (WELLINGTON) BOOTS, SILICONE RUBBER

↗ *These jars illustrate the preparation of synthetic rubber, isoprene being synthesized from turpentine.*

"A pencil and rubber are of more use to thought than a battalion of assistants."

Theodor Adorno, philosopher and musicologist

Fuel Injection (1910)

Adams-Farwell does away with the choke.

As anyone who has ever tried to start up an old automobile will tell you, a choke is a miserable thing to operate. Pull it out too far and you flood the engine, do not pull it out far enough and the engine will not fire. It is not surprising, therefore, that automotive engineers did everything in their power to eradicate the need for carburetors. How did they do this? They invented fuel injection.

Now commonplace on all production cars, fuel injection is an automatic, accurate way of keeping the engine's fuel-to-air ratio at suitable levels. Modern computer systems use precise sensors and gauges that react to very quick changes in operation, such as

---

"I put a new engine in my car, but forgot to take the old one out. Now my car goes 500 miles per hour."

Stephen Wright, comedian

---

sudden accelerations to decide how much fuel is needed at any one point. Then the fuel injector releases a spray of pressurized fuel into the air stream passing through the engine. The first fuel injection system, developed by the manufacturer Adams-Farwell of Dubuque, Iowa, employed this principle but was entirely mechanical.

Designed for use with automotive diesel engines, the idea of fuel injection sat on the shelf largely untouched for about thirty years before it was used in wartime aviation. Even after that, it still wasn't really seriously considered for use in spark-ignited gasoline engines until the mid-1950s. **CL**

SEE ALSO: CARBURETOR, DIESEL ENGINE, MOTORCAR, SPARK IGNITION, STARTER MOTOR

Rayon (1910)

American Viscose Co. produces artificial silk.

From the first recorded uses of flax as a textile fiber around 5,000 BCE, to the adoption of cotton, wool, and silk, people have always been ready to exploit any material that can be made into long flexible fibers and woven or knitted into fabric.

In 1655 the English scientist Robert Hooke first proposed creating artificial silk from a gelatinous mass, but it it took another 200 years for anyone to realize that ambition. In 1895 the Swiss chemist Georges Audemars made a mixture of pulp from the bark of a mulberry tree and rubber. By dipping a needle into this mess he was able to draw out fibers of artificial silk. His process took a long time for each fiber and was too slow to be of any practical use. In 1884 the French chemist, Comte Hilaire Bernigaud de Chardonnet, having refined the method used by Audemars, patented "Chardonnay silk." It was very flammable and expensive and, despite causing a sensation at the Paris Exhibition in 1899, it was never particularly successful.

Various companies tried to produce artificial silk, but it was not until 1910 that a commercially successful version was made by the American Viscose Company. By the 1920s it was possible to buy this man-made fiber for half the price of silk and, in 1924 the word *rayon* was coined to describe the fiber. Rayon is extremely versatile and, because it is made from naturally occurring polymers found in plant cellulose, it has properties similar to those of natural fibers. Crucially, however, unlike more modern fibers such as nylon, rayon does not insulate the body, and so it allows clothes to "breathe," making them more comfortable to wear during hot weather. **DHk**

SEE ALSO: CLOTHING, WOVEN CLOTH, NYLON

➡ *An analyst in July 1948 tests the cellulose content of a sample of the mixture that forms the base of rayon.*

Aluminum Foil
(1910)

J. G. Neher and Sons pioneer a new wrapping.

The production of aluminum foil via the process of the endless rolling of aluminum sheets cast from molten aluminum was pioneered at a foil-rolling plant at the foot of the Rhine Falls in the Swiss town of Kreuzlingen in 1910. The plant was owned by the aluminum manufacturing firm J. G. Neher and Sons.

The firm had experimented with sheets of pure aluminum, placing them between two heavy, adjustable rollers and filling the interior of the rollers with boiling water. Sheets were passed continuously through the rollers, which were gradually brought closer and closer together until the desired thickness of foil was achieved. Its earliest uses were as wrappers for various tobacco and confectionery products, and with its effectiveness as a barrier to oxygen and light, inhibiting the growth of bacteria, aluminum foil soon supplanted tin as the preferred metal in the wrapping and preservation of foodstuffs. Processes evolved to include the use of print and color.

Throughout most of the nineteenth century, aluminum was considered a precious metal. Despite ranking third on the list of the world's most abundant chemical element after silicon and oxygen, it is not naturally occurring in its pure form and had to be extracted using crude and expensive methods. The discovery of electrolytic reduction in 1886 enabled the isolation and separation of aluminum for the first time, dramatically lowering its cost and leading to the production of pure aluminum in all its forms.

In the 1940s, aluminum became popular in households throughout the United States with the marketing of Reynolds Wrap Aluminum Foil, developed after a company sales executive used aluminum to wrap a Thanksgiving Day turkey. **BS**

SEE ALSO: HALL-HÉROULT PROCESS, ANODIZED ALUMINUM, CARBORUNDUM, TETRA PAK, WELDING

Infrared Photography
(1910)

Wood opens up the electromagnetic spectrum.

Traditional photography relies upon the interaction of visible light with film or plates to produce an image. As visible light makes up only a small portion of the electromagnetic spectrum, it follows that images could also be produced by infrared (IR) or ultraviolet (UV) light, which flank the visible in the spectrum.

The first intentionally produced infrared images were taken by Professor Robert W. Wood (1868–1955) and displayed in 1910. Wood used a filter over the camera lens to remove all but the infrared light and a film that was sensitive to IR. The technique was used initially for landscapes, because of its long exposure

> *"Wood was credited with 'opening up two new worlds; the worlds at each end of the spectrum'…"*
>
> Professor Robin Williams, RMIT University, Australia

times. Chlorophyll reflects large quantities of infrared light, making foliage appear bright white, while clear blue sky appears almost black.

IR photography is used in aerial surveys and military reconnaissance as it enables the depth of water and the presence of underwater obstacles to be seen. It is used in forest surveys to distinguish coniferous and deciduous trees, while botanists can use it to observe changes in pigmentation and cellular structure as plant diseases progress. Astrophysicists use it for observing distant galaxies and discovering new stars. IR and UV photography is also used to examine damaged documents by comparing inks. **HP**

SEE ALSO: COLOR PHOTOGRAPHY, COMPOUND CAMERA LENS, KINEMATOGRAPH, PHOTOGRAPHIC FILM, PHOTOGRAPHY

Depth Charge

(1910)

Taylor develops the first effective antisubmarine weapon.

The first significant use of submarines in warfare occurred during World War I, and with that came the need for anti-submarine weapons. The idea of using the destructive shockwaves of a "dropping mine" against submarines was discussed by the British Navy in 1910. However, it was not until the Commander in Chief, Sir George Callaghan (1852–1920), requested their production in 1914 that they became a reality.

It is the "D" type, developed by Herbert Taylor in 1915 at HMS *Vernon* Torpedo and Mine School, Portsmouth, England, that is credited with being the first effective depth charge. This was essentially a steel barrel packed with high explosive that could be detonated at preselected depths.

Early depth charges were deployed simply by rolling them off racks on the stern of the attacking vessel. Although a simple and effective deployment method, it was not without its problems. Even the larger depth charges had to detonate within 16 feet (5 m) of the target submarine to rupture its hull. Most losses were due to accumulated damages acquired over sustained attacks. Due to the close proximity of the attacking vessels and submarines, attacks were often detected, allowing submariners to dive and maneuver to avoid attack. To overcome this, the depth charges were streamlined to enable them to sink faster, and propulsion devices were developed, as were charges that could be dropped from planes.

Although largely replaced by forward-throwing devices, such as the "Hedgehog" and homing torpedoes, the depth charge is still in use today. **RP**

SEE ALSO: SUBMARINE, SUBMERSIBLE CRAFT

↗ *The Russian Navy fires a depth charge at a German U-boat in November 1941 during World War II.*

"The depth charge was such a successful device that it attracted the attention of the United States ..."

Chris Henry, Museum of Naval Fire Power

Neon Lamp
(1910)

Claude lights up the world with noble gases.

Following the invention of the electric lightbulb by Edison and Swan in 1878, the race was on to improve its design and performance. French chemical engineer, Georges Claude (1870–1960) was working on an invention to extract oxygen from air, for use in hospitals and welding, and his experiments resulted in his discovery of the noble gases—helium, argon, krypton, xenon, radon, and neon—so called because they do not react with other elements.

Aware of the race for the perfect lightbulb, Claude experimented by passing an electric current through tubes containing different noble gases at low pressure. In 1902 he discovered that neon gas, with only a small current, produced an intense orange glow. He was unimpressed by the amount of light it produced, but Jacques Fonseque, an advertising agent, saw the potential for its application and the two men started to make neon signs, shaping the glass tubes and using mixtures of the other noble gases to produce different colors. In 1910, Claude displayed the first neon lamp to the public in Paris. In 1912, Claude and Fonseque sold the first neon sign to a barber's shop in Paris and, in 1919, just after World War I, they erected a huge neon sign over the entrance to the Paris Opera House.

However, it was in the United States that neon really took off. The first sign there, made by Claude's factory in 1923, was an advertisement for a car dealership in Los Angeles. By 1927, New York had 750 neon signs. During the years before the Great Depression of the 1930s, neon lighting had become a symbol of American opulence and extravagance. **EH**

> *". . . the flash of a neon light that split the night / And touched the sound of silence."*
>
> Paul Simon, "The Sound of Silence"

SEE ALSO: ARC LAMP, ARGAND LAMP, FLUORESCENT LAMP, GAS LIGHTING, INCANDESCENT LIGHT, HALOGEN LAMP

◰ *Neon light was first displayed to the public in Paris in 1910. This early neon light dates from around 1922.*

Starter Motor

(1911)

Kettering takes some labor out of motoring.

Before the invention of the starter motor, the motorist had to indulge in some energetic "cranking" of the engine with a starting handle before it would fire. Sometimes the engine would misfire, jerking the handle violently and injuring the hapless motorist.

Automatic starters had been proposed some years before—a patent for what turned out to be an impractical solution was granted to Clyde J. Coleman in 1903. Research into a practical system was revisited with some urgency when a friend of Cadillac's founder, Henry Leland, died following an injury sustained when a starting handle, thrown by a backfiring engine, struck him in the face. Cadillac called upon Charles F.

"The Cadillac car will kill no more men if we can help it. We're going to develop a fool-proof device. . . ."

Henry M. Leland, founder Cadillac Automobile Co.

Kettering's (1876–1958) company, Dayton Engineering Laboratories (DELCO)—an entity set up specifically to exploit Kettering's already proven success in automotive electrical design—for help.

Kettering realized that an automobile starter motor need only operate successfully for a few seconds at a time, so it need not be impossibly large. The unit he developed was a combined starter motor and generator. First installed in a Cadillac on February 27, 1911, it incorporated an over-run clutch and reduction gear and could thus provide enough torque to crank the engine quickly and disengage once the engine fired. Car ownership was revolutionized. **MD**

SEE ALSO: CARBURETOR, DIESEL ENGINE, FOUR-STROKE CYCLE, MOTORCAR, MOTORCAR GEARS, SPARK IGNITION

Knapsack Parachute

(1911)

Kotelnikov invents a safety device for pilots.

The Wright brothers' historic flight of 1903 had made aeronautics a worldwide phenomenon. Air shows became increasingly popular spectacles throughout the world. A Russian artillery school graduate, Gleb Kotelnikov (1872–1944), was attending such a show in 1910 when he witnessed the death of a pilot. He was so affected by the accident that he vowed to create a safety device to help prevent such deaths.

The parachute is not a particularly new idea—there are many accounts of rudimentary parachutes being used throughout the ancient world. Early prototypes of a Kotelnikov parachute contained within the pilot's helmet failed, yet Kotelnikov was

---

"Our greatest glory is not in never falling, but in getting up every time we do."

Confucius, Chinese thinker and educator

---

unperturbed and eventually devised a parachute within a knapsack that could be worn within the confines of a plane's cockpit.

Despite the dangers to their military pilots, many governments were reluctant to provide parachutes to their armed forces. However, in 1918 the German Army Air Service became the first to introduce the parachute as standard issue.

Nowadays, parachutes are so reliable that they are used not just to save lives but also for recreation and entertainment. Kotelnikov was honored in 1949 for his invention by having the town in Russia where he first tested his device renamed Kotelnikovo. **JB**

SEE ALSO: AIRSHIP, GLIDER, HOT-AIR BALLOON, KITE, POWERED AIRPLANE, SPACE PROBE

X-ray Crystallography (1912)

Von Laue and Ewald study crystal atoms.

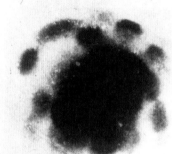

Crystals are solids—like salt, diamond, and quartz—that have their constituent atoms (or molecules) in regular orders. These patterns repeat in all directions. X-rays penetrate solids and are scattered by the clouds of electrons that surround the nuclei of each atom. Because the atomic arrays in crystals are strictly regular, the X-ray scattering is not random. Measuring the intensity of the X-rays in different directions and the specific angles at which the scattering occurs enables the separation of the arrays of crystal atoms to be calculated. The study of crystal atom spacing and ordering is known as crystallography. X-rays have wavelengths that are of the same order of magnitude as both the sizes of typical atoms, and also the spacing between solid arrays of atoms.

In 1912 Max von Laue (1879–1960) and Paul Ewald (1888–1985) suggested that the regular arrays of atoms in a crystal might act like the lines on a diffraction grating. Laue shone a beam of X-rays into a sphalerite crystal. Placing a photographic plate behind the crystal he noticed that the scattered X-rays produced a set of circular spot patterns around the larger spot formed by the central beam. He calculated the crystal array separations from the scattering angles.

Since then X-rays have been used to probe metals, chemicals, and biological samples. Huge advances have been made in the fields of organometallic and supramolecular chemistry. Dorothy Hodgkin used X-ray crystallography to calculate the atomic structure of such things as cholesterol, vitamin B_{12}, penicillin, and insulin. X-rays are widely used today in the pharmaceutical industry. **DH**

SEE ALSO: X-RAY PHOTOGRAPHY, X-RAY TELESCOPE, X-RAY TUBE, RADIOTHERAPY

> *"Crystals grew inside rock like arithmetic flowers … obedience to an absolute geometry."*
>
> Anne Dillard, American author

◩ *The first experiment of X-ray diffraction signed by Max von Laue in 1912.*

Cloud Chamber (1912)

Wilson discovers ionizing radiation.

In September 1894, Charles Thomson Rees Wilson (1869–1959) climbed to the summit of the Scottish mountain, Ben Nevis. He was so impressed by the effects of sunlight upon the clouds that he decided to try to reproduce them in his laboratory. His work was based on that done by John Aitken, the engineer who had created artificial clouds in a laboratory container. Aitken had found that if you put water vapor into a glass jar, a cloud would form if the air was unfiltered—that is, dusty. The water molecules in the air were treating the dust particles as tiny nucleation points onto which they could condense, forming a cloud.

Air pressure inside the container could be reduced by expanding the volume, which forced water vapor

"To those few weeks spent on the highest point of my native land I owe many happy years of work . . ."

C. T. R. Wilson, on receiving the Nobel Prize in 1927

to condense onto dust. Wilson also discovered that he was able to make airborne water droplets even when the dust had been removed. He theorized that condensation was occurring due to the presence of charged particles. To test this, he fired X-rays into his cloud chamber and also exposed it to uranium. In both cases he was able to form clouds from the radiation.

Wilson perfected his cloud chamber in 1912. He was able to track the movements of ionizing particles, which leave a track of ions through the chamber. This formed tracks of water droplets that he would photograph, allowing scientists to study the properties of the electrons and helium nuclei that formed them. **DHk**

SEE ALSO: X-RAY PHOTOGRAPHY, X-RAY TUBE,

ELECTRON MICROSCOPE

MDMA (Ecstasy) (1912)

Köllisch happens upon a psychotropic drug.

MDMA (3,4-methylenedioxy-methamphetamine) was originally synthesized by Anton Köllisch (1888–1916)—who was working for German pharmaceutical giant Merck—in 1912 as a side product to a drug intended to control bleeding. It was routinely patented in 1914, and Köllisch died two years later, oblivious to the impact that his discovery would have in years to come.

MDMA found its way onto the streets of America during the 1960s. Legend has it that the drug was deliberately developed as an appetite suppressor for use in World War I, but there is no documented evidence that human trials were carried out. It is also "unclear" whether experiments carried out in 1959 by Merck employee Wolfgang Fruhstorfer may have

"I hate to advocate drugs . . . to anyone, but they've always worked for me."

Hunter S. Thompson, journalist and author

involved tests on humans. The experiments were carried out with an unknown partner, creating even more mystery as to the appearance of MDMA on the streets, where it got its more familiar name, Ecstasy.

Today it is one of the most popular and most prevalent recreational drugs. Despite its illegality, production of the drug is a worldwide industry estimated to be worth over $65 billion (2003). Its production funds the organized crime required to smuggle the drug across borders. With increased availability and lower prices, the purity of the drug has decreased and users are increasingly exposing themselves to an unknown cocktail of ingredients. **JB**

SEE ALSO: HEROIN, LYSERGIC ACID DIETHYLAMIDE (LSD),

VALIUM

Rotavator (1912)

Howard develops a motorized rotary tiller.

Man has cultivated the Earth for thousands of years, and for a large portion of that time he has been "tilling"—turning the soil to bury weeds and mix in fertilizer—in order to grow crops. Tillage, and agriculture in general, took a big step toward modern intensive processes when Australian inventor Arthur Clifford (Cliff) Howard (1893–1971) created the motorized tiller—the Howard Rotovator—in 1912.

The son of a farmer, Howard studied engineering in Australia at Moss Vale, New South Wales. In 1912 he began experimenting on farming methods—primarily machines to improve tillage—on the family farm at Gilgandra, New South Wales. Howard noticed that regular plowing methods compacted the soil, making it more difficult to mix in fertilizer. Rotary tillage already existed, but was operated manually. Howard took a standard manual tiller and coupled it to his father's steam tractor. This proved superior to the standard plowing techniques, taking less effort to run, mixing the soil better and more evenly, and resulting in less crop residue being left on the surface.

Howard patented his creation, trademarked the name "Rotavator," and formed Austral Auto Cultivators Pty Ltd. in 1922 to market his invention. Howard went on to develop an extensive line of rotary tillers powered by internal combustion engines for many specific terrains, including orchards and vineyards. He also designed machines that could destroy weeds. Eventually, Howard marketed his machines around the world. Today rotavators are commonly used for soil preparation and are essential for maintaining the high yields of intensive modern agriculture. **SR**

SEE ALSO: MOLDBOARD PLOW, QUERNSTONE, SCRATCH PLOW, SEED DRILL, STEEL PLOW

⬆ *A man in Northmead, Australia, uses a rotavator like the one Howard conceived at the age of sixteen.*

Conveyor Belt (1913)

Ford introduces the continuous moving band for transporting objects.

Conveyers of various kinds have been in use since 250 BCE, the earliest example being the Archimedes screw used to raise water. The bucket conveyor, a simple chain of buckets used to move bulk materials, became an important technological innovation in the burgeoning mining industry of the fifteenth century.

Conveyor belts were a development of these simple machines. Early versions in the 1700s were nothing more than leather or canvas belts over flat wooden beds, used mostly for transporting sacks of grain and in the mining industry. But over the next couple of hundred years they developed, rubber replacing leather, canvas as the belt material and mechanization being introduced.

In 1913 Henry Ford installed conveyor belts in his factory in Michigan to create a production line. Combined with other factory manufacturing techniques and the principle of uniformity and interchangeability of parts, Ford revolutionized the motor industry and effectively created the standard for industry mass production. The idea of having a single or series of conveyors with your workers each contributing one small part of the end product at fixed points along the line is one that was quickly adopted by many factories. The conveyor belt quickly became used for transporting heavy and light objects in various stages of production in factories throughout the developed world.

Conveyor belts are still used in factories and in the mining industry; the longest conveyor belt in the world is over sixty miles long and is used in phosphate mining in the Western Sahara. **BG**

SEE ALSO: ARCHIMEDES SCREW, REAPER, COMBINE HARVESTER, GRAIN ELEVATOR

⊥ *An early conveyor belt used for haymaking in Sussex, England, circa 1912–1915.*

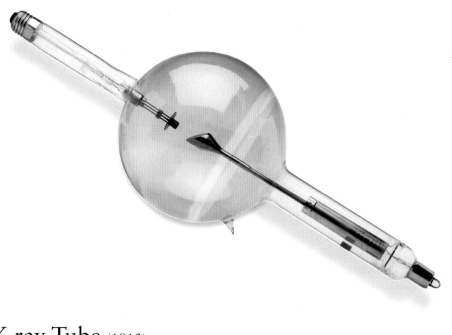

X-ray Tube (1913)

Coolidge develops a new diagnostic tool.

The first X-rays were produced by accident—a by-product emission from gas discharge tubes. The high voltage that is applied between the anode and cathode of the tube accelerates the ionized atoms of the gas into the metal cathode. This releases high-energy electrons, and when these electrons hit the glass wall of the tube, X-rays are emitted.

X-ray tubes use exactly the same physical processes but here the cathode is specially shaped so that a parallel beam of electrons is emitted through a window. Metal targets produce both a continuum of X-rays and a line spectrum. These "K lines" are formed by outer shell electrons falling into the inner K-shell. Metals such as molybdenum, copper, cobalt, iron, chromium, and tungsten are used for the cathode, each producing a K line of a different wavelength.

William Coolidge (1873–1975), an inventor and physicist working for General Electric in Schenectady, New York, had become interested in the properties of ductile tungsten when trying to improve the filaments of lightbulbs. As X-rays were another of his interests he decided to try using ductile tungsten as the target anode of a discharge tube. The Coolidge tube was gas-free and had a high-temperature tungsten cathode (tungsten melts at 6,170°F/3,410°C). These new tubes produced an intense, stable, and controllable beam of X-rays. They made the use of X-rays in medical diagnosis both safe and convenient. The tube was patented in 1913 and was used extensively in field hospitals during World War I. Doctors and dentists were introduced to this device and the subject of radiology blossomed. **DH**

SEE ALSO: MAGNETIC RESONANCE IMAGING, RADAR, RADIOTHERAPY, X-RAY PHOTOGRAPHY, X-RAY TELESCOPE

⬆ *The Coolidge X-ray tube used a heated cathode and was more consistent than earlier models.*

Zip Fastener (1913)

Sundbäck invents a hookless fastening.

The first person to attempt to reversibly connect two materials with a ziplike mechanism was the American inventor Elias Howe (1819–1867) in 1851 with his "Automatic, Continuous Clothing Closure." However, Howe devoted little time to his fastening invention. A short while later, fellow-American Whitcomb Judson invented the clasp locker, primarily to help a friend who had a bad back and couldn't do up his shoes. The design was based around a hook-and-eye mechanism and had little commercial success. One of Judson's employees, however, went on to hit the jackpot.

Gideon Sundbäck (1880–1954) worked for Judson's Universal Fastener Company. Because of his great skill he was appointed as head designer. He had been tasked with improving the Judson hook-and-eye fastener, which had an unfortunate tendency to come apart. Sundbäck's breakthrough design was based on the principle of interlocking teeth, and he called his invention the "Hookless Fastener." It consisted of two

rows of facing teeth that were interlocked with a slider, and he received a patent in 1913. Further improvements to his design resulted in the "Separable Fastener" in 1917. Sundbäck even developed a manufacturing machine for his new invention, which soon had the capacity to produce hundreds of feet of fastener every day. One of his first major customers was B. F. Goodrich, who used the fastener in his new line of shoes. Goodrich is credited with coining the term "zipper" for the device. Sundbäck's fastener also found great utility in tobacco pouches. But it was only two decades later that it entered the fashion industry and became the everyday object it is today. **RB**

SEE ALSO: **BLUE JEANS, BUCKLE, BUTTON, CLOTHING, SEWING MACHINE, WOVEN CLOTH**

⬆ *This zip fastener is a French design dating from the early twentieth century.*

Stainless Steel (1913)

Brearley develops a "rustless" alloy.

Stainless steel is an alloy of iron and chromium. It does not corrode in contact with air and water, stays bright, and can be polished. The chromium, which has a great affinity for oxygen, protects the iron by forming a molecular layer of chromium oxide at the surface, preventing contact between iron and oxygen.

Harry Brearley (1871–1948) was head of the research team at the Brown Firth company in Sheffield, England, when the firm was commissioned to develop an erosion-resistant metal for gun barrels. Brearley experimented with iron-chromium alloys, which were known to have higher melting points than steel. He varied the proportion of chromium between

> *"Brearley launched his 'rustless steel' (later renamed ... 'stainless') on the world with great gusto."*
>
> IP Review: Accidental Inventions

6 and 15 percent and also changed the carbon content until he developed an alloy with 12.8 percent chromium and 0.24 percent carbon. The alloy was impressively resistant to corrosion and Brearley recognized the potential for its use in cutlery.

Brearley was not the first to observe the properties of chromium-iron alloys. French metallurgist Pierre Berthier (1782–1861) had made such alloys, and other metallurgists were working on the same problem at the same time, notably Leon Guillet who developed an iron-nickel-chromium alloy in 1906. In 1908 the Krupp Iron Works in Germany manufactured chrome-nickel steel for the hull of a yacht, which sadly sank. **EH**

SEE ALSO: BESSEMER PROCESS, WROUGHT IRON, STEEL

Liquid Fuel Rocket (1914)

Goddard establishes modern rocketry.

Robert Goddard (1882–1945) gave serious thought to how we might get to the Moon. Despite being met with ridicule, what he gave to rocketry is still being used today. Using mathematics, Goddard worked out the energy-to-weight ratios of various fuels. This showed that gunpowder would never be powerful enough to lift a rocket into space. Also, to burn fuel in a vacuum, a rocket would have to carry its own oxygen supply. To get around these problems, Goddard used gasoline as a fuel and mixed it with liquid oxygen. Oxygen as a gas takes up a lot of space, but as a liquid it gives the rocket a lot of energy with much less weight and volume. In 1914 Goddard patented the first liquid-fueled rocket.

> *"It has often proved true that the dream of yesterday is the hope of today, and the reality of tomorrow."*
>
> Dr. Robert Goddard, American scientist

Twelve years later, after many failed experiments, Goddard took "Nell," a 9-foot-tall (3 m) rocket, out to a field. Getting an assistant to light the fuse with a blowtorch on a stick, he saw the rocket rest for a second, then shoot about 30 feet (12 m) into the air before tipping over and crashing into a field. The flight lasted two and half seconds, at an average speed of 60 miles per hour (96 kph). In addition to proving his theory to be sound, Goddard had laid the foundations for every space-going rocket since. **DK**

SEE ALSO: ARTIFICIAL SATELLITE, BALLISTIC MISSILE, POWERED AIRPLANE, REUSABLE SPACECRAFT, SUPERSONIC AIRPLANE

▣ *Goddard and a liquid oxygen-gasoline rocket in the frame from which it was fired on March 16, 1926.*

Traffic Lights (1914)

Morgan imposes a regulated control system on busy streets.

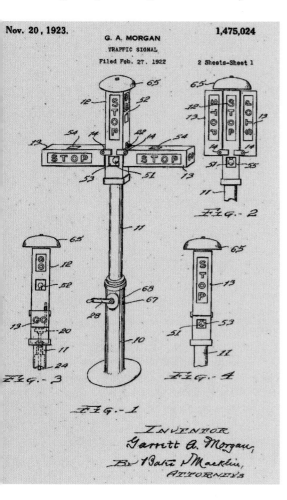

Henry Ford opened his first factory in America in 1903 and, with horse-drawn carts, pedestrians, bicycles, and cars sharing the treacherous and largely unlegislated roads, accidents were frequent.

Garrett Morgan (1877–1963), an African-American inventor, was inspired to design the traffic signal after witnessing a horrific crash between a horse-drawn carriage and a motor car. The invention was an indication of Morgan's perceptiveness; long before automobiles were as prevalent on the roads as they are today, he had the foresight to predict a problem and engineer a solution. It is a testament to his design that the signals are still in use today.

Morgan came from humble roots. The son of a former slave, his formal education ended after elementary school, but his natural talent for fixing and inventing soon earned himself a reputation as a master repairer, and he was quickly offered contracts around Cleveland, Ohio. In addition to the traffic signal, Morgan also invented the gas mask, a zigzag attachment for sewing machines, and a straightening comb for hair. He also published a newspaper for the black community called the *Cleveland Call* and stood as a candidate for the City of Cleveland government.

Morgan's traffic signal consisted of three arms in a T-shape, with hand-cranked "STOP" and "GO" signals. It could be alternated to stop traffic from single or multiple directions, allowing pedestrians to cross the road safely as well as preventing accidents and saving lives. The invention was patented in 1923, and Morgan eventually sold the rights to his traffic signal to the General Electric Company for $40,000. **KB**

SEE ALSO: **OMNIBUS, ELECTRIC CAR, MOTORCYCLE, MOTORCAR**

"An optimist is a person who sees a green light everywhere, while a pessimist sees only the red."

Albert Schweitzer, Alsatian philosopher

⬕ *Morgan's 1923 traffic light has arms that were raised and lowered to reveal Stop and Go signs.*

Military Tank (1914)

The British Army finds a way to break through defended obstacles.

In the early 1900s, the stalemate of trench warfare sparked military powers to look for alternative methods of breaking through enemy lines. Before World War I, motorized vehicles were still uncommon, and the current designs were unsuitable for combat. It was the British military, in 1914, that created the first tanks. They included tracks to make moving over muddy terrain easy and were fitted with internal-combustion engine, bulletproof casing, and mounted, revolving machine guns. Surprisingly it was the navy rather than the army that oversaw the deployment of the new war vehicles during World War I.

Before being put into service, the first tanks were demonstrated to two future British prime ministers—David Lloyd George and Winston Churchill. Amazed at the machines' ability to mow through barbed wire they set in motion the process of constructing more of what they called "landships." The criteria for a successful tank were a minimum speed of 4 miles per hour (6.5 kph) and the ability to climb 5-foot (1.5-m) high obstacles, span a trench 5 feet (1.5 m) wide, and, most important, be resistant to small-arms fire. The tank was developed remarkably quickly—in just under three years from idea to combat.

In November 1917, around 400 tanks penetrated almost all the way through a 7-mile (11 km) front at Cambrai, France. Unfortunately the infantry did not consolidate the tanks' gains and the incursion was not a complete success. However the British quickly honed their strategy and, in retaliation, all military powers began to develop their own version of the tanks in preparation for future warfare. **LS**

SEE ALSO: BALLISTIC MISSILE, BATTERING RAM, BAZOOKA, CANNON, IRON ROCKET, MACHINE GUN, REPEATING RIFLE

↗ *A British tank in November 1917. In World War I, tanks were used to resolve the war in the trenches.*

"It was effectively ironclad against bullets, and could at a pinch cross a thirty-foot trench ..."

H. G. Wells, *The Land Ironclads*, 1908

Oven-proof Glass (1915)

Sullivan and Taylor invent new cookware.

Nowadays, we take for granted that our oven doors and measuring cups will be see-through. But it was not always so. Normal glass will expand and shatter under the temperatures of the kitchen, so ceramics and metals were the only solution.

The magic material that changed all this was Pyrex, a borosilicate glass produced by Corning Incorporated. The heat-resistant glass was an immediate success and soon found its way into every cook's armamentarium. Oddly, its origins lie in the railway industry. Lanterns on trains would often crack because of large temperature differences between the hot lamp inside the glass and the cold weather outside. William C. Taylor and Eugene Sullivan at Corning's New York facility discovered that adding boron to the traditional glass mix improved its resilience to temperature extremes. The material was dubbed "nonex" and used with success on the railways for lanterns and telegraph battery jars. Corning went in search of other applications for the material.

Observing his wife using a nonex jar as an impromptu casserole dish, Corning scientist Jesse T. Littleton found that the glass could also withstand oven temperatures. With a little tweaking to remove lead and other undesirable chemicals to make it safe for food use, a new product line was born. Pyrex cookware went on sale in Boston in 1915, and laboratory equipment followed in the same year. The name Pyrex is somewhat arbitrary, following on the Corning affinity for the -ex suffix.

The use of this material does not end in the kitchen. Most glassware in chemical laboratories is made from Pyrex, and borosilicate glass is even used in telescope mirrors, starting with the famous Palomar Observatory in San Diego County. **MB**

SEE ALSO: CONTROLLED FIRE, OVEN, PRESSURE COOKER, AGA COOKER, ELECTRIC STOVE, MICROWAVE OVEN

Lipstick (1915)

Levy puts lip color into a sliding tube.

Women have added color to their lips for at least five millennia. The earliest evidence of a colored paste or lipstick comes from Mesopotamia in around 3000 BCE. There, it was made of crushed semi-precious jewels and then put on to the eyelids as well as the lips. Cleopatra, Pharoah of Egypt (69–30 BCE) used crushed carmine beetles in a base made of ants as lipstick. Some formulations would have resulted in serious illness or even death, such as the Ancient Egyptian concoction from 1400 BCE, which used a red dye extracted from seaweed, mixed with iodine and toxic bromine compounds.

In 1915, Maurice Levy invented the sliding tube that we know as lipstick. Levy's tubes were just 2 inches

> *"You can't keep changing your men, so you settle for changing your lipstick."*
>
> Heather Locklear, American actress

(5 cm) long. The sliding tube worked by a set of slide levers in the casing. In subsequent developments, Levy added a slide-and-twist mechanism, creating the lipstick tube as we know it today.

Lipstick, in its new and convenient form, caught the imagination of women in America and Europe from the early part of the twentieth century. Movie stars and other performers made it a relatively inexpensive luxury in an otherwise tough and unglamorous world of economic hardship and wartime. **LC**

SEE ALSO: BREAST IMPLANT, HAIR SPRAY, LIPOSUCTION

▣ *A photograph taken around 1920, showing a Valentine's Day card imprinted with a lipstick kiss.*

Gas Mask (1915)

Zelinsky filters toxic chemicals from air.

The need for a sealed face mask to protect against the inhalation of smoke and airborne toxins became apparent with the development of underground mining in the eighteenth century. A crude respirator was invented by the Prussian naturalist Alexander von Humboldt in 1799 while he was working as a mining engineer. Other designs included Garratt Morgan's "Safety Hood and Smoke Protector," developed in 1912, consisted of a cotton hood with two hoses reaching down to the floor, where cleaner air would often be found.

But the crucial imperative in the development of an effective gas mask was the use of poison gas, especially chlorine. This gas was used for the first time in World War I by the German Army at the Second Battle of Ypres in April 1915. Initial attempts to counteract the chlorine gas using cotton mouth pads and absorbent hoods proved either ineffective or cumbersome and thousands of soldiers perished.

In 1915, the Russian chemist Nikolay Zelinsky (1861–1953) invented the coal gas mask, which used a carbon filter to absorb the chlorine. The design was quickly adopted by the British, French, and Russian armies and has continued to play a crucial part in counteracting chemical warfare.

Carbon filters work by absorbing airborne particles and gases. Carbon powder is especially effective because of its very high surface area. Later gas masks were designed to protect against biological agents such as viruses and bacteria and radioactive fallout. However, masks alone are insufficient against many of these agents. **RBd**

SEE ALSO: IRON LUNG, AQUALUNG, VENTILATOR

⬆ *Gas masks like the one above were used by servicemen in World War II.*

Tungsten Filament (1916)

Langmuir improves the lightbulb's core.

In modern culture, the lightbulb is the symbol we often use to denote a sudden flash of inspiration. It is ironic, then, that the lightbulb itself is not the result of just one exceptional idea, but followed the contributions of many men over several decades. Thomas Edison is often credited with the design for the first practical lightbulb, but he was not responsible for the early developments. Without the input of American chemist Irving Langmuir (1881–1957), the lightbulb as we know it today might not have existed.

In the late nineteenth century, Edison filed numerous patents for improvements to electric lighting. His 1880 patent application for an electric lamp certainly depicts what we would immediately recognize as a lightbulb. But it was Langmuir's improvements to the filament at its core that gave us the lightbulb's modern design.

Langmuir was a researcher at Edison's General Electric Company, where he was studying the chemical reactions of tungsten inside lightbulbs. The company had replaced its old carbon filaments with tungsten, which lasted longer but was prone to vaporize very quickly, causing the bulbs to become black and dim. Langmuir slowed down this evaporation process by coiling the filament and adding nitrogen, a fairly unreactive gas, to the bulb.

The improved filament and lightbulb design was patented on April 18, 1916. It is essentially the design that is still used, although argon is now the unreactive gas most commonly used in lightbulbs.

Langmuir went on to win a Nobel prize in 1932 for his contributions to surface chemistry. **HB**

SEE ALSO: OIL LAMP, CANDLE, GAS LIGHTING, ARC LAMP, LIGHTBULB, NEON LAMP, HALOGEN LAMP, LAVA LAMP

⬆ *An early high-quality Mazda brand light bulb with a tungsten filament.*

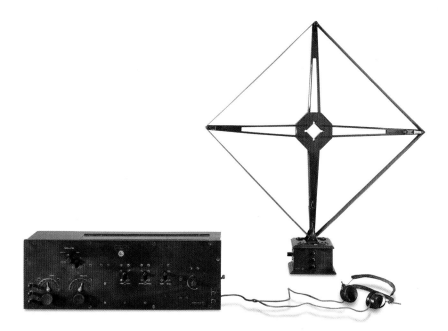

Superheterodyne Receiver (1918)

Armstrong improves radio technology.

It may not be the most eye-catching of names, but the superheterodyne receiver vastly improved radio technology in the early twentieth century by filtering out unwanted radio signals.

The "superhet," as it is commonly called, is the invention of Edwin Howard Armstrong (1890–1954). While working on ways to amplify radio signals during World War I, he hit upon a cunning idea. An incoming radio signal is mixed with an internal signal generated by the receiver (known as a local oscillator). The two signals are multiplied together. This creates two new signals: One is the sum of the two originals, and one is the difference. So, if you have an incoming radio frequency at 5 MHz and a local oscillator at 4 MHz, signals will be seen at 9 MHz and 1 MHz. This is known as heterodyning. A filter can be used to remove any signal that does not fit a narrow frequency range around the difference signal (here 1 MHz). The beauty of the system is that precise tuning can be achieved by varying the frequency of the local oscillator. For example, if you have a filter fixed at 1 MHz, and radio stations broadcasting at 5, 6, and 7 MHz, the three stations can be "tuned in" by turning the dial—and therefore the local oscillator—to 4, 5, and 6 MHz respectively.

Armstrong patented the technique in 1918, and it was quickly adopted by the military. The superhet principle was later used to tune TV and radio broadcasts and is still widely used today. Armstrong went on to invent FM radio, but he also spent many years fighting patent lawsuits and eventually committed suicide. **MB**

SEE ALSO: AMPLITUDE MODULATION, COHERER, RADIO TELESCOPE, RADIO TRANSMITTER

⊡ *A 1924 seven-valve "superhet" heterodyne radio receiver shown with its frame aerial and headphones.*

Mass Spectrometer (1918)

Dempster devises an instrument essential to analysis of the physical world.

The mass spectrometer sorts ionized atoms emitted by specific heated substances according to their individual masses. The analysis is usually performed by passing these ions through a vacuum chamber placed in a strong magnetic field. The path of individual ions is a function of their mass, velocity, and charge.

The foundation of the technique was the investigation of the ions produced by cathode ray tubes, research that led to the discovery of isotopes by J. J. Thomson in 1913. An effective mass spectrometer was produced after World War I by Arthur J. Dempster (1886–1950), who used an electrical collector in 1918 at the University of Chicago. One year later Francis W. Aston (1877–1945) used a photographic plate as a detector at the University of Cambridge. Aston discovered 212 naturally occurring isotopes, and was awarded the Nobel Prize for Chemistry in 1922.

By the 1940s mass spectrometers were being commercially manufactured and were being used by many experimenters—geologists, to measure the age of rocks by looking at their radioactive decay products; chemists, to analyze complex organic molecules; and space scientists, to analyze the composition of the upper atmosphere. Electron beams and lasers were used to excite the ions.

Miniature mass spectrometers were produced for portable medical applications, and versions were placed on spacecraft to analyze the atmospheres of Venus and Mars and the gases emitted from Halley's comet. Not only could the instruments measure minute samples, they could also measure the mass of an atom to the accuracy of one part in a billion. **DH**

SEE ALSO: CATHODE RAY TUBE, RADIOCARBON DATING

⬆ *Developing his work with isotopes, Aston devised this mass spectrometer with magnetic deviation in 1930.*

Hydraulic Brake (1918)

Lougheed uses fluid to improve braking.

Before the invention of hydraulic brakes, various systems of levers and pads were developed, but their main drawback was that they needed regular adjustment to maintain equal braking on all wheels. Hydraulic brakes, invented in 1918 by Malcolm Lougheed, addressed this problem and gave much more responsive braking.

Hydraulic brakes work by having a series of pistons connected to the brake pedal and the brake pads themselves. These are all interconnected by a central "reservoir" of non-compressible fluid, initially a mix of water and alcohol. The differences in diameter of the pistons means that the same pressure inside the system magnifies the force applied to the brake pedal.

> *"My hydraulic brakes stop on a dime—and with change left over."*
>
> Malcolm Lougheed, aviation and auto engineer

This greater force allowed the brakes to be applied much more firmly and so bring cars to a quicker halt.

The system was taken up by the Duesenberg motor company, and the first passenger car to use a hydraulic braking system—the "Model A"—went into production in 1921. Despite the system's superiority to previous braking systems, other car manufacturers were slow to follow, possibly because of early problems with leaks. Gradually, however, as technology for lines and pistons improved, hydraulic brakes were introduced universally, and in 1938, Ford—the last manufacturer to catch on—finally switched over on all its cars. **SB**

SEE ALSO: ACCELERATOR, AIR BRAKE, BAND BRAKE, DISC BRAKE, DRUM BRAKE, MOTORCAR, REGENERATIVE BRAKES

Anthrax Vaccine (1918)

Smith reveals a breakthrough medication.

Before the invention of an effective vaccine for anthrax, the disease was a major agricultural problem and economic burden. Anthrax is a potentially fatal disease that affects animals and humans and is spread via airborne spores. Before it was fully understood, the illness was referred to as "ragpickers' disease" or "woolsorter's disease" because it was mostly caught by people working closely with animal hides.

In 1877, the Prussian physician Robert Koch (1843–1910) finally made a link between anthrax infection and a spore-forming bacteria called *Bacillus anthracis*. In the late nineteenth century, renowned scientist Louis Pasteur (1822–1895) developed a two-dose vaccine for anthrax that he had tested on sheep. However, storage quickly reduced the efficacy of the vaccine, and its effect was sometimes fatal.

It was the Australian scientist John McGarvie Smith (1844–1918) and his research collaborator, John Gunn (1860–1910), who eventually developed a safe, single-dose vaccine. Though the the two men shared an amicable relationship, neither one ever conceded that the other was responsible for the invention. Following Gunn's death in 1910, Smith refused to make the formula public and even resorted to setting up a laboratory in his own home to start production of the vaccine.

After years of resisting pressure by Australia's Minister of Agriculture, W. C. Grahame, to divulge the secret, Smith's health was deteriorating and he finally agreed to make his work public. In 1918, he handed over the desperately sought formula to Dr. Frank Edgar Wall on behalf of the State government. Smith also donated $10,000 to set up the John McGarvie Smith Institute where his vaccine could be mass-produced and distributed. **HI**

SEE ALSO: BCG VACCINE, CHOLERA VACCINE, INOCULATION, POLIO VACCINE, RABIES VACCINE, RUBELLA VACCINE

Pivoting Bed (1918)

Murphy improves one-room living with an ingenious space-saving bed.

In early-twentieth-century San Francisco, William Murphy (1876–1959) found that he had a problem. He wanted to court a young opera singer from a good background, but he lived in a small one-room apartment with a bed taking up most of the floor space. Since it would not be proper to bring a young lady back to his bedroom, he began experimenting with pivoting beds to make his room respectable.

He applied for his first patent in 1900 and formed the Murphy Wall Bed Company that year. The company is still in business today, and is the second oldest furniture company in America.

In the earliest and most familiar Murphy beds, the bed flips up at the head to be stored in a closet. In 1918 Murphy invented the pivoting bed that pivoted on a swinging arm around the door jamb of a closet, then lowered into a sleeping position. The beds were at the height of their popularity in the 1920s and 1930s, but declined after World War II as the economy grew and single-family homes became more common. But with the recession of the 1970s, the beds came back into style as a way of making the most of limited space, and the business began to grow once more.

Murphy beds have provided countless opportunities for slapstick comedy throughout the years, regularly featuring in the films of Abbot and Costello, Laurel and Hardy, and Charlie Chaplin, always ready to spring from the wall to bonk someone on the head or suddenly fold up to leave someone trapped helplessly in the wall.

And Murphy's inventiveness paid off, too—he eventually married his musical sweetheart. **BO**

SEE ALSO: CHAIR, COILED SPRING, SOFA BED, CARPET SWEEPER, WATERBED

↗ *The 1931 comedy* Flying High *exploited the amusements to be had with a pivoting bed.*

"Believe me, you've got to get up pretty early in the morning if you want to get out of bed."

Groucho Marx, American comedian and actor

Flip-flop Circuit (1919)

Eccles and Jordan invent digital logic circuit.

In 1919, long before the invention of electronic computers, a pair of British physicists invented the circuit that would become their key building block. William Eccles (1875–1966), who had pioneered the development of radio communication and assisted Guglielmo Marconi, together with Frank Jordan worked with the leading-edge electronic technology of the time—vacuum tubes, the predecessor of the transistor.

Eccles was a radio pioneer, and his interest in vacuum tubes stemmed from their use in radio. In particular, the vacuum-tube diode was used to detect radio signals (the term "diode" was coined by Eccles).

Experimenting with vacuum tubes, Eccles and Jordan found a circuit with an interesting property—it had a memory. Unlike other circuits whose output would change depending on what their inputs were doing, this "trigger" circuit would cling on to the last state it had been put in. The circuit had two stable states: a brief pulse applied to one input would "flip" the circuit into one state, and it would stay there—until a pulse was applied to the other input, which would "flop" the circuit back into the other state.

With the "flip" and "flop" standing for 0 and 1, the fundamental element of an electronic digital memory had been created. With no moving parts, this vacuum tube memory was much faster than its mechanical predecessors.

Vacuum tube flip-flops were first pressed into service as computer memories in the build-up to World War II, enabling a new electronic generation of computers to be prepared for battle. Flip-flop circuits, now transistorized and crammed onto silicon chips, still flip and flop in their millions in the hearts of modern computers. **MG**

SEE ALSO: **DIGITAL ELECTRONIC COMPUTER, INTEGRATED CIRCUIT, RAM (RANDOM ACCESS MEMORY), WORD PROCESSOR**

Pop-up Toaster (1919)

Strite makes burned toast a thing of the past.

In 1919 Charles Strite, a factory worker in a manufacturing plant in Stillwater, Minnesota, became annoyed by the burned toast on offer in the factory canteen and set about trying to solve the problem.

Originally, toasting bread would have been carried out over a fire, but labor-saving devices to help with this procedure followed, and the first electric toaster was invented in 1893. It worked by passing electricity through coils of "Nichrome" (a nickel-chromium alloy), which caused them to give off heat, thus toasting the bread. Early toasters were sold as a status symbol, even before electricity was common in homes, with the power cord designed to connect to a light socket, the only electrical connection any house would have had.

> *"I'm quitting.... I'm going to open up an appliance store, I've always really been into toasters."*
>
> **Dane Cook, actor and comedian**

Strite's innovation was to add a clockwork timer and springs in addition to the heating coils. It was designed so that after a set amount of cooking time, the heating elements were switched off and the perfectly cooked toast was ejected.

Originally intended for sale to commercial kitchens, Strite's invention was adapted for home use, released in the mid-1920s, and dubbed the "Toastmaster." It soon became a global success and, by 1930, was selling at a rate of more than 1.2 million per year. **SB**

SEE ALSO: **SANDWICH, AUTOMATIC BREAD SLICER**

➡ *In 1926 the Waters-Genter Company began selling the Toastmaster, designed by Charles Strite.*

TOASTMASTER

utomatic Electric Toaster

You do not have to
Watch it~
The Toast can't Burn.

Study Carefully These
INSTRUCTIONS

Allen Key (1920)

Unbrako (in the United States) and Brugola (in Italy) devise similar hexagonal drivers.

In appearance it is a simple piece of metal with a hexagonal cross-section and a ninety-degree bend about three-quarters along its length. Called variously Hex key, Allen key, Alum key, Inbus, and Unbrako key, this uncomplicated device may date back to the 1920s. The Unbrako company developed a hexagonal-head key and screw in the 1920s which went on to become popular in the United States and Britain.

During the same decade it is claimed that Italian Egidio Brugola, founder in 1926 of Brugola manufacturing company, also created a hexagonal-head fastener, which was the foundation for a business that still thrives today. In Italy (unsurprisingly) the Allen key is called the "Brugola." A couple of decades later, in 1943, the Allen Manufacturing company took out a trademark on an "Allen Key"—a name that would become popular in the United States and the United Kingdom.

Whatever you choose to call it, however, it is undoubtedly one of the most ubiquitous tools on the planet. Its simplicity is undoubtedly one of the many things that makes it such a popular tool, plus the fact that it is small and lightweight, yet being cast from a solid piece of metal is hard-wearing. Either end of the key can be used, which adds versatility to its use.

Used extensively in the motor and bicycle industry because the working part of the hex-screw is protected from the elements, it has also recently found popularity with the rise of the "flat-pack" furniture trade, where, because they are relatively cheap to produce, manufacturers are able to include a hex-key with the furniture. **BG**

> *"It's essential to recognize that no tool . . . can approach the vastness of the universe and life itself."*
>
> Hanz Decoz, numerologist

SEE ALSO: METALWORKING, SCREW CUTTING LATHE, STANDARDIZED SCREW SYSTEM, ADJUSTABLE WRENCH

◨ *Allen keys are used to drive screws and bolts that have a hexagonal socket in the head.*

Jubilee Clip (1921)

Robinson firmly seals hoses and pipes.

The Jubilee Clip was invented by Royal Navy Lieutenant Commander Lumley Robinson (d. 1939) in 1921. This ingenious device consists of a stainless-steel band that is put around a hose or tube, then tightened and the fitting sealed by turning the screw on one end of the clip. The screw acts as a worm drive, and so these types of clips are sometimes called "worm-drive hose clips" or simply "hose clamps." The clips could be used for simple household uses, such as plumbing, or for larger applications—such as piping on ships. The Jubilee Clip was issued a patent and Robinson began marketing the clip commercially in 1923. Following his death, L. Robinson and Company Limited was established in 1948 by Robinson's son John in

"Arrest those drips with genuine Jubilee worm drive hose clips"

Jubilee Clip advertising poster

Gillingham, England. The company still manufactures the Jubilee Clip today and is considered one of the best companies producing these types of clips.

Even with its simple design and purpose, when the clip was introduced it was considered revolutionary. Warships, for example, had to use such items as wire whipping and split pins to keep their pipes together before the invention of the Jubilee Clip.

The Jubilee Clip is now used for tasks as varied as gardening, automotive needs, and aerospace. In 2005 L. Robinson and Company even donated Jubilee Clips to a car participating in the grueling Mongol Rally, a motor race throughout Europe and Asia. **RH**

SEE ALSO: SCREW CUTTING LATHE, STANDARDIZED SCREW SYSTEM, STAINLESS STEEL

Microelectrode (1921)

Hyde facilitates the study of single cells.

Being a woman in science in the early 1900s was difficult for American neurophysiologist Ida Hyde (1857–1945). Born in Davenport, Iowa, to German immigrant parents, she struggled to find a university that would accept her. She eventually earned a bachelor's degree at Cornell University and later became the first woman to earn a PhD in science at the University of Heidelberg, in Germany. Despite this, Hyde was still not acknowledged for her invention of the microelectrode until after her death. Since then, the microelectrode has revolutionized neurophysiology. Hyde's electrode was so small that at the time her methods were among the first capable of studying single cells.

Hyde's research focused on the breathing mechanism and nervous systems of a range of organisms, from grasshoppers to humans. During her research, she invented the microelectrode so that she could deliver electrical or chemical stimuli to a cell and record the electrical activity from an individual cell. This way she could find out exactly what happens when a cell or nerve conducts electricity.

It was not until twenty years later, when the microelectrode was reinvented, in ignorance of Hyde's pioneering work, that it was used in extensive studies. Today they are an essential piece of equipment needed to examine electrical impulses in the brain. If Hyde has been recognized at the time then perhaps the microelectrode would have been used by neurologists decades earlier. It is difficult to guess how many years research was delayed simply because of sexist views about women scientists.

Hyde's experiences made her very concerned about women's education and she spent her later years lecturing on women's issues. **LS**

SEE ALSO: COMPUTED TOMOGRAPHY (CT OR CAT SCANS), ELECTROENCEPHALOGRAPH (EEG)

BCG Vaccine (1921)

Calmette and Guerin fight tuberculosis.

Since it was first developed in 1921, the BCG, or Bacille-Calmette-Guerin, vaccine has been given to over a billion people worldwide to prevent tuberculosis.

Tuberculosis was a huge killer of adults in the nineteenth century. In 1882 Robert Koch proved that the bacterium *tubercle bacillus* was the cause. Using it in killed or treated form to protect people from infection did not work, however. French bacteriologist, Albert Calmette (1863–1933) and his colleague, veterinary surgeon Camille Guerin (1872–1961) made a significant step forward when they found that placing bovine tuberculosis in a glycerine-bile-potato mixture caused it to grow bacilli that were less virulent. By 1906, through further subculturing, Calmette and

> *"The launching of the BCG vaccine was … a gigantic dishonest commercial operation."*
>
> Dr. Jean Elmiger, Swiss doctor and homeopath

Guerin produced a strain of living bacilli that were so weakened they could not produce disease but could still be used as a vaccine. They first tested their BCG vaccine on humans in 1921, and in 1928 it was used to successfully inoculate 116,000 French children.

In recent years, more cases of tuberculosis are being reported alongside an increase in multi-resistant strains of the bacterium. This has led to trials of a new vaccine to complement the BCG vaccine. The MVA85A, contains a protein found in all strains of tuberculosis that aims to boost the response of T-cells already primed by the BCG vaccine. The new vaccine is being tested in South Africa's Western Cape. **LH**

SEE ALSO: CHOLERA VACCINE, INOCULATION, POLIO
VACCINE, RUBELLA VACCINE, VACCINATION

Lie Detector (1921)

Larson identifies untruthful responses.

Some lies roll off the tongue all too easily, while others can put the body under noticeable stress, inducing sweating palms, nervous twitches, contorted voices, and pounding heartbeats. Thanks to John Augustus Larson (1892–1983), signals such as these can also help to incriminate the sneakiest of liars.

Larson, a medical student at the University of California, invented the polygraph in 1921 and, in doing so, introduced one of the most contentious tools ever brought to the police officer's locker. Larson's lie detector worked by continuously and simultaneously monitoring physical responses such as changes to blood pressure, pulse rate, and respiration. Unfortunately, there is no known lie hormone that is

> *"One may sometimes tell a lie, but the grimace that accompanies it tells the truth."*
>
> Friedrich Nietzsche, German philosopher

secreted during acts of deception, and as the responses under surveillance can be triggered by any stressful situation or indeed be suppressed by any cunning criminal, there will always be question marks over the polygraph's credibility.

Despite the doubts, the lie detector is seen as one of the greatest inventions of all time, used most effectively in the earlier stages of an investigation when prime suspects need to be identified, or in civil investigations, which carry less severe penalties. **CB**

SEE ALSO: TACHISTOSCOPE, DNA FINGERPRINTING

➡ *A prisoner undergoes psychological testing with a primitive lie detector in 1915.*

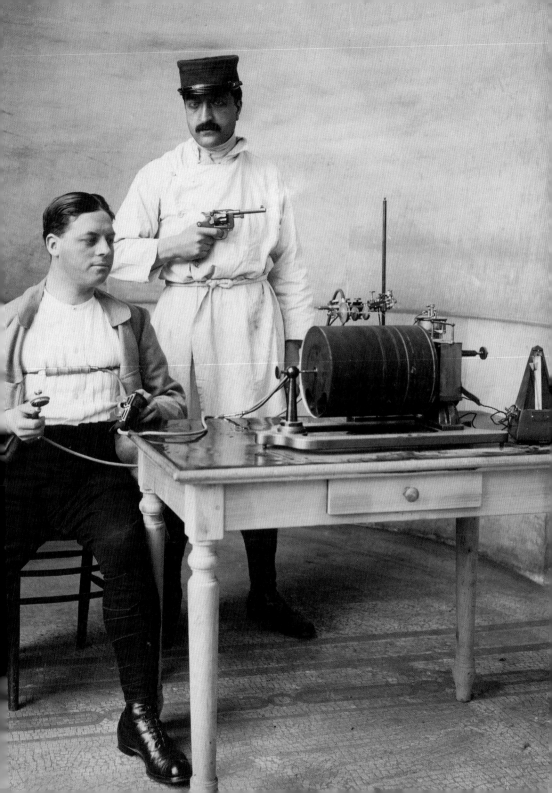

Refrigerator (1922)

Von Platen and Munters develop a cooling cabinet for preserving food.

The refrigerator is one of the key inventions of the twentieth century. Its use in food storage is vital, slowing the development of bacteria and keeping food edible for much longer. Before its invention, the only source of cold was blocks of ice, which could be bought in some places and used with a cool-box. Most homes had no means of chilling food.

Baltzar von Platen (1898–1984) and Carl Munters (1897–1989) were students at the Royal Institute of Technology in Stockholm, Sweden, when they collectively invented and developed the gas absorption refrigerator. Unlike modern fridges, the invention did not require electricity driving a compressor, but relied instead upon an ingenious process whereby a refrigerant gas is put through a series of changes of state. In von Platen's process, ammonia mixed with water is heated until the ammonia evaporates. This gas is then passed through a condenser, which conducts heat away from the pure ammonia until it becomes liquid at a much lower temperature than when mixed with water. This liquid is then passed through brine and cools it, which in turn chills the unit. The ammonia is then returned to a gas and reabsorbed into water so that the process can begin again.

The gas absorption refrigerator went into production in 1923 by AB Artic (later purchased by Electrolux), but it never truly caught on. The electric refrigerator, developed at the same time, gained much more investment and advertising and soon came to dominate the market. By the 1930s, the gas absorption refrigerator had ceased to be produced. **JG**

> *"When you hunt ... you may succeed or not. When you open the fridge, you succeed [all] the time."*
>
> Nora Volkow, National Institute on Drug Abuse

SEE ALSO: FREON, FROZEN FOOD, HABER PROCESS (MANUFACTURE OF AMMONIA), ICE-MAKING MACHINE

◩ *A "Monitor Top," electric compression domestic refrigerator made by General Electric in 1934.*

Blender (1922)

Poplawski shows the way to milkshakes and smoothies.

Although ubiquitous in modern kitchens, the blender started life as a tool for the soft drinks industry, designed to mix malted milkshakes. By the 1920s both milkshakes and malted milk drinks were extremely popular. However, malted milk powder was developed as a hot drink and so often clumped together when mixed with cold milk, resulting in a lumpy milkshake. This problem was overcome in 1922 when Stephen Poplawski (1885–1956) invented the blender. His design had small spinning blades, driven by an electric motor, at the bottom of a tall container into which all the ingredients would be added and mixed. The device also resulted in a lighter, frothy drink, forever changing the consistency of milkshakes and at the same time paving the way for the smoothie.

With modifications by Fred Osius and some nifty marketing by Fred Waring, the blender soon found its way into hotels and restaurants and, by the 1950s, was flying into the kitchens of many households.

The blender's uses do not end there. Wherever there is a need to mush things into a puree, you will find a blender. Hospitals, for example, were quick to employ blenders for the preparation of specific diets, and they do not even look out of place in a modern scientific laboratory—Dr. Jonas Salk famously used a blender whilst developing his polio vaccine. More recently, the blender has evolved into a source of entertainment, with the award-winning website, www.willitblend.com, testing the blender's blending limits.

Today the blender is as popular as ever; no health-conscious household would now be without their smoothie-making machine. **RP**

SEE ALSO: FOOD MIXER, FOOD PROCESSOR

⬀ *This Magimix food blender, with its mixing unit attached, was manufactured in 1951.*

> *"A two-year-old is kind of like having a blender, but you don't have a top for it."*
>
> Jerry Seinfeld, American actor and comedian

Snowmobile (1922)

Bombardier uses tracks to travel over snow.

Snowmobiles use a continuous, motor-driven ribbed rubber/Kevlar track to propel them across all types of snow and ice, with one or more skis at the front for steering. The idea for this caterpillar track system was devised by Canadian inventor Joseph-Armand Bombardier (1907–1964). This self-taught mechanical engineer ran a small garage in Valcourt, southeast of Montreal. The fact that the Quebec government did not plow the roads in winter and the inhabitants had to resort to horse-drawn sleighs challenged him to make a snow vehicle. By 1937 he was producing an enclosed snowmobile that would carry seven passengers. These were quickly adapted into ambulances, post vans, winter school buses, and forestry machines. They were also used by the army during World War II.

A further breakthrough came in the late 1950s with the introduction of small, lightweight, four-stroke gasoline engines. Bombadier invented a one- or two-person, ride-on, open machine called the "Ski-Doo" (it started off as the "Ski-Dog," but there was confusion at the printers).

Today more than 200,000 snowmobiles are manufactured every year and they have made a huge difference to winter tourism in Canada, northern United States, and Europe. High speed, impressive acceleration, and enabling easy access to remote areas and regions of deep snow have made them ideal winter recreational vehicles. Snowmobile races are held, one of the most famous being the 1,971-mile-long (3,172 km) "Iron Dog" race between Wasilla, Nome, and Fairbanks in Alaska. Better motors have been developed that make the machines less polluting. Some national parks, such as Yellowstone, allow only the use of four-stroke snowmobiles. **DH**

SEE ALSO: SLEDGE, SKI, ICE SKATE, FOUR-STROKE CYCLE, SKI LIFT, KEVLAR

Technicolor (1922)

Kalmus brings color to the movies.

Perhaps the ultimate accolade for an invention is for it to become a term used in everyday language. "Technicolor" is such an invention—a color film processing technique pioneered by Herbert Kalmus (1881–1963) and colleagues at Technicolor Motion Picture Corporation in 1922.

The first widely released film to use Technicolor was *The Toll of the Sea* in 1922. The technique involved the exposure of two adjacent frames of black and white film, one behind a green filter and the other behind red. Each set of frames was then printed onto separate strips of film and toned with the opposite color. The resulting parallel strips were then stuck together to create the final print.

> *"Technicolor makes me look like death warmed over."*
>
> Bette Davis, American actress

Although commercially successful, the process was far from perfect. In 1928, the introduction of a new process replaced the need for two separate strips to be stuck together and heralded a boom in color film production. Further refinements followed but producing color films was expensive and did not attract large enough audiences to make it economical.

Its revival owed much to Walt Disney who bought an exclusive contract to use it in his studios. The success of *Snow White and the Seven Dwarfs* (1937) confirmed the pre-eminence of Technicolor, which became the most widely used color film process in Hollywood for the next twenty years. **RBd**

SEE ALSO: CELLULOID, COLOR PHOTOGRAPHY, FILM CAMERA/PROJECTOR, KINEMATOGRAPH

Aga Cooker (1922)

Dalén produces an iconic kitchen appliance.

The Aga was the invention of the brilliant Swedish Nobel-prize-winning physicist, Gustaf Dalén (1869–1937). Blinded in an industrial explosion in 1912, he put his talents to improving his wife's cooker during his enforced convalescence at home.

Fortunately for subsequent generations of owners of large cold houses in the United Kingdom, Dalén was also the Managing Director of the Swedish Aktiebolaget Gas Accumulator Company (which most famously manufactured the Dalén-designed, acetylene-lit automatic lights in lighthouses and light buoys), and the Aga, patented in 1922, went into general production. Its cast-iron ovens and hotplates maintained a steady heat when fueled by readily available coke, and the cream enamel exterior was both attractive and easy to clean.

Agas were exported to the United Kingdom in 1929, initially sharing the domestic market with the then similar Esse. By the end of World War II sales of the Aga were taking off (by 1948 the Aga Heat Ltd Works had manufactured 100,000 in England under license) and an increasingly lively export market was mainly supplied from the United Kingdom. Modified models were gradually introduced in brightly colored enamels and running on varied fuels: oil in 1964, electricity in 1975, and, to date, kerosene, diesel, natural gas, and propane gas. There are even prototypes designed for biofuels and windmill power. The four-oven model was introduced in 1987. The Coalbrookdale foundry, in Shropshire, England, now casts all Agas that are sold anywhere in the world—currently to twenty-one different countries. **AE-D**

SEE ALSO: CONTROLLED FIRE, ELECTRIC STOVE, MICROWAVE OVEN, PRESSURE COOKER

↗ *A young woman places a kettle on the hob of an Aga solid-fuel cooker made around 1940.*

> *"… when you meet another Aga owner it is like discovering an instant friend."*

Mary Berry, cook book writer

Water Ski (1922)

Samuelson invents a glamorous new sport.

Messing about on the water provides many of life's great pastimes; swimming, fishing, and boating being just a few. In 1922 one more activity was added to this list—waterskiing. Its inventor, Ralph Samuelson (1903–1977), was already a keen exponent of aquaplaning—the art of being dragged behind a boat on a shaped piece of wood—but he wanted to replicate snow skiing on the water. After several abortive attempts with barrel staves and snow skis, he build his own skis out of 8-foot (2.5 m) lengths of plank, using his mother's copper kettle to boil the ends of the wood to shape them. When they were ready, his next task was to work out the correct technique for getting himself "launched."

For many years, the origins of the water ski were controversial, with several groups claiming they were the first, and a 1924 patent for "Akwa Skees" confusing matters. It was not until 1963, when a reporter accidentally stumbled across Samuelson and his story, that he finally became recognized as the "father of waterskiing," and inducted into the Water Ski Hall of Fame. Samuelson himself, despite touring and putting on exhibitions of his waterskiing, had not even realized he had done anything extraordinary.

Samuelson went on to be the first to achieve a water ski jump, launching himself off an angled diving platform greased with lard in 1925, and also being pulled behind a World War I flying boat at speeds of up to 80 miles (129 km) per hour. His second pair of skis (the originals having been broken by a particularly heavy landing) can still be seen in the museum at the Water Ski Hall of Fame. **SB**

SEE ALSO: SKI, MOTORBOAT, JET BOAT

"I decided that if you could ski on snow, you could ski on water. Everyone . . . thought I was nuts."

Ralph Samuelson, waterskiing pioneer

◩ *Samuelson on water skis in 1925, the year Fred Waller received a patent for Samuelson's invention.*

Cotton Buds (1923)

Gerstenzang invents a swab on a stick.

Leo Gerstenzang began to design a cotton swab after he saw his wife gluing cotton onto the ends of toothpicks to clean their baby's ears. He used cardboard material for the stem of the swabs to avoid any splinters harming the baby, found a way to attach equal amounts of cotton to each end of the swab, and ensured that the swab stayed put during cleaning. He created the Leo Gerstenzang Infant Novelty Company to supply his swabs and, in 1923, launched his refined product under the name "Baby Gays."

In 1926 Gerstenzang changed the name to Q-Tips Baby Gays, with the "Q" standing for quality, but eventually the product became known simply as the Q-Tips that we know today.

> "[C]otton-bud-related injuries are a common reason for attendances at ... clinics."

J. C. Hobson and J. A. Lavy

As well as supplying the baby accessory market, Q-Tips expanded into the cosmetic market in the 1950s. Hollywood makeup artists began using them as tools of their trade and a booklet, *Lessons in Loveliness with Q-Tips*, was produced.

In the early 1970s fears arose that damage could be caused to the ears by cleaning them with cotton buds, particularly perforation of the ear drum or impaction of the earwax. This led to manufacturers advising that cotton buds should no longer be used to clean ears. Q-Tips are still widely available and remain essentially unchanged from Leo's original design, although they are now very different in purpose. **HP**

SEE ALSO: DISPOSABLE RAZOR BLADE, PAPER TISSUES, SOAP, TOOTHBRUSH, TOOTHPASTE

Film Sound (1923)

De Forest ends the era of silent movies.

It was at the 1900 Paris Exposition that the first public demonstration of sound and vision in a movie theater took place. However, projecting volume into a large auditorium was difficult at a time when amplified public address systems did not yet exist; furthermore, the crude synchronization of sound and vision was simply a case of starting the movie projector and audio playback cylinder at the same moment and hoping for the best.

Low-key experimentation continued over the next two decades until, in 1919, electrical engineer Lee De Forest (1873–1961) developed the first sound-on-film technology: a system where a soundtrack "strip" was added to the movie film. Four years later, on April 23, 1923, De Forest's Phonofilms studio was responsible for the first public screening of a fully synchronized talking picture. A year later he made the first commercial dramatic talking picture, *Love's Old Sweet Song*, directed by H. Manning Haynes and starring John Stuart and Joan Wyndham.

There was anxiety within Hollywood, however, that this new technology would threaten their dominant position at the heart of what was now a multi-million-dollar industry. Furthermore, a number of major Hollywood studios began developing their own competing, incompatible technologies.

It took the success in 1927 of *The Jazz Singer*, directed by Alan Crosland, to prove that talking pictures could yield great profitability—even if this success was more down to its star, Al Jolson, being one of America's biggest celebrities than any public desire to experience synchronized sound and image. Nonetheless, the major studios gradually began to back the idea, and by 1930 the era of the silent movie was all but over. **TB**

SEE ALSO: LOUDSPEAKER, MICROPHONE, STEREO SOUND, SURROUND SOUND

Bulldozer (1923)

Cummings introduces an earth-moving tractor to replace mule power.

The steam shovel, invented in 1839 by William Otis, was used to dig the Suez Canal in 1869 and the Panama Canal in 1910. But eighty years after the invention of a digging machine, trenches were still being filled using mule power.

In 1923, American farmer James Cummings (1895–1981) saw mules being used to backfill oil pipeline trenches and realized that a machine could do the job more efficiently. He and draftsman J. Earl McLeod drew up plans, built the first bulldozer from junkyard parts, and won the contract to backfill the pipeline.

The purpose of the bulldozer is to move material from one place to another. Bulldozers have wide tracks to facilitate movement over mud and sand, and a heavy metal plate (or blade) to smooth, push, or carry rocks, sand, soil, or debris. They are the most frequently used earth-moving machines on construction projects today and are essential to quarrying, mining, constructing roads and buildings, and demolition.

Bulldozers were critical to the 1944 Allied invasion of Europe. British armored bulldozers cleared beaches and roads and filled in bomb craters. Some tanks were converted into bulldozers by removing the turrets and adding bulldozer blades. Today, armored bulldozers are the mainstay of combat engineers around the world, who use them for constructing earthworks, removing obstacles, clearing mines, and demolishing structures. Bulldozers are also important tools for responding to natural disasters such as earthquakes. A bulldozer can clear a collapsed high-rise building in a few days. Without bulldozers to clear rubble, a city hit by a major earthquake might never recover. **ES**

SEE ALSO: BACKHOE (JCB), CAST IRON PLOW, MOLDBOARD PLOW, STEEL PLOW, THRESHING MACHINE, TRACTOR

⬆ *Armored bulldozers were used during a battle at Ktima, Cyprus, in March 1964.*

Autogyro (1923)

De la Cierva paves the way for vertical flight.

In 1923 Juan de la Cierva (1895–1936) pioneered the first autogyro. These machines appear superficially similar to helicopters, but with a single unpowered rotor. Early autogyros were less maneuverable than helicopters and were unable to take off or descend vertically. The invention of the autogyro predated the helicopter and so paved the way for vertical flight. Autogyro rotors are not powered, unlike those of a helicopter, and thus work in a similar way to spinning "helicopter" seed pods such as those of the box elder tree, *Acer negundo*. These seeds are aerodynamically shaped to spin as they fall, allowing the seed to disperse much further; autogyro rotors autorotate in the same way.

The power or thrust of the autogyro comes from a powered propeller (or in later designs a jet engine) meaning that most do require some takeoff runway, but normally only tens of feet. As they can land in an equally small space, the autogyro had distinct advantages over airplanes as they are much more maneuverable and stable flying at low speeds, but can also fly faster than helicopters. The autogyro is also unable to "stall" in mid-air, making it considerably safer than other aircraft.

De la Cierva started developing his ideas for the autogyro around 1920, motivated partially by the frequent crashes that fixed-wing aircraft often suffered—especially at low speeds. The first successful autogyro flight was in 1923, in a machine called the C4. This was not a perfect machine by any means, but each time it stalled or there was a problem in mid-air, it was able to glide slowly back to earth on its autorotating blades. **JB**

SEE ALSO: AIRSHIP, GLIDER, HELICOPTER, HOT-AIR BALLOON, KITE, POWERED AIRPLANE, SPACE PROBE

⬆ *People inspect F. T. Courtney's autogyro at Farnborough Airfield, England, in 1925.*

Power Steering (1923)

Davis and Jessup banish heavy steering.

Power steering reduces the effort required to steer a car by using an external power source to assist in turning the wheels. The system was developed in the 1920s by Francis W. Davis and George Jessup in Waltham, Massachusetts.

Davis was the chief engineer of the truck division of the Pierce Arrow Motor Car Company, and saw first-hand how hard it was to steer heavy vehicles. He quit his job and got work developing the hydraulic steering system that led to power steering. Chrysler introduced the first commercially available power steering system on its 1951 Imperial, under the name "Hydraguide."

Most power-steering systems work by using a belt-driven pump to provide hydraulic pressure to the system. This pressure is generated by a rotary-vane pump driven by the vehicle's engine. As the speed of the engine increases, the pressure in the hydraulic fluid also increases, so a relief valve is needed to allow excess pressure to be bled away.

When the power steering is not being used, for example when driving in a straight line, twin hydraulic lines provide equal pressure to both sides of the steering wheel gear. When the wheel is turned, the hydraulic lines provide unequal pressures and hence assist in turning the wheels in the intended direction.

Electric power steering systems are now starting to replace hydraulic ones. In this system, sensors detect the motion and torque of the steering column and a computer applies assisted power via electric motors. This allows varying amounts of assistance to be applied depending on driving conditions—more at low speed and less at high speed. Electric systems do not require engine power to operate, so are an estimated 3 percent more fuel efficient than the original hydraulic system. **BO**

SEE ALSO: MOTORCAR, STEERING WHEEL

Wind Tunnel (1923)

Munk invents realistic wind simulation.

Early airplane engineers based their flying machines on the flight of birds. It soon became clear, however, that this method was limited. When a bird is in flight, air flows over its wings, and engineers realized that the flow of air over an airplane's wings would need to be simulated in order to uncover the secrets of flight.

Early simulation methods included the whirling arm in which a wing was attached to a pole and rotated. Shortly after, Frank Wenham (1824–1908) designed a crude wind tunnel in which a fan channeled air down a tube. This produced a controlled airflow, and harnessing this led to the first variable-

> *"Technical progress is made by integration, not differentiation."*

Max M. Munk, physicist and mathematician

density wind tunnel of Max Munk (1890–1986).

Munk moved from Germany to the United States in 1920 to work for the National Advisory Committee for Aeronautics. He decided to improve the modeling of air flow in early wind tunnel designs so they could recreate the conditions experienced by a full-sized plane at high altitude. The breakthrough came with his idea of increasing the density of the air in the tunnel by compressing it. This laid the groundwork for the first variable-density closed-circuit wind tunnel, which went into operation in 1923 and revolutionized aircraft and automobile design. **RB**

SEE ALSO: GLIDER, POWERED AIRPLANE

An engineer tests a Flying Flea aircraft in a wind tunnel at Farnborough, England, in 1936.

Xenon Flash Lamp (1923)

Edgerton illuminates photography.

Xenon flash lamps were pioneered by Harold Edgerton (1903–1990), a professor of electrical engineering at the Massachusetts Institute of Technology (MIT). He was an expert on the stroboscope as well as a creative photographer whose freeze-frame images are world famous.

Xenon flash lamps, which are commonly used in photographic strobe lights, can produce an intense flash of light that lasts for between a millionth and a thousandth of a second depending on the volume and amount of gas in each lamp. The flashes can also be repeated up to a few hundred times per second. At a pressure of between 1 and 10 percent of an atmosphere, the xenon is contained in a sealed tube, usually of fused quartz. The discharge is triggered by switching a radio frequency high-voltage on to a small cathode, thus ionizing the gas. Then a short, thousand ampere pulse of current (from a charged capacitor) is passed through the tube, exciting the xenon atoms. These decay immediately, emitting the flash of light.

Xenon is extremely efficient at converting electrical energy into visual radiation. The length of the flash is governed by the distance between the electrodes. Careful design of the electrode shapes and composition can ensure that the energy output of each flash is the same. The xenon flash produces a broad continuum spectrum topped with a host of spectral lines, but as these have wavelengths all through the visual spectrum, the light appears white to the human eye. Krypton can be used if more infrared radiation is required. **DH**

SEE ALSO: ARC LAMP, ARGAND LAMP, FLUORESCENT LAMP, HALOGEN LAMP, INCANDESCENT LIGHT BULB, NEON LAMP

"Life is not significant details, illuminated by a flash, fixed forever. Photographs are."

Susan Sontag, writer and critic

⬉ *Edgerton's strobe lamp was designed to produce an intense flash of light as the camera's shutter opened.*

Iconoscope (1923)

Zworykin paves the way toward television.

Though seen as a thoroughly American invention, television's roots are Russian. Vladimir Zworykin (1889–1982) studied electrical engineering at the St. Petersburg State Institute of Technology in St. Petersburg. Boris Rosing, a professor in charge of laboratory projects, tutored Zworykin and introduced his student to his experiments of transmitting pictures by wire. Zworykin and Rosing went on to develop a very rudimentary television system in the early 1900s.

Russia's revolution split up the duo and halted their research. Rosing died in exile, but his student, Zworykin, settled in the United States and continued their research as an employee of the Westinghouse Electric Corporation. By 1923 Zworykin had developed the first all-electric camera tube, which he called an "iconoscope" meaning "viewer of icons." The iconoscope was a modified cathode ray tube, a device developed by William Crookes.

Zworykin's device did not impress his supervisors at Westinghouse, but in his free time he developed a better iconoscope (the "kinescope") and in 1929 showed off his all-electric television to a radio engineer convention. David Sarnoff, an executive at the Radio Corporation of America (RCA), saw potential in television and set Zworykin to work as Director of RCA's Electronic Research Lab.

It would be twenty years before televisions were common to American households. Along the way there was enough drama—lawsuits, blown budgets, buy-outs, and World War II—to fill a television show. It all began, however, with a persistent inventor and the iconoscope he created. **RBk**

SEE ALSO: COLOR TELEVISION, CABLE TELEVISION, TELEVISION REMOTE CONTROL

⬀ *This "Emitron" camera tube developed for the BBC in 1935 is similar in design to Zworykin's iconoscope.*

"Television will be of no importance in your lifetime or mine."

Bertrand Russell, British philosopher, 1948

Fast-frozen Food (1924)

Birdseye freezes foods more quickly to preserve them in optimum condition.

In 1912 Clarence Birdseye (1886–1956) was working as a field biologist in northern Canada when he was taught by Inuit people how to preserve fish by freezing under very thick ice at around -40°F (-40°C). Frozen almost instantly, it tasted fresh when thawed days or weeks later. Birdseye realized that this fish was fresher than that sold in the fish markets in New York, which had also been frozen but slowly and at higher temperatures. He also saw that when food was frozen quickly, only small ice crystals formed in the cells, causing less damage to the texture. Frozen food will keep for several months, so long as it is stored at a constant temperature no higher than -0.4°F (-18°C).

Birdseye joined the Clothel Refrigerating Company in 1922 where he worked on freezing techniques. He set up his own company, Birdseye Seafoods Inc., in 1923, to supply fish fillets that had been frozen to -45.4°F (-43°C) using chilled air. The company failed to sell enough, however, and was bankrupt only a year later. Undeterred, in 1924 Birdseye patented a new process for quick freezing. Fish was packed into wax cartons and frozen quickly, under pressure.

Birdseye set up another company and turned his attention to marketing. His double-belt freezer, patented in 1925, used brine to chill a pair of stainless steel belts carrying the fish so that it froze extremely quickly. Meat and vegetables could be prepared and frozen using the same technique. Birdseye sold his company and its many patents in 1929 for $22 million to Goldman Sachs and the Postum Company, which became the General Foods Corporation and introduced the Birds Eye brand to the world. **EH**

"One of the marvels of the age is the new quick freeze process … time ceases to exist for foods."

Better Homes and Gardens, September 1930

SEE ALSO: FREON, ICE-MAKING MACHINE, REFRIGERATOR

◩ *Birdseye experiments with chopped carrots in 1943 to identify ways of improving the dehydration process.*

Paper Tissues (1924)

Kimberly-Clark creates paper hankies.

While the need to wipe your nose is as old as the need to sneeze, it took a surprisingly long time to come up with a solution that disposed of nasal mucus politely and hygienically. Although a seemingly simple solution, the key features of facial paper tissues are that they are cheap, soft, disposable, and absorbent, especially in comparison to other similar types of paper products. Even toilet paper is not as effective, being designed to break down in water.

In fact when the Kimberly-Clark Corporation first developed the material that would later make them household names, it was to use as bandages in World War I. Cellucotton, as it was called, was made from processed wood pulp and was five times as absorbent and half as expensive as cotton. As a result of army nurses using cellucotton as disposable sanitary pads, Kimberly-Clark introduced the first disposable feminine hygiene product in 1920. But it was not until 1924 that they developed the first facial tissues, under the brand name Kleenex. Even then they designed Kleenex as a means of removing face cream and other make-up. It was only in the late 1920s that they realized that the tissues were being used as disposable handkerchiefs, and quickly changed their marketing strategy.

The name Kleenex soon became synonymous with disposable facial tissues. In 1928, the characteristic carton box with a perforated opening was introduced, soon followed by colored and printed pattern tissues. Pocket packs were introduced in 1932. While the paper tissue has largely replaced the traditional cotton handkerchief in most industrialized countries, concerns have been expressed about their environmental impact, resulting in the more recent development of recycled and chlorine-free tissues. **RBd**

SEE ALSO: PAPER, TOILET PAPER, COTTON BUDS, TAMPON WITH APPLICATOR

Ultracentrifuge (1924)

Svedberg speeds up the centrifuge.

The centrifuge has been around since the mid-1800s. It is a device for separating a precipitate from a solution by spinning it at high speeds (the increased gravitational force means that anything suspended in the solution is forced to the bottom of the container much faster than if left to settle naturally). The first centrifuges were hand-driven, reaching speeds of 800 revolutions per minute (rpm). The first ultracentrifuge, developed by Swedish chemist Theodor Svedberg (1884–1971) in 1924, could rotate at speeds of up to 160,000 rpm, meaning it was capable of exaggerating gravity 1,100,000 times. In his first tests, Svedberg separated hemoglobin from blood in about six hours; using normal gravity it would have taken 180 years.

> *"Our age is ... a practical one. It demands of us all clear and tangible results of our work."*
>
> Theodor Svedberg, chemist

The device's speed makes it a powerful analytical tool. It can measure a molecule's weight, size, shape, and density. It measures the movement of molecules at a certain rotational speed and this "sedimentation rate" is used to calculate the Svedberg unit (the velocity of the molecule per unit of gravitational field); this is used to work out the molecule's weight.

Svedberg used his new technique to support the theories of Brownian motion put forward by Einstein and Von Smoluchowski, winning him the Nobel Prize in 1926. Nowadays, the ultracentrifuge proves useful for analyzing large molecules while still in solutions similar to those they naturally occur in. **JM**

SEE ALSO: CENTRIFUGAL PUMP, CENTRIFUGAL CLUTCH

The years just before and during World War II saw a rise in military inventions such as ballistic missiles, cluster bombs, and the Enigma machine. At the same time, the preservation of life gained even more importance, and the discovery of penicillin served to counter the losses sustained during war. Meanwhile, the invention of the helicopter and the supersonic airplane cut travel times between distant lands, and 35-mm cameras, vinyl records, and television offered light relief in difficult times.

◄ A line of nuclear reactor fuel rods signals the atomic age.

WAR *and* PEACE

1925 to 1951

35-mm Camera

(1925)

Barnack ushers in the age of photojournalism with a highly responsive camera.

Not everyone has their own award named after them. The Oskar Barnack award, given annually to photo journalists, was initiated in 1979 to mark the hundredth anniversary of the birth of the man who invented the 35-mm still camera. Barnack (1879–1936) had the idea for it back in 1905, but it was not until 1913–1914, while he was working as head of development at the German camera company Leitz, in Wetzlar, Hesse, that he was able to transform his idea into reality.

Traditional heavy plate cameras were cumbersome to use and required significant preparation before each shot. It was impossible to take a "quick snap" of anything. Barnack's camera was a tough metal box that could fit in a jacket pocket and used a new kind of film, adapted from Thomas Edison's 35-mm cine film. In 1914 Barnack took a picture of a soldier who had just put up the Imperial Order for mobilization. This was a new kind of picture—spontaneous and capturing a moment in history. Barnack had held up a strip of his new camera film and stretched his arms out. The length of film between his arms contained thirty-six frames, and this has been the number of negatives on a standard 35-mm roll of film ever since.

World War I put a halt to Barnack's progress, and it was not until 1925 that the Leica 1 camera was introduced (the name standing for **Lei**tz **ca**mera). According to one historian, old-school photographers regarded the new camera as toylike, but over the next seven years almost 60,000 of them were sold. **DHk**

SEE ALSO: PHOTOGRAPHY, SINGLE-LENS REFLEX CAMERA, PORTABLE "BROWNIE" CAMERA, DIGITAL CAMERA

↖ *The first commercially available 35-mm camera was the Leica I, manufactured by Leitz of Germany in 1925.*

← *This 1937 advertisement shows a model of the Leica III series, which included an integrated rangefinder.*

Fluidized Bed Reactor
(1925)

Winkler advances catalytic cracking.

In 1925 Fritz Winkler (1888–1950) patented a chemical reactor in which large particles could be "fluidized" by forcing an upward current of gas through the solid. Winkler used his reactor to extract gas from lignite, which was then piped directly to engines for compressing ammonia.

Catalytic cracking is a process in which the heavier molecules obtained from petroleum deposits are broken down into the more useful lighter molecules, such as gasoline, by heating them in the presence of a catalyst. The catalyst remains active for a short time and then becomes inactive as a layer known as coke is deposited on the surface. Catalysts can be removed

> *"[In] the history of fluid beds, the … speed of certain developments signal their intrinsic rightness."*

A. M. Squires, M. Kwauk, and A. A. Avidan

and regenerated by heating them in air. This takes a relatively long time and there is the problem of removing them without disturbing the products of the reaction.

Fluidized bed reactors were the ideal solution. A gas is forced up through the catalyst, enabling it to take on liquid properties, including running off for regeneration before being put back into the reaction mixture. Adapting Winkler's design to work with fine powders, the first U.S. reactor was installed in Baton Rouge's Standard Oil Refinery in 1940. Fluidized beds are still in use in the fuel industry and are also commonly used in the manufacture of polymers such as PVC and rubber. **HP**

SEE ALSO: OIL REFINERY, SYNTHETIC RUBBER, NEOPRENE

Aerosol
(1926)

Rotheim invents the spray canister.

The innocuous spray can has been the subject of some controversial press coverage over the years. Its use as a cheap means of getting "high," termed solvent abuse, involves inhaling the fumes of certain aerosols to achieve an effect a little like being drunk. The aerosol can is also popularly used to spray-paint graffiti and, during the 1970s, the increasing awareness that chlorofluorocarbons were causing damage to the ozone layer led to the use of CFCs being phased out in agreement with the Montreal Protocol. Despite the negative aspects of the product, the aerosol spray can is a ubiquitous invention that enjoys widespread use around the world today.

Aerosols date back to the late 1800s, with metal spray cans being tested as early as 1862. The real breakthrough in aerosol technology did not come until 1926 when Norwegian chemical engineer Erik Rotheim (1898–1938) discovered that a payload could be mixed with a propellant in a pressurized canister, allowing the product to be sprayed through a small directional nozzle.

The key to the aerosol is in the propellant. The gas used is stored in the can as a pressurized liquid vapor, at a temperature just below its natural boiling point (usually slightly below normal room temperature). When the liquid is released through the nozzle, it evaporates and turns into a gas, thus propelling the product in the form of a spray.

Despite its detractors, Rotheim's invention has brought improvements and greater convenience in many areas of activity, from spray paints that produce a satisfyingly even coat on refurbished cars to the improbable but appetizing arabesques of spray-administered whipped cream topping. **BG**

SEE ALSO: HAIR SPRAY

Enigma Machine (1926)

Scherbius's reflector-assisted encryption apparatus eclipses previous rotor machines.

In 1926 the German Army adopted a supposedly impenetrable electro-mechanical encryption device that eventually proved their undoing in World War II. Eight years earlier, electrical engineer Arthur Scherbius (1878–1929) had developed the first model of the Enigma machine, a bulky contraption incorporating a full-sized typewriter and three rotors to code messages. When typing a letter, the first one of these electrical discs rotated and caused the next one to do likewise, similar to the wheels in an odometer. Wires connecting the rotors provided an electrical path from the keys on the typewriter to the output end plate, with the various connections in between ensuring that the final product of the plaintext input was ciphered.

However, the army deemed Scherbius's initial 110-pound (50 kg) Enigma A and B to be insufficiently secure, and hence did not acquire them. To guarantee that messages sent from the machine were not decipherable, the inventor created a safety device for his third model that distinguished his by now smaller and lighter apparatus from all the other rotor encryption machines of the time: a reflector. Stuck behind the last rotor, the reflector gathered pairs of signals sent from the keyboard and then returned the currents as single, unified signals through the rotors.

The Nazis considered the reflector-assisted Enigma—whose messages were decrypted in the same way they were encrypted—undecipherable. However, the flaw of the reflector was that the machine's output was never the same as the input, that is, "a" could never be encrypted as "a." When British cryptologists became aware of this fact, and identified recurrent use of standard German phrases (such as "Heil Hitler") in the messages, they were able to decipher the codes and turn the war around. **DaH**

"The science of cryptography is very elegant . . . the ends for which it's used is less elegant."

James Sanborn, sculptor

⤒ *Introduced on Atlantic U-boats in early 1942, the MK 4 Enigma's code was broken in December that year.*

⤳ *German signal troops communicate by using an Enigma encoding machine during World War II.*

SEE ALSO: ODOMETER, WIRELESS COMMUNICATION, SONAR, RADAR, PUBLIC KEY CRYPTOGRAPHY

Fluorescent Lamp (1926)

Germer develops a cooler and more efficient light bulb.

If one individual can be said to have "invented" the fluorescent light, it was probably Edmund Germer (1901–1987), although many scientists and inventors, including Heinrich Geissler and Thomas Edison, contributed to its creation.

A fluorescent bulb works by passing an electric current through mercury vapor in a low-pressure inert gas. This causes the electrons in the atoms to "move," releasing ultraviolet light. Unfortunately we cannot see ultraviolet, so the inside of the tube is coated in phosphorous powder, which absorbs the ultraviolet light and re-emits it as visible light. This elaborate process leads to lights that are cooler and more efficient than incandescent ones.

Possibly because its process of manufacture was more complex than that of incandescent bulbs, the commercial use of the fluorescent bulb was not really realized until the 1920s, despite much experimentation throughout the first decades of the 1900s. In 1926,

Jacques Risler was granted a patent for a neon-based tube with a fluorescent coating, mainly used for advertising purposes, but apart from this the flourescent bulb languished in obscurity.

The patent received by Germer, Meyer, and Spanner was for a high-pressure mercury vapor lamp. Although this lamp did not go into production, it was similar to a patent application filed by General Electric, which was also trying to develop a fluorescent lamp at the time. After some legal wrangling, General Electric paid off Germer et al, and acquired the patent, thus becoming the dominant manufacturing force for fluorescent tubes. **BG**

SEE ALSO: OIL LAMP, GAS LIGHTING, ARC LAMP, INCANDESCENT LIGHT BULB, NEON LAMP, HALOGEN LAMP

⬆ *A circular tube lamp with magnifier (1967); a battery-operated hand lamp (1966); and a bench lamp (1957).*

Television (1926)

Farnsworth demonstrates an all-electronic television system.

A television system, by definition, transmits and receives live, moving half-tone images. Early versions, such as those invented by John Logie Baird in the 1920s, used crude, electromechanical, spinning, perforated, scanning discs to record and subsequently produce the images. The first transatlantic images were transmitted with this system in 1928.

Television relies on the fact that the human brain can convert a sequence of slightly different still images into a moving picture if more than fifteen frames are received every second. As soon as the number drops below fifteen, the motion looks jerky.

Today's televisions are a product of the invention of the cathode ray tube. This is coated with a phosphor that glows when an electron beam hits it. Behind the phosphor is a shadow mask that divides the image into picture elements (pixels). Television sets typically have 525 lines down the screen and these are raster scanned every sixtieth of a second. The scanning is interlaced so that odd-numbered lines are "painted" on one scan and even-numbered lines on the next.

In 1926 Philo Farnsworth (1906–1971) developed the world's first all-electronic system, where, like today, the cameras scan electronically and the television receiver is scanned electronically, too. By 1936 the BBC was producing 405-line, high-resolution images using this system. By 1949, ten million monochrome televisions had been sold in the United States, and now the average American spends between two and five hours per day "glued to the tube." The breakthrough in U.K. television watching was the broadcast of the coronation of Queen Elizabeth II in June 1953. **DH**

SEE ALSO: CATHODE RAY TUBE, IMAGE RASTERIZATION, COLOR TELEVISION, TELEVISION REMOTE CONTROL, PLASMA SCREEN

⬆ *Baird's basic mechanical television apparatus used thirty lenses on a disc that reflected light to a receiver.*

PVC (1926)

Semon forms a versatile plastic product.

Since it was first made, polyvinyl chloride (PVC) has been one of the most widely used plastics. The molecule was first discovered by accident in the 1800s by two chemists, each of whom left flasks of chemicals out in the sun and noticed that a white solid formed in them. However, it was not until 1926 that Waldo Semon (1898–1999) made the crucial breakthrough. He blended this material with additives, known as plasticizers, so that a useful product was formed, and the plasticized PVC, often shortened to "vinyl," was born.

Until this point, people had not seen any value in the product—in Semon's own words, "People thought of PVC as worthless back then. They'd throw it in the trash." The new plastic quickly became used for a

---

"The four building blocks of the universe are fire, water, gravel, and vinyl."

Dave Barry, author and humorist

---

myriad of different applications and is now the second most produced plastic in the world, generating billions of dollars of revenue each year.

First used as the soles of shoes and a coating for wires and tool handles, applications of this cheap-to-produce material soon became more widespread, particularly in the building industry, where it is used in everything from guttering and wiring insulation to window frames, floor tiles, and fencing. It also has innumerable other uses in everyday items, including food packaging, toys, car dashboards, vinyl records, and even clothing, being one of the fabrics of choice for pop stars, superheroes, and fetishists alike. **SB**

SEE ALSO: POLYSTYRENE, POLYPROPYLENE

Iron Lung (1927)

Drinker and Shaw devise a tank respirator.

Doctors treating polio patients found that while many sufferers were unable to breathe in the acute stage, when the action of the virus paralyzed muscles in the chest, those who survived this stage usually recovered completely. Such observations indicated the need to develop strategies to maintain respiration until the patient could breathe independently again.

In 1927, chemical engineers Philip Drinker (1894–1972) and Louis Agassiz Shaw, from Harvard University, devised a tank respirator to maintain respiration. In the device, the patient's head stuck out of the end of the tank, with a sponge rubber seal to make it airtight. Air was then pumped from the tank to produce negative pressure causing the chest to expand and thus produce breathing.

The first iron lung was installed in 1927 at Bellevue Hospital, New York, and in 1928 the first patient was an eight-year-old girl with polio, comatosed from lack of oxygen. One minute after the device was switched on, she regained consciousness and asked for ice cream. Further refinements included a garage mechanic's "creeper" that allowed patients to slide out of the tank, and boat "portholes" through which medical staff provided treatment. Equipment designer John Haven Emerson produced an iron lung that could vary the respiration rate, with the added advantage that it cost half as much to manufacture.

The iron lung helped to save thousands of lives during the polio outbreaks of the 1940s and 1950s. In 1959, 1,200 people were using tank respirators in the United States, but with the advent of the polio vaccine this figure had fallen to thirty by 2004. **JF**

SEE ALSO: HEART-LUNG MACHINE, VENTILATOR, ARTIFICIAL HEART

▶ *Drinker with his "life machine" in 1928, the year the iron lung was first used to resuscitate a person.*

Chain Saw

(1927)

Stihl designs the first electric saw.

The roar of a chain saw is a sound that is hard to separate from images of destruction and violence. Horror movies and wildlife documentaries have taught us that chain saws are, in general, a "bad thing." For maintenance workers and lumberjacks in the 1920s, the invention of the chain saw was undoubtedly a blessing.

It is possible that early chain saws were in use before World War I, although there is little solid evidence for this and it was not until World War II that Andreas Stihl's (1896–1973) "hand held mobile chain saw," invented in 1927, came into its own. German troops used the saws for making quick progress through wooded areas. When the Allies caught wind of this, they promptly dropped a bomb on the Germans' chain saw factory—but not before stealing a saw for themselves so that they could copy it.

Stihl's original chain saw was equal in weight to an average teenage boy and required two people to wield it. Having studied the stolen German chain saw, a U.S. company called Mercury developed an improved version that could be held by a single person. Stihl's company fought back and, by the 1970s, claimed to be the world's largest manufacturer of the machines.

In recent decades, chain saws have helped to bring about the destruction of some of the most diverse areas of life on Earth. The BBC series "Life of Birds" famously, and tragically, shows the lyre bird mimicking the sounds of chain saws destroying its natural habitat. As chain saws rip through the Amazonian rain forest—not to mention practically every other tree-studded landscape on the planet—it is hard to find sympathy with the loggers. **HB**

SEE ALSO: SAW, CIRCULAR SAW, BAND SAW

Anodized Aluminum

(1927)

Gower and O'Brien develop a tough coating.

A rusty automobile is a shame. A rusty piece of aluminum, however, is not only desirable, it is anodized. When exposed to oxygen, pure aluminum metal builds up a layer of aluminum oxide. The aluminum oxide has a significantly greater resistance to corrosion and abrasion and consequently serves as a sturdy shell to protect the rest of the aluminum.

Anodizing is a process right out of a mad scientist movie. In 1927 Charles Gower and Stafford O'Brien patented a sulfuric acid anodizing process that is now the most common way to anodize aluminum. The aluminum is first immersed in electrified sulfuric acid. Electric charges cause oxygen to build up on the

> *"The oxide on aluminum is naturally corrosion resistant, an insulator and very tenacious."*
>
> Mario S. Pennisi, consultant

surface of the aluminum, creating a thick coat of aluminum oxide. Next, the aluminum can be easily colored and used in countless applications. The metal's new coat is porous and thus can have additives like coloring dyes or lubricity aids easily infused into it. Finally, the aluminum is sealed. Sealing closes up holes at the surface of the coat and helps to reduce any color loss or scratching.

Anodized aluminum's protective oxide finish is one of the hardest naturally occurring substances. Applications include MP3 players, appliances, satellites, computer hardware, and buildings like the Sears Tower. **LW**

SEE ALSO: HALL-HÉROULT PROCESS, ALUMINUM FOIL

Electric Razor

(1928)

Schick frees shavers from the risks of the wet razor.

Shaving without soap or water had been the dream of men (and probably some women) for centuries. The traditional "cutthroat" razor had given way to the safety razor in the late nineteenth century, but shaving was still a wet, time-consuming, and delicate operation.

The electric, or dry, razor was patented by the U.S. inventor Jacob Schick (1878–1937) in 1928. Having dabbled with some very unwieldy devices, powered directly by household electricity and large external motors, Schick's most successful innovation was finding a way to house a small but powerful electrical motor inside a handheld shell. The motor drove a sharp, sliding cutter capable of slicing through a beard. All the parts were contained neatly within a Bakelite case.

After a slow start, the first successful Schick electric razor appeared in 1931. As the design improved, sales took off. Although it was doubtful whether the electric razor provided a closer shave than the safety razor, it had clear advantages with no need for water, soap, or cream and liberating the user from the sink.

Later developments included the introduction of a foiled shearing head by the Remington company in 1937 and rotating cutters invented by Alexandre Horowitz of the Philips Laboratories in the Netherlands. The first "Philishave" rotary electric razor appeared in 1939. Battery-powered, cordless razors were introduced in the late 1940s, followed by Remington's first "Lady Shaver" in 1947. The rechargeable battery-powered razor appeared in 1960. Ironically, some recent electric razors have been designed for use with shaving cream. **RBd**

SEE ALSO: SAFETY RAZOR, ELECTRIC RAZOR, BAKELITE

↗ This Schick electric razor from 1934 requires 110 volts. The Schick manufacturing plant closed in 1979.

"Many a man began to wonder how he had got along without one."

Time magazine

Bathysphere (1928)

Barton designs a deep-sea diving vessel.

Otis Barton's (1899–1992) famous marine exploration vehicle was reminiscent of a naval mine. A simple sphere made of steel, it once dangled from a 2-mile (3.5 km) cable deep into the ocean. Unlike a mine, however, the bathysphere was intended to hold two intrepid explorers, and at less than 5 feet (1.5 m) across, it was not exactly designed for comfort.

In 1928, Barton was just a student, but his blueprint for the bathysphere caught the attention of scientist and explorer William Beebe (1877–1962). A collector of rare species, Beebe had already consulted several professional engineers about building a deep-sea diving vessel. He was impressed by Barton's design and started making plans for a test dive.

"The longer we were in it, the smaller it seemed to get...."

William Beebe, deep-sea explorer

The secret to the bathysphere's success was its simplicity—its spherical form meant that pressure was evenly distributed across its surface. Underwater, pressure rapidly increases with depth. At just 328 feet (100 m) below the surface, pressure is equivalent to 142 pounds per square inch (10 kg/cm²). In 1934, after several previous descents, Barton and Beebe plummeted a staggering 2,950 feet (900 m) into the North Atlantic Ocean near Bermuda, breaking all previous records for manned submersibles. **HB**

SEE ALSO: SUBMERSIBLE CRAFT, SUBMARINE

⬅ *Inside the bathysphere, Barton and Beebe await release after their record-breaking dive in 1934.*

Freon (1928)

Midgley introduces CFCs in cooling systems.

Chlorofluorocarbons (CFCs) were for decades commonly used in the cooling systems of refrigerators and in aerosol cans. However, when these usually inert compounds get zapped by radiation in the upper atmosphere, they are energized and produce chlorine radicals that react with ozone. This has led to a significant depletion in the ozone layer, notably at the North and South Poles. The world was on the brink of losing its natural protection from harmful ultraviolet (UV) rays, mostly thanks to a household appliance used worldwide.

Ironically, CFCs started out as savior compounds, replacing nasty substances such as ammonia and sulfur dioxide in refrigeration equipment. Faulty units would cause unpleasant illnesses and even deaths. The American Thomas Midgley (1889–1944) first proposed CFCs as refrigerants in 1928, demonstrating their suitability with theatrical flourish—he proved their lack of reactivity and toxicity by inhaling the gas and then exhaling onto a lit candle. Developed with the help of Charles Kettering (1876–1958), the new compound was dubbed "Freon," a trademark held by the DuPont company.

Midgley's wonder compound powered a revolution in home refrigeration, and before long nearly every home in the Western world was able to store and enjoy chilled and frozen foods in safety. But the death knell rang in 1974, when research highlighted the harmful nature of CFCs. Restrictions were quickly put in place, and the 1987 Montreal Protocol implemented the gradual elimination of CFCs and similar gases. Although the quantities of CFCs in the atmosphere are now measurably lower, it will take many decades for the ozone layer to fully recover. **MB**

SEE ALSO: ICE-MAKING MACHINE, HABER PROCESS
(MANUFACTURE OF AMMONIA), REFRIGERATOR, FROZEN FOOD

Automatic Bread Slicer
(1928)

Rohwedder's machine slices and wraps bread.

U.S. inventor Otto Frederick Rohwedder (1880–1960) started working on the design of an automatic bread slicer in around 1912 and developed several prototypes, including one that held a sliced loaf together with metal pins. These early designs were not successful and Rohwedder faced a major setback in 1917 when his designs were destroyed in a fire at the factory at Monmouth, Illinois, that had agreed to build the first slicing machines.

Rohwedder, having earlier trained as a jeweler, was employed by a security firm while working on the development phase of his invention in his spare time. He continued improving his designs and realized that the main challenge he faced was keeping the bread fresh, because after slicing the loaf went stale more quickly. By 1927 he had devised a machine that both sliced and wrapped the bread. The timing for the launch of the bread slicer was good: the pop-up toaster, invented in Britain in 1919 by Charles Strite, was just becoming popular in the United States.

In 1928, the first of Rohwedder's bread-slicing and wrapping machines was installed at the Chillicothe Baking Co., which started to sell Kleen Maid Sliced Bread. Their customers loved sliced bread and demand for the machine from other bakeries had Rohwedder's production unit struggling to keep pace with orders. He established the Mac-Roh Manufacturing plant at Davenport in 1929 but was forced to sell the plant and his patents during the Great Depression.

Demand for sliced bread continued to grow, however, and by 1933 bakeries in the United States were selling a greater quantity of sliced than unsliced bread. Today, approximately 80 percent of all bread is sold sliced. **EH**

SEE ALSO: SANDWICH, POP-UP TOASTER

Audiotape Recording
(1928)

Pfleumer pioneers the use of magnetic tape.

At the end of the nineteenth century, Valdemar Poulsen developed the telegraphone as a means of recording sound on a magnetized wire. However, the sound quality of these machines was poor, and the wire itself was usually built into the machine, making it of little use for long-term audio storage. A breakthrough came in 1928 when German engineer Dr. Fritz Pfleumer (1881–1945) successfully fixed magnetic powder to a thin strip of paper. This was then able to record magnetic signals more effectively than magnetic wire.

In 1930, the AEG company of Berlin began work on the magnetophone, an audio recorder that would make use of the Pfleumer principle. To develop the

> *"Tape recording in your basement or bedroom used to be a freak thing. Now anybody can do it."*
>
> Les Paul, musician

tape itself, it collaborated with another illustrious name in German electronics, BASF, which used its expertise in plastics to create a new type of recording medium. The system BASF developed used a narrow band of cellulose acetate coated with a lacquer of iron oxide, thus enabling the tape to be magnetized in the recording process. The AEG K1 magnetophone and the BASF magnetic tape were exhibited together at the 1935 Berlin Radio Fair.

The AEG K1 was used extensively by the Nazis during World War II. After the war, a number of models were shipped to the United States, where they formed the basis of a revolution in audio recording. **TB**

SEE ALSO: MAGNETIC RECORDING (TELEGRAPHONE), AUDIOTAPE CASSETTE, EIGHT-TRACK AUDIOTAPE

Color Television

(1928)

Baird adds two electron beams to create colored moving images.

Color television is possible because the human brain can convert a grid of differently colored dots (usually known as pixels, short for picture elements) into a complete color image. Color television cathode ray tubes have three electron beams, as opposed to the single electron beam in a black and white TV. The screen is coated with red, green, and blue phosphor dots placed behind the holes of the tube's shadow mask. All the observed colors are combinations of the red, green, and blue signals (that is, if all dots are firing, the image appears white).

The fact that color television is essentially only three times more complicated than black and white television means that the basic invention processes for the two devices almost took place simultaneously. John Logie Baird (1888–1946) is recognized as a leading pioneer in the development of television, alongside Philo Farnsworth, and in 1928 Baird first demonstrated the transmission of color images.

One of the early important commercial considerations was that the signals that transmitted the color pictures should not only result in color images on "color TV" sets, but should also be rendered as black and white images on mono sets. Development was slow, and sales of shadow-masked RCA cathode ray tube color TV sets only started in 1954. The sets were extremely expensive and typically cost more than $1,000. The 1961 Sunday evening transmission of "Walt Disney's Wonderful World of Color" encouraged U.S. citizens to buy color TVs. Only by 1972 were color TV sales exceeding black and white. **DH**

SEE ALSO: CATHODE RAY TUBE, IMAGE RASTERIZATION, TELEVISION, TELEVISION REMOTE CONTROL, PLASMA SCREEN

⤢ *The cover of the August 1928 issue of* Television, *the world's first TV magazine, featured Baird.*

"Television has proved that people will look at anything rather than each other."

Ann Landers, journalist

Barometric Altimeter (1928)

Kollsman improves the accuracy of the altimeter.

The relationship between atmospheric pressure and height above sea level was known by Blaise Pascal and Edmond Halley in the mid-seventeenth century: one can be used to measure the other. In 1862 James Glaisher used an aneroid barometer as a height instrument on a 7-mile (11 km) high balloon flight.

The barometric (or pressure) altimeter uses an aneroid barometer to register pressure change. The higher you go, the less air there is above pressing down on the instrument. Near sea level, under normal conditions, the pressure changes by one millibar (0.03 inches of mercury) for every 27 feet (8 m) of altitude. Alas, the pressure also changes with the weather and the temperature. In addition, the pressure–height relationship is not truly linear but exponential.

Paul Kollsman (1900–1982), a German mechanical engineer, emigrated to the United States in 1923. His watchword was "accuracy." Previous altimeters determined airplane heights to a few hundred feet.

This was of no problem in perfect weather and daylight, but made flying in fog or clouds or at night rather too adventurous. Kollsman's instruments, when correctly set, measured height to within a few feet.

In 1929 army flyer 1st Lieutenant Jimmy Doolittle used the barometric altimeter, together with ground-based radio navigation systems and gyroscopic artificial horizons and compasses, to fly 15 miles (24 km) "by the gauges." These successful "instrument flights" ensured that few of the world's planes flew without a Kollsman altimeter thereafter. The ground pressure at the airport a flight is approaching is still referred to as the Kollsman number. **DH**

SEE ALSO: GROUND PROXIMITY WARNING SYSTEM

⬆ *The instrument supplies a simultaneous pressure reading for the altitude at which the pilot is flying.*

Cardiac Pacemaker (1928)

Lidwell discovers how to maintain a healthy heartbeat with an electronic device.

A cardiac pacemaker is a surgically implanted electronic device that regulates a slow or erratic heartbeat. The first artificial pacemaker was invented by Dr. Mark Lidwell, an Australian anesthetist, who developed an external device running on an alternating current that required a needle to be inserted into the patient's upper heart chamber (ventricle). In 1928 Lidwell used the device to resuscitate a baby born in cardiac arrest at the Crown Street Women's Hospital in Sydney. Lidwell reported the case to the Third Congress of the Australian Medical Society in 1929, but kept a low profile due to controversy at the time surrounding research into artificially extending human life.

In 1932 U.S. physiologist Albert Hyman (1893–1972) independently developed an electromechanical instrument, powered by a spring-wound, hand-cranked motor, that he referred to as an "artificial pacemaker." The first internal pacemaker was developed by Swede Rune Elmqvist and implanted into a patient in 1958 at the Karolinska University Hospital, Sweden, by surgeon Åke Senningn. These early models were powered by mercury-zinc batteries, lasting only two or three years. In 1973 a lithium-iodide fuel cell was developed that could last for around six years.

Today more than 100,000 pacemakers are implanted each year in the United States. Modern pacemakers have sophisticated programming capabilities and are extremely compact. The device contains a pulse generator, circuitry programmed to monitor the heart rate and deliver stimulation, and a lithium iodide battery, with a life of seven to fifteen years. **JF**

SEE ALSO: STETHOSCOPE, ELECTROCARDIOGRAPH (EKG), DEFIBRILLATOR, HEART-LUNG MACHINE, ARTIFICIAL HEART

⬆ *This battery-powered cardiac pacemaker of 1958 was the first to operate entirely inside the patient.*

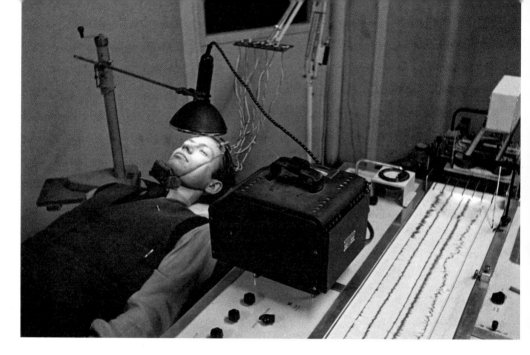

Electroencephalograph (EEG) (1929)

Berger records the electrical activity of the human brain.

We often think of thoughts as instantaneous, but in truth it stands to reason that they are limited by the speed of certain chemical reactions and electrical impulses in our brain. Given that these physical activities accompany thinking, it also stands to reason that if one looks hard enough, one should be able to measure the electrical activity, despite the seemingly fleeting nature of brain activity. The recording of these impulses—or electroencephalography—matured at something of a snail's pace until the work of Hans Berger (1873–1941).

In 1875, English physician Richard Caton figured out that he could measure brain activity in animals with a galvanometer. A Polish physician, Adolph Beck, also working with animals, advanced the topic further in the 1890s, going so far as to discover the location of some sensory impulses and noting a change in activity that took place with loud noises or bright light. The link between animal and human models was not well understood in this era, however, and it took a few more decades before anyone made this leap.

Hans Berger began experimenting with recording the electrical activity of the brain in the 1920s. Experimenting in secret, and oddly using lectures on telepathy as a cover, he refined his technique. Using his own son Klaus as a research subject, Berger recorded the first human electroencephalograph, or EEG, in 1924 and carried on his experiments for the next few years. Berger published his work in 1929 and revolutionized the world of neurology. EEG is used today to evaluate epilepsy, sleep disorders, and a host of other neurologic problems. **BN**

SEE ALSO: ELECTROCARDIOGRAPH (EKG), COMPUTED TOMOGRAPHY (CT OR CAT SCAN)

⬆ *An electroencephalograph is used to record the electrical activity within a volunteer's brain.*

Cyclotron (1929)

Lawrence accelerates particles to greater velocities than ever before.

It sounds rather like an exercise bicycle from the 1950s but the cyclotron is actually the grandfather of today's most powerful particle accelerators. Having originally studied chemistry, Ernest Lawrence (1901–1958) switched to physics and received his PhD from Yale University in 1925. At this time, scientific insights into the nature of matter were starting to yield interesting results. In Cambridge, England, Ernest Rutherford had been using atomic particles as projectiles with which to bombard atoms. By 1919 he had succeeded in bombarding the nucleus of a nitrogen atom and getting it to absorb a helium nucleus, creating oxygen.

This kind of work, however, was reaching a technical limit. The atomic particles from naturally radioactive materials were too few and did not have the energy required to pursue the experiments that Rutherford wanted to perform. In 1927 he issued a plea to physicists to find methods to produce a "copious supply" of high-energy particles.

Lawrence answered the call and, just two years later, in 1929, came up with the idea that would quickly become the cyclotron. He was inspired to experiment with a magnetic field that would force charged particles to travel in a circular trajectory. This would make the particles pass through the same accelerating magnetism over and over again. In 1931 his first model was ready for production. His first cyclotron was a relatively simple device, measuring just a few inches across, that could accelerate hydrogen ions to much higher energies than were achievable before by any method. The new field of high-energy physics had finally arrived. **DHk**

SEE ALSO: SYNCHROTRON, LARGE HADRON COLLIDER

⊡ *Students at Cambridge University, England, gather around the control board of a cyclotron in 1938.*

Antilock Braking System (ABS)

(1929)

Voisin designs a safer braking mechanism to prevent skidding.

The antilock braking system (ABS) was designed by Frenchman Gabriel Voisin (1880–1973), originally as a way of preventing planes from swerving on landing . He first installed a system to keep airplane brakes from locking up in 1920. Several decades later, after limited success with similar systems in cars, a breakthrough came in 1978, when Mercedes-Benz announced the installation of electronic ABS in its S-Class car.

Theoretically, ABS can stop a serious car accident from occurring by allowing a driver to maintain control in slippery conditions or during an emergency stop. On an icy road, a car's wheels can lock up, sending the vehicle into a spin. In old cars that did not have ABS, drivers had to try to pump the brakes to prevent this. With ABS, the brakes are automatically pumped—the system senses the change in conditions and alters the brake fluid pressure accordingly. Meanwhile, the driver keeps his or her foot firmly on the brake pedal. Unfortunately, there is not very much evidence to suggest that ABS reduces accidents on the roads, partly because many drivers do not realize how the system works or how to use it. Several studies have been carried out that suggest ABS has been of little practical use, one even concluding that ABS may actually increase instances of single-vehicle collisions.

Manufacturers are now fitting cars with advanced safety systems. The electronic stability program (ESP) —developed in the 1990s—works during both normal braking and skidding. Some manufacturers are already producing intelligent cars that can completely take control of the vehicle in an emergency situation. **HB**

"Antilocks help by preventing lockup; a vehicle with such brakes remains stable during hard braking."

Brian O'Neill, Highway Institute for Traffic Safety

SEE ALSO: AIR BRAKE, ACCELERATOR, BAND BRAKE, DISC BRAKE, DRUM BRAKE, HYDRAULIC BRAKE, REGENERATIVE BRAKE

◪ *The antilock braking system ensures that the wheels resume rotation in the event of a wheel-locked skid.*

Polarizing Filter
(1929)

Land filters out bright light and glare.

In 1854, an English doctor called William Bird Herapath recounted how unusual crystals formed when iodine was dropped into the urine of a dog that had been fed quinine. The crystals (later called herapathite) appeared dark alone, but light when they overlapped. This was an example of polarizing: the crystals formed a screen allowing light through in only one direction.

While still a teenager, U.S. scientist Edwin Land (1909–1991) became fascinated by polarizing effects, which he observed in button-sized crystals of the mineral tourmaline. He combed through past attempts to create polarizing filters, but the problem lay in creating crystals large enough for practical use. Land's

"Fifty years after the first synthetic polarizers, we find them the essential layer in digital liquid crystal."

Edwin Land, scientist

breakthrough was to create much smaller crystals and fuse them together in a syrupy suspension, drying into a solid sheet of film. He called it Polaroid film.

Land's main purpose for the filter was to filter out glare from car headlights, to avoid oncoming drivers from briefly blinding one another, but his filters found an important use in World War II, by filtering out bright sunlight in binoculars and enabling German submarines to be spotted more easily.

This naturally led to the use of his increasingly improved Polaroid filters in sunglasses, as well as for photography. Today's ubiquitous liquid crystal displays rely on his polarizing filters. **AC**

SEE ALSO: POLAROID SELF-DEVELOPING FILM CAMERA, LIQUID CRYSTAL DISPLAY

Tampon with Applicator (1929)

Haas offers an alternative to sanitary napkins.

The tampon with applicator was invented in 1929 by Dr. Earle Haas (1888–1981). The design, submitted for patent in 1931, consisted of a narrow tube nestling inside a bigger tube containing a cotton plug. When the narrow tube was pushed into the bigger tube, the tampon was guided into place in the vagina. Dangling from the end of the tampon was a piece of string that could be used for easy withdrawal.

The use of disposable plugs for menstrual flow dated back to the ancient Egyptians who invented tampons made from softened papyrus. Over the years women improvised with the materials at hand: in Rome it was wool, in Japan paper, in Indonesia vegetable fibers, and in Africa rolls of grass.

Haas registered the name Tampax as a trademark. In 1934, Haas's patents were purchased by a group of investors, leading to the birth of the Tampax Sales Corporation. The corporation's most significant contribution was to figure out how best to market the concept to women. The idea caught on quickly and women found tampons offered many advantages over conventional sanitary napkins: they were not visible under clothing and could be worn during activities like swimming.

Over the years tampons have been improved to fit inside women's bodies more efficiently. In 1976 an expandable tampon was patented by Dr. Kermit Krantz, a professor of anatomy at the University of Kansas. For women, tampons came to symbolize freedom and comfort, but the product has not been without controversy. There have been health concerns that tampons left in too long can cause toxic shock syndrome, where toxins from the *Staphylococcus aureus* bacterium are released into the bloodstream. **JF**

SEE ALSO: PAPER TISSUES, COTTON BUDS

Synchromesh Gears (1929)

Cadillac simplifies the way drivers change gear.

Changing gear while driving a car is something we take for granted; but in the early days of motoring it was a much more delicate operation that required a lot of skill and practice. With the old straight-cut gears, the rotational speed of the gears had to be the same before they could be meshed together to power the wheels. However good a driver you were, the result was often a terrible grating noise.

Drivers had to use a complicated procedure known as double de-clutching. When changing up a gear the driver had to disengage the clutch, switch to neutral, and let the engine run down to a slower speed. It was necessary to re-engage the clutch for a moment, which slowed down the gears, allowing you to shift into the new higher gear. Changing down a gear was even worse. Once you had disengaged the clutch and shifted to neutral, you had to briefly engage the clutch again and give the accelerator pedal a boost, which would spin up the gears to a higher rotational speed, allowing you to then engage the higher, faster gear.

When Cadillac introduced synchromesh gears in 1929, it was a blessed relief for drivers without three feet. The concept was a simple one. The rotation of gear wheels still had to match up if you were going to engage one toothed wheel with another, but synchromesh did it for you. As the rotating wheels approach each other, protruding bronze rings and grooves on the gear wheels come into contact before the teeth. The contact friction quickly makes sure the wheels are spinning at the same rate before the teeth on the gears actually meet. By the 1950s synchromesh gears had become practically universal. **DHk**

> *"The driver of a racing car is a component.... I changed gear so hard that I damaged my hand."*

Juan Manuel Fangio, Formula One World Champion

SEE ALSO: MOTORCAR GEARS, CLUTCH, AUTOMATIC TRANSMISSION, CENTRIFUGAL CLUTCH

◰ *Synchromesh gears have angled teeth to facilitate smooth engagement of gears within the gearbox.*

Coaxial Cable (1929)

Espenschied and Affel devise a new cable.

In the early 1920s it was clear to communications engineers that high-frequency transmission lines were paramount to the success of any further developments in communications, since ordinary wires and cables simply could not cope. Two engineers at Bell Laboratories, Lloyd Espenschied (1889–1986) and Herman A. Affel (1893–1972), came to the rescue. Together they created the coaxial cable, which is capable of carrying high-frequency (or broadband) signals successfully. Instead of having just single strands of copper covered by a jacket of a flexible plastic, they widened their working diameter to include an insulating spacer and a conducting shield, which gives the cable a very distinctive cross section.

"AT&T is proud to follow in the footsteps of Espenschied and Affel as we continue to drive innovation."

Dave Belanger, chief scientist at AT&T Labs

Running through the very center of the cable is the conductor, which carries the signal. Wrapped around this is the inner dielectric insulator and wrapped around that is a conducting shield that reduces electromagnetic interference from any external sources, meaning that the signal stays clear. The shield can be made from layers of braided wire (which allows flexibility, but creates gaps) or can be a solid metal tube (which is rigid, but more secure). Usually, the whole cable is coated in some sort of vinyl material.

The name coaxial means "sharing the same axis," which is what the conductor, the spacer, the shield, and the jacket all do. **CL**

SEE ALSO: INSULATED WIRE, OPTICAL FIBER, HIGH
TEMPERATURE SUPERCONDUCTOR

Polystyrene (1930)

I. G. Farben creates an important compound.

Many everyday objects are formed from polystyrene, such as pens, electrical equipment, and toys. This diversity of form comes from a relatively uninspiring molecular structure—a long chain of carbon atoms, each attached to a ring of six carbon atoms known as a phenyl group. When expanded with a gas such as pentane or carbon dioxide, it forms a light, foamlike structure ideal for packaging and insulation.

The compound was first identified in 1839 by Eduard Simon of Berlin. He isolated an oily substance from tree resin that he named styrol, which, over time, thickened into a jelly. It was later discovered by Hermann Staudinger that the substance was a monomer, a type of molecule that, with heat, combines with others to create a plastic polymer. The polymerization process was joining together single units of the styrol to make a long chainlike molecule.

The material found few applications until 1930 when Carl Wulff and Eugen Dorrer, working at BASF (under trust to I. G. Farben), patented an economical method for manufacturing the compound from crude oil. They used a heated tube to draw the polystyrene from the reaction vessel as pellets. Small-scale manufacture began in 1931. Early polystyrenes were brittle, but practical plastics were soon forthcoming with the use of additives.

Expanded polystyrene (Styrofoam) came along in 1954. As a waterproof insulator, it found many uses, most familiarly for drinking cups and food packaging. Although economical, concerns have been raised about the environmental impact of expanded polystyrene, and bans of its use are in place in several territories. Diverse applications for polystyrene show no signs of slowing, and the plastic is used to build everything from houses to Xboxes. **MB**

SEE ALSO: SYNTHETIC RUBBER, FLUIDIZED BED REACTOR,
NYLON, SILICONE RUBBER

Neoprene (1930)

Carothers makes a versatile rubber compound.

Neoprene—trade name for polychloroprene—is a synthetic rubber produced from chloroprene by polymerization (changing short chains of molecules to longer chains). Chloroprene is a liquid that when polymerized forms a solid, rubbery substance. When compared with natural rubber, it is lighter, does not perish, is a better thermal insulator, and is chemically inert. Neoprene was invented in 1930 by Wallace Carothers (1896–1937), and by 1931 it had become the first mass-produced synthetic rubber compound. It is commonly used in wetsuits, car fan belts, gaskets, hoses, and corrosion-resistant coatings.

When used for wet suits, the air spaces in the neoprene are filled with nitrogen to increase its insulation properties. This also makes the material more buoyant. More recently, it has become a fashionable material for lifestyle accessories, including laptop covers, iPod holders, pouches for remote controls, and even jewelry.

The development of synthetic rubber began at DuPont, the U.S. chemical company. It was based on the research of Father Julius Arthur Nieuwland, a professor of chemistry at the University of Notre Dame, in Paris. Nieuwland had successfully polymerized acetylene. DuPont bought the patent rights, and Carothers, who worked for them, collaborated with Nieuwland on the successful commercial development of neoprene.

Carothers was a professor at Harvard when he started his research into polymers, moving to DuPont to pursue his research in a commercial environment. He suffered from depression and carried a capsule of cyanide with him at all times as an escape route. **LC**

SEE ALSO: SYNTHETIC RUBBER, FLUIDIZED BED REACTOR, NYLON, SILICONE RUBBER

◀ *Carothers demonstrates the high tensile strength of the synthetic rubber developed by his team.*

Image Intensifier (1930)

Kubetsky devises the photomultiplier tube.

In 1930, the Soviet Russian physicist Leonid Kubetsky (1906–1956) proposed a method to amplify weak photoelectric currents. He exploited the photoelectric effect where the energy of photons is converted into the energy of moving electrons. These electrons can then be accelerated. When these electrons impact a fluorescent plate, several photons are then dislodged, converting the energy back into visible light. The device, a photomutliplier tube, can increase the illumination fifty-fold. Used in series, these tubes can create a cascade of photons, producing gains in excess of 50,000-fold illumination.

In the 1930s, Vladimir Zworykin (1889–1982) was working at RCA on what became the first commercially

> *"All technology can be used for bad or good. It's up to you how to use it."*
>
> Vladimir Zworykin, scientist

viable television. To overcome the problem of a weak signal, Zworykin employed a similar photomultiplier. It was Zworykin's design, produced in 1936, that became commercially successful, and Kubetsky is largely forgotten outside of the former U.S.S.R.

The military quickly saw the potential of these devices. Working with G. A. Morton, at RCA, Zworykin developed the first generation of night vision devices. These devices sensed infrared light, which is invisible to the unaided human eye. The photomultiplier tubes found use in astronomy in 1937 so that the weak visible light from distant stars could now be seen. The tubes were also employed in medical imaging. **SS**

SEE ALSO: X-RAY PHOTOGRAPHY, INFRARED PHOTOGRAPHY, COLOR NIGHT VISION

Tape (1930)

Drew invents a must-have product.

The earliest reference to a sticking tape goes right back to 1676, when lute makers used small pieces of paper with glue on to hold pieces in place while they built instruments. By the early twentieth century, much progress had been made in the area of surgical tape for bandages and plasters, but it was not until the 1930s that what we know as sticky tape appeared.

Bakers, grocers, and meat packers in the 1920s had started using cellophane to wrap their products. Unfortunately there was no way to seal it and prevent moisture getting in and spoiling the food. To solve this problem they turned to 3M, and 3M turned to engineer Richard Drew (1899–1980), the man behind the invention of masking tape. In 1930, after more than

---

"Tape is wonderful at preserving evidence . . . especially on the sticky side."

Michael Baden, forensic pathologist

---

a year of work, Drew and his team produced cellulose tape. It was made from four layers: first the sticky part, then a primer to hold the sticky part to the cellulose, next a layer of cellulose or any other flexible backing, and finally a layer called the release coat, which stopped the tape sticking to itself when rolled up. Shortly after the tape's invention, another company figured out a way to seal the cellophane by heating it.

Things were looking bad for cellulose tape, but other manufacturers found out about the tape and started using it to seal different types of packages. Soon the public discovered it as well and the tape found uses that Drew had never imagined. **DK**

SEE ALSO: GLUE, CELLOPHANE, SUPERGLUE

Radiosonde (1930)

Molchanov transmits meteorological data.

Modern meteorology depends hugely on radiosondes, and hundreds of these balloon-borne instrument packages are released daily from weather stations all over the world. The sonde collects data up to a height of 18 miles (30 km), while being tracked in position using radar. The real-time data is continuously transmitted back to Earth by radio and the position of the sonde as time passes indicates the wind speed. Rising at the rate of about 0.2 miles (0.3 km) per minute, the flight usually takes about two hours.

Silk weather balloons were first used to collect data in 1892 by the French scientist Gustave Hermite. In 1901 these were replaced by sealed rubber balloons. These burst when they reached a height of 12 miles (20 km), and the instrument package parachuted to the ground. The rest was left somewhat to chance because the flight records were picked up when, or if, someone happened to find them.

Military requirements during World War I stressed the need for more instant data. But the radiosondes' requirement of a cheap, low weight, reliable radio system and a transducer that could convert the recorded temperature, pressure, and humidity into radio signal modulations had to wait until 1930. The Russian Pavel Molchanov (1893–1941) is generally credited with the development of this system, and the instrument was quickly adopted by the meteorological offices of a number of countries.

Modern neoprene balloons burst when the ambient pressure falls to about 10 millibar, this being at a height of around 23 miles (37 km). The weight of the whole package is a few pounds. **DH**

SEE ALSO: RAIN GAUGE, ANEMOMETER, BAROMETER, HYGROMETER, ANEROID BAROMETER, WEATHER RADAR

⮕ *A polar researcher launches a radiosonde balloon from Kotelny Island in the Russian Arctic.*

Blind Rivet (1931)

Huck makes a more versatile fastener.

The rivet has existed for a very long time, in fact since the Bronze Age, and remains one of the best methods of permanently fastening two things (normally metal sheets) together. Rivets are commonly used when it is really rather important that whatever has been fastened stays that way, such as aircraft and ship hulls. The vibrations created by movement also have a habit of loosening nuts and bolts so the rivet is generally the preferred fastening.

Back in 1931, Louis Huck was looking for a method to speed up aircraft production, and the blind rivet was his answer. Unlike normal rivets, the blind rivet requires only one-sided access to your desired material, which is particularly useful in airplane construction because of ergonomically shaped hulls that are tricky to access.

Just as with normal rivets, a blind rivet is inserted into a predrilled hole, but the rivet has a mandrel—a cylindrical core made of tough material and with an engineered break notch. As the blind riveting tool draws the tough mandrel head back through the hole, it squashes the softer rivet head on the other side. Once enough force is applied, the mandrel head breaks, leaving it in the hole and a formed head on the blind side of the material.

A blind rivet is easier and faster to use than a traditional "solid" rivet. Although blind rivets are not necessarily as strong as normal rivets, the use of normal rivets is sometimes not a viable option. The basic design of rivets has changed little over time, but improvement in materials and riveting tools means that they are easier to fit and more reliable than ever. They are commonly used in the construction of anything from prefabricated houses to train carriages, buses, cars, and even sea containers. **JB**

SEE ALSO: RIVET, NAIL, SCREW, PHILLIPS SCREW

Radio Telescope (1931)

Jansky picks up emissions from outer space.

Bell Telephone Laboratories at Holmdel, New Jersey, was investigating the introduction of short-wave radio transatlantic telephone services, and was worried that static signals might interfere with voice transmission. In 1931 Bell physicist and engineer Karl Guthe Jansky (1905–1950) was instructed to find the source of the static. Using a high-quality 14.6 m (20.5 MHz) radio receiver and a quaint, wheel-mounted antenna system, he found three sources: nearby thunderstorms, distant thunderstorms, and a faint background hiss.

The intensity of the latter varied daily. After a few months' work, Jansky realized that the period was not the solar day of twenty-four hours but the twenty-three hours fifty-six minutes sidereal day. By 1932 he

---

"We live in a changing universe, and few things are changing faster than our conception of it."

Timothy Ferris, *The Whole Shebang*

---

had pinpointed the source of the hiss as being the Sagittarius region of the Milky Way galaxy. His antenna thus became the first radio telescope.

Grote Reber, a ham radio operator and radio engineer from Wheaton, Illinois, subsequently built the first fully steerable radio telescope (that is, one that could move in both altitude and azimuth). His parabolic reflector turned out to be the prototype of several generations of radio telescopes. Reber started to map the radio sky, discovering the radio sources of Cassiopeia A and Cygnus A (in 1940) and the Andromeda Galaxy (in 1944). Radio emissions from the sun were discovered during World War II. **DH**

SEE ALSO: COHERER (FOR DETECTING RADIO WAVES)

Electric Guitar (1931)

Rickenbacker crafts the instrument that dominated postwar popular music.

Although the guitar had existed in some form since the Renaissance, it was most commonly used as a parlor instrument. The nineteenth century saw it gradually move toward the concert hall, but the guitar still remained a solo or small-ensemble instrument. It played a formative role in the birth of jazz in the 1920s, but as bands became larger and brass sections became louder, the guitar struggled to make itself heard.

The solution was to amplify the sound. Around 1924, an engineer named Lloyd Loar, working for the Gibson guitar company, developed the idea of the magnetic pickup. Placed beneath the strings of the guitar, the pickup creates a magnetic field. The strings vibrate and disturb the magnetic field; these disturbances are converted to electrical current that is amplified and played back through a loudspeaker. Gibson, however, chose not to pursue Loar's idea.

The instrument most widely accepted as being the first electric guitar was Adolph Rickenbacker's (1886–1976) Frying Pan, conceived in 1931. Tapping into a brief for Hawaiian music, he produced a cast aluminum, lap steel guitar fitted with a large horseshoe-magnet pickup. Shortly afterward, Rickenbacker and his colleagues produced the Electro-Spanish guitar—an acoustic guitar with the same pickup attached. Some would argue that this was, in fact, the first true commercially produced electric guitar. Neither instrument achieved a great deal of success, however. Indeed, the electric guitar was not treated seriously until the late 1930s when virtuoso jazz musicians, such as Charlie Christian, began to demonstrate its potential as an instrument in its own right. **TB**

SEE ALSO: PIPE ORGAN, ELECTRONIC SYNTHESIZER

⬀ *The prototype Frying Pan had a wooden neck and body, but the production model was cast aluminum.*

"Patent examiners questioned whether [Rickenbacker's Frying Pan] was 'operative.'"

Monica Smith, Smithsonian Institution

Aerogel
(1931)

Kistler produces the lightest known solid.

Gels are jellylike colloidal substances that have a liquid body containing a network of interconnecting 2–5 nanometer nanoparticles surrounding 100-nanometer pores. If the liquid is carefully removed and replaced by a gas, you have an aerogel. These are light, low-density solid foams, sometimes called "frozen smoke."

Steven S. Kistler (1900–1975), a chemical engineer at the College of the Pacific in California, was investigating the gels produced by the acidic condensation of aqueous sodium silicate. He noticed that gels shrank and cracked as they dried, due to the high surface tension of the water they contained. Kistler managed to stop the volume reduction by

> *"When the tide of misfortune moves over you, even jelly will break your teeth."*

Persian proverb

replacing the water with low-surface-tension alcohol. The end-product, aerogel, was simply the unshrunk, microporous, solid component of the original gel.

In the late 1970s French rocket engineers considered aerogels—whose 99 percent air content makes them ideal thermal insulators—as a safe storage medium for rocket fuels. Their silica aerogel was cheap and easily produced, using a tetramethylorthosilicate base and methanol hydrolization. The Stardust space mission—which collected comet dust—also used aerogels. **DH**

SEE ALSO: LIQUID FUEL ROCKET, SPACE PROBE

🡐 *The aerogel used by NASA's Stardust program was manufactured at the Jet Propulsion Laboratory.*

Electron Microscope
(1931)

Ruska achieves greater magnification.

The first microscopes were made around 1590 by the father and son team of Hans and Zacharias Jansen. These Dutch spectacle-makers fashioned a microscope with a magnification of just twenty times.

In 1673 Dutch Antony van Leeuwenhoek discovered bacteria (animacules), blood cells, protozoa, and spermatozoa with a microscope that magnified objects by 300 times. By 1886 Ernst Abbe had advanced the technique quite considerably, and his microscope reached the limits of resolution with visible light—about 2,000 angstroms, or 0.0002 millimeters.

But to get better resolution you need something with a smaller wavelength. Ernst Ruska (1906–1988)

> *"A weak mind is like a microscope, which magnifies trifling things, but cannot receive great ones."*

Lord Chesterfield, English aristocrat

and his professor, Max Knoll (1897–1969), realized that if electrons were accelerated in a vacuum, their wavelength could be one hundred thousandths that of visible light. These electron beams could then be focused with magnetic coils to produce images. He built the first electron microscope in 1931. Ironically, it had worse resolution than the Jansens' microscope. But by 1933 Ruska had made an electron microscope with a resolution that exceeded that of visible light microscopes, and by 1939 the first commercially available electron microscopes were built. Ruska was awarded half of the 1986 Noble Prize in Physics for his development of the electron microscope. **SS**

SEE ALSO: LENS, MICROSCOPE, TELESCOPE, SCANNING TUNNELING MICROSCOPE

Stereo Sound (1931)

Blumlein channels sound across a spectrum.

For most people, "stereophonic" means listening to audio through a two-channel loudspeaker system. Different elements of the recording can be heard to come from different directions, just as the human ear is naturally able to pinpoint the location of a sound.

The man who invented stereophonic sound was an English electrical engineer named Alan Blumlein (1903–1942). While watching one of the early "talkies" in his local cinema in 1931, he became distracted by the disembodied effect of voices coming from a single location when the actors speaking were positioned across the cinema screen. The system he developed to enable the sound to "follow" the voice was called "binaural"—what we now know as stereo. His idea was

> *"Do you realize the sound only comes from one person? I've got a way to make it follow the person."*
>
> Alan Blumlein, electrical engineer

a simple one. Two microphones were set apart by a fixed distance, each microphone connected to an audio channel. The position of the actors' voices would then be positioned in the binaural "spectrum" according to how much volume each voice projected to each channel. If the recording was replayed over a two-channel system—with one loudspeaker placed on either edge of the screen—the voices would indeed reflect the position of the actors on the screen.

Blumlein filed a series of related patents at the end of 1931, including not only his binaural system, but a method in which two channels could be captured in the single groove of a gramophone record. **TB**

SEE ALSO: LOUDSPEAKER, MICROPHONE, FILM SOUND, SURROUND SOUND, DOLBY NOISE REDUCTION

Bourke Engine (1932)

Bourke thinks his engine will shape the future.

In 1932, Russell L. Bourke built an engine he thought was destined to change the world. It had only two moving parts (the pistons) and a fluid bearing connecting the pistons to a Scotch yoke (a mechanism Bourke used instead of a crankshaft to change the linear motion of the pistons to rotary motion). Four years later, Bourke applied for three U.S. patents for his engine; these were issued in 1938.

For twenty years, Bourke was unable to interest government or industry in his engine. Then, in 1957, his patents ran out, enabling anyone in the world to manufacture it. And yet, few were interested. Over the last half century, gas prices have skyrocketed, concern over greenhouse gases has dramatically increased,

> *"It will run on any fuel with a hydro-carbon base, needs no repair and the oil in it is good for life."*
>
> Russell L. Bourke, engineer

and the Bourke engine remains unused. Bourke thought the oil, engine parts manufacture, and repair industries were suppressing his invention. There are those who believe he was right and that the Bourke engine can decrease demand for oil, alleviate global climate change, and, not incidentally, make a fortune for investors in the engine.

Other have more mundane opinions. They believe the Bourke engine is not a reasonable replacement for current engines. They cite a number of problems, including the large engine weight, decreased efficiency due to shock waves and excess heat transfer, and excessive nitrogen dioxide emissions. **ES**

SEE ALSO: ATMOSPHERIC STEAM ENGINE, INTERNAL COMBUSTION ENGINE, TWO-STROKE ENGINE

Anglepoise® Lamp (1932)

Carwardine employs a simple principle to direct light to suit individual needs.

George Carwardine's (1887–1948) company was a car manufacturing factory and it was there that he came up with the idea of arranging springs on a metal arm that could be adjusted in orientation and yet stay in place when released. The design mimicked the movement of the human arm and was inspired by the constant tension principle. He patented the design in 1931, but it was not until the year after that the idea came to him to use it to angle temporarily the direction of a lamp.

To the moveable sprung arm, he attached a heavy base and a directional lamp, which allowed the lamp to be moved to face any direction but remain rigid in position. Carwardine found the lamps useful in his factory for illuminating the assembly process, but he soon realized there was nothing to stop them being used in offices and elsewhere. With the addition of a shade over the bulb, the beam could be focused in a particular direction, giving his design the advantage of using less energy than competing models. The shade also helped direct the beam away from the eyes and could be adjusted to stop light dazzling the person working.

The Anglepoise® lamp boasted to give as much light with a 25-Watt bulb as a conventional lamp would with a 60-Watt bulb. The directional action of the lamp turned out to be ideal for office use, especially for illuminating books and papers. The product took off straight away. Today Anglepoise® lamps are still commonplace in offices worldwide and over the years there have only been minor changes to Carwardine's original design. **LS**

SEE ALSO: OIL LAMP, CANDLE, GAS LIGHTING, ARC LAMP, INCANDESCENT LIGHT BULB, NEON LAMP, HALOGEN LAMP

↗ *An early Anglepoise® lamp from 1935, made by spring maker Herbert Terry & Son of Redditch, England.*

"We waste our lights in vain, like lamps by day . . ."

William Shakespeare, *Romeo and Juliet*

Folding Wheelchair (1932)

Jennings designs a market-leading product.

After breaking his back in a mining accident, Herbert Everest so disliked his unwieldy wheelchair that he enlisted the help of an engineer friend, Harry Jennings, to help design a new chair. The device that Jennings came up with revolutionized the wheelchair market.

Although wheelchairs had been in existence since the sixteenth century (the first one thought to be built for King Philip II of Spain), they had seen little development on their basic design until 1909. It was at around that time that the first lightweight models were made out of tubular steel rather than wood.

These models had some foldable features, but it was the introduction of the folding X-brace frame that was the secret to Jennings's successful design.

> *"Money cannot buy health, but I'd settle for a diamond-studded wheelchair."*
>
> Dorothy Parker, writer

Beforehand, any wheelchairs that had foldable features used a T-shaped or an I-shaped frame. The use of a collapsible X-shape not only created a greater rigidity, it also meant the wheelchair could be folded in by pushing the sides in toward each other with the wheels remaining fixed in place. In fact, this new wheelchair was so revolutionary that Everest and Jennings went into business together, launching the rather unimaginatively named Everest & Jennings Company. The company enjoyed massive success and market dominance in its early days, so much so that the U.S. Justice Department served it with an antitrust suit for attempting to monopolize the industry. **CL**

SEE ALSO: CHAIR, COLLAPSIBLE STROLLER

Sodium Thiopental (1932)

Volwiler and Tabern inject anesthetic safely.

Sodium thiopental was discovered in 1932 by Ernest H. Volwiler (1893–1992) and Donalee L. Tabern (1900–1974), two scientists on a quest to discover an anesthetic that could be injected directly into the bloodstream. Working for Abbott Laboratories, the pair spent three years screening hundreds of compounds to find one that could produce unconsciousness prior to surgery, with limited side effects.

Sodium thiopental was first tried in humans on March 8, 1934, by Dr. Ralph M. Waters in an investigation of its properties. It was found to induce anesthesia for ten to thirty minutes by depression of the central nervous system within sixty seconds of injection. It was also found to show surprisingly little analgesia. For this reason, it was commonly used to make it easier for doctors to administer longer lasting, inhalable anesthetics after patients had comfortably "gone under." Sodium thiopental was the first general anesthetic to be widely used intravenously and spawned an entirely new family of "short-acting" barbiturate drugs, including Brevital and Surital.

After the Pearl Harbor attack in 1941, sodium thiopental was associated with a number of anesthetic deaths because excessive doses were given to shocked trauma patients. This resulted in its temporary discontinuation of use as an anesthetic. As with nearly all anesthetic drugs, thiopental causes cardiovascular and respiratory depression resulting in low blood pressure and a reduced rate of respiration.

Later use of sodium thiopental was expanded. The CIA used it as a truth-inducing treatment during interrogations because small doses relaxed patients without producing unconsciousness. Some U.S. states have also used the drug as one of the ingredients in their lethal injection. **JF**

SEE ALSO: ANESTHESIA, NITROUS OXIDE ANESTHESIA, ETHER ANESTHETIC, CHLOROFORM ANESTHETIC

Parking Meter (1932)

Magee reduces traffic congestion in cities and provides local revenue.

In 1932, Carl Magee, lawyer, newspaper publisher, and newly appointed chair of the Oklahoma traffic committee, was asked to develop a solution to the problem of traffic congestion in the city. He observed that many people were driving into town and parking their cars all day, blocking up the streets. This slowed trade in shops, because people could not park nearby, and there was no turnover of custom. As a solution to this problem, Magee struck upon the idea of the parking meter. He designed a crude prototype and then joined with the Oklahoma State University to develop the idea. The result of this was the first coin-operated parking meter, dubbed the "Black Maria," and it was installed on July 16, 1935.

The Reverend C. H. North was the first recipient of a parking ticket for being parked by an expired meter. Like many since, he argued in court that he had only stopped for a moment to go to a shop, and like many since, his case was dismissed.

The invention soon spread from Oklahoma and was introduced in New York in September 1951. With a geared and sprung mechanism, early meters operated like wind-up clocks. In fact, the early meters had to be wound once a week, although later models had a self-regulated mechanism. As technology has improved, almost all meters have been replaced with digital models, which are more reliable and less easily vandalized. Recent developments include alternative payment methods and solar-powered meters. Although Magee's original clockwork technology has been replaced, his invention has continued to thrive and expand as car ownership has increased. **JG**

SEE ALSO: PAYPHONE, TRAFFIC LIGHTS, VENDING MACHINE

🡕 *Parking meters like this one were used in the United States when the device was introduced in the 1930s.*

> "[The parking meter is] just another way of getting money out of people..."

Sugar Ray Robinson, on New York's first meter

Vinyl Phonograph Record (1932)

RCA Victor markets the vinyl discs that will dominate the music industry for decades.

From 1932 until the early 1990s, the vinyl gramophone record was the single most popular medium for the reproduction of recorded music. Like many significant inventions, the line of evolution began with Thomas Alva Edison, who developed the cylinder recording system. It was Emile Berliner—already known as the man who invented the microphone used in the mouthpiece of Alexander Graham Bell's first commercial telephones—who, in 1887, developed the idea of using discs with a lateral groove cut in spiral.

The first discs were manufactured from rubber, but shellac was introduced in 1896, and the 10-inch, 78 rpm "single" became the norm. This system, however, had three notable problems: Shellac was extremely brittle, and records broke easily; it generated high levels of background noise; and it was simply not possible to get much music on a record—around four minutes on each side. Large-scale classical compositions had to be chopped up into movements and sold as a collection of discs bound in cardboard albums—indeed, this is where the term "album" originated.

In 1932, RCA Victor launched the first commercially available long-playing (LP) records using a new material—vinyl. These featured a low surface noise, were 12 inches in diameter, and played back at 33 rpm, and so could contain more music. That this new system failed to take off at first was a matter of timing: As the Great Depression kicked in, consumer confidence was unsurprisingly low. Nevertheless, Victor continued to back what was clearly a superior approach, and gradually, during the 1950s, the 12-inch LP and 7-inch single took over. **TB**

SEE ALSO: PHONOGRAPH RECORD, AUDIOTAPE CASSETTE, OPTICAL DISC, MINIDISC

⬆ *Vinyl records can contain thirty minutes of music on each side, about eight times more than earlier records.*

FM Radio (1933)

Armstrong pioneers a radio frequency that minimizes interference.

The first radio broadcasts made at the beginning of the twentieth century were based on the principles of amplitude modulation, or AM. Radio waves were broadcast at a specific frequency, and the receiver would detect amplitude variations before decoding the signal. Although poor in sound quality—primarily a deliberate technical limitation to deal with the sheer volume of AM radio stations—and prone to interference, it remained the dominant commercial mode for broadcast until the 1970s.

In 1933, Edwin H. Armstrong (1890–1954) patented an alternative method for making radio broadcasts using frequency modulation—FM. The basic principle has the carrier wave modulated so that its frequency varies with the audio signal being transmitted. The main benefit of FM over AM is that it enables broadcasts to be received with a minimum of interference: assorted atmospheric conditions, such as thunderstorms, or surrounding electrical activity, such

as car ignitions, can themselves create AM signals thus interfering with AM broadcasts but FM is not prone to these problems.

Armstrong presented his paper, "A Method of Reducing Disturbances in Radio Signaling by a System of Frequency Modulation," to the Institute of Radio Engineers in November 1935, and two years later the first FM radio station, W1ZOJ, began broadcasting. Until the advent of digital radio in the late 1990s, FM was the only practical means for making high-fidelity, stereo broadcasts. FM remains the most commonly used broadcast method worldwide and is still used in analog television broadcasting. **TB**

SEE ALSO: WIRELESS COMMUNICATION, IMPROVED RADIO TRANSMITTER, AMPLITUDE MODULATION, CRYSTAL SET RADIO

⬆ *FM radio became popular in Europe in the 1950s, while the United States only widely adopted it in 1978.*

Kitchen Extractor (1933)

Vent-a-Hood improves air quality in the home.

Before kitchens were ventilated, cooking was a smelly and smoky experience. The kitchen would quickly fill with smoke and nasty gases, stinky odors would spread throughout the house, and grease would stick to and damage the walls. However, in 1933, Vent-a-Hood set about making cooking in the home a more enjoyable and safe experience and invented the first kitchen extractor. Also known as an extractor hood, it was able to capture all the by-products of cooking, except grease.

Early kitchen extractors were considered, quite correctly, to be fire hazards because the grease they could not catch could easily ignite with the high temperatures generated by cookers. Later extractor designs, using wire mesh to catch grease, were incredibly inefficient and could only capture about 25 percent of the product. In 1937, Vent-a-Hood improved their original design and incorporated a special blower into their kitchen extractor, known as the Magic Lung®, which could pressurize the air blowing through it to liquefy grease and reduce the risk of fire in the kitchen.

Basic extractor hoods, now found in kitchens worldwide, comprise of a skirt to capture the rising gases, some kind of grease filter, and a fan to force ventilation. There are two types of kitchen extractor: the ducted hoods, which are vented and release cooking by-products outside of the home, and the ductless hoods, which rely on charcoal to clean the air by removing any odors and smoke particles before releasing it back into the kitchen.

Kitchen extractors have revolutionized home-living, making cooking a far safer, more pleasant, and smoke-free experience than it was at the turn of the twentieth century. **FS**

SEE ALSO: GAS STOVE, ELECTRIC STOVE

Catseyes (1934)

Shaw produces the design of the century.

One night in 1933 when the road mender Percy Shaw (1890–1976) was driving home in Yorkshire, he saw the light of his car headlamps reflected in the eyes of a cat beside the road. This gave Shaw the inspiration that by replicating this effect he could produce a practical way of helping drivers navigate poorly lit roads.

Shaw's challenge was to create a device bright enough to illuminate roads at night, robust enough to cope with cars constantly driving across it, and that also required minimum maintenance. Shaw came up with a small device that could be inserted into the road as a marker. It consisted of four glass beads placed in two pairs facing in opposite directions, embedded in a flexible rubber dome. When vehicles

> *"The Catseye is what great design is all about. Simple, functional, and beautiful."*
>
> James May, presenter of British TV show *Top Gear*

drove over the dome, the rubber contracted and the glass beads dropped safely beneath the road surface. The device was even self-cleaning. The cast-iron base collected rainwater and whenever the top of the dome was depressed, the rubber would wash the water across the glass beads to cleanse away any grime, just as the eye is cleansed by tears. The patent for the Catseye was registered in 1934, and in 2001 the product was voted the greatest design of the twentieth century, ahead even of Concorde. **JF**

SEE ALSO: TRAFFIC LIGHTS

▶ *Raised, reflective studs are laid at a pedestrian crossing at Wimbledon Park, London, England, in 1936.*

Perspex® (1934)

Hill introduces a viable alternative to glass.

Acrylic glass, or Perspex® under its commonly used trade name, is a transparent material that has been found to have many advantages over traditional glass. It is a polymer of methyl methacrylate, a simple organic compound, and is called by a number of monikers, including plexiglass, acrylic, and, properly, poly(methyl 2-methylpropenoate). It is a simple polymer, formed from chains of carbon, hydrogen, and oxygen.

Despite its simplicity, Perspex® is a versatile material. Durable, yet easily molded, it found early use as a safe alternative to glass, with particular utility during World War II in Spitfire canopies, and later in visors and shields. At about half the weight of glass, but seventeen times its strength under constant load, Perspex® is also useful in aquariums, skylights, and anywhere else a transparent material under pressure is required. Medical uses include contact lenses, dentures, and bone cement. Furthermore, when suspended in water, it forms acrylic paint.

The material was first produced by Imperial Chemical Industries (ICI), thanks to the work of two of its employees. Rowland Hill first described the polymer in 1931 and filed a patent for its preparation and that of the monomer. Meanwhile, John Crawford, working in ICI's explosives group in Scotland, was experimenting with synthetic materials for use in safety glass and found his ideal solution in Hill's polymer. Crawford's contribution was to find an economic way to synthesize the precursor, 2-methylpropenoate.

Over the next couple of years, ICI investigated ways to produce the material on an industrial scale. The first transparent sheet was produced in 1933 and the trade name Perspex® was born in 1934. Today, if you are indoors, you are unlikely to be far away from a Perspex® object, so numerous are its applications. **MB**

SEE ALSO: GLASS, BLOWN GLASS, FIBERGLASS

Phillips Screw (1934)

Phillips invents a screw for the assembly line.

The cross-shaped screw-head is ubiquitous today, yet its widespread use came only a lifetime ago.

Traditional flat-head screws, in use since the late seventeenth century, suffer from two disadvantages: one, the slot and screwdriver have to be precisely aligned; and two, the screwdriver can easily slip from its position thanks to centrifugal force. These problems are particularly pronounced in automated production lines, where robots are not able to compensate for slippage as easily as humans. The cross-head screw, with a pointed tip for self-centering, was designed to eliminate these problems and deliver more torque to boot.

The cross-head was popularized by Henry F. Phillips (1890–1958) from Portland, Oregon. Phillips built on

> *"The first off the assembly line were plated with gold and silver and made into a necklace. . . ."*
>
> Norman Nock, *Austin Healy Magazine* (1996)

the work of inventor J. P. Thompson, who had designed, but failed to capitalize on, a recessed-head screw. Phillips developed the concept and founded the Phillips Screw Company in 1933. He received five U.S. patents for the design between 1934 and 1936.

A means to mass manufacture the components was developed by the American Screw Company, but not without $500,000 of development costs. The investment began to pay off when General Motors adopted the screw on the 1936 Cadillac. The Phillips screw really came to the fore during World War II, where its use sped up production of tanks, jeeps, and other vehicles, including the Spitfire fighter plane. **MB**

SEE ALSO: NAIL, ARCHIMEDES SCREW, SCREW-CUTTING LATHE, SQUARE-HEADED SCREW

Trampoline (1934)

Nissen's bouncing frame finds application in the army as well as the leisure industry.

George Nissen (1914–2010) was a gymnastics coach and Larry Griswold (1905–1996) a gymnastic tumbler and acrobat. In the summer of 1935, they became acquainted with the great trapeze family, known as the Flying Wards, and often helped the Wards mend their nets at the local YMCA. Nissen thought a small rebound net may assist aerialists with tumbling practice and, with Griswold, set up a workshop in his parents' garage to develop a bouncing frame. They gathered a section of canvas, had it sewn up, and attached to it a series of springs along its outer edge. This was then joined to an angled iron frame scavenged from a nearby demolition yard.

The pair took their new bouncing frame to a University of Iowa summer camp, where it garnered much interest. After graduation, Nissen and two friends, collectively known as The Leonardos, developed an act and took it south into Mexico. There they found work as a hand-balancing and tumbling act at a small nightclub called the El Retiro in Mexico City. There Nissen gradually set about improving his Spanish. A *canguro* was a jack-knife, *aeroplano* meant swan-dive, and the Spanish word for spring board was *el trampolin*. Nissen anglicized the word by adding an "e," and the trampoline, known alternately as a magic carpet, *catro elastico*, or simply a "Nissen," was born.

In 1942 the Griswold-Nissen Trampoline & Tumbling Company was established. Their trampolines were used among the armed forces during World War II as a tool in the mental and physical conditioning of pilots and as an aid to the pilots orientating themselves while in the air. **BS**

SEE ALSO: METALWORKING, NYLON

⬀ *Trampolining on the beach has long been popular, as shown by this boy in Yorkshire, England, in the 1960s.*

"Every time you jump, there is two seconds of freedom . . ."

George Nissen, gymnastics coach

Electric pH Meter (1935)

Beckman makes the first precision instrument.

In 1934, Glen Joseph, a chemist at the California Fruit Growers Exchange laboratory, was trying to accurately measure acidity in citrus fruit products. The most common method at the time for testing pH, a measure of acidity or alkalinity of a substance, was to use litmus paper. Litmus paper turns different colors depending on the acidity of a substance. This was of no use to Joseph, however, because the sulfur dioxide used as a preservative in citrus juice bleaches out the paper.

Joseph tried using glass electrodes, but these were susceptible to breakage and gave a very weak signal. He finally called upon his old classmate Arnold Beckman (1900–2004), then employed as a professor at Caltech, who told Joseph that he needed to use vacuum tubes. Beckman later ended up making the instrument himself.

The instrument worked so well that Joseph soon ordered another one for his laboratory. Beckman realized he was on to something, and a patent was filed in 1934 for this "acidimeter." The following year, the National Technical Laboratories started selling this acidimeter for $195. Despite its steep cost (especially compared to dirt-cheap litmus paper), the company managed to sell eighty-seven instruments in the first three months of production. In 1939 Beckman left his professorship to run National Technical Laboratories full time, and went on to invent many more important laboratory instruments.

The pH meter was a revolutionary instrument in the science laboratory. It was essentially the first portable, precision instrument that could take accurate measurements immediately and reliably. This left the scientist time to concentrate on their research instead of building complicated equipment to make simple measurements. **RH**

SEE ALSO: PYROMETER, SI UNITS, ARTIFICIAL FERTILIZER

Radar (1935)

Watson-Watt invents device to detect enemies.

In 1935, with hostilities looming in Europe, the British government asked Scottish physicist Robert Watson-Watt (1892–1973) to develop a personnel-destroying "death ray" that could harm the opposition. Watson-Watt demonstrated that this was an impossible ambition. However, he theorized that a radio wave could be sent to bounce against a moving object, and monitoring its travel could then provide information about the target, such as its speed, direction of travel, and altitude. This could also determine the distance from the transmitter of fixed objects. A demonstration won favor with the government, and Watson-Watt subsequently gave the technology the name radar (radio detection and ranging).

> *"The use of radar in World War II … was a vital factor in the successful defense of Great Britain."*
>
> R. Hanbury Brown, Robert Watson-Watt, physicists

By the beginning of World War II, Watson-Watt had installed a chain of radar stations across the United Kingdom, and two other British physicists, Henry Boot and John T. Randall, developed his concept of the resonant-cavity magnetron. This electron tube was capable of generating high-frequency radio pulses with huge amounts of power. This in turn led to microwave radar, which is now used to measure atmospheric pollution and, more importantly, in modern forms of communications. **SD**

SEE ALSO: SONAR, WEATHER RADAR, GROUND PROXIMITY WARNING SYSTEM

➡ *The 61-foot (19 m) diameter antenna dish of a U.S. Air Force radar station, erected in California in 1945.*

Spectrophotometer (1935)

Hardy develops an instrument to measure the intensity of electromagnetic radiation.

Spectrophotometers are used to measure the intensity of electromagnetic radiation. Usually the measurements are confined by filters to a very narrow spectral range and the instrument is used to detect the change in brightness after the light radiation has either passed through a sample or been reflected off it. Early devices used the naked eye to determine the differences in intensity between two beams.

Arthur Hardy (1895–1977), a physicist at the Massachusetts Institute of Technology, decided to replace the eye with the new cesium photocells, and thus detect intensities electronically. The plan was to produce a spectrophotometer that automatically scanned through the visible spectrum and produced a pen-drawn spectrum showing how the light intensity varied with wavelength. Beam splitters and rotating polarizers were used and the two beams were compared by blinking quickly from one to another using a flicker photometer technique. Working in

collaboration with the firm General Electric, the first operational production machine was ready by 1935. Soon, the National Bureau of Standards was using Hardy's spectrophotometers to test pigments and dyes and to set paint color standards. The fully automated machine was extremely expensive, and soon cheaper versions were being produced that required manual fixed-point scanning of the spectrum.

Spectrophotometers found their way into industrial and scientific laboratories. Similar instruments have been used widely to monitor the seasonal and latitudinal variation of ozone, and a spectrophotometer led to the discovery of the ozone hole over Antarctica. **DH**

SEE ALSO: GEIGER COUNTER, ELECTRON SPECTROMETER

⬆ *Part of a spectrophotometer, circa 1938. In World War II they were used to measure vitamins in food.*

Nylon (1935)

Carothers's substitute for silk becomes a huge success.

Wallace Carothers (1896–1937) was an excellent chemist and, as an undergraduate, was made head of the chemistry department at Tarkio College, Missouri. After becoming a professor at Harvard, he was lured into industry by chemical company DuPont, which had opened a science laboratory to develop new products.

In 1928 he headed up a team looking into artificial materials. Carothers was interested in the hydrocarbon acetylene and its family of chemicals. After having developed the first synthetic rubber polymer—mass produced by DuPont as neoprene in 1931—he looked into creating a man-made synthetic fiber that could replace silk. Japan was the United States's main source of silk, but trade relations were breaking down as the political situation between the two countries worsened. Silk was becoming harder to source and very expensive.

In 1935, Carothers made significant steps toward a silklike fiber by developing a chemical reaction between chemical units called monomers. The reaction resulted in a polymeric fiber and generated water as a by-product. If the water was removed from the reaction as it formed, then an even stronger fiber resulted. This fiber is now known as nylon and was introduced to the world in 1938. During the war it replaced silk in the manufacture of parachutes, and it subsequently found a huge and lucrative market in nylon stockings. Nylon now underpins the multi-billion-dollar textile industry.

The highly ambitious Carothers suffered from severe depression and, in April 1937, despairing that he had no fresh inspiration, he committed suicide by taking a dose of potassium cyanide. **RB**

SEE ALSO: SYNTHETIC RUBBER, FLUIDIZED BED REACTOR, NEOPRENE, SILICONE RUBBER

⤴ *Nylon has far greater strength and resistance to heat and water than any previous man-made fiber.*

Squeegee (1936)

Steccone simplifies window cleaning.

Italian immigrant Ettore Steccone (1896–1984) liked to keep windows clean. When he moved across the Atlantic in 1922, window cleaners were using heavy, cumbersome tools, but he was about to change all that. In 1936 the patent for Steccone's squeegee was filed, and in 1938 it was published. But he was viewed as an uneducated Italian immigrant, and his attempts to sell his invention to dealers proved fruitless. Instead he approached fellow window cleaners, offering them the chance to try his tool for one day. Its simplicity and ergonomic design made it an instant hit and the Steccone family made a great deal of money.

The squeegee has been making the lives of window cleaners easier ever since, and the design has

---

> *"I never understood what people did with them— who's buying all these?"*

Diane Smahlik, daughter of Ettore Steccone

---

hardly changed in the process. The key lies in the rubber, as even the slightest imperfections will leave streaks on the glass. Despite the fact that Steccone's company—still based in Oakland—has evolved into "the world's squeegee empire," the workers still cut every single strip of rubber by hand, using Steccone's original wooden measuring system.

The squeegee looks just like a T-shaped device with a rubber bit across the front edge, but the formula for producing rubber that leaves windows squeaky clean is still carefully guarded to this day, even if imitations do exist. However, in the future self-cleaning windows may take some of Steccone's market share. **CB**

SEE ALSO: GLASS, ICE RINK CLEANING MACHINE, SELF-CLEANING WINDOWS

Epoxy Resin (1936)

Castan and Greenlee create strong adhesive.

If they ever create a hall of fame for materials, epoxy resin would be a shoe-in. This exceptional substance is the adhesive of choice when you really do not want two surfaces to come unstuck; holding bits of an aircraft together, for example, or the rotor blades of wind turbines. Epoxy resin is also resistant to heat and chemicals, while some epoxies are waterproof and even capable of curing underwater. They are also excellent electrical insulators.

Epoxy is a thermosetting plastic. That is to say, when it is mixed with a "hardener" or catalyst, it forms crosslinks with itself, curing into a robust material with the properties mentioned above. The raw compound comes in many forms, including a low-viscosity liquid and a powder. Because the hardener is also highly variable, a broad suite of cured polymers can be created with differing properties.

Swiss chemist Pierre Castan (1899–1985) and the American Sylvan Greenlee share the credit for this invention. In 1936, Castan, working on materials for denture repair, reacted the compounds bisphenol A and epichlorhydrin to produce an amber material with a low melting point. Meanwhile, across the Atlantic, Greenlee was investigating a similar reaction and produced his own resin, which varied from Castan's only in that it had a higher molecular weight. Both scientists filed patents at the same time, and went on to bag further intellectual property, exploiting the versatility of epoxy chemistry to develop alternative resins with different physical properties.

The biggest market for epoxy resins is as a protective coating. Here again, the ability to fix tightly to a surface and resistance to just about everything make epoxy ideal for coating washing machines, pipes, and the insides of tin cans. **MB**

SEE ALSO: GLUE, SUPER GLUE

Ski Lift (1936)

Curran designs the first chair and pulley lift.

The idea of using ropes to climb mountains is almost as old as rope itself, and evidence can be found from the 1600s for people using ropes to cross chasms or valleys suspended below a rope bridge (although some consider rope to have been used as far back as 15,000 BCE). Where this practice first began to be adapted to aid the sport of skiing (saving the time and tedium of having to climb the mountain before skiing back down) is somewhat uncertain, and also depends on your definition of "ski lift."

Broadly, ski lifts fall into a number of categories. The first of which is where skiers appropriate an existing lift system. A good example of this was on Gold Mountain (later named Eureka Peak) in the 1850s, when skiers used an existing network of gravity-powered, ore-mining buckets to climb the mountain to ski. A different type of lift, more appropriately called the ski-tow, was famously erected in 1934 in Vermont. Consisting of a long length of rope attached to a Model T Ford engine, the rope ran through pulleys, and skiers could hold on to be pulled along the slope up to the top of the hill.

The first purpose-built ski lift, where skiers were propelled along in chairs suspended from an aerial rope-way, was designed by Jim Curran and built for the Sun Valley resort in 1936. The idea, adapted from a system to load bunches of bananas onto boats, consisted of chairs suspended from a single rope that propelled skiers up the slope at 4–5 miles per hour (6–8 kph). Initially, there were reservations about safety, but when it was opened, the promise of the new system was obvious and was soon widely copied. **SB**

SEE ALSO: SKI, PULLEY, CABLE CAR

🔼 *T-bar lifts, like this one photographed in 1947, are provided on slopes used by relatively few skiers.*

"Curran revolutionized . . . skiing by designing an easy . . . method for skiers to ascend the mountain."

Press release from the Ski Hall of Fame

Suntan Lotion (1936)

Schueller opens up a new market for skin products by introducing the first sunscreen.

"Truth is like the sun. You can shut it out for a time, but it ain't goin' away."

Elvis Presley, singer

When the legendary fashion designer Coco Chanel inadvertently developed a suntan during a Mediterranean cruise in the 1920s, tanned skin immediately became synonymous with beauty, fashion, and a healthy lifestyle. When the Popular Front won the 1936 French general elections and legislated for annual paid holidays, people began to spend more of their time in the sun. In response to increasing French demands for a product to assist in the tanning process, the French chemist and founder of L'Oreal, Eugène Schueller (1881–1957), created "Bellis," the world's first sunscreen lotion. Its effectiveness in the prevention of skin cancers, however, was poor when compared with modern formulations, and many early attempts at tanning lotions amounted to nothing more than crude oil-based pastes. Twenty-six years would pass before the chemist Franz Greiter introduced the Sun Protection Factor (SPF) in 1962, the first real attempt to grade a sunscreen's effectiveness at blocking the sun's harmful ultraviolet radiation.

Schueller began experimenting at home with chemicals responsible for color pigmentation in 1903. He found the chemicals became permanently absorbed into human hair when mixed with ammonia and peroxide. In 1907 he successfully began to market his new synthetic hair-coloring formula to Parisien hairdressers under the name Aureole. The first mass-produced suntan lotion was a combination of jasmine and cocoa butter, mixed in an old granite coffee pot in 1944 by the Florida-based pharmacist Benjamin Green, a researcher with the Coppertone Company. **BS**

SEE ALSO: VASELINE, DEODORANT, HAIR SPRAY

⬉ *In 1953 Coppertone launched an ad campaign featuring the Coppertone Girl to promote its sunscreen.*

Disposable Diapers (1936)

Bruk is the first to make a single-use product.

It took a Swedish paper mill, a Connecticut housewife, and a Proctor & Gamble engineer to free parents from the shackles of diaper duty.

Paper mill Pauliström Bruk put pads of treated paper into rubber pants and produced the first diaper product destined for the bin in 1936. These disposable paper pads failed to impress the baby supply industry. A more commercially successful product came along more than a decade later, thanks to an inventive housewife named Marion Donovan (1917–1998).

Forgoing rubber pants for nylon and skipping paper entirely, Donovan created the "Boater" in 1946. Donovan's simple nylon diaper covers far surpassed rubber pants for one simple reasons—diaper rash.

> "This is the mystery of the modern disposable diaper: how does something so small do so much?"
>
> Malcolm Gladwell, journalist

"We couldn't say it, but it did cure diaper rash, and many doctors recommended it," said Donovan of her Boater. When it was launched on the market in 1949, the Boater was popular with parents. Donovan hit the jackpot when she sold the rights for $1 million in 1951.

Donovan set her sights on making a diaper that was disposable and leak-proof. She turned to treated paper, soon showing prototypes to various companies only to be shot down. By the late 1950s, one of those companies, Proctor & Gamble, had picked up the paper diaper idea and asked engineer Victor Mills (1897–1997) to perfect it. In 1961, Proctor & Gamble introduced the world to Pampers®. **RBk**

SEE ALSO: COTTON BUDS, ROTARY CLOTHES LINE, BABY CARRIAGE, COLLAPSIBLE STROLLER

Biodiesel (1937)

Chavanne transforms vegetable oil into fuel.

Biodiesel, technically described as mono-alkyl esters of vegetable oil or animal fat, is to many a new concept in man's quest to rely less on petroleum-based products in daily life. The first biodiesel, however, was produced in the lab decades ago and is still cited as an important technological advance in the area of alternative fuel sources.

The name *biodiesel* originates from the word *diesel*, a type of engine invented by Rudolph Diesel in 1892. Diesel first displayed his invention in 1900, at the Paris World's Fair. Rather unexpectedly, his diesel engine actually ran on peanut oil.

However, it was in 1937 that a Belgian scientist at the University of Brussels, G. Chavanne, was granted a patent entitled "Procedure for the transformation of vegetable oils for their uses as fuels." In the patent, Chavanne describes the use of palm oil as diesel fuel. The reaction was called alcoholysis (also known as transesterification), and involved mixing the vegetable oil with ethanol to produce glycerol and a vegetable oil ester, or biodiesel. Scientists still consider this to be the first account of the production of biodiesel.

A related report in 1942 by Chavanne described the use of palm oil ethyl esters as a fuel source, specifically used as a test fuel for a bus that ran between Brussels and Louvain during the summer of 1938. The article focused on the chemical and physical properties of the fuel, and reported "satisfactory" results on this use of palm oil.

This groundbreaking work, while at the time not focused on environmental or emissions implications, seems more important now than ever in the quest for renewable, clean and efficient fuel sources. Indeed, biodiesel is increasingly seen as a way to reduce dependence on traditional oil-based fuels. **RH**

SEE ALSO: DIESEL ENGINE, BIOETHANOL

Blood Bank (1937)

Drew separates blood into red cells and plasma and establishes the first blood bank.

Charles Drew (1904–1950) is widely credited as the father of the modern blood bank. In 1937, Drew made the key discovery that separating red blood cells from the plasma (the liquid part of blood that can be given to anyone), and freezing the two separately, allowed blood to be preserved for longer and reconstituted at a later date.

In February of 1941, Drew was appointed director of the first American Red Cross Blood Bank, and launched the "Plasma for Britain Project" where he collected thousands of units of plasma for the British war effort. From these samples the British Army established its own blood transfusion service, where dried and powdered plasma could be stored and turned into a liquid with the addition of sterile, distilled water. After the war, doctors who had seen the effectiveness of transfusion therapy in battle began to demand that blood be made available for treatment of civilian patients.

An earlier discovery in 1915 by Richard Lewishon proved that adding sodium citrate to freshly drawn blood prevented clotting, thus opening the way for the development of blood banks.

In 1950 breakable glass bottles were replaced with plastic bags, allowing the development of a system with multiple blood samples. The shelf life of stored blood was extended by the addition of an anticoagulation preservative, CPDA-1, in 1979, facilitating resource-sharing among blood banks.

Ironically Drew died in a car accident in North Carolina, and was too severely injured to benefit from his own invention. **JF**

SEE ALSO: ANTISEPTIC SURGERY, BLOOD TRANSFUSION, HYPODERMIC SYRINGE, SYNTHETIC BLOOD

⬆ *In 1944 San Franciscans were eager to donate blood for injured soldiers after being offered $4 for each pint.*

Shopping Cart (1937)

Goldman helps shoppers carry more—and buy more—and makes his fortune.

Sylvan Goldman (1898–1984) was the owner of the Humpty Dumpty supermarket chain in Oklahoma City when in 1937 he struck upon the idea of the shopping cart. Goldman had observed that shoppers would often struggle with the wire and wicker baskets once they became too full, and he realized that this would stop them from buying. His initial inspiration for the design came from a folding chair in his office. Along with employee Fred Young, Goldman designed and built the first shopping cart, which held two wire baskets, one above the other, in a metal frame with wheels at the base. When the carts were not in use, the frame would fold flat, like the chair that suggested the design.

Goldman founded the Folding Carrier Basket Co., while a mechanic called Arthur Kosted developed a production line process to mass produce the carts, and began to introduce them in his stores. They were initially unsuccessful as men found them effeminate, and women found them a little too close to strollers.

As one customer said to Goldman, "I have been pushing enough baby carriages." Goldman overcame this problem by employing models of both sexes and various ages to use the carts around the store. The carts took off, with a seven year waiting list by 1940, and Goldman made his fortune collecting royalties on every shopping cart until his patents expired.

Although his original design did well, Goldman improved it some years later by developing the "nest" carts, now ubiquitous across the world. With a single basket, and significantly more room, this design has to its credit that nothing has been found to replace it after more than seventy years of use. **JG**

SEE ALSO: CART, BABY CARRIAGE, COLLAPSIBLE STROLLER

⬆ *A woman loads a cart—modestly sized by modern standards—at an A&P store in New York, in 1942.*

Jet Engine (1937)

Whittle and von Ohain speed up air travel for both the public and the military.

A jet engine uses a fan to suck air into a cylindrical chamber and a second turbine fan to compress it. Then fuel is mixed with this high-pressure air and ignited. The hot expanding burning gasses then blast out of a nozzle at the rear of the engine thrusting the engine forward, and the aircraft it is usually attached to, with great force.

The independent coinventors were the English aviation engineer Sir Frank Whittle (1907–1996) and the German airplane designer Dr. Hans von Ohain (1911–1998). Ohain's engine was tested in the Heinkel He178, flying first in August 27, 1939. Whittle's first engine, the Whittle Unit (WU), was completed in 1937 and subsequently fitted to an aircraft called the Pioneer (E.28/39), built by the Gloster Aircraft Company. The first flight was on May 15, 1941.

World War II saw swift developments of the jet engine and airplane. The United States's Bell XP-59 flew in September 1942, and by 1944 both the Messerschmitt Me 262 and the Gloster Meteor were being mass-produced. Jet-to-jet dogfights were taking place during the Korean War in 1950. By 1952 BOAC were using the de Havilland Comet jetliner on their London to Johannesburg route.

Jet engines work most efficiently at altitudes of between 6 and 9 miles (10 and 15 km). Here, modern aircraft have cruising speeds of between 420 and 580 miles (680 and 900 km) per hour, this being about 80 percent the ambient speed of sound. Propeller powered aircraft have to fly at much lower altitudes and fly much more slowly. Also propeller engines are much more costly to manufacture and maintain. **DH**

SEE ALSO: AIRSHIP, POWERED AIRPLANE, TURBOPROP ENGINE, SUPERSONIC AIRPLANE, SCRAMJET

⬆ *A 1944 turbojet engine made by Jets Ltd., a company formed with the backing of the British Air Ministry.*

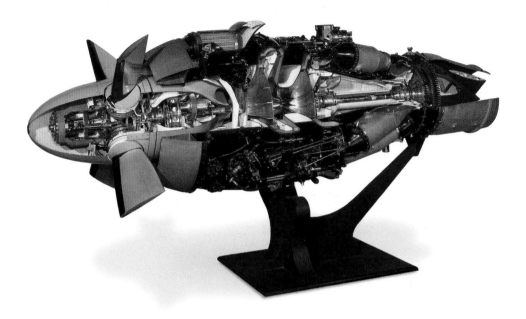

Turboprop Engine (1937)

Jendrassik designs suitable engine for slower-moving aircraft.

In 1937, Hungarian engineer György Jendrassik (1898-1954) designed and constructed a small turboprop engine. A year later he completed the larger "CS-1" engine, intending to use it on a military bomber, the RMI-1. Unfortunately, the CS-1 ran into combustion problems, keeping it from reaching its projected 1,000 horse-power. The RMI-1's designers were unable to implement an alternative engine before the RMI was annihilated in an air raid by the United States.

In spite of its turbulent birth the turboprop engine enjoyed subsequent success. It is primarily distinguished by its namesake, the turbine-driven propeller at the front. Whereas turbojet and turbofan engines generate thrust only at their rears, the thrust of the turboprop engine is generated mostly by the propeller.

The turboprop engine produces motion like any gas turbine engine. Air enters the engine and is compressed by a spinning, blade-covered cone called an axial compressor. The axial compressor pulls the air into a progressively smaller tube until it reaches a combustion area, where fuel injectors combine fuel and high-pressure air to create a powerful explosion that pushes exhaust out of the engine. As the exhaust leaves, it spins a turbine that rapidly rotates a drive shaft. The rotating drive shaft is then used to spin a gearbox, which in turn powers the plane's propeller.

Turboprop engines are most efficient at speeds below 500 miles (800 km) per hour and so are usually used in smaller, slower-moving aircraft that land and take off frequently. In order to increase efficiency at higher speeds, many modern turboprop engines now use smaller and more numerous blades. **LW**

SEE ALSO: AIRSHIP, POWERED AIRPLANE, JET ENGINE, SUPERSONIC AIRPLANE, SCRAMJET

⤒ *This engine from the 1940s powered the first turboprop airliner, the Vickers Viscount.*

Ballistic Missile (1938)

Dornberger initiates Nazi Germany's rocket program.

The history of rocketry dates back to around 900 CE, but the use of rockets as highly destructive missiles able to carry large payloads of explosives was not feasible until the late 1930s. War has been the catalyst for many inventions, both benevolent and destructive. The ballistic missile is intriguing because it can be both of these things; it has made possible some of the greatest deeds mankind has ever achieved, and also some of the worst.

German Walter Dornberger (1895–1980) and his team began developing rockets in 1938, but it was not until 1944 that the first ballistic missile, the Aggregat-4 or V-2 rocket, was ready for use. V-2s were used extensively by the Nazis at the end of World War II, primarily as a terror weapon against civilian targets. They were powerful and imposing: 46 feet (14 m) long, able to reach speeds of around 3,500 miles per hour (5,600 kph) and deliver a warhead of around 2,200 pounds (1,000 kg) at a range of 200 miles (320 km).

Ballistic missiles follow a ballistic flight path, determined by the brief initial powered phase of the missile's flight. This is unlike guided missiles, such as cruise missiles, which are essentially unmanned airplanes packed with explosives. This meant that the early V-2s flew inaccurately, so they were of most use in attacking large, city-sized targets such as London, Paris, and Antwerp.

The Nazi ballistic missile program has had both a great and a terrible legacy. Ballistic missiles such as the V-2 were scaled up to produce intercontinental ballistic missiles with a variety of warheads, but also the craft that have carried people into space. Ballistic missiles may have led us to the point of self-destruction, but without them man would have never been able to venture beyond our atmosphere. **JB**

> *"This third day of October, 1942, is the first of a new era in transportation, that of space travel."*
>
> Walter Dornberger

↑ *A V-2 rocket engine on a test stand; the engine was fueled by ethyl alcohol and liquid oxygen.*

→ *First fired in 1942, the V-2 rocket was the world's earliest successful large liquid-propellant missile.*

SEE ALSO: ROCKET, IRON ROCKET, TORPEDO, LIQUID FUEL ROCKET, LASER GUIDED BOMB

Teflon® (1938)

Plunkett discovers a useful new polymer.

In 1938 research chemist Roy Plunkett (1910–1994) was working at the DuPont Jackson laboratory in New Jersey. He had been trying to improve refrigerants to make them nontoxic and nonflammable. Plunkett and his technician Jack Rebok had produced 100 pounds (45 kg) of tetrafluoroethylene gas (TFE), storing it in cylinders on dry ice. When the time came to use the material, nothing came out of the cylinder, even though it weighed the same as before. The gas had turned into a white powder.

Plunkett and others at DuPont found that the substance was quite slippery and proved to be a good lubricant. It was resistant to chemicals and heat, and other substances would not adhere to it. The material

> "We recognized almost at once that the material was different and that it had potential . . ."

Lois Plunkett

was resistant to temperatures as high as 500°F (260°C). Plunkett and his colleagues realized the potential of this new polymer and DuPont set about marketing it.

At first Teflon® (the new trade name for this substance) was so expensive that no one seemed interested in purchasing it. However, this slowly changed as the material was used first in military and industrial applications and later in household use, most notably on nonstick pans. Awarded a patent in 1941, Teflon® is used as a coating for fabrics, wires, and metals—three-quarters of the pots and pans sold in the United States are coated with it—and also in industries such as aerospace and pharmaceuticals. **RH**

SEE ALSO: NONSTICK PANS

Fiberglass (1938)

Slayter and Thomas create a new insulator.

Fiberglass consists of extremely fine glass fibers, made from molten glass extruded at a specified diameter. Glassmakers have experimented with glass fibers throughout history, but mass manufacturing had to wait for the refinement of machine tooling before a practical product could be made possible.

The product commonly referred to as "fiberglass" was invented as a form of insulation by Russell Games Slayter (1896–1964)—he dropped the Russell early in life—and John Thomas of the Owens-Illinois Glass Company in the 1930s. In 1938 Owens-Illinois and Corning Glass formed Owens-Corning to make fiberglass using Slayter and Thomas's method.

Fiberglass was discovered—like many other scientific discoveries—by accident. While Thomas's assistant, Dale Kleist, was spraying molten glass for a project, tiny fibers formed. Thomas realized the process could be used to improve the production of fiberglass. Thomas and Slayter refined the process, leading to what is known as the steam-blowing method. As the molten glass is extruded through hundreds of tiny nozzles on the "brushing plate," it is hit by a blast of steam or compressed air, which draws out the filaments to extremely long lengths. The important part of Slayter and Thomas's discovery was the method of applying the blast of steam in a way that avoided breaking or disrupting the filaments.

The practical uses of fiberglass are virtually endless. Slayter himself held over ninety patents for fiberglass technologies. Besides insulation, it is used for medical equipment, fire-resistant textiles, electronics, and wall coverings. In 1953 Owens-Corning went into partnership with General Motors on the first car with a body made entirely of fiberglass-reinforced plastic, the Chevrolet Corvette. **BO**

SEE ALSO: GLASS, BLOWN GLASS, PERSPEX®

Walkie-talkie (1938)

Gross introduces the hand-held two-way radio.

Walkie-talkies are the portable two-way radios that paved the way for mobile phones by showing the public the joys of talking to faraway people while walking around. In World War II they allowed troops to communicate and, since then, the police, the coast guard, and even children playing games have used them to relay information.

Their exact origins are rather hazy, though. Once radios had been invented, the next big thing was making them smaller and more portable and there is much disagreement over exactly when a two-way radio became a walkie-talkie.

In 1937 a man called Don Hings, born in England and raised in Canada, built a waterproof two-way field radio. This radio weighed almost 12 pounds (5.5 kg) and was about the size of a toaster, but was definitely portable enough to count. Hings's radio was built for air crashes, so that survivors could guide in rescuers by transmitting to the manufacturer or the Canadian Signal Corps. Before this radio, most were portable only if someone carried it while another used it; but Hings' could be carried and operated by one person.

Engineer Alfred J. Gross (1918–2000) made his own lighter, smaller version in 1938 while still a teenager. His designs caught the attention of the U.S. Office of Strategic Services (now the CIA), who recruited him to design their radios. Gross's designs went on to play a large role in World War II and soon entered civilian use.

It was Gross's designs of wrist radios that found their way into Chester Gould's *Dick Tracy* comics, but it was walkie-talkies in general that have led to the many developments since. **DK**

SEE ALSO: WIRELESS COMMUNICATION, CITIZENS' BAND RADIO, CELLULAR MOBILE PHONE, DIGITAL CELL PHONE

↗ *The BC-611 walkie-talkie was made by Motorola and used by the U.S. military in World War II.*

BEFORE OPERATING
READ TM-11-235

"Gould [asked to use] the walkie-talkie idea . . . and he gave Dick Tracy that two-way wristwatch."

Alfred J. Gross

Lysergic Acid Diethylamide (LSD) (1938)

Hofmann discovers the world's most powerful mind-altering drug.

Lysergic acid diethylamide (LSD) is a powerful psychedelic drug. While now commonly associated with 1960s dropout youth culture, it was heralded as a wonder drug in the 1940s and 1950s and was used to treat thousands of psychiatric patients.

Swiss chemist Albert Hofmann (1906–2008) first synthesized LSD in 1938, expecting it to be useful as a medicinal stimulant. In 1943 he returned to studying it and after experiencing some pleasant sensations while working with the drug he took a dose of 0.25 mg. Hofmann bicycled home and began to experience its psychedelic effects, the world's first "trip." He reported that the morning after he felt entirely renewed and that his senses were "vibrating in a condition of highest sensitivity."

Today LSD is mainly taken as a recreational drug for its psychological effects. Common accounts are of colorful hallucinations, time distortions, loss of identity, and synesthesia. A trip can last up to around twelve hours, depending on dose. Physical effects include hypothermia, fever, increased heart rate, perspiration, tremors, and insomnia. LSD has been known to induce psychosis, and more commonly can result in intermittent flashbacks to the trip.

Due to LSD's extraordinary impact on the psyche, it has attracted a number of high profile users, who believed that it could unlock certain aspects of experience that are otherwise hidden. Preeminent among these were countercultural psychologist Timothy Leary, who urged Americans to "turn on, tune in, and drop out," and author Aldous Huxley, who chose to be injected with LSD as he died. **JG**

> "Wrong and inappropriate use has caused LSD to become my problem child."
>
> Albert Hofmann

SEE ALSO: HEROIN, MDMA (ECSTASY), VALIUM

◩ *Once a secret weapon of the CIA, LSD was hijacked for its psychoactive effect by 1960s counterculture.*

Bazooka (1942)

Americans develop an antitank weapon.

Tanks were a great problem for infantry during the first years of World War II. The thick armor was proof against small arms, and grenades powerful enough to penetrate it had to be placed directly on the tank.

Three Americans addressed this problem by combining a shaped-charge hand grenade with an electrically fired rocket launcher. Dr. Clarence Hickman had worked with Robert Goddard on tube-fired rockets during World War I. Starting in 1940, he helped U.S. Army officers Edward Uhl and Leslie Skinner to develop an electrically fired rocket launcher. When the user pulled the trigger, a battery in the stock sent a charge to ignite the rocket, which then fired through a steel tube. The first fielded version was officially called the M1 Rocket Launcher, but it was soon nicknamed the bazooka because it resembled a musical instrument of the same name.

By the end of 1942 U.S. troops in North Africa were equipped with the bazooka, and early the following year they used it against enemy armor for the first time. Not only was it effective against enemy tanks, it could also damage pillboxes, blast holes in barbed wire, and clear minefields. By the end of World War II the United States had produced almost half a million bazookas and more than fifteen million rockets.

German engineers analyzed bazookas captured in North Africa in early 1943 and developed their own version with a greater range and a larger warhead. The Soviet Union later incorporated features of the bazooka in their design for launchers of rocket-propelled grenades, many of which remain in use across the globe today. **ES**

SEE ALSO: SHRAPNEL SHELL

⬅ *U.S. troops destroy a German tank with a bazooka during the Allied invasion of Normandy in 1944.*

Super Glue (1942)

Coover stumbles on a powerful adhesive.

In 1942 Harry Coover (1917–2011), a chemist working for Eastman-Kodak, was seeking a way to manufacture ultra-clear plastic gunsights. The group of chemicals his team were investigating, the cyanoacrylates, proved not very useful. They were very sticky, and contact with even a tiny amount of water (such as is found on virtually every surface) caused them to bind.

It was not until several years later, when he revisited the cyanoacrylates while working on another project, that Coover realized they had stumbled upon something special. The prototype glue stuck together everything they tried, without requiring any heat or pressure. The substance, marketed as "Eastman 910" in 1958, became popularly known as super glue.

> *"The medics used the [super glue] spray, stopped the bleeding. . . . And many, many lives were saved."*
>
> Harry Coover

As well as being a powerful and useful adhesive, super glue has been put to a number of other uses. During the Vietnam War it was used extensively as an emergency medical intervention. Because it bonded skin and tissue so efficiently, it was used to seal wounds and stop bleeding quickly, without the need for time-consuming stitches.

Crime scene investigators also use it as a way of revealing fingerprints. The object to be printed is "fumed" by placing a few drops of the super glue on a heater inside a sealed tank. Gas produced by the heated super glue sticks to the oils left by the fingerprint, making it visible to the naked eye. **SB**

SEE ALSO: GLUE, TAPE

Cluster Bomb (1939)

Several countries simultaneously introduce a bomb that disperses smaller bombs.

The cluster bomb has courted controversy since its induction in modern warfare in 1939. A conventional bomb consists of a single container carrying an explosive charge that is designed to explode upon impact. The cluster bomb differs through the addition of an outer casing carrying dozens of small bomblets. The casing splits open in mid-air, releasing a shower of smaller bomblets that impact over a broad area.

Often dropped by parachute, cluster bombs are highly versatile, if not particularly accurate. They can wreak havoc on soft or unarmored targets such as airfields and formations of men; cluster bombs containing shrapnel are able to pierce armored tanks and penetrate concrete. Cluster bombs really came to the fore during the Vietnam War. U.S. forces carpet-bombed the dense forests of Vietnam, Laos, and Cambodia with cluster bombs carrying chemical weapons such as napalm. The bombs were designed to set fire to the foliage in order to unmask the enemy,

but such indiscriminate bombing led to much loss of civilian life, compounded by the roughly 5 percent of bombs that fail to detonate.

Despite the dangers from unexploded duds and conclusive evidence that cluster bombs cannot be targeted precisely even with guided circuitry, cluster bombs have been widely deployed during recent conflicts in Kosovo, Iraq, Afghanistan, and the Israeli-Hezbollah war. Pressure has been mounting from the International Committee of the Red Cross (ICRC) for a global pact to ban cluster bombs. To date the three major stockpilers, Russia, China, and the United States, have refused to commit to any treaty. **SG**

SEE ALSO: SHRAPNEL SHELL, BOUNCING BOMB, ATOMIC BOMB, HYDROGEN BOMB

⬆ *Two 500-pound incendiary cluster bombs fall toward Kiel, Germany, from U.S. aircraft in 1943.*

Helicopter (1939)

Sikorsky solves the problem of stabilizing helicopter takeoff and flight.

The first manned helicopter flight was achieved by the Frenchman Paul Cornu who lifted his twin-rotor craft off the ground for twenty seconds in 1907; his machine unfortunately broke up on landing. In 1909 Igor Sikorsky (1889–1972) built two helicopters but these could lift very little more than their own weight. The first practical helicopter was the German Focke-Wulf FW 61, which flew in 1936. By 1939 the British had built the two-seater Weir W.6, which was powered by a pair of rotors mounted independently, one on each side of the fuselage. The Weir W.6's prototype was the first helicopter in the world to carry three occupants

Many control problems had to be solved, the main ones being unsymmetrical lift, which caused the craft to flip over on takeoff, and the fact that the body's natural tendency was to spin in the opposite direction to the rotors. However, one big advance was the realization that changing the angle at which the rotor blades were set was much more effective for

stabilizing the helicopter in flight than trying to change the rate at which the rotors rotated.

The real breakthrough came with Sikorsky's VS-300 in 1939. As well as the horizontal main rotor the prototype had two smaller tail rotors, one ensuring horizontal stability and the other acting like a rudder and controlling the direction of flight. This was followed in the United States in 1945 by the highly successful mass-produced Bell 47.

Today there are more helicopters in military service than in civilian operations. The vision of city-center "vertiports" speeding passengers to local airports and between nearby airports is still to be realized. **DH**

SEE ALSO: AUTOGYRO

⤴ *This VS-300 has only one tail rotor, but the prototype had a second, horizontal rotor mounted above it.*

Semiconductor Diode

(1939)

Ohl paves the way to the transistor.

Russell Ohl (1898–1987) was a precocious talent who by the age of sixteen had already entered Pennsylvania State University. After a period in the Army Signal Corps and a brief career as a teacher, Ohl finally took a research position in U.S. industry.

Early radios were only able to receive low-frequency transmissions. At Bell Labs in Holmdel, New Jersey, Ohl worked to create an improved radio receiver for high frequencies. Here he experimented with semiconductor materials that he thought would outperform the electron tubes used in existing receivers. An expert in the behavior of crystals, Ohl investigated different materials for semiconductors, such as germanium and silicon. The crystals were heated, and when cool would be sliced for use.

In 1939 Ohl was working with a silicon sample that he noticed had a crack down the middle. When he tested the electrical resistance of the sample he noted that when it was exposed to light, the current flowing between the two sides of the crack jumped significantly. As a result of this accidental impurity in the silicon, atoms on one side of the crack were shown to have extra electrons; the crystallized silicon on the other side had a deficit. Ohl deduced that it was these impurities that made different areas of the silicon more or less resistant to electrical flow, and thus it was the "barrier" between these areas of differing purities that made the semiconductor work. Ohl named the two areas "p" (positive) and "n" (negative); the barrier between was called the "p-n junction." In the research that followed, Ohl was able to show that by super-purifying germanium crystals he was able to create semiconductor diodes that could behave in predictable and measurable ways. **TB**

SEE ALSO: VALVE DIODE, TRIODE, LIGHT-EMITTING DIODE (LED), TRANSISTOR, TRANSISTOR RADIO, MOSFET

DDT

(1940)

Müller patents a powerful insecticide.

In 1874 Austrian chemist Othmar Zeidler made one of the more famous chemicals of all time—dichlorodiphenyltrichloroethane (DDT).

Zeidler was more interested in making chemicals then determining their possible uses. It was not until 1935 that another chemist, Paul Hermann Müller (1899–1965) began a hunt for insecticides while working at the J. R. Geigy Corporation in Switzerland. By 1939 Müller had independently synthesized DDT, studied it for use as an insecticide, and found it was lethal for mosquitoes, potato beetles, body lice, and other pests. Müller's studies also revealed that in small doses DDT was safe for humans. In 1940, he was

> *"To only a few chemicals does man owe as great a debt as to DDT . . . "*
>
> National Academy of Sciences, 1970

granted a Swiss patent and by 1942 DDT products were commercially available.

Almost immediately, the power of DDT was recognized as unparalleled in combating infectious diseases such as yellow fever, dengue fever, and malaria, all spread by mosquitoes. Use of DDT actually eradicated malaria from whole islands and proved that insecticides were a way to combat disease. By the early 1970s, however, use of DDT was universally banned due to environmental concerns. **RBk**

SEE ALSO: PESTICIDE

→ *Parents smile indulgently as their child is sprayed with DDT delousing powder at a school in Germany.*

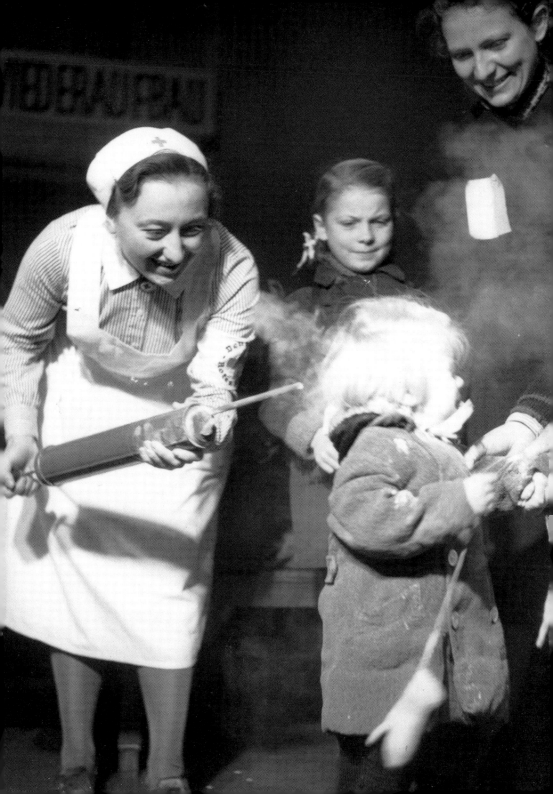

Silicone Rubber

(1940)

Rochow develops a rubber substitute.

In the 1930s alternatives to natural rubber were desperately needed. Uses for rubber were increasing, but the supply—trees grown mostly in Asia—was literally being tapped out. Soon World War II made it impossible to obtain natural rubber. However, synthetic rubber had been around for a few decades. Russian Sergey Lebedev made the first fake, butadiene rubber (BR), in 1910. Since then scientists had been in a race, either to make bulk quantities of BR faster and cheaper or to discover the next great fake.

That fake turned out to be the silicone rubber of American Eugene Rochow (1909–2002). Within five years of his first day at General Electric's (GE) Research

> *"The most important single experiment . . . in the history of the silicone industry."*
>
> Herman Liebhafsky, GE chemist

Laboratory, Rochow made one of the most important materials of the modern age—silicon rubber (SR). A unique fake, Rochow's rubber was the first with no carbon–carbon bonds. In their place were silicon–carbon bonds, giving SR unique properties.

Scaling up Rochow's SR recipe would not be easy. Disposal or recycling of its by-products was tricky, one ingredient was flammable, and another ingredient was controlled by a rival company. Within the year Rochow had a better SR recipe. From Rochow's work sprang the SR industry, which by the end of the twentieth century was producing nearly three million metric tons of SR and its derivatives per year. **RBk**

SEE ALSO: SYNTHETIC RUBBER

Penicillin Production

(1941)

Heatley initiates manufacture of penicillin.

In 1938, Howard Florey (1898–1968) and Ernst Chain (1906–1979), two pathologists working at the University of Oxford, read a paper published nine years earlier about a substance called penicillin. Its author, Alexander Fleming, recounted how spores of the mold *Penicillium notatum* had entered his bacterial culture dishes and killed some of the bacteria.

Florey and Chain recognized the significance of Fleming's observation and obtained a culture of the original mold. Initially they encountered difficulties in obtaining enough penicillin, but Norman Heatley (1911–2004), a biochemist on the team, devised ways of isolating penicillin without destroying it. Monitoring the extracted penicillin on mice infected with bacteria, they found animals treated with penicillin survived, while untreated animals died.

With World War II now underway, the group recognized penicillin's enormous potential to treat war wounds. In 1941 Heatley traveled to the United States to start the commercial production of penicillin. Working with a team at Northern Regional Research Laboratory, Peoria, Illinois, he increased the yield of penicillin thirty-four times by adding a by-product of cornstarch and lactose to the fermentation. By D-day in June 1944 enough penicillin was available to allow unlimited treatment of Allied troops.

In 1945 Florey, Chain, and Fleming shared the Nobel Prize. In his Nobel address, Fleming presciently predicted the problem of antibiotic resistance, where bacteria would develop resistance to penicillin and other antibiotics through their overuse. **JF**

SEE ALSO: ASPIRIN, ACETAMINOPHEN, BETA BLOCKERS, TETRACYCLINE, ANTIVIRAL DRUGS, HUMULIN, PROZAC®

➡ *Alexander Fleming examining petri dishes in his lab at St. Mary's Hospital, London, on December 18, 1943.*

Digital Electronic Computer (1941)

Zuse builds first electromechanical computer.

The two world wars led to many breakthroughs in all areas of science and technology. It was not, however, an easy time to get independently funded inventions off the ground, as German engineer Konrad Zuse (1910–1995) discovered.

In 1936 Zuse invented the Z1, an electromechanical binary computer, but it was completely obliterated by World War II bombing that left no trace of it or its blueprints behind. Work on the Z2 was difficult because the war made it impossible for Zuse to work with other computer engineers from Britain or the United States, but he still managed to complete it in 1940. The Z3, a more sophisticated version of the Z2, was finished in 1941, partially funded by contributions from the DVL (the German Experimentation Institution for Aviation). It was the first fully functional program-controlled electromechanical digital computer in the world. Sadly this too was destroyed in the war, but greater care was taken with the Z4, which was moved from country to country to ensure its survival.

One of Zuse's main motivations to create a computer was to make life easier for his fellow engineers and scientists. He had a passionate distaste for performing the long, time-consuming calculations that his profession was so often called upon to make. It was during the time that he was studying as a civil engineer that he began to wish for a machine that would take care of these irksome problems for him.

Although the original Z3 was destroyed, a working reconstruction was made in 1960. It is on permanent display at the Deutsches Museum, Munich. **CL**

SEE ALSO: MECHANICAL COMPUTER, PERSONAL COMPUTER, SUPERCOMPUTER, LAPTOP, PALMTOP COMPUTER

◄ *Zuse inspects a replica of the Z1, his first computer, in the Museum of Transport and Technology, Berlin.*

Night Vision Goggles (1942)

Spicer improves vision for night warfare.

In early World War II, U.S. engineer William Spicer (1929–2004) was aware of the visibility problem of conducting military operations at night and was looking into a solution based on photoemission. In photoemission, light is treated as packages of energy called photons, which strike a material to bounce out electrons. In 1942 Spicer developed the first night vision goggles using image enhancement.

At night, light is present in small quantities from various sources, but our eyes may not detect it. The photons from this light enter the goggle lens and strike a light-sensitive surface called a photocathode, releasing electrons. The electrons are accelerated

> *"When you see the little green pictures on CNN of people . . . at night, think of Professor Spicer."*
>
> Piero Pianetta

toward a microchannel plate that releases thousands more electrons through a cascade reaction. These electrons hit a screen coated with phosphor chemicals to emit visible light. As thousands of electrons are generated from the small quantity of original photons, a much brighter corresponding image is produced.

When light enters the goggles it can be of any color, but once the photons transfer their energy to the electrons this information is lost. The image should therefore be black and white but it is actually green. Green phosphors are used in the screen because the eye can differentiate more shades of green than any other color, gaining as much detail as possible. **RB**

SEE ALSO: SPECTACLES, INFRARED PHOTOGRAPHY, IMAGE INTENSIFIER, ELECTRON MICROSCOPE, COLOR NIGHT VISION

Bazooka (1942)

Americans develop an antitank weapon.

Tanks were a great problem for infantry during the first years of World War II. The thick armor was proof against small arms, and grenades powerful enough to penetrate it had to be placed directly on the tank.

Three Americans addressed this problem by combining a shaped-charge hand grenade with an electrically fired rocket launcher. Dr. Clarence Hickman had worked with Robert Goddard on tube-fired rockets during World War I. Starting in 1940, he helped U.S. Army officers Edward Uhl and Leslie Skinner to develop an electrically fired rocket launcher. When the user pulled the trigger, a battery in the stock sent a charge to ignite the rocket, which then fired through a steel tube. The first fielded version was officially called the M1 Rocket Launcher, but it was soon nicknamed the bazooka because it resembled a musical instrument of the same name.

By the end of 1942 U.S. troops in North Africa were equipped with the bazooka, and early the following year they used it against enemy armor for the first time. Not only was it effective against enemy tanks, it could also damage pillboxes, blast holes in barbed wire, and clear minefields. By the end of World War II the United States had produced almost half a million bazookas and more than fifteen million rockets.

German engineers analyzed bazookas captured in North Africa in early 1943 and developed their own version with a greater range and a larger warhead. The Soviet Union later incorporated features of the bazooka in their design for launchers of rocket-propelled grenades, many of which remain in use across the globe today. **ES**

SEE ALSO: SHRAPNEL SHELL

◁ *U.S. troops destroy a German tank with a bazooka during the Allied invasion of Normandy in 1944.*

Super Glue (1942)

Coover stumbles on a powerful adhesive.

In 1942 Harry Coover (1917–2011), a chemist working for Eastman-Kodak, was seeking a way to manufacture ultra-clear plastic gunsights. The group of chemicals his team were investigating, the cyanoacrylates, proved not very useful. They were very sticky, and contact with even a tiny amount of water (such as is found on virtually every surface) caused them to bind.

It was not until several years later, when he revisited the cyanoacrylates while working on another project, that Coover realized they had stumbled upon something special. The prototype glue stuck together everything they tried, without requiring any heat or pressure. The substance, marketed as "Eastman 910" in 1958, became popularly known as super glue.

> *"The medics used the [super glue] spray, stopped the bleeding … And many, many lives were saved."*
>
> Harry Coover

As well as being a powerful and useful adhesive, super glue has been put to a number of other uses. During the Vietnam War it was used extensively as an emergency medical intervention. Because it bonded skin and tissue so efficiently, it was used to seal wounds and stop bleeding quickly, without the need for time-consuming stitches.

Crime scene investigators also use it as a way of revealing fingerprints. The object to be printed is "fumed" by placing a few drops of the super glue on a heater inside a sealed tank. Gas produced by the heated super glue sticks to the oils left by the fingerprint, making it visible to the naked eye. **SB**

SEE ALSO: GLUE, TAPE

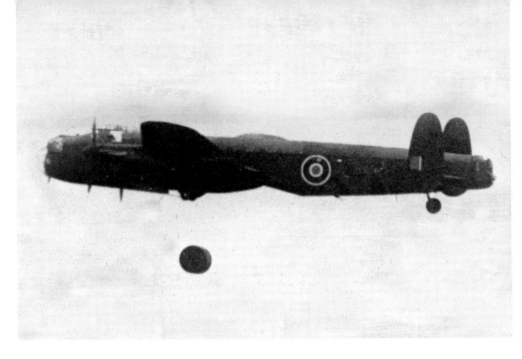

Bouncing Bomb (1942)

Wallis builds the dambusting bombs.

Conceived at the height of World War II by English aeronautical engineer Barnes Neville Wallis (1887–1979), the bouncing bomb was a weapon with a unique purpose; namely the destruction of Hitler's Ruhr Valley-based hydroelectric plants.

Codenamed "Upkeep," the cylinder-shaped bouncing bomb had its origins among naval gunners of two centuries before, who had increased the range of their cannons' projectiles by literally skimming or bouncing them off the water.

The effect had also been noticed by Allied pilots attacking ships. Forced by enemy fire to drop their bombs prematurely, the bombs had, under certain conditions, similarly bounced their way onward to the target. Reasoning that a bomb system able to do this by design would enable the destruction of targets otherwise requiring an impossibly heavy load of explosives, or a suicidal pilot, Wallis set about creating one. Numerous designs of scaled-down "bombs" were

tested before Wallis concluded that a casing striking the water at an angle of around seven degrees would produce a reliable bounce. Backspin improved the results greatly.

Successful testing against a disused Welsh dam confirmed that, with a depth-sensitive fuse, and the tamping effect of the water, 6,500 pounds (2,950 kg) of explosive would suffice for each dambusting bomb. But the bombs would need to be dropped from the dangerously low height of 60 feet (18 m), and with a precision previously unheard of in combat warfare. Success was finally achieved with the aid of special sights in each aircraft. **MD**

SEE ALSO: CLUSTER BOMB, ATOMIC BOMB, HYDROGEN BOMB

⊞ *An Avro Lancaster testing a bouncing bomb in 1943 before attacking the dams of the Rhur Valley, Germany.*

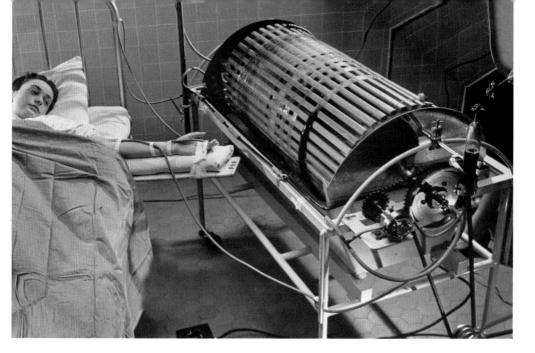

Kidney Dialysis (1943)

Kolff removes toxic waste products from the blood of kidney patients.

Willem J. Kolff (1911–2009), a doctor working in occupied Holland during World War II, cobbled together the first kidney dialysis (hemodialysis) machine.

When kidneys are not functioning correctly, waste products accumulate in the blood, and can be fatal. Kolff was aware of experiments showing that when two solutions of different chemical concentrations are separated by a permeable membrane an exchange of molecules takes place from the area of greater concentration to the area of lower concentration.

Kolff's machine consisted of 66 feet (20 m) of cellophane tubing wound around a wooden drum that was suspended horizontally inside a tank filled with saline. As the drum was rotated by a motor, the patient's blood was forced through the tubing and its waste products crossed the membrane into the saline. A severe shortage of materials due to the war forced Kolff to improvise; he used cellophane that came from sausage casings. Nevertheless, he was able to treat his first patient in 1943.

After the war Kolff moved to the United States, where he continued to improve the design of the artificial kidney, collaborating with a producer of intravenous and saline solutions. The first commercial artificial kidney was marketed in 1956. Initially dialysis was restricted to acute cases (drug intoxication, third-degree burns, and mismatched transfusions), with the idea of keeping people alive long enough for their kidneys to recover or the poison to be eliminated. By the early 1960s long-term dialysis was made possible by the development of a Teflon U-shaped shunt that was permanently attached to the patient. **JF**

SEE ALSO: SYNTHETIC BLOOD, LITHOTRIPTER, HEART-LUNG MACHINE, VENTILATOR, ARTIFICIAL HEART, ARTIFICIAL LIVER

⬆ *A kidney dialysis machine from 1947 purifies blood flowing from a tube inserted into the patient's wrist.*

Aqualung

(1943)

Cousteau and Gagnan breathe at depth.

The aqualung, invented in 1943 by Frenchmen Jacques Cousteau (1910–1997) and Émile Gagnan (1900–1979) transformed underwater exploration, engineering, and marine biology. For the first time, divers could move around freely, and undertake lengthy dives. The invention consisted of compressed air in cylinders, worn on the diver's back and connected to a mouthpiece through a regulated "demand" valve. This Self-Contained Underwater Breathing Apparatus—SCUBA—is very similar to the diving equipment we use today.

Before this aqualung was invented, divers used various techniques for underwater exploration. Snorkeling, using a short tube connected to the mouth with one end above the surface, allows exploration in shallow waters. Attempts to explore deeper using a longer snorkel tube were limited by the diver's need to stay fairly close to the air source. A rebreather system, in which the diver's exhaled gases were filtered, was unreliable and often caused deaths.

Gagnan was an engineer with expertise in gas valves, while Cousteau, a commander in the French Navy during World War II, realized the potential value of free-moving "frogmen" who could attach explosives to enemy ships. Even with their new scuba equipment, there was a lot to learn. Cousteau's first dive was nearly disastrous, as he did not know that oxygen becomes toxic at depths below 30 feet (9 m).

After the war, Cousteau continued to lead the development of underwater exploration and became known for underwater filming and photography. **SC**

"Man carries the weight of gravity on his shoulders. But he has only to sink below the surface to be free."

Jacques Cousteau

SEE ALSO: STANDARD DIVING SUIT, REBREATHER, SUBMARINE, BATHYSPHERE

◹ *Melbourne engineer Ted Eldred's Porpoise scuba set had a single hose, so avoiding Cousteau's patents.*

Disposable Catheter

(1944)

Sheridan improves health care with a tube.

Half a century ago, urinary catheters were made of laminated, braided cotton. Rather than being disposable, they were cleaned and reused. Far more than just turning the stomachs of the patients who needed them, the catheters posed an increased risk of infection—with potentially fatal consequences.

That all changed with the dream of David Sheridan (1908–2004). While working as a floor refinisher, this American son of Russian immigrants decided that he could make a better catheter. That dream, coupled with the fact that World War II threatened to cut off U.S. hospitals from their French catheter suppliers, caused Sheridan to act. Despite having only an eighth-

> *"I always had the idea that a catheter shouldn't be used more than one time."*
>
> David Sheridan

grade education, Sheridan invented a machine that allowed hollow rubber tubes to be made into catheters that could be used a single time, and then discarded. A painted strip on the catheter ensured that it would show up on an X-ray so that the catheter's correct placement could be determined.

Though he started out with the humble urinary catheter, Sheridan soon advanced to other, more sophisticated tubes. He invented an endotracheal tube, which allows patients to breathe during surgery, and also several catheters for use in heart surgery. This all started out, however, with a hollow tube that enabled people to urinate. **BMcC**

SEE ALSO: BALLOON CATHETER, INTRAVASCULAR STENT

Electron Spectrometer

(1944)

An MIT team measure electron energy.

The first spectrometer was devised by Martin Deutsch (1917–2002) and Robley D. Evans (1907–1995) at the Massachusetts Institute of Technology (MIT). Electrons are a by-product of nuclear reactions and the first electron spectrometers were used to monitor the radiation from the nuclear tests that took place toward the end of World War II. They have since become a "must have" instrument on scientific space missions.

The fourth state of matter is plasma, where some of the outer electrons of the atoms have been knocked away, and move off freely in space. Electrons are relatively simple fundamental particles. Their mass, charge, and collision cross-section are well known.

> *"I couldn't reduce the explanation to a freshman level. That means we really don't understand it."*
>
> Richard Feynman on the behavior of electrons

That leaves their speed and direction of motion as unknowns. An electron spectrometer measures their kinetic energy by registering the way in which the trajectories of the electrons are bent as they pass through either an electrostatic or a magnetic field.

The Earth's upper ionosphere is plasma, as is the wind of particles ejected by the sun. Interesting auroral physical processes occur when this solar wind hits a planetary ionosphere; electron spectrometers are used to measure the distribution of electron energies. Placed on spacecraft, they can also measure how the electron energy spectrum varies with the direction in which the electrons are moving. **DH**

SEE ALSO: GEIGER COUNTER, SPECTROPHOTOMETER

Atomic Bomb (1945)

Scientists of the Manhattan Project complete the bomb destined to end World War II.

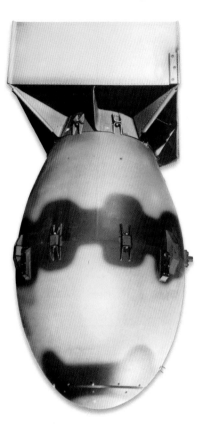

"It was a paradise for scientists. In Los Alamos, whatever you wanted, you got."

Joseph Rotblat, physicist

⬆ *A reproduction of the bulbous "Fat Man" bomb used against Nagasaki, Japan, in 1945.*

➡ *In an explosion codenamed "Trinity," the world's first atomic bomb detonates at Alamogordo.*

During World War II the United States used an unprecedented $2 billion to feed an ultra-secret research and development program, the outcome of which would alter the relationships of nations forever. Known as the Manhattan Project, it was the search by the United States and her closest allies to create a practical atomic bomb: a single device capable of mass destruction, the threat of which alone could be powerful enough to end the war.

The motivation was simple. Scientists escaping the Nazi regime had revealed that research in Germany had confirmed the theoretical viability of atomic bombs. In 1939, in support of their fears that the Nazis might now be developing such a weapon, Albert Einstein and others wrote to President Franklin D. Roosevelt (FDR) warning of the need for atomic research. By 1941 FDR had authorized formal, coordinated scientific research into such a device. Among those whose efforts would ultimately unleash the power of the atom was Robert Oppenheimer (1904–1967), who was appointed the project's scientific director in 1942. Under his direction the famous laboratories at Los Alamos would be constructed and the scientific team assembled.

An atomic bomb initiates a nuclear chain reaction, thereby releasing truly vast amounts of energy. An initial problem was the production of enough "enriched" uranium to sustain such a reaction. Instrumental in the solution was Italian physicist Enrico Fermi. In a former squash court under Chicago University, he and other scientists created the first ever controlled, self-sustaining nuclear chain reaction. The project ultimately created the first, man-made nuclear explosion, which Oppenheimer called "Trinity," at Alamagordo, New Mexico, on July 16, 1945. **MD**

SEE ALSO: BOUNCING BOMB, CLUSTER BOMB, HYDROGEN BOMB

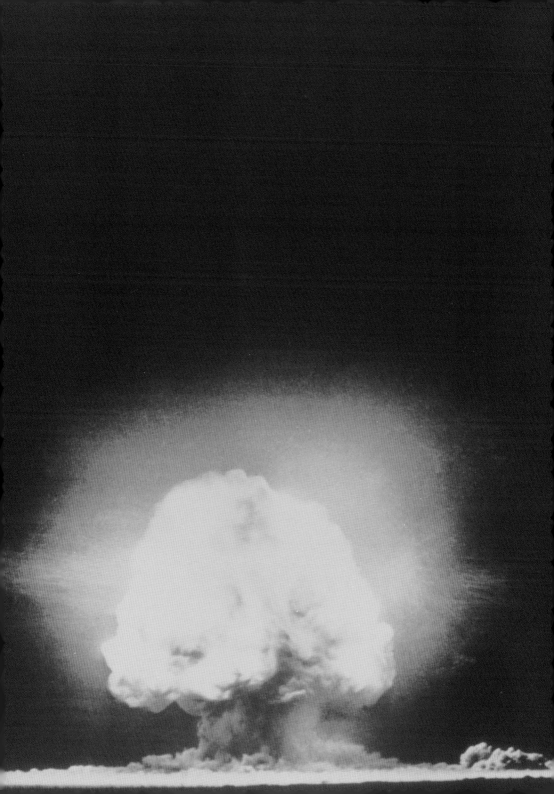

Cruise Control (1945)

Teetor automates speed control in cars.

Ralph Teetor (1890–1982), a prolific—and blind—inventor, was inspired to invent cruise control one day while taking a ride in a car driven by his lawyer. The lawyer had the habit of slowing down while talking and speeding up while listening. The car's jerky, rocking motion so annoyed Teetor that he became determined to invent a speed-control device.

Teetor received his first patent on a cruise-control device in 1945 after a decade of tinkering, and it was first offered commercially on Chrysler's Imperial, New Yorker, and Windsor models in 1958. Early names for his invention included Controlmatic, Touchomatic, Pressomatic, and Speedostat, before people settled on the familiar Cruise Control.

With cruise control, the driver sets the speed and the system then takes over the vehicle's throttle to maintain that speed. The cruise control gets its speed signal from a rotating driveshaft, the speedometer cable, a speed sensor on the wheels, or from the engine's revolutions per minute (rpm). The car maintains its speed by pulling the throttle cable with a solenoid or a vacuum-driven servomechanism.

Buttons, usually located on the steering wheel, allow the driver to set the speed, accelerate, and decelerate, and return to the set speed after braking. A tap from the driver on the clutch or brake will turn off the cruise control, while pressing the accelerator will allow the car to speed up; releasing the accelerator returns the car to the set speed. Most cruise control systems will not engage at speeds lower than 15 miles per hour (25 kph).

More advanced adaptive cruise-control systems are currently being developed that will automatically adjust a car's speed at a safe following distance, using radar installed behind the front grill. **BO**

SEE ALSO: MOTORCAR, ACCELERATOR

Sofa Bed (1945)

Castro eases life in a restricted space.

In Depression-era America, Sicilian immigrant Bernard Castro (1904–1991) spotted a niche in the market. Noting that New Yorkers did not have much money to spend on luxuries and that their apartments in the city were cramped, he proposed a bed that could be folded into a sofa during the day, significantly increasing available living space.

Naming the beds "Castro's Convertibles," he sold them from his loft store on Twenty-first Avenue. The only convertible sofa otherwise available was known as a davenport, which was difficult to open and close and did not look like a sofa when folded. One night Castro and his wife found that their four-year-old daughter had managed to open their convertible;

> *"Mr. Castro [was a] genius not only at designing a piece of furniture … but at marketing his product."*
>
> Dennis Hevesi, journalist

they realized that this could be a major selling point.

Castro's other masterstroke was recognizing the power of advertising on television. He was the first person to buy a local advertising spot from the single television station in New York. The commercial was also the first to feature a child, his daughter Bernadette, who showed how easy it was to open the bed. His slogans bragged that his invention was "the first to conquer living space" and that conversion between sofa and bed was "so easy a child can do it."

For a time all sofa beds were referred to as Castro's and he became a millionaire. In total Castro sold more than 5 million of his convertibles before retiring. **HP**

SEE ALSO: HAMMOCK

Synchrotron (1945)

McMillan and Veksler increase the energy of charged particles.

The cyclotron was a subatomic particle accelerator that used a magnetic field to force the particles to move around a circular path in a thin, doughnut-shaped vacuum chamber, with a fixed frequency electric field for acceleration. This was then developed into the synchrotron, in which both the magnetic and the electrical fields could be varied. By decreasing the frequency of the applied voltage as the electrons move faster, the accelerating voltage and the orbiting particles could be synchronized. This phase stability ensured that the particles that were going too fast were accelerated less than those going too slowly, and the result was a stable cloud of particles that were gradually accelerated together. In 1945 two proposals for a synchrotron were put forward, one by Edwin McMillan (1907–1991) in the United States and the other by Vladimir Veksler in the U.S.S.R.

Soon a host of synchrotrons were being built. Collisions between the particles that they accelerated could be used to study their structures, and the synchrotron radiation that these particles emitted could be used as a source for investigation in the far ultraviolet and X-ray regions of the spectrum.

Completed in 1983, the Tevatron at the Fermi National Accelerator Lab in the United States is 3.9 miles (6.3 km) in diameter and it accelerates protons and antiprotons to enormous energies and then lets them collide. A sevenfold increase in final energy is obtained by the newer Large Hadron Collider (16.6 miles/26.7 km in diameter) at the European Laboratory for High Energy Physics (CERN) on the Switzerland/France border. **DH**

SEE ALSO: CYCLOTRON, LARGE HADRON COLLIDER

↗ *McMillan (left) stands with Edward Lofgren on the shielding of the Bevatron, a 1954 particle accelerator.*

"To pinpoint the smallest fragments of the universe you have to build the biggest machine in the world."

The Guardian

Microwave Oven (1945)

Spencer ushers in the age of near-instantaneous cooking.

The microwave oven is an invention that arrived almost entirely by accident. Its inventor, Percy Spencer (1894–1970), was an electronics whizz, working on designing radar equipment. He paused next to a "magnetron," one of the power components of the machinery, and was amazed to discover that the chocolate bar in his pocket had melted.

Understandably curious, he tried placing other objects near the magnetron. Some unpopped popcorn popped successfully (with Spencer standing further away, so as not to start cooking himself), and the next morning an egg was cooked, demonstrating for the first time that eggs in their shells explode if cooked in the microwave.

Spencer realized the potential of his discovery and set about designing a more efficient food-cooking device. He filed for a patent in 1945, and by late 1946 a prototype device was being tested in a Boston, Massachusetts, restaurant, and soon commercial models became available. These early models were not received well by consumers, possibly because they were over 6 feet (1.8 m) tall, cost $5,000, and required special plumbing to cool the magnetron apparatus. Gradually, however, units became more practical and affordable, and became safe and reliable enough to be used by the average consumer. By 1975, microwave sales were exceeding those of gas cookers.

The microwave oven does have its shortcomings, however. Meat does not brown on the outside, and the quick cooking time means that food can end up cooked unevenly. Despite these issues, many people regard their microwave as indispensable. **SB**

SEE ALSO: CONTROLLED FIRE, OVEN, PRESSURE COOKER, GAS STOVE, ELECTRIC STOVE

⬆ *A 1968 commercial microwave oven model HN 1102, by Philips, Eindhoven, Netherlands.*

Electronic Synthesizer (1945)

Le Caine pioneers electronic music.

It is hard to imagine how popular music might have evolved without the use of the synthesizer and other such electronic innovations. Although Dr. Robert Moog is a household name for his pioneering efforts in this field, important groundwork was done several decades earlier by a Canadian physicist and instrument designer named Hugh Le Caine (1914–1977). One of his instruments in particular, the electronic sackbut, is now widely recognized as being the first voltage-controlled synthesizer.

Developed at the University of Toronto, Canada, in 1945, Le Caine's sackbut featured a piano-type keyboard built into an old desk. Unlike most of the early commercial synthesizers that appeared at the start of the 1970s, Le Caine's instrument was touch-sensitive; the characteristics of the sound altered according to how lightly or forcefully the keyboard was played, giving the sackbut much the same potential expressiveness as a real acoustic instrument.

Two other particularly innovative design features also stand out: the use of adjustable wave forms to create basic sounds, and the use of voltage control to alter certain characteristics of the sound, such as pitch and modulation. Sound quality was also altered electronically by manipulating special filters. Voltage control would be used on most subsequent synthesizers produced until the early 1980s.

Le Caine would revisit the sackbut in 1971, but although he intended to produce a commercial instrument, the project failed. Despite his public obscurity, Hugh Le Caine's instruments had a great deal of influence on the world of electronic music. **TB**

SEE ALSO: PIPE ORGAN, ELECTRIC GUITAR

⬆ *Le Caine mounted his prototype electronic sackbut of 1948 on a crude three-legged wooden stand.*

Air-cushioned Sole
(1945)

Maertens creates healthy subcultural icon.

While many inventions are deemed "accidental discoveries," the air-cushioned sole was, in fact, the result of a real mishap rather than a metaphorical one. In 1945 the young Bavarian doctor Klaus Maertens had a skiing accident, and in order to speed up his recovery he devised a shock-absorbing shoe. Its sole has cushions of trapped air that constantly stimulate the muscles and tendons of the person wearing it. The principle of pressure and counter pressure mimics the effects of an elastic insole, thereby protecting the joints.

Two years later, Maertens and a friend, engineer Dr. Herbert Funck, set up a company to sell the shoes. But while his innovation—marketed as orthopedic footwear—was well received by the medical world, it was not considered fashionable. That changed when Maertens and Funck placed ads in international trade magazines. Bill Griggs—a producer of army and working boots in Northamptonshire, England—spotted one of them and bought the global rights to the new sole in 1959. The following year, on April 1, 1960, he launched his new 1460 boot (the name deriving from the date of the launch), which incorporated Maertens's sole as well as the newly designed yellow stitching that was to become the hallmark of the anglicized Dr. Martens AirWair brand.

Initially popular with postmen, dustmen, and factory workers in England, the shoes were soon adopted by subcultures such as skinheads and punks, who considered the brand's working class image ideal for their protest attitudes. After a sales slump in the late 1990s, the shoes have become popular again. **DaH**

SEE ALSO: CLOTHING, SEWING, SHOE

← *Maertens's air-cushioned sole only became a fashion accessory after Bill Griggs created the AirWair brand.*

Citizens' Band Radio
(1945)

Gross develops the first CB radio.

In 1945 the U.S. government allocated radio frequencies, called the Citizens' Band (CB), for personal radio services. In response, Alfred J. Gross (1918–2000) set up a company to produce two-way CB radios; in 1948 his radio was the first to receive federal approval.

Citizens' Band radios are used for short-distance communication, generally within the 27 MHz band. Unlike amateur radio, CB can be used for commercial communications. CB radios first became popular with small businesses and truck drivers, but during the 1970s their popularity around the world soared—a popularity that was bolstered by film and television with *Smokey and the Bandit* and "The Dukes of

> *"If I still had the patents on my inventions, Bill Gates would have to stand aside for me."*
>
> **Alfred Gross**

Hazzard." Governments around the world often released the CB frequencies only after CB radio equipment had been imported and used illegally.

After the release of the frequencies, to use a CB radio legally a license had to be obtained and a call sign assigned. Despite this, many people chose to continue illegally, using nicknames or "handles" to avoid identification. A whole new culture grew within the CB fraternity and even had its own slang language. Police were referred to as "bears," speed cameras as "flash for cash," and the CB radio itself became "ears." They also incorporated and modified the ten-code, a call of nature becoming "10-100." **RP**

SEE ALSO: WIRELESS COMMUNICATION, WALKIE-TALKIE, PAGER, CELLULAR MOBILE PHONE, DIGITAL CELL PHONE

Radial Tire
(1946)
Michelin creates an improved pneumatic tire.

With the transition from horse and buggy to more modern modes of transportation, faster vehicles capable of longer distances of travel became the norm. The wheels on such vehicles obviously became quite important—the wood and metal constructs that initially performed well on wagons were not as well suited to automobiles, motorcycles, and bicycles.

Needing a material with added durability and also cushioning, early tire manufacturers turned to rubber. Initially composed of solid rubber, the first tires were durable, but also heavy and rough on roads.

Though it had been invented earlier, the pneumatic tire found a niche again in the early 1900s in the form of bicycle tires. The idea soon spread to cars, where the new inflatable tires were lighter and provided better shock absorption, allowing for a smoother ride. These early tires consisted of an inner inflatable tube paired with an outer tire that provided protection and traction. Not completely composed of rubber, these outer tires had alternating layers of rubberized cords, or plys, imbedded within them to provide some degree of reinforcement. Because these plys ran at an angle from the outer tire rim to the inner tire rim, they were referred to as bias-ply tires.

The Michelin company hit upon the next major advancement in the 1940s by producing the first radial-ply tire. In contrast to the bias-ply tire, the plys on a radial tire run at a ninety-degree angle to the tire rims, like the spokes of a bicycle wheel. Michelin also embedded a belt of steel mesh in the tire to further reinforce it. The new tires lasted longer, enhanced steering, and improved mileage, but were more costly to manufacture. Nevertheless, they have become the gold standard of the tire industry. **BMcC**

SEE ALSO: VULCANIZATION, PNEUMATIC TIRE, TIRE VALVE

Tupperware®
(1946)
Tupper improves the plastic container.

In the early 1940s plastics were still relatively new compounds and their practical applications had not yet been fully realized. Early plastics were brittle, greasy, and had a rather unpleasant odor. It would take an ex-tree surgeon by the name of Earl Tupper (1907–1983) to come up with the perfect plastic.

Tupper worked at the chemical company DuPont, where he learned about the design and manufacture of plastics. In 1938 he founded the Earl S. Tupper Company, which manufactured parts for gas masks during World War II. After the war, he turned his attention toward creating a peacetime product.

Tupper discovered a way of turning polyethylene slag—a by-product of crude oil refinement—into a strong, resilient, grease-free plastic that he called Tupperware®. By 1946 Tupperware® was on the market in an array of brightly colored incarnations: cigarette cases, water tumblers, and food storage containers. In 1947 he patented the Tupperware® seal, which was based on the lid of paint tins and was both airtight and watertight. Though Tupperware® was undoubtedly innovative, sales were sluggish and Tupper enlisted the help of the Stanley Home Products company; it came up with the idea of selling the containers at in-home parties.

The "Tupperware® party" was born, and by 1951 sales were so impressive that the product was taken off the store shelves to be sold exclusively in this way. Competitors launched products in an attempt to cash in on Tupper's success, but by then the Tupperware® brand had secured its place in the vocabulary of virtually every Western household. **HI**

SEE ALSO: POLYPROPYLENE

▶ *Tupperware® parties like this one in 1963 appealed to the domesticated lifestyle of the female consumer.*

Chemotherapy (1946)

Gilman and Goodman treat lymphoma.

Chemotherapy's effectiveness against cancer was discovered from mustard gas, a lethal weapon used in the World War I trenches. Autopsy observations of soldiers exposed to the gas revealed destruction of lymphatic tissue and bone marrow. Scientists at the time reasoned that mustard gas might destroy cancer cells in lymph nodes, but nothing was done.

Early in 1942 Alfred Gilman (1908–1984) and Louis S. Goodman (1906–2000), two pharmacologists at Yale University, were recruited by the U.S. Department of Defense to investigate potential therapeutic applications of nitrogen mustard (a derivative of mustard gas) on lymphoma. After establishing lymphomas in mice and rabbits, they went on to show that they could treat them with mustard agents.

The next step was to inject mustine (the prototype anticancer agent) into a patient with non-Hodgkin's lymphoma, whose cancer had become resistant to radiation. Initially the patient responded well, with doctors noting a softening of the tumor masses within two days, but unfortunately the beneficial effect lasted only a few weeks. In 1946 the government gave the team permission to publish the first paper on the use of nitrogen mustard in cancer treatment, providing the first "proof" that cancer could be treated by pharmacological agents.

Nitrogen mustard became a model for the discovery of further classes of chemotherapy, drugs that damage the cell control centers that make cells divide, thereby preventing further division. Chemotherapy gradually became a standard treatment for many cancers, often in combination with surgery and radiation, and nitrogen mustard was incorporated into today's multidrug chemotherapy for Hodgkin's disease. **JF**

SEE ALSO: RADIOTHERAPY

Acrylic Paint (1947)

Bocour markets a new paint for artists.

Otto Röhm studied acrylic plastics for his 1901 Ph.D. thesis. Six years later, he co-founded the Röhm and Haas Company, which in 1936 began selling shatter-proof acrylic glass (more commonly known by trade names such as Plexiglas®, Perspex®, and Lucite®). Sales were slow until World War II, when the United States began manufacturing tens of thousands of aircraft each year, all with Plexiglas® canopies. Chemists at Röhm and Haas had worked with acrylics for decades, but they were not the ones who invented acrylic paint. Instead, the inventor was an artist turned paint-maker.

In 1941 Leonard Bocour (1910–1993) was making oil paint and selling it to artists when he was shown a sample of acrylic and was impressed by how white it

> *"Some guy walked into the shop … with something like white syrup. He said, 'It's an acrylic.'"*
>
> Leonard Bocour

was. After the war ended, Bocour worked with Röhm and Haas to produce an acrylic that could be used in paint. In 1947 Bocour began selling "Magna," the first acrylic paint and the first significant change in paint technology for artists since the fifteenth century.

Acrylic paint has many advantages over oil paint. It does not peel, crack, or fade, and can be applied to many different surfaces. It also dries quickly, which can be an advantage or a disadvantage, depending on the artist's goal. In 1953 Röhm and Haas introduced water-based acrylic paint for interior walls. The new paint was easier to apply, dried faster, and was easier to clean up than the solvent-based paint it replaced. **ES**

SEE ALSO: PERSPEX®

Transistor (1947)

Bardeen, Shockley, and Brattain replace the problematic valve triode.

The development of the transistor was one of the true landmark inventions of the twentieth century. Before their existence, almost all electronic circuits made use of cumbersome and unreliable valves.

The transistor was developed in the United States at the Bell Telephone Laboratories in New Jersey. Scientists John Bardeen (1908–1991), William Shockley (1910–1989), and Walter Brattain (1902–1987) were researching the behavior and suitability of germanium crystals for use as semiconductors that could replace valve diodes. When this was successfully achieved, the same group turned their attention to the considerably more demanding task of creating a solid-state germanium triode—one that could replace the ubiquitous valve equivalents. Since its development and evolution during the early years of the twentieth century, the valve triode had been at the very heart of every piece of electronic equipment. However, valves consumed large amounts of power, created a sometimes unacceptable level of heat, and, like the electric light bulb from which they evolved, they had a short lifespan.

On December 16, 1947, Bardeen, Shockley, and Brattain created the first transistor—so named from the "trans" of transmitter and "sistor" of resistor. A week later they gave their first public demonstration: December 23, 1947, is usually cited as the birthday of the transistor. The transistor was tiny, consumed very little power, and gave off no heat whatsoever. It was also supremely predictable in its behavior. Within a decade, valve technology in most commercial applications was rendered all but obsolete. **TB**

SEE ALSO: VALVE DIODE, TRIODE, TRANSISTOR RADIO

↗ *Metal contacts are separated by an insulator pressing onto germanium, itself supported by a metal plate.*

"Brattain decided to try dunking the entire apparatus into a tub of water. It worked . . . a little bit."

Ira Flatow, *Transistorized!*

Defibrillator (1947)

Beck uses electricity to restore a regular heart rhythm after a cardiac arrest.

A defibrillator is a device that delivers an electric shock to the heart through the chest wall, with the aim of restoring a regular rhythm. It is used to treat ventricular fibrillation, a condition where the heart muscle is no longer contracting in a coordinated fashion, thus preventing blood from being pumped around the body. If untreated, death frequently results.

American cardiac surgeon Claude Beck (1894–1971) performed the first successful defibrillation procedure in 1947 at Case Western Reserve University in Cleveland, Ohio. Beck had been operating on a fourteen-year-old boy with a congenital heart problem, and had just closed the boy's chest when he suffered a cardiac arrest. Beck immediately reopened the chest and, after unsuccessfully massaging the heart by hand, tried out the defibrillator device that he was in the process of developing. Beck's device consisted of silver paddles (the size of large tablespoons) laid over the heart and connected to a 60-Hz alternating current .

In 1954 William B. Kouwenhoven, working with William Milnor, demonstrated the first closed chest defibrillation on a dog. Two years later Paul Zoll went on to perform the first successful external defibrillation of a human heart. Today defibrillators are widely available in medical settings, and automated external defibrillators (AEDs) are even located in places such as shopping malls and sports stadiums for use by the general public. People judged at extreme risk of ventricular fibrillation are often fitted with implantatable defibrillators that automatically deliver an electric shock if a change in rhythm is detected. **JF**

"Defibrillators should be as common as fire extinguishers."

Michael Tighe, saved by a portable defibrillator

SEE ALSO: ELECTROCARDIOGRAPH (EKG), CARDIAC PACEMAKER, HEART-LUNG MACHINE, ARTIFICIAL HEART

◩ *Beck developed his defibrillator after a colleague successfully revived research animals with electricity.*

Polaroid Self-developing Film Camera (1947)

Land introduces the photograph that develops almost instantly.

The "Polaroid" camera became an instant classic following its conception, more than sixty years ago. Although the technology behind self-developing film was already present at the time, it was Edwin Land (1909–1991), founder of the Polaroid Corporation, who designed and produced the first commercially available self-developing camera in 1946, an invention that won its creator many accolades.

Land formed his company in 1937 to produce and sell the polarizing filters he had patented eight years before, and soon the company was making filters for the United States in World War II. Land was on vacation with his daughter in 1943 when, after snapping a photo of her, she asked why she had to wait so long to see the image. He soon visualized a system of "one-step dry photography," whereby the image would develop within sixty seconds inside the camera.

In the years following World War II, when demand for polarizing filters sagged, Land turned his "instant camera" idea into reality. He unveiled his finished invention at the Optical Society of America in 1947 by taking a photo of himself and revealing it to the audience after just a minute. The first of Land's cameras, the Model 95, became available in November the next year, and was an instant sellout.

Polaroid continued to produce redesigned cameras and film—including the release of an X-ray film for radiography in 1951—right up until February 2008, when the company stopped producing all instant film after falling sales due to the advent of digital cameras. It seems that this once revolutionary photographic invention has had its day. **SR**

SEE ALSO: PHOTOGRAPHY, PORTABLE CAMERA, POLARIZING FILTER, DIGITAL CAMERA

↗ *Like the Kodak "Brownie," the Polaroid camera led to a more spontaneous approach to photography.*

"[It was on this day] I suddenly knew how to make a one-step dry photographic process."

Edwin Land

Radiocarbon Dating

(1947)

Libby ascertains the age of organic materials.

Before the 1940s, scientists wanting to determine the age of fossils and other organic materials relied on relative dating techniques, grouping objects by the estimated date of surrounding rock strata. This method could produce large inaccuracies.

Strangely, the ability to date objects more accurately was to originate from outer space. High-energy cosmic rays are constantly bombarding our atmosphere; when they hit, they break up atoms in the stratosphere. A consequence of this is the production of an unstable radioactive form of carbon called Carbon-14 (C-14) from atmospheric nitrogen.

American scientist William Frank Libby (1908–1980) reasoned that, as all living creatures are made up of carbon and are constantly replacing this carbon from the atmosphere (plants take it in as carbon dioxide and animals absorb it from plants), all living organisms should have an equal amount of C-14 in their bodies in relation to the proportions occurring in the atmosphere. However, when an organism dies, it stops replacing this C-14. Being unstable, its C-14 breaks down and is converted back to normal nitrogen at a fixed rate (approximately half every 5,700 years). The time since the death of an organic sample can be worked out by comparing the proportions of C-14 in the atmosphere to what is in the sample. The method should be accurate up to about 50,000 years ago (after which all the C-14 will have broken down).

Libby presented his radiocarbon dating method in 1947 after testing it on fir trees (where the age had already been ascertained by counting their rings). It was soon widely adopted and has been used to date such diverse objects as the Dead Sea Scrolls and relics from the end of the last Ice Age. **JM**

SEE ALSO: GEIGER COUNTER, MASS SPECTROMETER

Supersonic Airplane

(1947)

U.S. Air Force pilots break the sound barrier.

Supersonic airplanes fly faster than the velocity of sound, this being about 770 miles per hour (1,230 kph) at ground level. During World War II certain fighter planes, such as the Mitsubishi Zero and the Supermarine Spitfire could approach this speed in a dive but the near supersonic air passing over the plane produced disruptive shockwaves and turbulence. The propellers became much less efficient and chaotic effects amplified pressure perturbations, producing an increase in drag and a loss of lift and control. These effects became known as the "sound barrier," which was "broken" by introducing the much more powerful jet engine and by strengthening the airframes and

"I was always afraid of dying. It was my fear that … kept me … always alert in the cockpit."

Chuck Yeager

wings. Fortunately, flight became smooth again when the aircraft moved faster than the speed of sound.

There is still a slight controversy over who flew the first supersonic airplane. Chuck Yeager broke the sound barrier on October 14, 1947, over the Mojave Desert in his rocket-powered Bell X-1 plane, by flying at Mach 1.06. But shortly before that, ground witnesses saw George Welch, testing the prototype of the F-86 Sabre jet fighter, go into a steep dive and heard the ba-boom of the sound barrier being broken. **DH**

SEE ALSO: AIRSHIP, POWERED AIRPLANE, JET ENGINE, SCRAMJET

➡ *The Bell X-1 was modeled on a .50-caliber Browning bullet for its known stability in supersonic flight.*

Multitrack Audio Recording (1947)

Paul combines eight performances into one.

When Thomas Edison first captured sound, his audio recordings, made using a wax cylinder, were a simple reflection of a single moment in time. During the 1940s, when a number of experimenters began looking at different approaches to audio recording, guitarist and inventor Les Paul (1915–2009) worked on the idea of capturing a number of single events and playing them back simultaneously to produce something completely new—a synthetic recording.

In 1947 Les Paul's experiments came to fruition. Capitol Records issued his instrumental solo "Lover (When You're Near Me)," featuring Paul playing eight different guitar parts at the same time. Paul made the recording using two wax-cylinder recording machines; after making one recording on the first cylinder, the second would be of himself playing along with the first recording. He continued in this way until he had built up all eight parts. This pioneering recording technique is now referred to as "sound on sound."

Les Paul was also highly important in the evolution of true multitrack tape recording. In 1948 he received a gift from his friend, the singer Bing Crosby, of the world's first commercially produced magnetic tape recorder, the Ampex Model 200. Paul experimented with the machine, modifying the record and playback heads, and in 1953 he commissioned Ampex to build the first eight-track recorder, a machine that allowed tracks to be recorded in parallel on magnetic tape. By the end of the decade most of the major recording studios in the United States were equipped with multitrack tape recorders. **TB**

SEE ALSO: MAGNETIC RECORDING (TELEGRAPHONE)

← *Les Paul, here in the studio in 1940, was frustrated by the limitations of recording in a single take.*

Barcode (1948)

Silver and Woodland improve data tracking.

The development of barcodes stemmed from comments made by a food-chain president to a university dean at the Drexel Institute of Technology, Philadelphia, in 1948, and overheard by graduate student Bernard Silver (1924–1962). The company wanted some sort of system to collect product information at the checkout automatically, but the dean had little interest in initiating such research. Silver decided that he and his friend Norman Woodland (1921–2012) should pursue a solution. Eventually the pair turned to a combination of movie soundtrack technology invented by Lee De Forest in the 1920s and Morse code dots and dashes; "I just extended the

> *"[In] 1974, the first product with a barcode [chewing gum] was scanned at a checkout counter."*
>
> Russ Adams, author

dots and dashes downward and made narrow lines and wide lines out of them," said Woodland.

De Forest's film included a varying transparency pattern on its edge. When a light was shined through it, sensing equipment on the other side converted the changes in brightness into electronic waveforms that translated into sound. Woodland adapted this system by reflecting light off his wide and narrow lines and coupling the sensor to an oscilloscope. They had invented the first electronic reader of printed data. The advent of affordable, low-power, laser light sources, and "compact" computers to process the captured data, did not occur until several years later. **MD**

SEE ALSO: RADIO FREQUENCY IDENTIFICATION (RFID), CHARGE-COUPLE DEVICE (CCD)

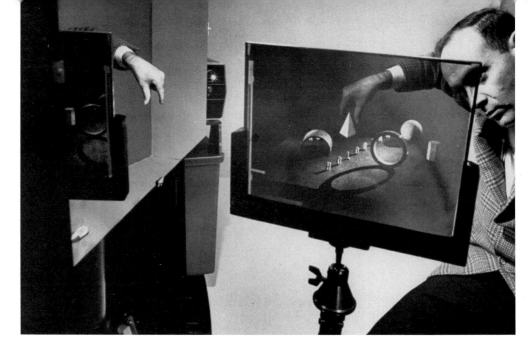

Holography (1948)

Gabor creates three-dimensional images.

Holography, coming from the Greek words *hòlòs* (whole) and *grafè* (writing), is a form of photography that allows an image to be recorded in 3D. Discovered in 1948 by Hungarian born Dennis Gabor (1900–1979) while working for the Thomson-Houston company in Rugby, England, its early development was hampered by insufficient light sources.

Gabor's holography stored 3D images by encoding them within a beam of light, but the mercury arc lamp he used produced variable results. The invention of the laser by a team of Russian and U.S. scientists in 1960 provided a pure, intense light that was ideal for creating holograms. The pulsed-ruby laser emits a very powerful burst of light that lasts a few nanoseconds and effectively freezes movement.

The development of the laser enabled the first experiments in optically storing and retrieving images. The first laser-transmission hologram of 3D objects, a toy train and a bird, occurred in 1962. Stephen A. Benton invented white-light-transmission holography In 1968, and this form finally enabled mass production. Benton's technique meant that, when viewed in ordinary white light, the hologram created a three-dimensional "rainbow" image from the seven colors that make up white light.

The use of holograms in every form is widespread. Rainbow holograms are used on credit cards, artists such as Salvador Dali employed holography artistically, and holograms also appeared in some of the movies that defined the latter half of the twentieth century. Where would *Star Wars* be without a 3D Princess Leia begging for help, or the swoosh of a light saber? **SD**

SEE ALSO: **PHOTOGRAPHY, HOLOGRAPHIC MEMORY**

⬆ *Emmett Leith at University of Michigan creates an image in three dimensions using a laser beam.*

Weather Radar (1948)

Ex-World War II radar is directed toward weather systems.

The combatant nations of World War II developed radar to detect enemy airplanes. They also noticed that they were picking up signals from raindrops, hailstones, and snowflakes. After the war these signals were used for weather forecasting. From 1948 weather radar was carried on aircraft to help the crew detect potentially hazardous cumulonimbus cloud systems.

Weather radar stations transmit a directional, narrow beam of pulsed microwaves up toward the clouds, using an antenna that rotates and scans the sky. These microwaves typically have wavelengths of 1–10 centimeters, these lengths being about ten times the size of the raindrops, hailstones, and snowflakes they are aiming to detect. The falling particles scatter the microwaves and a return pulse is then picked up back at the radar station.

Five things are measured. The time taken for the pulse to travel to and from the scattering body gives the distance to the precipitation. The Doppler shift of the return pulse gives the direction and speed in which the cloud is moving. Variability in the Doppler signal can indicate the turbulence of the rain drops and thus gives clues as to whether thunderstorms or even tornadoes might develop. The strength of the return pulse is a function of the amount of rain in the specific cloud, and the shape of the return pulse distinguishes between rain, snow, and hail. To get a complete 3D moving weather picture, together with cloud heights, requires an interlinked series of weather radars dotted over the land. Meteorological weather radar usually provides a complete radar rain picture every half hour. **DH**

SEE ALSO: RAIN GAUGE, ANEMOMETER, BAROMETER, HYGROMETER, ANEROID BAROMETER, RADIOSONDE, RADAR

⬆ *A staff member at the Meteorological Office in London uses a radar scanner to track storms.*

Cable Television (1948)

Walson invents the community antenna television system.

Nothing brings a community together like the collective glow of its televisions. In the spring of 1948, American John Walson (1914–1993) installed community antenna television, bringing the wonders of cable television to his customers.

Walson and his wife Margaret, owners of the Service Electric Company of Mahanoy City, Pennsylvania, came up with cable television as a way to help their customers pick up signals blocked by nearby mountains. Walson decided to take his service literally to new heights by climbing to the top of a mountain and planting an antenna. Using cables and signal boosters, he connected the antenna to his appliance store. Along the way he dropped the signal directly off at his customers' homes, thus creating the first community antenna television system.

Community antenna television, now known as cable TV, is found in nearly 60 percent of U.S. homes and throughout Europe. The first cable systems consisted of a large antenna to capture the signal and a grid of signal amplifiers arranged throughout the service area. As the signal moved along the cable it weakened, forcing the designers to implement amplifiers at regular intervals. But amplification added noise and distortion to the signal and created a potential place for the signal to be completely lost.

Soon thereafter, cable was fortified as more and more channels were added. Microwave transmitters and receivers were used to capture distant signals, giving cable providers more channel choices. The wave for sofa-bound channel surfers was fast approaching tidal proportions. **LW**

SEE ALSO: CATHODE RAY TUBE, TRANSFORMER, TELEVISION, COLOR TELEVISION, REMOTE CONTROL, PLASMA SCREEN

⬆ *When cable television arrived in 1948, television sets were still expensive; this one incorporates a radio.*

Robot (1948)

Walter creates the first electronic autonomous robots.

Robots, in one guise or another, had been suggested as far back as 1495, when Leonardo da Vinci created his mechanical knight robot. However, the first significant robot prototypes, built in 1948, were a pair of tortoiselike robots named Elmer and Elsie.

Created by United States-born neurophysiologist and inventor Dr. William Grey Walter (1910–1977), the tortoise robots were remarkable in their ability to mimic lifelike behavior. These experimental robots incorporated sensors for light and touch, as well as motors for propulsion and steering, and a two-vacuum tube (valve) analog "computer."

With the aid of simple circuitry, these electromechanical, three-wheeled robots were capable of phototaxis (an automatic movement toward or away from light), and could thus find their own way to a recharging station when they ran low on "food"—a precursor of the technique used in the popular Sony Aibo® robot dog some sixty years later. Using a combination of light-level and motor-power settings, four modes of operation were possible. This produced a variety of unique behavior patterns in the tortoises. In one experiment, Dr. Walter placed a light on the front of a tortoise and watched as the robot considered itself in a mirror.

Because of their speculative, exploratory tendencies, Dr. Walter called his tortoises "Machina Speculatrix." Designed to aid the study and testing of theories of behavior arising from neural interconnections, these small robots with reflexes were hugely influential in the birth and development of the sciences of cybernetics and robotics. **MD**

SEE ALSO: AUTOMATON, INDUSTRIAL ROBOT, BIPEDAL ROBOT, SURGICAL ROBOT

↑ *A photoreceptor allowed the "tortoise" to approach moderate light but made it avoid bright illumination.*

Ice Rink Cleaning Machine (1948)

Zamboni automates a time-consuming chore.

The Zamboni name is synonymous with ice rink resurfacing in the United States, although few people outside of ice-skating would recognize it.

Frank Zamboni (1901–1988) was born just after the turn of the twentieth century to Italian immigrant parents. As a young man he worked on the family farm and as a mechanic in a local garage. Frank opened an ice-making plant, with his younger brother Lawrence, producing blocks of ice for refrigeration. When electrical refrigerators were invented in the mid 1930s, making the production of ice blocks redundant, the Zambonis transformed their ice-making equipment and expertise into an ice rink. "Iceland" proved a popular local attraction, with 150,000 visitors a year.

Cleaning, or resurfacing, of the rink was a time-consuming business, requiring three men and taking an hour and a half to complete. Zamboni started to think about ways in which this process could be made more efficient, and in 1948 he succeeded in producing a working ice rink resurfacer. The "Model A" was born and instantly transformed the high-pressure world of ice rink resurfacing, a job that now could be performed by one man in ten minutes.

The genius of Frank's machine is that it performs all the multiple tasks needed to resurface a rink simultaneously. Blades shave off a thin layer of the surface ice and a series of conveyors transport the waste to the back of the vehicle where the waste ice (or "snow") is stored in a bucket. Water is blasted over the surface of the ice at high pressure to remove dirt and refill holes in the ice. Excess water is then removed with the help of a giant squeegee. **BG**

SEE ALSO: SQUEEGEE, SELF-CLEANING WINDOWS

> *"… people like to stare at:*
> *a flowing stream, a crackling fire,*
> *and a Zamboni clearing the ice."*

Charlie Brown, *Peanuts* character

Richard Zamboni drives a descendant of the machine that his father developed to resurface his own ice rink.

Synthetic Zeolites (1948)

Barrer synthesizes zeolites for industrial use.

In 1948 New Zealander Richard Barrer (1910–1996) was the first to manufacture a type of zeolite that had not been found in nature. He became the professor of physical chemistry at Imperial College, London.

Natural zeolites are a form of microporous stone that have such a regular and tiny pore structure that they can be used as molecular sieves. Up to 50 percent of the volume of the stone is open space, or air.

Zeolites are produced when layers of fine volcanic ash react with alkaline water; the mineral natrolite is a typical example. Open-cast mines yield about 4 million tons of zeolites per year and these are mainly used in the concrete industry.

In agriculture zeolites can be added to soil to act as a water trap. The zeolite is hygroscopic and can absorb up to half its own mass of water from the morning dew, only to release the water slowly later. Potassium and ammonium can be added to zeolites for use as a slow-release fertilizer.

Synthetic zeolites can be formed by slowly crystallizing a silica alumina gel, held initially in a mixture of alkali and water, onto an organic substrate. The advantage of manufacturing zeolites synthetically is that the end product is pure and uncontaminated by traces of volcanic metals, quartz, and minerals. Also the controlled temperature of the production process leads to a structure that is more uniform than would ever appear in nature.

Synthetic zeolites are used as ion filters in water-softening and purification systems. As sieves they can remove certain molecules from industrial flue gases. They can also filter out unwanted fission products in the nuclear fuel industry. As they act as traps for large molecules, they can speed up chemical processes such as alkylation in petrochemical engineering. **DH**

SEE ALSO: SYNTHETIC DIAMOND, SYNTHETIC RUBBER, SYNTHETIC BLOOD

Acetaminophen (1948)

Axelrod and Brodie discover a safer painkiller.

Despite earning a degree in biology, American biochemist Julius Axelrod (1912–2004) was rejected from all the medical schools he applied to and so got a laboratory technician job in New York's public health department, looking at vitamins in food. He attended night school to get a master's degree in science in 1942, then worked under scientist Bernard Brodie (1909–1989) at Goldwater Memorial Hospital in 1946.

During the 1940s the use of nonaspirin-based pain analgesics was being linked to the blood-poisoning condition methemoglobinemia. Axelrod and Brodie wanted to find out why. Their detailed research found acetanilde, the main ingredient in many of these products, degraded into the blood toxin aniline. They

> *"I soon learned that it did not require a great brain to do original research."*

Julius Axelrod

suggested this was the most likely cause. They also discovered that one of its breakdown products also worked to kill pain in the same way as acetanilde.

In a 1948 paper they discussed the idea that this product, acetaminophen or paracetamol (Tylenol) as it became known, could be used as an alternative to acetanilde. From the 1950s it was being recommended as safer than aspirin and marketed in preparations for children. Today it is widely used to relieve pain, headache, fever, symptoms of common viral conditions like colds and influenza, as well as period, joint, and dental pain. Acetaminophen has few side effects and reacts with few other drugs. **LH**

SEE ALSO: ANESTHESIA, DISSOLVABLE PILL, ASPIRIN, MODERN GENERAL ANESTHETIC

Ventilator (1949)

Emerson helps incapable patients to breathe.

After working to improve the iron lung in 1931, American biomedical device inventor John Emerson (1906–1997) went on to develop a mechanical ventilator.

A ventilator is a machine that automatically moves air into and out of the lungs of a patient who is unable to breathe. Emerson worked on his mechanical assistor with colleagues in the department of anesthesia at Harvard University in 1949. Polio epidemics of the 1940s had increased the demand for such machines, as had an increased use of muscle relaxant drugs during surgery; these would paralyze the patients' respiratory muscles so that they could not breathe.

Other models followed Emerson's through the 1950s, such as the Bird ventilator, driven only by gas. In

---

"Emerson developed a respirator selling for less than $1000 and solved ... breathing-rate changes."

Charles C. Smith, Jr., *Special Air Mission: Polio*

Britain the East Radcliffe and Beaver models were used; the Beaver ventilator had an automotive wiper motor that operated bellows to inflate the lungs. But such motors were dangerous in surgical environments where flammable anesthetics were used. The Manley ventilator of 1952 was another gas driven model and so got around this problem. The Manley Mark II became popular and was made and sold in the thousands. Manley's ventilators led to positive-pressure ventilation techniques being used widely across Europe.

In the 1980s high-frequency percussive ventilators, combining mechanical ventilation and high-frequency oscillatory ventilation, were introduced. **JF**

SEE ALSO: IRON LUNG, HEART-LUNG MACHINE

Atomic Clock (1949)

The race for the perfect timekeeper begins.

Most types of clocks rely on the oscillation of a solid body, be it a pendulum, a balance-wheel, or a quartz crystal, but each suffers from the effects of temperature, pressure, and gravity. Time measuring devices have also depended on the spin of the Earth, but these suffer from seasonal effects and tidal friction. Atoms, however, vibrate a fixed number of times per second. Both the U.S. National Bureau of Standards and the United Kingdom's National Physics Laboratory tried to take advantage of these vibrations.

In 1949 the Americans built a quartz clock that was synchronized by the 24-GHz vibrations of low-pressure gaseous ammonium molecules. The British, under the leadership of physicist Louis Essen (1908–1997), used the oscillations of an electrical circuit synchronized to the vibrations of caesium atoms, the first caesium clock being built in 1955. The caesium was kept in a tuneable microwave cavity and the clock relied on the fact that there were 9,192,631,770 transitions between two hyperfine ground-state energy levels every second. This number defined the second, as opposed to the old definition of there being 86,400 seconds in one day. A good atomic clock was accurate to one part in 1,014, and therefore would take about 3 million years to lose or gain a second.

Four atomic clocks are used in each of the many satellites of the Global Positioning System and comparisons of electromagnetic-wave travel times enable positions on Earth to be measured very precisely. The clocks are also used by geophysicists to monitor variations in the spin rate of Earth, and the drifting of the continents. **DH**

SEE ALSO: SHADOW CLOCK, WATER CLOCK, TOWER CLOCK, POCKET WATCH, DIGITAL CLOCK, QUARTZ WATCH

→ *Built by the British in 1955, this atomic clock depends on the vibrations of caesium exposed to radio waves.*

Centrifugal Clutch (1949)

Fogarty invents an effective clutch for lightweight applications.

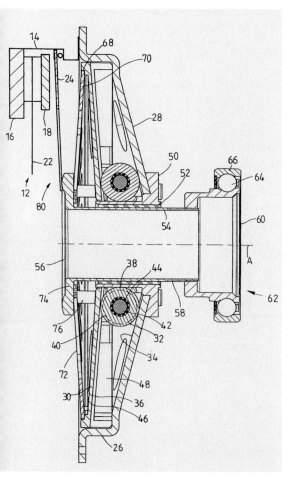

The centrifugal clutch owes its existence to the frustration Thomas Fogarty (*b.* 1934) experienced with the gears of his motor scooter.

A traditional car clutch works via longitudinal mechanical motion, disconnecting the driveshaft for the wheels from the motor by moving a pressure plate away from the clutch disc.

The centrifugal clutch works differently, using the rotational motion of the motor to engage and disengage the driveshaft. The clutch is cylindrical, with the crankshaft from the engine rotating in its center. Attached to this shaft is a pair of weights that are held in place by springs. The crankshaft and weights rotate together, at the same number of revolutions per minute (rpm) as the engine. When the engine increases in rpm, the revolution speed of the weights also increases and they swing outwards. When a certain rpm is reached, the weights swing out far enough to come into contact with the clutch's outer casing, causing this to begin spinning also. This outer casing or hub is attached to whatever the motor is designed to power, causing it to begin to move also.

The weights inside the clutch slip against the hub, allowing it to rotate at a slower rate than that of the engine. As the rpm increases, the weights swing out even further and stop slipping. Eventually this drives the hub at the same rate as the engine, and the maximum speed for the equipment is achieved.

Centrifugal clutches are used in lawnmowers, chainsaws, motor scooters, mopeds, and racing go-karts, as they are effectively automatic, avoid stalling, and last a very long time. **HP**

SEE ALSO: MOTORCAR GEARS, CLUTCH, AUTOMATIC TRANSMISSION, SYNCHROMESH GEARS

"If . . . you had to shift into low gear, the motor scooter would jump about ten yards ahead . . ."

Thomas Fogarty

◹ *Opening the throttle causes the clutch to expand and grip the inside of the clutch's casing, driving the wheel.*

Intraocular Lens (1949)

Ridley inserts the world's first artificial lens.

As recently as the 1950s, people with sight clouded by cataracts would slowly go blind with no hope of a cure. Today, in most cases, a cataract sufferer's eyesight can be restored to what it was when they were a teenager in an operation taking just thirty minutes. The man responsible for this incredible breakthrough was British ophthalmologist Harold Ridley (1906–2001), although he had to battle for this achievement to be recognized by his peers.

During World War II Ridley treated pilots with injuries caused by shards of Perspex® from their cockpits being lodged in the eye. He noticed that the Perspex® did not react with the eye and realized that, its inert quality, combined with its lightness and optical properties, made it ideal for the construction of replacement lenses for damaged eyes. He confided this information to optical scientist John Pike, who helped design and make the first-ever intraocular lens. Ridley implanted it into a forty-five-year-old female patient in a two-step operation in 1949.

Despite being a breakthrough, the procedure was not seen as such by a skeptical medical community. Previous to this, eye surgeons only ever removed objects such as foreign bodies or tumors from the eye, and never inserted anything into it. Eventually, after staying resolute in the face of scornful ridicule, his genius was recognized. The invention marked a major change in ophthalmologic practice and helped to pioneer artificial device implementation in medicine.

Ridley's procedure is now a routine operation and surgeons annually implant more than 6 million lenses. More than 60 million people with otherwise disabling cataracts have benefited from his invention. Since 1999 intraocular lenses have also been used to correct serious problems with focusing the eye. **JM**

SEE ALSO: LENS, SPECTACLES, BIFOCALS, CONTACT LENS,
LASER EYE SURGERY

Pager (1949)

Gross improves hospital communication.

Alfred J. Gross (1918–2000), inventor of the pager, had a lifelong passion for communication devices. As a boy he built his own radio set from junk materials, and, by the age of sixteen, he had earned his amateur radio operator's license, the beginning of a career in the radio communications industry.

The pager was originally developed as a tool for the medical profession, and was first introduced in a New York hospital. The devices responded to specific high-frequency radio signals and beeped to alert doctors to an emergency. They were not well received in the medical profession, however, with some doctors complaining that the beeping of the device would annoy patients, or even interrupt their golf swing. But

> *"I don't own a cell phone or a pager. I just hang around everyone I know, all the time."*
>
> Mitch Hedberg, comedian

the pager soon proved to be a useful device and became an essential tool for the medical profession.

The idea gradually spread outside of the medical community, with the first commercial pager available in 1974, and it became a must-have gadget. While the earliest pagers simply beeped to alert the user, later ones had a small display, allowing a message to be sent. However, the short message format meant that users had to rely on codes and shorthand.

Unfortunately for Gross, his ideas were ahead of their time, becoming popular only long after their patents had run out, meaning he never earned any money from their widespread popular use. **SR**

SEE ALSO: WIRELESS COMMUNICATION, CITIZENS' BAND
RADIO, WALKIE-TALKIE, CELLULAR MOBILE PHONE

Crash Test Dummy (1949)

Alderson creates a dummy to help assess potential injury.

In the late 1940s, the U.S. Air Force wanted data on how deployment of their newly designed ejection seats would affect the pilots who were strapped into them. For the first time, a crash test dummy was created to obtain the information. This very smart dummy was named "Sierra Sam" and was built in 1949 by American Samuel Alderson (1914–2005) in partnership with the Sierra Engineering Co.

Prior to the arrival of the crash test dummy, human cadavers were used to guide safety design. Working with corpses was of course highly unpleasant, but also the human bodies were very limited in terms of the information they could convey to researchers. It was also impossible to use them repeatedly to any useful purpose, and although they gave limited information on what injuries might be sustained, they gave no clue how the living would react in the scenario under investigation. But Alderson's carefully designed fake humans could. While crash test dummies were made famous by their well-publicized work in car safety, their use in the auto industry began more than twenty years after dummies were first created.

Alderson remained involved in crash test dummy research after his first invention, and in 1968, responding to the dangers of driving, he designed a car-specific dummy that he dubbed "V.I.P." This very important dummy and its modern descendents have been used to test seat belts, airbags, reinforced doors, antilock brakes, as well as influence total car designs. Without complaining they have endured endless simulated collisions to measure velocity of impact, crushing force, deceleration rates, bending, folding, and torque of the body. We owe modern automobile safety features to "anthropomorphic test devices." For dummies, they turned out to be very smart. **RBk**

"Any of us who [have walked] away from an automobile accident is likely to have a dummy to thank."

Jack Jensen, General Motors

⬆ *Only one "Sierra Sam" dummy was made; it appears here in a boiler suit with a recorder in its leg pouch.*

➡ *Two Irving Air Chute Co. employees unload dummies fitted with parachutes for use in skydiving research.*

SEE ALSO: MOTORCAR, AIRBAG, THREE-POINT SEAT BELT, INFLATABLE AIRCRAFT ESCAPE SLIDE

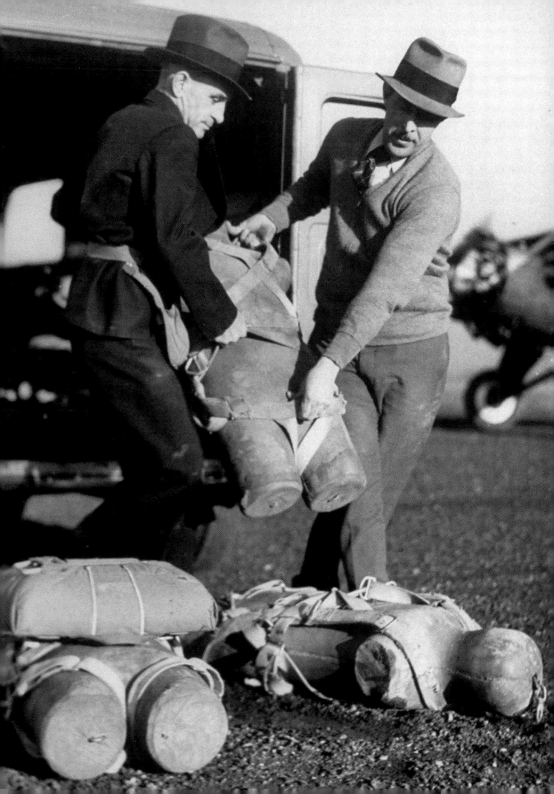

Magnetic Core Memory (1949)

Wang and Woo store data permanently.

Nowadays we trust computers to store more and more of our important records and information. However, before the 1950s no one would have even considered trusting a computer to store data—there simply was not the technology for long-term storage.

The first computers were not reliable because they stored data using such means as acoustic waves in mercury-filled tubes or complicated circuits of vacuum tubes. Then, in 1949, a new way of storing binary data was invented: "magnetic core memory."

Computers store data as binary information, where each bit of data is stored as a zero or a one (off or on). With core memory, metal cores are magnetized in one

> *"Success is more a function of consistent common sense than it is of genius."*
>
> An Wang

of two directions, and this determines whether they store a zero or a one. These cores are threaded together by wire in a flat lattice formation and data is recorded, or read, by sending pulses of current through the wires to the cores.

The technology behind this core memory was created by the Shanghai-born U.S. physicists An Wang (1920–1990) and Way-Dong Woo. Its nature explains why computer memory is called "memory," as magnetic cores, like magnetic disks, keep a record of their content even after the power supply is cut off. For this reason it is still sometimes used in specialized applications in military and space vehicles. **JM**

SEE ALSO: COMPUTER PROGRAM, DIGITAL ELECTRONIC COMPUTER, C PROGRAMMING LANGUAGE, FLASH MEMORY

Credit Card (1950)

McNamara devises a new way to pay bills.

The concept of credit is not a new one. The word "credit" comes from the Latin word *credo*, to believe or trust, and we know that the Egyptians were using credit methods over 3,000 years ago.

The credit card is a system of payment and is named after the plastic card issued to people who subscribe to the system. Although the term was used several times in Edward Bellamy's 1887 novel *Looking Backward*, and Western Union issued charge cards to users from 1914, the modern credit card was not invented until 1949. It is said that Frank X. McNamara, a New York banker, came up with the idea after forgetting his wallet when out entertaining clients. He distributed 200 Diners Club cards, which were essentially charge card accounts as the bill had to be settled entirely on issue.

Modern credit cards, where the issuer lends money to the consumer to be paid to the merchant, evolved quickly. In 1958 the Bank of America created the first credit card to be made available to the mass market. Bank Americard evolved into the Visa system, and MasterCard was established in 1966 by a group of credit-issuing banks. Arriving in the postwar boom period of growth, the credit card system was embraced wholeheartedly by developed countries. However, the perceived unreliability of the banking system in certain developing countries has meant that adoption there is still slow, even now.

Recently, alarm has begun to grow that the credit system in developed countries is out of control. The United Kingdom alone was facing a consumer debt of one trillion pounds sterling in 2004. McNamara's system, more than just a convenience, has changed the way we interact with money forever. **SD**

SEE ALSO: CASH REGISTER, CASH MACHINE (ATM)

Television Remote Control

(1950)

Zenith Radio Corporation eliminates exertion from television viewing.

By the 1950s the television set was beginning to establish itself as a firm feature of the family living room and viewers soon grew tired of constantly having to leave the comfort of their favorite armchair to change the channel.

One of the first incarnations of the remote control was the 1948 Garod "Telezoom." This small, round, single-button remote was connected to the televison by a wire, but its only function was to enlarge the picture on the screen. Thus it was Zenith Radio Corporation (now Zenith Electronics Corporation) that pioneered the modern remote control. The founder-president, Eugene F. McDonald, Jr., was worried that advertisements would kill television, and so challenged his engineers to invent something that would "tune out annoying commercials." In 1950 the rather aptly-named "Lazy Bones" was born. The remote control activated a motor in the television set that operated the tuner. However, this device was still connected to the television via a cable.

Five years later Zenith engineer Eugene Polley came up with the "Flashmatic," which used a beam of light rather like a flashlight to change channel, cut off the sound, and activate the television. Unfortunately the light sensors in the television were so sensitive that a beam of sunlight could turn it on or off randomly.

In 1956 Zenith's Dr. Robert Adler suggested using ultrasonics (high-frequency sound) and the "Zenith Space Command" went into production. This model set the technological trend until the 1980s, when newly developed infrared technology arrived. **HI**

SEE ALSO: TELEVISION, COLOR TELEVISION, CABLE TELEVISION, PLASMA SCREEN

↗ *Since the earliest remote controls, broadcasters have had to work much harder to retain viewer loyalty.*

"One might wonder how we would be controlling our television sets today if it wasn't for... Zenith."

James Fohl, writer

Tokamak (1950)

Sakharov and Tamm conceive nuclear fusion.

In 1950 Andrei Sakharov (1921–1989) took a break from designing nuclear weapons to study plasma physics with Igor Tamm (1895–1971) for a year, during which time they developed the concept for a machine that could produce energy through nuclear fusion. Tamm went on to win the 1958 Nobel Prize in Physics; Sakharov was awarded the 1975 Nobel Peace Prize. Progress on fusion power has been slow.

Nuclear fission produces energy by splitting highly radioactive atoms of uranium or plutonium. However, fission also produces waste that is dangerously radioactive for 10,000 years. Fusion, on the other hand, produces energy by combining deuterium (heavy hydrogen) into helium, with no dangerous waste. Sea

> *"We have never succeeded in slowing down our nuclear fusion reactors."*
>
> Wilson Greatbatch, inventor

water has one atom of deuterium for each 6,500 atoms of hydrogen. Deuterium atoms are twice as massive as hydrogen atoms, so they are relatively easy to separate. The energy thus promises to be inexpensive and clean.

Tamm and Sakharov's invention was called a TOKAMAK, the Russian acronym for a toroidal chamber with an axial magnetic field. The donut-shaped device uses magnetic fields to compress charged particles to the temperatures and densities required for fusion. In 1968, a team led by Lev Artsimovich used a tokamak to produce the first terrestrial thermonuclear fusion reaction. It achieved a temperature of 8 million degrees Kelvin for 0.02 seconds. **ES**

SEE ALSO: GAS TURBINE, ELECTRICAL GENERATOR, NUCLEAR REACTOR

Nuclear Reactor (1951)

Zinn uses nuclear fission to make electricity.

Harnessing the power of the atom has been a major goal of both science and science fiction, capturing the public imagination with the promise of cheap, clean energy. The initial idea was developed until December 20, 1951, when a switch was flicked and the Experimental Breeder Reactor 1 (EBR1) was switched on, becoming the first nuclear reactor to generate electrical energy, and therefore become the first nuclear power plant.

The heat generated from the reaction was used to turn water into steam, turning turbines to generate electricity. On its first successful run the reactor produced enough power to run just four lightbulbs. The next day it produced enough to power the entire research facility, and today nuclear power stations, many of them based on this original design, provide some 16 percent of the world's electricity.

Walter Zinn (1906–2000), the chief scientist behind the work, started his career in nuclear engineering in 1939. Just three years later, in 1942, he became the first to produce a self-sustaining nuclear reaction as part of the Manhattan Project, which was developing the nuclear bomb. After World War II ended in 1945, the Atomic Energy Commission (AEC) assigned resources to research peaceful use of the power of the atom, and develop nuclear power for electricity generation.

The reactor EBR1 ran until 1963, continuing research into nuclear energy. Zinn himself continued to refine the reactor design. His boiling-water reactor, among many of his later designs in the field, became the prototype for commercial nuclear power plants across the globe. **SB**

SEE ALSO: GAS TURBINE, ELECTRICAL GENERATOR, TOKAMAK

➔ *The condensation chamber at the heart of the Phénix nuclear reactor, opened in France in 1973.*

Hydrogen Bomb (1951)

Teller and Ulam build an awesome weapon.

Despite the continuing secrecy surrounding the development of the atomic bomb, it is public knowledge that Edward Teller (1908–2003), a Hungarian physicist, worked on the Manhattan Project to produce the first atomic bomb based on uranium fission. Teller had long been interested in a hydrogen fusion bomb, but secrecy and the lack of access to computers contributed to slow progress.

Stanislaw Ulam (1909–1984), a Polish mathematician, realized that a fission bomb could be used as a trigger for a fusion reaction. It is believed that Teller seized on this for what became, in 1951, the "Teller-Ulam" design.

Most sources agree that the H-bomb works in a series of stages, occurring in microseconds, one after the other. A narrow metal case houses two nuclear devices separated by polystyrene foam. One is ball shaped; the other is cylindrical. The ball is essentially a standard atomic fission bomb. When this is detonated, high-energy radiation rushes out ahead of the blast. It is believed that this radiation is "reflected" by the outer metal case toward the cylindrical secondary nuclear device. The cylinder, made of uranium 235, is crushed by the pressure of the radiation. Inside, deuterium (an isotope of hydrogen) is compressed and heated to the point where it undergoes nuclear fusion. This in turn releases more energy as well as neutrons, which start a further fission chain reaction in the outer uranium 235 of the cylinder and in an inner core of plutonium.

The fission bombs used in World War II had a yield equivalent to 20,000 tons of TNT; Ivy Mike, the first U.S. H-bomb test in November, 1952, is thought to have had the power of 10,400,000 tons of TNT. Albert Einstein commented, "I don't know what weapons countries might use to fight World War III, but wars after that will be fought with sticks and stones." **AKo**

SEE ALSO: CLUSTER BOMB, BOUNCING BOMB, ATOMIC BOMB, LASER-GUIDED BOMB

Liquid Paper (1951)

Graham improves correction of typing errors.

The correction fluid used worldwide to cover mistakes on paper began life in the United States in a single mother's North Dallas kitchen. In 1951, Bette Graham (1924–1980) was a young divorcée bringing up a small son, Michael (later Mike Nesmith of the Monkees pop group) and working as a secretary at Texas Bank & Trust. Bette and her colleagues appreciated the new speedy electric typewriters, but their carbon-film ribbons made fixing mistakes messy. The only answer was to type the whole sheet again.

Bette harbored artistic ambitions that had been thwarted by an early marriage. The story goes that her artist's eye spotted workers who were decorating the bank's windows and covering mistakes with an extra

> *"I put some tempera water-based paint in a bottle and took my watercolor brush to the office . . ."*
>
> Bette Graham

paint layer. Soon Bette was doing the same with a pale water-based tempera paint. Her boss never noticed and she began giving out bottles of her miracle "Mistake Out" fluid to colleagues.

Bette enlisted help, including a school chemistry teacher, to improve her formula. When, in 1958, she was fired for using her company headed paper for a bank letter, she was able to focus on promoting her product—now called Liquid Paper. She transferred operations to a factory in 1968, and by the 1970s was running a Dallas headquarters producing 25 million bottles a year. Bette sold out to Gillette in 1979 for $47.5 million, but was to die shortly afterward. **AK**

SEE ALSO: INK, PAPER, GRAPHITE PENCIL, CARBON PAPER, FOUNTAIN PEN, BALLPOINT PEN, FELT-TIP PEN

Tetra Pak (1951)

Rausing and Wallenberg revisit packaging.

Tetra Pak, a multinational company, revolutionized the food and drink industry with its unique cardboard carton production in the 1940s. It was founded in 1951 by Ruben Rausing (1895–1983) and Erik Wallenberg (1915–1999) in Sweden. Work began in 1943 with development of a new storage medium for milk, which had previously been sold only in glass bottles. The challenge was to provide a hygienic container using minimal materials. Tetra Paks are made from paper, polyethylene, and aluminum foil, arranged in seven layers to create a lightweight product.

The original design, the "Tetra Classic," was launched in 1952. This was a four-sided pyramidal milk container sold in Sweden. Initial responses were very positive and the company continued to refine its designs with the more familiar rectangular "Brik" carton produced in 1959.

In 1961 the simple addition of an aluminum layer provided the first aseptic package. Tetra Pak also used a short-term, high-temperature sterilization technique that dramatically increased the shelf lives of dairy products and juices, removing the need for refrigeration. It also allowed easier and cheaper transportation of goods across the globe.

Developments continue at Tetra Pak, and many containers of different shapes and sizes are now available. In 2003 the Sensory Straw was launched, a novel take on the drinking straw with a unique four-holed design. The straw causes the drink to squirt into the mouth in four directions, providing a "change in the consumer's drinking experience." Tetra Pak has also branched out into nonliquid packaging. **JG**

SEE ALSO: CANNED FOOD, SELF-HEATING FOOD CAN, ALUMINUM FOIL, FROZEN FOOD

↗ *Tetra Pak continues to work on cap designs that combine seal effectiveness with ease of use.*

"Lactomangulation, n. [Badly] manhandling the 'open here' spout on a milk carton . . ."

Rich Hall, comedian

The 1950s and 60s saw a changing of the guard, as driving a car lost its innocence and computers became the new high-tech products. The introduction of three-point seat belts, breathalyzers, air bags, and speed cameras put an end to careless driving, while hard-disk drives, video game consoles, hypertext, and the computer mouse hinted at technological developments to come. Man (and dog) left the planet and entered space for the first time, while money could now also be obtained from a cash machine.

◁ A starboard solar array onboard the Space Shuttle *Endeavour*.

Going
GLOBAL

1952 to 1968

Optical Fiber (1952)

Kapany pioneers the use of fiber optics for medicine and telecommunications.

Optical fiber, made from glass or plastic, is used to guide light from a source to another location. First used in medicine to examine internal organs, the technology has since been developed for many applications, including telecommunications.

Indian-born Narinder Singh Kapany (1926–2020) is the father of fiber optics. While undertaking research at Imperial College in London in 1952, Kapany drew out fine filaments of optical-quality glass and found that when he shone a light in at one end, it emerged unchanged from the other, even if the fiber was twisted. The concept behind optical fibers was first shown by Irish inventor John Tyndall, who used the principle to illuminate water fountains in the 1850s.

Kapany discovered that light was guided by total internal reflection within the glass fiber and that each fiber could carry many different wavelengths of light simultaneously over considerable distances. Adding a glass coating to the fiber increased its effectiveness by increasing the fiber's internal reflectivity. These qualities make optical fiber better than copper wires for many applications.

The concurrent development of laser technology was combined with optical fiber technology in the mid-1960s. While working for Standard Telephones and Cables of Essex, United Kingdom, Charles Kao realized that the attenuation of signal in optical fibers could be reduced by eliminating impurities in the glass. His work paved the way for today's use of optical fibers in telecommunications, where they form the majority of long-distance telephone and computer communication cabling worldwide. **LC**

"A teacher told Kapany that light could travel only in a straight line. Kapany set out to prove him wrong."

Fortune Magazine (1999)

SEE ALSO: INSULATED WIRE, COAXIAL CABLE, HIGH TEMPERATURE SUPERCONDUCTOR

⬈ *Optical fiber illumination can be used for decorative applications like signs and artificial Christmas trees.*

Bubble Chamber (1952)

Glaser advances the study of particle physics.

The story goes that Donald Glaser (1926–2013), a faculty member at the University of Michigan, was drinking a nice cold beer when he got the inspiration for what would eventually earn him the 1960 Nobel Prize in Physics. In 1952 he invented the very first bubble chamber, which was no bigger than a thumb. The idea was that subatomic particles, accelerated by a particle accelerator, could pass through a chamber containing a liquid. Under the right conditions a trail of tiny bubbles would be created as a particle passed through. By photographing and analyzing the bubbles, physicists could gain precious information about the nature of particles.

In 2006 Glaser denied that the bubbles in beer inspired him, though he did use beer as an early experimental liquid. Irrespective of how he got the idea, Glaser's concept worked, and bubble chambers sprang up all over the world to be used in atom smashers. The best and largest grew to sizes that could contain 706 cubic feet (20 cu m) of liquid hydrogen, stored just below its boiling point. As the particles enter the chamber, the internal pressure is reduced slightly using pistons, putting the liquid hydrogen into what can be described as a superheated metastable phase. This forces bubbles to form in the wake of the particles and, by decreasing the internal pressure further, these bubbles will expand until they are large enough to be imaged in 3-D by multiple cameras.

Bubble chambers dominated particle physics until the advent of wire-chamber detectors in the mid-1980s. However, the bubble chamber is making a comeback as a tool to look for dark matter. **DHk**

SEE ALSO: GEIGER COUNTER, CYCLOTRON, SYNCHROTRON, LARGE HADRON COLLIDER

⬈ *This bubble chamber, built for Imperial College London, England, contained liquid hydrogen.*

"In a way I represent the third generation of Nobel laureates in physics."

Donald Glaser, on receiving his Nobel Prize

Air Bag (1952)

Hetrick's invention saves lives in automobile accidents.

One day in 1952, John W. Hetrick was driving, with his wife and daughter in the front seat, when he had to swerve and brake quickly to avoid an obstacle. Instinctively, he and his wife put their arms out to shield their daughter in case of a crash. This inspired him to provide automobiles with air bags to protect people during accidents.

Hetrick had been an engineer in the U.S. Navy during World War II, and he recalled a compressed-air torpedo accidentally turning itself on, causing its canvas cover to shoot "up into the air, quicker than you could blink an eye." In 1952 Hetrick proposed using compressed air to inflate air bags rapidly during car crashes. He received a patent for his invention in 1953, but car manufacturers in the 1950s were more concerned with style than safety. Air bag technology improved and consumers became more safety conscious—the first air bags were offered as optional, but by the 1990s they had become standard.

An air bag system has three main components: the bag itself, a sensor for measuring the severity of a crash, and gas to inflate the bag in a severe crash. Air bags cushion sudden forward movement by spreading the impact over a greater area and slowing the impact so that it occurs over a longer time. Both decrease the damage done to bodies in a crash.

Although air bags have saved thousands of lives, they are not always sufficient to prevent death and injury during crashes. Travelers must also wear lap and shoulder seat belts, and automobiles must have padded dashboards and steering columns that can absorb energy during the impact of car crashes. **EH**

SEE ALSO: MOTORCAR, CRASH TEST DUMMY, THREE-POINT SEAT BELT, SPEED CAMERA

⬆ *Ford's president Donald Petersen shows an air bag to U.S. Transportation Secretary Elizabeth Dole in 1983.*

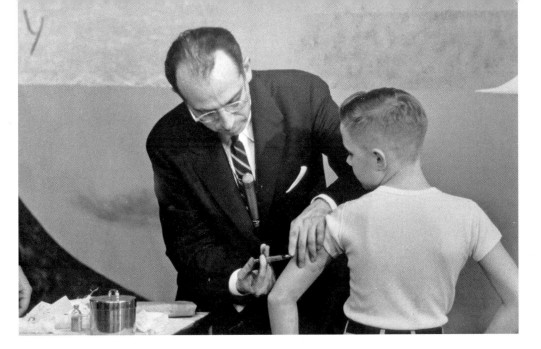

Polio Vaccine (1952)

Salk develops a life-saving injectable vaccine.

Today the deadly and debilitating poliomyelitis virus is only endemic in four countries—Afghanistan, India, Nigeria, and Pakistan. This is thanks to the groundbreaking research undertaken by U.S. medic and biologist Jonas Salk (1914–1995).

In 1947, at the University of Pittsburgh, Salk combined his work on the influenza vaccine with searching for a vaccine to protect against polio. The virus was deadly in 5 to 10 percent of cases where patients became paralyzed, and thus were unable to breathe. Medical opinion at the time held that only a live virus could prompt complete immunity, but Salk disproved this. In 1952 he used formaldehyde to inactivate the polio virus and developed a vaccine still capable of triggering an immune response in a host. Initially tested on monkeys, then patients at the D. T. Watson Home for Crippled Children, Salk's success convinced him to test it on himself, his family, his staff, and other volunteers.

In 1954 one of the earliest double-blind, placebo-controlled trials ever gave the vaccine to one million children aged between six and nine years, and a placebo to another million of the same age. A year later, the children treated with vaccine had developed immunity to the live disease. In 1952 there were 57,628 cases of polio recorded in the United States. Following vaccine use, this fell 85 to 90 percent in two years.

An oral vaccine replaced Salk's injectable version in 1961 after Albert Sabin developed a live dose administered on a sugar lump. More recently, the United States and United Kingdom have reverted to Salk-type injectables of the inactivated virus. **JF**

SEE ALSO: INOCULATION, VACCINATION, CHOLERA VACCINE, ANTHRAX VACCINE, BCG VACCINE, RABIES VACCINE

⬆ *Salk is photographed inoculating a young boy with his polio vaccine following extensive trials.*

Cloning (1952)

Briggs and King genetically copy organisms.

When most people think of cloning, their thoughts inevitably turn to science fiction movies or a sheep named Dolly. However, cloning has been around a lot longer than either of these would suggest. The word itself comes from the ancient Greek word for twig and was initially used to refer to plant grafting in the early 1900s. The word's meaning as it is used today, originated in the 1950s.

The scientists responsible for what is now thought of as cloning did not set out to create a genetically identical model of another organism—as clones are defined these days. They were simply trying to understand how an embryo develops into an adult. Nobel Laureate Hans Spemann began experimenting with salamander embryos in the 1930s. Using a piece of hair as a tool, he manipulated a dividing nucleus from a salamander embryo in such a way that he had two different nuclei in two different cells. The eventual result, two identical salamanders, caused Spemann to wonder if a nucleus from an adult creature could be placed in an embryo and divided in the same way.

Robert Briggs (1911–1983) and Thomas King (1921–2000) took up the quest several decades later. Briggs, apparently unaware of Spemann's work, also speculated that transplanting an adult nucleus into an enucleated fertilized egg would ultimately lead to a fully formed adult animal. He approached King about the microsurgical aspects of manipulating such minute amounts of matter, and soon the pair managed to implant nuclei from adult Leopard frogs into an enucleated egg. In 1952—forty-four years before Dolly became the first mammal to be cloned—they succeeded in cloning an organism for the first time, when the eggs they used went on to form tadpoles. **BMcC**

SEE ALSO: **INTERFERON CLONING**

Roll-on Deodorant (1952)

Diserens invents easily applicable deodorant.

The development of the roll-on deodorant is a perfect example of how ingenious lateral thinking can link one seemingly unrelated invention to another and result in a useful product. "Mum" deodorant—first developed in the late nineteenth century—was a rather sticky, hard-to-apply substance. When Helen Barnett Diserens (1918–2008) joined the product team of Bristol-Myers as a researcher in the late 1940s, one of the company's skills was working creatively on a select list of consumer-based toiletries. A member of Diserens's team suggested that she take a look at another recent marvel, the ballpoint pen (actually based on an idea from the late 1800s) to get some inspiration for improving Mum's applicator.

> *"Your bath took care of the past, but for future freshness, make Mum your next step."*
>
> Mum advertisement, 1946

Diserens's new roller applicator—made of glass with a rolling ball at its tip—was tested in the United States in 1952. After further testing, the product was launched globally and to wide acclaim in 1955. However, the roll-on was soon eclipsed by aerosol antiperspirant deodorants, which held a massive market share until the 1970s. Then fortune turned once more. With increasing strictures on the chemical ingredients of the aerosol product, the roll-on has come into its own once again. **AK**

SEE ALSO: **TOOTHPASTE, SOAP, TOOTHBRUSH, DEODORANT**

→ *Mum deodorants are still widely used in parts of Europe as well as South America and Asia.*

Another case for Mum Rollette

THE NEW DEODORANT THAT **ROLLS ON!**

Speech Recognition (1952)

Bell creates computers that recognize speech.

Her voice was once described as sounding like two robot tobacco auctioneers fighting over a cigar butt. However, Audrey's synthesized speech was state-of-the-art back in 1952. A clumsy acronym for Automatic Digit Recognition, Audrey was an analog computer at Bell Laboratories. But it was not her primitive voice that she was renowned for. Scientists at Bell Laboratories had tried for many years to devise a technology that could recognize human speech and Audrey was their first working solution.

The potential applications for a computer that could convert words spoken by humans directly into digital text were obvious. However, the sheer number of variations in the qualities of people's voices, along with their different intonations and pronunciations, means that foolproof speech recognition is a huge task. Audrey could only recognize spoken numbers from one to ten, and used flashing lights to illustrate what she had heard—or thought she had heard. She compared the spoken sounds to the sound patterns stored in her memory and simply decided which were the most similar.

The technology was full of promise, and Bell Laboratories wanted to turn her into an operator that could dial telephone numbers for you. The problem was accuracy. Audrey had to be "tuned" for specific people's voices, and then she could only guarantee an accuracy of 98 percent, which is nowhere near good enough for commercial application.

Since the 1950s, the rapid increase in computing power has enabled many groups to experiment with their own speech recognition systems. One can even buy them for a home computer and program them with specific phrases in order to play computer games without touching the keyboard. **DHk**

SEE ALSO: DICTATING MACHINE, SPEECH SYNTHESIS,
OPTICAL CHARACTER RECOGNITION

Hovercraft (1953)

Cockerell invents a craft that floats on air.

A hovercraft—also known as an Air Cushion Vehicle (ACV)—floats on a cushion of air, the air being pumped under it by a large ducted centrifugal fan. It is amphibious and can be driven across relatively flat land for loading purposes, and then across reasonably calm water. As the cushion of air greatly reduces drag, the craft can travel across water at high speed.

Sir Christopher Cockerell (1910–1999) designed the first commercial hovercraft, the Saunders Roe Nautical One (SRN1). He conceived the idea in 1953, but it was not until June 1959 that his three-man craft—traveling at about 29 miles (46 km) per hour—was first tested. Cockerell decided to pump air into a narrow tunnel around the circumference of the craft. The air then

> *"Fair is foul and foul is fair; Hover through the fog and filthy air."*
>
> William Shakespeare, *Macbeth*

escaped towards the center, where it built up a high-pressure cushion that supported the weight of the craft, making it stand clear of the ground.

Today, large commercial hovercrafts can cross the English Channel in less than half an hour. Military forces use them as personnel and weapons carriers, whereas oil companies employ them to transport drilling equipment to inaccessible places. They can also travel across wetlands, reeds, and swamps with very little impact on the water life near the surface. **DH**

SEE ALSO: ROWBOAT, SAIL, RUDDER, STEAM BOAT,
SUBMERSIBLE CRAFT, SUBMARINE, MOTORBOAT, JET BOAT

➡ *In July 1959 Cockerell's hovercraft made its first trip across the English Channel, going from Calais to Dover.*

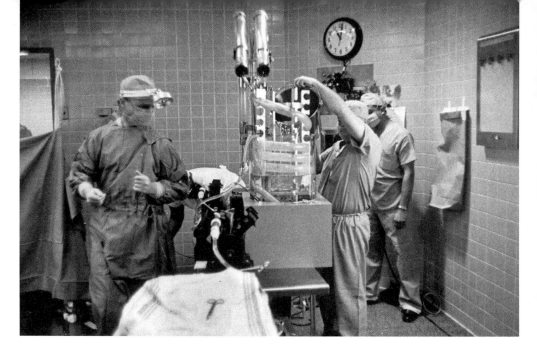

Heart-Lung Machine (1953)

Gibbon's machine is a major breakthrough for heart surgery.

In October 1930, a young surgical resident sat vigil as a patient with a blood clot in her pulmonary artery labored to breathe. The surgery she needed had never been successfully performed in the United States. Developed in Germany, the "Trendelenburg operation" had a 6 percent survival rate. After seventeen hours, it was clear the patient was not going to survive without surgery, so, with nothing to lose, the procedure was successfully carried out, but the patient died.

For the next twenty-three years, Dr. John Heysham Gibbon (1903–1973) and his wife Mary worked to produce a machine that could supply oxygenated blood while the heart was stopped. In 1935 he used a prototype heart-lung bypass machine to keep a cat alive for twenty-six minutes. Venous blood was fed into the machine where it was spun over a cylinder to provide oxygen, and then pumped back into an artery.

Many improvements were made, and in 1951 the first coronary artery bypass on a human being was performed, although the patient died. Fifteen months later, on May 6, 1953, an eighteen-year-old girl's heart function was replaced for twenty-six minutes with Gibbon's machine while a large hole in her heart was repaired. She is known to have survived for at least another thirty years.

Although this version was only used a few times, improved Gibbon machines and other models have enabled surgeons to perform many operations using cardiopulmonary bypass to correct heart defects, replace heart valves, and repair aortic aneurysms .

Gibbon, the pioneering heart surgeon, died of a heart attack in 1973. **SS**

SEE ALSO: IRON LUNG, CARDIAC PACEMAKER, KIDNEY DIALYSIS, ARTIFICIAL HEART

⬆ *A heart-lung machine is readied for use at the University of Minnesota Hospital in 1957.*

Adaptive Optics (1953)

Babcock comes up with an optical system to aid astronomers.

Astronomers observe light from stars many light years away and yet the thin atmosphere that surrounds Earth can play havoc with their results. Small volumes of the atmosphere have different temperatures and densities; they move around and this turbulence causes incoming light to change direction. The change in direction shows up as distortions in the data and can either render measurements useless or make their interpretation difficult. Observatories are often built on high mountains to limit the thickness of atmosphere the light needs to travel through. However, for the most detailed projects the thinnest atmosphere is still enough to cause problems.

U.S. astronomer Horace Babcock (1912–2003) devised an optical system in 1953 that could adapt to the changes and correct the errors in real time. Although his design gave hope to toiling astronomers, it was not used until the 1990s when computers could keep up with the speed of the fast changing atmosphere.

The technique is called adaptive optics and works by measuring the incoming distortion and quickly sending this information to deform a mirror. The incoming light from a star is reflected from the deformed mirror and measured as what it would have been without an atmosphere. The computer must be able to analyze the information about the changing atmosphere and send signals to the mirror to change its shape many times in a second. Implementing the technique has been difficult—materials for the bendable mirror and computing power have held the technology back, but when successful it has made massive improvements to the quality of images. **LS**

SEE ALSO: TELESCOPE, BLINK COMPARATOR, SPACE OBSERVATORY, HUBBLE SPACE TELESCOPE

⬆ *Babcock examines a paper inside the Mount Palomar Observatory near San Diego, California.*

WD-40® (1953)

Larsen invents product with unimagined uses.

There can be few people who have not encountered this remarkable product. Assumed by some to have been a spin-off from wartime, military technology, and therefore named "War Department—1940," the title actually relates to the number of attempts it took the inventors of this water displacing (WD) chemical to perfect the product.

It was at a lab in San Diego, California, in 1953, that the Rocket Chemical Company and its three personnel embarked on their mission to create a range of rust-prevention and degreasing products for use in the aerospace industry. After thirty-nine "almosts," they succeeded. The aerospace contractor Convair bought the chemical for use as a corrosion inhibitor on its Atlas missiles, and other wholesale orders soon followed.

Rocket Chemical Company employees had for some time been taking small amounts of the petrochemical-based product home for their personal use. Company founder and chemist Norm Larsen (1923–1970), suspecting that the general public might find the product useful, experimented with putting WD-40® into aerosols. The brand made its retail debut in stores in San Diego in 1958 and sales, along with new and unimagined uses, steadily grew. In 1969, the company was renamed for its famous product.

These "unintended" uses have remained a theme throughout the life of the product. Initially devised as a moisture repellent, WD-40® has attracted an army of loyal users, who delight in finding new applications. So, whether using it on the edges of plant pots to keep slugs at bay, to remove chewing gum from the carpet, applying it to unstick a jammed zipper, or as a cure for arthritis (although the manufacturers are at pains to point out that they do not endorse this), WD-40® is likely to be around for a long time to come. **MD**

SEE ALSO: AUTOMATIC LUBRICATION, AEROSOL

Synthetic Diamond (1953)

Von Platen achieves diamond alchemy.

The idea of creating a man-made diamond is a very appealing one—a kind of alchemy (the process of turning base metal into gold), but one that is, remarkably, achievable. Synthetic diamonds actually come in two kinds—simulant (which are diamond-like in looks and structure) and synthetic, where the chemical structure of the stone is also the same. The latter is a "true" synthetic diamond.

Simulant diamonds made from silicon carbide were first discovered by Henri Moissan in a meteorite crater in 1893. They were later reproduced by him and others in the lab, most famously Willard Hersey, whose diamond can still be seen at the McPherson Museum in Kansas. It was not until 1953, however, when a team led by Baltzar von Platen (1898–1984), working in secret for the Swedish electrical company ASEA, actually succeeded in generating sufficient heat and pressure to create the first synthetic diamond. Their machine generated 83,000 atmospheres of pressure and ran for two hours. Initially, it was so dangerous that von Platen ordered all nonessential personnel from the room when the machine was switched on.

The discovery was kept secret for fear of competitors stealing their methods, and it was a team working separately for General Electric a year later that first publicly claimed to have created a synthetic diamond. Commercial production followed and man-made diamonds have since found a wealth of uses. In addition to jewelry, they are used as hard edges on cutting tools and have various uses in electronics on complex machinery such as particle accelerators and precision measurement devices. **SS**

SEE ALSO: CARBORUNDUM

→ *Scientists at the General Electric Company operate a 1,000-ton press used to make diamonds in 1955.*

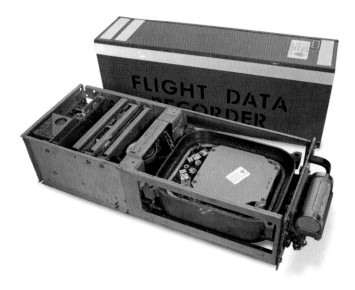

Black Box Flight Recorder (1953)

Warren's box is a major step forward for the safety of air travel.

The black box flight recorder, now a required feature on all aircraft, was the brainchild of Australian aviation scientist David Warren (1925–2010), whose own father was killed in a plane crash in 1934.

After World War II, there was a massive rise in commercial air travel, but after several planes came down in unexplained circumstances, the public's confidence in flying was understandably shaken. In 1953, Warren was part of the team investigating the crash of the world's first jet-powered passenger plane—the Comet. He thought how useful it would be to have an account of the events inside the plane during the last moments before it came down, and so he set out to develop a crash-proof device that could record sound and instrument readings in the cockpit. He built a prototype called the "ARL Flight Memory Unit" that could record up to four hours of speech onto steel wire. Little interest was aroused in his native Australia, but the British first took up the design in 1958.

After another unexplained crash in Queensland in 1960, Australia became the first country in the world to make black box recorders mandatory on all its aircraft. Modern planes now have two black box devices (which are actually bright orange): the cockpit voice recorder (CVR) and the flight data recorder (FDR), both of which are located in the aircraft's tail. They record everything from cockpit and radio communications, to airspeed, altitude, and engine temperature. The recorders are encased in materials such as titanium, and are insulated to survive the huge impact of a crash so that they may help investigators piece together the events prior to an accident. **HI**

SEE ALSO: RADAR, BAROMETRIC ALTIMETER, GROUND PROXIMITY WARNING SYSTEM

⬆ *Black boxes are designed to withstand intense flames and the pressure of the deepest oceans.*

Backhoe Loader (1953)

Bamford invents a powerful, versatile piece of construction equipment.

The backhoe loader, or JCB®, was invented by Joseph Cyril Bamford (1916–2001) of Staffordshire, England. In 1945, using only scrap metal, war surplus Jeep axles, and a cheap welding kit, he produced an hydraulically operated tipping trailer that he later sold for £45 ($180 at the time). With the dump truck-like tipping trailer created, Bamford's business grew steadily as he designed and built new machines using hydraulic power. When the backhoe loader was created in 1953, the JCB® logo was used for the first time, based on its inventor's initials.

Typical backhoe loaders are made up of three components: a tractor, a loader, and a backhoe. The tractor enables the backhoe loader to traverse difficult terrain of all sorts. At the front of the tractor, the loader can scoop, smooth, and push great quantities of material. Meanwhile, the back of the tractor sports a backhoe, which is a large maneuverable arm with a deep bucket. The backhoe can dig up material above

or below the tractor, lift heavy loads, and—in the wrong hands—cause serious structural damage. The backhoe and the loader are controlled and powered by a complex system of bidirectional hydraulic pistons called hydraulic rams. A diesel engine pumps oil into the backhoe loader's various hydraulic rams to move the backhoe and loader appendages.

JCB® began painting their construction machinery its distinctive yellow and black colors in 1951. The sporty yellow and black backhoe loader is now among the most popular pieces of construction equipment in the world. Since its inception, more than 325,000 JCB® backhoe loaders have been sold worldwide. **LW**

SEE ALSO: TRACTOR, BULLDOZER

⊡ *As one of the world's top three makers of construction equipment, JCB® sells products in 150 countries.*

Felt-tip Pen
(1953)

Rosenthal's pen leaves an indelible mark.

The fiber or "felt-tip" pen is a wondrous development in the history of writing. It allows the writer to scrawl their message to the world on almost any surface, safe in the knowledge that their pearls of wisdom will be read for the rest of time.

Primitive felt-tip pens date back to the 1940s. They were crude devices—not unlike a fountain pen in design—with a reservoir for ink, but with a piece of felt or some other porous material for the ink to slowly run through, instead of the traditional nib or stylus.

Sidney Rosenthal dramatically improved on this in design in 1953. He took a squat glass bottle of ink with a wool-felt wick and writing tip. He called this new device the "Magic Marker" due to its ability to write on any surface. It represented a significant development in the technology of felt-tip pens, which started to be used more and more.

In 1962 Yukio Horie of the Tokyo Stationery Company in Japan invented the modern felt-tip pen, which had a smaller tip than previous incarnations and was therefore more suited to writing on paper. Horie's invention was the result of his desire to develop a writing brush (commonly used in Japan) that had the convenience of a pen. He created a firm narrow point by binding together and shaping acryl fibers. It was the first pen to use dye rather than ink, which meant than many different colors could be used. Gravity caused the dye to be channelled to the little fibers in the tip of the pen.

Some people predict that soon all of our writing will be done using digital media instead of paper, making pens a thing of the past. But others believe that whenever people want to express themselves in writing, the convenient felt-tip pen will still rule. **BG**

SEE ALSO: QUILL PEN, FOUNTAIN PEN, BALLPOINT PEN

Medical Ultrasonography (1953)

Edler and Hertz advance medical imaging.

Since its tentative beginnings, the diagnostic imaging technique of ultrasound—which uses sound waves to visualize the body's internal organs—has found its way into most areas of medicine. Early experiments were made by Dr. George Ludwig at the Naval Medical Research Institute in Bethesda, Maryland, in the late 1940s. Ludwig successfully used ultrasonics to detect human gallstones implanted in a dog.

Inge Edler (1911–2001), a Swedish cardiologist, took the idea further. He was frustrated with the inadequacies of techniques then available to examine the heart. While making decisions on heart disease prior to surgery at University Hospital Lund, he found that cardiac catheterization and contrast X-rays did not tell clinicians enough about the heart's mitral valve. He asked Carl Hertz (1920–1990), who was working in nuclear physics at Lund University, if radar might be the answer. Hertz said it was not, but he suggested ultrasonography might work instead.

After borrowing an ultrasonic reflectoscope, the men managed to get some good echoes on a screen that moved in sync with a heartbeat on October 29, 1953. Six weeks later, they used the technique to generate an ultrasonic probe of the brain. Meanwhile, in Scotland, physician Ian Donald of the Glasgow Maternity Hospital was also developing the use of ultrasound for obstetric applications.

Today, ultrasound uses probes with acoustic transducers that send pulses into an organ. It is good for imaging muscle and soft tissue, and has no known long-term side effects. **LH**

SEE ALSO: X-RAY PHOTOGRAPHY, MAGNETIC RESONANCE IMAGING, COMPUTED TOMOGRAPHY (CT OR CAT SCANS)

➡ *This ultrasound scan shows a human fetus in the first trimester of pregnancy.*

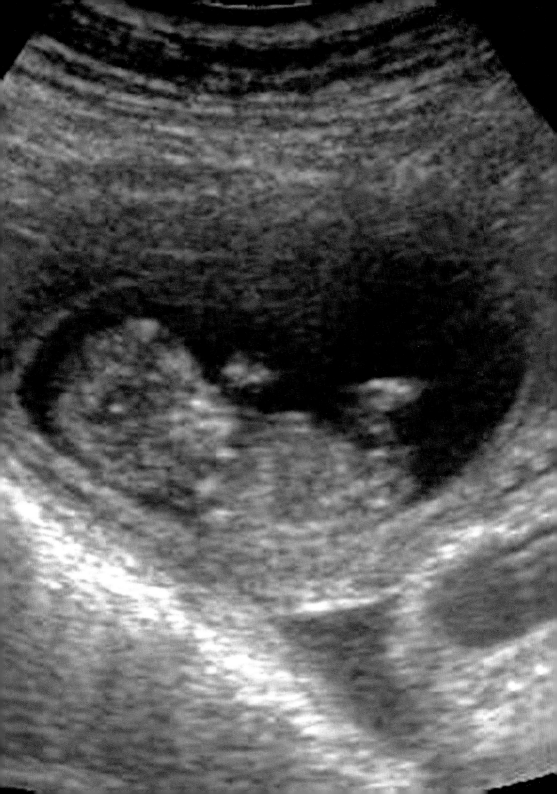

Geodesic Dome (1954)

Fuller creates an architectural form from patterns of reinforcing triangles.

Richard Buckminster Fuller (1895–1983) was a true polymath, dipping into fields as diverse as poetry and architecture. It is for the latter that he is best known and, in particular, for the geodesic dome. Many geodesic domes have been built around the world, including the Epcot Center in Disneyworld, Florida, and the Eden Project in Cornwall, England.

Geodesic domes come in many shapes and sizes, but they're basically a sphere made from self-reinforcing triangular sections. The dome has a number of advantages: It maximizes volume to surface ratio, it is very strong for its weight, it is quick to assemble, and it is aerodynamic, allowing it to stand up to strong winds. Although Fuller received the U.S. patent on the concept, he was actually building on earlier work by the German Walther Bauersfeld, who constructed a planetarium along these principles in 1922.

Fuller's dome had a remarkable renaissance in 1985, when Harold Kroto and his coworkers discovered a third stable form of carbon (after diamond and graphite) comprising sixty atoms of carbon. Kroto, pondering the possible structure of such a compound, recalled a geodesic dome he had seen in Montreal, and subsequent analysis suggested that carbon cages of sixty atoms are arranged just as in one of Fuller's geodesic domes. The structures were named fullerenes in his honor. Buckyballs, as they are also known, have been touted as delivery devices for medicines, and structural components in nanotechnology applications. Even though large-scale production remains a challenge, this unusual structure looks to have a promising future. **MB**

SEE ALSO: BUILT SHELTER, DOME

⬆ *Fuller, posing in front of one of his geodesic domes in 1960, also built prototypes of a fuel-efficient car.*

Automatic Doors (1954)

Horton and Hewitt introduce a new door system to the world.

In 1954 in Corpus Christi, Texas, Horton Glass Company employees, Dee Horton and Lew Hewitt, had just finished replacing yet another customer's wind-damaged glass door. Powerful, gusting South Texas winds wreaked havoc with traditional glass, push-pull doors, and ensured a steady demand for such repairs. Worse, the unpredictable winds could cause a door to blow open and shut in someone's face just as they were trying to walk through it. So the duo decided to invent a better door system.

Initially, the system used a simple, electrically activated sliding door that opened only when a mat actuator in front of the door was stepped upon. Not only did this solve the problem of wind-blown accidents, it also enabled visitors or customers to leave, and delivery persons to enter a shop or business with their hands full. Having installed a test unit, for free, at the city's utilities department, sales of the door system proper began in 1960. The first commercial unit found service in a hotel restaurant in Corpus Christi. They patented the invention in 1964, and their pioneering company, Horton Automatics, took off. Hewitt later invented the automatic sliding window.

The sliding door remains, to this day, the most widely used automatic entry system for offices and public buildings. Boasting simplicity of operation and user-friendliness, the basic design has been developed over the years to an extremely high level of technical sophistication. Automatic doors are now so commonplace that some people can experience a momentary confusion when a large glass door does not open automatically. **MD**

SEE ALSO: BUILT SHELTER, REVOLVING DOOR

⬆ *Automatic doors can be activated by a sensor that detects approaching traffic, or by pushing a button.*

Non-Stick Pans (1954)

Marc and Colette Grégoire create an ingenious kitchen utensil.

No one can dispute the value of the non-stick pan in today's kitchen. It is one of the best-selling utensils in kitchen history. Its story begins with the invention of Teflon® by the DuPont company in 1938. DuPont was researching refrigerants when it accidentally produced PolyTetraFluorEthylene or PTFE. The slippery substance was resistant to chemicals, temperature, electricity, mold, and fungus. DuPont used it to coat bread pans.

In the early 1950s, French engineer Marc Grégoire heard about Teflon® from a colleague who used it to coat aluminum for industrial applications. Grégoire managed to find a way to make Teflon® adhere to aluminum, and he then used it as a coating to prevent his fishing gear from tangling. His wife Colette then asked him if he could coat her cooking pans. He did, and it worked so well and safely that he was awarded a patent in 1954.

The couple began producing the non-stick pans and selling them in 1955. Marc made the pans in the kitchen and Colette sold them on the street. Even traditional French cooks were buying them, and in 1956 Marc and Colette formed the now famous Tefal Corporation. The word *Tefal* comes from combining Teflon® and aluminum, the two components of non-stick pans. By 1958 the couple was selling one million non-stick items from their factory.

Thomas Hardie began manufacturing the pans in the United States in 1960 under the name T-fal after DuPont decided Tefal was too close to their brand name of Teflon®. The companies have continued to sell non-stick products for decades and they are considered staples of the kitchen worldwide. **RH**

SEE ALSO: TEFLON®

"Franklin D. Roosevelt was a 'Teflon® president' long before Teflon® was invented."

Robert S. McElvaine, historian

◤ *Teflon® coatings are applied to steel or aluminum pans, while cast iron pans become non-stick through use.*

Transistor Radio (1954)

Texas Instruments produce the first radio in miniature.

When the transistor first appeared in 1947, few could have guessed how it would quickly transform the world of consumer electronics. The transistor was the size of a fingernail, weighed practically nothing, and was a direct replacement for the bulky, delicate electricity-powered glass valve. Suddenly, electronic circuitry was able to shrink to a fraction of its former size and be powered by small DC (direct current) batteries.

The first popular consumer product to take advantage of this miniaturization was the transistor radio. Although there had been a number of small portable models appearing at trade fairs in Europe and Japan, the 1954 Regency TR-1 was the first to go into mass production. The technology behind the Regency transistor radio came from an innovative U.S. electronics corporation, Texas Instruments. They had designed and built a prototype model, but sought an established radio manufacturer to develop and market their circuits. Now viewed as something of an icon of 1950s design, the TR-1 went on sale for just under $50, which at today's value is approximately $400 dollars—making it by no means a cheap product.

The transistor radio became hugely popular in the 1960s, by which time cheap imports from the Far East meant it could be bought for well under $10. It created the fashion for portable audio—now you could listen to music on the beach, in the garden, or walking down the street. Most "trannies" also came with an earpiece, so you could listen in private. It was the first step along a path that would lead to the Sony Walkman personal stereo and the Apple iPod, both essential accessories for their respective generations. **TB**

SEE ALSO: WIRELESS COMMUNICATION, IMPROVED RADIO TRANSMITTER, TRANSISTOR, PERSONAL STEREO

⬀ *The Regency TR-1 was unveiled in October 1954 and went on sale a month later, selling 150,000 units.*

"Whatever happened / To Tuesday and so slow / Going down the old mine / With a transistor radio."

Van Morrison, "Brown Eyed Girl"

Polypropylene (1954)

Natta and Ziegler create one of the world's most important plastics.

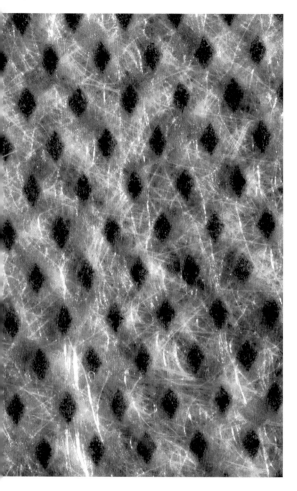

> *"There's so much plastic in this culture that vinyl leopard skin is becoming an endangered synthetic."*
>
> Lily Tomlin, comedian and actress

Polypropylene is one of the modern world's most important plastics. It is used as an artificial fiber in carpets, upholstery, and industrial ropes, as well as in food and toiletry bottles, toys, furniture, and car components. Polymers are very common in nature because they are the major components of hair, bones, muscles, and plant fibers. In the 1920s, scientists were working on how to make polymer chains longer and heavier, and therefore more useful. One of the first breakthroughs came in England in 1931, when polyethylene—the first man-made polymer with a high molecular weight—was made. Polyethylene dominated the global plastics market, but scientists were disappointed by its material weakness. The race to develop a better plastic was on.

Italian scientist Giulio Natta (1903–1979) had been working with the German scientist Karl Ziegler (1898–1973) on developing catalysts to help create new polymers. Based on titanium, these new Ziegler-Natta catalysts opened up a whole new way of creating polymers, and Natta was quick to exploit them. In 1954 he managed to polymerize propylene, but, in what is seen by many as an act of betrayal, he applied for a patent before telling Ziegler about his discovery.

This heralded a legal fight, during which it became clear that other groups around the world had also managed to make polypropylene using Ziegler-Natta catalysts. Eventually, the lawyers decided that, while Natta had the right to claim polypropylene, he must pay 30 percent of all royalties to Ziegler. The rift between the two men lasted until 1963 when they jointly shared a Nobel Prize in Chemistry. **DHk**

SEE ALSO: POLYSTYRENE, PVC, TUPPERWARE, PLASTIC BOTTLE (PET)

⬉ *This example of polypropylene is mainly used in automotive interior trim, such as car seats.*

SI Units (1954)

The Committee for Weights and Measures introduces a universal system of measurement.

Scientists and engineers constantly need to measure distances, masses, times, temperatures, densities, velocities, electrical currents, and so on. All these quantities are then expressed as a number, and this means that units are absolutely vital.

The idea of implementing common bases for all the units began with the British Association for the Advancement of Science in 1874. They suggested the centimeter for length, the gram for weight, and the second for time, which was known as the c.g.s. sytem. Prefixes such as *mega-* and *micro-* could then be used to indicate decimal multiples and submultiples.

Unfortunately, this c.g.s. system was rather inconvenient in the field of magnetism and electricity. In 1889 the Conférence Générale des Poids et Mesures decided that the meter, kilogram, and second (m.k.s.) might be more appropriate. In 1946 the ampere, a unit of electrical current—named for André-Marie Ampère—was added.

By 1971 this system had grown, and at the fourteenth Conférence Générale des Poids et Mesures, these four base units were joined by the Kelvin (for temperature), the candela (a unit of luminous intensity), and the mole (a unit quantifying the amount of substance, or atoms and molecules).

Le Système International d'Unités—the SI unit—was thus established. The international system of units has replaced many traditional measurement systems around the world. Today, most countries employ the SI, although the United States, Liberia, and Myanmar have not officially adopted the SI unit as their primary system of measurement. **DH**

SEE ALSO: STANDARD WEIGHTS AND MEASURES, METRIC SYSTEM

⤢ *The Bureau's platinum-and-iridium standard 1-kilogram weight is stored under three glass bells.*

"The SI is not static but evolves to match the world's increasingly demanding requirements."

International Bureau of Weights and Measures

Breathalyzer (1954)

Borkenstein invents a device to deter driving under the influence of alcohol.

In 1954 Robert Borkenstein (1912–2002), of the Indiana State Police, invented the breathalzyer, a portable device that provided scientific evidence of alcohol intoxication. After the consumption of alcohol, blood flows through the lungs and some of the alcohol evaporates and moves across the membranes into the air sacs. The breathalyzer shows a direct relationship between the concentration of alcohol found in the air sacs and that in blood. Subjects blow up a balloon (ensuring a deep air lung sample is taken), the contents of which are released over a chemical solution, resulting in a color change detected by a monochromatic light beam. The extent of the color change relates to the percentage of alcohol in the breath.

The breathalyzer replaced the drunk-o-meter, an earlier device developed by Rolla Harger and patented in 1936. It relied on the sampled breath being analyzed in specialist laboratories. Prior to this, police had to rely on observation of a suspect's physical condition, evidence that did not stand up well in court. The breathalyzer met the demands of legal evidence, although there were criticisms that the results varied between individuals consuming identical amounts of alcohol due to gender, weight, and metabolic rates.

In 1964, Borkenstein published a milestone study in which he argued that the 0.08 percent blood-alcohol content should be the standard, above which any driver should be considered impaired.

By the mid-1980s, chemical-based devices had given way to infrared technology where a narrow band of infrared light, of a frequency that is absorbed by alcohol, is passed through a breath sample. How much of the light makes it through to the other side of the sample without being absorbed gives the precise concentration of alcohol. **JF**

"Drinking and driving:
there are stupider things,
but it's a very short list."

Author unknown

⤒ *This breathalyzer from the early 1960s was made by the Stephenson Corporation in New Jersey.*

⤷ *This portable drunk-o-meter was exhibited at the National Safety Congress in Kansas City in 1937.*

SEE ALSO: MOTORCAR, PARKING METER, SPEED CAMERA, AIR BAG, THREE-POINT SEAT BELT

Ventouse (1954)

Malmström invents a device to assist birth.

The ventouse is a vacuum device that was developed to assist in the delivery of a baby when labor is not progressing well. It was developed in Sweden by obstetrician Tage Malmström (1911–1995) as an alternative to forceps. Malmström had the idea of using a bicycle pump to create the vacuum pressure. A bowl-like apparatus is attached to the baby's head in the birth canal. Air between the baby's head and the apparatus is then sucked out to create a vacuum. The doctor, who controls the amount of suction applied, synchronizes the pressure applied with the mother's contractions, enabling the baby to be delivered. The suction cups can be either flexible (soft plastic or silicon) or rigid (hard plastic or metal).

"The vacuum cup may save many mothers from difficult and dangerous forceps deliveries."

Time magazine (1960)

The ventouse may cause less maternal trauma than forceps, but is thought to be less reliable at achieving delivery. Ventouse delivery was slow to gain acceptance and was not widely used until the 1990s. Although it continues to be popular, it is no longer considered a totally benign procedure. In 1998, the United States Food and Drug Administration issued a warning that ventouse delivery can cause serious or fatal complications, including bleeding between the muscles of the scalp and bleeding within the skull. This advisory was prompted by reports of twelve vacuum-related deaths and nine serious injuries between 1994 and 1998. **JF**

SEE ALSO: INCUBATOR

Gel Electrophoresis (1955)

Smithies discovers a molecular sieve.

Oliver Smithies (1925–2017), a scientist working at the University of Toronto in the 1950s, discovered a way for scientists to separate proteins quickly and easily by their size, using potato starch to create a type of microscopic sieve.

Smithies, who was one of three scientists sharing the 2007 Nobel Prize in Physiology or Medicine, was looking for a way to separate insulin from its precursor when he developed starch-gel electrophoresis. Using potato starch, Smithies created a gel that allowed proteins to be separated by size. He found that the best potato starch came from a powder produced in Canada. Later, when Smithies worked at the University of Wisconsin, he made many trips from Madison, Wisconsin, to Toronto to obtain this special powder.

The powder was used as a matrix in the gel. The proteins were applied to the gel and exposed to an electric charge, which allowed them to migrate through the matrix of potato starch. The starch acted as a sieve and enabled scientists to see the size separation of the proteins that they were studying.

Although electrophoretic methods had been developing since the late 1930s, this method of size separation by a sieve created from powder was an important advance in protein science. Smithies' initial device for starch-gel electrophoresis proved to be quite bulky. The equipment used to run the electric current did not produce a constant current, and the gels used, which were about 0.4 inches (1 cm) thick, were excessive by today's standards.

Improvements to the process were developed over the next few decades. Today, gel electrophoresis equipment is much more streamlined, but the simple sieve idea that Smithies discovered still forms the basis of protein separation. **RH**

SEE ALSO: MASS SPECTROMETER

Velcro (1955)

Mestral invents a unique fastening material inspired by nature.

Velcro is the brand name for a lightweight, durable, and washable fastening, widely used instead of zips in clothing and luggage. It consists of two strips of nylon fabric, one densely covered in small, strong hooks and the other containing small loops. When pressed together, the two strips form a strong bond that can be peeled apart again, making a characteristic ripping noise, but will not open if pulled in any other direction.

Swiss engineer George de Mestral (1907–1990) had the idea for velcro in 1941 after getting his microscope out to study the burrs stuck to his dog's fur and his own clothing, following a hike in the Alps. Burrs—seedheads of the burdock plant—have lots of strong little hooks that fix onto passing animals (and walkers) and stay tenaciously attached until the animal cleans them off, usually depositing them some distance from the parent plant. De Mestral saw the potential for a two-sided fastener, using strong hooks on one side and a looped fabric into which the hooks became entangled, and called it *velcro* from the French words *velour* (velvet) and *crochet* (hook).

Although De Mestral's idea was initially greeted with skepticism and laughter, he set up Velcro Industries in Switzerland in 1952, and worked with a weaver from a French textiles company to perfect the design; he patented the invention in 1955. By the early 1960s velcro was becoming a household name throughout the world. Velcro is a registered trademark in many countries, but the brand name has also become the generic term. **EH**

SEE ALSO: CLOTHING, METALWORKING, SHOES, BUTTON, BUCKLE, TEFLON

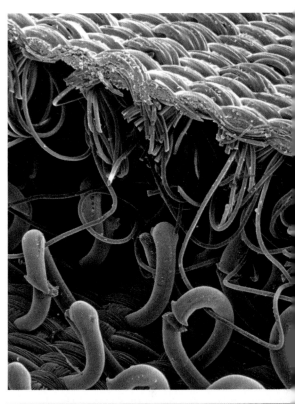

↗ *Velcro consists of a smooth surface (top) that forms a strong bond with the hooked surface at the bottom.*

⮕ *The silvered Teflon® cover of NASA's Long Duration Exposure Facility was held in place by velcro pads.*

Speed Camera (1955)

Gatsonides's device reduces traffic accidents.

Rather ironically, it was a Dutch rally driver, Maurits Gatsonides (1911–1998), who invented the speed camera. Gatsonides enjoyed most of his driving successes in the 1950s, and it was during this period that he came up with a device—known as the Gatso—to measure his speed while cornering in a bid to improve his driving, that is, to make him drive faster.

The camera works by using radar to measure the speed at which a vehicle passes the device, photographing those that break the limit. Two photographs are taken and, should the initial measurement be questioned, the position of the vehicle relative to the white lines painted on the road indicates the average speed that the vehicle traveled

---

"Speed has never killed anyone; suddenly becoming stationary … that's what gets you."

Jeremy Clarkson, TV presenter

---

at during a set time interval.

Fixed speed cameras have been used widely in the United Kingdom, Australia, and France, but only marginally in the United States. Despite being billed as life-saving devices, speed cameras remain unpopular with large portions of society. Some people consider speeding fines an unethical source of revenue for the local law enforcement agencies or the private organizations that operate them. However, a research study conducted in 2006 estimated that "Gatsos" and other traffic enforcement measures have reduced road fatalities by about a third in the United Kingdom. **CB**

SEE ALSO: MOTORCAR, PARKING METER, AIR BAG, THREE-POINT SEAT BELT, BREATHALYZER

Hair Spray (1955)

Chase Products' spray holds hair in place.

After the straight hairstyles of the 1920s, waves and curls became the fashion, and a new product was needed to hold hair firmly in place. Women had been using natural compounds such as clays and gums to hold their hair in place for centuries, but it was the invention of the aerosol can that led to the development of the first hair spray. During World War II, the United States government was looking for a way to spray insecticides to kill malaria-carrying bugs. In 1943, two Department of Agriculture workers designed an aerosol can pressurized by liquefied gas.

Soon, hair spray was produced using the same principle, with a debate still raging over whether it was Chase Products of Broadview, Illinois, in 1948, or Helene Curtis of Chicago seven years later who came up with the idea. Early hair sprays contained polymers (long-chain chemical compounds) that when dry form tiny gluelike spots at the junctions between hairs, holding them in place. These polymers are combined with solvents such as water, or alcohol, to help propel them onto the hair. Originally, hair sprays used chlorofluorocarbons as propellants, but during the 1990s a link was made between these gases and the depletion of the ozone layer, and they were subsequently banned from use. Also, environmental legislation was introduced that called for a reduction in the amount of volatile organic compounds (such as ethanol) in products.

Modern hair sprays now use hydrocarbons such as butane or propane. These products also contain a much more sophisticated mixture of compounds to hold hair in place in unfavorable weather conditions. **HI**

SEE ALSO: AEROSOL, BLOW-DRY HAIR DRYER

➡ *Images of Helene Curtis magazine ads like this one from the early 1960s are sold as posters these days.*

Tetracycline (1955)

Conover develops a broad spectrum antibiotic.

The world existed for a long time without antibiotics, but as soon as they were discovered, the race was on to find more. Penicillin and streptomycin had both been isolated from some form of fungus, and the search continued for products that came from naturally occurring bacteria.

Research groups looked everywhere but one of the more promising avenues of investigation began to emerge from groups studying certain fungi that lived in soil—organisms known as actinomycetes. Lederle Laboratories found the now forgotten antibiotic aureomycin in dirt samples. Pfizer soon followed suit with the equally obscure antibiotic terramycin. Both aureomycin and terramycin functioned as broad-

---

"An antibiotic for all seasons, it's used for a dozen reasons . . . there is no proxy, for our antibiotic doxy."

James McCallum, from the poem "Doxycycline"

---

spectrum antibiotics, covering a range of both gram-positive and gram-negative bacteria. This sparked a great deal of interest in figuring out how they worked.

Lloyd Conover (1923–2017) made a breakthrough while studying the actinomycete-derived antibiotics. He figured out that they both shared a common structure that would prove to be the active component. The new chemical, dubbed tetracycline, represented the first time a naturally occurring substance was modified to produce an enhanced drug. The resulting chemical and its derivatives doxycycline and minocycline, are still used today to treat a variety of ailments. **BMcM**

SEE ALSO: PENICILLIN PRODUCTION

Beta Blockers (1956)

Black finds a drug to relieve stress on the heart.

Beta blockers are drugs that block the stimulating action of noradrenaline—the "fight or flight" hormone—thereby reducing the force of the heartbeat and the workload of the heart. Today they are widely used to treat angina, high blood pressure, and irregular heart rhythms, and to improve heart muscle function in cardiomyopathies.

Beta blockers were developed in 1956 by Sir James Black (1924–2010), a Scottish doctor working for ICI in the United Kingdom. Black had personal as well as professional reasons for his interest in cardiovascular disease. His father had a fatal heart attack following a car accident, making Black ponder the role of stress in producing adrenaline, angina, and heart attacks.

At the time, many of the drugs used to treat angina were vasodilators (causing dilation of the blood vessels), in particular nitrites, which increased the blood supply, and therefore the amount of oxygen to the heart. However, these caused unpleasant side effects such as flushing of the face and headaches. Black hypothesized that instead of treating angina by increasing the oxygen supply to the heart, it might be possible to do so by reducing the demands made on the heart.

Black acknowledges the influence of the 1948 paper by U.S. scientist Raymond Ahlquist, which suggested that the effects of noradrenaline of speeding up and slowing down the heart are mediated by different receptors in the target organ, which he called alpha and beta receptors. Black and his colleagues substituted different chemicals to find structures that blocked the beta receptor without stimulating it. Eventually they synthesized propranolol, the first beta blocker to be marketed. Since then many different beta-blocking agents have been produced. **JF**

SEE ALSO: HEART-LUNG MACHINE, PROZAC®

Catalytic Converter (1956)

Houdry's device cleans up car exhaust fumes.

French mechanical engineer Eugene Houdry (1892–1962) is probably best known for inventing a process for "cracking" crude oil and turning it into high-octane gasoline—otherwise known as the Houdry process. However, he is also credited with inventing the first catalytic converter for cleaning up car exhaust fumes.

Houdry had an avid interest in all things automotive and enjoyed car racing. This made him acutely aware of the need for high-performance fuels in engines, and he developed a way to make high-octane fuels from petroleum using what is known as a catalyst. A catalyst is a substance that can instigate or speed up a chemical reaction, without itself being changed during the process.

In the 1950s, early studies about smog in Los Angeles were published and Houdry became concerned about the effect of harmful chemicals in exhaust fumes. The main products from car exhausts are carbon dioxide, nitrogen, and water vapor, but they also produce toxic gases such as carbon monoxide and hydrocarbons. Houdry made a device that used catalysts such as platinum, rhodium, and palladium to reduce the amount of toxic gases released in the fumes. Unfortunately, catalytic converters were not used until after Houdry's death.

In 1970, the U.S. Environmental Protection Agency set strict emission-control standards, and in 1974 the first cars with catalytic converters were produced. The devices were not introduced in Europe until 1985, but now most new cars have a catalytic converter. **HI**

SEE ALSO: FOUR-STROKE CYCLE, MOTORCAR, DIESEL ENGINE

↗ This catalytic converter from 1992 was a finalist for the 1996 Environmental Award for Engineers.

➔ The expanding mat on the outside provides an unbreakable enclosure for the underlying layers.

Hard Disk Drive (1956)

Johnson and IBM give computer users direct access to their data.

For most of the twentieth century, the primary medium for data entry, storage, and processing was the punched card. In the 1930s, IBM hired teacher and inventor Reynold B. Johnson (1906–1998) to develop the IBM 805 test-scoring machine to convert pencil marks on forms into punched cards. Twenty years later, Johnson led the team that developed the technology that made the vast majority of punched cards obsolete—the hard disk.

Unlike punched cards and magnetic tape, in which data must be accessed sequentially, hard disk drives provide access to all data almost simultaneously. Some computers in the late 1940s stored data on the outside of magnetic drums, but this left most of the internal space unused. Johnson and his team sought to store data on a stack of spinning disks, vastly increasing the data storage per volume. The main problem they had was preventing the read-write heads from hitting and damaging the disks. Johnson et al. solved this problem by supporting the heads on thin layers of air.

Their 350 Disk Storage Unit, proudly unveiled by IBM in September 1956, held five megabytes of data on fifty 23 ½-inch (60 cm) diameter platters rotating twenty times a second. By modern standards, the disk drive was a crude device. Fifty years later, IBM introduced the System Storage DS8000 Turbo, which could hold up to 320 terabytes (more than sixty million times more than the IBM 350).

Today, people routinely take pictures with digital cameras, listen to music from a library of thousands of songs on digital audio players, and watch long movies on their computers, all of which depend on sufficient digital storage. Over the past fifty years, storage capacity has doubled every two years or so, a trend that looks set to continue into the future. **ES**

> *"The first magnetic slurry coating on the first disk drive was poured ... from a Dixie cup."*
>
> Barry Rudolph, IBM vice president

⬆ *Technology's progress is evident when comparing a 1984 hard disk to its smaller counterpart from 1999.*

➡ *The spinning magnetic disks that formed the core of IBM's RAM 650 Data Processing System hard drive.*

SEE ALSO: PUNCHED CARD, COMPUTER PROGRAM, PUNCHCARD ACCOUNTING, FLOPPY DISK

Synthetic Blood (1956)

Chan invents a lifesaving alternative to blood transfusion.

Synthetic blood is a product that acts as a substitute for red blood cells, designed with the purpose of transporting oxygen and carbon dioxide around the body. The development of artificial blood is desirable because of the problems associated with blood transfusions, particularly the risk of transmitting viral diseases such as HIV and hepatitis. There are also difficulties with transporting and storing blood (synthetic blood is kept in powder form), as well as a perpetual shortage of blood donors.

In 1956 Thomas Chan (*b.* 1933), working on an undergraduate research project at McGill University, Montreal, created the first artificial blood cells. Turning his dormitory room into a makeshift laboratory, Chan used improvised materials (including perfume atomizers) and cellulose nitrate solution (a material used to coat wounds) to create a permeable sack that could transport hemoglobin. Hemoglobin can be extracted from old donor blood, cow's blood, plants, and fungi. It is modified to ensure it is stable before being used in the human body.

Synthetic blood has not yet become an alternative to real blood, but it has been used outside of research. There are ethical questions about its use and whether people will accept artificial blood. Europe has explored hemoglobin-based oxygen carriers (HBOCs), whereas the United States has focused on perfluorocarbons (PFCs), a Teflon-type group of synthetic liquids. But HBOCs and PFCs still lack two ingredients of real blood: white blood cells to fight infection and platelets to help blood clot. Finding a substitute that performs all of blood's functions still eludes scientists. **JF**

SEE ALSO: BLOOD TRANSFUSION, BLOOD BANK

⬆ *The oxygen-carrying blood substitute PolyHeme® has been developed at Northfield Laboratories since 2006.*

Videotape Recording (1956)

Ampex and Ginsburg develop the technology to record videos.

Prior to videotape, film was the only practical medium via which television programs could be recorded, which was a problem for U.S. television executives whose audiences were spread across several time zones. For a viewer on the East Coast to see a show on the same night as someone on the West Coast, the live broadcast had to be filmed, sent for processing, returned to the studio, and then retransmitted a few hours later.

A team at Ampex, including the enemy of hiss, Ray Dolby, and led by Charles Paulson Ginsburg (1920–1992), developed the first solution in the form of a Videotape Recorder (VTR), a unit that could capture live images from television cameras, convert them into electrical signals, and save the information onto magnetic tape. In an audiotape recorder, the information is recorded linearly, the tape traveling past the record head at, for example, 3–7 inches (7.6–17.7 cm) per second (ips). But TV signals can contain 500 times more information than regular audio, so linear

video recording, even with a much larger tape area, would require speeds of several feet per second. The Ginsburg/Ampex solution was to use multiple recording heads that were both slanted and rotated rapidly to create a transverse, rather than linear, recording pattern. This meant that the tape itself could pass the heads relatively slowly.

Ampex unveiled the first commercial VTR in April 1956. It used 2-inch (5 cm) wide 3M tape traveling at 15 ips. The monochrome machines were offered at around $50,000 each. Within four days, Ampex had received orders worth $5 million. Radio stars everywhere shed a tear. **MD**

SEE ALSO: TELEVISION, OPTICAL DISC, CAMCORDER, DVD (DIGITAL VERSATILE DISC), BLU-RAY/HD DVD

⬆ *In late 1956 CBS became the first TV station to use videotape recorders like the one above.*

Artificial Intelligence
(1956)

McCarthy's research makes little progress.

Since the dawn of computers, people have wondered if they can be made to show intelligence—to think in the way that humans think. Charles Babbage and Ada Lovelace first debated the question when they worked together to create the first computer in 1835.

By 1950, U.S. mathematician Claude Shannon was busy trying to figure out how computers could play a good game of chess. On the other side of the Atlantic, Alan Turing published his paper "On Computing Machinery and Intelligence," which considered the thorny problem of how you could actually tell if a machine was intelligent or not.

In 1955, John McCarthy, of Dartmouth College, New Hampshire, proposed a conference to study the issue of intelligence research. In his proposal, he used the phrase "artificial intelligence" for the first time, and an entire field of study was born. The 1956 Dartmouth Conference is now known as the defining moment of artificial intelligence (AI) research. The conference set the path for research for years to come, asking questions that remain unanswered to this day. Many of the great minds who attended, such as Harvard's Marvin Minsky, devoted the rest of their careers to a subject that had only just been given a name.

While there was great optimism at the conference, with many attendees expecting intelligent machines to appear within a decade, artificial intelligence remains elusive. Although there have been notable successes—IBM's Deep Blue computer beating chess champion Gary Kasparov in 1997, for example—they have been in very narrow fields. **MG**

SEE ALSO: MECHANICAL COMPUTER, COMPUTER PROGRAM, HYPERTEXT, INTERNET, TOUCH SCREEN

⏎ *In 1997 Deep Blue became the first machine to win a chess match against a reigning world champion.*

Metered Dose Inhaler
(1956)

3M invents a medicine for asthma sufferers.

Around 300 million people have asthma across the planet. For more than fifty years, many of these have benefited from the metered dose inhaler. But the idea to create a device that delivers measured particles of aerosol medicines into the lungs did not come from a scientist, but from a thirteen-year-old girl. Susie Maison asked her father why her asthma spray could not be put into a hairspraylike device instead of the bulky, glass nebulizer she struggled with. Mr. Maison, then president of Riker Laboratories (acquired by 3M Pharmaceuticals in 1970), took the idea to the company where a team looked into making it a reality.

They examined devices used for aerosol perfumes and tried mixing the medications, isoprotenerol or epinephrine with alcohol, ascorbic acid, and chlorofluorocarbon propellants. Early trials showed the effectiveness of this method of drug delivery and the first pressurized metered dose inhalers were launched in March 1956. An amyl nitrate inhaler for angina was launched the same year. With suspension formulations came the first nasal inhaler in 1957. Refinement of both oral and nasal inhalers led to better designs, and by 1970 the first breath actuated inhaler was launched. Use of chlorofluorocarbon propellants switched to hydrofluorocarbon ones with laws to protect the ozone layer in 1987. This was a challenge to the industry as all materials had to be reworked and retested for efficacy and safety.

Metered dose inhalers are still being improved to become more effective and easier to use. Increased understanding of how particles work has meant products that were less receptive to conventional metered dose inhalers, such as insulin, have at last found their way into inhaled formulations. **LH**

SEE ALSO: AEROSOL

Birth Control Pill (1956)

Pincus and Rock invent female contraception.

In 1952 Gregory Pincus (1903–1967), a biologist working at the Worcester Foundation for Experimental Biology in the United States, demonstrated that a synthetic form of the hormone progesterone, known as norethidrone, inhibited ovulation in rabbits and rats. Norethidrone had been developed a year earlier by Carl Djerassi, a chemist working at the Syntex company in Mexico City. It had initially been created with the aim of producing high concentrations of progesterone to treat menstrual disorders. It had the advantage of being more active than the human hormone, and also of being effective when taken orally. Margaret Sanger, founder of the American Birth Control League, saw the potential of Pincus's work. She enlisted the help of heiress Katharine McCormick, who agreed to fund research to develop a contraceptive pill.

Pincus developed a pill with the aid of gynecologist John Rock (1890–1984), a devout Catholic aiming to improve conception among infertile couples. The resulting pill, combining estrogen and progesterone, worked by fooling the pituitary gland into thinking that a woman is pregnant. The pituitary gland then cuts the output of the FSH and LH hormones, which are essential for egg release. Clinical trials started in Puerto Rico in 1956, and the following year the U.S. Food and Drug Administration approved Syntex's norethidrone. It was passed for contraceptive use in 1960.

"The Pill" has been used by millions of women worldwide, but it has been cloaked with controversy from the start. Critics said it interfered with women's natural cycles, encouraged sexual freedom in unmarried women, and even undermined the social order. Supporters countered that it had the potential to stem poverty, reduce unnecessary deaths, and free women from fear of unwanted pregnancy. **JF**

SEE ALSO: CONDOM

Digital Clock (1956)

An electronic form of timekeeping is born.

Clocks usually have two jobs. One is to display the right time; the other is to give a measurement of a time interval. Digital clocks are numeric. Hours, minutes, and seconds are represented by numbers and the display can be made small and linear. Although German inventor Josef Pallweber patented a digital watch as early as 1883, the development of the digital clock proper is closely associated with the history of digital displays. The earliest examples of these were the glowing end of valve tubes that could indicate numbers. These were much loved by the nuclear physics instrumentation industry in the 1950s.

The modern digital clock relied significantly on the development of the light emitting diode (LED) and

---

"The hours of folly are measured by the clock; but of wisdom, no clock can measure."

William Blake, poet

---

liquid crystal display (LCD). The first commercially usable LEDs were developed in the 1960s, and the first active matrix LCD panel was produced in 1972. Film director Stanley Kubrick famously showed a futuristic digital clock in his 1968 science fiction masterpiece, *2001: A Space Odyssey*.

A problem with digital alarm clocks is that they flash to a default setting when switched off and can fail to alarm after a power surge or outage. Small and cheap digital clocks have been used in many other devices from microwave ovens to cell phones. While digital watches were very popular in the 1980s, the analogue display is arguably more fashionable again today. **DH**

SEE ALSO: LCD (LIQUID CRYSTAL DISPLAY), DIGITAL WATCH

Artificial Satellite (1957)

The Soviet Union takes the first important step in the space age.

The Soviet Union launched the first artificial satellite, *Sputnik 1*, on October 4, 1957, thus triggering the space race with the United States.

Sputnik 1 was a nitrogen-filled sphere about the size of a beach ball—23 inches (58 cm) across—which orbited the Earth every ninety-six minutes. It had four long, whiplike aerials that transmitted information back to Earth. In November the same year, *Sputnik 2* carried a living passenger, the dog Laika, into space. (It is thought that Laika only survived a few hours rather than the intended ten days because of stress and overheating.) By August 1960, when *Sputnik 5* was launched, two dogs, forty mice, two rats, and a collection of plants had been sent into orbit. The goal was the manned exploration of space.

The United States were taken horribly by surprise by the Soviets' achievement and responded by pumping money into space research and founding NASA, the National Aeronautics and Space Administration. Soon near-Earth space was being crisscrossed by a variety of artificial satellites.

Around forty countries have since manufactured and launched their own artificial satellites. About 3,000 useful satellites are thought to be orbiting Earth, along with 6,000 pieces of "space junk" (such as empty fuel tanks and rocket boosters). Although artificial satellites have orbited the moon, the sun, asteroids, and planets, most are operated orbiting the Earth. They are used to study the universe, forecast weather, transmit telephone calls and television broadcasts, and assist in sea and air navigation as well as military activities. **DH**

SEE ALSO: GEOSTATIONARY COMMUNICATIONS SATELLITE, SATELLITE RADIO BROADCASTING

⤤ *This photograph of Sputnik 1 shows its various parts separated. The satellite transmitted signals for 21 days.*

"Sputnik 1 fascinated and frightened vast numbers of people."

Don Mitchell, "Sputnik: 50 Years Ago"

Bubble Wrap® (1957)

Fielding and Chavannes invent new material.

Most people are aware of the usefulness of Bubble Wrap® for packaging china, glass, and other precious items to prevent them from breaking. However, what is not quite so well-known is that this versatile polythene packaging was originally intended as a unique type of wall decoration.

In 1957, U.S. engineer Alfred Fielding and Swiss inventor Marc Chavannes were trying to produce a textured plastic wallpaper that would be easy to clean. Their early designs did not work, but thinking quickly they transformed their accidental product into a new kind of packaging. On the basis of this new invention, they then founded Sealed Air, a global corporation that now turns over $4 billion a year. Not bad for an invention based on thin air. Sealed Air's Bubble Wrap® is superior to other cellular polythene packaging because it has a barrier layer giving extra protection.

During the manufacturing process, a sheet of polythene is wrapped tightly around the outside of a drum that has a regular arrangement of holes in it. Suction draws the film through the holes to create the bubbles, which become fixed when a second sheet is added to trap the air. The result, of course, is a lightweight, flexible packaging that also makes satisfying pops when you stamp on it.

In 2007, Sealed Air held a competition for young inventors to create products made out of Bubble Wrap®. As a result of the resourcefulness of thousands of school children, there are now plenty of new applications for bubble wrap—as well as a few old. The winner created a bubbly blue wallpaper intended to engage children with autism. **HB**

SEE ALSO: CELLOPHANE, ALUMINUM FOIL

◧ *Popping the bubbles of Bubble Wrap® is considered to have a cathartic effect on people as a stress relief.*

Photolithography (1957)

Lathrop and Nall make silicon devices.

Photolithography is a modern development of an older process used in printing. In lithography, smooth pieces of limestone had oil-based images burned by acid into their surface. The non-oily portions were then sealed by gum arabic. Oil-based ink then only adhered to the unsealed areas so that complicated pictures and typed regions could be reproduced.

Photolithography is used in the mass production of transistors and electronic components. In April 1957 Jay Lathrop and James Nall of the U.S. Army's Diamond Ordnance Fuse Laboratories in Maryland produced the first electronic components that did not require manual soldering. The required design is often pre-formed on a photo-mask, consisting of a series of opaque chromium lines on glass. The aim of the process is to transfer the pattern of the mask onto the flat surface of a wafer of silicon. The silicon is cleaned and covered with a photo-resistant substance. Ultraviolet light is then shone through the mask, and this light chemically changes the photo-resistant substance. A developing chemical then washes away the chemically altered material leaving the underlying material exposed. The whole process is similar to that of producing a photographic negative. Projection printing photolithography is only capable of producing a small final image, but images can be stepped over the surface of the wafer. The final step bakes the photoresistant material to the surface. Then a new layer of material is added and the process is repeated. The whole series of electrical components and the connecting wiring is built up layer by layer.

Photolithography has advanced to the smallest scale in microlithography and nanolithography, where it is used in the production of today's highly complex microchip cores for computers and cell phones. **DH**

SEE ALSO: INTEGRATED CIRCUIT, MICROPROCESSOR

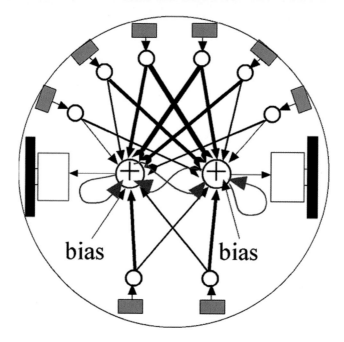

Artificial Neural Network (1957)

Rosenblatt creates the first computer to simulate the human memory.

The study of human memory was greatly changed in 1943, when Warren McCulloch and Walter Pitts wrote a paper on how neurons might work. (Neurons are the cells that make up tissue in the part of the nervous system involved in learning and recognition.) Six years later, D. O. Hebb described the strengthening of neural connections that occurred each time they were used.

In the early days of artificial intelligence research, Frank Rosenblatt (1928–1971)—a computer scientist at the Cornell Aeronautical Laboratory in New York—was studying how the eyes of a fly work. Rosenblatt observed that when a fly perceives danger, its reaction occurs faster than the information can be processed by its brain. He then produced the perceptron, the first computer to learn new skills using a neural network mimicking human thought processes. The perceptron had a layer of interconnected input and output nodes. Each connection is "weighted" to make it more or less likely to stimulate another node. Rosenblatt's Mark 1

perceptron followed in 1960, the first machine to "learn" to recognize and identify optical patterns.

Practicality grew with the advent of "backprop" perceptrons, which include "hidden" layers that greatly increase the complexity. John Hopfield then introduced his model of neural networks, which can store memories of patterns so that when the network is presented with even partial information, it can retrieve the full pattern—rather like humans.

Today, neural networks are the foundation for the optical character recognition employed in scanners, weather forecasts, bomb detectors, and even financial market predictions. **SS**

SEE ALSO: ARTIFICIAL INTELLIGENCE

⬆ *Just like the structure of the human brain, an artificial neural network is an interconnected group of nodes.*

Jet Boat (1957)

Hamilton invents the first propeller-free boat.

Sir William Hamilton (1899–1978) already had some experience of working with water-based mechanics when he invented the jet boat. In 1954, he had built a jet pump, the first of its kind. The pump was essentially a system for water propulsion. It used a propeller to create a centrifugal force that caused a forward thrust action underwater, drawing water through and back inside the pump.

The jet boat works in a similar way to the jet pump. Simply put, a traditional screw propeller accelerates a large volume of water by a small amount and, in agreement with Newton's Third Law of Physics—for every action there is an equal and opposite reaction—a large amount of thrust is created. This thrust is used to drive the boat and allows it to move speedily through the currents.

Hamilton lived and worked in New Zealand. There his boat was able to power quickly through the fast-flowing, shallow waters and was particularly adept at avoiding obstacles, such as rocks. The maneuverability of the jet boat made its design highly marketable.

Hamilton was not the first person to come up with the idea for a jet boat. Italian inventor Secondo Campini had devised a remarkably similar jet-powered boat as early as 1931, but he did not have the foresight to patent his design. Well before that, Greek scholar Archimedes had dreamed of a jet boat when he devised his water screw, an early propeller, in the third century BCE.

Hamilton has nonetheless been credited with "revolutionizing the conventional world of boating," although as he himself has commented, he probably has Archimedes to thank, at least in part, for this. **KB**

SEE ALSO: HOVERCRAFT

⬆ *Mr. and Mrs. Hanning-Lee test an early jet boat prototype on Lake Windermere, England, in 1952.*

Magnetic Stripe (1958)

Parry invents the swipe card.

Odds are that if you check your purse or wallet right now, you'll find an invention that owes its existence as much to a desire for neatly pressed clothes as a need for portable data storage—the magnetic stripe card.

Conceived by IBM engineer, Forrest Parry, as part of a government security system project, the technique of attaching a strip, or "stripe," of magnetic tape to a card facilitated a revolution in portable, personal data retrieval. Until the advent of the chip-based smart card during the 1980s, the magstripe ruled—from club membership cards, through to phone, credit, and debit bank cards.

Parry had experienced several frustrating failures with trying to fix the magnetic material to the card with adhesive. When a forlorn-looking Parry returned home from the laboratory, his wife, who was ironing at the time, stepped in and suggested she try using the heat of the iron to bond the magstripe to the plastic. After a period of technical development, international standardization, and testing, the magstripe went into mass production. By the 1960s, they were being used by organizations as diverse as the London Underground transport system and the Central Intelligence Agency (CIA).

Today, the stripe is usually made from tiny magnetic particles secured in resin, which are either applied directly to the card or produced as a plastic backed stripe that is then attached to the card. Particles may be standard iron oxide, or, if a higher "coercivity" stripe is needed, barium ferrite. Information is written to a stripe by changing the polarity of its particles; as the coercivity is increased, the chances of accidental erasure are reduced—a factor of prime importance to card issuers and users alike. **MD**

SEE ALSO: MICROPROCESSOR, FLIP-FLOP CIRCUIT, PERSONAL
COMPUTER MODEM, LAPTOP NOTEBOOK COMPUTER

LEGO® (1958)

Christiansen develops an iconic children's toy.

Ole Kirk Christiansen was a humble carpenter from the small Danish town of Billund when he began making wooden toys for children in 1932. With Europe still in the grip of a depression and work as a carpenter hard to find, Christiansen's finely crafted toys proved hugely popular. In 1934 he formed his own company and named it Lego, a contraction of the Danish phrase *leg godt*, meaning "play well."

In 1947 the company bought Denmark's first plastic injection-molding machine and began to focus on plastic toys. By 1951 plastic toys made up half of the company's sales, although it was another eleven years until Ole, working with his son Godtfred, created the plastic blocks that became officially known as LEGO®.

> *"The challenge is to make toys that allow kids to create, experiment, and explore."*
>
> Mitchel Resnick, MIT Media Lab

Christiansen had first made plastic interlocking blocks in 1949. These "Automatic Binding Bricks" were based on the Kiddicraft blocks made in 1947 in the United Kingdom, but were only sold in Denmark. By 1958 they had evolved into rectangular blocks made of cellulose acetate with grooves down each side. They were topped by a new stud-and-tube coupling system that gave the blocks greater stability when stacked, with the LEGO® name visible on the underside of each block. A legendary toy had arrived. **BS**

SEE ALSO: TOY ELECTRIC TRAIN SET, MECCANO®, CRAYONS

▣ *Six-year old Philippa Smith plays with LEGO®
at the Selfridges department store in London in 1962.*

Integrated Circuit (1959)

Kilby and Noyce pave the way for the personal computer.

When, in 1958, fifteen-year-old Bobby Fisher became the youngest Grandmaster in chess history, few onlookers would have believed that, one day, a machine would be capable of beating him.

After the invention of the transistor, electronic equipment became ever more complex. Thousands of different sized components had to be soldered together to make the circuits and were being crammed into less and less space. This was time-consuming, expensive, and unreliable. The Micro-Module program undertaken by the U.S. Army's Signal Corps made pre-wired building block components of a standardized size that could be snapped together. This still did not solve the core problem, however.

Texas Instruments were working with the Micro-Module project when Jack Kilby joined them in 1958. Before long he saw a better solution. As passive components such as capacitors and resistors could be made from the same semiconductor material as active

devices such as transistors, it should be possible to manufacture them in one process from a single block, or "monolith" of that material, thereby creating a complete or integrated circuit (IC). His germanium-based prototype—comprising a transistor, capacitor, and three resistors, connected by fine gold wire—was the world's first integrated circuit. A patent was granted in February 1959. Meanwhile, Robert Noyce, of Fairchild Semiconductors, was working on a unitary circuit, and his ideas regarding depositing the component connections directly during manufacture were key to the IC's development. Ultimately, the IC would lead to the first personal computer. **MD**

SEE ALSO: MICROPROCESSOR, FLIP-FLOP CIRCUIT, PERSONAL COMPUTER MODEM, LAPTOP NOTEBOOK COMPUTER

⬆ *A close-up from 1984 showing the integrated circuits and the ribbon connector wiring of a circuit board.*

In Vitro Fertilization (1959)

Chueh Chang develops a reproductive procedure to help infertile couples.

In vitro fertilization involves clinicians collecting eggs from a woman's ovaries and allowing sperm to fertilize them outside of the womb. The fertilized eggs are then put back into the woman's womb in the hope that a successful pregnancy will follow. Arguably one of the most significant developments in reproductive medicine, it has been used for infertile couples the world over with more than a million babies born via the technique. Today, in vitro fertilization continues to be controversial in medical, legal, and moral terms, but it has gained wider social acceptance.

The story begins with Chinese reproductive biologist Min Chueh Chang (1908–1991), who went to the United States to work with the eventual pioneer of the oral contraceptive pill Gregory Pincus after World War II. Pincus had claimed to have successfully used in vitro fertilization with rabbits in 1935. His claims were met with disbelief and neither scientist was able to successfully repeat the procedure.

However, Chang did discover that sperm must incubate inside the female before it has fertilizing capacity. It took until 1959 for Chang, in competition with others, to finally demonstrate in vitro fertilization in rabbits by transferring fertilized eggs from black rabbits into the uterus of a white rabbit, which gave birth to black young. Further work by Chang and others led to knowledge of the conditions needed for in vitro fertilization in other species.

In England, Patrick Steptoe and Robert Edwards applied this knowledge to humans. After years of trials and opposition, they produced the world's first "test-tube" baby in 1978 with the birth of Louise Brown. **LH**

SEE ALSO: BIRTH CONTROL PILL, INTRACYTOPLASMIC SPERM INJECTION (ICSI)

⊡ *IVF research is conducted in Sheffield, England, in 1970, eight years before the birth of the first "test-tube baby."*

Float Glass (1959)

Pilkington invents a cost-effective process to make high-quality flat glass.

It is likely that you will not have given much thought to how glass is made. In fact, it is very difficult to make flat glass to the standard needed for uses such as shop windows, cars, and mirrors. Or at least it was before Sir Alastair Pilkington (1920–1995) created the float process in 1959 in the United Kingdom. He was a technical engineer at the glass manufacturers Pilkington Brothers (no relation) and it took him seven years to perfect his technique. Before this, the plate process, first developed by English engineer Henry Bessemer in 1848, was used. But this process was expensive because it resulted in distortion and marking of the glass, which had to be corrected by polishing.

In the float process, the melted glass mixture, traditionally made of a combination of silica, sodium carbonate, calcium oxide, magnesium oxide, and aluminum oxide, is poured onto a bath of molten tin. This forms a floating ribbon of glass that is perfectly smooth on both sides and evened out to a uniform thickness of ¼ inch (6.8 mm)—although thinner glass can be made by stretching this glass ribbon and thicker glass can be made by not allowing the glass pool to flatten. It travels along the tin bath, cooling gradually, and then undergoes a heat treatment in a long furnace called a lehr. This is necessary because if the glass cools over quickly, the stress will be too great and it will break under the cutters.

A float glass line can be nearly 1/3 mile (0.5 km) in length and can produce up to 3,728 miles (6,000 km) of glass a year. Thanks to this innovative method, 970,000 tons of flat glass is created every week by around 370 float plants all over the world. **CL**

SEE ALSO: GLASS, BLOWN GLASS, GLASS MIRROR, AUTOMATIC GLASS BOTTLE-MAKING MACHINE

⬆ *The research into float glass began in 1952. Sales became profitable in 1960, a year after its launch.*

Three-Point Seat Belt (1959)

Bohlin's invention encourages drivers to buckle up and saves thousands of lives.

It may come as no surprise to some that the three-point seat belt in widespread use today was invented by a Swede working for Volvo. Nils Bohlin (1920–2002) was an aircraft designer for Saab, where he developed ejector seats, before joining Volvo as its first safety engineer in 1958. Seat belts at that time involved just one belt across the lap, a design that risked injuries to internal organs in high-speed crashes.

Bohlin sought to find a simple, comfortable alternative that would protect both the upper and lower body. His three-point solution allowed occupants to buckle up with one hand, using one strap across the chest and lap with the buckle placed next to the hip. The design spread out the forces of a crash more evenly across the body, resulting in fewer injuries. It was effective at restraining the body and preventing ejection from the vehicle in high-speed crashes. Bohlin said his design worked because it was comfortable and easy to use, as well as being safer.

Volvo began including the new seat belt in its cars in Sweden in 1959, and other car manufacturers soon followed suit. In 1970 the state of Victoria in Australia passed the first laws requiring drivers and front seat passengers to wear seat belts. Since then most countries have introduced seat belt laws. Volvo first put three-point seat belts in the rear of their cars in 1972.

The National Highway Traffic Safety Administration estimates that seat belts reduce the risk of car crash fatalities by 45 percent and prevent more than 100,000 injuries a year in the United States alone. Meanwhile in the United Kingdom, 370 still die in car crashes every year from not wearing seat belts. **BO**

SEE ALSO: MOTORCAR, CRASH TEST DUMMY, AIR BAG

⊼ *Lap belts can cause intra-abdominal tears, but three-point seat belts like this one from 1960 prevent them.*

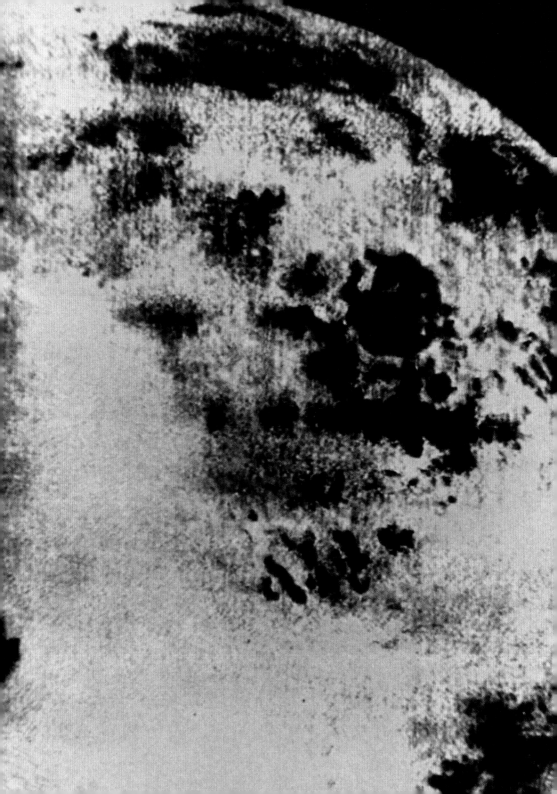

Space Probe

(1959)

The Soviets boost our knowledge of space.

A space probe is a satellite that leaves the grip of Earth's gravity and moves off to flyby, or orbit, another solar system body. The first target was the moon. There were three U.S. attempts in 1958, *Pioneer 1, 2,* and *3*, which all failed. The U.S.S.R. had the first success in January 1959 with the *Luna 1* probe, which was launched with the objective of impacting the lunar surface. It flew by, missing by 3,700 miles (5,950 km), but at least it got there. The U.S.'s *Pioneer 4* flew past at a miss distance of 37,300 miles (60,000 km).

The greatest early successes were *Luna 2* (September 1959) and *Luna 3* (October 1959), the first hitting the moon just east of the Sea of Serenity and

> *"We have your satellite. If you want it back send twenty billion in Martian money."*
>
> Graffiti at NASA laboratory, Pasadena, California

the second imaging the far side of the moon for the first time. The 1960s saw a host of lunar missions that was to culminate in the first moon landing.

The first successful planetary space probe reached Venus in 1962, and the U.S.S.R.'s *Venera 7* probe soft landed there in 1970. In June and August 1976 the U.S.'s *Viking 1* and *2* probes achieved the first successful landings on Mars and returned images from the planet's surface. *Voyager 1* and *2* have since become interstellar space probes and are on their way to the stars. **DH**

SEE ALSO: ARTIFICIAL SATELLITE, LIQUID FUEL ROCKET, SPACE STATION, REUSABLE SPACECRAFT

⬅ *This image, sent by the Russian space probe Luna 3 in 1959, is the first photo of the far side of the moon.*

Computer-assisted Instruction (1960)

Bitzer kicks off e-learning.

Today computers can be used to teach young children how to add up, government employees how to speak a foreign language, and even medical students how to dissect a human body. The idea of using a computer as an educational tool, or e-learning, is often seen as a recent innovation, whereas it has existed, at least in concept, for quite some time. By the end of the 1950s, computers were being used educationally by such early innovators as IBM and the University of Illinois.

In 1960, Donald Bitzer was concerned about the number of illiterate students coming out of schools. When a professor posed a simple question in class one day about the use of computers for teaching, Bitzer

> *"Computers offer the possibility of truly individualized instruction."*
>
> Donald Bitzer

took up the task. He designed one of the first learning systems specifically for computers, which used graphics and touch-sensitive screens. Dubbed "Programmed Logic for Automatic Teaching Operation," or PLATO, for short, it was an ideal medium for teaching, as well as for games and social networking. Along with an associated computer language, known aptly as TUTOR, it allowed simultaneous teaching to a great many users at the same time. As if the creation of computer-assisted learning were not enough, Bitzer also coinvented the plasma screen display panel in 1971, for which they won an Emmy Award for technical achievement in 2002. **BMcC**

SEE ALSO: MECHANICAL COMPUTER, COMPUTER PROGRAM, DIGITAL-ELECTRONIC COMPUTER

Optical Character Recognition (1960)

Rabinow invents a machine that can read.

Jacob Rabinow (1910–1999) left Russia with his family during the Russian Revolution, arriving eventually in New York. After studying electrical engineering, he worked for the American National Bureau of Standards, where he began churning out inventions, which led to a staggering 230 U.S. patents. His first inventions were military and involved missile guidance systems, but he is probably best known for his "Reading Machine."

Optical character recognition (OCR) had been around for some time. A commercial system was installed at the offices of Reader's Digest in 1955, but the results were often unpredictable. In 1960 Rabinow incorporated a new principle into the idea that would

> *"In 1918, I think it was that I made my first invention. I built a machine to throw rocks."*

Jacob Rabinow

greatly reduce the errors made by other reading systems. His machine was the first to include a "Best Match Principle," which compared the information it scanned with a set of standards in a matrix, consisting of letters, numbers, and other symbols.

This new method of scanning documents soon became popular and is still the basis of all modern OCR systems. The optical recognition of Latin-based languages is now 99 percent accurate; however, the race is still on to perfect a method for electronically reading handwriting. OCR has been used to sort mail; a less popular application has been in car number plate recognition by speed cameras. **DHk**

SEE ALSO: FACSIMILE MACHINE, SPEECH RECOGNITION, COMPUTER SCANNER

Laser (1960)

Maiman invents a versatile optical light device.

Lasers (standing for Light Amplification by Stimulated Emission of Radiation) are now part of everyday life. They form a key component of CD and DVD players and the scanners used at supermarket checkouts. Their ability to direct energy with pinpoint precision give them a wide spectrum of uses as dental drills, metal cutters, welders, and scalpels for eye surgery.

The laser process begins with a population of atoms that has been optically pumped so that there are more atoms at a higher energy level than at a lower level. Decay then produces photons of light that stimulate the emission of other decays and so a cascade of photons is produced. The lasing substance is contained in a resonant cavity that has mirrors at each end. Photons travel back and forth between the two mirrors stimulating feedback and the system "lases" if the gain due to stimulated emission exceeds the losses due to absorption and scattering. One of the mirrors is half silvered and the radiation that passes through this forms the beam of laser light.

Albert Einstein discussed stimulated emission as early as 1916, but the actual phenomenon was first produced in 1954 using microwaves (therefore MASER). In May 1960 Theodore Maiman (1927–2007), who was working at the Hughes Research laboratories in California, used a synthetic ruby crystal to produce the first operating pulsed laser.

There are now many different kinds of lasing materials. Some are solid, such as ruby and garnet, whereas others are gases such as helium and liquid solutions of organic dyes. **DH**

SEE ALSO: CARBON DIOXIDE LASER, LASER GUIDED BOMB, COMPUTER LASER PRINTER, ATOM LASER

➡ *This is the first commercially available laser capable of producing a continuously intense beam.*

Halogen Lamp (1960)

Moby advances the quest for a brighter, more efficient light bulb.

Ever since Irving Langmuir's invention of the tungsten filament lamp in 1916, researchers at the General Electric Company (GEC) have been trying to produce more effective light bulbs. One of the problems with the tungsten filament (or incandescent) light bulb is that the tungsten evaporates during operation. This not only produces an absorbing coating on the inside wall of the bulb but it also weakens the filament, eventually leading to breakage. The manufacturers' dilemma is that the hotter filaments produce greater efficiency, but the greater evaporation shortens their lifetime.

GEC research engineer Fredrick Moby made a breakthrough in 1960 when he placed an electrically heated, high temperature tungsten filament inside a compact fused quartz envelope filled with a halogen gas (usually iodine or bromine). His halogen lamp fitted into a standard light bulb socket. Not only did this bulb have a higher luminous efficiency, it also had

about twice the lifetime (2,000 to 4,000 hours) of previous tungsten lamps. With a normal filament bulb, 98 percent of the power used is emitted as heat and only 2 percent as light. A halogen bulb changes these figures to 91 percent and 9 percent, and this saves money. Moby was working parallel to other engineers involved in the GEC project. Elmer Fridrich and Emmett Wiley also patented an improved type of incandescent lamp in 1959.

Getting more light for a specific power expenditure makes halogen bulbs ideal for car headlights. Also in offices much less air conditioning power is needed to counteract the effect of heating from lighting. **DH**

SEE ALSO: INCANDESCENT LIGHT BULB, TUNGSTEN FILAMENT, HALOGEN COOKTOP

⬆ *The tungsten wire in lamps like this one is enclosed in a quartz envelope that contains halogen vapor.*

Industrial Robot (1961)

Devol and Engelberger invent a machine to perform tasks for humans.

Ever since science fiction first described futuristic machines that could perform unpleasant, dangerous, or boring tasks for people, inventors and designers have sought to make such dreams a reality. In 1961, following prototype trials, a robotic manipulator called Unimate heralded the dawn of this new exciting era when it began employment on a General Motors assembly line. A stationary industrial robot, Unimate spent its working day moving hot die castings from machines and welding vehicle bodies. Operating from sequential commands stored on a magnetic drum, the robot's arm, weighing around two tons, was versatile enough to perform any number of different tasks.

Unimate was conceived in the late 1950s by American engineers George Devol (1912–2011) and Joseph Engelberger (1925–2015). Its development was undertaken by Engelberger's company, Unimation Inc. The fledgling industry of industrial robotics grew rapidly, and soon a variety of other mundane, tedious,

or dangerous jobs were being carried out by robots of various types. As the primary use of such robots was initially to move objects from one point to another, less than a few feet away, they were also referred to as "programmable transfer machines." Using hydraulic actuators, they are programmed—in what is called the "learning phase"—by positioning and recording the angles of the various joints. The resulting sequences are then replayed to perform the required task.

The progeny of Unimate may still be encountered reliably picking and moving different items, but also performing tasks as diverse as bomb disposal, sorting mail, and—building other robots. **MD**

SEE ALSO: AUTOMATON, COMPUTER-AIDED MANUFACTURING (CAM), BIPEDAL ROBOT, SURGICAL ROBOT

⬆ *Robots weld the bodies of cars in Japan, the only job left for humans being the inspection of the machines.*

Optical Disk (1961)

Gregg conceives a precursor to the DVD.

The DVD disk has its roots way back in 1958 when engineer David Paul Gregg, while working for the U.S. Westrex corporation, had the inspiration to create a new format for audio and video. He had seen an article that showed the results from an early scanning electron microscope, which had lines drawn by the electron beam that were less than a tenth of a micrometer wide. Gregg imagined a plastic disk with tracks written on it that could be read with an inexpensive optical reader—the equivalent of a stylus reading the tracks on a traditional vinyl LP.

Gregg's original concept, which he patented in 1961, was of a transparent disk, through which a concentrated beam of light would be shone and

---

"The inspiration for the optical disk was an illustration in a technical news magazine … "

David Paul Gregg

---

picked up on the other side by a reader. Like an old LP, his "videodisk" was analog, rather than digital, and included two audio tracks as well as the video.

In 1965 Gregg formed his own company, Gauss Electrophysics, and continued to work on refining his invention. He quickly caught the attention of various large corporations including MCA, which bought Gregg's company in 1968.

An extra patent that Gregg received in 1969 became the basis for the Philips Compact Disk, the Sony MiniDisc, and the Pioneer DVD LaserDisc. Today the DVD and its successor, the Blu-ray Disc, are among the fastest growing media formats in history. **DHk**

SEE ALSO: DVD, BLU-RAY/HD DVD

Space Observatory (1962)

NASA launches the first space telescope.

Only the visual and radio bands of the electromagnetic spectrum pass through Earth's atmosphere relatively unhindered. For everything else, gamma-ray, X-ray, ultraviolet, and infrared, observations of space are much better from above Earth's atmosphere.

The first real telescope in space was Orbiting Solar Observatory–1 (OSO-1), which was launched into a low Earth orbit on March 7, 1962, from Cape Canaveral. The lower portion rotated every two seconds, and the upper portion was fixed in space, using sun sensors and servomotors to point instruments accurately at the solar disc. The main aim was to observe solar flares—extremely violent explosions that occur near sunspots on the solar surface—in the gamma-ray, X-ray, and ultraviolet wavelengths. The spinning part of the spacecraft searched the whole sky for stellar gamma-ray sources. After three months and over 1,000 orbits the effectiveness of OSO-1 was greatly reduced when the United States tested a nuclear device at high altitude.

OSO-1 demonstrated that delicate instrumentation could be flown successfully for long periods of time on stabilized orbiting spacecraft. The National Aeronautics and Space Administration (NASA) went on to build a series of space observatories, including the Hubble Space Telescope (launched in 1990), the Compton Gamma-Ray Observatory (launched 1991), the Chandra X-ray Observatory (launched 1999), and the Space Infrared Telescope Facility, now called the Spitzer Space Telescope (launched 2003). Space observatories either search the sky for sources or collect data from specific objects. **DH**

SEE ALSO: TELESCOPE, SPACE PROBE, X-RAY TELESCOPE, REUSABLE SPACECRAFT, HUBBLE SPACE TELESCOPE

➔ *The satellite OSO-4, launched in 1967, was used to measure the spectral emissions of solar flares.*

Light-emitting Diode (LED) (1962)

Holonyak creates the first LED to emit visible light.

The LED (light-emitting diode) is a semiconductor device. All semiconductors have a variable ability to conduct electric current because of impurities (caused by trace chemical additives) in their structure. An N-type impurity adds an extra electron to the semiconductor, and a P-type impurity provides an electron hole. Electrons, negatively charged particles, naturally move from areas with many electrons (negative) to areas with few electrons (positive).

In a diode, an N-type material is placed next to a P-type one, and the two are sandwiched between electrodes. This setup only allows electric current (a stream of electrons) to flow in one direction, from the N-type side's electrode to the P-type side's electrode.

When an electron drops into an electron hole, it releases energy in the form of a photon. As a result, when electrons move from one side of the diode to the other, light is released. Depending on the types of materials used in the semiconductor, different wavelengths of light are produced.

In 1962 Nick Holonyak (*b.* 1928) created a diode made of his synthetic gallium arsenide phosphide crystals; this produced visible light, making it the first visible-spectrum LED. Using similar principles, he also constructed a semiconductor laser prototype that was a precursor to the CD-reading lasers of today.

LEDs are used in digital clocks, watches, televisions, traffic lights, and display screens for many electronic devices. Infrared LEDs are used in remote controls. Because LEDs create less waste heat energy than conventional bulbs, they are also used in power-efficient lighting systems, lamps, or flashlights. **LW**

"I wanted to work in the visible spectrum . . . and everybody else was working in the infrared."

Nick Holonyak

SEE ALSO: VALVE DIODE, INCANDESCENT LIGHT BULB, TRIODE, SEMICONDUCTOR DIODE

◹ *The amount and direction of light emitted by an LED is determined by the type of semiconductor used.*

MOSFET (1962)

Hofstein and Heiman devise a revolutionary transistor.

Some words, such as *laser* and *radar*, are so common that most of us are unaware of their humble origins as acronyms. Others, despite being worthy of recognition, have not made the leap to public fame; the MOSFET is definitely within this category.

A MOSFET (metal oxide semiconductor field effect transistor) has no moving parts. We can roughly understand its form and function by thinking of a sandwich composed of three types of material, generally referred to by the letters N P N. If wires are connected to the two N components of our sandwich, the current does not flow. This is due to the different electrical properties of the P layer.

How do we get the current to flow? Strangely we first have to coat one side of the P layer with an insulating nonconductive material and then place a plate of metal on top of this insulator. This still does not permit a current, but when the metal plate (called a "gate") has a voltage applied to it, an electric field is generated through the nonconductive material and into the underlying P layer. The electric field will repel positive charges in the P layer, causing them to move away. This creates a "channel" in the P layer that allows conduction between the two N layers in the NPN sandwich. The current flows.

The MOSFET may be small and sound insignificant but it was built for great things. It could act as a switch or amplifier, it proved suitable for miniaturization, it had low heat properties, and it was cheap to produce. Basically, it had all the properties needed to make it a standard component of the integrated circuits used in the modern computer. **AKo**

SEE ALSO: TRANSISTOR

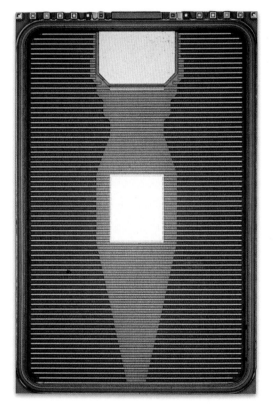

⬈ *A light micrograph shows the surface containing the integrated circuit of a typical field effect transistor.*

> *"If you succeed, you'll be a hero. If you fail, so what? You'll just go to work on a different project."*

Tom Stanley "motivating" the MOSFET team

Hip Replacement (1962)

Charnley transforms the prospects of people with hip disorders.

More than 800,000 hip replacement operations are carried out globally every year. These enable their recipients to live more mobile and pain-free lives after their worn out, damaged, or diseased hip joints are replaced with artificial prostheses.

It was the pioneering work of English surgeon John Charnley (1911–1982) that led to low-friction arthroplasty becoming the gold-standard procedure for hip replacement. Charnley moved from a method of compression fixation of fractures to considering actually replacing the joint during his investigations into the best way to treat osteoarthritis and other conditions limiting hip movement. In a lecture to the East Denbigh and Flint division of the British Medical Association in 1959, he said, "In orthopedics, surgeons yearn for an easy hip operation, or, if a good operation is difficult, . . . [that] it should be universally applicable."

Charnley met with opposition from colleagues but, working in Wrightington Hospital in Wigan, England, he persisted. His design included a steel femoral stem-and-ball section and a polyethylene hip socket, both attached to the bone with acrylic bone cement. He performed the first successful operation in 1962.

Charnley reduced infection in theater by using a clean-air enclosure, suits that covered the entire bodies of surgical staff, and an instrument-tray system. He also followed up on the effectiveness of each operation by persuading patients to bequeath their joints back to him so that he could examine each bone-cement interface and improve his technique.

Today hip replacement continues to evolve as people live longer and outlast the ten-year duration of their prostheses. Research into better materials will lead to more durable designs that should reduce the need for artificial hips to be replaced. **LH**

"The only type of operation that could ever be universal would be an arthroplasty."

John Charnley, lecture in 1959

↑ *In this hip-replacement prosthesis, a steel femoral head fits into a polyethylene liner, or "socket."*

→ *An X-ray reveals a prosthesis successfully implanted, with the steel cemented into the patient's thighbone.*

SEE ALSO: ARTIFICIAL LIMB, JOINTED ARTIFICIAL LIMB, ARTIFICIAL HEART, POWERED PROSTHESIS, ARTIFICIAL SKIN

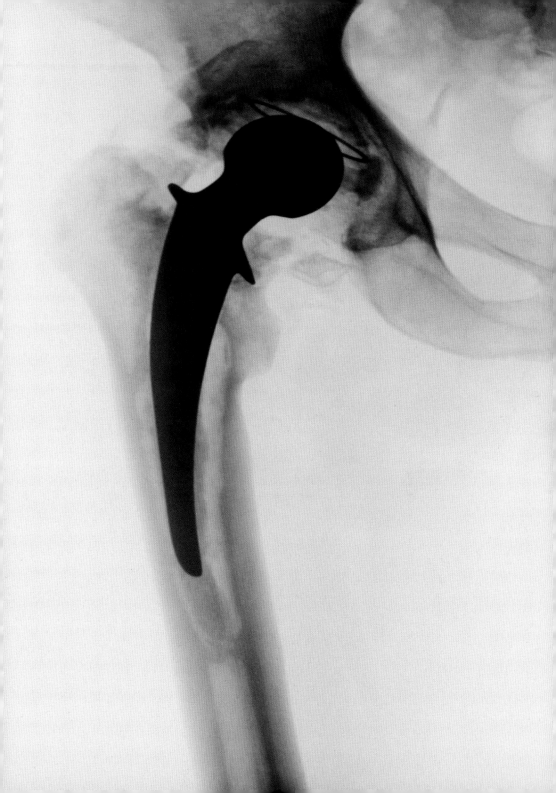

Ring Pull (1963)

Fraze makes it easy to open a drinks can without a special tool.

The canned drink is one of the most familiar and practical inventions of the twentieth century. Until Ermal Fraze (1913–1989) came along, the problem was how to open them. Before his ring pull, cans had to be opened with a "church key," a tool similar to a bottle opener but with a sharp point at either end. One end was used to make a hole in the top to drink from, and the other to make a smaller hole for air to enter, allowing the liquid inside to escape.

Fraze struck upon the idea of the ring pull when he was at a picnic and had forgotten his church key. Like most people of that time, he was aware that it was easy to injure yourself while trying to open a can with a sharp object. In the event he managed to improvise by using the car fender to open his drinks, but he realized that it would be much simpler if cans could always be opened without a separate tool.

Fraze's invention went through several stages that culminated in the "pull-top" can. Similar to today's cans, this had a small tab of metal, the edge of which was scored, attached by a rivet to a ring. By pulling the ring, the tab would be pulled out of the can, opening a hole that was large enough to let both air in and liquid out. Fraze's innovation was the rivet, and his invention proved a hit. It was first picked up by the Pittsburgh Brewing Company in the early 1960s, and by 1965 more than 75 percent of brewers in the United States were using his pull-top, although the patent was not awarded to him until 1967.

While it certainly solved one problem, the pull-top did generate a significant amount of litter. This led in the 1970s to the development of the nonremovable tops in use today. Fraze's invention has been superseded, but the idea is the same, and his pull-tops are still used widely in some areas of the world. **JG**

> *"I ... did not invent the easy-open can end. What I did was develop a method of attaching a tab."*

Ermal Fraze

⬆ *Without the riveted tab, the scored section of the can's end would be impossible to lift from the can.*

➔ *Soft-drink makers were among the first to adopt the ring-pull; the pulls are now inseparable from the can.*

SEE ALSO: CANNED FOOD, CAN OPENER, SELF-HEATING FOOD CAN

Artificial Heart (1963)

Winchell devises a means to keep blood circulating during open-heart surgery.

The artificial heart is a machine that pumps blood around the body and is designed to replace the natural heart when it no longer works efficiently due to conditions such as heart failure. Paul Winchell (1922–2005), a U.S. television ventriloquist, was the unlikely inventor of the artificial heart.

At a cast party, Winchell met surgeon Dr. Henry Heimlich, inventor of the Heimlich maneuver for choking. After observing Heimlich in his operating room, Winchell thought that an artificial heart could keep blood pumping in during difficult open-heart procedures. With Heimlich's advice, Winchell designed an artificial heart and built the first prototype. He filed for a patent in 1956, which he received in 1963.

Winchell donated the rights to his design to the University of Utah, allowing Robert Jarvik and others to build an artificial heart, dubbed the Jarvik-7. Jarvik introduced an ovoid shape to fit inside the human chest, and used a more suitable polyurethane material. The Jarvik-7 had two pumps (like the ventricles), each with a disk-shaped mechanism that pushed the blood from the inlet valve to the outlet valve. On December 2, 1982, Dr. Willem DeVries implanted the first artificial heart into retired dentist Dr. Barney Clark, who survived 112 days with the device in place.

Artificial hearts, or left ventricular assist devices as they are commonly called, are now used as a bridge to keep alive patients with heart failure until donor hearts become available. Modern devices are much smaller; they do not need to store blood because constant flow-impeller pumps are used to keep the blood permanently circulating around the body. **JF**

"The valves and chambers were not unlike the moving eyes and closing mouth of a puppet."

Paul Winchell

SEE ALSO: ARTIFICIAL LIMB, HIP REPLACEMENT, POWERED PROSTHESIS, ARTIFICIAL SKIN, ARTIFICIAL LIVER

◨ *The Jarvik-7 was powered by compressed air and electricity, supplied by a 6-foot (1.8 m) lifeline.*

Valium (1963)

Sternbach discovers a drug to calm anxiety.

At its peak in 1978, Valium was the most widely prescribed drug for tension and anxiety. Valium was discovered in 1963 by Leo H. Sternbach (1908–2005), a Polish chemist working in the United States for Hoffmann-La Roche. He wanted to create a better "chill" pill, after barbiturates were found to cause dependence and toxicity in overdose.

Sternbach started by fiddling with compounds he had cast aside twenty years earlier and found that one, Ro-5-0690, had hypnotic and sedative effects in mice. Hoffmann-La Roche named the drug Librium, which was the first of the new benzodiazepine class of drugs. Benziodiazepines work by depressing activity of the reticular activating system (RAS) that controls mental

> *"She goes running for the shelter/ Of a mother's little helper/ And it helps her on her way . . ."*

The Rolling Stones, "Mother's Little Helper," (1967)

activity in the brain. In 1963 Sternbach synthesized a simplified version of the Librium molecule. Five to ten times stronger than its predecessor, the drug was called Valium and used for treating insomnia and panic and phobia disorders.

Valium was alluded to in "Mother's Little Helper," a Rolling Stones' song about a housewife using Valium to help cope with a demanding family. Barbara Gordon's 1979 memoir, *I'm Dancing as Fast as I Can*, led to recognition of the risk of physical dependence in regular users, and of the withdrawal symptoms caused if Valium is withdrawn suddenly. For this reason Valium is now prescribed for only two weeks or less. **JF**

SEE ALSO: ANESTHESIA, DISSOLVABLE PILL, ASPIRIN, HEROIN, ACETAMINOPHEN, TETRACYCLINE, BETA BLOCKERS, PROZAC®

Audiotape Cassette (1963)

Philips announce a new recording medium.

Like SMS (short message service) text messaging three decades later, the audiotape cassette (or "compact cassette," to give it its official name) is a classic case of an innovation created for one purpose that finds unexpected success for another. Although the cassette (derived from the French word meaning "little box") was an audio storage medium, Philips saw little potential for its use within the high-fidelity music market. It had, in fact, been designed primarily for use in dictation machines and cheap portable recorders.

Introduced in 1963, the cassette slowly established itself in the decade that followed. Its success was largely due to Philips's decision to license aspects of their technology free of charge. The tape used in the

> *"In the mid-1980s . . . music networking through things called audio cassettes was at its peak."*

Carl Howard, alternative music network pioneer

early cassette cartridges was thin, low-quality, and only half the width of standard reel-to-reel tape. Since cassettes were reversible—designed to record and play back in both directions—recordings for each "side" had to be squeezed into 1/16th of an inch. The cassette was therefore highly unsuited to capturing the more extreme frequencies.

The compact cassette was never taken seriously by audiophiles, even when Dolby® noise reduction and high-quality chromium dioxide tapes made high-quality recordings possible. Nevertheless, up until the mid-1990s, it was an important format, rivaling, and even surpassing, the vinyl record in some regions. **TB**

SEE ALSO: VINYL PHONOGRAPH RECORD, AUDIOTAPE RECORDING, DIGITAL AUDIOTAPE (DAT), MINIDISC

Lava Lamp

(1963)

Walker devises a fascinating lamp for the Peace Generation.

Englishman Edward Craven Walker (1918–2000) had the idea for the lava lamp while enjoying a drink in a bar. Looking at a homemade lamp, made from a cocktail shaker and some cans, Walker realized the potential for a glass containing fluids that did not mix and had different densities. Back home, he started work on a novelty lamp, using an incandescent bulb to heat the contents of a glass bottle containing a mixture of water, translucent wax, and carbon tetrachloride. In the lamp the wax heated up, melted, and rose in the bottle—when it reached the top, it cooled and fell back to the bottom. Molten wax would have floated on water at any temperature, but the carbon tetrachloride increased its density.

Walker started a company called Crestworth and in 1963 launched a range of what he called "Astro" lamps. Invited to inspect the lamps, retailers in England thought them unattractive, but Walker presented them at a Brussels trade show, where they were spotted by U.S. entrepreneur Adolph Wertheimer. Wertheimer and his business partner, Hy Spector, bought the U.S. rights and their company, Lava Simplex International, began to manufacture the lamp. They called it the "Lava Lite."

The U.S. company carried on developing the lamp, using a range of shapes for the glass bottle and adding bright colors. The Lava Lite was a huge success during the 1960s, becoming an icon as a standard item of psychedelic paraphernalia—not least because the lamp's brightly colored blobs were thought to recall hallucinations caused by drugs such as LSD. **SC**

"[The company name of] 'Lava brand motion lamp' hasn't caught on with American consumers."

James P. Miller, *Chicago Tribune*

SEE ALSO: ARGAND LAMP, ARC LAMP, FLUORESCENT LAMP, ANGLEPOISE® LAMP, HALOGEN LAMP

◹ *Watching the rise and fall of the lava lamp's colored blobs was a favored pastime of the Love Generation.*

Unmanned Aerial Vehicle (UAV) (1963)

U.S. Military spearheads new pilotless aircraft.

It is impossible to put an exact date on the invention of unmanned aerial vehicles (UAVs), commonly known as drones, as the development of the technology stretched over many decades. Remotely controlled airplanes first took off in the 1910s, and full-size pilotless aircraft were developed as practice targets for anti-aircraft gunners from the 1920s onward. By the end of World War II, some radio controlled aircraft had the ability to be flown beyond the operators' sight, because they could transmit information such as the aircraft's speed and altitude via radio signals.

In 1960, the U.S. Air Force engaged Ryan Aeronautical to produce long-range reconnaissance drones after two spy planes had been shot down and their pilots and crew captured. By 1963, the jet-powered Ryan Model 147B, codenamed "Lightning Bug" and based on Ryan's earlier "Firebee" target drone, was fully operational. With a range of 1,200 miles (1,900 km), these drones saw action over China and Vietnam over the next decade. Today, sophisticated drones are commonly used for surveillance and sometimes for targeted strikes.

Small, electrically powered drones have a wide range of non-military applications. For example, hobbyists and professionals alike use drones for aerial videography; farmers use them to monitor their fields or spray crops; police forces use them for surveillance; and groups of coordinated, illuminated drones are sometimes used in public displays. These small drones typically take the form of remotely controlled "quadcopters"—flying machines with four electrically powered rotors. Trials are ongoing involving smaller drones that autonomously deliver online purchases and even fast food orders. **JC**

SEE ALSO: WIRELESS COMMUNICATION, POWERED AIRPLANE, MISSILE, LASER-GUIDED BOMB

Carbon Dioxide Laser (1964)

Patel creates a specialized light source.

The carbon dioxide laser is considered the most useful and versatile type of laser. It was invented in 1964 by Kumar Patel (b. 1938) while he was working in the United States at Bell Laboratories, New Jersey.

Carbon dioxide lasers emit infrared radiation with a wavelength between 9 and 11 micrometers. The active medium in the laser is a mixture of carbon dioxide, nitrogen, and helium. The nitrogen molecules, made to vibrate by an electric current, cannot lose this energy by electron emission, so they in turn excite the carbon dioxide molecules, which produce the laser light. The helium plays two roles; it assists the transfer from the gas of heat caused by the electric discharge,

> *"The atoms become like a moth, seeking out the region of higher laser intensity."*

Steven Chu, physicist

and it helps the carbon dioxide molecules to return to their ground state after excitation. The gas mixture is generally contained in a sealed chamber, with a reflecting mirror of polished metal at one end and a partially transmitting one of coated zinc selenide at the other, through which the laser beam escapes.

Patel found many uses for his device. As a result, the carbon dioxide laser has more practical applications today than any other type of laser. It has improved high-resolution and saturation spectroscopy, contributed to laser-induced fusion and nonlinear optics, and is even used for the optical pumping that has made possible newer types of lasers. **BO**

SEE ALSO: LASER, LASER-GUIDED BOMB, LASER EYE SURGERY, ATOM LASER, LASER TRACK AND TRACE SYSTEM

Eight-track Audiotape (1964)

Lear introduces a new audio format for in-car entertainment.

The eight-track cartridge (or "Stereo 8") was developed by a consortium of U.S. businesses lead by Bill Lear (1902–1978) of the Lear Jet Corporation. Their aim was to produce a convenient magnetic tape playback system that could be used in cars.

The eight-track cartridge contained a continuous loop of 1/4-inch (0.6 cm) tape that played back at 3.75 inches (9.5 cm) per second. It was so named since the music was recorded as four parallel pairs of stereo tracks on a single piece of tape. Switching between tracks was achieved automatically by mechanically altering the height of the playback head so that it aligned with the correct piece of music.

While Lear touted the system as being superior to reel-to-reel because more music was available in a smaller package, the audio quality was inevitably compromised because the four parallel stereo tracks were squeezed into the 1/4-inch tape. Furthermore, the design of the transport mechanism meant that rewinding tapes was not physically possible. A greater problem existed, however: long player (LP) releases comprised two distinct "programs" of music—one on each side of the record; the configuration of the eight-track tape meant that music content had to be divided into four such programs. In some cases this meant album track listings being reorganized to fit in with the length of the tape, long passages of silences between programs as the mechanism shuffled and the heads were realigned, or songs fading out at the end of one program and fading back in at the start of the next.

In spite of these drawbacks, Stereo 8 became the first popular in-car music format, aided in no small way by the fact that throughout the 1970s most mainstream album releases were also available on eight-track cartridge. **TB**

> *"Players have been installed in powerboats and airplanes, as well as in funeral limousines . . ."*

Time magazine (August 5, 1966)

⬆ *Until the early 1980s, eight-track stereo systems were coveted, despite a lack of material relative to vinyl.*

➡ *Cars were fitted with eight-track players, although storage of the bulky cartridges could be problematic.*

SEE ALSO: MULTI-TRACK AUDIO RECORDING, AUDIOTAPE CASSETTE, DIGITAL AUDIOTAPE (DAT)

Geostationary Communications Satellite (1964)

The Hughes Aircraft Company facilitates phone calls between continents.

In 1945, in an article entitled "Extra-Terrestrial Relays," British novelist Arthur C. Clarke described a way to bounce information off orbiting satellites so one side of the earth could communicate with the other almost instantly. Although the idea had been put forward previously by the Russian scientist Konstantin Tsiolkovsky, it was Clarke's detailed description that caught the attention of Harold Rosen of Hughes Aircraft Corporation. In 1961 the project, called the Synchronous Communications Satellite program (or Syncom), was given funding to make it happen.

A mere seventeen months later the satellite *Syncom I* was launched, but it stopped sending signals before it reached orbit. *Syncom II*, which followed in 1963, achieved a geosynchronous orbit (it traveled at an inclined angle, so was not stationary above one spot) but nevertheless proved the concept with a two-way satellite call between President Kennedy in the United States and Prime Minister Abubakar Balewa in Nigeria. *Syncom III* finally achieved a true geostationary orbit in 1964 and transmitted live television coverage of the Tokyo Olympic Games to North America and Europe.

Today, due to the proliferation of satellite-building worldwide, establishing a satellite in geostationary orbit is not simply a matter of launching spacecraft as required. Satellites traveling at the same speed as Earth in geostationary orbit must all occupy a single ring 22,300 miles (35,800 km) above the equator. The satellites have to be spaced apart, so the number in geostationary orbit is naturally restricted. Those countries wishing to maintain satellites in the skies above their longitude, as well as those wanting to control airspace above the equator, are governed by an international allocation mechanism. **JM**

> " . . . In the long run . . . the most daring prophecies seem laughably conservative."

Arthur C. Clarke, novelist and writer

⬆ *Olympic athletes Rafer Johnson (U.S., right) and C. K. Yang (China) discuss Syncom III in Japan, in 1964.*

➡ *Since the Syncom launchings, numerous satellites have been placed in geostationary orbit above Earth.*

SEE ALSO: GLOBAL POSITIONING SYSTEM (GPS), SATELLITE RADIO BROADCASTING

AstroTurf (1964)

Faria and Wright invent easily maintainable substitute for grass.

James Faria and Robert Wright were researchers at Monsanto, Inc., in the United States when they invented AstroTurf. Originally called Chemgrass, this artificial grass for sports and playing fields was first installed in 1964 at Moses Brown School in Providence, Rhode Island. Monsanto adopted the name AstroTurf after the product was installed at Houston Astrodome stadium in 1966. It was patented in 1967.

The first AstroTurf, a short-pile carpet of nylon, was harder than real grass and caused serious injuries to players. Several English League soccer teams adopted AstroTurf in the 1980s, but the clubs quickly returned to natural grass. Recent developments have largely overcome these problems. The addition of sand or rubber infill, superior backing, and nylon yarn fibers have made it as safe as normal grass.

The newest version of AstroTurf, the so-called "third-generation" product, consists of green blades of polyethylene that resemble grass, mixed with small pieces of black rubber that make the playing surface very springy. Despite the bouncing of the rubber pieces around the players' feet, playing on third-generation AstroTurf is closely comparable to playing on grass—but without the mud.

AstroTurf has transformed the maintenance of sports grounds throughout the world. Unlike natural grass, it requires no watering, making it suitable for hot climates, and obviously there is no need to mow or roll it. Its development has also meant that many sports that were only playable outside—such as baseball— could be brought into indoor stadiums for occasions such as professional-level matches. **SC**

SEE ALSO: REINFORCED CONCRETE

"I don't know whether I prefer AstroTurf to grass. I never smoked AstroTurf."

Joe Namath, American football quarterback

◤ *Freshly applied yellow marker paint traverses AstroTurf made of green and brown polyethylene.*

Plasma Screen (1964)

Three Americans create a huge screen.

The plasma screen was invented in 1964 by Donald Bitzer, Gene Slottow, and Robert Wilson at the University of Illinois. It was an alternative to the traditional television set that used an electron gun inside a glass tube to excite atoms of phosphorous coated on the inside of the screen, making them glow. The need for the electron gun and tube meant that a normal television set required depth, making it bulky.

The plasma screen uses different technology. Just behind the screen are hundreds of thousands of tiny cells containing xenon and neon gas, with electrodes behind them. An electric charge from these electrodes can make the gas temporarily become a glowing ionized gas—a plasma. The same physics underlie the

---

"On their new 150-inch screen: 'Can you imagine watching the Olympics on this baby?'"

Toshihiro Sakamoto, Panasonic president

---

tendrils in a plasma ball, as well as the aurora.

Unlike traditional televisions, plasma screens do not flicker, and this was one of the reasons behind their creation. People using computers for long periods would be much less likely to suffer headache and eyestrain. However, the most obvious benefit of plasma screens is that, because they do not need an electron gun and tube, the display unit can be thin and light enough to mount on a wall.

For the true television aficionado, plasma screens with a diagonal measurement of 150 inches (380 cm) are available, but the purchaser needs a room at least 30 feet (9 m) long to get the most out of one. **DHk**

SEE ALSO: TRANSFORMER, TELEVISION, COLOR TELEVISION, CABLE TELEVISION, TELEVISION REMOTE CONTROL

SQUID (1964)

Jaklevic, Lambe, and Silver check magnetism.

At the heart of a SQUID (Superconducting Quantum Interface Device) are two Josephson junctions. These consist of two superconductors separated by an insulating barrier that is so thin that electrons can "tunnel" through. If a voltage is applied, the current across the barrier starts to oscillate at a high frequency, this being influenced by the ambient magnetic field. In the process, the electrical resistance of the SQUID changes. These changes can be used to measure very small and weak magnetic fields.

The direct-current SQUID was invented by Robert Jaklevic, John Lambe, Arnold Silver, and James Mercereau in 1964, one year after the production of the first Josephson junction. The first models worked

---

"Science cannot solve the ultimate mystery of Nature . . . because . . . we ourselves are part of [it]."

Max Planck, physicist

---

only at the temperature of liquid helium, -452°F (-269°C). The discovery of higher-temperature superconducting ceramics in 1987 meant that devices could be made to work at the temperature of boiling liquid nitrogen, at around -321°F (-196°C).

Magnetic fields that are a hundred billion times weaker than those exerted by a fridge magnet can easily be quantified. Magnetic fields produced in the brain, muscles, and nerves can be measured. SQUIDs also play a role in the testing of structures that contain metal. As non-natural objects affect their magnetic surroundings, SQUIDs can be used in delicate military surveillance tasks, as well as airport security. **DH**

SEE ALSO: MAGNETOMETER, ELECTROENCEPHALOGRAPH (EEG)

Dolby Noise Reduction

(1965)

Dolby takes the hiss out of tape recordings.

For most of the second half of the twentieth century, magnetic tape was used in the making of most audio recordings, but there was always some background noise present on them. This tape hiss, or "white noise," was most noticeable in quieter musical passages.

In 1965 electrical engineer Dr. Ray Dolby (1933–2013) proposed the first magnetic tape noise reduction system. The task was to improve the signal-to-noise ratio of the recording, reducing the level of hiss without affecting the quality of the sound. His approach was to "compand" (compress, then expand) the sound. During recording, an encoding circuit was inserted between the recording source and the tape

---

". . . developments start with the desire of the developer to get what he . . . wants so that he can use it."

Dr. Ray Dolby

---

recorder; this compressed the dynamic range of the recording. During playback, a decoding circuit was inserted between the tape recorder and the playback amplifier, expanding the dynamic range once again. The quieter sounds in the spectrum received a greater proportional boost; the signal level dropped and was filtered in the higher frequencies (where the tape hiss existed). On playback the white noise was less audible.

Dolby's first system, Dolby A, was widely adopted within the professional recording market. When the lower-fidelity Philips cassette replaced the reel-to-reel machine, Dolby produced a simpler version and licensed it to cassette recorder manufacturers. **TB**

SEE ALSO: LOUDSPEAKER, STEREO SOUND, SURROUND SOUND

Collapsible Stroller

(1965)

Maclaren makes life easier for parents.

From the Victorian era, right up until the 1960s, mothers struggled with huge and heavy baby carriages. These days, however, a stroller is the item at the top of every new parent's baby list. Not only do parents appreciate the practicalities of having a stroller that they can fold up and carry under one arm, they also want to distance themselves from the unstylish and cumbersome older styles of baby buggy.

Unexpectedly, the inventor of the collapsible stroller was Owen Maclaren (1907–1978), a retired test pilot who had previously designed landing gear and protective seals for Spitfire aircraft. His departure into strollers may seem odd, but building a practical stroller required knowledge of strong, lightweight structures, and Maclaren's aeronautics experience had given him exactly that.

At around 6½ pounds (3 kg), Maclaren's first model cost about $10 (£7) and weighed less than the child it was intended to seat—his granddaughter. It had an aluminum frame and could be collapsed with one hand, an important design feature for a parent trying to keep control of a toddler while doing so. After patenting his design in 1965, Maclaren began manufacturing the strollers from his home in Northamptonshire, England. By 1976 he was selling more than a half a million of the strollers a year, with a large proportion of these going overseas.

Maclaren was awarded the rank of Member of the British Empire (MBE) for his achievements in both aeronautical engineering and transportation design, which, of course, included the beloved stroller. **HB**

SEE ALSO: CHAIR, CART, BABY CARRIAGE, FOLDING WHEELCHAIR

➡ *These drawings of an updated stroller appear in Maclaren's U.S. patent documents, dated 1987.*

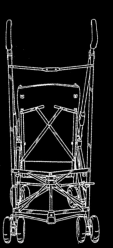

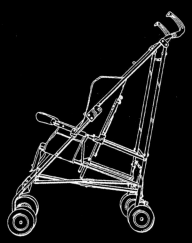

Inflatable Aircraft Escape Slide (1965)

Grant speeds emergency exits from aircraft.

If an airplane crashes, an exit strategy needs to be in place. In fact, aviation authority rules state that it must be possible to completely evacuate an airplane within ninety seconds, under conditions of pitch black darkness and with half the exits blocked.

In 1965 Jack Grant, who was working at Quantas Airlines as a safety superintendent, invented a superior inflatable escape slide that could double up as a life raft in the event of a crash landing at sea. His design was tried and tested in Sydney, Australia, with great success. In the 1960s, aviation authorities suggested that inflatable slides would only be useful if they could be fully deployed within twenty-five seconds, in

---

"[Developing an escape slide] is like trying to balance a sheet of plywood on the head of a pin."

Mark Robertson, engineer

---

moderate weather. The slide met these requirements and was also light and compact; it fit inside the aircraft door or below the emergency exit window.

In an emergency, the slide mechanism is activated to release the slide. It then begins to inflate, initially drawing air from a cylinder of compressed carbon dioxide and nitrogen gas. The cylinder provides approximately a third of the volume required to inflate the slide, after which it sucks in ambient air, channeled through aspirators, to reach full inflation. **RB**

SEE ALSO: PARACHUTE, KNAPSACK PARACHUTE, THREE-POINT SEAT BELT, CRASH TEXT DUMMY, AIR BAG

⬅ *People use an inflatable escape slide in an illustration from a safety instruction sheet for airline passengers.*

Rubella Vaccine (1965)

Meyer and Parkman trial a new vaccine.

In 1814 German researchers first described the disease "German measles"—later known as rubella from the Latin *rubellus* meaning "reddish." Rubella is a single-stranded RNA virus spread from person to person via respiratory droplets. It usually causes a mild illness (symptoms include low-grade fever and swollen lymph nodes followed by a generalized rash), but in pregnant women it is an altogether different matter. In fetuses, it can result in congenital rubella syndrome, a condition characterized by deafness, mental retardation, cataracts, heart defects, and diseases of the liver and spleen.

Between 1963 and 1964 a rubella epidemic in the United States resulted in 30,000 babies being born with permanent disabilities as a result of exposure to the virus. The tragedy prompted the National Institutes of Health to launch a campaign to find a vaccine. Two pediatricians, Harry Martin Meyer (1928–2001) and Paul Parkman (b. 1932), isolated the rubella virus and then went on to develop the first vaccine against rubella. The team grew the virus in cultures of kidney cells taken from African green monkeys, seeding each crop from the preceding crop. Finally, after two years spent growing seventy-seven crops, they inoculated rhesus monkeys with what they called High Passage Virus 77 (HPV-77). The vaccinated monkeys showed no signs of rubella but developed antibodies against the virus, while their cage mates, who were unvaccinated, remained free of infection. In 1965 the team started the first clinical trials in women and children, and again showed that the virus did not spread and that subjects developed antibodies.

The rubella vaccine was later refined into a vaccine known as MMR for mumps, measles, and rubella. **JF**

SEE ALSO: INOCULATION, VACCINATION, CHOLERA VACCINE, BCG VACCINE, RABIES VACCINE, POLIO VACCINE

Kevlar®

(1965)

Kwoleck discovers a highly tough polymer.

A fiber with half the density of fiberglass and five times stronger, weight for weight, than steel, Kevlar® is now globally recognized and widely used. It is best known for its use in bulletproof and stabproof vests, where it has saved thousands of lives.

After graduating from the Carnegie Institute of Technology in 1946, American chemist Stephanie Kwoleck (1923–2014) was employed by DuPont to research high-performance chemicals. Her work on polymers yielded a number of successful discoveries. Kwoleck, who holds twenty-eight U.S. patents, specialized in developing polymers at low temperatures and in the 1960s discovered a new group called liquid crystalline polymers.

Kwoleck's invention of Kevlar fibers in 1965 stemmed from an interest in the chemicals produced during the process of polymer synthesis. These substances are sensitive to moisture and heat and easily undergo hydrolysis and self-polymerization. She discovered that, in cool conditions, these chemicals created an aramid polymer that was a cloudy liquid. Kwoleck's research team spun it out and the resulting fibers were much stiffer and stronger than any others previously created.

The Pioneering Lab at DuPont was put to work finding a commercial use for these fibers, and now Kevlar has many important applications; it replaced asbestos brake pads, and it is used for racing sails and standing rigging on performance boats. Fiberoptic cables, spacecraft shells, radial tires, and suspension-bridge cables are now frequently made of Kevlar. It is widely used in bulletproof vests for military and police personnel. One clothing firm has even incorporated Kevlar into fabric for children's school uniforms. **LC**

SEE ALSO: CLOTHING, SHOE, HELMET

Powered Exoskeleton

(1965)

General Electric creates a superstrength suit.

The powered exoskeleton is a good case of life imitating art. Robert A. Heinlein's 1959 novel *Starship Troopers* described warriors in powered suits. The idea was used again in the Marvel Comic *Iron Man*, with a man inside a powerful homemade iron suit. The idea struck a chord, and General Electric took up the task of turning it into reality. By 1965 they had produced "Hardiman," the first powered exoskeleton.

The idea behind the device was to produce a robot that reacted to the natural muscle movements of the wearer. It was designed to act like a "second skin," albeit one that weighed as much as a car. Hardiman was a ¾-ton monster, designed to lift 1,500 pounds

> *"… you don't have to think about it … you just wear it and it takes orders directly from your muscles."*
>
> Robert A. Heinlein, novelist

(680 kg). Unfortunately the team never managed to get Hardiman working. Any attempt to power up the full frame caused a "violent and uncontrolled movement," and as such, the machine was never fully turned on with a person inside. One arm of the behemoth did work, lifting 750 pounds (340 kg) as predicted, but this was as far as the project went. The challenges in making such a monster work in a controlled way, without crushing the human inside, made it impossible to build a practical product. **SB**

SEE ALSO: CHAIN MAIL, MILITARY CAMOUFLAGE, POWERED PROSTHESIS, SEGWAY PT

▣ *An operator demonstrates the Hardiman's arm—the one part of the behemoth that worked reliably.*

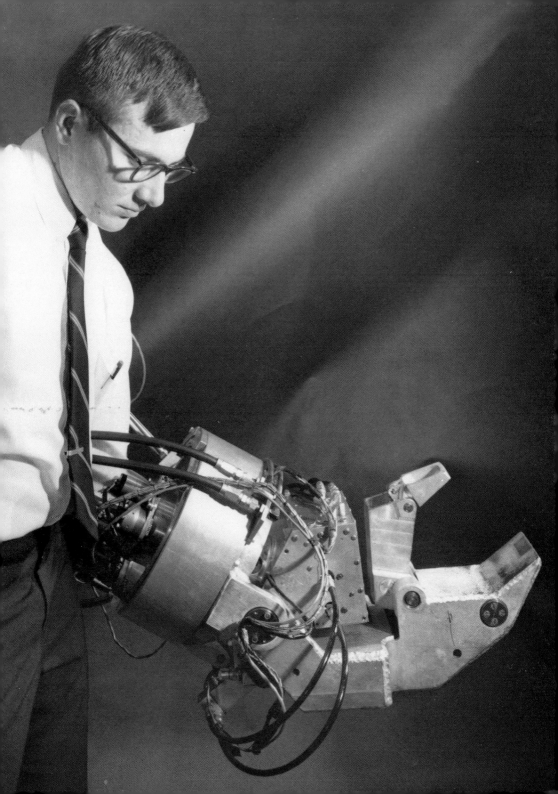

Laser-guided Bomb (1967)

The U.S. Air Force uses lasers to direct bombs onto pinpoint targets.

> *"Our scientific power has outrun our spiritual power [with] guided missiles and misguided men."*

Dr. Martin Luther King, Jr.

⬆ *BOLT-117 bombs are loaded onto a weapon station aboard USS Truman in 2005, during the Gulf War.*

➡ *The Paveway II laser-guided bomb featured pop-out rear wings to enhance its glide performance.*

In 1962 the United States Air Force (USAF) began research into developing a laser-guided weapon that could be used to target stationary targets accurately. During World War II, unguided bombs dropped by aircraft caused immense damage, but many bombs were often required to hit one target successfully. From a military perspective the risk to bomber pilots from antiaircraft fire had been great, and there had been a terrible toll of deaths and injuries among civilians at some distance from the actual targets.

By 1967 the USAF had produced the first laser-guided bomb, the BOLT-117, and were able to use it during the Vietnam War the following year. This breakthrough in ordnance capability turned unguided, "dumb" bombs (which simply fall to the ground) into precision guided, "smart" bombs.

Laser-guided bombs rely upon the target being illuminated by a laser beam from the releasing aircraft, an accompanying aircraft, or a unit operating on the ground. The laser light reflects off the target to form an inverted cone, or "basket." After releasing the bomb, personnel in the aircraft direct it to the cone by means of signals sent to the bomb's control fins. However, early use of laser-guided bombs proved to be literally hit or miss. If pilots were faced with bad weather, the weapon became virtually redundant.

Refinements in laser and computer technology led to more sophisticated and deadly laser-guided missiles. First deployed with huge success during the Gulf War in 1991, aircraft acting alone were able to direct a laser that could emit a heat signature on the target. Once locked on to the target the missile follows that heat, even if the target is moving. Although still prone to error, laser-guided weaponry has gone some way to reducing the loss of innocent lives. **SG**

SEE ALSO: ROCKET, CANNON, CLUSTER BOMB, BOUNCING BOMB, ATOMIC BOMB, HYDROGEN BOMB

Ground Proximity Warning System (1967)

Bateman devises a safety aid for aircrew.

Following a series of controlled-flight-into-terrain (CFIT) crashes in the 1960s, where aircraft had crashed simply because their pilots did not realize how close to the ground they were flying, Donald Bateman (b. 1932) designed a Ground Proximity Warning System.

The system works by collating data from the radar altimeter (which measures the height above the ground), the barometric altimeter, the sensor that detects the glide path, and sensors attached to the flaps and landing gear. The data are then used to predict trends in flight pattern. The pilot is warned if the system registers an excessive descent rate, excessive terrain closure rate, altitude loss after take-off, unsafe terrain clearance, or below glide path deviation.

Each of the five danger modes has its own audio and visual warnings to alert the pilot. Within each mode, there are also different warnings depending on the severity of the deviation. For example, a sink rate of 4,000 feet (1,200 m) per minute will trigger the warning "Sink Rate" at 1,500 feet (460 m), but "Pull Up" at 500 feet (150 m).

The introduction of Bateman's system in 1967 reduced the fatal CFIT crash rate for large passenger aircraft to fall from 3.5 per year to two per year and was made mandatory on such planes by the U.S. Federal Aviation Administration in 1973. The systems were referred to as "Bitching Betty" or "Bitching Bob" by U.S. pilots, depending on whether the voice was female or male. An improved version of the system, Enhanced Ground Proximity Warning System, is linked to a global terrain map, enabling a plane to pinpoint exactly where it is and to predict sudden changes such as steep slopes, giving more timely warnings to pilots and enabling them to fly in poor conditions. **HP**

SEE ALSO: POWERED AIRPLANE, RADAR, BAROMETRIC ALTIMETER, BLACK BOX FLIGHT RECORDER

Regenerative Brakes (1967)

AMC tries out a new power technology.

Regenerative brakes were designed by the American Motors Corporation (AMC) in the 1960s as a way to increase the potential range of electric cars and match the performance of their fossil fuel-burning contemporaries. The resulting car, the Amitron, built in 1967, never made it into full production.

A brake is a device that converts kinetic energy (movement) into another type of energy. Traditional braking systems use the friction of a brake pad to slow down the vehicle, turning its kinetic energy into heat. This system is effective but not energy-efficient—it would be much more resourceful to try to recover some of the kinetic energy back in the form of fuel.

"Regenerative braking improves overall efficiency and prolongs brake component life."

Industrial equipment advertisement

The problem with vehicles powered by the internal combustion engine and fossil fuels is that it is impossible to turn inertia back into petroleum. In an electric vehicle, however, the problem can be solved with regenerative brakes. Motion is normally provided in an electric vehicle by electricity passing through coils, creating a magnetic field that repels magnets on a movable part, the rotor. If, during braking, that process is reversed, the magnets generate electricity in the coils, which can then be stored.

Regenerative brakes have proved very useful in other electric vehicles, particularly trains, where they typically improve efficiency by about 15 percent. **JB**

SEE ALSO: BAND BRAKE, DISC BRAKE, DRUM BRAKE, HYDRAULIC BRAKE

Cash Machine (ATM) (1967)

Shepherd-Barron speeds the process of cash withdrawal.

There was a time not so long ago when there was no such thing as a cash machine. If you wanted to withdraw some money, you had to go into a building and speak to a teller. Now, of course, it is possible to get cash from one of over 1.6 million automated telling machines (ATM) worldwide, in stores, cinemas, and even the southern rim of the Grand Canyon.

Exactly who we have to thank for this stroke of technological banking genius is a matter of some controversy. Luther George Simjian, a prolific inventor of his time, devised the very first "cash-point" in 1939. Installed by the City Bank of New York this cash machine saw little use except with "...prostitutes and gamblers who didn't want to deal with tellers face to face." The machine was removed.

There followed a lull in the history of the cash-point that lasted nearly thirty years. Then, in 1967, John-Shepherd Barron (1925–2010), an inventor of Scottish descent, had an idea in the bath for a machine that would give you money, anywhere in the world, and the ATM was reborn. The first one was installed in Enfield, North London, in 1967. This early cash machine was operated by a "token" resembling a check impregnated with radioactive material, which was verified against a four-digit personal identification number (PIN code). Why four digits? Because that is the most the inventor's wife could remember.

The first plastic card-operated ATM was invented by Texan Don Wetzel a short time later and some people (including the Smithsonian Institution) credit him with being the inventor of the ATM. **BG**

SEE ALSO: CASH REGISTER, CREDIT CARD

⬈ *A customer counts her money after using a London bank's pioneering cash-point machine in 1967.*

"I hit upon the idea of a chocolate bar dispenser, but replacing chocolate with cash."

John Shepherd-Barron

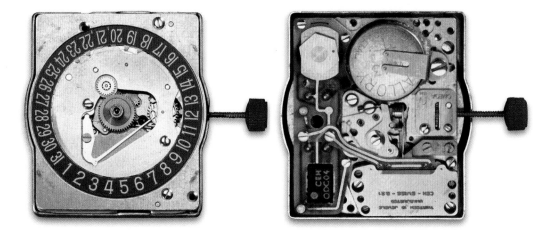

Quartz Watch (1967)

The Centre Electronique Horloger makes a major breakthrough in timekeeper accuracy.

At the heart of a quartz watch is a 4-mm bar of quartz piezoelectric crystal that is made to vibrate by applying a small voltage. The crystal is laser trimmed so that it oscillates exactly 32,768 times per second. Higher-frequency crystals would need too large a driving current and quickly drain a watch battery, and lower-frequency ones would be physically too large. The signal, one cycle per second, either drives a second hand or triggers an LCD (liquid crystal display).

Quartz is used because it has a very low coefficient of thermal expansion and thus is not affected by changes in the weather. A fairly standard mass-produced quartz watch typically gains or loses less than one second per day.

The first quartz oscillator was produced in 1921. By 1927 Warren Marrison, a telecommunications engineer at Bell Laboratories in Canada, had made the first quartz clock. Unfortunately, its valve-driven counting electronics were bulky and unreliable. The great breakthrough came in the 1960s with the introduction of robust inexpensive digital logic systems that used simple semiconductors. In 1967 the Centre Electronique Horloger (CEH) in Neuchâtel, Switzerland, developed the famous Beta 21, the world's first analog quartz wristwatch, and two years later Seiko produced the world's first commercial quartz wristwatch, the Astron. Soon clocks, timers, and alarm mechanisms were routinely being fitted with quartz crystals, as opposed to "old-fashioned," inaccurate, and high-maintenance mechanical oscillating balance wheels.

For purposes requiring extreme accuracy, the atomic clock is preferred to the quartz clock. **DH**

SEE ALSO: TOWER CLOCK, PENDULUM CLOCK, ATOMIC CLOCK, DIGITAL WATCH, LCD (LIQUID CRYSTAL DISPLAY)

⬆ *The date wheel (left) and movement of CEH's pioneering Beta 21 quartz wristwatch of 1967.*

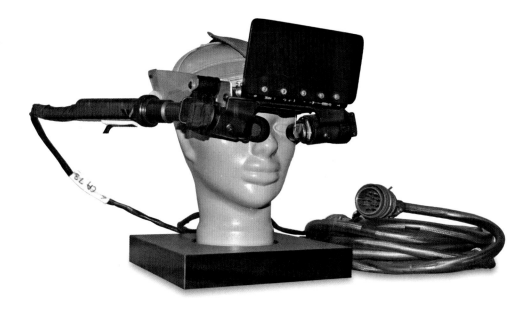

Virtual Reality Headset (1968)

Sutherland introduces the world to a wholly simulated alternative world.

Virtual reality (or VR) is a system that allows individuals to interact with a computer-simulated environment. There is much debate about the origin of the term, although it seems to have come into popular usage in the 1970s. Ivan Sutherland (*b.* 1938), an American engineer, was one of the first to explore the potential of computers to enable people to have experiences that are unavailable to them in real life.

In 1960 cinematographer Morgan Heilig built a single-user console called the Sensorama, which stimulated all the senses of the user in an all-surrounded environment. Heilig's concept involved only passive viewing, but many of his initial ideas were used by Harvard postgraduate student Sutherland. Sutherland wanted to develop a head-mounted display that enabled the wearer to look into a virtual world that would appear completely real. The user could fully interact with this virtual world, which would be maintained by a computer.

In 1968, aided by his student Bob Sproull, Sutherland built a primitive head-mounted display system that was tethered to a computer. The initial invention was so heavy that it had to be suspended from the ceiling, and so intimidating that others nicknamed it the Sword of Damocles. The system displayed images that were relayed from the computer, presenting the images in stereo, which gave the illusion of three dimensions. It also tracked the user's head movements so that the field of view could be constantly updated.

Sutherland's invention has seen applications in flight simulators and movies such as *The Matrix*. **SD**

SEE ALSO: DIGITAL ELECTRONIC COMPUTER, VIDEO GAME CONSOLE, 3D COMPUTER GRAPHICS

⬆ *Sutherland's first headset of 1968 is now an exhibit at the Computer History Museum in California.*

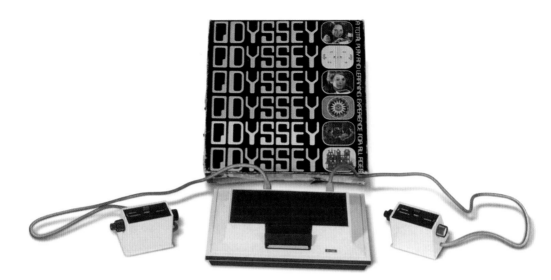

Video Game Console (1968)

Baer produces the first game console to interact with a television set.

While working for the U.S. defense contractor Sanders Associates, Ralph H. Baer (1922–2014) had for some years been pondering possible ways of playing games through a television set. In 1966 he sat down and, in four short pages, produced a document that he calls the "Magna Carta of video games." This document was to form the basis of the very first video game console. At first he produced his prototypes in his spare time, but as he developed his ideas his employer realized that there might be big money to be made, so they started to support his project.

Baer's "brown box" appeared two years later in 1968. It featured seven games: table tennis, volleyball, handball, soccer, golf, checkers, and even target shooting, using the very first prototype light gun. The console made use of clear plastic "overlays" that were placed in front of the television to simulate scenery and in-game obstacles. The system was licensed to Magnavox and released in 1972 as the Odyssey system.

Despite being a pioneering console offering an enjoyable and new form of entertainment (Baer himself is reported to prefer these early games to today's more advanced systems), it was not a big seller. Ineffective marketing put off prospective buyers, and the high price of $100 was prohibitive.

The video game did not really take off in the public consciousness until another inventor, after playing the Magnavox ping-pong game, built an arcade version called Pong, which became extremely popular. From then on, video games have become a billion-dollar industry and an integral part of leisure time—all of this stemming from Baer's little brown box. **SB**

SEE ALSO: TELEVISION, DIGITAL ELECTRONIC COMPUTER, VIRTUAL REALITY HEADSET, 3D COMPUTER GRAPHICS

⬆ *The Magnavox Odyssey system of 1972 included a boxful of accessories for use with its seven games.*

Graphical User Interface (GUI) (1968)

Engelbart presents computer processes visually on screen.

Television is traditionally vilified as being a vacuous, flashing goggle box—"chewing gum for the eyes." However, in the twenty-first century people are much more likely to be goggling at a computer screen than a television set.

That is strange because, unlike the television, the computer did not originally include a screen. When they were first created, computers simply processed data from manually inserted punch cards. Save for some mechanical whirring, they gave no real visual clues as to what they were up to. To have even the most basic understanding of what a computer was doing at any given point in its processing, a person needed a fairly advanced degree in mathematics.

Since the invention of the Graphical User Interface (GUI), that privilege has been accessible to everyone. The man who paved the way was Douglas Engelbart (1925–2013). Inspired by an essay by Vannevar Bush that he read in *The Atlantic Monthly* magazine, Engelbart led

the development in a pioneering new human/computer interaction system known as the "oN-Line System," or NLS. The first to employ a display screen, it used vector graphics, clickable hypertext links, and screen-windowing, all controlled by a cursor. When demonstrated in 1968, it caused a huge sensation.

Engelbart's radical ideas were further developed by Alan Kay (b. 1940) of the Palo Alto Research Center, who introduced the idea of graphical representations of computing functions. The folders, menus, and overlapping windows with which we are all familiar today grew out of Kay's pre-eminent work. Together, they transformed how we viewed computers. **CL**

SEE ALSO: MECHANICAL COMPUTER, COMPUTER PROGRAM, HYPERTEXT, TOUCH SCREEN, 3D COMPUTER GRAPHICS

⬆ *Apple's Lisa computer was the first to have a GUI, but its $10,000 price daunted most potential customers.*

Computer Mouse (1968)

Engelbart and English hugely improve the computer/user interface.

The 1968 Fall Joint Computer Conference at San Francisco in the United States presented a remarkable number of "firsts." Among them was the first video teleconference; the first use of hypertext (the foundation of today's web links); and the first presentation, by the Stanford Research Institute (SRI), of NLS, short for oNLine System, the revolutionary ancestor of modern computer server software. Such dazzling displays likely distracted people from another important first, moved by the hand of SRI researcher Douglas Engelbart (1925–2013): the computer mouse.

Far from the sleek ergonomic devices of today, the first computer mouse was a wooden box with wheels and a thick electric cord. Engelbart and colleague Bill English (1929–2020) first came up with the idea in 1963 and created the device as a very small piece of a much larger computer project. They were looking for something that allowed computer users to easily interact with computers. The first prototype had a cord to the front, but this was so cumbersome it was moved to the back, becoming a "tail," which gave rise to the device's name. "It just looked like a mouse with a tail, and so we called it that in the lab," commented Engelbart.

Neither Engelbart, English, nor SRI ever marketed the mouse. The next lab to work on it, Xerox's Palo Alto Research Center (PARC), gave it some modern touches but failed to bring it to the masses. That job was done by Steve Jobs, founder of Apple, Inc., in the 1980s. Jobs's company polished up the mouse, making it affordable, available, and an integral part of the personal computer. Apple may have made the mouse famous, but Engelbart and English were there first. **RBk**

SEE ALSO: MECHANICAL COMPUTER, COMPUTER PROGRAM, TOUCH SCREEN, TOUCHPAD, WEBCAM

⬆ *Moving the two wheels of Engelbart's "X-Y position indicator" moved a pointer on the computer screen.*

Water Bed (1968)

Hall designs a bedroom artifact for adventurous sleepers.

A water bed consists of a heated, water-filled mattress inside either a "hard-sided" rectangular wooden frame or a "soft-sided" robust foam frame. Both types rely on a strong metal platform to bring the mattress to a convenient height. Early models had only one water chamber within the mattress, making it very bouncy, but later designs incorporated fiber blocks and several interconnected water chambers to reduce the wave action. Some modern water beds comprise a mixture of water and air chambers. Thermostatically controlled electrical heating pads usually maintain the water at body temperature, at around 86°F (30°C).

In 1883 Dr. William Hooper of Portsmouth, England, obtained a British patent for a water bed designed for patients at risk of developing bedsores. However, his model lacked the technologies to ensure that the beds were watertight, and he was unable to control the water temperature, so his invention was not a commercial success.

Charles Prior Hall (b. 1943) designed the modern water bed in 1968, when he was a student of design at San Francisco State University in California. He originally planned to design a chair, using a vinyl bag filled initially with liquid cornstarch and then jelly, but the resulting structure was not particularly comfortable with either filling. Hall then turned his attention to designing a water-filled bed. He was unable to obtain a patent for his design, though, because similar water beds had already been described in considerable detail in the science fiction novels of Robert A. Heinlein. However, Heinlein never attempted to construct his design. **LC**

SEE ALSO: SOFA BED, PIVOTING BED

⬆ *The water bed became associated with shameless luxury, evidenced by this "Pleasure Island" creation.*

Jacuzzi® (1968)

Jacuzzi develops the world's first hot tub.

As with so many inventions, the Jacuzzi®—the best-known brand of hot tub—was invented to fulfill a very practical need: in this case, the healthcare of a family member. The Jacuzzi family were Italian immigrants to the United States in the early 1900s who developed a strong company in the aviation industry, and then flourished by designing irrigation pumps. But it was not until 1948 that they began developing the technology that would make them world famous.

Candido Jacuzzi (1903–1986) had a son who contracted rheumatoid arthritis and received hydrotherapy treatment in hospital. Wanting to bring his son home, he developed a hydraulic pump to replicate the boy's therapy, and for some years the

> *"The Jacuzzi is to this generation what the drive-in movie was in the fifties."*
>
> Mike Darnell, U.S. television executive

pump was marketed as the J-300 therapeutic device.

In 1968 third-generation family member Roy Jacuzzi developed the first version of the modern hot tub. Roy had joined the company as head of research and was looking for new products. He took Candido's J-300 device and developed it to sell to the leisure industry as a whirlpool bath. Marketed as the "Roman," Roy's design was the first self-contained consumer model to be sold. Previously an external pump had been lowered into the bath, but the Roman's pump was fully incorporated. The Roman pumped a 50/50 mix of water and air into the bath, a patented formula that the Jacuzzi brand still uses. **JG**

SEE ALSO: WATER BED

Hypertext (1968)

Nelson and Van Dam ease computer access.

As soon as hypertext was unveiled to the public in 1968 at the Convention Center of San Francisco in the United States, technology experts knew it was unique. The demonstration, now known as the "Mother of All Demos," showed how this tool for data organization enabled the user to read information, not just in the linear way we read regular text, but for the first time in a dynamic and interactive way. Hypertext later became the fundamental language of the Internet in Hyper Text Markup Language (HTML), and has revolutionized the way information is accessed.

Whereas standard text is read linearly (for instance, Western scripts are read left to right, top to bottom), hypertext allowed the user to retrieve information by "clicking" on links that shifted the page, opened further texts, and activated video and audio. The forerunner of this breakthrough was called the Memex system (from **MEM**ory **Ex**tender), imagined by U.S. engineer Vannevar Bush in his 1945 article "As We May Think" in *The Atlantic Monthly*. Bush envisaged an individual at a mechanical desk able to access information in the form of linked microfilm rolls.

The article is said to have inspired the creation of hypertext by two young computer scientists, Ted Nelson (*b.* 1937) and Andries van Dam (*b.* 1938). While working at Brown University, Nelson coined the word "hypertext" in 1963 to describe his vision of a fully indexed information system. Nelson and van Dam went on to develop the "Hypertext Editing System," the research project that led to the formation of the standard hypertext language and the historic "Mother of all Demos" three years later. **SR**

SEE ALSO: COMPUTER MOUSE, GRAPHICAL USER INTERFACE (GUI), TOUCH SCREEN, PERSONAL COMPUTER

➡ *The development of hypertext led to HTML code, the markup language for pages on the World Wide Web.*

```
ar lplg=(window.navigator.plugins['Sho
 (lplg) {
 (lplg.charAt(lplg.indexOf('.')-1)>=InM) in
else if (window.navigator.userAgent.ind
ocument.write('<scr'+'lpt language=vbs
extincap=(IsObject(CreateObject("Sho

        if (ad_jsl && document.getElement
        ad_el('l_fl').previousSibling.style.vi
        if (!lcap&&typeof(LAMP)=='undefin
        ad_el('l_fl').parentNode.style.margi
        ad_el('l_fl').parentNode.style.margi

                   hodObi('swf','11',1 ff.","ad_para
```

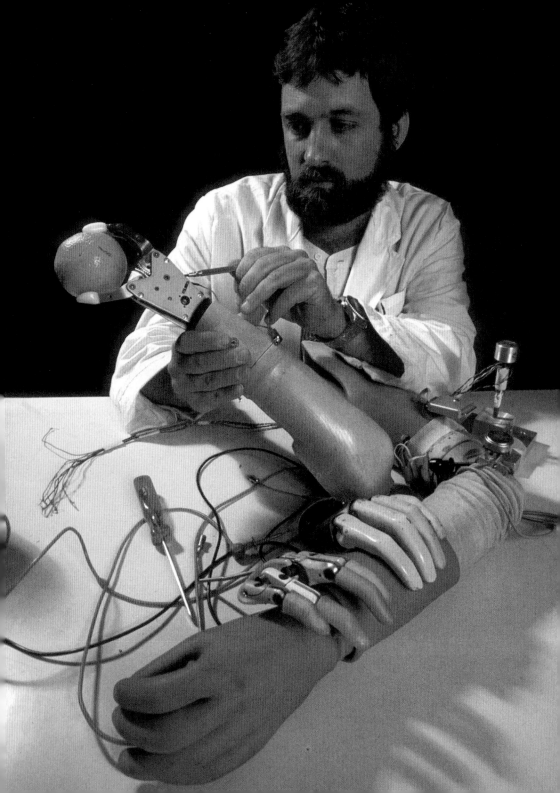

Powered Prosthesis
(1968)

An MIT team creates a powered artificial arm.

Samuel Alderson developed the first working model of an electrically powered artificial arm, which appeared in 1949. Designed for factory workers who had suffered amputation, the device was very bulky and was plugged into an external power source.

Reinhold Reiter, a physics student at Munich University, patented the first myoelectric prosthetic arm. Also requiring an external power source, it used muscle-contraction signals from the remaining biceps to control the opening and closing of the hand. It was bulky and used vacuum tubes. The transistor would have made the technology more feasible, but this was not invented until 1948. By then the German currency was revalued, and the project lost its funding.

In 1958 a Russian team led by A. E. Kobrinski had developed a myoelectric hand controlled by signals from surviving wrist muscles. Both Otto Bock Orthopaedic Industry in Germany and Viennatone in Austria marketed versions of the "Russian hand."

The first successful myoelectric arm is the "Boston Elbow." Mathematician Norbert Weiner, orthopedist Melvin Glimcher, Amar Bose, Robert Mann, and others from the Massachusetts Institute of Technology (MIT) created working prototypes in 1968 and a viable prosthetic arm by 1974.

The device worked by having sensors in the socket that sensed the currents generated by muscle contractions: these small signals were amplified and moved the prosthesis with the aid of battery-driven motors. Future advances are expected to include transmission of temperature and tactile sensation. **SS**

SEE ALSO: ARTIFICIAL LIMB, JOINTED ARTIFICIAL LIMB, HIP
REPLACEMENT, ARTIFICIAL HEART, POWERED EXOSKELETON

⬅ *A laboratory technician addresses the complex mechanics of an artificial forearm with hand.*

Controlled Drug Delivery (1968)

Zaffaroni pioneers slow-release medications.

Anyone taking a medication only once a day should thank Alejandro Zaffaroni (1923–2014). It was his pioneering attitude that brought about slow-release medications, including drugs that are absorbed through the skin and five-year reversible birth control.

In 1949 Zaffaroni received a PhD from the University of Rochester in New York after his thesis on quantitative analysis of natural steroids. His work had taught him that organisms generally released steroids in small amounts over relatively long periods of time. This was in stark contrast to most medications of the 1940s, which involved relatively large doses in pill-like forms.

In 1968 he founded Alza (an acronym of his own name) to pursue his concept of improving medical treatment through controlled drug delivery. He had seen the side effects that many medicines produced when they were sent to the bloodstream all in one massive dose, and knew there had to be a better way. By studying endocrinology, where glands deliver very small amounts of hormones that have a powerful effect, he was convinced that delivering drugs in small, steady doses would be more appropriate.

Today's controlled drug-release mechanisms include implanted pumps for the delivery of insulin or pain medications, and transdermal patches, which release medicine slowly into the skin. The latter are used for certain pain medications, motion sickness medicine, and birth control, as well as nicotine withdrawal where success depends on a gradual reduction of the dose. Controlled drug delivery not only allows steady, predictable drug levels but also enhances the duration of effect of short-acting medications and decreases their side effects. **SS**

SEE ALSO: HYPODERMIC SYRINGE, DISSOLVABLE PILL, ASPIRIN,
ANTIVIRAL DRUGS, TRANSDERMAL PATCH

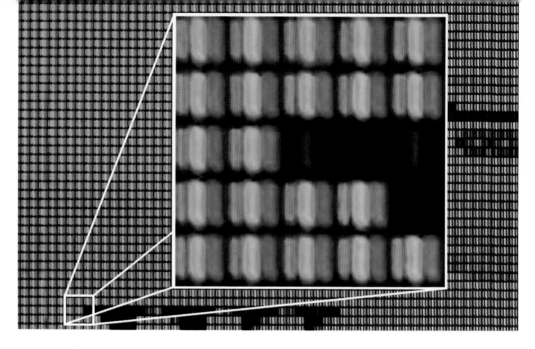

Liquid Crystal Display (LCD) (1968)

Heilmeier exploits the two melting points of crystals.

LCDs (liquid crystal displays) are used in televisions, laptop computers, and many portable electronic devices. The properties of liquid crystals were first discovered in 1888 by Friedrich Reinitzer. He was measuring the melting point of a cholesterol-based substance and noticed that it had two melting points: it melted at 293°F (145°C) to give a cloudy, gluelike liquid, then again at 352°F (178°C) to give a clear liquid.

Otto Lehmann, an expert in crystal optics, studied these phases and found that the cloudy liquid had similar properties to the solid crystal. In the solid crystal, the molecules are lined up neatly and in parallel. In the cloudy liquid, the molecules can move around. However they tend to line up like in the solid crystal, reflecting light to appear cloudy. Lehmann named the liquid *fliessende Kristalle*, or liquid crystal.

In 1968 George Heilmeier (1936–2014) led a group at the Radio Corporation of America to develop the first LCD. They used the dynamic scattering method, in which an electrical charge is applied to a liquid crystal, causing the molecules to rearrange and scatter light.

The LCD was made up of a liquid crystal substance sandwiched between two polarized filters. When electricity is applied to the LCD, the electric field causes the molecules to twist. Light passes through the first filter, rotates around the liquid crystal, and passes through the second filter to produce a light spot on the reflecting screen. When no electric field is present, the molecules cannot twist, and light does not pass through, resulting in a dark spot on the screen. How much light can pass through depends on the degree to which the molecules are twisted. **RB**

SEE ALSO: DIGITAL CLOCK, ELECTRONIC PAPER, CATHODE RAY TUBE, TELEVISION

⭱ *The red, green, and blue elements of an LCD monitor are activated individually to produce moving images.*

Random Access Memory (RAM) (1968)

Dennard combines a transistor with a capacitor in a revolutionary memory cell.

RAM, or Random Access Memory, the short-term, high-speed "working" memory of a computer, has existed since the invention of magnetic core memory in 1949. Modern RAM, though, owes its invention to Texan Robert Dennard (b. 1932).

In 1966 Dennard was working at IBM's Thomas J. Watson Research Center. IBM knew that magnetic core memory was too bulky, power-hungry, and slow, and that transistors would be the answer to replacing it. They had reduced the problem of storing a single bit of memory to a cell that used only six transistors. Added to a silicon chip, this cell was already tiny compared to a magnetic core. Dennard, though, simplified the memory cell even more, to a single transistor and a capacitor, a component that can hold an electric charge. The memory was stored as charge on the capacitor, and the transistor was used to read and write it. Capacitors "leak" charge, though, so the memory had to be continuously refreshed, many times a second. Because of this constant forgetting and refreshing, Dennard's system is called "dynamic" RAM, or DRAM.

Despite its need to be refreshed, DRAM had a world-beating advantage. With only two components, which could be placed side by side in the thousands on a single silicon chip, it was the smallest memory ever made. The computer industry quickly took advantage of Dennard's invention, and fledgling company Intel released the first commercial DRAM chip in 1970. Magnetic core memory became the technology of yesteryear almost as soon as Intel began to ship its new chip. **MG**

SEE ALSO: DIGITAL ELECTRONIC COMPUTER, MAGNETIC CORE MEMORY, C PROGRAMMING LANGUAGE, FLASH MEMORY

⬆ *A false-color scanning electron micrograph shows part of a DRAM integrated circuit or silicon chip.*

Despite the fact that most of us had never heard of the Internet before the 1990s, it was first conceived decades earlier. Similarly, and surprisingly, the first cellular mobile phones were around in 1970—the year before the first e-mail was sent. In fact, many other products such as personal computers and laptops were developed long before their launch as consumer products. And in an era of permanently manned space stations, the things yet to come are the most exciting prospect of all.

⬅ A cluster of servers pave the road toward tomorrow's inventions.

The INTERNET AGE

1969 to Present

Internet (1969)

Advanced Research Project Agency (ARPA) develops the first computer network.

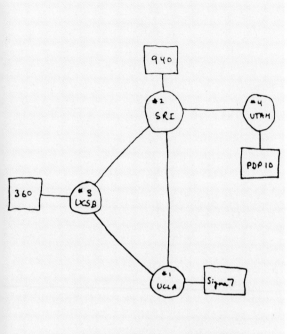

THE ARPA NETWORK

DEC 1969

4 NODES

"Getting information off the Internet is like taking a drink from a fire hydrant."

Mitchell Kapor, software designer

In 1963, the Advanced Research Project Agency (ARPA) unit, set up by the U.S. Defense Department, began to build a computer network. Driven by fear of the Soviet nuclear threat, it aimed to link computers at different locations, so researchers could share data electronically without having fixed routes between them, making the system less vulnerable to attacks—even nuclear ones.

Data was converted into telephone signals using a modem (**mo**dulator-**dem**odulator), developed at AT&T in the late 1950s. In the 1960s, key advances were made, including "packet switching"—the system of packaging, labeling, and routing data that enables it to be delivered across the network between machines. Paul Baran (1926–2011) proposed this system, which broke each message down into tiny chunks. These would be fired into the network, which would then route ("switch") the various pieces to the desired destination. So, if chunks of a message were traveling from Seattle to New York via Dallas, but Dallas suddenly went offline, the network would automatically route via Denver instead. Different parts—or "packets"—of a message would go by different routes, before being reassembled back into the original message at their destination, even if they arrived in the wrong order. Baran published his concept in 1964, and five years later the new network—called ARPANET—went live.

As the threat of nuclear war receded in the early 1970s, ARPANET was renamed the Internet and effectively opened to all users. Since then, the development of e-mail, the creation of the World Wide Web, and browser technology has enabled the Internet to become a rich communications facility. **SC**

SEE ALSO: CABLE MODEM, E-MAIL, INTERNET PROTOCOL (TCP/IP), SEARCH ENGINE, WEBCAM, WORLD WIDE WEB

◩ *Sketching out the future: A diagram of the ARPANET created by one of its key architects—Larry Roberts.*

Balloon Catheter (1969)

Fogarty invents a noninvasive clot-busting device.

Thomas Fogarty (b. 1934) was working as a scrub technician at the Good Samaritan Hospital, Cincinnati, Ohio, when he noticed the difficulty surgeons had in removing blood clots that formed in arteries and veins. The operation, which often took nine to twelve hours to perform, necessitated opening up the entire length of the vessel and often resulted in the patient dying or having their limb amputated.

Fogarty devised a scheme that could be used to overcome the need for invasive surgery. It involved using a urethral catheter, which is flexible and strong enough to be pushed through a blood vessel and penetrate the blood clot.

Working in his attic, Fogarty had the further inspiration to use the fly-tying skills he had learned as a fisherman to attach the "fingertip" of a latex glove to the catheter, which could then be inflated with saline once it was past the clot. The idea is that the balloon expands to the size of the artery and is then pulled back out, bringing the clot with it.

In 1961, Fogarty's balloon embolectomy catheter— named for the clot-removal procedure—was used for the first time on a human patient. A small incision was made, and the catheter was threaded up through the patient's blocked artery. When inflated and pulled back out, it did indeed bring the clot out with it.

Today Fogarty's balloon catheter (patented in 1969) is still the most widely used technique for blood clot removal. The technology has also been extended for use in angioplasty, where balloons are inflated to widen narrowings in the heart arteries that cause symptoms of angina. **JF**

SEE ALSO: DISPOSABLE CATHETER, INTRAVASCULAR STENT

⤴ *Manufactured in c. 1990 by USCI and Bard, these instruments reduce the risks of vascular surgery.*

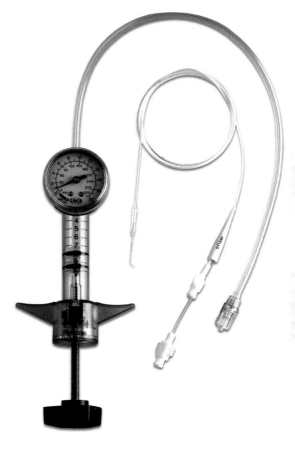

"Fogarty's ... procedure was the first successful example of 'less invasive'... vascular surgery."

Inventor of the Week archive

Charge-coupled Device (CCD) (1969)

Boyle and Smith pioneer a form of memory.

Charge-coupled device (CCD) technology is the bedrock of digital cameras and video but it started out as a new form of memory. One day in 1969 William Boyle (1924–2011) and George Smith (b. 1930) were brainstorming at Bell Labs in New Jersey and decided to play around with merging two of the new technologies that were being worked on—semiconductor bubble memory and the video phone.

The pair worked on a new principle of handling small pockets of electrical charge on a silicon chip that was similar to the work being done on moving microscopic "bubbles" of magnetism around on various materials. They called their invention the

> *"Well we've got a new device here. It's not a transistor, it's something different."*
>
> William Boyle, coinventor

charge-coupled device. It soon became clear that the small packets of charge on the CCD could be put there using the photoelectric effect, which meant that incoming photons could be "captured." By the end of 1969 Boyle and Smith were able to use their new device to take electronic images at Bell Labs.

Various companies began developing the CCD, and in 1974 the first commercial device was released by Fairchild Semiconductor, capable of taking an image that measured 100 x 100 pixels. Today CCDs are widespread in astronomical telescopes, scanners, and bar-code readers, as well as robotic vision and, of course, in everyday digital cameras. **DHk**

SEE ALSO: BAR CODE, COMPUTER SCANNER, DIGITAL CAMERA, HUBBLE SPACE TELESCOPE, SURGICAL ROBOT

Far-ultraviolet Camera (1969)

Carruthers enables observations of the stars.

Ultraviolet (UV) light is found beyond the violet end of our visible spectrum of light, toward the X-rays. It is given off by the sun and is harmful to living things, which is why we need to wear sunscreen when we go out in the sun. Fortunately for us, most of it is absorbed by Earth's atmosphere.

Ultraviolet light from places other than the sun can tell us a great deal about the universe—specifically about stars that are between twice and ten times the temperature of the sun. Because Earth's atmosphere gets in the way, astronomers find it hard to see them. By the mid-1960s, however, humans were journeying beyond Earth's atmosphere…

Normal cameras pick up only light around the visible spectrum, but on November 11, 1969, astrophysicist Dr. George Carruthers (1939–2020) was granted a patent for an "Image converter for detecting electromagnetic radiation especially in short wave lengths." The far-ultraviolet camera was a 3-inch (7.62 cm), 48.5-pound (22 kg), gold-plated apparatus that could see stars that are a hundred times fainter than those that can be seen with the human eye. The camera was sent up with the *Apollo 16* mission in 1972 and placed on the Moon's surface, allowing researchers to examine Earth's atmosphere for concentrations of pollutants. It recorded nearly 200 images, giving astronomers data on over 550 stars, nebulae, and galaxies, as well as providing new views of Earth. The camera looked into the ionosphere (the highest part of our atmosphere) and gave us some of our earliest solid data on the concentrations of man-made pollutants. **DK**

SEE ALSO: COLOR PHOTOGRAPHY, INFRARED PHOTOGRAPHY, PHOTOGRAPHIC FILM, PHOTOGRAPHY, SPACE OBSERVATORY

➡ *This image, taken by a NASA ultraviolet detector, shows the double-star system Z Cam (white object, center).*

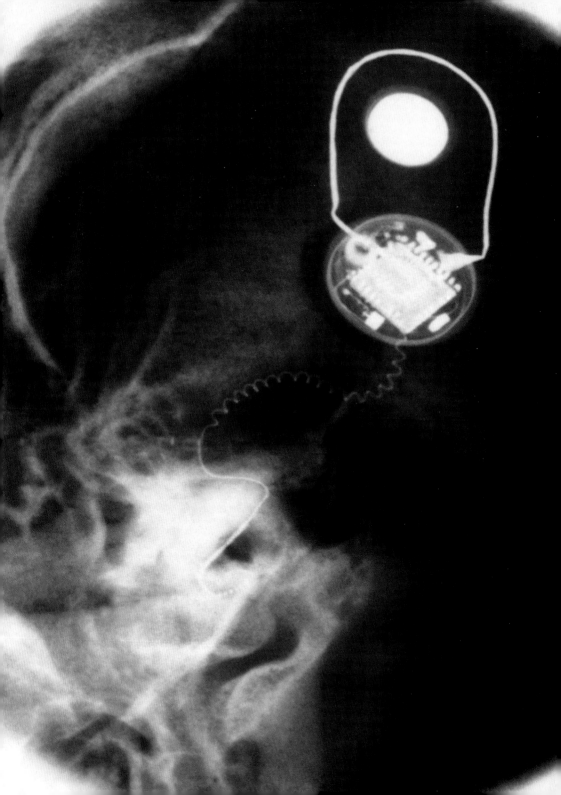

Cochlear Implant (1969)

House's implant stimulates auditory nerves.

A cochlear implant is a surgically implanted electronic device that provides a sense of sound to a person who is profoundly deaf. It works by directly stimulating the auditory (hearing) nerves with electrical impulses.

American William House (1923–2012) is credited with being the first surgeon to implant a cochlear-type device. In 1957 House saw an article by two French surgeons who had inserted an electrode into the auditory nerve of a deaf man and shown that he could perceive sounds when the nerve was stimulated. In 1961 House placed cochlear implants in three patients, who gained some benefits. After research into the best positioning of the electrodes, House created the first wearable implant in 1969.

Despite hostile criticism and fears that electrical stimulation of the cochlea might destroy brain tissue or spread infections, by December 1984 cochlear implants had the U.S. Food and Drug Administration stamp of approval, and were no longer deemed experimental. In the device, sound is picked up by a microphone worn near the ear and transmitted to a speech processor worn on the body. The sound is then analyzed and converted into electrical signals that are transmitted to a surgically implanted receiver behind the ear. The receiver sends the signal through an electrode array (wire) into the inner ear, where the electrical impulses are transmitted to the brain.

Throughout the 1990s further improvements were made, particularly the miniaturization of speech processors that could be incorporated into hearing-aid-like devices. Finally, in 2005 the first fully implantable devices were developed. **JF**

SEE ALSO: SPEECH RECOGNITION, SPEECH SYNTHESIS, VISUAL PROSTHETIC (BIONIC EYE)

◄ *This X-ray shows a cochlear implant—an electronic device to restore hearing—in place in the inner ear.*

Smoke Detector (1969)

Pearsall makes a battery-powered alarm.

Installing at least one smoke detector in your house is estimated to halve the chances of a fatal fire, which means that thousands of lives have been saved worldwide since home devices were introduced in the late 1960s.

The forerunner of the modern detector was invented by the British electrical engineer George Darby in 1902. His device, which detected heat rather than smoke, consisted of two electrical plates with a wedge of butter between them. As the room temperature rose, the butter melted, causing the two plates to fall onto each other triggering the alarm, and presumably dripping butter everywhere.

The most common smoke detectors now use an ionization chamber, a device developed by the Swiss physicist Ernst Meili in 1939 to detect poisonous gases in mines, although not specifically smoke. In the ionization chamber, a radioactive material produces ions (electrically charged atoms). In the presence of smoke, the flow of these ions between two electrodes is disturbed, triggering the alarm.

Although ionization chamber smoke detectors were available in the 1950s, they were expensive and used only in factories and other large buildings. The American Duane Pearsall (1922–2010) is credited with the first practical home smoke alarm in 1967, achieved by adding a battery to power a more compact ionization chamber. Two other Americans, Kenneth House and Randolph Smith, may have been the first to patent the battery-powered smoke detector in 1969. Even the National Aeronautics and Space Administration (NASA) is thought to have created the smoke detector as a spin-off from the space program. In fact NASA developed a type of detector for the Skylab project but did not invent the detector itself. **RBd**

SEE ALSO: AUTOMATIC FIRE SPRINKLER, CONTROLLED FIRE, ELECTRIC FIRE ALARM, FIRE EXTINGUISHER

Digital Watch (1970)

Hamilton Watch Company develops a space-age timepiece.

The first wrist-worn timepiece to tell the time digitally was the Hamilton Watch Company's "Pulsar." This 18-carat gold-cased device used a red LED display to tell the user what time it was in clean, crisp, twentieth-century digits at the push of a simple button, and retailed for a cool $2,100. Teething problems meant the Pulsar did not become commercially available until 1972, but when it was released, it caused many to think that the end had come for conventional dial face watches with mechanical movements. Hamilton claimed that their inspiration for developing a digital watch was the futuristic digital clock that they had created for the 1968 film *2001: A Space Odyssey*.

The only problem with the Pulsar was the hefty accompanying price tag—although many would argue that $2,100 was cheap for the opportunity to look like James Bond wearing a swanky gold digital watch. The answer for those who did not want to pay so much for a watch came from Texas Instruments, who introduced a plastic-strapped version that retailed for just $20 in 1975 and soon after dropped to $10. This spelled the end for Hamilton and led to them becoming a subsidiary of Seiko.

Developments in digital watches continued over the next three decades and included the use of liquid crystal displays to replace light-emitting diodes (which could not always be left on due to their high level of power consumption) in the early 1970s. Throughout the 1980s many amazing innovations were incorporated into the digital display, including thermometers, language translators, calculators, and even miniature televisions. **BG**

"You guys have really come up with somethin'..."

Dr. Flood, *2001: A Space Odyssey* (1968)

SEE ALSO: ATOMIC CLOCK, PENDULUM CLOCK, QUARTZ WATCH, SHADOW CLOCK, TOWER CLOCK, WATER CLOCK, DIGITAL CLOCK

◪ *A Tiffany and Co. version of the Hamilton "Pulsar," the first electronic digital watch, produced in 1970.*

Cellular Mobile Phone (1970)

Joel opens up a portable communication system.

In 1970, when he was working as an electrical engineer at Bell Labs, in Murray Hill, New Jersey, Amos Joel (1918–2008) came up with an idea that worked so well that many of us use it today without noticing that anything has happened at all. Joel had invented the idea of the cellular mobile phone.

Cell phones existed prior to 1970, but they had problems. Since each call was made on a single channel, the number of simultaneous calls was limited to the number of available channels. Additionally, a cell phone user could not leave the base station area of coverage in which the call was initiated or the connection to the network would be lost.

Joel's cellular mobile communication system proposed providing phone service to a geographic area by dividing it up into many small, low-powered base stations or "cells." This network of cells could deal with a greater number of simultaneous calls by allocating a radio channel to a call in one cell and reusing the same radio channel in a number of other cells separated by enough distance to prevent interference. When a user traveled from the service area of one cell to another, the user's phone call would not be disrupted as each cell would "hand-off" to the other, switching cellular base stations and radio channels without anyone noticing.

The key ingredient? A microprocessor that could make all the decisions, identify the phone, find cellular base stations, and control the connection as users moved through the network. Today's cell phones support a number of additional services, such as text messaging, access to the Internet, and camera. **AKo**

SEE ALSO: DIGITAL CELL PHONE, SMARTPHONE, WALKIE-TALKIE, WIRELESS COMMUNICATION

⬈ *Motorola produced the first commercially approved cellular phone in 1983; it weighed 28 ounces (794 g).*

"What got me into this business was the curiosity about how the dial telephone system worked."

Amos Joel

Holographic Memory (1970)

Caulfield, Soref, and McMahon develop a new form of data storage.

While holography is commonly associated with three-dimensional images, some of the most important developments have been in using holograms to store and retrieve information in optical form.

Holographic memory is created by interference patterns between two sources of light, called the reference and signal, with lasers being the most precise. Instead of recording an image of an object, as with normal holograms, a set of data can be captured instead. Binary data in the form of 1s and 0s can be represented by a pattern of light and dark.

Holographic memory is still very much in development, but the pioneering work was done almost forty years ago. In 1970, Henry Caulfield (1873–1966)—then principal scientist at the Sperry Rand Research Center in the United States—along with electrical engineers Richard Soref and Donald McMahon filed the first patent for "holographic data storage." In it they proposed holography as a means of

recording and playing back information. In principle, a hologram can store as much as four gigabits per cubic millimeter, although practical limitations make that figure lower. In addition to potentially huge capacity, holographic memory promises almost instant retrieval of the whole data set at once.

In 2006 the company InPhase Technologies set a new record for data storage using holographic techniques, with 515 gigabits per square inch. The company has since launched the world's first holographic storage product range. Thanks to the principles established by Caulfield and his colleagues, DVDs and hard disks look set to become obsolete. **AC**

SEE ALSO: HOLOGRAPHY, LASER

⬆ *Tapestry™ is the latest storage device from InPhase; at 300 GB it meets the commercial need for high capacity.*

Pocket Calculator (1970)

Busicom Corporation develops the first truly portable calculator.

Today, when nearly every device is available in a portable, pocket-sized version, it is hard to imagine a time when a simple, handheld calculator was the stuff of science fiction. In the early 1960s, calculators were the size of modern-day desktop computers but not nearly as powerful. Personal calculators were nonexistent, and workplace desktop calculators were limited to four simple arithmetic functions. In the workplace, complex math was left to humans.

In 1965, mathematicians and engineers at Texas Instruments (TI) set to work shrinking calculators, using integrated circuit technology that had been invented in-house. By 1967, they had a battery-powered prototype capable of the four simple arithmetic functions on six-digit numbers. Dubbed "Cal-Tech," the calculator was the size of a large paperback book—4.25 x 6.15 x 1.75 inches (11 x 16 x 4 cm)—and weighed nearly 3 pounds (1.3 kg)—hardly a "pocket calculator" and not commercially available or viable. TI partnered with Canon and by 1970 they had marketed the Pocketronic®, which was only slightly smaller.

In late 1970 the first truly pocket-sized calculator was introduced by Japan's Busicom. The Busicom LE-120A® was about half the size of the Pocketronic®, the first with a light-emitting diode (LED) display, and the first handheld to use a special integrated circuit specifically designed for calculators. The 1970s saw a calculator war, with a variety of electronics companies fighting for market supremacy. Texas Instruments would win, dominating the market and becoming nearly synonymous with the word "calculator." **RBk**

SEE ALSO: **ABACUS, DIGITAL ELECTRONIC COMPUTER, MECHANICAL CALCULATOR, TALLY STICK**

⬆ *Busicom's "Handy-Le" went on sale in 1971; it was the first pocket-sized calculator and the first to use LED display.*

Computer Scanner (1970)

Bell Labs develops an imaging technique.

Computer scanners (or "optical scanners") take images and turn them into signals that a computer can understand. Although image scanners did exist prior to the invention of the charge-coupled device (CCD), most modern scanners now capture images in this way.

A scanner houses an array of CCDs for capturing light. When a photograph or book is scanned, a lamp illuminates the image, which is then reflected by a series of mirrors and focused through a lens to reach the CCD array. The CCDs create an analog signal, which is then converted to a digital signal that can be read by a computer and stored. More recently, manufacturers have started to use a different technology—the contact image sensor (CIS) method, which reflects

---

"Man lives by images. They lean at us from the world's wall, and Time's."

Robert Penn Warren, poet

---

light from LEDs (light-emitting diodes), instead of from a normal lamp, and receives it via a sensor. Despite producing lower quality imagery, CIS technology is cheaper and is used in portable scanners.

Scanning has moved up a dimension since the 1970s. Artists and architects can now use sophisticated scanners, some incorporating the latest laser technology, to generate 3D computer models of objects that they want to integrate into their designs. Film production companies also use 3D scanners to help them create special effects from computer-generated imagery, taking their inspiration from nature's physical forms. **HB**

SEE ALSO: CHARGE-COUPLED DEVICE (CCD), OPTICAL CHARACTER RECOGNITION

Car Alarm (1970)

Stereo company makes a mechanical alarm.

The car alarm is far from being a universally popular invention: a British television poll listed it as one of the United Kingdom's ten least favorite inventions (just behind the cell phone), and as mayor of New York Rudy Giuliani commented, "Noise pollution is a major problem. Even in a city as exciting as New York, people should be able to sleep without being disturbed by car alarms . . . " There have been petitions to ban the devices, and there have even been claims that car alarms, far from helping reduce instances of car theft, actually make the problem worse!

The first recorded instance of car theft occurred in 1896. Cars are valuable and simple to "disguise," so thieves regard them as easy targets. Reputedly the product of car stereo manufacturers in California in 1970, the car alarm seems to offer the perfect solution; an automated alarm that will emit a loud blaring noise when activated, scaring potential thieves away and alerting people to the fact that a car is being stolen. The problem with the device has traditionally been the sensitivity of the activation mechanism.

Modern car alarms consist of an array of sensors, spread around the car, wired up to activate the alarm when the door opens, or when the car is moved suddenly. The frequent accidental activation of these motion sensors has led people to ignore the alarm, like a mechanical version of the boy who cried wolf.

In place of the typical car alarm, modern stolen vehicle recovery systems, such as LoJack, are becoming more prevalent. Once a car has been reported stolen, a silent wireless device hidden in the vehicle relays a radio signal to the police, who, using corresponding computer programs, can then track and recover the car (often apprehending the thief in the meantime). **BG**

SEE ALSO: BURGLAR ALARM, ELECTRIC FIRE ALARM, MOTORCAR, COMBINATION LOCK

Optical Tweezers (1970)

Ashkin uses lasers to trap and manipulate tiny particles.

Radiation pressure is the force exerted by a beam of light when it is reflected from or absorbed by a body. Normal light beams are wide and brutal, but a focused laser beam can apply extremely delicate forces.

If a small object has a mass less than about 1 gram and is dielectric, it can diffract a laser beam in such a way that the difference between the momentum of the radiation entering the object and that leaving it can be made to produce a force that traps the object in a specific position. A single laser beam can thus act like a pair of tweezers and individual atoms, molecules, and biological cells can be micro-manipulated using the beam. In 1970, Arthur Ashkin (1922–2020) of Bell Laboratories detected optical scattering and showed that particles can be trapped in this way. It took another fifteen years before he created a workable process, derived from the laser-cooling techniques achieved by Steven Chu of Bell Laboratories. Ashkin was trapping very small biological bodies by 1987.

When investigating strands of DNA, bacteria, or viruses, these have to be biochemically attached to a dielectric sphere of glass or polystyrene that is being positioned by the "tweezers." Care must be taken, in choosing the wavelength and power of the laser light, so that the specimen does not heat up or become damaged by the radiation. The positioning of the trapped particle is usually controlled using acousto/electro-optical systems that are computer-controlled.

Minute details and movements on the nanometer scale can be studied. The whole system is, however, extremely complicated and expensive and needs skillful operation. **DH**

SEE ALSO: LASER, LASER COOLING OF ATOMS, MAGNETIC ATOM TRAP

⬈ *Optical tweezers handle the cell nuclei of transgenic plants at France's National Center for Scientific Research.*

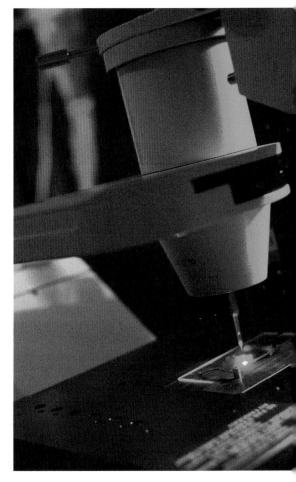

"[The optical trap] turned out to be a pretty important discovery. It led to Steve's [Chu] Nobel Prize . . . "

Arthur Ashkin

Radio Frequency Identification (1971)

Cardullo invents an identifying device.

There is a good chance you have never heard of Radio Frequency Identification (RFID), but there is every likelihood that you have used it on more than one occasion. Whether it is through a name tag that allows a person to enter a secure part of a building, or a device that monitors library books, RFID has become very much a part of modern society.

The concept was first employed in military aircraft and had a single purpose—to broadcast a signal indicating if it was a friendly aircraft or an enemy. A similar rudimentary device was also in use in the security industry, to determine if an object being protected was indeed where it was supposed to be. Mario Cardullo was aware of both systems in 1969, when a chance meeting with an engineer lamenting the difficulties of using bar codes to track railroads caused him to advance RFID dramatically.

What Cardullo actually conceptualized was a relatively simple device. He conceived of an RFID that would consist of some form of tag that could be read electromagnetically by a reader. One of the earliest applications was in an unmanned toll booth for the New York Port Authority. At about the same time, a competing inventor, Charles Walton, figured out how to make a door lock that could be opened only when a certain radio frequency was broadcast—in other words a door opened by a key card.

Current applications have also expanded to fill a significant retail need. Although the bar code remains the gold standard for identification on inventory at present, RFID devices are in development and may supplant them, given enough time. RFIDs are already being implanted into household pets as a way to track them when lost. **BMcC**

SEE ALSO: BARCODE, LASER TRACK AND TRACE SYSTEM

Computer-aided Manufacturing (1971)

Bézier's program works the production line.

People often voice disappointment that robots do not quite have the pervading influence on modern-day life that was once envisaged. They are not making us breakfast, cleaning our cars, or putting our kids to bed. In fact, it is easy to think that they have all but forsaken us. This is not the case, however—robots are constantly working for us. They are just a lot more humble then we ever expected them to be, since the work they do is all behind the scenes.

French engineer Pierre Étienne Bézier (1910–1999) first developed the UNISURF program while working at Renault. UNISURF was created to use computers to aid with the design and manufacture of automobile bodies and other similar systems. Computer-aided manufacturing (CAM) makes use of software tools to communicate with the machines in charge of producing components.

People were initially concerned that CAM would do away with the need for skilled engineers and technicians. In fact, the use of CAM has arguably improved their position, since they can now get on with the more desirous work and the tasks they have chosen to do, rather than getting bogged down with the repetitive manual labor of recreating part after part after part.

Although many different varieties of CAM software now exist, Bézier left a lasting mark on the industry, not only for pioneering the technology but also for putting his name to the Bézier curve and the Bézier surface—both of which are used extensively in computer graphics and design modeling. **CL**

SEE ALSO: AUTOMATON, BIPEDAL ROBOT, INDUSTRIAL ROBOT, SURGICAL ROBOT

➡ *This computer display shows a simulation of manufacturing using industrial robots.*

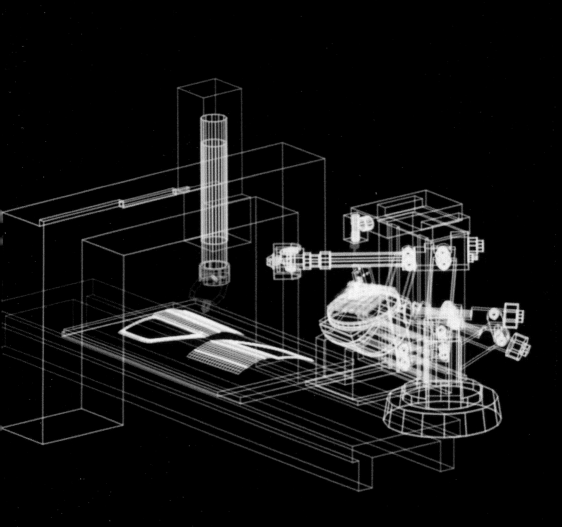

Computer Laser Printer (1971)

Starkweather invents a process for printing high-quality text and graphics.

In 1969, while everyone else was experimenting with LSD, Gary Starkweather was hard at work at Xerox's research facility in Webster, New York, experimenting with the laser printer. Two years later, Starkweather, who has since won an Academy Award for his work with Pixar on film input scanning, constructed the first working laser printer system.

The laser printer relies on the concept that—like magnets—opposites attract. The laser printing process begins with the laser applying the desired pattern on the revolving printer drum. The beam reverses the drum's positive charge, making certain areas negatively charged. These negatively charged areas attract toner, a positively charged powder.

After a wire applies a strong negative charge to the printing paper, a rolling belt feeds the paper past the drum. The toner is pulled from the drum by the paper. The paper is then zapped to neutralize its charge. Before the paper can be ejected, it needs to go through the fuser. This is a pair of heated rollers that melt the toner powder, thus fusing it onto the paper. The only thing that saves the paper from incineration at the hands of these rollers is the speed at which it flies through them.

Xerox failed to see the potential of the laser printer, however, discouraging Starkweather from pursuing his invention. Even when they finally began producing commercial models, Xerox missed out on the business prospects abounding in the sale of toner and paper. Hewlett-Packard began selling the world's first personal laser printer in 1980, leaving Xerox in their toner dust. **LW**

SEE ALSO: INKJET PRINTER, LASER, PRINTING PRESS, PRINTING TELEGRAPH (TELEPRINTER), ROTARY PRINTING PRESS

⭱ *The interior of a Xerox laser printer from 1986; laser printers were not commonly used until the 1990s.*

Floppy Disk (1971)

Noble and IBM develop flexible data storage.

In 1967, IBM was looking for a better way of sending software to its customers. Their popular System/370 mainframe computers "booted up" from big, heavy magnetic tapes, which were slow and expensive to ship. Engineer David L. Noble (1918–2004) tried all sorts of improvement schemes, from better tape systems to vinyl records, just like those used for music.

None of them were right for the job, so Noble proposed a new system based on a flexible disk of magnetic material. Developed by IBM over the next few years and finally released commercially in 1971, IBM's 8-inch (20 cm) "floppy" disk was made from flexible plastic. After a hunt for a package for mailing the new disks, the engineers had the idea of making a protective envelope part of the design of the disk itself. The disk was sandwiched into a square jacket that included a fabric liner—a built-in cleaning cloth—and the final design of all future floppy disks was born.

These cheap, light disks really caught on and over the years they shrank from 8 inches (20 cm) to 5.25 inches (13.3 cm), then to the 3.5-inch (8.89 cm) standard that you may still see occasionally today. Untold billions of floppy disks were manufactured over the next thirty years.

The floppy disk is now almost obsolete, replaced by networks, flash memory, or optical media such as DVDs. These innovations can store much more information than their predecessors—to match the storage capacity of a single Blu-ray DVD, you would need 600,000 of IBM's original 80 kilobyte, 8-inch (20 cm) floppies. **MG**

SEE ALSO: FLASH MEMORY, HIGH-DENSITY COMPUTER STORAGE, MAGNETIC CORE MEMORY

⊞ *Floppy disks measuring 5.25 inches (13.3 cm) came in when IBM introduced personal computers in 1981.*

Genetically Modified Organisms (1971)

Chakrabarty bypasses an evolutionary process.

Humans have sought to improve the qualities of other species since time immemorial. Selective breeding led to faster horses, hardier crops, and the diverse pool of dog breeds. With the advent of molecular science, a process that had hitherto relied upon chance and breeding for many generations could be achieved quickly and by design.

It works like this. First, identify a trait that you would like to introduce into the organism you are interested in. Next, find a gene or set of genes that code for this trait in another organism. With favorable circumstances, you should be able to introduce the genes into the recipient organism where they will be passed on to future generations.

An early pioneer of the technology was microbiologist Ananda Chakrabarty (1938–2020). He worked with bacterial plasmids—small rings of DNA that can be transferred between bacteria. Chakrabarty wanted to produce bacteria that could degrade crude oil to help clean up spills. Four oil-metabolizing *Pseudomonas* species were known, but no combination would yield a helpful result. By transferring plasmids between species, Chakrabarty united four genes in one species, producing a new bacteria with much improved oil metabolism. This became the first organism to be subject to a patent.

Today, genetic modification of crops, animals, and microorganisms is a flourishing and diverse field of research. Getting bacteria to produce insulin, as first marketed by Genentech in 1982, is one of the most successful uses of this technology. However, the ability to manipulate life at the molecular level raises many objections, and applications such as pest-resistant crops have received dogged opposition. **MB**

SEE ALSO: GENE THERAPY, STEM CELL THERAPY

Vacuum Forming (1971)

Japanese scientists form plastic parts.

Yogurt pots, boat hulls, dashboards, and children's lunch boxes all have something in common. Not only are they made of the same material, but they are also manufactured through vacuum forming—a process that dates back to 1971 when Japanese scientists devised a way to form complicated shapes out of plastic. The researchers credited are Yoshimasa Kubo and Nataka Kunii.

The process works by first producing a former mold in the shape of the desired final product, constructed from wood, structural foam, or aluminum. This is placed within a chamber that can induce a vacuum. A sheet of plastic (typically high-impact polystyrene

> *"I love Hollywood. They're beautiful. Everybody's plastic, but I love plastic. I want to be plastic."*
>
> Andy Warhol, U.S. artist

sheeting or acrylic) is then placed above the former in the upper half of the chamber where it is heated from above. Once the plastic has become flexible enough to be molded, the heater is switched off and the former is raised into the plastic from below.

With the components in place, the vacuum is then activated beneath the plastic, sucking the sheet down to form a tight fit with the former. The vacuum is maintained until the plastic has cooled. At this point, the sheet of plastic is removed, and its edges are trimmed to leave a finished product in the correct shape. The process is usually restricted to forming plastic parts that are quite shallow in depth. **CB**

SEE ALSO: PLASTIC BOTTLE (PET), POLYSTYRENE, TUPPERWARE®, VACUUM FLASK, VACUUM PUMP

Food Processor

(1971)

Verdan takes the labor out of slicing, chopping, and mixing.

The creation of what is perhaps one of the most popular modern kitchen accessories—the food processor—was actually the brainchild of a salesman rather than a scientist. Pierre Verdan was a French catering services salesman who saw the amount of time his clients spent chopping, slicing, and grating in their kitchens and spotted a niche in the market.

Verdan set to work to develop a machine that could take on the laborious task of preparing vegetables and save the kitchen staff significant time. He realized that his idea could save catering businesses hundreds of hours a year, as well as giving kitchen staff a method of creating smaller and more uniformly sliced ingredients that were more pleasing, aesthetically and texturally.

Verdan began building his prototype and, by all accounts, it was a pretty basic design. His device consisted of a large plastic bowl with a fixed revolving blade, attached to a shaft, in the base. The blade was able to rotate rapidly around the shaft and slice the contents evenly and swiftly. The cover to the bowl had a feed tube through which foods are added. This basic design was later developed to include a rather faster and more powerful motored blade that was even more adept and speedy. Food processors can chop, slice, dice, shred, and puree most foods, as well as knead dough. Verdan also designed a smaller, handheld version of his device, which has been extremely popular due to its ease of use and convenient size.

Other manufacturers developed versions of Verdan's processor, including Cuisinart in the United States. The invention has revolutionized food preparation, not only in large commercial kitchens but also in the home. **KB**

SEE ALSO: BLENDER, FOOD MIXER, PRESSURE COOKER

⬀ *The original Magimix processor, made by Robot Coupe, has changed little over time and remains popular.*

"Although a food processor is not an essential piece of equipment, it saves you time and energy."

Delia Smith, British cookbook writer

Computerized Telephone Exchange (1971)

Hoover advances telecommunications.

When telephonic communication systems were first invented, and few people actually owned a telephone with number buttons. Telephone calls would be connected via a human operator. You would lift a receiver, the switchboard operator would ask you for the number you wished to call, and he or she would plug a wire into the requisite part of the board to have you connected.

However, as telephones became more popular, it became increasingly impractical to have humans connecting the calls by hand and so electromechanical switchboards were invented. Soon communications companies found that even this was not enough—there was so much telephonic traffic to deal with that they started to overload and seize up.

It was Erna Schneider Hoover (b. 1926), a computer programmer with a PhD in mathematics and a specialist in symbolic logic, who struck upon the solution when she was working for Bell Laboratories in New Jersey. She created a computerized switching system that monitored the frequency of incoming telephone calls at various times and rearranged the call acceptance rate at peak times to prevent overloading. This software was patented in 1971, and is renowned for being one of the very first software patents ever issued.

Not only a telecommunications pioneer, Dr. Hoover also successfully juggled her family life with a high-flying career. The story goes that she drew the very first sketches for the system when she was in hospital after having given birth to the first of her three children. It was the success of this software that led her to take up the first position as a female supervisor in a Bell Laboratories technical department. **CL**

SEE ALSO: TELEPHONE, AUTOMATIC TELEPHONE EXCHANGE, CELLULAR MOBILE PHONE

Magnetic Resonance Imaging (1971)

Damadian investigates living cells with scans.

Although Raymond Vahan Damadian (b. 1936) is credited with the idea of turning to nuclear magnetic resonance to look inside the human body, it was Paul Lauterbur (1929–2007) and Peter Mansfield (1933–2017), who carried out the work most strongly linked to Magnetic Resonance Imaging (MRI) technology. The technique makes use of hydrogen atoms resonating when bombarded with magnetic energy. MRI provides three-dimensional images without harmful radiation and offers more detail than older techniques.

While training as a doctor in New York, Damadian started investigating living cells with a nuclear magnetic resonance machine. In 1971 he found that the signals carried on for longer with cells from tumors than from healthy ones. But the methods used at this time were neither effective nor practical, although Damadian received a patent for such a machine to be used by doctors to pick up cancer cells in 1974.

The real shift came when Lauterbur, a U.S. chemist, introduced gradients to the magnetic field so that the origin of radio waves from the nuclei of the scanned object could be worked out. Through this he created the first MRI images in two and three dimensions. Mansfield, a physicist from England, came up with a mathematical technique that would speed up scanning and make clearer images.

Damadian went on to build the full-body MRI machine in 1977 and he produced the first full MRI scan of the heart, lungs, and chest wall of his skinny graduate student, Larry Minkoff—although in a very different way to modern imaging. **LH**

SEE ALSO: COMPUTED TOMOGRAPHY (CT/CAT SCAN), X-RAY PHOTOGRAPHY, INFRARED PHOTOGRAPHY, ULTRASOUND

➡ *A sagittal (side view) MRI scan of the human brain, showing cerebrum, cerebellum, and brainstem.*

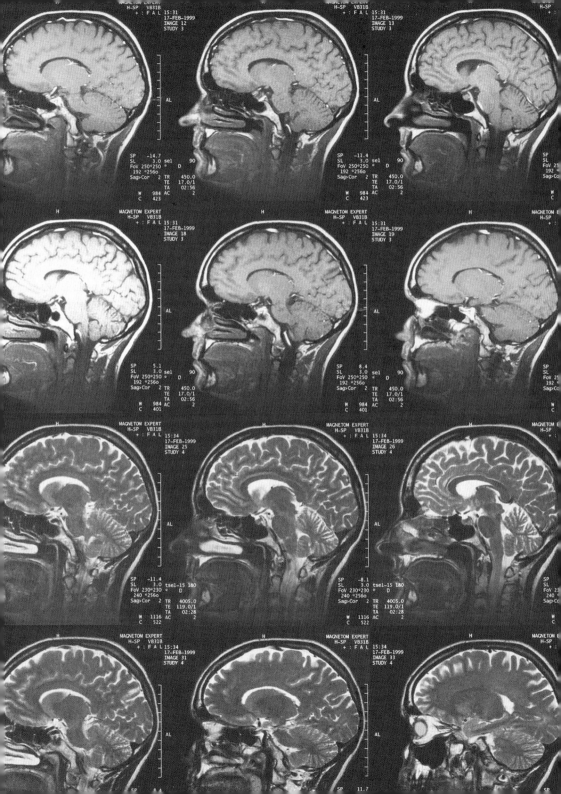

Space Station (1971)

The Soviets put the first permanently manned craft into space.

Space stations are permanently manned orbiting crafts that are designed to stay in space continuously. The first one, *Salyut 1*, was launched by the U.S.S.R. on April 19, 1971. It was about 65 feet (20 m) in length and 13 feet (4 m) in diameter. The docking mechanism on the first crew's *Soyuz 10* spacecraft failed, so it was first occupied by the *Soyuz 11* three-man team who stayed in orbit for just under twenty-four days. Sadly all three men died when their capsule depressured on reentry to Earth's atmosphere. Later generations of Soviet space stations, such as *Mir*, had two docking ports and water regeneration facilities. They were designed to be manned permanently.

The first U.S. space station, *Skylab*, was three times the size of *Mir* and was launched in 1973. In 1984 President Ronald Reagan announced that the United States planned to build a huge space station called *Freedom*. After many budget cuts the plans changed, and in 1993 the U.S. decided to go into partnership with Russia, Japan, Canada, and the European Space Agency (ESA) to build the International Space Station (ISS). The first element was launched in 1998.

The ISS—a joint project between space agencies from the United States, Russia, Canada, and eleven European countries—has been continuously occupied since November 2000. It makes 15.77 orbits of Earth each day, at a height of between 189 and 248 miles (350 and 460 km). It is serviced by the Space Shuttle and Russia's *Soyuz* and *Progress* spacecraft. When complete there will be six astronauts onboard, each staying there for a few months at a time. They will concentrate on microgravity, biology, biomedical, and fluid physics experiments. One of the main goals is to assess the long-term effects of space exposure on the human body. **DH**

"Poet Yevgeny Yevtushenko told a television interviewer that 'the price they had to pay was not fair.'"

Time *magazine on* Soyuz 11

⬆ *The crew of* Soyuz 11 *in the command module before takeoff in June 1971; the mission ended in tragedy.*

➡ *View of the* Mir *space station in orbit taken in 1995. The Russian station was originally launched in 1986.*

SEE ALSO: POWERED AIRPLANE, LIQUID FUEL ROCKET, BALLISTIC MISSILE, SPACE PROBE, REUSABLE SPACECRAFT

Computed Tomography (CT or CAT Scan) (1971)

Hounsfield develops a new imaging technology.

Conrad Röntgen (1845–1923) was the first person to take X-ray photographs of a person, winning the first Nobel Prize in Physics in 1901. X-ray films are sometimes referred to as Röntgenograms in his honor. For the first time, surgeons could see shrapnel and bullets contained within the body. However, X-rays were two-dimensional. In order to see how deep an object was, a second X-ray picture, usually perpendicular to the first, had to be taken. X-rays also fail to image the body's soft tissues very well.

Many techniques had been tried to improve the images produced by X-rays, but it was not until computer-assisted tomography (CAT) was developed that these problems were solved. Godfrey Hounsfield (1919–2004) devised the CAT scan in 1968, and by 1971 a prototype scanner was installed at Atkinson Morley's Hospital, Wimbledon, England, for use in clinical trials. In computer-assisted tomography, the X-ray tube is moved so that many images are taken from different angles, allowing depth to be viewed. The X-ray film is replaced by sensitive detectors, and a computer reconstructs the images. Because the detectors are over a hundred times more sensitive than film, subtle variations in tissue density can be elicited.

The first use of the new technology was to distinguish normal brain from diseased tissue. By 1975, larger scanners that could image the entire body were being marketed. In the 1960s, Allan Cormack (1924–1998) had independently begun work with the mathematical technique necessary to reconstruct the images. Cormack and Hounsfield shared the 1979 Nobel Prize in Medicine or Physiology. **SS**

SEE ALSO: ELECTROCARDIOGRAPH, ULTRASONOGRAPHY, ELECTROENCEPHALOGRAPH, X-RAY PHOTOGRAPHY

⬆ *During a CAT scan, patients are placed in a tunnel-like cylinder and the X-ray detectors rotate around them.*

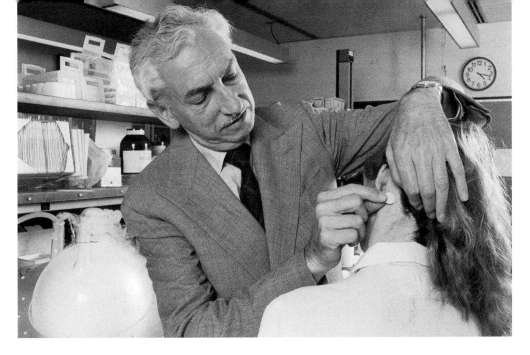

Transdermal Patches (1971)

Zaffaroni's invention improves drug delivery.

Needle and pill phobia sufferers must have cheered when the U.S. Food and Drug Administration (FDA) approved the first transdermal patch in 1979. This new mode of drug delivery promised all the benefits of shots and pills but with no downside.

Patient comfort was not the reason biochemist Alejandro Zaffaroni (1923–2014) developed transdermal patches. Zaffaroni wanted to mimic the body's timed release of hormones and thought available drug delivery methods were not sophisticated enough. In 1969 he started his company, ALZA, and by 1971 had been awarded a U.S. patent for a "bandage for administering drugs." Big pharmaceutical companies thought the patch was the path to nowhere. "I thought the industry would look at what we were doing and say, 'Gee, it makes a good deal of sense. But they didn't,'" said Zaffaroni in an interview.

The pharmaceutical industry, however, soon realized that Zaffaroni's patches made good sense and would produce good business. By the early 1980s the first transdermal patch, delivering a motion sickness drug, went on sale. Patches for the heart medication nitroglycerin followed closely, and now we have patches for nicotine addiction, pain management, contraception, hormone replacement therapy, and many other applications.

Transdermal patches are a multibillion-dollar industry, worth $3 billion in the U.S. market alone. It all began at ALZA, Zaffaroni's start-up, which was sold in 2001 to pharmaceutical giant Johnson & Johnson (of Band-Aid® fame) for $10 billion. Not bad for an idea that industry thought would go nowhere. **RBk**

SEE ALSO: ANTIVIRAL DRUGS, BIRTH CONTROL PILL, CONTROLLED DRUG DELIVERY, DISSOLVABLE PILL, HUMULIN®

⬆ *The pioneer of transdermal patches, Alejandro Zaffaroni, attaches a patch to a patient in 1980.*

E-mail (1971)

Tomlinson develops a program allowing communication between computer networks.

In 1969 a company called Bolt Barenek and Newman won the contract to develop a communication network called ARPANET that would enable scientists and researchers to use each other's computer facilities. During its development, an engineer named Ray Tomlinson (1941–2016) started to experiment with the coding of two programs. SNDMSG allowed members of the same network to exchange messages among one another, whereas CPYNET allowed file transfers to occur between two separate networks. It occurred to Tomlinson that by combining the two he could create a system that would make message transfer possible between different users of independent networks.

One of the most significant decisions made by Tomlinson was his choice of the @ symbol to separate the user's name from the host network name. It was a fairly logical choice, but one that revived the rather esoteric symbol and saved it from the brink of linguistic extinction.

Unaware of the global significance that the 200 lines of code that made up the e-mail program would have, Tomlinson neglected to note what he wrote in the first e-mail ever sent (he claims it was something banal like "QWERTYUIOP" or "TESTING 1 2 3 4").

Allegedly, when Tomlinson first demonstrated his program to a coworker, the latter told him not to show the system to anyone because it was not part of their job description. Tomlinson has since said that even though there was no direct stated objective to create e-mail, the ARPANET project was in fact a giant and worthwhile investigation into the multifarious uses of computer communication. **CL**

SEE ALSO: COMPUTER PROGRAM, INTERNET PROTOCOL (TCP/IP), WORLD WIDE WEB

⬆ *E-mail led to the introduction of a host of universally recognizable symbols onto computer screens.*

Microprocessor (1971))

Hoff's invention brings computers to the masses.

At the end of the 1960s, Intel's Ted Hoff (b. 1937) was asked to design several different calculators for a Japanese client. The traditional way would have been to develop several different integrated circuits— silicon chips—to do the work. Even though these were small enough to be put into handheld calculators, programmable computers, which could do a variety of jobs, were still huge devices.

Combining the small size of integrated circuits with the power of programmable computers was an inevitable idea. Hoff decided that he would make a single integrated circuit that could be programmed to do many different things. Joined by fellow engineers Stan Mazor (b. 1941) and Federico Faggin (b. 1941), Hoff squeezed an entire computer onto a single silicon chip, pairing it with a small memory to give it its instructions. His range of calculators all used the same chip, but each one had different instructions to instruct it how to behave.

Intel quickly realized that they had, quite literally in the palms of their hands, a programmable, general-purpose computer with the power of machines that a decade before had taken up entire rooms. Making a deal with the calculator manufacturer, Intel kept the rights to sell the chip to other people and released the Intel 4004 processor in 1971.

The 4004, the first commercial microprocessor, was also the first step in the revolution that would sweep the world during the 1970s and 1980s, taking computers from their air-conditioned industrial servitude and bringing them to homes, cars, and even washing machines. **MG**

SEE ALSO: COMPUTER PROGRAM, MECHANICAL CALCULATOR, MECHANICAL COMPUTER, RANDOM ACCESS MEMORY (RAM)

⊡ *By 1974, Intel had developed the 8080 microprocessor; it was used in one of the first personal computers.*

Touch Screen
(1971)

Hurst makes computer interaction easier.

During the rapid rise of the computer in the second half of the twentieth century, people were always searching for the next best way to interact with them. The early days of punched cards and paper tape became too cumbersome as computers advanced and keyboards became the input device of choice.

In the 1960s, U.S. inventor Douglas Engelbart invented the computer mouse, which represented a milestone in computer interaction. The next big leap forward came in 1971 when Dr. Samuel C. Hurst invented the electronic touch screen interface. While teaching at the University of Kentucky, he was faced with the daunting task of reading a huge amount of data from a strip chart. Realizing that this work would normally take graduate students at least two months to complete, he decided to work on an easier method.

What he came up with was the Elograph coordinate measuring system. It was an input tablet that could measure where the user was pressing a stylus. Hurst quickly formed the Elographics company (now Elo TouchSystems) to make and sell the device. Working furiously to develop their concept, Hurst and his team took just three years to make a proper transparent version that could sit over a screen. Four years later, in 1977, they came up with what was to become the most popular technology for touch screens today. The five-wire resistive touch screen contains transparent layers that are squeezed together by the pressure of a finger touching them. Easily translated into electrical resistive data, this modern touch screen is durable and offers high resolution. **DHk**

SEE ALSO: COMPUTER MOUSE, INTERNET COMPUTER SCANNER, TOUCH PAD, VIRTUAL REALITY HEADSET

◄ *John K. Lee of Hewlett-Packard demonstrates touch screen technology on a personal computer (HP 150).*

Public Key Cryptography (*c.* 1972)

Ellis, Cocks, and Williamson codify identities.

Public Key Cryptography (PKC) is a technological tool that enables participants to confirm their identity with each other electronically. Traditional signatures have been around for thousands of years, originally being used to mark artwork such as pottery with the identity of the creator. However, as the concept of currency and contracts spread across the globe, so did the use of signatures. Although signatures were adequate for society's needs, and still are as a whole, they clearly did not satisfy the demands of electronic security.

When governments began to learn of the potentials of PKC, they endeavored to keep the technology to themselves. In the early 1970s, while working for the British government, James Ellis (1924–1997), Clifford Cocks (*b.* 1950), and Malcolm Williamson (1950–2015) contributed to its development. It was only in 1997, under a new government "openness" policy, that it was revealed to the world that Britain had developed PKC twenty-four years previously. Numerous text books had been written on the subject in the intervening years, and, frustratingly for Ellis, Cocks, and Williamson, the texts had credited Martin Hellman, Ralph Merkle, and Whitfield Diffie of Stanford University with the discovery of PKC.

To accurately describe PKC would be a long and complicated process. However, the fundamental point to PKC is that it allows two computers or people to communicate in impenetrable privacy (there is no known way to maliciously decipher the code). Exactly who should be credited with the mathematical trick for this encryption process seems irrelevant now, given that it is used so widely across the Internet. Many people will have used it online via a transaction that was secured through PKC. **CB**

SEE ALSO: MAGNETIC STRIPE, RADIO FREQUENCY IDENTIFICATION (RFID)

C Programming Language (1972)

Ritchie develops a more flexible language.

Dennis Ritchie (1941–2011) is idolized by computer programmers all over the world. Why? Because he wrote what is, without a doubt, the most widely used programming language in the world.

After gaining undergraduate and graduate degrees from Harvard, Ritchie went to work for Bell Laboratories in 1968. It was there, alongside Ken Thompson (b. 1943), that he created the UNIX operating system. At the time, Bell Labs was using a programming language called "B," which was used to write UNIX. Building on this operating system, Ritchie, in his own words, "added data types and new syntax to Thompson's B language, thus producing the new language 'C.'"

This new language, designed to be used with the UNIX operating system, is general purpose and, critically, was written to allow it to be "ported," or transferred from one type of computer to another. At the time, Ritchie and Thompson had been working with their B language on a PDP-7 computer and, when Bell Labs acquired one of the new PDP-11s, they understandably wanted to switch over to the more powerful machine. It was B's inability to take advantage of the new features of the PDP-11 that led Ritchie to come up with early versions of C, which he used to rewrite their UNIX operating system so it could be ported to the new computer.

Thus the bulky PDP-11, complete with magnetic tape drive, became the first major computer to use C. The flexibility and simplicity of the language means it is still used, along with UNIX, on modern PCs and by real programmers. Ritchie and Thompson received the U.S. National Medal of Technology from President Bill Clinton in 1999 for their work on UNIX and C. **DHk**

SEE ALSO: COMPUTER PROGRAM, JAVA COMPUTING LANGUAGE, MAGNETIC CORE MEMORY, MICROPROCESSOR

Insulin Pump (1972)

Kamen evens out drug delivery for diabetics.

The insulin pump is a small, battery-powered device that releases varying amounts of insulin into the bloodstream of diabetics. Diabetes is a disease that affects the body's ability to break down sugar, caused by an absence or insensitivity to the hormone insulin. Until the invention of the insulin pump, the only way for diabetics to control their disease was to inject themselves daily with insulin. The first insulin pump was invented by Dr. Arnold Kadish in the 1960s, but it was so large it had to be worn like a backpack.

While Dean Kamen (b. 1951) was at college, his brother, then a medical student, approached him with a problem. He complained that there was no way to

> " . . . if you are not working on important things, you are wasting time."
>
> Dean Kamen

provide patients with steady doses of drugs, such as insulin. In response, Kamen constructed a circuit that controlled a small pump of insulin connected to a syringe. He used a new form of microchip—which did not require much power—to control the circuit. The device was wearable and programmable and delivered small, precise doses of insulin over a long period of time, evening out the peaks and troughs of insulin levels associated with injections. Kamen's brother showed this device to his colleagues and they were immediately impressed. In 1976, Kamen founded his first company, AutoSyringes Inc., to market and manufacture the pumps. **HI**

SEE ALSO: HYPODERMIC SYRINGE, DISSOLVABLE PILL, ASPIRIN, BETA BLOCKERS, TRANSDERMAL PATCHES, INSULIN, HUMULIN®

Personal Computer

(1973)

Xerox PARC's Alto inspires the home-computing industry.

There are many contenders for the title of first ever "personal computer." However, it was Xerox PARC in 1973 that was responsible for creating perhaps the most innovative design in computer history—a personal computer as we would recognize it today. The Alto, named after the Californian Palo Alto Research Center (PARC) where it was created, was made up of a cabinet (containing a 16-bit custom-made processor and disk storage), a monitor, a keyboard, a mouse, and even the first what-you-see-is-what-you-get graphical user interface featuring windows and clickable icons.

The Alto was designed primarily for research and had to be compact enough to fit in an office, but powerful enough to support a user interface while being able to share information between machines. This led it to feature groundbreaking innovations that would not be common until a decade later and would still be cutting edge in the 1990s. These included an object-oriented operating system and the first Ethernet networking cards. It also came with that most essential of research tools, a pinball game. Xerox donated its Alto machines to various research institutions, and they quickly became the bar by which all future personal computer design would be judged. The Alto is also believed to have inspired others, including Apple's cofounder Steve Jobs and his team who were impressed with the Alto's sharp graphics and user interface. The team who produced the Xerox Alto was honored with the prestigious Draper engineering prize in 2004 as the catalyst of a "golden age" of computer research. **JM**

SEE ALSO: COMPUTER MOUSE, COMPUTER PROGRAM, MECHANICAL COMPUTER, WINDOWS-BASED COMPUTING

☑ *The Alto cost $32,000; the design of the monitor allowed it to display a whole page of text.*

" . . . the luminous Alto display, covered with images and graphical fonts, was a revelation."

John Markoff, *New York Times* (April 3, 2003)

Ethernet (1973)

Metcalfe and Boggs link computers in a network.

In 1973, Bob Metcalfe (*b.* 1946) of Xerox's Palo Alto Research Center (PARC) faced a problem. Increasing numbers of computers were springing up around him, all of which needed to be connected to each other. Just down the hallway the world's first laser printer, invented at PARC in 1971, was hungry for documents.

Computer networking was in its infancy. The hardware was expensive, and the wiring at PARC looked like an explosion in a spaghetti factory. Any glitch in the computers or the cabling would bring down the whole system. Metcalfe was given the job of building a simpler, more reliable computer network.

Desperate for inspiration from any source, he stumbled across the University of Hawaii's ALOHAnet, a radio network. Unlike most computer networks, which were carefully regulated so only one computer could talk to another at any given moment, ALOHAnet was a free-for-all. If several computers tried to talk at the same time, each computer would "back off" for a little while before trying again.

Grabbing graduate student David Boggs from PARC's basement to help him, Metcalfe set out to build a wired network based on ALOHAnet's ideas. Metcalfe's "Ethernet" system expected network collisions and glitches and worked around them. Connecting all the computers together on a single long wire hugely simplified the cabling.

Constantly developed ever since, Ethernet is now the most popular standard for local networks. If you send a document to your office printer today, it is likely to travel down the wires of an Ethernet-based network, directly descended from Metcalfe's work. **MG**

> *"Today, Ethernet stands as the dominant networking technology . . ."*
>
> The Economist (2003)

SEE ALSO: INTERNET, INTERNET PROTOCOL (TCP/IP), WORLD WIDE WEB

◰ *Computers are linked to the network via a patch panel—a system still in operation today.*

Plastic Bottle (PET) (1973)

Wyeth invents a new plastic to contain carbonated drinks.

The humble plastic bottle is now one of the most commonly recycled household objects. But it was the product's cheapness and durability that led to its popularity over glass bottles. Nathaniel Wyeth (1911–1990), a U.S. engineer, worked on the invention for almost a decade. When he asked a colleague if plastic could be used to store carbonated beverages such as Coca Cola, he was told that they would explode. A series of early experiments proved that carbonated beverages caused the plastic to expand. Obviously, the plastic was too weak, but the plastic could be strengthened if the long strands of molecules that form plastic were woven together. Wyeth knew that nylon gets stronger when its molecules are stretched and aligned, and developed a pre-formed mold that forced the nylon threads to interweave when plastic is extruded into it.

Molten plastic was lowered into the bottle-shaped mold and air shot through the plastic to spatter it all over the mold and create a form through layers of weaving. Initial results were far from successful, and Wyeth took nearly 10,000 attempts to solve what he termed the "pop bottle problem." He finally replaced the nylon and polypropylene material he had been using with polyethylene-terphathalate (PET), whose superior elastic properties guaranteed a transparent, resilient, wholly recyclable plastic bottle.

The bottle was quickly taken up by the booming soft drink industry, and it was estimated that by 1999 ten billion bottles per year were being produced. Almost half of the polyester carpet made in America today comes from recycled plastic bottles. **SD**

SEE ALSO: GLASS, GLASS BOTTLE-MAKING MACHINE, SODA WATER, POLYPROPYLENE, TETRA PAK®

↗ *The trade name of PET is Terylene; it is popularly used in the manufacture of sportswear and as drinks bottles.*

"After months of frustration, we had grown used to seeing blobs of resin caked on the mold … "

Nathaniel Wyeth

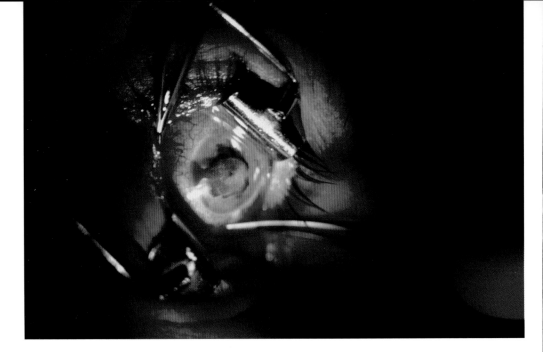

Laser Eye Surgery (1973)

Bhaumik improves eyesight with an excimer laser.

Millions have turned to laser-assisted in situ keratomileusis (LASIK) to correct myopia, hyperopia, and astigmatism, and do away with wearing glasses.

In the 1950s, Spanish ophthalmologist Jose Barraquer devised a method for surgically changing the shape of the cornea. His technique was developed further by Russian ophthalmologist and eye surgeon Svyatoslav Fyodorov, who created radial keratotomy while treating a boy whose glasses had smashed into his eye after a fall. He made several radial incisions from the pupil to the edge of the cornea to remove the glass. Once the cornea had healed, the boy's eyesight was much better. Fyodorov's success fueled much interest in refractive surgery.

In 1968, work was going on at the University of California, where Indian physicist Mani Lal Bhaumik and colleagues developed an excimer laser. This creates new molecules when xenon, argon, or krypton gases are excited. In 1973, Bhaumik told the world of

the new technique at a meeting of the Denver Optical Society of America and later went on to patent it. Seven years later, Rangaswamy Srinivasan, an Indian chemist, found that an ultraviolet excimer laser could accurately etch living tissue without damaging the surrounding tissues.

Building on all this, LASIK surgery was finally put together by Italian Lucio Buratto, and Greek Ioannis Pallikaris in 1990. It was faster, more accurate, and had lower risk of complications than older techniques. Newer technologies have taken LASIK further performance-wise, but its limitations have also led to a host of other developments. **LH**

SEE ALSO: LASER, LASER CATARACT SURGERY, VISUAL PROSTHETIC (BIONIC EYE)

⬆ *Precision laser surgery is performed on the eye of a patient using an excimer laser.*

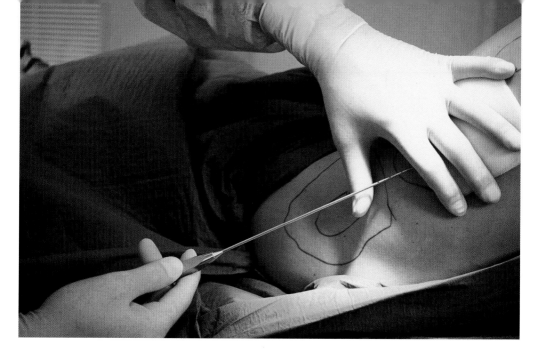

Liposuction (1974)

Fischer develops a technique for removing excess abdominal fat.

Liposuction is a cosmetic technique in which excess fatty tissue is suctioned from beneath the skin. It was first developed in 1974 by Italian gynecologist Giorgio Fischer (b. 1934) who found that he could remove fat through tiny incisions with an electrical rotating scalpel connected to a cannula that was attached, in turn, to a suction device. The procedure was initially developed to remove excess fat to make abdominal surgery easier. A major drawback was that patients often suffered considerable blood loss.

Four years later, French plastic surgeon Yves-Gerard Illouz was the first to recognize the potential of liposuction as a cosmetic procedure. He used a blunt-tipped cannula, which resulted in fewer complications and a shorter recovery time. In the early 1980s the procedure was introduced to the United States, but the high failure rate damped enthusiasm.

In 1985 Dr. Jeffrey Klein, a Californian dermatologist, solved the problem with the development of tumescent anesthesia, a technique that uses large volumes of the local anesthetic lidocaine, in combination with the vasoconstrictive drug epinephrine. The combination of drugs helped to reduce the risk of bleeding and reduced the requirement for a general anesthetic.

With today's prevailing obsession with the body perfect, liposuction has never been more popular. It is particularly suitable for people who are more or less of normal weight but who have isolated pockets of fat that cause their bodies to appear disproportionate. The areas most often treated include the abdomen, hips, thighs, and knees. **JF**

SEE ALSO: HYPODERMIC SYRINGE, BREAST IMPLANT, LASER EYE SURGERY, TRANSDERMAL PATCHES

⬆ *A female patient undergoes liposuction; the technique allows pockets of fat to be targeted.*

Hybrid Vehicle (1974)

Wouk develops a car with reduced carbon dioxide emissions and improved fuel efficiency.

Hybrid vehicles have two different types of engines running in parallel—usually a gasoline or diesel engine to provide most of the motive force and an electric motor, powered by batteries, for additional assistance.

The main advantages of hybrids are a reduction in carbon dioxide emissions and increased fuel efficiency due to the use of regenerative braking to recharge the battery (these batteries are sometimes referred to as onboard rechargeable energy storage systems). Fuel can also be saved by switching off the gasoline or diesel engine when coasting or stationary in traffic.

The major breakthrough was made by Victor Wouk (1919–2005), the brother of the novelist Herman Wouk. Victor, an electrical engineer, was spurred into action by the Federal Clean Car Incentive Program of 1970. After founding a company called Petro-Electric Motors Ltd. with a chemist friend, Charles Rosen, in 1974 they adapted a Buick Skylark as a gasoline/electric hybrid that doubled the miles per gallon of similarly sized cars.

The twin-engine approach overcame the sluggish performance of all-electric cars. Unfortunately, the major car manufacturers in the United States and the U.S. Environmental Protection Agency showed little enthusiasm. The two partners eventually dissolved the company, and Wouk returned to consulting work.

The term hybrid can be applied more generally. Bicycles fitted with small electric motors, diesel-electric trains, and diesel-powered ships with masts and sails are also referred to as hybrid vehicles. Today the sale of hybrid cars in the United States is more than 200,000 a year. Unfortunately for the United States, about 95 percent of these are made in Japan. **DH**

SEE ALSO: AIR CAR, ELECTRIC CAR, MOTORCAR

⬆ *Wouk pictured with his hybrid Buick Skylark at the test site of the U.S. Environmental Protection Agency.*

Electronic Paper (1974)

Sheridon paves the way to the paperless office.

Electronic paper has many of the properties of paper; thin, flexible, and readable from a wide angle, but it also has the distinct advantage of being reusable.

In the 1970s Xerox PARC had developed a personal computer, and Nicholas K. Sheridon had the task of developing a display that improved on the then very dimly lit cathode ray displays. In 1974 Sheridon developed the Gyricon, which—although it was never used as a monitor because the PC project was dropped—formed the foundations of e-paper.

The Gyricon, Greek for rotating image, consists of a thin, flexible sheet of plastic peppered with oil-filled wells containing small beads. Each bead, colored white on one half and a contrasting color (usually black) on the other, is charged with positive and negative ends that corresponded to the two colors. When an electrical current is applied to the Gyricon, the beads rotate in a predicted manner displaying their black sides where marks are to be made on the paper. A major advantage

of this type of display is that once the initial current is applied no further energy input is required to maintain or refresh the image.

Today several types of e-paper have been produced, such as the bistable LCD (liquid crystal display), cholesteric LCD, and the electrophoretic display, which works on a similar principle to the Gyricon and has also been developed to display color. Initially highly expensive, e-paper hit the mass market in 2004 when it was used by Sony in the first e-book reader. It is now utilized in a wide range of devices, including wristwatches, cell phones, smart cards, and traffic signs. **RP**

SEE ALSO: PAPER, CARBON PAPER, KRAFT PROCESS, LIQUID PAPER, LIQUID CRYSTAL DISPLAY (LCD)

⊡ *At Xerox's research center, California, Sheridon shows each bead of Gyricon is smaller than a grain of sand.*

Digital Audiotape (DAT) (1975)

Stockham's Soundstream, Inc., leads the way in an alternative recording system.

Until the 1980s, most commercial audio recordings were made according to analog principles, with the original sound modulated onto another medium, the physical characteristics of which are directly related to the original sound. In contrast, digital recording sees the original sound converted to digital information and stored as a series of 1s and 0s—known as "bits."

Although the principles of digital recording were already in place in the late 1930s, it was not until 1975 that a usable commercial system was developed, when Dr. Thomas Stockham (1933–2004) established Soundstream, Inc., the first dedicated digital recording company. The original audio was passed through an Analog-Digital-Converter (ADC), converted to 16-bit audio, and stored on a 1-inch (2.54 cm) Honeywell tape deck. To play back the sound, the digital information was passed through a Digital-Analog-Converter (DAC). The system offered the highest quality sound without any of the problems of analog recording. Mechanical deficiencies commonly associated with analog equipment were also radically improved.

Competing systems, such as those produced by 3M, Sony, and JVC appeared in quick succession, but for the rest of the decade it was Stockham's system that dominated. From the early 1980s, the Sony DASH and Mitsubishi ProDigi formats rendered previous systems obsolete. At the same time, Sony's DAT cassette format became the de facto standard for making digital stereo recordings. All of these formats remained in use until the early 1990s, after which hard disk recording began its gradual domination. **TB**

SEE ALSO: VINYL PHONOGRAPH RECORD, AUDIOTAPE RECORDING, MAGNETIC RECORDING, MINIDISC

⬆ *DAT cartridges are similar in appearance to audio cassette tapes but are roughly the size of a credit card.*

Digital Camera (1975)

Sasson eliminates the need for photographic film.

They are so commonplace today that many people even have them in their cell phones; some companies have even added them to digital watches. Digital cameras are now taken for granted, but it was only in 1975 that the first prototype was produced.

Steve Sasson (b. 1950) had recently graduated in electrical engineering before taking a job with Kodak. His assignment was a broad one: Was it possible to make a camera using solid-state electronics? Starting from scratch, he gathered together various pieces of electronic equipment including a movie camera lens, an analog-to-digital converter, and, most importantly, charge-coupled devices (CCDs).

By December 1975 Sasson's rough prototype was ready for initial testing. Weighing in at 8 pounds (3.6 kg) and the size of a toaster, the camera was hardly portable, but Sasson convinced a lab assistant to pose for a test shot. It took twenty-three seconds for the image to be recorded onto a cassette and then an additional twenty-three seconds to be read from the tape before appearing on a television screen.

Over the next year, Sasson and his colleagues presented his invention to Kodak officials. The response, however, was often less than encouraging. The employees of a company best known for making camera film were understandably perplexed at the notion of a camera that required no film. Why would anyone want to view their photos on a television? How could you store these digital images? Ultimately, the rapid increase in computer capabilities would solve these issues, but it was still many years before digital cameras started to replace traditional film. **CA**

SEE ALSO: CHARGE-COUPLED DEVICE (CCD), PHOTOGRAPHY, PORTABLE CAMERA, SELF-DEVELOPING FILM CAMERA

⬆ *Sasson's first digital camera used a CCD image sensor and the technology was patented in 1978.*

3D Computer Graphics (1976)

Catmull creates the first 3D computer-generated image in the movies.

The idea of 3D graphics—just like painting—is making an image that tricks the brain into thinking it is looking at something with three dimensions rather than two. To do this, you must consider the effect of lighting on the object, as well as depth, perspective, texture, and many more qualities, which the computer then has to project on a two-dimensional surface in a realistic way.

The advance in computer graphics started in the early 1960s, with the first commercially available graphics terminal, the IBM 2250, hitting the market in 1965. Three years later, Ivan Sutherland (b. 1938) created the first computer-controlled head-mounted display (HMD). The wearer of this helmet was able to see a computer scene in stereoscopic 3D, because separate images were displayed for each eye.

Sutherland subsequently joined what was then the world's leading research center for computer graphics at the University of Utah. One of his students was Edwin Catmull (b. 1945), a wannabe animator who conceived the idea of texture mapping, which is based on the fact that the majority of real-life objects have detailed surfaces. He was convinced that you could apply similar patterns to computer-generated items by taking a flat 2D image of an object's surface and placing it onto a 3D computer-generated object.

Using this method, he created an animated version of his left hand. Following the world's first computer animation in movies—in the Canadian short film *The Hunger* from 1974—an animated face as well as Catmull's hand became the first 3D computer-generated images in movies when they featured in *Futureworld* in 1976. **BG**

SEE ALSO: COMPUTER PROGRAM, COMPUTER-AIDED DESIGN (CAD), VIRTUAL REALITY HEADSET

⬆ *Photographed in 1985, Catmull shows off the detail of a CAT scan displayed in 3D on a Pixar monitor.*

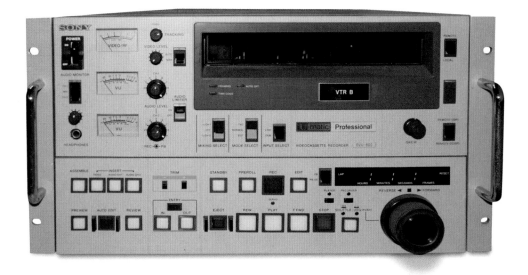

Electronic News Gathering (1976)

Sony's invention ousts film from the newsrooms.

Years before the format war between VHS and Betamax, and decades before DVDs came onto the scene, there was film. But film was hard to use and eventually videotape was invented. It was much faster to edit than film and speeded up the news business. But these first video machines were so big that they were stationary, and the roving reporters of the world stuck to film until portable versions eventually appeared. The first portable broadcast recorder, which had a huge backpack in which the tape sat, allowed news to be filmed on video outside a studio, although at a lower quality than film.

The videotape was stored on reels, just like film, but this changed in 1971 when Sony released the U-Matic video recorder. The U-Matic had been designed for home use, and the tape was in cassette form, which made it easy to handle, although the units themselves were heavy and expensive—so heavy and expensive that after trying to sell them to consumers

in Japan, Sony had a go at selling them to businesses in the United States. Unfortunately the quality was so much lower than professional standards that broadcasting companies could not use them.

In 1974 CBS, a U.S. broadcaster, asked Sony to make a version of the U-Matic for broadcasters. After more than a year of development, Sony delivered the Broadcasting Video series in 1976. Combining the shooting, recording, and editing into the one unit, it was built to match what the customer needed and this is what made it successful. Before the end of the decade, film was disappearing from newsrooms and electronic news gathering had emerged. **DK**

SEE ALSO: FILM CAMERA/PROJECTOR, CAMCORDER, OPTICAL DISK, VIDEOTAPE RECORDING

⬆ *The Broadcasting Video incorporated the shooting, recording, and editing functions of news gathering.*

GORE-TEX® (1976)

Gore family creates a new waterproof material.

Wrapping up warm to toil against the elements has one problem: you can get too warm. Donning a fleece or heavy overcoat might protect you from the wind and rain, but any serious physical exertion will leave you hot and sweaty. Step forward GORE-TEX®, a synthetic fabric that keeps you dry while allowing the skin to breathe.

GORE-TEX® is produced by expanding polytetra-fluoroethylene (PTFE)—a polymer comprising long carbon-fluorine chains—under high temperature. The material contains many pores, allowing perspiration from the wearer to evaporate and escape. The pores are, however, much too small to allow passage of water droplets, rendering the material impenetrable to the elements. For comfort and protection, the special membrane is normally sandwiched between several fabric layers.

The invention of GORE-TEX® was a family effort; it was developed by Wilbert L. Gore (1912–1986) and his son Robert W. Gore, plus Rowena Taylor. In 1958, Gore Senior and his wife Vieve set up a company in the basement of their Delaware home to produce insulating material from PTFE for electronics. Around ten years later, Robert Gore discovered that rapidly stretching PTFE produces a strong, waterproof material, which he christened GORE-TEX®. The first fabric application came in 1976, when the material was used as a tent covering. Coats, boots, and other outdoor equipment made of the material soon followed—even dental floss.

The merits of this breathable fabric are such that many armies dress their soldiers in outdoor gear made from the material. It has also found numerous surgical applications, providing the material for more than 25 million implants in plastic and heart surgery. **MB**

SEE ALSO: CLOTHING, PVC (POLYVINYL CHLORIDE), GAMOW®
BAG, PLASTIC BOTTLE (PET)

Supercomputer (1976)

Cray speeds up computing.

Imagine if you could revolutionize the design of computers and leave the competition standing; U.S. inventor Seymour Cray (1925–1996) made a habit of doing just that. In 1972, with a long history of extending the reach of computer technology already behind him, Cray set up the Cray Research company to concentrate on building a powerful computer. His design for the Cray 1 was the first major commercial success in supercomputing. It was essentially a giant microprocessor capable of completing 133 million floating-point operations per second with an 8-megabyte main memory. The secret to its immense speed was Cray's own vector register technology and its revolutionary "C" shape, which meant its integrated

> *"He [Seymour Cray] is the Thomas Edison of the supercomputing industry."*
>
> Larry L. Smarr, physicist

circuits could be packed together as tightly as possible. It produced an immense amount of heat and it needed a complex Freon-based cooling system to prevent it from melting.

The Cray 2, introduced in 1985, was six to twelve times faster, with ten times the memory. Cray's supercomputers have been of inestimable value to science and have been used to predict the weather, design airplanes, explore for oil, and even provide computer simulations of nuclear tests. **JM**

SEE ALSO: MECHANICAL COMPUTER, COMPUTER PROGRAM,
DIGITAL ELECTRONIC COMPUTER, MICROPROCESSOR

▣ *The second incarnation of the Cray supercomputer, located at NASA's Ames Research Center, California.*

Inkjet Printer (1977)

Endo invents a new printing method.

Most useful technologies take time to mature; usually there are at least a couple of years between an idea's conception and a final working model. When it comes to inkjet printing, however, it took much longer—the patent for directing ink onto paper using electrostatic forces was granted to Lord Kelvin in 1867.

Before the 1980s, printing from a computer was a slow, unrewarding task. The mechanisms behind those early printers used moving parts, pumps, and bladders that made them expensive, clumsy, and inefficient. The modern inkjet was to change all this by using heat or electrostatic forces to produce uniform droplets and precision results. In Japan in the 1970s, Canon and Hewlett-Packard were competing with each other to produce the first reliable inkjet printer.

Hewlett-Packard was beaten by a Canon researcher named Ichiro Endo who invented the first thermal inkjet printer in 1977. He was inspired when he saw a syringe full of ink accidentally being touched with a hot soldering iron. The heat caused the ink to increase in volume and spurt out. Endo realized this was the solution to delivering controlled spurts of ink, and within days he produced a working model that later became the Canon Bubblejet printer.

Both Hewlett-Packard and Canon put in patents within months of each other, and although there is much discussion about who invented what first, the two companies ended up sharing a lot of printing technologies with each other. The winner, with high-quality cheaply produced printers available to all, was the home consumer. **JM**

SEE ALSO: COMPUTER LASER PRINTER, INK, PRINTING PRESS, PRINTING TELEGRAPH, ROTARY PRINTING PRESS

⬆ *Hewlett-Packard introduced the ThinkJet to the market in 1984; it was quiet, fast, and cheap ($495).*

Personal Stereo (1977)

Pavel patents the first portable cassette player.

Given its market domination from the moment it first appeared in 1979, many people might imagine that the Sony Walkman was the original personal cassette player. Its iconic status is beyond question—it all but created the vogue for listening to music on the move and is a direct antecedent of today's ubiquitous iPod. And yet seven years earlier, a lone inventor with little expertise in the field of electronics came up with a concept that was almost identical.

The story begins in Brazil in 1972 when a German-born former TV executive named Andreas Pavel (b. 1945) sought a way of listening to high-quality music while going about his everyday business. His idea was for a tiny portable cassette player—not that much larger than the cassette itself—that played back audio through a small pair of headphones. He called his novel idea the Stereobelt.

Having left Brazil and moved to Switzerland, Pavel made approaches to many of the leading electronic manufacturers, but none were interested in his idea: They believed that few would be prepared to wear headphones in public in order to listen to music. Although Pavel failed to find a backer for his idea, his faith was unshaken and, during 1977, he filed patents for his invention across the globe.

One year after Sony launched their renowned Walkman—to immediate acclaim—Pavel set out on what turned out to be a marathon legal battle taking up most of the next twenty-five years. It was eventually resolved in 2003, with an out-of-court settlement in which Sony is believed to have paid Pavel in excess of $10 million. **TB**

SEE ALSO: AUDIOTAPE CASSETTE, MINIDISC, TRANSISTOR, TRANSISTOR RADIO

⬆ *Also known as the Sound About and the Stowaway, Sony's Walkman was launched on July 1, 1979.*

Conductive Polymers
(1977)

Chemists create a plastic to conduct electricity.

While studying science in school we may have been taught that electrical conduction occurs in metals and in some liquids containing ions. That particular lesson changed when chemists Alan MacDiarmid (1927–2007) and Hideki Shirakawa (b. 1936), and physicist Alan Heeger (b. 1936) created the first polymer that could conduct electricity.

Shirakawa was working on the polymerization of acetylene when a mistake was made in the amount of a catalyst used in the process (a thousand times too much!). Instead of the usual black powder, a "ragged film" of polyacetylene was formed. This film had a metallic luster, and Shirakawa began investigating its properties. Alan MacDiarmid at the University of Pennsylvania invited him to collaborate, and together with Alan Heeger, the three experimented on modifying the polyacetylene by oxidization with iodine vapor. When one of Heeger's students was examining this iodine-doped version, it was discovered that the electrical conductivity had increased by a staggering ten million times. The three scientists published their work in 1977, launching the new field of polymer-based electronics. They were jointly awarded the Nobel Prize for Chemistry in 2000.

Today the best conductive polymers have an electrical conductivity close to that of copper. Since their discovery, a wide range of applications has arisen. Just as metal wire can glow via electricity, so can conductive polymers. This electroluminescence is much more efficient than traditional light bulbs. There are also organic LEDs (light-emitting diodes), which are being developed into tiny highly efficient screens for use in cell phones and cameras. **DHk**

SEE ALSO: ELECTROACTIVE POLYMERS (ARTIFICIAL MUSCLES)

Laser Cooling of Atoms
(1978)

Wineland develops a process to cool atoms.

If you can make the photons of a laser light that hits atoms have more energy than those that leave the atoms, then the atoms get colder. The trick is to tune the energy of a laser photon to a value that is slightly below that of the energy of an electronic transition in the atom. Due to the Doppler red-shift of the photons, those atoms moving toward the beam absorb more photons than those moving in the opposite direction. Emitted photons leave the atom in random directions, and the result is a general loss of momentum and kinetic energy. The atoms get colder because temperature is proportional to the kinetic energy.

The atoms need to be at very low concentrations. The idea was first suggested in 1975 by Theodore Hansch and Arthur Schawlow at Stanford University in California. Ten years later, Steven Chu of AT&T Bell Labs put it into practice. Sodium was cooled down to about 250 Kelvin using six lasers that provided three pairs of beams along each coordinate axis. By using optical pumping, the sodium was cooled even further to a temperature of 35 K and caesium to 3 K.

Physicist Dr. David Wineland pioneered the application of laser cooling to the next generation of even more accurate atomic clocks. Extremely cold atoms form themselves into a Bose-Einstein condensate, this being a pure quantum form of matter. In this state, atoms can be studied with greater accuracy. Interestingly, the main perturbing effect on laser cooled material is Earth's large gravitational field. The next generation of experiments will have to take place in the microgravity environment of nearby space. **DH**

SEE ALSO: ATOMIC BOMB, ATOMIC CLOCK, ATOM LASER, CARBON DIOXIDE LASER, LASER

⮕ *At AT&T Bell Labs, yellow laser beams surround the steel atom trap, cooling atoms to near absolute zero.*

Cyclone Vacuum Cleaner (1978)

Dyson designs an iconic new cleaner for the home.

In 1978, designer James Dyson (*b.* 1947) noticed that the suction of his vacuum cleaner diminished as the bag started to fill with dust. He realized that a system in which dust is siphoned into a bin, rather than retained by a filter, would maintain the suction of the appliance. Five years and some 5,127 prototypes later, Dyson had invented the technology that would be the cornerstone of the iconic Dyson vacuum cleaner.

James Dyson studied at the Royal College of Art, where he began a career in design that would lead him to develop many unique ideas. However, his big breakthrough came seven years after he began his research into developing filtering processes for vacuum cleaners.

The cleaning power of a Dyson vacuum cleaner lies in the process of cyclonic separation. By spinning a column of air, dust particles are sucked up and drawn to the insides of the container, as the centrifugal forces exerted increase their mass by hundreds of thousands of times. These particles sink to the base of the container and are collected in a "bin." The air, free of dust particles, is released back into the environment. Importantly, the cyclonic system means that in theory the cleaner never loses suction and never needs any new bags or filters.

In 1986, Dyson secured an agreement with a Japanese firm to market his product as the G-Force, the design of which won him many awards. Within seven years, the first Dyson-produced DC01 model reached U.K. shelves and became the quickest selling vacuum cleaner, not to mention an iconic design. Today, Dyson has branched out to offer a variety of cleaning products based around the original concept, including handheld vacuum cleaners and two-drum washing machines. **SR**

> *"I made 5,127 prototypes of my vacuum ... There were 5,126 failures. But I learned from each one."*
>
> James Dyson, *Fast Company* magazine (May 2007)

⬆ *Dyson sold the G-Force in 1986 in a Japanese catalog retailing for the equivalent of $3,500 (£2,000).*

➡ *Inventor James Dyson tinkering with one of the latest iterations of his bagless vacuum cleaners.*

SEE ALSO: CARPET SWEEPER, ELECTRIC VACUUM CLEANER, VACUUM PUMP

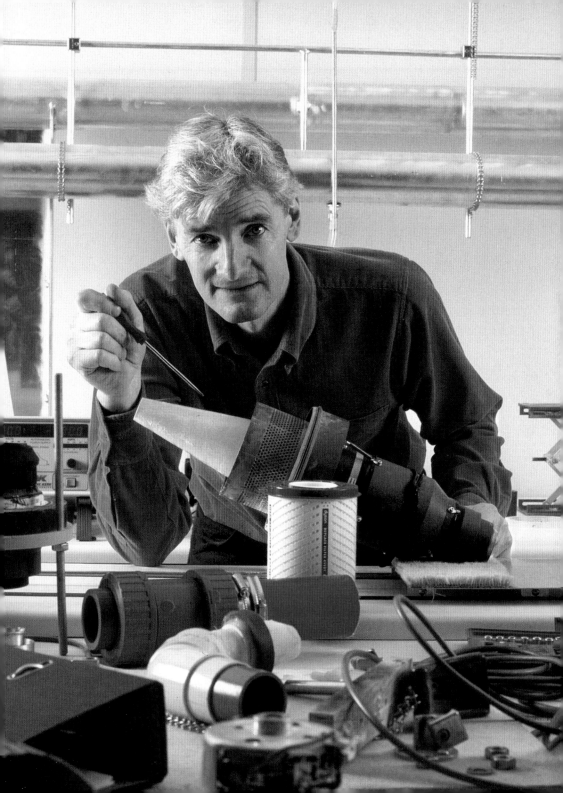

Word Processor (1978)

Rubinstein and Barnaby write the first commercially successful word processing software.

Unlike a great deal of the software that the first few major waves of computer programming produced, the word processor was slightly anomalous in that it was created to satisfy the needs of writers rather than those of mathematicians and engineers. After the introduction of the Graphical User Interface (GUI) in the late 1960s, there was some lofty talk about computers soon being able to take the place of books. The idea of using computers to do the work of typewriters also began to gain momentum.

Specialist computer hardware designed to act as a word processor already existed. It had a keyboard, a screen, and a printer all housed in one stand-alone unit that utilized modern technology to create printed, processed documents. But one young entrepreneur saw the chance to create a software program that would not be tied to any one particular machine.

Seymour Ivan Rubinstein (b. 1934) had been working as a director of marketing for the IMSAI

Manufacturing Corporation—a company that made microcomputers—but left in 1976 with $8,500 in cash to set up his own company, MicroPro International, Inc.

Rubinstein brought a number of IMSAI employees with him to staff his company, most notably Rob Barnaby. Within a few months, Barnaby had completed two programs: WordMaster and SuperSort.

Barnaby's magnum opus, however, came with WordStar, the first commercially successful word processing software. Feature-rich and easy to use, it fought off all of the emerging competition on the market to become the most popular word processing software in the world. **CL**

SEE ALSO: COMPUTER PROGRAM, C PROGRAMMING LANGUAGE, JAVA COMPUTING LANGUAGE, MICROPROCESSOR

⊡ *Barnaby's resume in WordStar. The program's most popular features was its integrated printing facility.*

Spreadsheet Program (1978)

Bricklin and Frankston speed up data sorting and calculations.

Modern offices rely heavily on spreadsheets for calculations and sorting data. The idea of transferring what used to be a large sheet of accounts that had to be calculated by hand to a computer was first proposed by Professor Richard Mattessich in 1961. In 1969, a spreadsheet-type application was developed by Rene Pardo and Remy Lamau for use on mainframe computers and was used for budgeting in some major companies. The spreadsheet in its modern incarnation, however, was conceived in 1978 by Dan Bricklin (b. 1951), a student at Harvard Business School.

Bricklin imagined a computer program that would show the data it contained and enable easy manipulation and calculations. It would also recalculate instantly if the initial data changed. The prototype spreadsheet was not very powerful, so Bricklin recruited Bob Frankston (b. 1949), from Massachusetts Institute of Technology (MIT), to improve it. Frankston added better arithmetic and the

ability to scroll to the application. Importantly, he also kept the program size down to 20 kilobytes, making it ideal for running on the personal microcomputers of the day. The software was named VisiCalc.

VisiCalc became an almost instant success and was instrumental in convincing businesses to invest in computers (in this case the Apple II). It paved the way for other spreadsheet programs such as Lotus 1-2-3, which contributed to the popularity of the IBM PC, and Microsoft Excel, which is the most commonly run spreadsheet today. Bricklin and Frankston never patented their invention because software was not eligible for a U.S. patent until 1981. **HP**

SEE ALSO: COMPUTER PROGRAM, C PROGRAMMING LANGUAGE, MAGNETIC CORE MEMORY, MICROPROCESSOR

⬆ *Dan Bricklin (right) with Bob Frankston; they worked in Bob's attic to develop the electronic spreadsheet.*

X-ray Telescope (1978)

NASA launches a super-sensitive, orbiting telescope.

"For NASA, space is still a high priority.... If we don't succeed we run the risk of failure."

Dan Quayle, U.S. vice-president (1989–1993)

⤒ *The Einstein Observatory satellite carrying two X-ray imaging telescopes, shown prior to its 1978 launch.*

⇥ *A bright X-ray image of quasar 3C 273 taken by the Einstein Observatory.*

X-rays are absorbed by Earth's atmosphere so one has to climb above the atmosphere to see them. NASA's Uhuru (1970) and the United Kingdom's Ariel V (1974) were spin-stabilized satellites that discovered around 400 bright X-ray sources. Astronomers realized that X-rays provide vital clues to the death-throws of stars, specifically supernova explosions and the final transitions to white dwarf, neutron star, and black hole states. X-rays are a vital component of the radiation coming from energetic events such as solar flares.

Two technical advances helped the development of space telescopes. One was the construction of advanced confocal mirror systems and the other was the development of two-dimensional X-ray imaging gas scintillation proportional counters. Using these, the United States launched the first orbiting satellite containing a fully imaging X-ray telescope, the High Energy Astrophysical Observatory-2 (HEAO-2), in November 1978.

HEAO-2 was renamed "Einstein" when it was in orbit and operating correctly. Einstein discovered that nearly all astronomical bodies emit X-rays. Also the angular resolution (a few seconds of arc) and the sensitivity was such that accurate maps could be made of objects such as the Cygnus loop supernova remnant. The Einstein instruments were a thousand times more sensitive than those on Uhuru.

Einstein remained operational until April 1981. Other space telescopes have followed, such as Exosat, Rosat, Chandra, and XXM Newton, and all have continued the quest for ever higher detail and sensitivity. In addition to new sources being discovered, it has been found that many X-rays are being emitted as material falls into the black holes at the center of active galactic nuclei. **DH**

SEE ALSO: GLASS, GLASS MIRROR, TELESCOPE, LENS, RADIO TELESCOPE, HUBBLE SPACE TELESCOPE

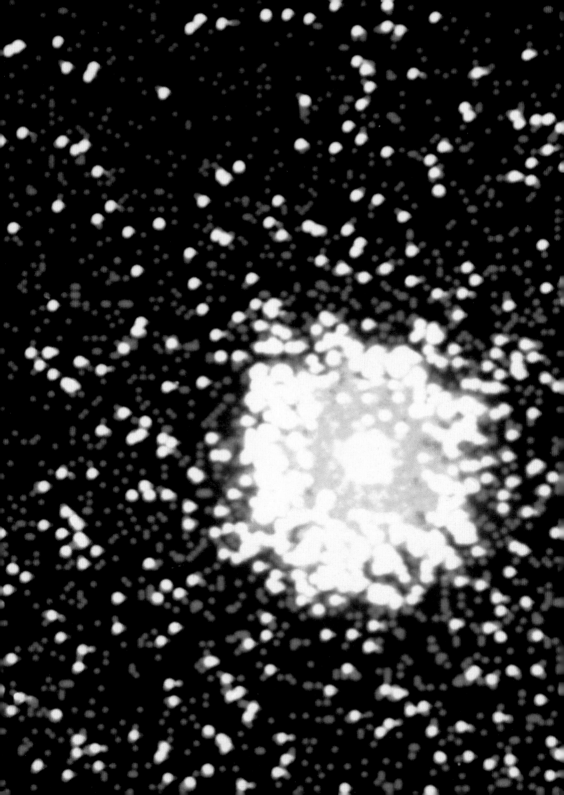

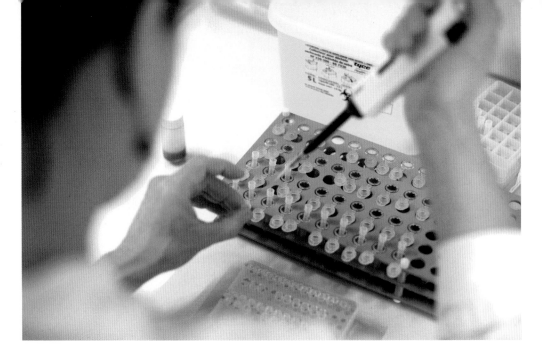

Gene Therapy (1978)

Zamecnik halts the replication of certain genetic materials.

Modern medicine has come a long way. We can treat pneumonia, hypertension, diabetes, and even heart failure—but certain genetic diseases have proved more difficult to deal with. That may change in the near future with the application of techniques pioneered by Paul Zamecnik (1912–2009).

The use of genetic manipulation has had a place in medicine for quite some time in the form of recombinant DNA. This technique has been the source of many medicines, but is not really manipulation of genetic material in an individual. Zamecnik, though, may have revolutionized gene therapy, not by inserting new genes into an individual, but by blocking the genes that were already there. While studying a virus prone to causing certain cancers in chickens, Zamecnik realized that rather than adding a gene, he could manipulate the viral RNA so that it would not be able to reproduce. He figured he could take advantage of the fact that most genetic material, be it DNA or its cousin RNA,

exists as a series of nucleotides that at some point pair up with other nucleotides. Zamecnik believed that if he could determine how to block that step, he might be able to prevent any number of bad outcomes.

In a study published in 1978, Zamecnik showed how one could prevent the normal function of a gene using a short strand of DNA that contained the base pairs opposite to the ones on the first strand. These strands, called oligonucleotides, tend to attach to the first strip and block any translation of genetic material. The technology, called antisense, is now in development for the treatment of everything from leukemia to malaria. **BMcC**

SEE ALSO: GENETICALLY MODIFIED ORGANISMS, STEM CELL THERAPY, TISSUE ENGINEERING

⊞ *A technician transfers DNA samples into Eppendorf tubes at a research institute in Brest, France.*

Air Car (1979)

Miller develops a compressed air vehicle.

In 1979, American Terry Miller showed that it was possible for a car to run on compressed air alone. After developing his Air Car One, which he built for $1,500, Miller patented his method in 1983.

Instead of burning fuel to drive pistons with hot expanding gas, air cars used the expansion of compressed air to drive the pistons. Initially energy is involved in compressing the air, and this is usually done with electricity, but it is still a more environmentally friendly process than that of gasoline cars. Many companies are developing air cars, though they are yet to be released for the public; this is likely to happen in the very near future.

There are a few drawbacks that the air car must overcome before it hits the market big time. When air expands from its compressed state, the engine is cooled and this can encourage icing. Also, in the event of an accident, the compressed air tanks are liable to explode. However, the air car certainly has advantages over gasoline-powered cars and other designs for the future, not least of which that it costs 20 percent less than a current car. Without a combustion engine, the wear and tear on internal parts is minimal and, with zero harmful emissions, the air car is an attractive design for the next step in car manufacturing.

Even with its low maintenance costs, the air car is arguably still the underdog in the race for the next generation of cars. Fuel cell and hydrogen-based models, as well as various hybrid designs, are ahead in the running, but as none of these have hit the commercial market yet, the top design for the future of the car is still to be decided. **LS**

SEE ALSO: ELECTRIC CAR, MOTORCAR, HYBRID CAR, DRIVERLESS CAR

⬆ *A CAD image of the engine of an air car; fully charged the vehicle can cover 93 miles (150 km).*

Antiviral Drugs (1979)

Scientists develop targeted drug therapy.

It is a medical practitioner's dream—a drug that will affect only the metabolism of a virus or cancer, with no side effects for the patient—the so-called "magic bullet." U.S. scientists George Hitchings (1905–1998) and Gertrude Elion (1918–1999) produced not one but a whole line of these type of medicines.

Hitchings hired Elion to work in his lab in 1944, and together they conducted experiments to study the differences between how DNA is synthesized in normal human cells, cancerous cells, bacteria, and viruses. They created new compounds similar to nucleic acids—the building blocks of DNA—that would interfere with the virus (or cancer) cell's ability to reproduce, but would not affect normal healthy human cells. Their work went on to revolutionize the way drugs were developed because they came up with the idea of "rational" drug design. Instead of the usual, time-consuming, trial and error method of hoping to find a chemical that would affect a disease, Hitchings and Elion actively designed molecules that would be accepted by the cells of foreign bodies and be actively absorbed by them, delivering the drug direct to its target.

The pair worked together for forty years, perfecting one medicinal compound after another. Their work led to the development of new drugs for leukemia, gout, malaria, herpes, organ transplant rejection, rheumatoid arthritis, and the first-ever treatment for AIDS. Hitchings and Elion were awarded the Nobel Prize in Medicine or Physiology in 1988, together with British scientist, James Black (1924–2010), who discovered beta blockers for high blood pressure and H-2 antagonists for ulcers. Together they had definitively proved that chemistry could be used to fight infection and cancer. **JM**

SEE ALSO: ANESTHESIA, ASPIRIN, DDT, ACETAMINOPHEN, PENICILLIN, TETRACYCLINE, STATINS, HUMULIN®

Bioethanol (1979)

Fiat runs an engine entirely on ethanol.

Bioethanol is a high-octane alcohol produced from sugar or starch and is considered an important alternative fuel to petroleum-based products. One of the first fuels used in the automobile business, it was used extensively during World War II in Germany, the United States, Brazil, and the Philippines. After the war, bioethanol was generally replaced by cheaper petroleum-based fuels.

Brazil has been the worldwide leader of bioethanol development since the 1973 oil crisis, with the initiation of the Brazilian Alcohol Program (PROALCOOL), whose aim was to produce bioethanol to combine with gasoline. Sugar cane was to be used as the source of bioethanol, and in the first phase of the project (from 1975 to 1979), distilleries were attached to existing sugar mills. In the later phase of the project, autonomous distilleries were built for bioethanol production.

The 1979 crisis, when the price of crude oil soared following the Iranian Revolution, reinforced the need for Brazil to become independent of foreign oil, and the government ordered automobile manufacturers to shift production to new engines that would run exclusively on ethanol. The first of these was a Fiat model 147, produced by Fiat's local subsidiary.

The PROALCOOL program lost momentum in the 1980s due to a drop in oil prices, and again in the 1990s due to an ethanol supply shortage. However, the market has turned around again since the late 1990s and sales of cars powered by bioethanol have risen.

A question still remains about the carbon footprint of bioethanol production. It appears that the labor and energy requirements of producing bioethanol are high and that it may someday be replaced by other alternative fuel sources. **RH**

SEE ALSO: BIODIESEL, DIESEL, THERMAL CRACKING OF OIL

Maglev Train (1979)

Danby and Powell develop a new technology for train transport.

Magnetically levitated (maglev) trains glide above a track, propelled by superconducting electromagnets. The principle is more than a century old, but initially the huge electrical currents needed to provide a sufficiently strong magnetic field were impractical. The breakthrough came when two physicists, Gordon Danby and James Powell, at Brookhaven National Laboratory decided to use high temperature superconductors as electromagnets. They obtained a patent for the technology in 1968, and by 1979 visitors to a transportation exhibition in Hamburg, Germany, were enjoying a short test run on a Transrapid maglev train.

Maglev trains need a guiding track and the carriages float just above it. Changing the field produced by the electromagnets in the guideway pulls the train along. The only friction is due to air resistance, so extremely high speeds are possible. Changes in field strength can produce very high accelerations, and much more variable track gradients can be accommodated than with normal trains, so cuttings and embankments would not be needed.

There are serious disadvantages to maglevs. The guideways are extremely expensive, and the trains cannot be diverted from the maglev track to normal railways to take in existing inner city termini.

Maglev trains were introduced in the United Kingdom in 1984, linking Birmingham airport to the nearby railway station. In China, another system links Shanghai to Pudong International Airport. The success of the TGV trains that run on normal track, however, has reduced the appeal of maglev. **DH**

SEE ALSO: **ELECTRIC TRAM, LOCOMOTIVE, CABLE CAR, ELECTROMAGNET**

⤤ *The maglev train route in Shanghai will be extended 106 miles (170 km) to reach Hangzhou.*

"Learn to deprive large masses of their gravity and give them absolute levity, for the sake of transport."

Benjamin Franklin

Post-it® Notes (1980)

Silver and Fry invent a new stationery item.

The Post-it® is a small reminder note, stuck temporarily to documents, computers, and other prominent spots. Launched commercially in 1980 by Arthur Fry (*b.* 1931) and Spencer Silver (*b.* 1941), employees of 3M in the United States, the notes are available in a wide range of shapes and colors, although the original yellow, three-inch (7.5 cm) square note is still the most popular.

In 1968 Silver developed a "low-tack" reusable adhesive made of tiny, indestructible acrylic spheres. Sticking to a given surface at a tangent rather than flat against it, the adhesive was sufficiently strong enough to hold papers together, but weak enough to separate them without tearing. Silver envisaged its application as a spray, or as a surface for notice boards. He spent

> *"[Give people] a Post-it® note and they immediately know what to do with it and see its value."*

Arthur Fry

five unsuccessful years promoting his idea within the company. However, in 1974 a colleague, Arthur Fry, attended one of his seminars, and a few days later, during a sermon at his church, he realized that the low-tack adhesive would solve a frustrating problem: He sang in the choir and saw that if his bookmarks were temporarily and lightly stuck into his hymnal, they would not fall out.

3M eventually agreed to invest in the product, taking five years to develop, design, and build the machines needed for manufacture. In 1980, after a sustained marketing campaign, the notes took off and were soon to be found all over the world. **EH**

SEE ALSO: GLUE, POSTAGE STAMP, TAPE, SUPER GLUE

Lithotripter (1980)

Dornier develops a remedy for kidney stones.

The lithotripter gave the world a new, noninvasive way to treat kidney stones with little to no pain. Discoveries in aerospace engineering and research on shock waves in the 1970s led to the technology's birth.

Early studies at the Dornier research group (founded by German engineer Claude Dornier) focused on aerospace technology. One new phenomenon noted by Dornier scientists was the pitting effect that occurred in airplanes as they approached the speed of sound. It was discovered that this was caused by shock waves created in front of droplets of moisture. This finding in 1974 was the beginning of a collaboration between Dornier engineers and hospitals, and led to the invention of clinical extracorporeal shock wave lithotripsy (ESWL).

In 1980 this new technology was used to treat its first patient, using the Dornier HM1 lithotripter. The treatment of kidney stones begins with an X-ray to identify the localization of the stone, followed by ESWL's high-frequency shock waves, which eventually break up and pulverize the stone. The process takes up to forty minutes, causing minimal collateral damage. The power level is slowly increased, so that the patient can get used to the feeling, with the final power level depending on the patient's pain threshold. The patient then passes the fragments of stone with the urine, usually over the next three weeks. The first-time success rate of the ESWL procedure is around 95 percent.

The lithotripter is used for all types of stone removal in the urinary tract. Shock waves are also used in orthopedics for the treatment of conditions such as heel spurs, tennis elbow, and calcinosis. It enables doctors to treat patients in a safe, effective, and noninvasive way. Millions of people are estimated to have benefited from this technology. **RH**

SEE ALSO: KIDNEY DIALYSIS, HEART-LUNG MACHINE, VENTILATOR, ARTIFICIAL LIVER

Liquid Soap (1980)

Minnetonka Corporation markets an old product in a new dispenser.

It is hard to believe that U.S. patent number 49,561 took more than a century to become widespread, but that is exactly what happened with liquid soap. It was New Yorker William Shepphard who discovered in 1865 that, by mixing large amounts of ammonium bicarbonate with normal soap, he could produce a soap that had the consistency of treacle.

Although his invention slowly began to appear in dispensers in public, it was not until 1980 that the U.S. Minnetonka Corporation began selling it widely to the general public. Under the brand name "Softsoap®," the liquid hand cleaner was an instant success, partially due to the way it was dispensed. On the top of its bottle was a small plastic pump that, when depressed, issued a fixed quantity of soap onto the hand. Minnetonka had entered a market that was dominated by large global corporations. Rather sneakily, Minnetonka bought up all the available plastic pumps from suppliers, putting a stranglehold on the market to stop anyone from threatening their survival.

Eventually, in 1987, the company sold its liquid soap brand to the Colgate Company, which continues to market and sell Softsoap® all over the world—still in its characteristic bottles complete with plastic pumps.

The creation of liquid soap itself was not rocket science and, as with many stories of successful innovations that lead to big business, it was the combination of inventions that led to the widespread use of the product. It was the use of the soap and the pump together that was new in 1980, and the humble plastic pump has since migrated on to dispense toothpaste and shaving cream. **DHk**

SEE ALSO: TOOTHPASTE, SOAP, DEODORANT

↗ *Initially disregarded, liquid soap became popular after a plastic pump became part of the product.*

"It's going to be a pump war out there . . . [Pump dispensers are] a real trend—not a fad."

Jack Salzman, industry analyst

Flash Memory
(1980)

Masuoka finds a new way to save data.

Since its infancy, modern computer memory, namely RAM, has suffered from a fundamental problem. It remembers things when it has power, but pull the plug out and its carefully held pattern of 0s and 1s fades and dies. The traditional solution is to use disks or tapes to save the contents of the computer's memory more permanently, but these have disadvantages. They are slower than RAM, and their motors and other moving parts use more electricity and do not like being shaken around.

In 1967 Simon Sze and Dawon Kahng invented a transistor that could remember a programmed state even without power. They programmed their "floating gate transistor" by forcing electrons onto a part of the transistor that was normally electrically isolated—the floating gate. When the power was turned off, this electrical charge was trapped, potentially for years.

While memory chips based on Sze and Kahng's invention were produced, they were complicated and expensive. In 1980 Toshiba's Fujio Masuoka (b. 1943) made a crucial improvement. Realizing that a lot of the complexity was in the clearing of the memory, where each individual memory cell was erased one by one, he came up with a new design. His method connected blocks of memory cells together, and allowed them all to be erased at once—in a "flash."

Masuoka's "flash" memory was cheaper to produce and is now used in any portable application where batteries might fade or devices might be dropped, such as mobile phones, digital cameras, and music players. As its capacity increases, and price comes down further, it is also replacing hard disks and tapes in video cameras and laptop computers, making them lighter and more reliable than ever. **MG**

SEE ALSO: COMPUTER PROGRAM, MAGNETIC CORE MEMORY, MICROPROCESSOR, RANDOM ACCESS MEMORY (RAM)

Hepatitis B Vaccine (1980)

Blumberg creates the first anticancer vaccine.

There are few people who can be said to have saved the lives of millions, but American scientist Dr. Baruch Blumberg (1925–2011) is one of them. In the 1960s he and his colleagues were screening aboriginal blood for diseases when they found a rare protein. They named it the "Australian antigen" and investigated whether it also occurred elsewhere in the world. It turned out to be uncommon in Americans but much more prevalent in Asians, Africans, and some Europeans. They discovered it was also found in leukemia sufferers who were receiving regular blood transfusions.

Further population studies pointed to the antigen being part of a relatively unknown virus that caused a

> *"'Australian antigen' was the Rosetta stone for unraveling the nature of the hepatitis viruses."*
>
> Robert H. Purcell, National Institute of Health

particularly virulent form of hepatitis—hepatitis B. Hepatitis B is a serious disease that attacks the liver causing cirrhosis (scarring), often leading to liver cancer and liver failure.

The fact that they knew this protein was linked to the virus meant they could test for it, which proved especially useful in screening donated blood and reducing the risk of hepatitis B being transmitted by blood transfusion. Blumberg and his team then started work on harvesting the outer coat of the virus from the blood of chronic carriers; this was a key step toward developing a working vaccine. Blumberg was awarded the Nobel Prize in 1976. **JM**

SEE ALSO: VACCINATION, CHOLERA VACCINE, BCG VACCINE, RABIES VACCINE, POLIO VACCINE, RUBELLA VACCINE

Artificial Skin
(1981)

Burke and Yannas offer hope of renewal for victims of severe burns.

Human skin is a marvel of engineering. It is tough yet stretchy and pliable, and acts as an impermeable barrier against water loss, infection, and cell damage from the sun's ultraviolet (UV) rays. With this range of properties, it is a very difficult material to duplicate.

John F. Burke, a surgeon at Massachusetts General Hospital in the United States, was looking for a reliable skin replacement for the treatment of burn victims. Skin is usually grafted from other parts of the patient's body, but in cases where the burns cover 50 percent or more of the body, often there is not enough healthy skin to cover the damaged area.

In the 1970s Burke teamed up with Ioannis V. Yannas, a chemistry professor at Massachusetts Institute of Technology (MIT), who was studying a stretchy protein called collagen that naturally occurs in animal tendons. Burke and Yannas combined collagen fibers taken from cowhide with long sugar molecules from shark cartilage to create a gridlike polymer membrane. They dried this membrane and stuck it onto a layer of viscous plastic.

The two layers, about as thick as a paper towel, offer a barrier against infection and dehydration while acting as a scaffold on which new skin cells can grow. As the patient's skin grows back, the artificial membrane breaks down naturally and can be peeled away. The new skin is not scarred and looks like normal skin, albeit without sweat glands or hair follicles.

In 1981 Burke and Yannis proved that their artificial skin worked on patients with 50 to 90 percent burns, vastly improving their chances of recovery. **JM**

SEE ALSO: ARTIFICIAL LIMB, ARTIFICIAL HEART, POWERED PROSTHESIS, ARTIFICIAL LIVER, CHITOSAN BANDAGES

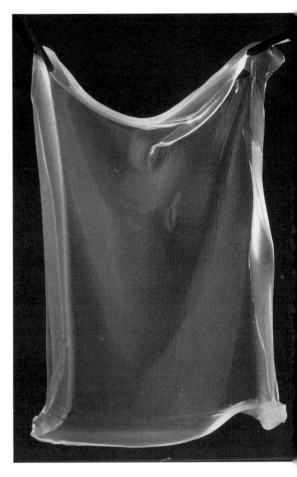

↗ Grown in the laboratory, epidermal strips like this one can be used to treat skin diseases such as vitiligo.

"[The artificial skin] is soft and pliable, unlike other substances used to cover burned-off skin."

John F. Burke

Musical Instrument Digital Interface (MIDI) (1981)

Smith suggests a common language for all electronic musical instruments.

During the early 1970s, sounds generated by the electronic synthesizer became increasingly popular in recordings and at rock concerts. Gradually, these once prohibitively expensive musical instruments became commonplace and affordable. In addition to a great variety of user-friendly electronic keyboards on offer, there were related devices, such as sequencers, that could "trigger" sounds from a connected keyboard, as well as drum machines with a variety of sounds.

One problem that arose with this electronic proliferation was that devices produced by different manufacturers tended not to be compatible with each other. American audio engineer Dave Smith sought a way forward when, at a 1981 meeting of the Audio Engineering Society, he presented a paper that proposed the first universal communication standard for musical equipment. He called it MIDI—an acronym for Musical Instrument Digital Interface. In essence, MIDI is a digital "language" that enables synthesizers,

MIDI recorders (whether hardware sequencers or computer-based software), drum machines, and other similarly equipped devices to "talk" to one another by sending and receiving messages via interconnecting MIDI cables. The simplest use of MIDI would be two synthesizers connected in such as way as to enable the sounds of both instruments to be played and controlled from just one of the keyboards.

Smith's paper was immediately adopted by manufacturers. Indeed, it would be no exaggeration to suggest that without MIDI most of the programmed and sampled electronic music of the past twenty-five years simply could not have been made. **TB**

SEE ALSO: PIPE ORGAN, METRONOME, ELECTRIC GUITAR, ELECTRONIC SYNTHESIZER

⬆ *MIDI enables synthesizers to be linked to—and even controlled by—other pieces of digital equipment.*

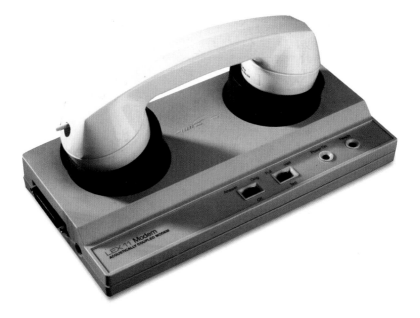

Personal Computer Modem (1981)

Hayes simplifies the link between computers and the telephone system.

The first modems date back to the cold war and 1958, when the North American Air Defense transmitted data over telephone wires to hundreds of radar stations in the United States and Canada. A modem has one essential function, which is to translate the digital language of computers into the analog language of the telephonic system and back again.

In 1981 Dennis Hayes (b. 1950) launched the Smartmodem (originally named the Hayes Stack Smartmodem). This was an automatic modem that for a time led the rapidly emerging personal computer market. Earlier modems were not adaptable to a variety of computers, were expensive to produce, and also were cumbersome to operate, requiring manual connection to telephone lines.

The brilliance of the Smartmodem lay in its ability to "think for itself" and program itself into the telephone networks. It did this by using its own data language to instruct itself to engage and disengage with other computers' phone lines according to the requirements of its own operator. It also opened the way to cheaper and smaller designs because, needing only data instructions from its host computer, it could be connected up to any computer by an easily accessible port.

By 1985 Hayes's company held nearly half the personal computer market and the term "Hayes compatible" had entered the language of computers as a benchmark against which to measure rival modems. The term was used to describe any modem that could claim to recognize the command sequence devised for Hayes's Smartmodem. **AE-D**

SEE ALSO: INTERNET, PERSONAL COMPUTER, CABLE MODEM

⊞ *A modem from the 1980s with an attached telephone receiver. Modems convert digital signals to analog.*

Reusable Spacecraft (1981)

NASA deploys the first Space Shuttle, designed to make multiple trips from Earth.

Booster rockets—such as the *Saturn V* that launched the *Apollo* astronauts toward the Moon—are extremely wasteful. They can fly only once, parts are thrown away after use, and 97 percent of the mass is consumed in the first few minutes. Clearly, what was really needed for advanced and economic space exploitation was a spacecraft that could take off and return, to be reused time after time.

In 1972 NASA in the United States decided to build the Space Shuttle. Rockets were to be used to assist the launch. Crew facilities were to be provided for up to eight people, and the huge cargo bay would be used to take satellites and sections of the International Space Station (ISS) into orbit. New instruments would also be taken up to existing spacecraft, and in-orbit repairs would be carried out.

The Space Shuttle would glide back to earth through the atmosphere, the nosecap and underwing tiles reaching temperatures of 2,600°F (1,430°C) as it came in. The craft would then land on a normal runway. *Columbia*, the first Shuttle, blasted off on April 12, 1981. *Challenger*, *Discovery*, *Atlantis*, and *Endeavor* followed. Everything went fine until the twenty-fifth flight, in January 1986. One of the solid rocket boosters blew up, leading to the death of the *Challenger* crew.

The Russian shuttle, *Buran*, had a test flight in 1988 but was soon abandoned. In January 2003 the U.S. craft *Columbia* broke up during re-entry. After a hiatus of two-and-a-half years, the U.S. launches were restarted. Six were planned for 2008, in an attempt to finish the construction of the ISS.

No U.S. Space Shuttle launches are scheduled beyond 2010. Even though *Discovery* has made thirty trips, the cost and the long turnaround time have never been satisfactory. **DH**

"The Space Shuttle is the most effective device known to man for destroying dollar bills."

Dana Rohrabacher, U.S. Congressman

⬆ *The Space Shuttle* Atlantis *is readied for launch prior to its mission to the Hubble Space Telescope in 2008.*

➡ *In 1998 the crew of the Russian space station* Mir *took this picture of the U.S. Space Shuttle* Discovery.

SEE ALSO: POWERED AIRPLANE, LIQUID FUEL ROCKET, BALLISTIC MISSILE, SPACE PROBE, SPACE STATION

Scanning Tunneling Microscope (1981)

Binnig and Rohrer make it possible to "see" at the atomic level.

As recently as thirty years ago the idea of being able to "see" bumps and grooves on substances at an atomic scale seemed unachievable. Then, in 1981, Gerd Karl Binnig (*b.* 1947) and Heinrich Rohrer (1933–2013) created a microscope capable of doing just that.

Their scanning tunneling microscope (STM) bears little resemblance to a conventional microscope. Operating at a low temperature and in a vacuum, it consists of a very sharp needle (its end is the width of a single atom), which can be brought very close to the sample being examined. It is a tool that "senses" rather than "sees" because it deals with sizes that are smaller than the wavelength of light.

The STM exploits a phenomenon of quantum mechanics known as electron tunneling where electrons can jump (or "tunnel through") small gaps between atoms. In electricity-conducting materials, such as metals, electrons flow from one atom to another. If the STM needle tip is projected close enough, this electron flow can be detected as the electrons tunnel from the metal surface to the tip. The needle tip can then follow the contours of the surface to reveal its shape. The result is a picture of the molecule in the nanometer range (one nanometer is a billionth of a meter, or the width of three atoms).

The STM's needle also allows scientists to move individual atoms by applying an electric field to the sample. The tip attracts the electrons making up the atoms of the sample and pulls them around. In 1990 this was famously used to make the smallest branding in the world when the IBM logo was created out of thirty-five xenon atoms on a nickel plate. **JM**

SEE ALSO: LENS, MICROSCOPE, TELESCOPE, ELECTRON MICROSCOPE

"As a research scientist you are driven by the desire to find things out with all your might."

Gerd Karl Binnig

◩ *This scanning tunneling microscope from 1986 was the first to be produced commercially.*

Humulin® (1982)

Eli Lilly produces human insulin.

The year 1982 marked a huge step in diabetes care and ushered in a new era of drug production, thanks to the development of Humulin®, the first fully human insulin product. Until then, diabetes patients were given insulin derived from animal sources, mostly cattle and pigs. At around the same time, advancements in gene technology finally allowed for the manufacturing of fully human insulin.

The molecule insulin was discovered in the early 1920s, and the first injections of insulin from cattle into humans quickly followed. Although this early insulin was extremely impure and had numerous side effects, it certainly saved the lives of many diabetics.

In the following decades there were further advancements in the development of insulin, including improvements in purity and the chemical synthesis of human insulin. In 1978 researchers at Genentech, Inc., in San Francisco, California began working on producing fully human insulin from recombinant DNA in the bacterium *Escherichia coli* (*E. coli*). The development and manufacture of this type of insulin was quickly picked up by the Eli Lilly company in 1982, when Humulin® production was approved. Two types were produced—Humulin® R (for rapid effect) and Humulin® N (a longer-lasting version).

Humulin® was the first nonanimal-based insulin, as well as the first approved genetically engineered pharmaceutical product. DNA was made synthetically, put into *E. coli* bacteria for processing into protein, and the resulting product purified. Humulin® was free of any animal contaminants and could be produced in huge quantities. This was welcome news to diabetes patients who depended on animal-based insulin. Four types of Humulin® are produced by Eli Lilly today, for use by approximately 4 million diabetes patients. **RH**

SEE ALSO: HYPODERMIC SYRINGE, DISSOLVABLE PILL, INSULIN PUMP

Halogen Cooktop (1983)

Schott creates a heat-resistant cooking surface.

Schott, the German specialty glass producer, originally made glass ceramics for use in astronomical telescopes. In 1983 they introduced CERAN®, a glass-ceramic cooktop panel for halogen cooking. Hot and long-lasting, halogen bulbs were perfect for cooking, but their application in the kitchen had been limited by available cooktop materials. CERAN® brought the power of halogen bulbs to chefs everywhere. The cooktops contain reflectors beneath the bulbs to direct maximum heat to the cooktop surface.

Everyday incandescent bulbs consist of a tungsten filament encased in blown glass that contains an inert gas, such as argon. Halogen bulbs are similar except that, instead of an inert gas, a halogen gas—usually

> *"If all these cooktop[s] were laid out in a line, they would cover about three quarters of . . . the earth."*
>
> Schott

iodine or bromine—is used. Using such gases ramped up the bulbs' light output and radiant heat, but also demanded a special cooktop. CERAN® is translucent, thermally stable, and highly heat-resistant.

Halogen cooktops provide a flat cooking surface, but flat cooktops had been around long before CERAN®. Electrical coils placed under glass-ceramic cooktops had first brought flat cooktops to kitchens, but these were no match for the heat of halogen bulbs. Schott's new CERAN® halogen cooktops proved highly popular; the company has sold over 50 million of them, incorporated into appliances produced by many different manufacturers. **RBk**

SEE ALSO: OVEN, PRESSURE COOKER, GAS STOVE, ELECTRIC STOVE, HALOGEN LAMP, MICROWAVE OVEN

Camcorder (1983)

Sony releases the first handheld video recorder.

Today mobile video-recording technology must fit into the palm of the hand, or be integrated into the back of a mobile phone, before anyone would consider paying any money for it. But before the camcorder was invented, anyone wanting to capture moving footage on film had to use an incredibly unwieldy two-part machine. Worse still, the camera itself was all that one person could reasonably carry, so a partner had to be persuaded to carry the video cassette recorder (VCR) alongside.

The older equipment also had no playback screen, so whenever it was necessary to watch recently recorded material, a television screen had to be available nearby in which to plug the VCR.

The cumbersome equipment was troublesome, particularly for broadcast journalists, movie makers, students, and others working in the field. However, advances in technology and design meant that soon various pioneers of video-recording equipment were shrinking down the basic component parts and creating a one-piece camera/recorder combination—what is now known as the camcorder.

The first commercially available camcorder was designed by Sony in Japan. Making use of Betamax video technology, the Betamovie model was unleashed upon the general public in May 1983. Even then there were no playback functions on the camera, and anyone wanting to watch or rewind a section of footage still had to eject the tape and insert it into a home Betamax player and watch it on a separate television screen. Really, though, the roving reporter's lot had improved immeasurably. **CL**

SEE ALSO: VIDEOTAPE RECORDING, OPTICAL DISC, ELECTRONIC NEWS GATHERING

⬆ *The Betacam system used the same cassette shell as the domestic Betamax but with a different tape inside.*

Laptop/Notebook Computer (1983)

Compaq releases the first commercially successful laptop.

Today's laptop computer has evolved over decades from different types of portable computers, but the Compaq was the most successful early model.

Alan Kay of the Xerox Corporation proposed the Dynabook concept in 1971. His idea was to create a portable, networked personal computer. However, at the time there was no market for it so the idea was shelved. In 1981 Alan Osborne of the Osborne Computer Corporation invented the Osborne 1, the first fully portable personal computer. The size of a small suitcase, it weighed about 24 pounds (11 kg).

The first clamshell design was the GRiD Compass 1101, invented by Bill Moggride and released in 1982. Galivan Computer released what is considered to be the first true "laptop" computer in 1983; it was the smallest and lightest portable computer to date.

However, it was Compaq Computer Corporation that stole the market from these rivals in 1983, with the Compaq Portable. Rod Canion, Jim Harris, and Bill Murto founded Compaq in 1982 after leaving Texas Instruments. The idea for the Compaq Portable was supposedly sketched out on a placemat from a Houston pie shop. The computer was "reverse-engineered" using IBM BIOS source code to create a new version of a system that operated like IBM's. This was important considering IBM's huge success in the computer market during this time.

Compaq enjoyed record-breaking revenue in 1983, the year it first released the Portable. Following the model's success in the market, laptops have evolved even further, developing into the smaller and faster models of today. **RH**

SEE ALSO: MECHANICAL COMPUTER, DIGITAL-ELECTRONIC COMPUTER, PERSONAL COMPUTER, PALMTOP COMPUTER

⬆ *The Osborne 1 portable microcomputer of 1981 paved the way for the Compaq Portable.*

Internet Protocol (TCP/IP) (1983)

Kahn and Cerf enable computers to "talk."

When we send files or e-mails to others over the Internet, few of us care how our messages arrive, almost instantaneously, at a distant terminal, with no direct connection between the two. But without the decentralized magic of Transmission Control Protocol (TCP) and Internet Protocol (IP), these remarkable events would not be possible. TCP and IP are separate networking protocols, but so intimately linked in use that they are generally referred to as a single entity.

Developed by Robert Kahn (b. 1938), Vinton Cerf (b. 1943), and others, and initially used on the U.S. Government's Advanced Research Projects Agency (ARPA) packet-switching network in 1983, TCP and IP

> *"The Internet … allows [us] at each level of the network to innovate free of any central control."*
>
> Vinton Cerf

permitted an ever growing number of remote networks to connect to each other and ultimately mature into the Internet. In essence, TCP and IP is "layered." The higher (TCP) layer concerns itself with dividing files or messages into smaller chunks, or "packets," for transmission, and reassembling received packets into their original form. The lower (IP) layer deals primarily with addressing and routing each packet so that it gets to the proper destination. The routes via which the parts of each message travel may well be different to each other. It is a bit like sending the pages of a book separately, via different routes, to an address at which they are then reassembled. **MD**

SEE ALSO: INTERNET, E-MAIL, ETHERNET, DIGITAL
SUBSCRIBER LINE (DSL), WORLD WIDE WEB

Stealth Technology (1983)

Lockheed Martin builds an "invisible" plane.

After years of development, the Lockheed Martin Corporation's first viable stealth fighter aircraft, the F-117 Nighthawk, finally came into operation in 1983. The technology it contained made it practically invisible to conventional radar at the time, giving it a major military advantage over conventional aircraft.

The rounded shape of conventional aircraft improves their aerodynamics but also makes them highly visible to enemy radar. The F-117's designers found that an aircraft of sharp angles and triangles reflected back less radar. The surface of the plane is also coated in paint that absorbs radar signals, making the plane even more invisible. Other adaptations

> *"This is a strategic weapon that really reshaped how the Air Force looked at strategic warfare."*
>
> Lt. Col. Chris Knehans, 7th Fighter Squadron

include an unconventional tailfin and reduced turbulence on the engine outlets. However, the aircraft is inherently unstable and would be impossible to fly without its extensive computer systems.

The U.S. military denied the existence of the stealth aircraft until as late as 1988, when the first grainy photographs appeared. It first saw military action in 1989 and quickly became the weapon of choice for some operations. Its scheduled retirement in 2008 is due to replacement by more advanced aircraft. **SB**

SEE ALSO: AIRSHIP, POWERED AIRPLANE, LIQUID FUEL ROCKET,
JET ENGINE, SUPERSONIC AIRPLANE, SCRAMJET

➡ *The designers of the F-117A Nighthawk stealth fighter minimized its cross-section signature on enemy radar.*

3D Printing

(1984)

A new technology, built up layer by layer.

3D printing is a computer-controlled process that builds physical, three-dimensional objects. There are three ways of "printing" three-dimensional objects: stereolithography, selective laser sintering (SLS), and fused deposition modeling (FDM). In the first two, laser or ultraviolet light is used to harden polymers, from either a vat of liquid or a powder. In FDM, a thin filament of polymer is extruded in a molten state from a heated nozzle, and hardens in the air. In all three cases, the object being made is built up layer by layer.

Although people had explored the concept of 3D printing since the 1950s, 1984 was an important year, because three important patents were filed that year.

> "…When you get enough smart people working on something, it always gets better."
>
> Chuck Hall, co-founder of the 3D Corporation

American entrepreneur Bill Masters outlined a system in which droplets or particles of material (such as ceramic) could be deposited at precise points to build up 3D objects. Later that year, a team of French inventors filed a patent for a stereolithography system. Later still, American engineer Chuck Hall filed a U.S. patent for a stereolithography system.

Today, FDM (developed at the end of the 1980s), is the most common form of 3D printing, since affordable machines are now available. 3D printing is found not only in "rapid prototyping" and small-scale manufacturing, but also in building medical implants and even in the construction of buildings. **JC**

SEE ALSO: COMPUTER LASER PRINTER, LITHOGRAPHY, LASER, PRINTING PRESS WITH MOVABLE TYPE, ELECTRIC MOTOR

Self-driving Cars

(1984)

Artificial intelligence learns to drive.

Early proposals for driverless vehicles relied on external control. The most visionary example was created by American engineer Norman Bel Geddes. In 1939, he suggested that, by 1960, roads and cars would work together to "prevent the driver from committing errors." In Bel Geddes' system, dispatchers in control towers would control lane changes and other movements of cars and trucks on highways via radio control. A full scale test track based on this idea was trialed in the 1950s, albeit with cables installed in the road rather than radio signals, but the extremely high cost of the system meant it was never rolled out. A series of similar test systems were trialed in the 1960s and '70s, with cables or devices in the roadway.

In true self-driving vehicles, driving decisions are taken by the vehicle itself. Such systems could not be realized until the advent of powerful computers and artificial intelligence (AI), relying on other technologies, including lasers, radar, satellite navigation, and motion sensors. In 1984, Carnegie Mellon University began its NavLab program, which produced eleven autonomous and semi-autonomous vehicles and blazed the trail for modern driverless cars. In 1995, semi-autonomous Navlab 5 (a 1990 Pontiac Trans Sports) completed a coast-to-coast traverse of the United States.

In 2004, the U.S. Defense Advanced Research Projects Agency (DARPA) held the first of three Grand Challenges, offering substantial prizes to engineers whose vehicles could succeed at autonomous tasks. The head of the Stanford University team, which won in 2005, went on to lead a self-driving car project at Google, which began in 2009. Since then, many motor manufacturers and also taxi companies such as Uber have invested hugely in the technology. **JC**

SEE ALSO: MOTORCAR, DIGITAL ELECTRONIC COMPUTER, ARTIFICIAL INTELLIGENCE, GLOBAL POSITIONING SYSTEM (GPS)

Manned Maneuvering Unit (MMU)

(1984)

NASA frees astronauts from the tether with a powered individual transporter.

The advent of the Space Shuttle and space stations made it obvious to the National Aeronautics and Space Administration (NASA) that some method of extravehicular space locomotion was needed. The space agency called on the Lockheed Corporation (now Lockheed Martin) to build a device that would allow astronauts to manipulate satellites, repair and build structures, and perhaps even rescue colleagues engaged in extravehicular activities (EVAs).

In 1984 the Manned Maneuvering Unit (MMU) became reality. The backpack-like unit was designed to work alongside the life-support systems already in place on a space suit. It was stowed in the payload bay of the Space Shuttle when not in use and simply attached to the standard Extravehicular Maneuvering Unit (EMU) spacesuit when astronauts needed to work on a part of the space station or ship that they could not easily access when walking with a tether. The unit used twenty-four different thrusters to allow astronauts to maneuver with some degree of agility, but it was still small enough to allow precision work. Astronauts moved using a handheld control unit, and an autopilot could hold them at a constant attitude.

The MMU was deployed on three separate Space Shuttle missions, but it has not been used since the explosion of Space Shuttle *Challenger*. For the most part, tethered space walks have proved adequate to replace the device, but a Simplified Aide For EVA Rescue (SAFER) unit is present on the International Space Station; its primary function is to assist in the rescue of an accidentally untethered astronaut. **BMcC**

SEE ALSO: LIQUID FUEL ROCKET, BALLISTIC MISSILE, SPACE PROBE, SPACE STATION, REUSABLE SPACECRAFT

⬈ *The MMU was first flown in 1984 by astronaut Bruce McCandless from Space Shuttle* Challenger.

"Ever since I've been an astronaut I knew I wanted to do a spacewalk."

John L. Phillips, astronaut

Apple Macintosh (1984)

Apple launches the first successful personal computer with a graphical user interface.

Although computers were in use long before the 1970s, they were incredibly difficult to operate. Early computers had barely enough processing power to solve the problems they were given. Further, nobody had taken time to look into how the computers were given the problems: their user interface.

The first big inroads were made at Xerox's legendary Palo Alto Research Center (PARC). PARC's "Alto" computers drew on earlier innovations like Doug Engelbart's computer mouse of 1968. The Alto was the very first computer where a click with a mouse on a file would open it. It also had the earliest "what you see is what you get" word processor, showing documents on its screen just as they would look if printed out.

Adding menus and icons to allow the user to make choices easily, and putting programs in different windows on the screen, PARC developed the first graphical user interface. Suddenly, people with only minimal training could write out a letter, click a mouse button, and have it printed out on PARC's laser printer.

Xerox developed their user interface throughout the 1970s, and in 1979 PARC gave a demonstration of their ideas to a group of engineers from a small company called Apple Computer. Apple had been working on similar concepts for their top-secret "Lisa" computer, and they were amazed by what they saw. Apple cofounder Steve Jobs left PARC buzzing with enthusiasm. Apple integrated Xerox's user interface ideas with their own, extending them and developing a new way of working with computers. In 1984 they launched the first successful personal computer with a graphical user interface: the Apple Macintosh. **MG**

SEE ALSO: COMPUTER PROGRAM, GRAPHICAL USER INTERFACE (GUI), 3D COMPUTER GRAPHICS, C PROGRAMMING LANGUAGE

⊡ *The Apple Macintosh model M001 was designed to be unpacked, plugged in, and used immediately.*

Widget In-can System (1984)

Forage and Byrne replicate in cans the characteristics of draft Guinness.

When beer is stored in large quantities—in the cool cellar of a bar or pub, for example—it can be retained in the casks in which it was delivered, where its continued fermentation improves the taste and strength of the drink. When the cask beer is eventually drawn out by a pump, it mixes with nitrogen in the air and gains its characteristic texture and head. Guinness Irish stout is unusual in that its distinctive head is caused by special taps that ensure a precise mix.

When brewers started putting beer in bottles so that it could be drunk away from the cask, they found that the fermentation in bottles caused a much higher level of carbonation. All this carbon dioxide is good because it leaves no room for oxygen in the bottle, which would sour the beer, but it creates a much fizzier drink. Without the taps to pour the beer, the texture of the beer is significantly different also.

Beer was first sold to the public in cans in 1935. However, many drinkers, perhaps even the majority,

prefer the taste and texture of draft beer to the canned product. And if people demand that, then companies such as Guinness want to sell it to them. Guinness has therefore spent a fortune in finding a way to transfer the taste, texture, and head of their stout into cans.

In the 1980s Alan Forage and William Byrne found a way to simulate the nitrogen mix that a Guinness tap gives the beer. A small ball full of nitrogen and beer called a "widget" is put in the can along with the beer. When the can is opened, the internal pressure suddenly drops, and the nitrogen escapes into the beer, giving it the required texture and head. **DK**

SEE ALSO: ALCOHOLIC DRINK, CANNED FOOD, RING PULL, GLASS BOTTLE-MAKING MACHINE

⬆ *Nitrogen from a single widget produces the thick, creamy head preferred by drinkers of stout.*

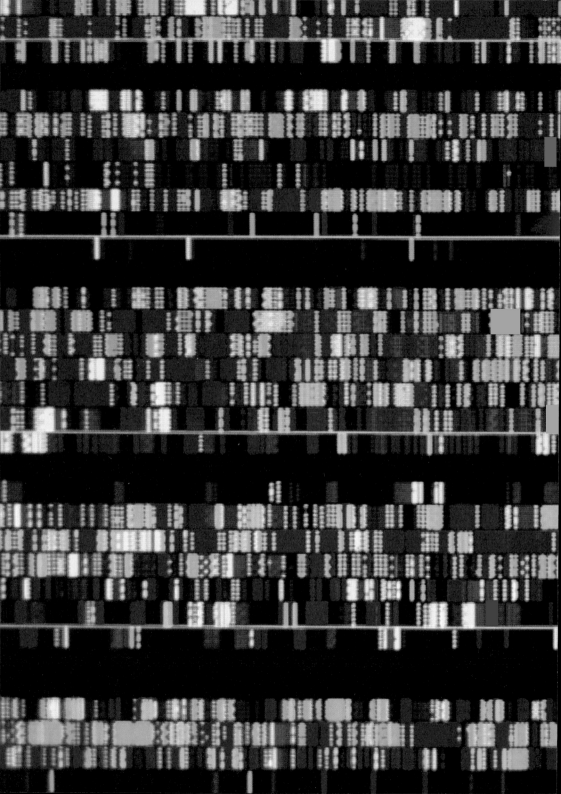

Automated DNA Sequencer (1985)

Hood and Smith speed up DNA analysis.

Imagine inventing a process that made a task 3,000 times faster than it used to be, not to mention substantially less dangerous and labor intensive. Now imagine that process as the key to unlocking and mapping the human genome. That is what Leroy Hood (b. 1938) and Lloyd Smith (b. 1954) did in 1985 when they invented automated DNA sequencing.

DNA sequencing was initially invented in the 1970s, but it involved a long, laborious process in which the nucleotide base pairs that make up DNA were tagged radioactively and then attached to existing single strands of DNA. The resulting strands of DNA were run through a separation gel and painstakingly examined manually, base pair by base pair, strand by strand, and the sequence recorded by hand.

Hood and Smith recognized that this process was at best impractical, and along with their colleagues Tim and Michael Hunkapillar set out to streamline it. In the mid-1980s they determined that the radio-labels on each base pair could be replaced by orange, red, blue, and green fluorescent dyes (one for each nucleotide). These dyes were not only more stable and less of a health risk than radio-labels, but they also could be illuminated with a laser. The resulting glows could then be counted by a light sensor attached to a computer, which could catalogue the strands automatically. Furthermore, the sequencing of all the bases could be done on a single gel.

The process made sequencing of large amounts of DNA practical and viable for research. It ultimately led to the Human Genome Project. **BMcC**

SEE ALSO: GENETICALLY MODIFIED ORGANISMS, GENE THERAPY, DNA FINGERPRINTING, DNA MICROARRAY

← *The genes that make up human DNA appear on a computer screen as colors representing specific bases.*

Edible Mycoprotein (1985)

Rank Hovis McDougall introduces Quorn™.

Meat substitutes are in increasing demand as more people reduce their meat intake or switch to vegetarianism. Edible mycoprotein, popularly known by the brand name Quorn™, offers a high-protein alternative to those consuming little or no meat. But vegetarianism was not on the minds of the U.K. pioneers of this food, who were seeking a ready source of protein against an anticipated global shortage.

The solution came from the least exotic of locations—a field near the small English town of Marlow where the mold species *Fusarium venenatum* was discovered in 1967; it was soon identified as a potential source of mycoprotein.

> *"To really enjoy fake meat, you might have to forget ... the true origins of the stuff you're [eating]."*
>
> Farhad Manjoo, *Wired* Magazine (2002)

Fungal extracts were assessed for human consumption by Rank Hovis McDougall (RHM) throughout the 1970s, and by 1980 large-scale production techniques had been mastered. Mycoprotein received permission for U.K. sale in 1985, and Quorn™ was launched by Marlow Foods. The fungus is grown in oxygenated fermentation tanks. Mycoprotein is then extracted, textured, and shaped, using chicken egg albumin as a binding agent.

Today the range of products based on Quorn™ is enormous, extending even to flavored "streaky bacon." Now, half a million Quorn™ meals are consumed in the United Kingdom each day. **MB**

SEE ALSO: SANDWICH, POWDERED MILK, SACCHARIN, BREAKFAST CEREAL

Polymerase Chain Reaction (PCR) (1985)

Mullis proposes a way to duplicate DNA.

Kary Mullis (1944–2019) was working at Cetus Corporation in Emeryville, California, in 1985 when he worked out a way to duplicate a single piece of DNA into as many copies as were wanted. The technology that resulted from his idea was named polymerase chain reaction, or PCR. The technique could be done in a test tube with the aid of enzymes and temperature changes.

After overcoming initial challenges in the laboratory, PCR took off as a huge technological advance in the study of molecular biology. Previously, 5,000 papers had been published on the subject. Not only did the technology provide scientists with a seemingly unlimited amount of DNA derived from

---

"[No one has] affected the current good or the future welfare of mankind as much as Kary Mullis."

Ted Koppel, on ABC's "Nightline"

---

something as small as a single strand, it also sped up the process by which they could analyze, clone, and modify DNA. Cetus eventually sold the technology to LaRoche for $300,000,000.

The uses of PCR today seem endless. Among many other applications, the technique is used to diagnose diseases, detect bacteria and viruses, amplify DNA from ancient fossils, analyze DNA for criminal cases, and compare DNA from different species. A profound and influential invention, PCR stands out as one of the most important innovations of our time. Few inventions have had the widespread impact of Kary Mullis's polymerase chain reaction. **RH**

SEE ALSO: GENE THERAPY, AUTOMATED DNA SEQUENCER, DNA FINGERPRINTING, DNA MICROARRAY

Surgical Robot (1985)

Kwoh refines robotically assisted surgery.

In 1954 George Devol created the first programmable industrial robot. It consisted of a multijointed manipulator arm and a magnetic storage device to hold and replay instructions. More advanced versions worked on assembly lines in the 1960s. In 1978 the PUMA (Programmable Universal Machine for Assembly) was introduced by Victor Scheinman and quickly became the standard for commercial robots.

Dr. Yik San Kwoh (*b.* 1946) invented the robot-software interface that allowed the first robot-aided surgery in 1985. "Ole" was a modified PUMA that could perform a type of neurosurgery. In the surgery, a small probe traveled into the skull, a linked CT scanner provided a 3D picture of the brain, and the robot plotted the best path to the lesion. "Ole" was used for biopsies of deeply located suspected tumors.

Before his device could be used on humans, Kwoh needed to test it. Small metal objects were inserted into four watermelons. The robot quickly located the objects and inserted an instrument to remove them.

Robots have since grown more complex and can now assist and even perform surgeries. In 1998 Dr. Friedrich-Wilhelm Mohr used a Da Vinci surgical robot to perform the first robotically assisted coronary artery bypass graft (CABG) at Leipzig in Germany.

In 1999 the world's first surgical robotics "beating-heart" CABG was performed at the London Health Sciences Center in Ontario, Canada, using a ZEUS surgical robot. In this type of surgery, the sternum of the patient is not opened, and the heart is not stopped as it is in conventional bypass surgeries. **SS**

SEE ALSO: AUTOMATON, INDUSTRIAL ROBOT, BIPEDAL ROBOT, COMPUTER-AIDED MANUFACTURING (CAM)

➲ *Guided by images from an endoscope, the surgeon controls surgical tools at the ends of the robot's arms.*

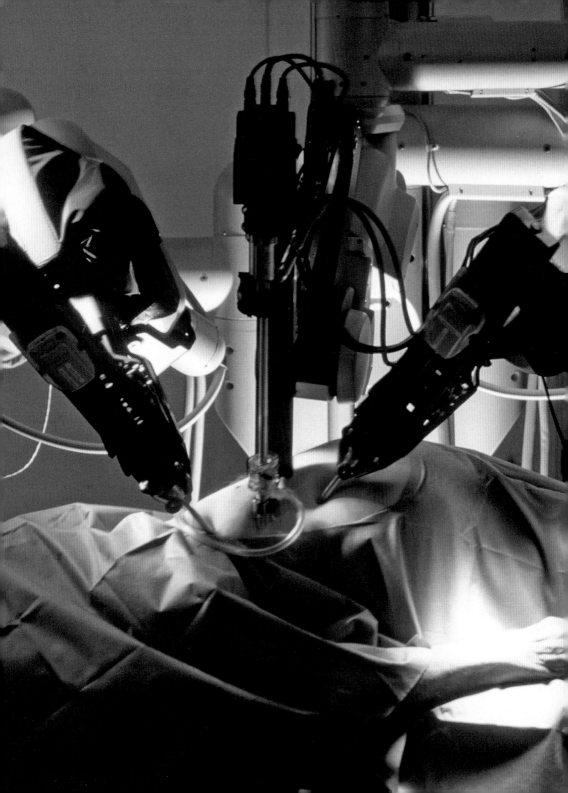

Bipedal Robot

(1985)

Hitachi creates a successful walking robot.

Before the Waseda-Hitachi Leg II (WHL-11) biped robot was unveiled at the 1985 Tsukuba Expo in Japan, robots could roll, drag, or crawl their way around. But here was a real breakthrough. This one could walk.

Developed by Hitachi in collaboration with Waseda University, Japan, and with an onboard computer and hydraulic pump, the WHL-11 was capable of static walking on a flat surface at around thirteen seconds per step. The robot was also able to turn.

If a robot could walk like a human, it had been reasoned, it would be able to maneuver around objects and go up and down stairs—the applications of such a machine would be almost limitless. To do so,

> "... an alien race that looked like insects ... would build robots to look like themselves ..."

Kevin J. Anderson

however, it had to have two legs, just like a person. It would also need to be able to walk on uneven ground.

Initial walking systems for experimental biped robots all emphasized static balance walking: the walking robot passed through successive states of static equilibrium. When walking with dynamic balance, the center of mass is allowed outside of the area prescribed by the feet, and the robot may actually be "falling" during parts of the gait cycle—like a human. Thus a foot must be moved at just the right instant in order to break the fall and achieve an effective transitional movement. Dynamic balance would be the next major milestone. **MD**

SEE ALSO: AUTOMATON, INDUSTRIAL ROBOT, SURGICAL ROBOT

DNA Fingerprinting

(1985)

Jeffreys identifies individuals by their DNA.

While some fear the invasion of privacy contingent on DNA databases produced from DNA fingerprinting, it is undeniable that the technique has had a positive impact in areas such as forensics, paternity testing, and animal classification.

After studying at Oxford University, biochemist Alec Jeffreys (*b.* 1950) became a professor in 1977 at the University of Leicester, where he worked on DNA variation and genetic evolution in families. He studied inheritance patterns of disease, specifically in what are called "mini-satellites," or areas of great genetic variation that occur in the human DNA sequence outside of core genes.

In 1984, while studying mini-satellites in the DNA of seals, Jeffreys tested a probe made of DNA on samples from various different people using X-ray film. When he developed the film, he saw what he described as a "complicated mess." Upon closer examination, however, he realized that certain patterns occurred that varied greatly from person to person. The mini-satellite DNA pattern from each person was unique, just like a fingerprint.

Jeffreys quickly realized the implication of his discovery, especially in the field of forensic science. He obtained patents in 1984 and published a series of papers in the journal *Nature* in 1985. The technique was first used in a U.K. immigration dispute, and in 1986 it was used in a criminal case in Leicestershire. DNA fingerprinting is now standard practice in criminal cases, and newer techniques have automated the process to produce quick and reliable results. **RH**

SEE ALSO: GENETICALLY MODIFIED ORGANISMS, AUTOMATED DNA SEQUENCER, DNA MICROARRAY

▣ *DNA synthesizers build short stretches of synthetic DNA that are used in the genetic fingerprinting process.*

Capsule Endoscopy
(1985)

Mullick devises a camera that is swallowed for investigation inside the body.

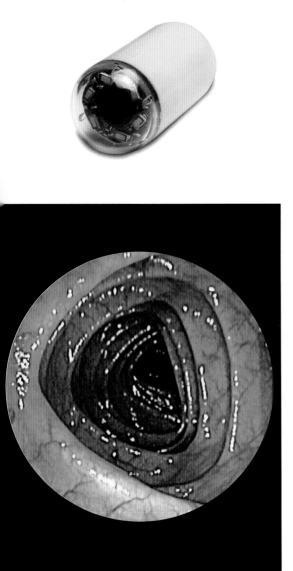

The human torso contains up to 26 feet (8 m) of intestines. When someone suffers a disorder of the gastrointestinal tract, it can be a difficult task to locate the problem in such an expansive length of tissue.

A traditional endoscopy involves a thin fiberoptic tube being inserted into the patient and images of the walls of the subject's innards being relayed to a television screen. It is a minimally invasive procedure and can cause the patient some mild discomfort.

However, as the technology of digital cameras has become smaller and more compact, an alternative has appeared. Created by a team of doctors led by Dr. Tarun Mullick in Baltimore, Maryland, the first wireless capsule endoscopy unit came into being in 1985. The camera-in-a-capsule is useful for spotting things such as vascular lesions, tumors, celiac disease, and Crohn's disease in areas that other noninvasive methods fail to reach. The capsule is roughly the size of an ordinary vitamin pill and is swallowed with water.

Much like the endoscopic tube, the capsule contains a light source and lens system, but it also contains equipment that acquires and transmits images to a receiver that the patient wears on a belt. After being swallowed, the camera takes an eight-hour trip through the entire length of gastrointestinal tract. On its journey it takes hundreds of color images of the patient's esophagus, stomach, and small and large intestines, enabling a diagnosis to be made based on comprehensive visual information. **CL**

SEE ALSO: ENDOSCOPE, FILM CAMERA/PROJECTOR, ZOOM LENS, LAPAROSCOPE

↖ *LEDs around the lens (black) of this battery-powered capsule illuminate the body part under investigation.*

← *An interior view of the colon as taken by the camera on an endoscope*

Driverless Car
(1986)

Dickmanns creates a fully automated vehicle.

As professor at the Universität der Bundeswehr München, Ernst Dieter Dickmanns (*b.* 1936) and his team designed a fully automated, driverless vehicle. They equipped a van with a series of cameras and sensors that processed images of the changing scenery and relayed the information to a mechanism that controlled the steering, accelerator, and brakes.

This was not the first autonomous vehicle to drive unmanned. Nearly a decade earlier, the Tsukuba Mechanical Engineering Lab in Japan created an automobile that could travel around a specially designed and clearly marked course at speeds of up to 18 miles (30 km) per hour. But in 1986 Dickmanns's van

> *"Today, we are in a state where a car can drive 100 miles . . . before human assistance is necessary."*
>
> Dr. Sebastian Thrun, Stanford University

did a lot better, navigating its way around ordinary (albeit empty) roads, attaining a top speed of about 60 miles (100 km) per hour the following year.

The European Commission began to fund a research and development project called PROMETHEUS. Dickmanns made huge advances in the eight years it was active. The final demonstration of the project involved two re-engineered Mercedes 500 SEL models driving 620 miles (1,000 km) on the multilane autoroutes at Paris. The VAmP and its twin, Vita-2, reached speeds of 80 miles (130 km) per hour and were successfully programmed to move into the fast lanes automatically to overtake slow vehicles. **CL**

SEE ALSO: AUTOMATON, ELECTRIC CAR, MOTORCAR, HYBRID CAR, AIR CAR

High-temperature Superconductor (1986)

Bednorz and Müller transform conductivity.

Superconductors are materials that have no electrical resistance, so electricity can flow through them without any loss. The superconductivity phenomenon was first discovered in 1911 by researchers in Germany who used solid mercury as their conducting material. Superconductivity was at first seen only in certain substances when supercooled to temperatures close to absolute zero, or -459°F (-273°C)—the coldest temperature theoretically possible.

In 1986 Georg Bednorz (*b.* 1950) and Alex Müller (*b.* 1927), both researchers at IBM, discovered a new type of superconducting material, copper oxide perovskites, that could superconduct at -396°F (-238°C). Paul Chu, at the University of Houston, improved on this by bringing the superconducting temperature up to a relatively balmy -296°F (-182°C).

For the first time, superconductivity could be made to occur at temperatures in the range of liquid nitrogen. The discovery quickly led to a huge meeting of physicists in New York, a meeting that became known as the "Woodstock of Physics."

In 1987 Bednorz and Müller were awarded the Nobel Prize in Physics and, in the same year, U.S. president Ronald Reagan declared that the United States was about to enter a new era of technology, thanks to high-temperature superconductors.

Although the technology has yet to take over the world, high-temperature superconductors have found applications in MRI medical scanners and in special superconducting wires, cooled by a sheath of liquid nitrogen. Japan uses coils of this super wire for their experimental maglev train, and the U.S. Navy is researching the use of the wire for their next generation of ships. **DHk**

SEE ALSO: INSULATED WIRE, COAXIAL CABLE, OPTICAL FIBER

Prozac® (Fluoxetine Hydrochloride) (1986)

Eli Lilly revolutionizes the drug treatment of depression.

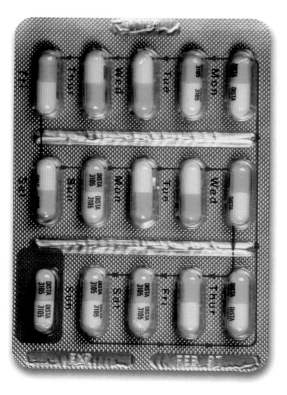

Prozac® is the registered trademarked name for fluoxetine hydrochloride, the world's most widely prescribed antidepressant. It was the first in a new class of drugs for depression called selective serotonin re-uptake inhibitors (SSRIs), which work by increasing brain levels of serotonin, the neurotransmitter thought to influence sleep, appetite, aggression, and mood. Prozac® works by inhibiting re-uptake of the neurotransmitter (where it is either destroyed or retrieved into the cell), thereby amplifying its levels.

At Eli Lilly and Company, the team of inventors behind Prozac® included Bryan Molloy, Ray Fuller, and David Wong. In the early 1980s it was known that the antihistamine diphenhydramine showed some antidepressant-like properties. As a starting point, the team took 3-phenoxy-3-phenylpropylamine (which is a compound structurally similar to diphenhydramine) and synthesized dozens of its derivatives. Fluoxetine hydrochloride was found to be most effective in mice.

Fluoxetine hydrochloride was first tested as an anti-obesity agent, and on people hospitalized with depression, but neither indication proved successful. Finally, Eli Lilly tested the drug on mild depressives, and all five of the subjects instantly cheered up.

Prozac® first hit the market in Belgium in 1986 and has had the fastest ever acceptance for a psychiatric drug. Within three years, 65,000 prescriptions per month were being written in the United States alone, and by the early nineties 4.5 million Americans had taken it. Since then, Prozac® has had a mixed reputation, with reports of the drug changing aspects of personality and also triggering suicide. **JF**

"Prozac® enjoyed [a] career of renown . . . rumors . . . scandal . . . and finally a quiet rehabilitation."

Peter D. Kramer, *Listening to Prozac* (1993)

SEE ALSO: ACETAMINOPHEN, VALIUM, VIAGRA® (SIDENAFIL CITRATE), HUMULIN®

◩ *Before its side-effects were realized, Prozac® seemed the way forward for millions of unhappy people.*

Palmtop Computer (PDA) (1986)

Psion offers the market a handheld computer for personal organization.

It is 1984. Economies are booming, *Newsweek* magazine declares it "the year of the yuppie," and in the U.K. the Filofax personal organizer is the must-have accessory for all young urban professionals. In London, though, Dr. David Potter is planning to make the leather-bound, paper-based Filofax obsolete.

Dr. Potter's company, Psion—the name comes from "Potter's Scientific Instruments"—had been in business for a few years already, making games and other software for early home computers like Sinclair's ZX Spectrum. In 1984 Psion entered the computer hardware market, releasing a new kind of handheld computer, the Psion Organiser. It was a hefty device, a rectangular slab of plastic with a small screen at the top and a keyboard protected by a sliding sleeve. It had a clock, a small memory, and it was powered by a 9-volt battery that could keep it running for months.

By today's standards, the Organiser was primitive. You could type in things on its tiny buttons, and it would remember them. And that was about it. Its drawbacks included a "write-once" memory that would eventually fill up with information and could only be erased by an ultraviolet lamp, and the fact that the keyboard was arranged in alphabetical order.

In 1986, though, Psion released a new version of the Organiser, increasing the size of the screen and adding a diary and alarm clock, among other handy applications. This, the Psion Organiser II, was the first computer truly worthy of the description "personal digital assistant" (PDA)—even though that phrase arrived only in 1992, when Apple's John Sculley used it to describe Apple's "Newton" MessagePad. **MG**

SEE ALSO: POCKET CALCULATOR, PERSONAL COMPUTER, SUPERCOMPUTER, LAPTOP/NOTEBOOK, SMARTPHONE

⬈ *The Psion Organiser II, released in 1986, soon became the must-have accessory for Yuppies worldwide.*

"… shaped like a small brick … and could hold the equivalent of about two-and-a-half pages."

Astrid Wendlandt, *Financial Times*

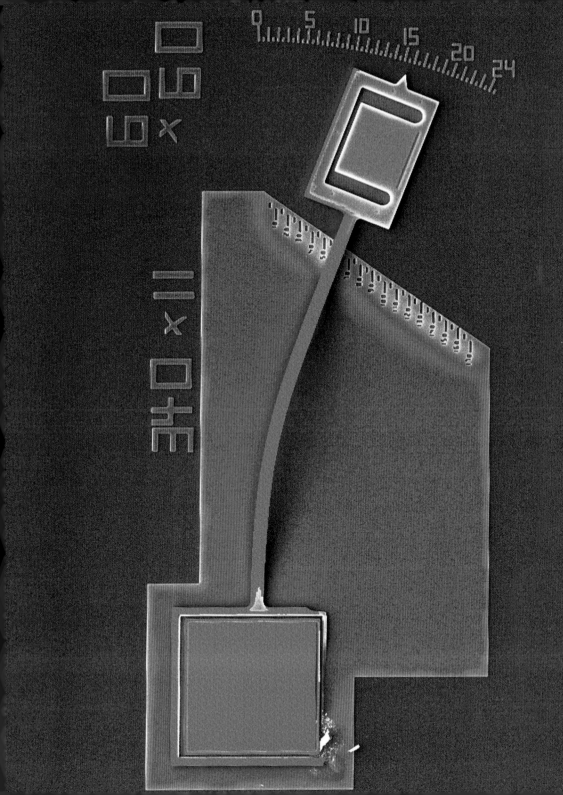

Micro-Electro-Mechanical Systems (1987)

Silicon chips gain sensory functions.

Micro-Electro-Mechanical Systems, or MEMS, are specialized silicon chips that incorporate not only miniaturized electronic circuits (the "electro" component) but also a "mechanical" part, such as tiny arms, gears, or springs. These chips, therefore, may possess not only the ability to process data, but also, etched into its surface, the ability to collect data in the form of some kind of sensor. Needless to say, the sensors are tiny, going down to the scale of millionths of a meter, or micrometers.

The term MEMS is itself thought to have arisen as a consequence of a 1987 gathering, called the Micro Robots and Teleoperators Workshop, conducted by the International Electrical and Electronics Engineers. Later, in 1988, a reliable method of bonding silicon components onto silicon chips was demonstrated.

Controllers in some video games, for example, utilize tiny accelerometers within their control pods to measure tilt, shock, and acceleration. While this is obviously fun, accelerometers are also used in the automotive industry, most commonly to detect a sudden change in inertia and deploy an airbag safety system at the appropriate moment. Another common MEMS application is in the nozzles of inkjet printers.

Future uses being researched for MEMS are far-ranging, from new computer displays to biochips with the ability to monitor glucose levels in diabetic patients. MEMS is quickly growing as a technology showing great potential to deliver affordable and effective solutions in a wide variety of situations. Many of them are now only at the planning stage. **AKo**

SEE ALSO: RANDOM ACCESS MEMORY (RAM), INDUSTRIAL ROBOT, MICROPROCESSOR

◧ *This scanning electron micrograph shows a MEMS with a sensor that gives readings from two scales.*

Massively Multiplayer Online Games (1987)

Flinn and Taylor pioneer graphic MMO games.

Few people today remember the simpler multiplayer creations that started the craze for shared gaming. The likes of *Everquest*, *World of Warcraft*, and even *Second Life* owe their existence to those who saw how much fun it would be for gamers to be able to interact with many others in an alternate reality on the Internet.

Two of the pioneers of Massively Multiplayer Online (MMO) gaming were Kelton Flinn and John Taylor, who together founded the games company KESMAI. Although several text-based multi-user dungeon (MUD) games had existed since the late 1970s, they worked on a huge step forward—a graphic-based massively multiplayer game: *Air Warrior*.

> *". . . if you had a multiplayer game that exceeded sixteen, you might as well call it massive."*
>
> Raph Koster, Sony Online Entertainment

Released in 1987, *Air Warrior* was a World War II flight-combat simulator that many people, hundreds at a time, could play together online (using a pay-for-use service called Genie). Its virtual environment was perpetual—while the players could enter and leave as they chose, the game continued. Players from around the world could log on, choose their own type of aircraft, and fly missions with and against each other.

Air Warrior was a major milestone of its time. Although the graphics were crude by today's standards, in those days anything capable of accommodating the activities of over sixteen players was a staggering achievement. **JM**

SEE ALSO: VIDEO GAME CONSOLE, VIRTUAL REALITY HEADSET

Statins (1987)

Merck releases a cholesterol-reducing drug.

With the knowledge that high cholesterol (blood fat) increases a person's risk of suffering a heart attack (or stroke), the quest was on to discover drugs that could be used to lower cholesterol. In 1959 scientists working at the Max Planck Institute in Heidelberg, Germany, identified the HMG-CoA reductase enzyme as a major contributor to production of internal cholesterol. The discovery inspired scientists worldwide to start searching for drugs to inhibit the enzyme and thereby reduce levels of cholesterol.

By 1976 Japanese researcher Akira Endo, from Sankyo, isolated the first inhibitor (Compactin, ML-236B) of HMG-CoA reductase enzyme from the fungus *Penicillium citrinium*. Endo had chosen to begin his

---

"… maybe people should be able to have their statin … with their drinking water."

Dr. John Reckless, consultant endocrinologist

---

search in molds and fungi for the sole reason that this was his own particular area of expertise. In 1979 Carl Hoffman and colleagues working for Merck & Co. in the United States isolated MK-733 (later to be named Simvastatin) from a strain of the fungus *Aspergillus terreus*. But in 1980 clinical trials with Compactin were discontinued after rumors it caused cancers in dogs, and further setbacks to the drug followed.

In July 1982 the Federal Drug Administration (FDA) allowed Merck special permission to give Simvastatin to patients with severely high cholesterol. Simvastatin produced dramatic results with very few side effects and was approved by the FDA in September 1987. **JF**

SEE ALSO: SYNTHETIC BLOOD, INTRAVASCULAR STENT

Tissue Engineering (1987)

Vacanti and Langer grow cell tissue.

Dr. W. T. Green, a pediatric orthopedist at Children's Hospital in Boston, Massachusetts, made one of the first attempts at tissue engineering. He tried to grow cartilage in laboratory mice. Although unsuccessful, his work set the stage for later attempts by suggesting that, once suitable materials were invented, cells would grow on configured scaffolds.

Joseph Vacanti, a transplant surgeon, and Robert Langer (*b.* 1948), an engineer, created an engineered, biosynthetic, biodegradable scaffold in 1987. The scaffolds they developed provided access to nutrients as well as waste removal for the growing cells. The final structure can resemble a natural organ.

A famous development was the "auriculosaurus," a mouse with a human-shaped ear growing on its back. The "ear" was a biodegradable scaffold seeded with bovine cartilage cells. The mouse was a nude animal specifically bred not to reject foreign proteins. The auriculosaurus, filmed by a BBC video crew, became the image of tissue engineering worldwide.

In 1998 a factory worker was presented to the University of Massachusetts Medical Center after the last bone in his thumb was ripped off in machinery, leaving just the flesh. Vacanti ground a piece of coral into the shape of the missing bone. The coral was then seeded with bone cells, and eventually implanted. As the coral dissolved, the bone replaced the structure.

These two examples were grown inside organisms. Engineered tissue is more commonly grown in bioreactors—vessels designed to provide nutrients and remove waste more efficiently than could be done in a petri dish. **SS**

SEE ALSO: GENETICALLY MODIFIED ORGANISMS, GENE THERAPY, ARTIFICIAL SKIN, STEM CELL THERAPY

➡ *Cut human tendons like these are placed between the ends of a broken tendon.*

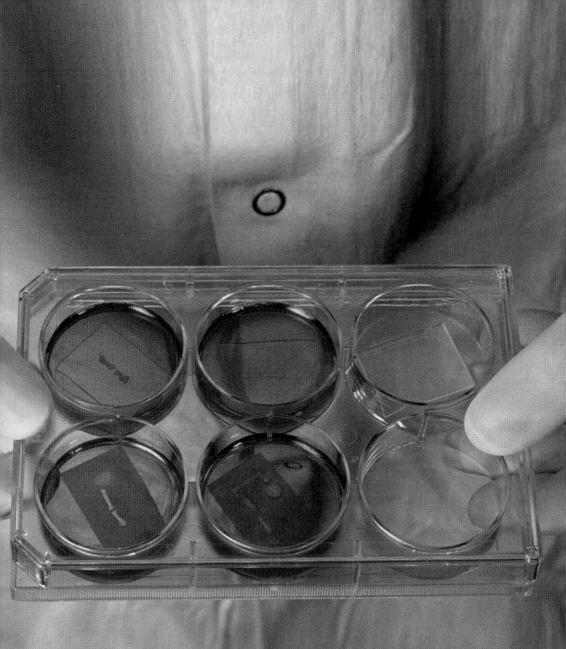

Digital Subscriber Line (DSL) (1988)

Lechleider transforms Internet connections.

When the Internet was a relatively new part of people's lives, getting connected was difficult. Dialing up and connecting to the server was slow, the connection itself would be sluggish, and the connection would block the phoneline, so no one could call.

Aware that cable communication companies had a head start over them in terms of high-frequency, broadband provision, the telephone companies were desperate to provide high-speed Internet access. This was the commercial consideration that spurred them to create DSL—the Digital Subscriber Line—a technology that used the as yet untapped potential of the copper wires already connecting the world.

> *"[On the Internet we can now] get very large amounts of data to very large numbers of people."*
>
> Frank James, *Chicago Tribune*

The man now recognized as the father of these technologies is Joseph Lechleider (1933–2015) of the U.S. firm Bellcore. He was the first to demonstrate that broadband signals could be sent through the copper lines along both high- and low-frequency bands. He also went on to develop the idea of an asymmetrical connection—known as an ADSL. As the typical residential Internet user downloads much more data than he or she uploads (partly due to the vast amount of multimedia content available on the Internet), the possibility of using more of the bandwidth and connection speed to download, rather than upload, was a brilliant, yet practical, inspiration. **CL**

SEE ALSO: INTERNET, CABLE MODEM, INTERNET PROTOCOL (TCP/IP), WORLD WIDE WEB

Intravascular Stent (1988)

Palmaz reopens blocked blood vessels.

The stent, a mesh tube device designed to hold open blood vessels, has revolutionized management of coronary artery disease. The first successful stent was invented by Argentinian doctor Julio Palmaz (b. 1945). Palmaz had heard that blood vessels had a tendency to close up after balloon angioplasty, in which narrowed heart vessels are opened with a catheter. Palmaz had the idea of putting a "scaffold" inside the vessels to prevent them from closing over.

Palmaz began working on creating prototypes of an implantable stent, using simple materials such as copper wire and a soldering iron. He modeled the mesh with a structure of staggered openings he just happened to find lying on his garage floor. The design proved perfect—the structure was collapsible but remained rigid when inserted into the blood vessel.

After testing his device on pigs and rabbits, Palmaz secured funding from the unlikely partnership of Phil Romano, a restaurant entrepreneur, and cardiologist Richard Schatz from the Brooke Army Medical Center. Calling themselves the Expandable Graft Partnership, the trio patented the device in 1988.

One complication was restenosis, the development of a blockage from scar tissue arising from implantation. The next stage was to produce drug-eluting stents to deliver therapeutic agents, such as sirolimus, to prevent that scarring from occurring. However, in a few cases these can lead to blood clots that cause heart attacks. The way forward currently being explored is to have biodegradable stents made of metals or polymers that slowly dissolve. **JF**

SEE ALSO: DISPOSABLE CATHETER, BALLOON CATHETER, LITHOTRIPTER, STATINS

▶ *After the narrowed artery is widened by the catheter, the stent is implanted and the catheter withdrawn.*

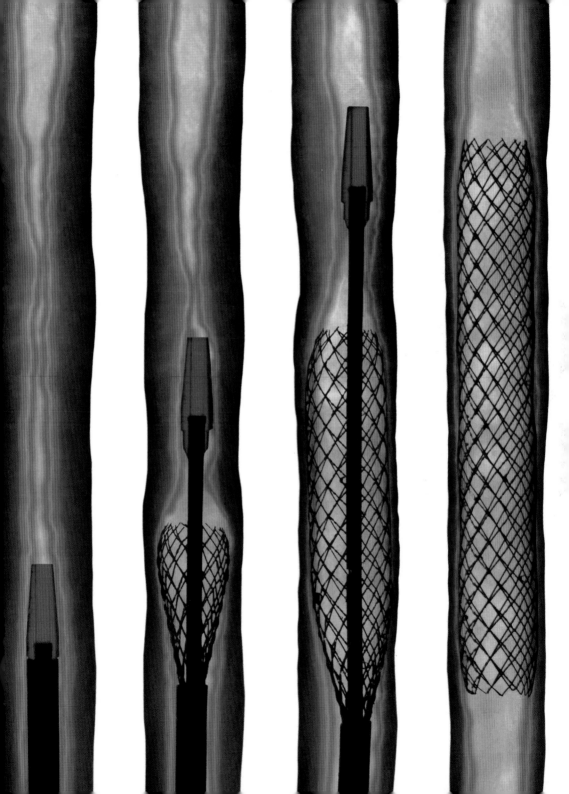

Laser Cataract Surgery

(1988)

Bath devises a laser instrument to facilitate lens removal from the eye.

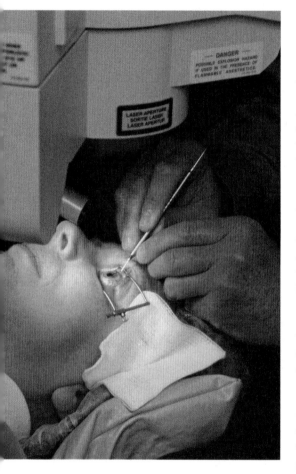

"When I talked to people about it, they said it couldn't be done . . ."

Dr. Patricia Bath

In nearly all circumstances, pointing a laser at your eye is a bad thing, but if you have cataracts it may just restore your sight. Cataracts are a leading cause of blindness. The disorder occurs when the part of the eye that focuses light—known, not surprisingly, as the lens—turns cloudy. This is a process that occurs in almost all of us if we live long enough. Unfortunately, currently there is no viable way to make a cloudy lens transparent again, and so ophthalmologists are forced to resort to other means to alleviate the problem.

The current method of dealing with a cataract is to remove the lens of the eye. One problem is that taking out the lens in its entirety requires a rather large incision into the eye in order to remove it. Ophthalmologists have long felt that it was more appropriate to find a method to break up the lens, which allows them to remove it through a smaller incision. This process, known as emulsification of the lens, was at one time done simply by grinding up the lens, but this method was subsequently replaced by a tool that transmits ultrasonic waves and uses sound energy to break up the lens. The resulting fragments are then vacuum-extracted from the eye.

All this may change with a tool invented by Dr. Patricia Bath (1942–2019). Dr. Bath postulated that lasers could be used to emulsify the lens of the eye as well, and she developed a model for a laser instrument to be used in removing cataracts. She received a patent for her invention in 1988. After much trial and error it turns out that Dr. Bath is correct, and her system may yet benefit cataract patients everywhere. **BMcC**

SEE ALSO: INTRAOCULAR LENS, LASER, CARBON DIOXIDE LASER, LASER EYE SURGERY

◩ *At present laser cataract surgery is in its infancy, but the technique promises to be important in the future.*

Touchpad

(1988)

Gerpheide finds a replacement for the mouse.

The introduction of the laptop computer presented a problem: namely, how to operate a computer's cursor without a mouse. The solution most widely adopted was created by American inventor Dr. George Gerpheide (b. 1952). His capacitive touchpad, invented in 1988, could detect a user's finger movement and transfer it to the on-screen cursor. Interestingly, Gerpheide developed this technology before point and click was the standard method of operating computer interfaces. This may explain why it was not until 1994 that Apple Computers bought the first license to use his technology (which first appeared on the Apple Powerbook 520).

His touchpad works by employing several layers of material. At the top is a protective layer, about 3 inches (8 cm) square, that the user touches. Underneath are successive layers of electrodes arranged in horizontal and vertical rows, each separated by a thin layer of insulation. The electrodes are all connected to a circuit board that provides them with a constant supply of alternating current. When the finger touches the pad, it interrupts the current, and the locality of this interruption is registered by the circuit board. Any further movements of the finger and new interruptions are then compared to the initial touch point and so the finger's movement can be recorded and reproduced on screen.

Nowadays touchpads are increasingly found on mobile devices, from personal digital assistants (PDAs) to cell phones. The reasoning behind the innovation is simple, as Gerpheide himself said: "When you're talking about mobile appliances, the mouse is not suitable. Nobody wants a mouse dangling from their net-connected cell phone." **JM**

SEE ALSO: COMPUTER MOUSE, TOUCH SCREEN

Preimplantation Genetic Diagnosis (1989)

Handyside tests cells for genetic disorders.

Preimplantation genetic diagnosis (PGD) was a pioneering genetic test developed in the late 1980s to enable concerned parents to test for genetic disorders before they even got pregnant.

British researcher Alan Handyside and his colleague Robert Winston (b. 1940) reported their new technique in 1989. It involved checking fertilized eggs for genetic disorders before they were implanted. Unaffected embryos were then implanted through conventional in-vitro fertilization (IVF) techniques.

Before PGD was introduced, parents likely to pass on genetic disorders to their children had few options to prevent this. They could remain childless, adopt, or

> *"What's important is that they are not designer babies. They are not perfect babies."*
>
> Lord Robert Winston

get pregnant with the risk of having to terminate a pregnancy if genetic disorders were discovered. PGD allowed them to pick only unaffected embryos.

The technique of PGD involves stimulating a woman's ovaries with hormones to increase egg production. The eggs are then fertilized by the father's sperm. After three to five days, cells from the embryos (which at this point are called blastocysts) are removed and tested for the particular genetic condition.

Critics of PGD claimed that screening out disorders is unethical. Others claimed that the technique might increase the risk of other genetic complications, but a recent study showed that not to be the case. **RH**

SEE ALSO: GENETICALLY MODIFIED ORGANISMS, GENE THERAPY, IN VITRO FERTILIZATION

MP3 Compression

(1989)

Fraunhofer speeds up music downloading.

From the early 1990s, as Internet usage first began to proliferate, users quickly saw the potential for sharing music. But a combination of basic connection speeds and large file sizes made uploading and downloading a painfully slow process. As early as 1987, Germany's prestigious Fraunhofer Institut had been engaged in researching high-quality, low bit-rate audio coding: in short, how an audio file can be compressed in size without affecting its sound quality. The format they came up with in 1989 was called MPEG (Moving Picture Experts Group) Audio Layer III, or MP3.

MP3 compression is a simple concept to understand—even if the process itself is highly complex. A compact disc (CD) stores its information digitally, in binary digits (bits); every second of stereo music contained on a CD consists of 1,411,200 bits. MP3 compression reduces the number of bits in a recording by taking out "unnecessary" information. It does this by using "perceptual noise shaping"; characteristics of the human ear are taken into account in its compression algorithm.

So, for example, certain sounds that cannot be heard by the human ear, or are masked by other louder sounds, may be removed without significantly altering the overall sound. In fact, compressing a song "ripped" from a CD can typically reduce its size by a factor of 12—the MP3 version of the song can therefore be downloaded twelve times more quickly than the uncompressed version.

The speed at which files could be transferred immediately made MP3 the format of choice for moving digital music around on the Internet and spawned a whole new phenomenon: the downloading of music—legal or otherwise. **TB**

SEE ALSO: PODCAST, JPEG COMPRESSION, HIGH DENSITY COMPUTER STORAGE

DNA Microarray

(1989)

Fodor simplifies the study of genetic activity.

There are about 30,000 different genes in human DNA. Different cells in the body, although having identical DNA, switch on and off different genes, depending on what is needed to build that particular cell. Studying which genes are active in a cell is a useful way to find out what makes it function, and helps identify what has gone wrong when it is not functioning properly.

In 1989 U.S. scientist Stephen Fodor presented a technique that was to revolutionize DNA analysis. He created a DNA microarray—a glass slide with up to 500,000 different strands of DNA attached to it. When a gene is switched on in a cell, a complementary copy of that gene's information (called messenger

---

"By being able to see … all the genes, all the genetic variation, we can readily pick out answers."

Eric Lander, a human genome project leader

---

ribonucleic acid, or mRNA) is produced by the cell. This is like a mirror image of a particular stretch of DNA. Matching pieces of mRNA and DNA will stick together. To find the active genes, a cell is treated with a dye that attaches itself to mRNA, and then the cell contents are added to the microarray. The mRNA strands will stick to any matching DNA sequences, and the dye will show them up. The microarray has already proved to be a powerful tool for learning about many different illnesses, from heart disease to cancer. **JM**

SEE ALSO: GENE THERAPY, DNA SEQUENCER, DNA FINGERPRINTING, PREIMPLANTATION GENETIC DIAGNOSIS

➡ *Using a DNA microarray, thousands of genes can be tested at once, providing a cell's entire genetic profile.*

World Wide Web (1989)

Berners-Lee creates the world's first website.

"The world can only be changed one piece at a time. The art is picking that piece."

Tim Berners-Lee

⬆ *Berners-Lee used this computer and server at CERN to create the world's first web browser and editor.*

➡ *A screen shot taken from a NeXT computer running Tim Berners-Lee's original World Wide Web browser.*

Tim Berners-Lee (*b.* 1955) knows a lot about changing the world. There was a time when his web pages were the only pages on the World Wide Web.

Born to parents who had met while developing one of the earliest computers, Berners-Lee studied for a physics degree from Queen's College, Oxford, then headed straight for the computer industry. By 1989 he was working in Geneva, Switzerland, at CERN, the European particle physics laboratory. CERN was interested in finding a way by which groups of researchers could share information more easily. Berners-Lee saw the potential for marrying hypertext—a technique for linking documents together using clickable words—with the Internet, which was already heavily used by CERN. Working with colleague Robert Cailliau, he proposed a system he called the World Wide Web.

The first web server was up and running by the end of 1990. At the time, Berners-Lee's "NeXT" computer did not have a color display, and the earliest web pages were simple black-and-white text. Now, less than twenty years later, there are more than a hundred million websites on the World Wide Web. The Web certainly continues to help people in their physics research, but it also helps them to do their shopping, listen to music, read their morning newspapers, and get back in touch with old friends.

Berners-Lee, who without doubt picked the right piece of the world to change, is now Sir Timothy Berners-Lee, Director of the World Wide Web Consortium, Senior Researcher at the Massachusetts Institute of Technology (MIT), and Professor of Computer Science at the University of Southampton. The significance of his work in shaping the early twenty-first century is beyond calculation. **MG**

SEE ALSO: INTERNET, WEBCAM, VOICE OVER INTERNET PROTOCOL (VOIP), WEB SYNDICATION (RSS)

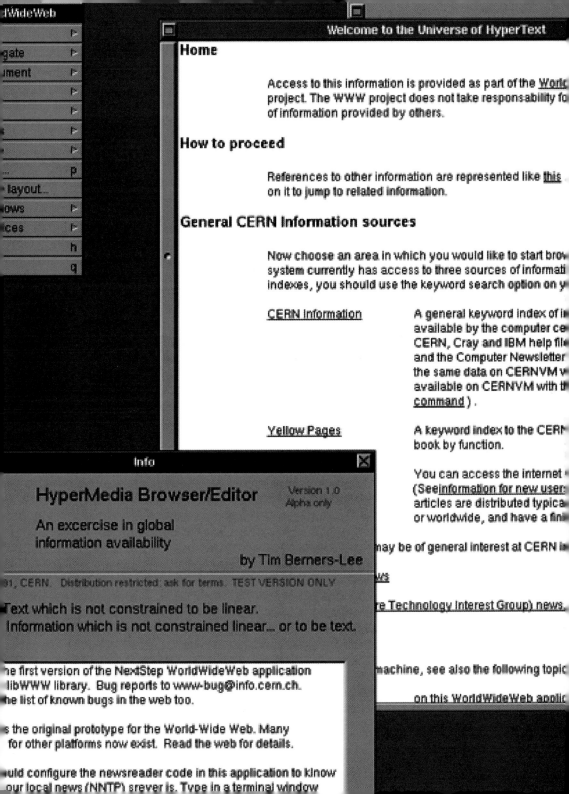

WideWeb

gate ▷
ment ▷
▷
▷
▷
▷
... p
- layout...
ows ▷
ces ▷
h
q

Welcome to the Universe of HyperText

Home

Access to this information is provided as part of the Worl
project. The WWW project does not take responsability fo
of information provided by others.

How to proceed

References to other information are represented like this
on it to jump to related information.

General CERN Information sources

Now choose an area in which you would like to start brow
system currently has access to three sources of informati
indexes, you should use the keyword search option on y

CERN Information

A general keyword index of i
available by the computer ce
CERN, Cray and IBM help fil
and the Computer Newsletter
the same data on CERNVM v
available on CERNVM with th
command) .

Yellow Pages

A keyword index to the CERN
book by function.

You can access the internet
(See information for new user
articles are distributed typica
or worldwide, and have a fin

may be of general interest at CERN i

ws

re Technology Interest Group) news.

Info ✕

HyperMedia Browser/Editor
Version 1.0
Alpha only

An excercise in global
information availability

by Tim Berners-Lee

91, CERN. Distribution restricted: ask for terms. TEST VERSION ONLY

Text which is not constrained to be linear.
Information which is not constrained linear... or to be text.

he first version of the NextStep WorldWideWeb application   machine, see also the following topic
libWWW library. Bug reports to www-bug@info.cern.ch.
he list of known bugs in the web too.   on this WorldWideWeb applic

s the original prototype for the World-Wide Web. Many
for other platforms now exist. Read the web for details.

uld configure the newsreader code in this application to klnow
our local news (NNTP) srever is. Type in a terminal window

Gamow Bag® (1990)

Gamow invents a lifesaving sleeping bag.

Descriptions of Dr. Igor Gamow (b. 1936) depict him as a cross between Indiana Jones and Albert Einstein, a bold adventuring spirit driven by an enquiring scientific mind. Gamow's father was the physicist and cosmologist George Gamow, and his mother was a famous ballet dancer. The young Gamow held down jobs as diverse as horse-breaker and karate instructor before he finally succumbed to the lure of science and enrolled at the University of Colorado. He eventually gained a PhD in Microbiology and Biophysics and went on to lecture in these subjects, although his passion for adventure and the outdoors did not wane.

It was while Gamow was investigating training at altitude that the idea for his sleeping bag came about. He envisaged a bubble that athletes living at high altitude could use to get the most out of their training. Altitude sickness occurs when a person ascends to a height in excess of 8,200 feet (2,500 m). The resulting lack of oxygen can have a severe detrimental effect on the human body, causing the lungs or brain to fill with fluid. Gamow's low-altitude simulation bubble was not successful because it was bulky and overheated. However, he incorporated his idea into an innovative pressurized, nylon sleeping bag that could reduce the effective altitude. The Gamow Bag®, now sold by DuPont, acts as a hyperbaric chamber in which the symptoms of altitude sickness are alleviated. The collapsible bag, which is pressurized by a foot pump, is easily carried by mountain climbers and trekkers. In addition to the rescue bag, Gamow has developed a "Gamow tent" that can accommodate two patients and has also patented an altitude bed. **BG**

SEE ALSO: GAS MASK, REBREATHER, GORE-TEX®

← *The Gamow Bag® effectively treats altitude sickness by increasing the air pressure around the patient.*

Sulfur Lamp (1990)

Ury and Wood reconsider the light bulb.

One of the goals of good lighting is to produce a radiation that has an energy distribution very similar to the sensitivity of the human eye. Michael Ury and Charles Wood decided that the ideal radiator would be ionized molecular sulfur (S_2), which produces a continuous spectrum as opposed to a line spectrum. About 73 percent of the light emission is in the visual spectrum, with only 1 percent in the ultraviolet.

Sulfur plasma is extremely corrosive so normal tungsten electrodes could not be used. Ury and Wood decided to use a magnetron power source rather like those used in microwave ovens, the sulfur being contained in a golf ball-sized quartz bulb. However, the sulfur inside the bulb gets extremely hot and the

---

> *"We've run these bulbs almost 10,000 hours in test cases, and there's no wear and tear."*
>
> Michael Ury

---

bulb has to be continually rotated and cooled by a fan to keep it from melting. Another disadvantage was that it was impractical to produce bulbs dimmer than 1,000 watts, and so light pipes or parabolic reflectors had to be used to distribute the luminous energy. The production of commercial sulfur lamps suffers from the fact that the cooling fans are noisy and the magnetrons produce radiation, which can interfere with Wi-Fi, cordless phones, and satellite radio.

A lamp that completely mimics sunlight remains a dream of the future, but Ury and Wood have continued to develop their bulbs, producing a brighter, more energy efficient, and environmentally friendly light. **DH**

SEE ALSO: CANDLE, OIL LAMP, ARGAND LAMP, GAS LIGHTING, ARC LAMP, INCANDESCENT LIGHT BULB, ANGLEPOISE® LAMP

Cable Modem (1990)

Yassini connects personal computers to the Internet.

Imagine a world where broadband did not exist. Without this high-speed data transfer, would the Internet still be the hub of information, pictures, movies, and business opportunities it is today? The fast connection speeds required the invention of the cable modem, and the man who did that was Iranian-born American electrical engineer Rouzbeh Yassini (b. 1958).

Yassini worked for General Electric in 1981 building television receivers. To understand how the signals flowed, he took home television sets and dismantled them to see how they functioned. This knowledge proved useful when in 1986 he joined Proteon, a data-networking company that used a network cable called "twisted pair" to carry data.

Despite being told that video and data did not mix, Yassini realized right from the beginning that he could employ the same coaxial wire that carried cable television into people's homes to deliver other information as well. In 1990 he created a new company called LANcity Corp. He and his thirteen-strong team started to build a device that would provide an interface between a data network on one side and a cable television network on the other—the first cable modem. The initial model retailed at a staggering $15,000 and took three months to install. Just five years later, however, the third-generation LANcity modem was "plug and play" and cost just $500.

The company was bought very soon after this for $59 million by Bay Networks. Yassini went on to spearhead the development, implementation, and certification of DOCSIS, a standard for carrying data over cable modems. **JM**

SEE ALSO: PERSONAL COMPUTER MODEM, WEBCAM, VOICE OVER INTERNET PROTOCOL (VOIP), WEB SYNDICATION (RSS)

⬆ *Yassini's local area network (LAN) cable modem connected computers to the Internet worldwide.*

```
                        Archie
  Server:  archie.ncu.edu.tw                    ▼

  Find:    bacteri

  ● Sub-string (dehqx)          ☐ Case sensitive
  ○ Pattern (dehqx*.hqx)
  ○ Regular Expr (dehqx.*\.hqx)   Matches:  100

     ( Cancel )        ( Save )      ( Find ▶ )
```

Name	Size	Date	Zone	Host
27.cyanobacteria_1.ps	188k	16/6/95	2	ftp.sunet.se
28.cyanobacteria_2.ps	123k	16/6/95	2	ftp.sunet.se
34.uncertain_bacteria_1.ps	124k	16/6/95	2	ftp.sunet.se
35.uncertain_bacteria_2.ps	108k	16/6/95	2	ftp.sunet.se
36.uncertain_bacteria_3.ps	116k	16/6/95	2	ftp.sunet.se
bacteria-11.hqx	5k	21/7/91	2	ftp.rrzn.uni-hannover.de
bacteria-11.hqx	5k	22/7/91	2	info.nic.surfnet.nl
bacteria-11.hqx	5k	21/7/91	2	ftp.sunet.se
bacteria.gif	13k	9/12/94	1	micros.hensa.ac.uk
bacteria.gif	13k	9/12/94	5	nctuccca.edu.tw
Bacteria.gz	30k	30/12/94	2	ftp.fu-berlin.de
bacteria.zip	45k	2/3/94	2	power.ci.uv.es
bacteria1.1.sit.hqx	5k	6/12/92	2	ftp.sunet.se
bacteria1.1.sit.hqx.gz	3k	6/12/92	2	gigaserv.uni-paderborn.de
CRS-Bacteria.lha	595k	13/6/95	2	gigaserv.uni-paderborn.de
CRS-Bacteria.readme	2k	13/6/95	2	gigaserv.uni-paderborn.de

(window title: **bacteri from archie.ncu.edu.tw**)

Search Engine (1990)

Emtage creates a search tool for the Internet.

Before the arrival of the search engine, computers were linked together simply to let people transfer files between themselves. Those who had files to share set up a server, and those who wanted the files would come and get them. In time these servers clumped together, and having lots of files in one place made them easier to find. But even with clumping, files were spread out over the Internet. If you did not know the location of a file, it was very hard to track it down.

This was the problem facing Alan Emtage (b. 1964), studying at McGill University in Montreal. With funding for software limited, it was Emtage's job to find free applications on the Internet for the university to use. At first he searched by hand, building a database of the software he had found, but eventually, being a computer scientist, he made a program to do the job.

In 1990 the first search engine was born. Emtage's program was built to archive, but the UNIX world standard for program names required them to be short and cryptic, so he dropped the "v" in "archive" and named the program "Archie." The software was a long way from modern search engines, but if you knew the name of the file you wanted, it could help you find it, which was a massive leap forward.

Archie searched file names, but in 1991, Gopher was created—this could search the text contained within files. Search engines then began to use statistics to aid the search. Yahoo added descriptions of pages and Lycos analyzed the closeness of words and gave you sites by relevance. By 1995, AltaVista had appeared, additionally capable of searching for pictures, music, and videos. **DK**

SEE ALSO: INTERNET, PERSONAL COMPUTER, INTERNET PROTOCOL (TCP/IP), WORLD WIDE WEB

⬆ *By typing a keyword, users could access the names of files held in Archie's database of archive sites.*

Hubble Space Telescope

(1990)

NASA expands knowledge of the universe.

Often known simply as Hubble, the Hubble Space Telescope was named for Edwin Hubble, the U.S. astronomer who showed the existence of other galaxies outside the Milky Way. Hubble is an orbiting reflecting telescope designed to study the distant universe in visual light. Funded by the National Aeronautics and Space Administration (NASA) since 1977, its launch was planned for 1986, but the *Challenger* Space Shuttle tragedy put it back until April 1990.

The four instruments attached to the telescope focal plane were designed to be modular, and it was intended that they should be replaced by other instruments during the mission. This attribute was

---

"We in astronomy have an advantage in studying the universe, in that we actually see the past."

Lord Rees, Astronomer Royal for England

---

extremely useful because it was soon found that the main mirror had a design fault. A space shuttle rescue mission in 1993 corrected the problem.

Apart from capturing spectacular images of cosmic objects, Hubble can expose its detectors to the same region of sky for several days, and thus image objects that are much more distant from Earth. By searching for supernova explosions in distant galaxies, Hubble has discovered that the expansion of the universe is actually accelerating. **DH**

SEE ALSO: TELESCOPE, VERNIER SCALE, CHARGE-COUPLED DEVICE (CCD), X-RAY TELESCOPE

← *Four servicing missions have added to the Hubble Space Telescope's capabilities since its 1990 launch.*

Webcam

(1991)

Live images join the world of cyberspace.

The first webcam in the world was born from the desire of computer science students at Cambridge University, England, for fresh coffee. Having only a single coffee pot situated at some distance from the computer labs meant that a freshly brewed supply soon ran out. To overcome this difficulty Quentin Stafford-Fraser and Paul Jardetzky had the idea of focusing a camera on the coffee pot.

A computer with a simple frame grabber was connected to a camera, which was focused on the coffee pot. Jardetzky wrote a server program that collected images from the camera every three minutes, while Stafford-Fraser developed the software to run on the computers of all the members of the "Trojan Room coffee club." Connecting to the server then provided an up-to-date, icon-sized image of the pot on-screen. The camera was connected to the Internet in 1993 and became a popular symbol of the early World Wide Web. When the webcam was finally switched off in August 2001, the international media covered the event and the original Krups coffee pot was auctioned for a significant sum on eBay.

The current longest running webcam in the world is Fogcam, which has been broadcasting from San Francisco State University since 1994. The camera was set up by students Jeff Schwartz and Dan Wong to capture images of daily life on campus, and still relays images from the front of the humanities building.

Webcam technology took off across the Internet after the pornography industry began to show an interest. The industry employed a Dutch developer to write software that could provide live images without the need for any web plug-ins and so the "live streaming webcam" was developed. **HP**

SEE ALSO: PERSONAL COMPUTER, INTERNET, COMPUTER MOUSE, TOUCH SCREEN

JPEG Compression (1992)

JPEG improves compression of digital images.

Before the early 1990s, the Internet was primarily a text-based medium as bitmap (binary data) images were too large to easily download or distribute. This all changed with the introduction in 1992 of the JPEG—a new standard for compressing the size of images.

These days the term JPEG is well known, but surprisingly it was not meant to be the name of a picture format. The term comes from the Joint Photographic Experts Group, the international organization set up in 1986 that came up with the standard compression algorithm responsible for reducing the amount of memory JPEG pictures take up. Technically, the JPEG format is called JFIF, for JPEG File Interchange Format, but the name never stuck.

JPEG works by converting the pixels of a color image into blocks of pixels, and then taking an average of the values of the brightness and color of these blocks so that less data needs to be saved. Further compression is done by quantization (a mechanism that removes high-frequency noise) and a complicated process called Discrete Cosine Transformation. The compression is done intelligently, so, depending on the level of compression, the loss of detail is not obvious to the human eye. Despite the fact that each time an image is resaved more data is thrown away and the quality is reduced, this has not prevented its ubiquitous use on the Internet.

The JPEG organization has also come up with a new format called the JPEG 2000, a new improved lossless form of the JPEG. It will have to do well to replace the original, which, due to its widespread adoption on the Web and on digital cameras, is likely to be the image format of choice in standardized digital archives, meaning its future is assured. **JM**

SEE ALSO: INTERNET, MP3 COMPRESSION

Global Positioning System (GPS) (1993)

The U.S. Defense Dept. updates navigation.

The Global Positioning System (GPS) is a system of satellites that transmit microwaves over specific wavelengths. GPS receivers pick up signals from these satellites and define a location from the information they obtain. The system was developed by the U.S. government in 1993, but similar mechanisms have been set up in other countries including the Russian GLONASS (incomplete), China's COMPASS system, and the upcoming Galileo system in Europe. The GPS system costs the U.S. government approximately $750 million a year and is used worldwide for navigation.

The system is extremely useful and there are a huge range of commercial and domestic applications.

"The most significant development for [navigation and surveillance] since . . . radio navigation."

National Aeronautical Association

In military terms, GPS is used to track targets, locate positions in unknown territory, and project missiles, and it is also used in search and rescue and reconnaissance missions. In the civilian world, GPS units are used for in-car navigation systems as well as on domestic seacraft. Today, many cell phones include internal GPS systems that allow for tracking and can be used to locate users over wide areas. GPS also brings improved accuracy to surveys of tectonic activity such as earthquakes and tremors. **JD**

SEE ALSO: GEOSTATIONARY COMMUNICATIONS SATELLITE

➡ *In 2004 a demonstrator of the PDA Navigation System refers to a map of the Athens Olympic Games.*

Wind-up Radio
(1993)

Baylis invents the battery-free radio.

The evolution of the battery-free radio is a curiously British success story. The tale begins in 1991 with part-time inventor Trevor Baylis (1937–2018) watching a television documentary about the spread of AIDS in Africa. It suggested that the epidemic could only be halted through education. A major problem, however, was that poverty and a lack of basic technology made communication to remote parts of Africa difficult.

Baylis saw an immediate solution—a simple and cheap radio set that required no household current or battery power to operate. Crudely cobbling together parts from an old transistor radio, a small electric motor from a toy car, and the clockwork mechanism

> *"The key to success is to risk thinking unconventional thoughts. Convention is the enemy of progress."*
>
> Trevor Baylis

from a music box, he created a prototype powered by a clockwork wind-up mechanism that drove a tiny internal electrical generator. Fully wound, the spring provided fourteen minutes of uninterrupted power.

Baylis patented the idea in 1993, but failed to convince potential investors of its worth. National interest was stirred in 1994, however, with an appearance on the BBC's program *Tomorrow's World*, and within a year his invention—now known as the Baylis Freeplay Radio—was in mass production in South

SEE ALSO: WIRELESS COMMUNICATION, CRYSTAL SET RADIO, FM RADIO, TRANSISTOR RADIO

◄ *In 1996 Nelson Mandela (with Baylis) emphasized the wind-up radio's role in African communications.*

Java Computing Language (1995)

Sun Microsystems boosts the World Wide Web.

In 1991 Sun Microsystems formed "the Green project" to create a new computer programming tool for the next generation. Trying to anticipate what would happen next in the computing world, the team led by James Gosling agreed that a likely future trend would see the convergence of digital consumer products with computer technology. Perplexed by the number of different types of computer platforms that existed, they decided to create a "write once, run anywhere" programming language that would work on any device. In order to interest the digital cable television industry, they mocked up an interactive home entertainment controller. However, the technology—

> *"The Internet is an opportunity for the best products to win. Java is great technically and people want it."*
>
> Bill Joy, cofounder, Sun Microsystems

featuring animations, a touchscreen, and a networking ability similar to the Internet—was so advanced that the companies could not see a use for it.

But the team realized that by allowing media and small programs to be distributed over a network, their programming language was a perfect match for the World Wide Web. In 1994 they created a demo that brought to life animations, moving objects, and dynamic executable content inside a web browser. In 1995 it was incorporated into the Internet browser Netscape Navigator. The Java platform's versatility, efficiency, and portability means that it is now found in devices ranging from PCs to credit cards. **JM**

SEE ALSO: COMPUTER PROGRAM, DIGITAL ELECTRONIC COMPUTER, C PROGRAMMING LANGUAGE

USB Connection

(1995)

USB Implementers Forum provides a standardized interface socket to increase connectivity.

Long-term computer users will be all too familiar with the frustration of having to switch off and reboot their machines. Thanks to USB (universal serial bus) connectivity, however, this scenario has become, largely, a distant memory. These days, almost every device you plug into a computer, such as a printer or scanner, comes complete with a USB connector, instead of a card that the machine has to learn to recognize in a lengthy installation process.

The impetus behind the USB was toward a future where you could connect any device to any computer, using any port—because all the ports and plugs would match. The universal, three-pronged "trident" symbol is used on all plugs and sockets to indicate USB functionality. The reality, of course, is that there are still a few rogue devices that do not conform, but now all PCs are made with USB ports as standard.

The first USB interconnect (USB 1.0) appeared in the mid-1990s. Today's USB 2.0 connections allow data transfer to occur at ten times the speed users expected of older types of connectors.

The formation of the USB Implementers Forum (USB-IF) by Intel in 1995 marked an important point in computing history. For the computing industry, it was a way to show their commitment to increasing connectivity. The next step will be to do away with physical connections altogether. In 2007 the USB-IF announced significant progress in wireless USB communications. Wireless USB will work like a small-scale WiFi network, so that your printer can be placed unconnected anywhere you like in the room. **HB**

SEE ALSO: FIREWIRE/IEEE 1394 CONNECTION, WLAN STANDARD

"'Cannot access printer'? Why can't you . . . access printer? I've plugged you in!"

Eddie Izzard, comedian

◩ *Plugging in and unplugging USBs does not disrupt a computer's operation, so rebooting is unnecessary.*

FireWire/IEEE 1394 Connection (1995)

Apple introduces a new digital interface.

During the mid-1980s, engineers at Apple, Inc., began working on a new, high-speed data transfer medium to exchange large amounts of data on computers. They dubbed this technology "FireWire" in view of the increased speed, and released it in 1995.

Apple's engineers produced their first specification sheet for the setup in 1987. The company realized that USB devices were fine for keyboards and mice, but faster speeds were required for high-memory applications, such as video cameras, which had many gigabytes of data to be exchanged. They approached the Institute of Electrical and Electronics Engineers (IEEE) with the intention of making the technology the

"The S3200 standard will sustain the position of IEEE 1394 as the absolute performance leader ..."

James Snider, 1394 Trade Association

standard for all computers, both Macintosh and Windows-based, which the IEEE endorsed in 1995.

The first instance of the interface system was called IEEE 1394 and had data exchange speeds almost forty times that of existing USB speeds. From conception, however, the technology was constructed for large packets of data rather than smaller devices such as keyboards, and so the formats rarely competed.

In 2002 IEEE 1394b doubled existing FireWire speeds and increased the distance over which it could be used. In 2008, the 1394 Trade Association announced a new version known as S3200, which will quadruple the speed of the current format. **SR**

SEE ALSO: USB CONNECTION, WLAN STANDARD

Grid Computing (1995)

Computers join together for greater power.

U.S. computer scientists Carl Kesselman, Ian Foster, and Steve Tuecke had the bright idea that computers could be loosely coupled together to provide massive computing ability, just as power stations can be linked to supply extra electricity. With computer grids and power grids the consumer does not have to worry about the source or the location of the input. Different types of computers can be incorporated and these may be located all over the world. Your own computer can be used in a grid when you are asleep, during lunch breaks, or even at random moments during the day when the computer is waiting for input. High-speed interconnections are usually not available and so the system works best on problems where independent calculations can be carried out without the participating processors having to communicate.

Needless to say, software has to be carefully designed to check for untrustworthy, malfunctioning, and malicious nodes. It also has to accommodate nodes going off-line at random times. The participating nodes also have to trust the central system not to interfere with their individual programs, security, and data storage.

Grid computing is widely used for solving extremely complicated mathematical problems and for specific computing applications where huge amounts of data are involved. The latter include the interrogation of molecular modeling data for pharmaceutical drug design, the analysis of electrical brain activity, the searching of radio telescope receiver output for messages from extraterrestrial civilizations, and the investigation of the outputs of high-energy physics machines, such as the Large Hadron Collider while it looks for new elementary particles. **DH**

SEE ALSO: INTERNET, ETHERNET, LARGE HADRON COLLIDER

DVD

(1995)

A consortium upgrades video recording.

After the futuristic-looking compact disc (CD) took the audio market by storm—consigning the humble cassette tape to the back of a billion cupboards—it was only a matter of time before technological wizards set their sights on abolishing the VHS tape.

Although the technology for LaserDisc already existed, it never really took off in the way that CD technology did, and so the market for a compact digital video disc was still very much open. The first proposals for a high-density CD were put forward in 1993, leading to the creation of two competing formats. Electronic powerhouses Sony and Philips led their collected investors forward with the MMCD format, going head to head with industry giants Toshiba, Masushita, and Time Warner's effort, the SD. Then, in 1995, a combined effort—known as the DVD—was officially announced and consequently developed by a consortium of ten companies.

The DVD was capable of storing two hours of high-quality digital video, eight tracks of digital audio, and thirty-two tracks of subtitle information, as well as offering the practical benefits of being lightweight, compact, easily rewindable, and durable. Dual-layer DVDs later doubled this capacity, and two-sided DVDs (which can be flipped over like vinyl LPs) doubled it again without creating needless bulk.

Although DVD is often cited as being an acronym for digital video disc or digital versatile disc, the official line on it—as stated in 1999 by the 250 company members in the DVD Forum—is that it is simply a three-letter name. So, in short, DVD stands for DVD. **CL**

SEE ALSO: VIDEO TAPE RECORDING, OPTICAL DISC, BLU-RAY/HD DVD

◀ *A DVD is manufactured at the factory of the Polystar Digidisc Company in Beijing, China.*

Voice Over Internet Protocol (VoIP) (1995)

VocalTec introduces cheap Internet calls.

In 1973, researcher Danny Cohen's Network Voice Protocol was first used on ARPANET (Advanced Research Projects Agency Network), where it allowed research sites to talk with each other over the computer network. For many years afterward, however, sending your voice over the Internet was the preserve of researchers, geeks, and early computer gamers.

But in 1995 a company called VocalTec released a piece of software it called Internet Phone. Designed for Microsoft Windows, it turned the speaker's voice into computer data, compressing it enough to send it in real time over a modem connection to another computer on the Internet.

> *"The advantage is obvious: I can call my mate in Sydney and chat for the price of a local London call."*
>
> John Diamond, journalist, *The Times*

Many people suddenly became interested in Internet Telephone, for one simple reason—it was cheap. In the United States, for example, the local call to connect to the Internet was often free, whereas long-distance calls were costly.

As Internet speeds improved and other companies started to offer similar services, making telephone calls across the Internet gained a generic name: Voice over Internet Protocol, commonly known as VoIP. VoIP is now extremely popular. Skype, one of the best-known VoIP companies, has clocked up more than a hundred billion minutes of calls between its users since 2003 when the service started. **MG**

SEE ALSO: TELEPHONE, INTERNET, ETHERNET, INTERNET PROTOCOL (TCP/IP)

WLAN Standard (1996)

Hayes regulates wireless networking.

Today's wireless networks owe much to one of the earliest computer networks, the University of Hawaii's ALOHAnet. This radio-based system, created in 1970, had many of the basic principles still in use today. Early wireless networks were expensive, however, and their equipment was bulky. They were used only in places where wired networks were awkward, such as across water or difficult terrain. It was not until the 1980s, with the arrival of cheaper, more portable equipment, that wireless networking began to go mainstream.

There was a problem, however. By the end of the 1980s several companies were selling wireless networking equipment, but it was all incompatible. What was needed was some joined-up thinking. Step

---

"Anything that needs communications or control will be wirelessly connected."

Vic Hayes

---

forward the Institute of Electrical and Electronic Engineers (IEEE) and in particular Vic Hayes (b. 1941). Hayes did not invent any new technology, but he took charge of the IEEE's wireless standards committee, and fostered cooperation between the manufacturers. In 1996 they released the first standard wireless local area network (WLAN); its IEEE designation was "802.11."

Adopted in 1999 by a group of like-minded industrial leaders, who gave it the more catchy name of "Wireless Fidelity," or Wi-Fi, the standard lets us take our laptops around the world, confident that, without wires, we will be able to browse the Web anywhere, from an airport in Australia to a zoo in Zanzibar. **MG**

SEE ALSO: INTERNET, WORLD WIDE WEB, USB CONNECTION

Atom Laser (1996)

Ketterle develops a working atom laser.

The idea of an atom laser has been around for many years, and its principle is based on the more conventional optical laser. A normal laser emits light, but, unlike a normal lamp, the laser light is "coherent," so that it can focus to a pinpoint, and also travel a long distance without spreading out like a flashlight beam. By the time the optical laser was introduced in 1960, scientists were already familiar with the wavelike properties of matter, and the atom laser was under consideration as a theoretical possibility.

But it was not until 1997 that reports of the first rudimentary working model were released. A bizarre form of supercooled matter called a Bose-Einstein condensate made it all possible. This strange stuff, in which individual atoms "lose their identity" and coalesce into a single "blob," is in some ways like the photons of light in a laser. It was Professor Wolfgang Ketterle (b. 1957) and his colleagues who first managed to produce a Bose-Einstein condensate in 1995.

Not long afterward, in November 1996, Ketterle and his team cheered as their atom laser worked for the first time. They had successfully used a Bose-Einstein condensate as a source of coherent atoms to create a "matter wave." Described like a dripping faucet, it emitted pulses of droplets of atoms, each containing up to several million atoms.

Practical uses of the atom laser have yet to materialize, and it is confined to research at present. However, it is likely that atom lasers will be used in the future to directly deposit atoms onto computer chips, enabling the creation of much smaller, finer patterns and more powerful computers. **DHk**

SEE ALSO: LASER, CARBON DIOXIDE LASER, LASER COOLING OF ATOMS, LARGE HADRON COLLIDER

➔ *Ketterle and two others received the Nobel Prize for Physics for their work on Bose-Einstein condensates.*

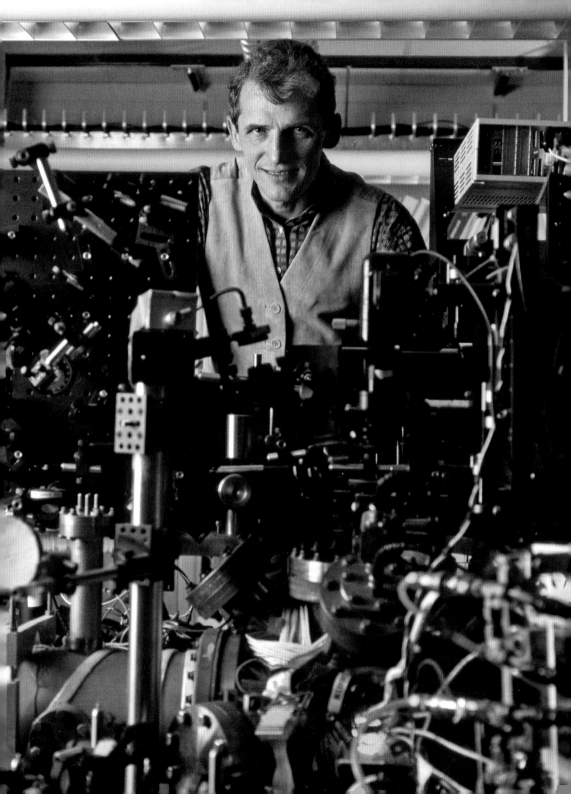

Stem Cell Therapy

(1998)

Thomson and colleagues advance a revolutionary medical therapy.

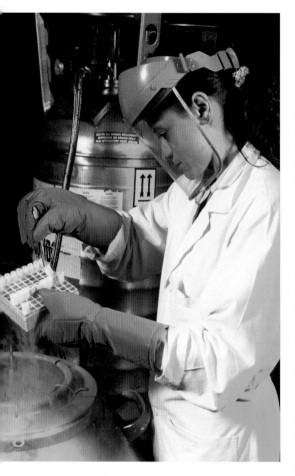

Stem cells are cells that have the ability to differentiate into a diverse range of cell types, creating the potential for the cells to be used to grow replacement tissues. American developmental biologist James Thomson (b. 1958), from the University of Wisconsin School of Medicine, won the race to isolate and culture human embryonic stem cells. On November 6, 1998, the journal *Science* published the results of Thomson's research, describing how he used embryos from fertility clinics (donated by couples who no longer needed them), and developed ways to extract stem cells and keep them reproducing indefinitely.

With the ability to develop into any one of the 220 cell types in the body, stem cells hold great promise for treating a host of debilitating illnesses, including diabetes, leukemia, Parkinson's disease, heart disease, and spinal cord injury. They also provide scientists with models of human disease and new ways of testing drugs more effectively in living organisms. But for all the hopes invested, progress has been slow. It has not helped that stem cell research has been steeped in controversy, with different groups questioning the ethics of harvesting stem cells from human embryos.

In 2007 Thomson and Shinya Yamanaka, from Kyoto University, Japan, both independently found a way to turn ordinary human skin cells into stem cells. Both groups used just four genes to reprogram human skin cells. Their work is being heralded as an opportunity to overcome problems including the shortage of human embryonic stem cells and restrictions on U.S. federal funding for research. **JF**

SEE ALSO: GENE THERAPY, DNA MICROARRAY, PREIMPLANTATION GENETIC DIAGNOSIS

"... embryonic stem cells were a researcher's dream. Now they're a political hot potato."

Frederic Golden, commentator

◤ *A researcher removes phials containing stem cells from storage in liquid nitrogen at -296°F (-182°C).*

Viagra® (Sildenafil Citrate) (1998)

Pfizer offers hope for impotent men.

Since Viagra® (sildenafil citrate)—the first oral drug to treat erectile dysfunction—went on sale ten years ago, more than 27 million men in 120 countries have been prescribed it for impotence. Initially the drug was designed to treat high blood pressure, but in clinical trials this use proved disappointing. One side effect, shyly reported by the healthy volunteers, was that the drug produced super-charged erections.

It was known that sexual arousal messages from the brain spark the production of cyclic guanosine monophosphate (cyclic GMP), a chemical that relaxes the pelvic muscles and allows the penis to become engorged with eight times its normal supply of blood. Sildenafil suppresses an enzyme (phosphodiesterase type 5), whose normal role is to break down cyclic GMP and cause the erection to subside.

Pharmaceutical company Pfizer conducted twenty-one randomized, placebo-controlled clinical trials involving more than 3,700 participants aged from nineteen to eighty-seven suffering from varying degrees of impotence. Results showed that Viagra® restored sexual function in seven out of ten men. Fortunately, the drug worked only when the men were sexually aroused, and not at other times also.

The drug was approved by the U.S. Food and Drug Administration in 1998. Previous treatments for erectile dysfunction included injections into the penis, urethral suppositories, surgery, and vacuum devices. Within fourteen weeks, two million Viagra® prescriptions had been written in the United States alone. The side effects of Viagra®, now also sold under other names, include headaches, flushing, indigestion, and a temporary color-vision change. A small number of users have also suffered a heart attack or stroke. **JF**

SEE ALSO: CONDOM, BIRTH CONTROL PILL, IN VITRO FERTILIZATION, INTRACYTOPLASMIC SPERM INJECTION (ICSI)

Web Syndication (RSS) (1999)

Guha tracks news on the Web.

Computer science is renowned for its use of confusing TLAs—Three Letter Abbreviations. Even in computing, however, RSS is notable for being a tricky example. During its history, RSS has stood for Rich Site Summary, RDF Site Summary, or Really Simple Syndication.

RSS is a way of describing websites, especially sites with fast-moving content such as news. The original RSS was created by Ramanathan V. Guha (b. 1965) in 1999, for the popular My.Netscape.Com site. The "portal" site enabled browsers to customize news on a single page. This was only possible because each of the different source websites agreed to publish a description of their news in Guha's common format.

> *"It takes a while before people realize they can do other things with a new medium."*
>
> Ramanathan V. Guha

Since then, RSS, which grew from Guha's earlier work at Apple's Advanced Technology Group, has been through several incarnations at the hands of many different organizations and people. Dave Winer, another podcast pioneer, added much to the standard, including the ability to handle audio files and other enclosures in RSS, paving the way for the arrival of podcasting in 2003.

Today RSS—currently standing for Really Simple Syndication—can gather news, blog entries, stock market prices, podcasts, pretty much anything in fact, into one place, often through the use of news reading software such as the popular Google Reader. **MG**

SEE ALSO: INTERNET, INTERNET PROTOCOL (TCP/IP), PERSONAL COMPUTER MODEM, WORLD WIDE WEB

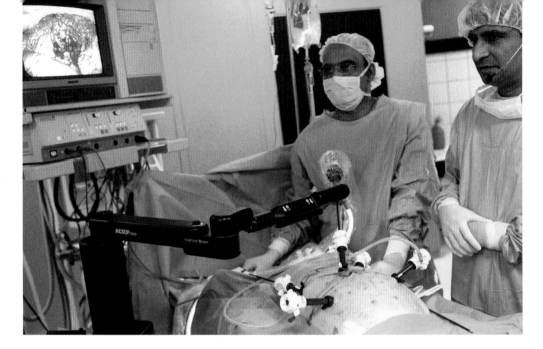

Telesurgery (2001)

Gagner performs remote surgery with the help of robots.

Surgeons using keyboards, joysticks, and the like can make movements that control advanced surgical robots—such as "Da Vinci®" and "Zeus®"—in the performance of long-distance, unassisted surgery.

On September 7, 2001, Jacques Marescaux of the University of Strasbourg, France, and the IRCAD European Institute of Telesurgery, and Michael Gagner, chief of the Department of Laparoscopic Surgery at Mount Sinai Medical Center in New York, removed the gallbladder of a sixty-eight-year-old woman in Strasbourg, more than 3,700 miles (6,000 km) from the surgeons. Other medical staff stood by in Strasbourg but did not have to intervene.

"Operation Lindbergh," as it was dubbed, required a high-speed optical-fiber network to relay the information. The time lag between the surgeons initiating a movement and watching the result on their display was one eighth of one second. The operation was performed laparoscopically: a camera and instruments were inserted into the patient through small incisions. The operation took fifty-four minutes, similar to conventional gallbladder surgery.

Such advances allow specialized surgery to take place virtually anywhere, even in such remote places as space stations, with the human surgeon far away.

In May 2006 a surgical robot performed an unassisted procedure in Milan, Italy. Carlo Pappone, the head of Arrhythmia and Cardiac Electrophysiology at Milan's San Raffaele University, monitored the operation from Boston, Massachusetts. The fifty-minute surgery was carried out to control the irregular heart rhythm of a thirty-four-year-old man. **SS**

SEE ALSO: ANTISEPTIC SURGERY, ENDOSCOPE, LAPAROSCOPE, SURGICAL ROBOT

⬆ *Surgeons undergo training with the surgical robot "Aesop®," which is operated by vocal commands.*

Artificial Liver (2001)

Matsumura invents a potential lifesaver for patients with acute liver failure.

People become quite excited about artificial hearts. But, when you think about it, the heart is basically a pump, the same kind of pump that people have been using for thousands of years. Described in these terms, it does not sound quite so advanced. By contrast, an artificial liver is a complex achievement.

Far from the one trick pony that the heart is, the human liver has to undertake many tasks simultaneously. Among other functions, it helps to break down food into usable substances, detoxifies harmful chemicals, stores energy in the form of glycogen, and manufactures any number of substances from bile to the proteins that make cuts stop bleeding. But how do you combine all those different functions into an artificial liver?

Numerous ways have been tried to treat liver failure, from replacing the entire blood volume in a person's body with new blood to hemodialysis. All have met with little success. However, in 2001,

Dr. Kenneth Matsumura and his team were among the first to produce functioning artificial livers. Matsumura and his team decided the best way to approach the problem was to take living liver cells and place them in sequence with a series of charcoal filters. The resulting device performed most of the activity that a normal liver would carry out because it was partially made of normal liver cells. It is mostly used as a bridge until a new liver is available for transplant rather than as a full replacement, but further test results are promising.

British scientists have since created the first artificial liver tissue from stem cells. In time it is hoped that it will provide whole organs for transplant. **BMcC**

SEE ALSO: HEART-LUNG MACHINE, ARTIFICIAL SKIN, ARTIFICIAL HEART, KIDNEY DIALYSIS, CONTROLLED DRUG DELIVERY

⬆ *In this Petri dish is fleece populated with liver cells, to be used in a temporary, external, artificial liver.*

Segway PT
(2001)

Kamen produces a gyroscopic scooter.

When the Segway PT (Personal Transporter), produced by Dean Kamen (*b.* 1951), was finally unveiled in December 2001, there was a sense of disappointment. Rumors had circulated that the Segway was an antigravity device, and influential figures had described it as "maybe bigger than the Internet."

In fact, the PT—developed over a decade at a cost of $100 million—is a 5,000-dollar gyroscopic scooter. Standing on a platform between two wheels and holding onto a T-shaped handlebar, its users can reach speeds of up to twelve miles per hour (20 kph). With a range of fifteen miles on a single battery, the PT's originality lies in its array of microprocessors and

---

"If we're serious about our environment, we've got to get serious about the Segway."

Lembit Öpik, British politician

---

sensors that register the slightest movements. The eco-friendly electric motors in the wheels instantly act upon these deviations from the norm to attain constant perfect balance. Hence, the PT—which does not have conventional brakes but decelerates in order to stop—moves in whatever direction you move.

However, the PT's prohibitive price, its clunky design, and its stigma as a gadget for geeks all account for the fact that the device currently has little success beyond the commercial sector. **DaH**

SEE ALSO: AUTOMATON, POWERED EXOSKELETON, BIPEDAL ROBOT, DRIVERLESS CAR

◁ *Mounted on Segway PTs, Chinese armed police demonstrate an anti-terrorist drill in 2008.*

Satellite Radio Broadcasting (2001)

XM revolutionizes radio broadcasting.

How often have you been listening to your favorite radio station in your car, only to have it slowly dwindle away as you drive out of range of the transmitter? The answer to this problem came in 2001 when a new way to receive radio made its debut—a national service beamed from outer space that, in return for a subscription fee, offered 100 different channels, none of which would be interrupted by poor reception.

The coverage, provided by two or three high-orbit satellites, came as a strong signal requiring no satellite dish, just an antenna the size of a matchbox. Although the broadcast could be dampened by skyscrapers or long tunnels, the signal was bolstered by transmissions from ground-based towers.

Two companies were originally granted licenses to provide satellite radio in 1997: XM and Sirius, with XM getting off the mark first in the U.S. in September 2001 and Sirius following in 2002. For the monthly subscription fee of $10 the XM service offered a wide choice: some channels were dedicated to specific topics such as news, sports, traffic, and weather reports; music channels had no commercials, and even no DJs, and talk shows were less censored than the free terrestrial radio broadcasts. What is more, satellite radio receivers displayed the name of each artist and song as it played. Soon receivers were being made that could be taken out of cars and carried into the home.

Initial uptake was slow. At launch it was estimated that there were only four listeners per channel, but by 2008 a total audience of over 16 million had been garnered. Despite legal wrangles over music copyright and complaints about the content of some chat shows, it seems that satellite radio is here to stay. **JM**

SEE ALSO: ARTIFICIAL SATELLITE, GEOSTATIONARY COMMUNICATIONS SATELLITE

Optical Camouflage (2002)

Tachi creates a see-through coat.

Thanks to Japanese research, the twenty-first-century soldier may soon be blending invisibly into the background. The man behind optical camouflage, Susumu Tachi (*b*. 1946), is a professor at Tokyo University, where he works on the "science and technology of artificial reality." Ironically, since he now works to make things invisible, Tachi previously developed a robotic guide dog for the blind.

The optical camouflage developed by Tachi and his research team works by filming the background environment and projecting it onto a coat worn by the test subject. However, this is no average coat. It is covered in thousands of tiny beads that reflect light

---

"I wanted to create a vision of invisibility. . . . This is a kind of augmented reality."

Susumu Tachi

---

back to its source, therefore rendering the coat invisible. This is the theory, but in reality the system is still far from perfect and in great need of cutting back on the volume of equipment required.

Tachi patented the "Method and Device for Providing Information" in 2002, and since then the technology or variations of it have been appearing in U.S. military prototypes; monitored surgery may also benefit from the technology. Modern warfare may soon be turning into one big game of hide and seek. **CB**

SEE ALSO: MILITARY CAMOUFLAGE, NIGHT VISION GOGGLES, COLOR NIGHT VISION

↩ *"Invisibility" is achieved by projecting the background onto a coat made of a special light-reflective fabric.*

Laser Track and Trace System (2002)

Drouillard traces product movements.

It is amazing how advancing technology has influenced what would seem to be the simple task of selling apples in supermarkets. Not so long ago, the checkout cashier was expected to know the difference between apple varieties such as Granny Smith and Golden Delicious, and charge the customer accordingly. When, for various reasons, this became too arduous, each apple was provided with a little sticker (known as a P. L. U., or price look-up). But these were time-consuming and expensive to apply, and provided too little information. Further, if they were too sticky they were difficult to remove, and if they were not sticky enough they fell off.

In 2002 the American laser expert Greg Drouillard offered a solution, patenting his idea of tattooing the skin of the apple with a laser-scannable barcode. The barcode would inform a computer as to the type of apple, where it came from, when it was picked, who picked it, how many calories it contained, and whether it was organic, genetically modified, or Fair Trade.

With the aid of the barcode, the apple could be tracked and traced on its whole journey, from tree to consumer. Laser tattoos can also be burned into a host of other fruits and vegetables, including pears, peaches, lemons, oranges, cucumbers, and peppers. Only soft and easily perishable foods such as strawberries are unsuitable. However, there remain a few questions, such as whether a fruit is still organic if it has been laser tattooed.

On the other hand, the laser track and trace system can be applied to many other manufactured items, and has been extremely useful in the pharmaceutical industry with its time-sensitive blood and vaccination products passing along the distribution pipeline. **DH**

SEE ALSO: RADIO FREQUENCY IDENTIFICATION (RFID), BARCODE, LASER

Scramjet (2002)

The University of Queensland and QinetiQ develop hypersonic travel.

"We … believe that a hypersonic airplane could be a reality in the not too distant future."

Dr. Steven Walker, DARPA Tactical Technology Office

In the twenty-first century, speed seems to be of paramount importance. As well as improving the simple but inefficient turbine-based engine systems that drive rockets and planes, scientists have developed Supersonic Combustion **Ramjet** engines (scramjet) to allow much faster travel.

Scramjets improve on ordinary engines by eliminating the need to carry a fuel oxidant. Instead, they use oxygen from the atmosphere to burn the onboard fuel, making them lighter, more efficient, and extremely fast. Scramjets have long been a theoretical possibility, but in 2002 scientists at the University of Queensland, Australia, and at the U.K. defense company QinetiQ successfully completed the first flight of a scramjet vehicle. Although the test simply demonstrated the technology and not a practical engine system, the vehicle reached Mach 7, which is seven times the speed of sound.

Since then, the National Aeronautics and Space Administration (NASA) has been working on its Hyper-X program to develop scramjets into a practical technology, able to provide the thrust to propel a craft. It is believed the technology could eventually allow vehicles to reach Mach 15, reducing an eighteen-hour flight from New York to Tokyo to just two hours.

In 2007 a joint project between the U.S. Defense Advanced Project Agency (DARPA) and the Australian Defence Science and Technology Organization (DSTO) launched a flight that succeeded in reaching the hypersonic speed of Mach 10. The imminent arrival of these hypersonic airplanes will make the world a much smaller place. **FS**

SEE ALSO: POWERED AIRPLANE, JET ENGINE, SUPERSONIC AIRPLANE, STEALTH TECHNOLOGY

⬦ *In 2004 the NASA X-43A scramjet achieved a record Mach 9.6, or nearly 7,000 miles per hour (11,265 kph).*

Blu-ray/HD DVD (2003)

Nakamura optimizes optical storage.

High-definition (HD) media was originally developed in 1998, three years after DVDs became commercially available. Despite the DVD having six times the storage space of CDs, it was not sufficient to store all the information in the new HD media. New formats were needed and, eventually, in April 2003, the first Blu-ray disk and HD DVD-compatible devices went on sale.

Blu-ray and HD DVD store exactly six times more information than a DVD. Members of the general public, disgruntled at having to watch poor-quality images on old-fashioned DVD players, now require these HD storage devices to watch movies on their new big-screen televisions.

There was a gap of several years between the development of HD media in 1998 and that of the HD DVD storage device in 2003. Although it was well known that lasers with shorter wavelengths could solve the problem, it was not until Dr. Shuji Nakamura (b. 1954), a professor at the University of California, Santa Barbara, invented blue laser diodes that Blu-ray and HD DVD became a reality. Nakumura had already invented blue, green, and white light-emitting diodes (LEDs) in the 1990s. The name Blu-ray originates from the blue laser used to read and write the disk.

DVDs were introduced as a standard format, the manufacturers having learned painful lessons from the costly war between the VHS and Betamax formats. When it came to the HD format, however, the manufacturers decided to ignore this precedent. Sony's Blu-ray Disc and Toshiba's HD DVD began to compete for dominance in the market. The clash of technological titans was won by Sony, with Toshiba announcing on February 19, 2008, that it was ceasing production of the HD DVD format. By that time, close to 1 million HD DVD players had been sold. **JB**

SEE ALSO: OPTICAL DISC, LASER, DVD, HIGH-DENSITY COMPUTER STORAGE

Podcast (2003)

Lydon and Winer pioneer digital media files.

The first decade of the twenty-first century has seen a proliferation in new communications technology. It is now possible to watch a favorite television show on a laptop, read a newspaper on a cell phone, or listen to a radio broadcast on an MP3 player. The evolution of the podcast is one more important development.

A podcast is a digital audio or video file, distributed automatically to a subscribed user. That user can then listen to or view the file on a mobile device such as a personal computer, MP3 player, or cell phone. The podcaster first creates a "show"—usually a video or MP3 audio file—and then an RSS (Really Simple Syndication) feed file that points to where the podcast can be found. The receiver uses "aggregator" software

> *"[The] question is how to be ready for a new world of reporting and commentary by Internet rules."*
>
> Christopher Lydon

to subscribe; this periodically checks the RSS feed to see if content exists or new content has been added, and then downloads that content automatically.

The term itself was coined by British technology writer Ben Hammersley in 2004, from the words "iPod" and "broadcast." Although former *New York Times* reporter and radio broadcaster Christopher Lydon (b. 1940) is often cited as creating the first true podcasts, much credit also belongs to background innovators such as Dave Winer (b. 1955), who developed the RSS syndication feed that enabled Lydon's blogs—a series of cutting-edge interviews—to achieve automated widespread distribution. **TB**

SEE ALSO: WEB SYNDICATION (RSS)

High-density Computer Storage (2005)

IBM boosts computer memory dramatically.

A computer has two fundamentally different types of memory. The main memory is Direct RAM (DRAM), which represents the active storage components of a computer's memory. DRAM is fast, but it stores relatively little information. The other type of memory is the hard drive, which uses a solid, locally magnetized, metal disc to store information. (For this reason, a powerful magnet should never be placed on top of a laptop.) The hard drive uses a single "pick-up" to read the information while the disc spins beneath it. This process takes a long time by computer standards.

In 2005 IBM demonstrated a possible alternative in the form of the "millipede"—so-called because it

> *"Supercomputers will achieve one human brain capacity by 2010, and [PCs] by about 2020."*
>
> Raymond Kurzweil, inventor and futurist

looks like one. The millipede uses nanotechnology to store information in tiny depressions on the surface of a silicon-treated polymer. This information is read by a series of atomic force probes, which can read and write information much more quickly than conventional hard drives.

The consequence is that the millipede is much faster than the hard drive, almost as fast as some DRAMs, and is capable of storing more information than some hard drives, as much as 1 terabyte per square inch (1 gigabyte per square millimeter). This capacity is four times greater than the magnetic storage currently available. **BG**

SEE ALSO: PUNCHED CARD, MAGNETIC CORE MEMORY,
FLASH MEMORY, JPEG COMPRESSION, MP3 COMPRESSION

Color Night Vision (2005)

TNO refines the technology of night vision.

There are a number of situations in which you might want to be able to see in the dark, but it was for military purposes that night vision goggles were originally developed by American Wiliam Spicer in 1942. In later years, cameras on the goggles would take tiny amounts of light and amplify it, producing a gray or green, albeit fuzzy, image of what was happening. Alternatively, in pitch black, where there was no light to amplify, infrared sensors showed the heat coming off things, enabling you to see everything in colors that depended on the degree of heat.

A big problem with a night-vision world represented in gray or green is that distances are hard to judge. Without the range of colors that are seen in daylight, grayish objects can be hard to distinguish from a background of similar color. The Dutch military approached the TNO research team led by Alex Toet in Soesterberg, Netherlands, to develop goggles that would show the night world in color.

The goggles—whose workings were revealed in 2005—are given a sample picture of the world during daytime from which all the colors can be identified. The goggles map this color picture onto a gray night vision picture and compare the two. After this they can instantly translate the grays of the night vision into colors whenever the goggles are used at night.

Initially the color night vision goggles have a range of settings based on urban, rural, desert, or coastal environments. Eventually, using GPS (Global Positioning System) data, they can be programmed to adjust the color ranges automatically for the specific location where they are being used. In addition to the military, the emergency services are likely to find the goggles invaluable in their night operations. **DK**

SEE ALSO: NIGHT VISION GOGGLES, GLOBAL POSITIONING
SYSTEM, IMAGE INTENSIFIER

Surface Computing

(2006)

Microsoft takes the personal computer off the table—to become the table itself.

In 2006 Microsoft announced a new way in which a human can interact with a computer. The mouse and keyboard are thrown away and replaced with a table-top containing an embedded rear-projection touch screen. Behind the screen, inside the table, are five cameras with overlapping fields of view. These can look through the screen and be programmed to recognize or read items that are placed on the screen. These cameras can also recognize physical objects, track hand gestures and the movement of pens and brushes, and read credit cards and "loyalty" cards.

The software supports a multitude of touch points so many people can use the computer at once. A single user can also multitask. For example, if a digital camera is placed on the computer table, the computer immediately recognizes that it is a camera, downloads the stored images, and shows them on the table-top screen. Enlarging an image is done simply by touching two opposite corners of the image and moving the fingers apart. Dragging songs into and out of music players is done in a similar operation. The aim is to replicate everyday intuitive manipulative interactions.

At the present time Microsoft Surface is expensive and aimed at hotels, public entertainment venues, stores, and restaurants, but soon people could be browsing though music and book lists and down-loading specific items, playing video poker, and ordering beer and food. Computer game technology could be revolutionized by these devices, and they could also be used with a real brush and a virtual palette to paint computer pictures in real time. **DH**

SEE ALSO: TOUCHPAD, TOUCH SCREEN, PLASMA SCREEN

↗ Images on the table-top screen can be moved around or resized by touching them with one or two fingers.

"A table, a chair, a bowl of fruit, and a violin; what else does a man need to be happy?"

Albert Einstein

Printing On Demand
(2006)

Books on Demand cuts short-run print costs.

Printing on demand (POD) has revolutionized the publishing world. The process, established in 2006 by Books on Demand, uses digital technology to print copies of a book once an order has been placed. POD is infinitely quicker and simpler to set up than conventional printing because the pages are stored electronically, eliminating the need for lithographic films or printing plates. It is also relatively cheap. Unit costs are higher per copy than with traditional printing methods, but POD offers lower costs per copy for small print runs after set-up costs are taken into account. POD books can be made available in either a traditional printed format or as an e-book.

POD digital technology is rescuing many older and relatively obscure titles from oblivion by enabling them to remain available and in print. Cambridge University Press (CUP) has been using the process since 1998 and now has more than 10,000 titles in its catalog. Their POD company, Lightning Source, churns out roughly one million books per month, on subjects as diverse as eighteenth-century Spanish music or the origins of meteor showers.

The printing-on-demand process also allows publishers to exploit the short-lived hype associated with celebrity biographies, television tie-ins, and reality TV programs. With consumer attention span being so brief, publishers can minimize the risk of unsold stacks of books in the warehouse by making a low estimate of demand, printing traditionally, and then using POD to meet any additional demand

POD is superior to conventional printing methods in its eco-friendliness. Less paper and ink is used and little waste is generated—a crucial aspect of new technology in a heavily industrialized world. **FS**

SEE ALSO: ETCHING FOR PRINT, LINOTYPE MACHINE, MONOTYPE MACHINE, MOVABLE TYPE, LITHOGRAPHY

iPhone
(2007)

Apple adds capabilities to the cell phone.

The first smartphone was developed by IBM in 1992, with features including a calendar, an address book, a calculator, a notepad, email, and fax capabilities. Announced in January 2007, the Apple iPhone took things to a whole new level—it could play music, take photographs, browse the internet, play movies, and store up to 8 gigabytes of information. And it could make a telephone call.

With its large screen, lack of buttons, and sleek appearance, the iPhone stood out in design terms alone. What is more, the touch screen was designed to be operated by a finger, so it was no longer necessary to find the stylus required by other models.

"We think the iPhone is a 'game changer'… it will change how people think about… handsets."

Randall Stephenson, CEO of AT&T

However, the iPhone did not perform as well as expected, with Apple taking only a 5 percent share of the worldwide smartphone market. This was in large part due to the initial retail price of $599, later reduced to $399, much to the annoyance of early purchasers.

In July 2008 Apple released its new, improved, and cheaper iPhone 3G, which could exploit the faster 3G networks. As for the inventor of the iPhone, Apple's CEO Steve Jobs heads the author list of the patent, although many Apple employees were involved. **RP**

SEE ALSO: CELLULAR MOBILE PHONE, DIGITAL CELL PHONE

➡ *Slim in profile, the iPhone presents its menu of functions on a liquid-crystal-display touch screen.*

Graphene Transistor (2008)

IBM researchers improve transistors made from single-atom-thick sheets of carbon.

The new era of post-silicon and semiconductor electronics is dawning, and graphene—a graphite-based, two-dimensional atomic-scale sheet of carbon atoms joined in a honeycomb-like lattice—is one of the materials that lies at its heart. Hundreds of times stronger than steel, it is believed that graphene transistors might enable the production of computer chips that are hundreds of times faster than the current generation of silicon chips.

Graphene was first measurably isolated in 2004, by researchers at the University of Manchester in the UK. Its electronic properties have always been known, but they have proved problematic to harness in transistors because the high conductivity of graphene makes it difficult to switch current on and off to create circuits. In 2008, researchers at IBM made the discovery that by placing two layers of graphene on top of each other, they were able to significantly reduce this problem and decrease the "noise" that prevents the current flowing.

The electronics industry is abuzz over graphene's potential. Not only is it the strongest, thinnest, and most pliable material we have, but it is also waterproof and could be the precursor to a whole new era of waterproof devices. In late 2015, a team of scientists at the University of British Columbia in Vancouver, Canada, succeeded in coating graphene with lithium atoms—thus creating the world's first example of superconducting graphene with the ability to conduct electricity without dissipating energy. Revolutionary developments such as this make it inevitable that graphene will one day be transforming the world of advanced circuitry. **BS**

SEE ALSO: MICROPROCESSOR, HIGH-TEMPERATURE SUPERCONDUCTOR, MICRO-ELECTRO-MECHANICAL SYSTEMS

⬆ *A graphene transistor held by tweezers in the palm of a person's hand.*

Cloud Computing (2009)

Web 2.0 enables shared computing resources.

Cloud computing is the practice of using a network of remote internet-hosted servers to store and access data and programs rather than a local server or a PC hard-drive. The idea of the cloud, like many computing concepts, has its origins in the 1960s. The American psychologist and computer scientist J.C.R. Licklider foresaw the era of interactive personal computing, and envisioned an "intergalactic computer network" that would link all users and be accessible anywhere in the world. As this became a reality with the growth of the Internet in the late 1990s, the term "cloud" was used to represent the computing space between the provider and the end user.

Amazon led the early charge to cloud technology with its Web Services in 2002 and Elastic Compute Cloud (EC2) in 2006, which allowed individuals and small businesses to "rent" computers on which they ran their own applications. It was in 2009, however, with the coming-of-age of Web 2.0 and a new emphasis on user-generated content and dynamic interfaces, that cloud computing truly took off. Google were at the forefront of the revolution, with the introduction of browser-based cloud enterprise applications through their Google Apps service.

There is, of course, a lot more to the cloud than it simply being a vehicle for apps and music streaming. There are private clouds, public clouds, and hybrid clouds of two or more existing clouds, linked to offer a multitude of deployment models. With security issues that once bedevilled its use for businesses now largely addressed, the cloud is fast achieving the potential it always promised. **BS**

SEE ALSO: INTERNET, ETHERNET, WORLD WIDE WEB, GRID COMPUTING

⬆ *The cloud allows users to access files remotely, as well as sharing them with friends and colleagues.*

Synthetic Genome (2010)

The first synthetic organism is created.

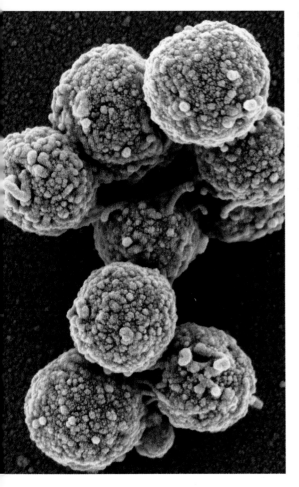

On May 20, 2010, a press release was issued by the J. Craig Venter Institute, a genomic research organization with facilities in California and Maryland. It read, in part, that its researchers had succeeded in constructing the first self-replicating, synthetic bacterial cell. Named *Mycoplasma mycoides* JCVI-syn1.0, its genome was designed on a computer, chemically made in a laboratory, and transplanted into a recipient cell to create a new self-replicating cell under the control of the synthetic genome.

It was a work fifteen years in the making, a dream that consumed the lives of those closest to it. To produce the synthetic cell the researchers had to learn how to sequence, synthesize, and finally transplant the genomes. All was done with the complete absence of natural DNA. Initial efforts to synthesize the genome failed to result in the creation of any viable cells, and new methods had to be developed to test each synthetic segment with other (natural) segments to establish if they were candidates for transplantation. The use of DNA sequencing ultimately unlocked the problem of rejection: one single base pair deletion in a certain gene was the reason for the unsuccessful transplants.

The researchers at the J. Craig Venter Institute believe the creation of the synthetic genome will lead to the creation of bacteria that can be engineered to produce biofuels and medicine. In 2014, Venter established Human Longevity Inc. to look at extending our human lifespans. The company also plans to sequence tens of thousands of genomes, focusing on cancer genomes and those of cancer patients. **BS**

"Venter is creaking open the most profound door in humanity's history . . . peeking into its destiny."

Julian Savulescu, Oxford University

SEE ALSO: GENE THERAPY, AUTOMATED DNA SEQUENCER, DNA MICROARRAY

◩ *A colored scanning electron micrograph of* Mycoplasma mycoides *JCVI-syn1.0 cells.*

Tablet Computer (2010)

Apple pioneers a new genre of computer.

When the iPad first appeared in 2010, it launched a whole new genre of devices. But this new cache of technology also ushered in a lot of questions. Larger than a smart phone yet smaller than a notebook, how was one to define these new gadgets that were quickly filling the grey area between those two established bookends? Was a tablet to be measured solely on its screen size, or were features and specifications also meant to be considered?

The concept of the tablet can be traced back to the 1960s and the prototype Dynabook (1968) created by computer scientist Alan Kay, which included a screen with an integrated keyboard. Then came the Apple Graphics Tablet (1979), the MS-DOS GriDpad (1989), the Apple Newton MessagePad (1993) —hated by Apple Inc. CEO Steve Jobs—and the Windows stylus-based XP Tablet PC (2002), among others. All had either poor battery life, were too heavy, had a lack of standalone capability, poor handwriting recognition software, or an exorbitant price tag.

Finally, in April 2010, Apple got it right with its iPad—a reasonably priced tablet with Wi-Fi, 3G options, a screen resolution of 1024 x 768, a choice of 16, 32, or 64GB of storage, and, significantly, Apple's very first branded processor, the Apple A4. It also came with an ambient light sensor that adjusted to different lighting conditions and an enlarged, two-columned iPhone interface that allowed for the inclusion of a whole new class of apps. With a surprisingly low cost of $499, this new device—the product of Steve Jobs' vision—gave the world a new form of mainstream computer. **BS**

SEE ALSO: TOUCH SCREEN, PERSONAL COMPUTER, LAPTOP/NOTEBOOK COMPUTER, SURFACE COMPUTING

↗ *Apple Inc. CEO Steve Jobs unveils the iPad at an Apple Special Event on January 27, 2010.*

"So much more intimate than a laptop, and so much more capable than a smartphone."

Steve Jobs

Intelligent Personal Assistant (2011)

Apple develops a computerized secretary.

An intelligent personal assistant is a computer software application that combines information about its user with data gleaned from the Internet to provide a vast range of services. It can perform the same function as a conventional diary, filling in vacant time slots, and providing reminders of imminent appointments and obligations. It can use the global positioning system (GPS) to provide the latest weather forecasts, and advise on traffic and train and plane schedules—everything that in the pre-computer age would have been done by a travel agent. Another of its uses is as a sort of personalized personal trainer or medical assistant: it can monitor calorific intake and output, heart rate, and the number of paces walked, and prescribe the healthiest option in each case. It responds to voice commands, too: you can ask it, "What's the capital of South Sudan?" and the answer will come back almost instantly, "Juba."

And, perhaps most amazing of all, it can perform most of these tasks autonomously. It can read your e-mail, and act on its contents—but only if you authorize it to do so. You won't need to tell it that you're going for a run, it'll work that out for itself, and take the appropriate readings while you're pounding the pavement.

The first intelligent personal assistant was Siri, which was launched as an app in 2011 by Apple Inc. Its voice recognition function enabled users to issue instructions—book restaurants, cancel theater tickets, order taxis, change airline seats—in natural language. Other intelligent personal assistants are now available from numerous suppliers, including Amazon, Baidu, BlackBerry, Facebook, Google, HTC, IBM, LG, Microsoft, Samsung, and SILVIA. **GL**

SEE ALSO: SPEECH RECOGNITION, PALMTOP COMPUTER (PDA), GLOBAL POSITIONING SYSTEM (GPS), IPHONE

CRISPR Gene Editing (2012)

A revolution in medical science.

CRISPR gene editing is one of the most important innovations in the history of biology, and it won its inventors a Nobel Prize in 2020. A gene is a section along the length of the DNA molecule that carries coded instructions for making a specific protein. Some proteins form an organism's physical structure; others are enzymes, which enable life's chemical reactions; still others are signaling molecules such as hormones. Together, the entirety of an organism's DNA, including all the genes (and non-coding sections of DNA) make up that organism's genome. The genome determines an organism's characteristics, and since genes are passed down from generation to generation, the gene is the basic unit of inheritance, as well as the instruction set for building a specific protein.

CRISPRs (clustered regularly interspaced short palindromic repeats) are regions that lie along the DNA of life-forms such as bacteria, part of a simple immune system. The enzyme Cas9 (CRISPR-associated protein 9) identifies harmful viruses by recognizing CRISPRs in the viruses' genome—and cuts the viruses' DNA at the matching points. In 2012, French microbiologist Emmanuelle Charpentier and American biochemist Jennifer Doudna found a way to utilize the CRISPR-Cas9 system to edit an organism's genome to order.

CRISPR gene editing can be used to develop new medicines or to create cell lines with specific diseases, including cancers—so those diseases can be studied more effectively in the laboratory. The ability to edit genes is a powerful tool, which raises ethical and moral questions. Most controversially, some researchers are looking to use the technique to engineer the genomes of human embryos—making changes that would be inherited down the generations. **JC**

SEE ALSO: GENE THERAPY, GENETICALLY MODIFIED ORGANISMS, AUTOMATIED DNA SEQUENCER

Mind-Controlled Prosthetic

(2015)

Ossur unveils a bionic leg that can be controlled by subconscious intentions.

"Over-the-skin" prosthetics are typically composed of a stretchable polymer infused with an array of sensors that are capable of sensing heat, moisture, and reacting to pressure. The hope is they might be able to bring sensitivity—feeling—to artificial limbs, and that their sensors, properly configured, will provide a level of tactile touch similar to that of a human hand, and be capable of sending that information to the user's brain. There have been some solid advances in recent years, but now there is something that lifts the concept of controlling artificial limbs into an entirely new realm—the mind-controlled prosthetic.

In 2015, the Icelandic orthopedics company Ossur unveiled a mind-controlled prosthetic leg, after successfully trialling it with two patients for over a year. The bionic limb uses tiny implanted myoelectric sensors (IMES), which are surgically inserted into the patient's residual muscle tissue in a simple 15 minute operation. These sensors receive electronic impulses from the brain, which they transmit wirelessly to a receiver located inside the prosthesis, instantaneously triggering the desired movement. The process occurs subconsciously and in real time, enabling faster, more natural responses and movements. It also has the side benefit of promoting muscle growth and reducing muscle atrophy—a common problem among amputees—by forcing the user to actively use muscles that were previously ignored. Furthermore, Ossur believe that the sensors should last a lifetime—as there are no integrated batteries, there is no need to replace the IMES unless they fail. **BS**

SEE ALSO: ELECTROACTIVE POLYMER (ARTIFICIAL MUSCLE), ARTIFICIAL LIVER, VISUAL PROSTHETIC (BIONIC EYE)

↗ *A "Power Knee" active prosthetic limb sits in a display case at the Ossur manufacturing plant.*

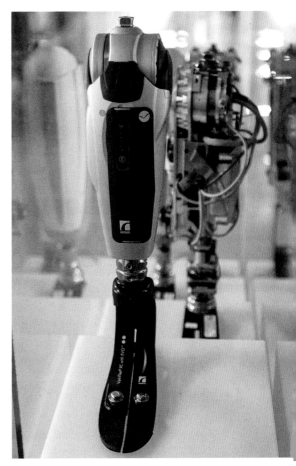

"The user's experience with their prosthesis [becomes] more intuitive and integrative."

Thorvaldur Ingvarsson, Ossur

Artificial Womb (2017)

A viable alternative to the female uterus.

In 2017, after three years of testing, researchers at the Children's Hospital of Philadelphia revealed the world's first artificial womb, a fluid-filled container designed to mimic the female uterus. The hope is that it will improve the survival odds of premature babies, born so small that today's technology—ventilators and incubators designed to assist ill-equipped tiny lungs with breathing—is of limited usefulness.

Premature babies would be placed inside the womb and take in its substitute for amniotic fluids, just as they would in the natural womb. A mechanical placenta has been designed to provide babies with oxygen, delivered through their existing umbilical

> *"It's promising research that will help us better understand how to support pre-term babies."*
>
> David Tingay, neonatal researcher

chords. The enclosed "Bio Bag" system, with its watertight ports, can house a variety of medical instruments. It is the result of years of research into how to recirculate sterile fluids, enhanced with nutritional electrolytes and proteins to avoid infections.

The technology has been trialled successfully on lambs, which are developmentally similar to humans, but human tests are a few years away. Hoping to end high rates of mortality and the spectre of cognitive impairment and chronic lung problems common to many "preemies," the artificial womb was described by Jay Greenspan, chair of pediatrics at Nemours duPont Pediatric, as "heroic and monumental." **BS**

SEE ALSO: ARTIFICIAL LIMB, COCHLEAR IMPLANT, KIDNEY DIALYSIS, INCUBATOR

Simultaneous Machine Translation (2017)

Eliminating language barriers with an app.

Waverly Labs, a small start-up lab in New York, has created Pilot, a smart earpiece that promises to deliver on the idea of simultaneous machine translation. Put simply, if your only language is English, and someone with a Pilot speaks to you in Spanish, your Pilot delivers the words to your ear via a computerized voice in English. Your reply will then be translated and the other person will hear it in their language.

This simple-looking device, the product of computational linguistics, complex machine learning algorithms, and speech synthesis technologies, may finally lead to the demise of the human translator, who first came under assault in the 1930s with IBM's first

> *"The Pilot promises a life untethered, free of language barriers."*
>
> Andrew Ochoa, Waverly Labs founder

generation of simultaneous interpretation equipment.

The device follows on from previous translation apps, and similarly is not without its drawbacks. As it functions wirelessly through a phone app and the cloud the device burns through mobile data. The first Pilots only work with the English language's close Latin cousins, and in all languages context will continue to cause the usual linguistic errors. The Pilot's Holy Grail is a lofty one—nothing less than a world no longer separated by the barriers of language. **BS**

SEE ALSO: C PROGRAMMING LANGUAGE, SPEECH RECOGNITION, OPTICAL CHARACTER RECOGNITION

▣ *Packed full of complex algorithms, Pilot has used science to breach the language threshold.*

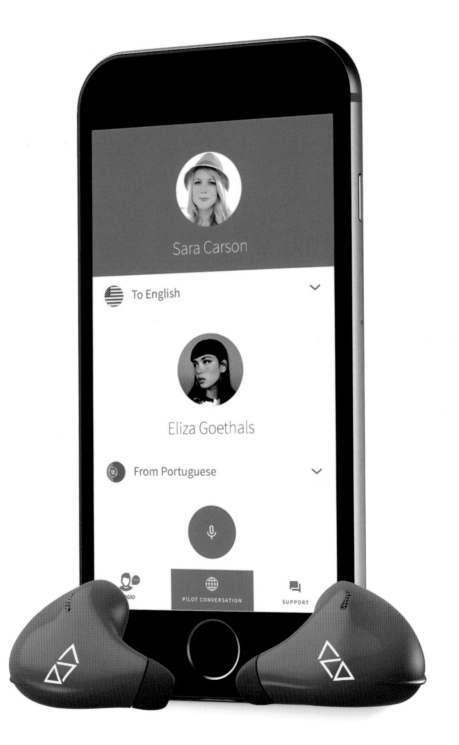

Jet Pack (2018)

Gravity Industries makes a science fiction dream come true.

The dream of strapping on a personal jet pack and speeding through the air has long been a fantasy, in science fiction and in real life. Many people have tried, with varying levels of success. Prototypes have been developed, powered either by jet or rocket engines. Whatever the fuel and type of engine used, jet packs create thrust in one direction by allowing a typically hot exhaust gas to escape through a nozzle in the opposite direction. Rocket engines have the disadvantage that the entire weight of the exhaust gas is present from lift-off. The most commonly used fuel is hydrogen peroxide (H_2O_2), which decomposes into oxygen and steam extremely quickly when it comes into contact with a catalyst. Flight times of rocket-propelled jet packs are typically less than thirty seconds. Jet engines, on the other hand, react fuel with oxygen from the air, the oxygen making up part of the weight of the exhaust. As a result, jet-propelled jet packs can have longer flight times.

Small gas turbine jet engines have been used on jet packs since the mid-1960s. Many inventors have succeeded in making short controlled flights, including some using wing suits that create extra lift in horizontal flight. In 2018, British inventor Richard Browning perfected the Daedalus Flight Pack—a jet pack capable of achieving high-speed controlled flight, hovering, and powered jumps. Browning's pack is an exoskeleton holding two small jet engines, and the pilot also has one small jet engine attached to each arm. This innovative approach led to commercial interest from the Great North Air Ambulance Service, who were interested in using the system in mountain rescue. **JC**

SEE ALSO: JET ENGINE, LIQUID FUEL ROCKET, IRON ROCKET, FIREWORKS, POWERED AIRPLANE

⬆ *Richard Browning makes a record-breaking flight over Brighton Pier, UK, in his Daedalus jet pack.*

Quantum Computer (2019)

IBM points to a stable quantum future.

Conventional computers work by manipulating numbers, text, and images represented by groups of binary digits, or bits. Those digits (0 and 1) can be represented inside the computer by anything that can be in one of two states—for example, the absence or presence of an electric current inside a transistor. A quantum computer manipulates information in the form of qubits (quantum bits). Each qubit can also only take the value 1 or 0 at the beginning and the end of a calculation. In between, however, as long as a qubit is not disturbed or its state observed, it can be 1 and 0 at the same time, thanks to the laws of quantum physics. Utilizing quantum effects gives quantum information processing a huge advantage over conventional information processing, and certain tasks that would take conventional computers millions of years are trivial with a quantum computer.

Originating in the early twentieth century, quantum physics is the study of very small particles, such as electrons, and how they interact with electromagnetic radiation, such as light or X-rays. Mathematicians and physicists became interested in the idea of quantum computing in the 1970s. Since then, computer scientists have been working on the theoretical aspects of handling quantum information, while engineers have been addressing the challenges involved in representing qubits and keeping them undisturbed, in shielded, cryogenic chambers. There were several breakthroughs in the 2010s, most notably perhaps IBM's Q System One, launched in 2018. IBM's device was the most stable quantum computer to date, and was the first commercially available quantum computer. **JC**

SEE ALSO: DIGITAL ELECTRONIC COMPUTER, LASER COOLING OF ATOMS, SQUID, SCANNING TUNNELING MICROSCOPE

⤒ *Quantum computers like the IBM Q System One will revolutionize computing and cyber security in the future.*

Glossary

Acid
A solution with a pH of less than 7; a chemical compound that releases hydrogen ions when dissolved in water. *See* **pH**

Alchemy
The medieval forerunner of chemistry, characterized by an erroneous belief in the transmutation of matter, such as turning base metals into gold.

Algorithm
The rules to be followed in mathematical calculations or other processing operations, such as computing.

Alkaline
A chemical compound with a pH of more than 7 that dissolves in water. *See* **pH**

Alloy
A metal produced by the combination of two or more metallic elements, usually to achieve greater strength or resistance to corrosion.

Altitude
The height of an object in relation to a known level, such as sea or ground level. In astronomy, altitude describes the apparent height above the horizon of an object in space, measured as an angle.

Ammonia
A colorless gas that dissolves in water to create a highly alkaline solution. Used as a cleaning fluid.

Analog
The use of continuous signals or information that are analogous to another set of signals, such as voltage. *See* **Digital**

Armillary Sphere
Made of brass rings and hoops, a model representing the Earth in relation to the Sun and Moon and the movement of stars and planets. First developed by the Greeks in the third century BCE, it was refined by medieval astronomers, eventually becoming heliocentric rather than geocentric as knowledge of the solar system increased.

Atom
The basic unit of a chemical element, made up of varying numbers of protons, neutrons, and electrons.

Azimuth
A concept measured as an angle between a reference plane, such as true north, and a point in space such as a star or planet. For example, using true north at 0 degrees as a reference, a point south would have an azimuth of 180 degrees.

Battery
Invented by Alessandro Volta in 1800, a device that stores chemical energy for conversion into electricity. Originally, batteries were created by linking up cells of liquid electrolytes, but now batteries are made of dry cells containing a paste.

Bose Einstein condensate
A prediction that under supercooled conditions most atoms would collapse and coalesce into a single state or "super atom." Named after a work published by Albert Einstein and Satyendra Nath Bose.

Bronze
An alloy of copper and tin.

Cantilever
A beam or platform supported at only one end, so that it projects outward without any other support. Cantilever bridges and balconies are common examples.

Catalyst
A substance that provokes or speeds up the rate of a chemical reaction without itself undergoing any permanent chemical change.

Celestial equator
An imaginary projection of Earth's equator into space.

CFCs
Chlorofluorocarbons (CFCs) are compounds containing chlorine, fluorine, and carbon. For a time they were used in refrigeration and as cleaning solvents, but they are now prohibited because of their effects on the ozone layer.

Chloroform
Also known as trichloromethane, a colorless, volatile liquid used as a solvent. In the nineteenth century it was also used as a general anesthetic.

Circuit
A closed path that enables the flow of an electric current, or a system of electrical conductors that together form such a path.

Cleat
A piece of metal or wood that prevents the slippage of ropes by acting as a stop.

Coaxial
Literally "sharing a common axis," but also describes coaxial cable in which a conducting medium is sheathed in several other media.

Combustion
Burning, or the sequence of reactions that take place when a fuel and an oxidant are heated.

Composite
A composite material is made from two or more materials with significantly different physical or chemical properties. These properties remain separate within the finished structure. Examples are plywood, wood-plastic composites, and some laminates.

Compound
A substance consisting of two or more different elements that are chemically bonded. Unlike the materials in a composite, which remain distinct, a compound's constituents react chemically to become an entirely new substance.

Compressor
A mechanism that supplies gas at increased pressure to power a turbine. The word also refers to an electrical amplifier that inhibits the range of a signal.

Conductor
A material or device that permits the transmission of heat, electricity, or sound. Some materials are better conductors than others—for example, copper is a good conductor of electricity, while wood is not.

Cryptography
The art of writing and solving codes.

Declination
In astronomy, declination refers to the angular distance of a point north or south of the celestial equator. It also describes the angular deviation between the magnetic north pole and the geographic north pole.
See **Celestial Equator**

Diameter
The measurement of a line passing through the center of a circle from one side to the other.
See **Radius**

Digital
The use of signals or information converted into digits of a physical quantity such as voltage, and used to represent arithmetic numbers.
See **Analog**

DNA
Deoxyribonucleic acid (DNA) is a nucleic acid that contains the genetic coding for all living organisms. DNA usually exists as a pair of DNA molecules twisted together in a double helix.
See **Helix**

Dynamo
A device that converts mechanical energy into electrical energy; an early form of electrical generator.

Electrode
An electrical conductor through which electricity leaves or enters a substance or device.

Electromagnetism
The interaction of electricity with magnetic fields.

Electron
A subatomic particle that carries a negative electric charge.

Frequency
The rate of vibration, usually measured per second, of sound, radio, or light waves. A measure of repeating events, such as sound waves, per unit of time. "High frequency" refers to many waves and a short wavelength.
See **Wavelength**

Fulcrum
The pivotal point on which a lever turns; the position of a fulcrum in relation to a lever is crucial to moving objects of great weight.

Fuselage
The body of an aircraft that holds the crew, passengers, and cargo.

Gamma-ray
High-energy electromagnetic radiation, or light emission of high frequencies, produced by sub-atomic particle interactions. Gamma rays easily penetrate and damage living cells.
See **X-ray**

Generator
A device that generates energy, such as electricity, or a mechanism that produces gas or steam.

Genetic
Relating to genes—the code in each cell that determines how an organism develops.

Geocentrism
The belief that the Earth is the center of the universe or the solar system, a view commonly held until the seventeenth century.

Gigawatt
Equal to one billion watts and used to described the output of large power plants or power grids.

Graphite
A form of carbon that occurs as a mineral in some rocks. Used in pencils and in nuclear reactors.

Heliocentrism
The Sun's position as the center of the solar system. Until the time of Galileo, the Earth was believed to be the center of the universe.
See **Geocentrism**

Helioscope
An instrument used to observe the Sun by projecting an image of the Sun onto a surface. This prevented direct observation of the Sun, which causes eye damage.

Helix
A curve in three-dimensional space in the shape of a spiral or corkscrew. Atoms in a protein, nucleic acid, or polymeric molecule are in the shape of a helix, as in the double helix of DNA. *See* **DNA**

Holograph
Originally, a document written in the author's own hand; this now has the additional meaning of a photographed image that appears three-dimensional.

Horsepower
The power of an internal combustion engine. The term originated with a comparison of horses versus steam engines, but is now an inexact measure of engine output. *See* **Kilowatt, Megawatt, Gigawatt, Terrawatt**

Hydraulics
The control of the flow of liquids through pipes, pumps, and turbines, such as the channeling of water for irrigation or via an aqueduct.

Hydroponic
A method of growing plants without soil that instead uses nutrients dissolved in water.

Hydropower
The use of the energy of flowing water to power mechanisms such as water wheels or watermills.
See **Hydraulics**

Infrared
Electromagnetic radiation that has a wavelength just greater than that of the red end of the visible light spectrum. It is emitted by heated objects and can be picked up by infrared cameras.

Insulator
A material that resists the flow of electricity, such as glass or plastic.

Ion
An atom or molecule that is positively charged because of the loss or gain of one or more electrons.

Iron
A hard, magnetic, silvery-gray metallic element commonly found as an ore, such as hematite. Iron is believed to be the chief constituent of Earth's core. It is also the main component of steel, cast iron, and wrought iron.

Kerosene
Originally a trade name for a lamp oil, the term is now generic and refers to a combustible hydrocarbon liquid used to propel jet engines and fuel domestic heating. Usually referred to as "paraffin" in the United Kingdom and South Africa.

Kilowatt
Equal to 1,000 watts and equivalent to about 1.34 horsepower.
See **Horsepower**

Kinetic energy
The energy acquired by an object or body when it is in motion. The kinetic energy of an object is directly proportional to the square of its speed.

Laser
An acronym for **L**ight **A**mplification by **S**timulated **E**mission of **R**adiation. A laser generates a beam of coherent light by stimulating the emission of photons from atoms or molecules. Lasers are used in laser printers, in fiberoptic cables, in players that read CDs and DVDs, and in manufacturing to shape metals and other materials.

Latex
A natural rubbery substance found in many plants and trees. Latex from the rubber tree is one of the substances used to make rubber for many commercial applications.

Mechanical escapement
A device in a watch or clock that regulates the movement of gears so that the swing of the pendulum or balance wheel is regulated.

Megawatt
Equal to one million watts of electricity and used to describe the huge output of lightning strikes or electrical generators.

Modem
A device for modulation and demodulation: for example, to convert analog signals from a telephone line into digital data for use in a computer.

Nitroglycerin
A heavy, oily liquid obtained by nitrating glycerol. It is unstable and highly explosive and has been used since the 1860s in explosives such as dynamite.

Optics
The study of light, and also the study of the eye and sight.

Ore
Naturally occurring material, such as rock, that contains valuable minerals or metals.

Oxidation
The chemical reaction of oxygen and another substance. Carbon, for example, oxidizes to produce carbon dioxide.
See **Redox**

Paraffin wax
Flammable waxy solid of saturated hydrocarbons distilled from petroleum. Used in candles, cosmetics, and polishes, and also for sealing and waterproofing.

Parchment
A thin writing material made from animal skin, such as goat's skin.
See **Vellum**

Patent
The filing of a right or title to an invention that asserts the sole right to make, use, or sell the invention.

pH
The measure of the concentration of hydrogen ions in a solution, which quantifies its acidity or alkalinity. Pure water has a pH of 7. A pH of less than 7 is acidic, a pH of more than 7 is alkaline.
See **Acid, Alkaline**

Phenol
A white crystalline solid, which is extracted from coal tar; when in a dilute solution, it is called carbolic and used as a disinfectant.

Photon
An elementary particle or wave that carries electromagnetic radiation. It differs from other elementary particles, such as electrons, in that it has zero rest mass.

Pixel
Abbreviation of "picture element," meaning the smallest unit of information in an image displayed on a screen. Pixel counts are now used to describe the definition of an image, and therefore photographic equipment, as in a 5-megapixel digital camera.

Plumbago
An old-fashioned word for graphite.
See **Graphite**

Pneumatic
Describes a device operated by air or gas under pressure, such as a pneumatic drill.

Polymath
A person of wide-ranging learning, adept at many subjects.

Polymer
Commonly used as a term for plastic, it refers to materials with a molecular structure consisting mostly of a large number of similar units, usually synthetic, such as nylon and PVC, although there are also naturally occurring polymers, such as amber and cellulose.

Port
The side of a ship that is on the left-hand side when facing forward.
See **Starboard**

Radar
"Radar" is an acronym for **Ra**dio **D**etection **a**nd **R**anging. It is a system that bounces electromagnetic waves off of things such as moving objects or weather systems and processes the returning signal to establish their range, altitude, direction, and speed.

Radiation
Energy emitted as electromagnetic waves when an atom or other body changes from a high-energy state to a lower-energy state. Radiation usually refers to ionizing radiation, which is an unstable and dangerous form of radiation. Radio waves and microwaves are examples of non-ionizing radiation, which is more stable and therefore not dangerous.

Radioactivity
The emission of ionizing radiation or particles from an unstable atomic nucleus. *See* **Radiation**

Radius
The measurement of a straight line drawn from the center of a circle to its perimeter. *See* **Diameter**

RAM
RAM is an acronym for **R**andom **A**ccess **M**emory. It is the short-term working memory (data storage) of a computer.

Ratchet
A device that permits rotary motion only in one direction: examples are the turnstile, jack, hoist, and ratchet screwdriver.

Redox
A chemical process that changes oxidation number, either decreasing the number of electrons attached to a molecule, atom, or ion (reduction) or increasing the number (oxidation). *See* **Oxidation**

Right Ascension
The distance of a point east of the zero point of longitude, measured along the celestial equator and expressed in hours, minutes, and seconds.

RNA
Ribonucleic acid (RNA) is a nucleic acid that is essential to the synthesis of proteins. Unlike DNA, RNA is a single-stranded molecule. *See* **DNA**

Scuba
Scuba is an acronym for **S**elf-**C**ontained **U**nderwater **B**reathing Apparatus, and it refers to equipment that enables divers to breathe underwater.

Semiconductor
Made of solid materials such as silicone, semiconductors both conduct and insulate against electricity. They are used in devices such as computers, cell phones, and digital televisions.

Smelting
The extraction of metal from ore by heating and melting the ore.

Solenoid
Wire coiled into a cylinder that acts as a magnet when an electric current is applied to it.

Solvent
A liquid or gas used to dissolve other substances, and which yields a solution. Common solvents are nail varnish remover, paint thinner, and organic solvents used in dry cleaning.

Starboard
The side of a ship that is on the right-hand side when facing forward. *See* **Port**

Steel
An alloy of iron with carbon and usually other elements such as manganese, chromium, vanadium, or tungsten.

Taxonomy
The classification of living organisms such as flora and fauna.

Terawatt
Equal to one trillion watts of electricity.

Thermodynamics
The study of the transformation of heat into power.

Tillage
The cultivation of land for the purpose of growing crops by weeding and turning the soil in order to loosen it.

Torque
The tendency of force to rotate an object about an axis. The torque will depend on the amount of force applied and on how long a lever is from its fulcrum.

Truss
A load-bearing structure used in architecture and bridge design. Usually triangular, trusses can be simple, two-dimensional shapes or more complex and three-dimensional in form.

Ultraviolet
A form of electromagnetic radiation in the form of light with shorter wavelengths than X-rays. It is invisible to the human eye but just beyond violet in the color spectrum. A certain amount of ultraviolet light is needed by the human body to stimulate the production of Vitamin D, though too much causes sunburn.

Vellum
A thin, fine-grained parchment made from calf skin. *See* **Parchment**

Volt
A measure of the force of electricity, named for Alessandro Volta, the Italian physicist who developed the first electric cell in 1800.

Watt
A measure of the rate of energy converted by an appliance: for example, the electrical energy that a light bulb converts into light. Named for James Watt, the nineteenth-century inventor of the steam engine.

Wavelength
The distance between the repetitive peaks of measured waves, with reference to light, sound, water, or electromagnetic waves. *See* **Frequency**

X-ray
A form of electromagnetic radiation in the form of light with longer wavelengths and lower frequencies than gamma rays, and therefore not so harmful. As a form of ionizing radiation, X-rays can be dangerous, but when used with care they have invaluable applications in medicine and public security.

Contributors

Simon Adams (SA) is a historian and writer living and working in London. He has contributed to a number of history and reference books and is the author of more than fifty books for children on subjects as varied as ancient Rome, the two world wars, and the sinking of the *Titanic*.

Clare Ashton (CA) was born in Dublin, Ireland. She lived there until moving to Scotland to study at the University of Edinburgh, where she received a first class honors degree in mechanical engineering. She currently lives in Aberdeen, where she works as a mechanical engineer for a major oil and gas company.

Katherine Ball (KB) grew up in Hampshire, and gained a BSc in English and environmental science at University of the West of England, Bristol. Today, she is still in the area, working in public relations.

Jim Bell (JB) earned a degree in zoology at the University of Leicester then a master's in science communication at University of the West of England, Bristol. He works as a presenter in science centers, online, and on local radio.

Stewart Bell (SB) was born in London and studied at the University of Bath and the University of the West of England, gaining a degree in applied biological science. He is currently a trainee science teacher based in Cambridge, England.

Riaz Bhunnoo (RB) completed a master's degree and PhD in chemistry at the University of Southampton. He currently lives in Swindon, where he works for the Research Councils UK. He loves writing and is on a mission both to enthuse the public with his passion for science and to demonstrate that not all scientists are geeks. He is a contributor to *Null Hypotheis: The Journal of Unlikely Science*.

Hayley Birch (HB) is a freelance writer and editor based in Bristol, UK, and holds degrees in life sciences and communications. Hayley has written for a number of books, magazines, and websites on a broad range of science and technology subjects. Her work has appeared in *Nature*, *Chemistry World*, *The Little Black Book of Science*, and *The Daily Telegraph*.

Richard Bond (RBd) has degrees in biology and history/philosophy of science from the University of London. He has written widely on science policy, and on science and society, and provides the humorous (or so he likes to think) Other Lab column for *Null Hypothesis*.

Christopher Booroff (CB) studied physics with space science at University College London and was elected as a Fellow of the Royal Astronomical Society in 2005. As a child he dreamed of being in the "A-Team" and building rickety tanks out of tractors and scrap metal; he now works in finance.

Matt Brown (MB) is a science writer and editor of the website Nature Network, a community site for scientists. He lives in London with his wife and his iPhone.

Raychelle Burks (RBk) is a native Californian studying in one of America's many land-locked states, Nebraska. Currently, she is finishing up both a master's in forensic science and a doctoral in chemistry. She devotes her free time to *Law & Order*, speed reading any book she can find, and her Labrador, Watson.

Terry Burrows (TB) is an author, musician, and academic. He has had more than fifty books published in such diverse fields as music, business, technology, education, psychology, media and modern history. He is based in London.

A keen sailor, **Lucy Cave (LC)** has a practical interest in the developments that have transformed her sport over time. She is studying the sciences and writes on a wide variety of subjects.

Stephen Cave (SC) is studying sciences, design, and technology and is particularly interested in the impact of effective design on our daily lives. His other passions include reading, rugby, rowing, and sailing.

Jack Challoner (JC) has a degree in physics from Imperial College, London, and worked for several years at the educational department of the Science Museum, also in London. He is now a science writer and editorial consultant and has published more than 20 popular science books. He lives in Bristol, England.

Andrew Chapman (AC) has been a freelance writer for more than fifteen years, and has published books on subjects as diverse as anthropology, genealogy, employment, and acting. He is also a professional designer and editor, as well as a publishing and Internet consultant, and additionally runs a popular website that recommends books.

Josh Davies (JD) is a struggling geochemist presently situated in Alberta, Canada, but open to offers from warmer climes.

Michael Davis (MD) has spent many years traveling and researching the origins of various inventions. He currently resides in Kent and works in London as a freelance wrangler of words.

Sarah Drinkwater (SD) is a freelance writer and journalist. She writes for magazines and newspapers and has an MA from UCL in Renaissance Literature. Her first novel was published in 2009.

Amanda Elsdon-Dew (AE-D) (née Batten) has a degree in English and history from York University. She worked for many years for general book publishers, carrying out a variety of minor editorial jobs. Now, with nearly grown children, she works as a freelance writer.

Janet Fricker (JF) has a degree in physiology and writes about medicine for a number of publications ranging from medical publications *BMJ* and *The Lancet* to the *Daily Mail*.

Margaret Fricker (MF) has worked in industry, technical colleges, schools, and adult education and recently was an honorary visiting research fellow. She has a BSc and MSc—both in physics—as well as an MPhil and PhD—both in education.

Matt Gibson (MG) has been programming computers since the age of eight. A Warwick University graduate, he's worked in the IT industry all his adult life. Lured into part-time journalism by the BBC, his writing now includes everything from book reviews to celebrity interviews. He is a native Londoner, but wound up in Bristol by mistake.

James Grant (JG) is an English language and literature graduate of King's College, London. His writing includes arts criticism and theater pieces. He currently lives in Surrey.

Reg Grant (RG) has written more than twenty books on a wide variety of historical subjects. His works include *The Rise and Fall of the Berlin Wall; Flight: 100 Years of Aviation; The Visual History of Britain from 1900 to the Present Day; Assassinations;* and *Soldier: A Visual History of the Fighting Man*.

Simon Gray (SG) studied media at Sheffield Hallam. He currently works as a freelance writer and travel planner.

Barney Grenfell (BG) worked as a science-communicator at the London Science Museum and at Bristol Science Centre before taking the plunge into the murky waters of freelance. He wears many hats: science, communication, new media, writing, and a purple one with a feather.

David Hawksett (DHk) writes, edits, consults, and presents on all science and technology topics, although originally he was an astrophysicist. He is senior contributing consultant to Guinness World Records and also works freelance for organizations including the BBC, Lucasfilm, and Starchaser Industries.

Rebecca Hernandez (RH) studied biochemistry and worked for eight years in Seattle's biotech sector before moving to the UK in 2006. In addition to bench science she dabbles in science writing and education, and spends her free time traveling, playing drums, and flamenco dancing.

Lisa Hitchen (LH) is a freelance science and medical journalist. Her writing has appeared in *New Scientist, Nursing Times,* and *British Medical Journal*, among other publications. She is also a short-story writer and performance poet.

Elizabeth Horne (EH) studied sciences at university and is particularly interested in the impact of scientific and technological developments on the way people live. She writes on a wide range of subjects and works in educational publishing.

David Hughes (DH) has degrees in physics and astronomy and spent 42 years lecturing to students at the University of Sheffield, UK—where he was Professor of Astronomy—and researching into the minor bodies of the solar system and the history of astrophysics. He retired in 2007.

David Hutter (DaH) was born in Germany and briefly lived in both England and France during his teens. After finishing high school he moved to Italy for a lacklustre and ultimately futile attempt at studying economics. Two years later, he moved to England to study creative writing and religious studies at Middlesex University. Having graduated, he now works as a freelance writer and editor in London.

Andrew Impey (AI) graduated from the University of Bristol in 2005, having completed a PhD studying the relationship between water nutrients and diving ducks. Today, he is a freelance tour guide and conducts natural history tours all over Europe. Dr. Impey is also the co-founder of *Null Hypothesis, the Journal of Unlikely Science*, an award-winning magazine and online resource that has been described by *The Daily Telegraph* as the "*Private Eye* of the science world." Its aim is to use humor to increase public awareness and understanding of science.

Hannah Isom (HI) studied biomedical science at the University of Manchester. She is now living in Leeds studying for a postgraduate diploma in print journalism.

Ann Kay (AK) is a writer and editor of many years' experience. Having tackled most subjects in her time, she has a particular fascination for how the things we use every day came into being. She is currently undertaking postgraduate studies at Bristol University.

Douglas Kitson (DK) studied math and physics at Warwick, before heading to Bristol to study science communication. He loves movies and science, and has very scruffy hair.

Andrey Kobilnyk (AKo) is a science journalist and former editor of FirstScience.com

Jane Laing (JL) is an editor and writer of more than 20 years' experience in the area of illustrated non-fiction. Specializing in western art and history, she is author of *Cicely Mary Barker and Her Art*, and was a major contributor to *The Visual History of the Modern World*.

George Lewis (GL) writes articles and encyclopedia entries for a wide range of publications, from *Cosmopolitan* to *The Times* (London).

Chris Lochery (CL) studied creative writing at Leeds University. After graduating he worked as a science-based children's entertainer to pay off his student debt. He now lives in London, working as a writer, and occasionally plays piano for theater companies in Dubai and Peterborough.

B. James McCallum (BMcC) holds degrees in biology from Davidson College and medicine from the University of South Carolina School of Medicine. He is an Academic Hospitalist and Assistant Professor of Clinical Internal Medicine in Columbia, South Carolina, U.S.A.

Jamie Middleton (JM) has a degree in biochemistry and a passion for all things scientific. He makes his living as a freelance writer and editor, and currently lives in Bath.

Brian Owens (BO) was born in Canada where he earned a degree in biology, before moving to London for his master's in science communication. He is currently News Editor of two science policy newsletters, *Research Fortnight* and *Research Europe*.

Tamsin Pickeral (TP) studied history of art before furthering her education in Italy. She has traveled extensively and divides her time between writing about art, science and inventions, and horses. Her publications include *The Horse: 30,000 Years of the Horse in Art* and *1001 Historic Sites*.

Becky Poole (RP) is currently working as a postdoctoral research scientist at the University of Bristol, where she is investigating the effects of the environment on gene expression in bread wheat. In between lab work, looking after her wheat plants and analyzing data, Becky enjoys writing about all aspects of science and has contributed a number of articles to thenakedscientists.com.

Helen Potter (HP) was born in England and lived in Singapore and Guyana, before returning to gain a BA and MSc in natural sciences from the University of Cambridge, where she is currently studying for her PhD in biological chemistry. She is a voracious reader, keen dancer, and plays the bad guy in pantomimes. She is also a contributor to *Null Hypothesis: The Journal of Unlikely Science*.

Steve Robinson (SR) is studying journalism at Oxford and also works as a freelance writer for science publications.

Having escaped from the depths of Glasgow at the tender age of sixteen, **Leila Sattary (LS)** emigrated to the island nation of St. Andrews, where she spent five years completing a master's degree in physics. She has carried out research at Glasgow and Oxford Universities and currently works for the Engineering and Physical Sciences Research Council.

Eric Schulman (ES) is an American PhD astronomer and the author of the 1999 popular science book *A Briefer History of Time*. His work has appeared in magazines, newspapers, radio, and television in North America, Europe, and Asia. He sits on the board of *The Annals of Improbable Research* (AIR), once memorably described as "The *Mad Magazine* of science," to which he has contributed a number of articles, including "The History of the Universe in 200 Words or Less."

Faith Smith (FS) was born in England, where she completed a degree in zoology. She wrote her own column for a local newspaper for several years and is currently studying for a PhD in veterinary parasitology.

Stuart Smith (SS) was born in the United States, where he completed a degree in biology—and wouldn't leave until they gave him a medical degree as well. Stuart continues to practice and teach medicine, and occasionally write.

Barry Stone (BS) was born in Melbourne, Australia, where he completed a diploma in travel writing and photography at the Chisholm Institute. He currently writes on travel, architecture, and modern history for a wide variety of newspapers and magazines.

Francisca Wiggins (FW) is originally from southern Lincolnshire, but having graduated in mechanical engineering from Edinburgh University, she now works as a ship surveyor in ports around the world.

Logan Wright (LW) lives in Canada and is currently pursuing a degree in engineering. He walks everywhere but thinks the combustion engine is better than the *Mona Lisa*.

Thomas Zeller (TZ) is associate professor of history at the University of Maryland, College Park, USA. He is the author of *Driving Germany: The Landscape of the German Autobahn, 1930-1970* (2007) and has coedited the volumes *The World Beyond the Windshield: Roads and Landscape in the United States and Europe* (2008), *Rivers in History: Perspectives on Waterways in Europe and North America* (2008), *How Green Were the Nazis? Nature, Environment, and Nation in the Third Reich* (2005), and *Germany's Nature: Cultural Landscapes and Environmental History* (2005).

Picture Credits

Every effort has been made to credit the copyright holders of the images used in this book. We apologize for any unintentional omissions or errors and will insert the appropriate acknowledgment to any companies or individuals in subsequent editions of the work.

20–21 © Jesper Jensen/Alamy 22 John Reader/Science Photo Library 25 The Art Archive/Musée Guimet Paris/Dagli Orti 26 The Art Archive/Egyptian Museum Turin/Dagli Orti 27 The Bridgeman Art Library 29 Science Museum 31 The Bridgeman Art Library 32 Alamy 34 Cintract Romain/Age Fotostock 36 The Art Archive/Musée Guimet Paris/Dagli Orti 37 The Art Archive/Private Collection/Dagli Orti 38 The Art Archive/British Museum/Dagli Orti (A) 41 The Art Archive/Musée du Louvre Paris/Dagli Orti 42 The Bridgeman Art Library 43 The Art Archive/Dagli Orti 45 akg-images/Erich Lessing 46 Corbis 47 Catalhoyuk Research Project 49 The Art Archive/Topkapi Museum Istanbul/Dagli Orti 50 The Art Archive/Dagli Orti 51 The Bridgeman Art Library 53 The Art Archive/Dagli Orti 54 The Art Archive/Bardo Museum Tunis/Dagli Orti 57 The Bridgeman Art Library 59 The Art Archive/Diozesanmuseum Trier/Dagli Orti 61 Gianni Dagli Orti/Corbis 63 The Art Archive/Dagli Orti 65 Kimbell Art Museum/Corbis 66 Västerbottens museum 67 Frederico Formenti/University of Oxford 69 The Art Archive/Museo Nazionale Reggio Calabria/Dagli Orti 71 The Art Archive/Dagli Orti 73 The Art Archive/Dagli Orti 75 The Art Archive/Dagli Orti 77 The Art Archive/Musée du Louvre Paris/Dagli Orti 78 Dinodia 79 Photo Scala, Florence 81 akg-images/Erich Lessing 82 The Art Archive/Egyptian Museum Cairo/Dagli Orti 85 Hip/Scala, Florence 87 Chris Cheadle/Alamy 88 Science Museum 89 The Art Archive/Musée du Louvre Paris/Gianni Dagli Orti 91 The Bridgeman Art Library 93 Mary Evans Picture Library 2008 94 The Art Archive/Dagli Orti 95 Christel Gerstenberg/Corbis 97 The Art Archive/British Museum/Dagli Orti 99 The Bridgeman Art Library 100 Hip/Scala, Florence 101 Louvre, Paris, France/The Bridgeman Art Library 102 akg-images/John Hios 103 akg images 104 The Art Archive/Musée Romain Nyon/Dagli Orti 105 Science Museum 107 Werner Forman Archive/Scala, Florence 108 Bibliotheque Nationale, Paris, France/Archives Charmet/The Bridgeman Art Library 111 akg-images/Erich Lessing 112 The Bridgeman Art Library 115 The Art Archive/British Museum/Eileen Tweedy 117 The Art Archive/Archaeological Museum Cividale Friuli/Dagli Orti 119 David Lees/Corbis 120 Dragon News/Rex Features 122 Science Museum 123 The Bridgeman Art Library 124–125 © Hans Delnoij/Alamy 127 The Hammer Museum 129 The Stapleton Collection/The Bridgeman Art Library 131 akg-images 132 Ben Plewes Travel Photography/Alamy 133 Mary Evans Picture Library 134 The Art Archive/Historiska Muséet Stockholm/Alfredo Dagli Orti 135 Hugh Threlfall/Alamy 137 akg-images/British Library 139 North Wind Picture Archives/Alamy 140 Fitzwilliam Museum, University of Cambridge, UK/The Bridgeman Art Library 142 The Bridgeman Art Library 143 Interfoto Pressebildagentur/Alamy 144 Science Photo Library 145 akg-images 146 The Bridgeman Art Library 148 akg-images 149 Ladi Kirn/Alamy 150 The Art Archive 152 The Art Archive 154 akg-images 155 The Art Archive 157 akg-images 158 The Bridgeman Art Library 159 Scala 160 The Art Archive/Museo Correr Venice/Alfredo Dagli Ort 163 Science Museum 164 Gutenberg-Museum, Mainz 165 Scala 167 Science Museum 168 © Bettmann/Corbis 170 Science Museum 173 Wellcome Images 174 Scala 176 Science Museum 177 Heidelberg University, Germany 178 Photo Scala, Florence 2004 179 Scala, Florence—courtesy of the Ministero Beni e Att. Culturali 180 Scala Florence/Hip 182 Dave King/Dorling Kindersley, Courtesy of The Science Museum, London 185 Science Museum 186 Science Museum 188 The Bridgeman Art Library 190 The Art Archive/Eileen Tweedy 191 Science Museum 192 Science Museum 194 Dave King/Dorling Kindersley, Courtesy of The Science Museum, London 195 Science Museum 197 Anna Clopet/Corbis 197 The Bridgeman Art Library 198 Corbis 199 Science Museum 200 DK Limited/Corbis 203 The Bridgeman Art Library 204 Getty 206 Science Museum 207 Science Museum 208 DK Limited/Corbis 211 Science Museum 212 Science Museum 215 Science Museum 215 Wellcome Images 216 The Art Archive/Collection J. M. Demange Paris/Gianni Dagli Orti 218 George Bernard/Science Museum 219 Archives Charmet/The Bridgeman Art Library 220 Science Museum 221 Science Museum 222 Canadian War Museum in Ottawa, Canada 224 Mary Evans Picture Library/Alamy 224 Science Museum 226 Giraudon/The Bridgeman Art Library 227 Courtesy of the Warden and Scholars of New College, Oxford/The Bridgeman Art Library 228 Mary Evans Picture Library 230 Science Museum 230 Science Museum 231 Science Museum Pictorial 232 Science Museum 232 Science Museum 235 Science Photo Library 236 Bridgeman Art Library 237 Wellcome Library, London. Wellcome Images 239 Science Museum 239 Science Museum 240 akg-images 241 Alamy 242 Hip/Scala, Florence 243 Science Museum 244–245 © Colin Garratt; Milepost 92 ½/Corbis 246 Wellcome Images 247 Bettmann/Corbis 248 The Bridgeman Art Library 249 Bettmann/Corbis 250 Interfoto Pressebildagentur/Alamy 252 Mary Evans Picture Library 253 Lordprice Collection/Alamy 255 Science Museum 256 Mary Evans Picture Library 258 McCord Museum 259 Bettmann/Corbis 260 Science Museum 261 Schenectady Museum; Hall of Electrical History Foundation/Corbis 263 Science Museum 264 Science Museum 265 Science Museum 267 Science Museum 269 Getty Images 270 Science Museum 271 Wellcome Images 273 Science Museum 274 The Bridgeman Art Library 275 Clive Streeter © Dorling Kindersley, Courtesy of The Science Museum, London 276 Vic Fowler/Rex Features 277 NMeM Science Museum 278 Science Museum 279 Mary Evans Picture Library 280 Mary Evans Picture Library 281 Beverly Wilgus 282 Bettmann/Corbis 283 akg-images 284 Mary Evans Picture Library 285 Getty Images 286 Bridgeman Art Library 287 Science Museum 289 © Bettmann/Corbis 290 The Bridgeman Art Library 291 © Swim Ink 2, LLC/Corbis 292 © Chad Davis/Alamy 292 © Bettmann/Corbis 295 The Bridgeman Art Library 295 Science Museum 296 Roger-Viollet/Topfoto 296 DK Limited/Corbis 298 National Museum of American History/Smithsonian 299 Science Museum 301 Science & Society Picture Library/Science Museum 301 Science Museum 303 Science Museum 304 Getty 304 Science Museum 305 The Print Collector/Alamy 305 Dave King © Dorling Kindersley, Courtesy of The Science Museum, London 307 Science Museum 308 Geoff Brightling © Dorling Kindersley 309 Gady Cojocaru/Alamy 310 Science Museum 311 Science Museum 312 Time & Life Pictures/Getty Image 313 Getty Images 315 Science Museum 315 Science Museum 316 akg-images 317 Getty Images 319 Antiques & Collectables/Alamy 319 Hip/Scala, Florence 320 Science & Society Picture Library/Science Museum 321 Science Museum 323 Schenectady Museum; Hall of Electrical History Foundation/Corbis 324 Science Museum 325 Sam Ogden/Science Photo Library 328 Bettmann/Corbis 329 DK Images 330 Science Museum 332 Ronald Pearson 334 Science Museum 335 Advertising Archives 336 Corbis 337 Sarah Quill/Alamy 339 Science Museum 340 The Advertising Archives 343 © Syd Brown 344 Science Museum 345 Science Museum 346 Mary Evans Picture Library 348 Science Museum 350 Science Museum 351 Science Museum 353 Science Museum 354 Historial de la Grande Guerre © Yazid Medmoun 356 Science Museum 357 Science Museum 358–359 Science Museum 361 Courtesy the National Firearm

Museum, Fairfax, VA **362** NMeM, Science Museum **363** Science Museum **365** Science Museum **367** U.S. National Archives and Records Administration **368** Hulton Archive/Getty Images **370** Imperial War Museum **371** akg-images **373** The Art Archive/Culver Pictures **374** Topical Press Agency/Getty Images **376** Three Lions/Getty Images **378** Mary Evans Picture Library 2008 **379** Boyer/Roger Viollet/Getty Images **381** Shutterstock **382** Courtesy Stanley **383** Science Museum **385** Vintage Power And Transport/Mark Sykes/Alamy **386** DK Images **387** Harry Vos/www.the-canopener.com **389** Corbis **390** Dave King © Dorling Kindersley **391** Bettmann/Corbis **393** Bettmann/Corbis **395** The Advertising Archives **396** Tyco Fire & Building products **397** Science Museum **399** Science Museum **400** Science Museum **401** Science Museum **402** Fox Photos/Getty Images **403** Mansell/Time & Life Pictures/Getty Images **404** Mary Evans Picture Library 2008 **406** Science Museum **407** Hulton Archive/Getty Images **409** Stuart Forster/Alamy **410** Science Museum **411** Jan Willem Bech **412** Science Museum **413** Mary Evans Picture Library **414** Boyer/Roger Viollet/Getty Images **416** Science Museum **417** David Kilpatrick/Alamy **419** Corbis **420** Science & Society Picture Library/Science Museum **424** Stockbyte/Getty Images **425** Clive Streeter © Dorling Kindersley, Courtesy of The Science Museum, London **426** Hulton Archive/Getty Images **427** Schenectady Museum; Hall of Electrical History Foundation/Corbis **428** The Art Archive/Dagli Orti **431** Science Museum **433** U.S. National Archives and Records Administration **434** Otis Elevator Company **435** Bettmann/Corbis **436** Science & Society Picture Library **437** Science & Society Picture Library **439** Fox Photos/Getty Images **440** Science & Society Picture Library **441** Science & Society Picture Library **443** Daimler **443** Daimler **445** Daily Herald Archive/Science & Society Picture Library **446** Science Photo Library **447** Hip/Scala, Florence **448** Science Museum **451** Fox Photos/Getty Images **453** Shawn Flock **455** Mary Evans Picture Library 2005 **456** Clive Streeter, Dorling Kindersley, Courtesy of The Science Museum, London **457** Hulton Archive/Getty Images **459** www.vintageinternetpatent.com **460** Underwood & Underwood/Corbis **461** Bettmann Corbis **463** Science Photo Library **464** National Museum of Photography & Corbis **464** Corbis **467** Mary Evans Picture Library 2005 **468** akg-images **471** Corbis **472** Corbis **475** National Media Museum/Science Photo Library **477** Otis Elevator Company Archives **478** Wenger SA **479** SNA Europe **481** ABI Europe **482–483** © David Muscroft/Alamy **484** Sheila Terry/science Photo Library **485** Science Photo Library **487** Getty Images **488** Dorling Kindersley **489** GEA WestfaliaSurge Inc. **491** Universität Heidelberg **492** Bettmann/Corbis **493** National Media Museum/Science Photo Library **494** Kodak Collection/NMeM/Science Photo Library **496** Science Photo Library **497** Science Photo Library **500** The History of Advertising Trust **501** Science Photo Library **503** The Art Archive **504** Getty Images **505** General Photographic Agency/Hulton/Getty Images **506** Science Museum **506** Mary Evans Picture Library 2008 **507** Science Museum **507** NMeM - Kodak Collection/Science Museum **509** Science Museum **510** Popperfoto/Getty Images **511** Bettmann Corbis **512** Science Photo Library **513** Science Photo Library **517** Carrier Corporation **518** Shutterstock **519** Motoring Picture Library/Alamy **521** Dave Rudkin © Dorling Kindersley **522** FPG/Hulton Archive/Getty Images **523** Science Photo Library **524** Crayola **528** Science Museum **529** Underwood & Underwood/Corbis **530** Getty Images **531** Getty Images **532** Science Museum **533** Science Museum **534** Science Museum **536** PhotoPower/Alamy **538** Bernard Hoffman/Time Life Pictures/Getty Images **539** Science Museum **540** Corbis **542** Mike Dunning © Dorling Kindersley, Courtesy of The Science Museum, London **543** MoMA, New York/Scala, Florence **545** Science Museum **546** Science Museum **547** Science Museum **549** Science Museum **551** Science Museum **553** Fox Photos/Getty Images **555** Time Life Pictures/Pix Inc./Time Life Pictures/Getty Images **556** Science Museum **560** DeAgostini Picture Library/Scala, Florence **561** John Martin, Science Museum **562** Science Museum **563** Science Museum **565** Science Museum **567** Hulton Archive/Getty Images **569** General Photographic Agency/Getty Images **570** The Bridgeman Art Library **572** Science Museum **573** Dave King © Dorling Kindersley, Courtesy of The Science Museum, London **575** Utilisation Presse/Edition uniquement, akg-images **578** Photodisc/Alamy **580** Bettmann Corbis **582** Science Museum **583** Science Museum **585** General Photographic Agency/Getty Images **586** Minnesota Historical Society **588** Hulton-Deutsch Collection/Corbis **589** Bettmann Corbis **591** Hulton-Deutsch Collection/Corbis **592** Bettmann Corbis **593** National Media Museum/Science Museum **594** Bettmann Corbis **596–597** Corbis RF/Alamy **598** Antiques & Collectables/Alamy **598** Kodak Collection/NMeM/Science Museum **600** Science Museum **601** Corbis **602** Science Museum **603** Science Museum **605** Daily Herald Archive/NMeM/Science Museum **607** Science Museum **608** Bettmann Corbis **611** Mary Evans Picture Library 2008 **612** De Agostini Picture Library, Scala, Firenze **613** Bettmann/Corbis **614** Hulton-Deutsch Collection/Corbis **615** George W. Hales/Fox Photos/Getty Images **616** Dave Rudkin © Dorling Kindersley **618** JupiterImages/Brand X/Alamy **620** Photo Researchers/science photo library **623** akg-images/RIA Nowosti **625** Ray Tang/Rex Features **626** NASA **629** Science Museum **631** Time Life Pictures/Getty Images **632** Redfx/Alamy **633** Gerard Tel **635** Harry Todd/Fox Photos/Getty Images **637** Hip/Scala, Florence **639** R. Gates/Hulton Archive/Getty Images **640** Science Museum **641** Science Museum **643** Dmitri Kessel/Time Life Pictures/Getty Images **644** Peter Stackpole/Time Life Pictures/Getty Images **646** Bettmann Corbis **647** Alfred Eisenstaedt/Pix Inc./Time & Life Pictures/Getty Images **648** Science Museum **649** Science Museum **650** Detlev Van Ravenswaay/Science Photo Library **651** Popperfoto/Getty Images **653** Motorola, Inc. **654** John Loengard/Time Life Pictures/Getty Images **656** The Art Archive/Culver Pictures **657** Dmitri Kessel/Time Life Pictures/Getty Images **659** George Konig/Keystone Features/Getty Images **661** Daily Herald Archive/NMeM/Science Museum **662** Utilisation Presse/Edition uniquement/akg-images **664** Fred Ramage/Getty Images **666** Matthew Richardson/Alamy **667** Courtesy of the Collection of the Australian National Maritime Museum, Darling Harbour, Sydney **668** Fritz Goro/Time Life Pictures/Getty Images **670** Science Museum **671** Los Alamos National Laboratory/Science Photo Library **673** Lawrence Berkeley National Lab **674** Science Museum **675** Canada Science and Technology Museum Corporation **676** The Advertising Archives **679** Daily Herald Archive/NMeM/Science Museum **681** Science Museum **682** Dittrick Medical History Center, Case Western Reserve University **683** National Media Museum/Science Museum **685** Smithsonian Institution/Corbis **686** Michael Ochs Archives/Getty Images **688** Fritz Goro/Time Life Pictures/Getty Images **689** Hulton-Deutsch Collection/Corbis **690** Science Museum **691** Science Museum **692** Neal Preston/Corbis **695** Science Museum **698** Courtesy of Dr. Gary Enever/Royal Victoria Infirmary **698** Margaret Bourke-white/Time & Life Pictures/Getty Images **701** NMPFT/Daily Herald Archive/Science Museum **703** Jean Gaumy/Magnum Photos **705** Tetra Pak **706-707** © Corbis Sygma **708** Michael Rougier/Time Life Pictures/Getty Images **709** Science Museum **710** Bettmann Corbis **711** Al Fenn/Time Life Pictures/Getty Images **713** The Advertising Archives **715** Daily Herald Archive/NMeM/Science Museum **716** Al Fenn/Time Life Pictures/Getty Images **717** J. R. Eyerman/Time Life Pictures/Getty Images **719** Bettmann Corbis **720** Edward Kinsman/Science Photo Library **721** Dorling Kindersley **722** Dr. Najeeb Layyous/Science Photo Library **724** Hulton Archive/Getty Images **725** Photodisc/Alamy **726** David Murray, Dorling Kindersley **727** Bettmann Corbis **728** Alfred Pasieka/Science Photo Library **729** AFP/Getty Images **730** Science Museum **731** Bettmann Corbis **733** NASA **733** Science Photo Library **735** The Advertising Archives **737** Document General Motors/Reuter R/Corbis Sygma **737** Science Museum **738** Science Museum **739** The Advertising Archives **740** Tim Boyle/Getty Images **741** Science Museum **742** Louie Psihoyos/Corbis **745** Ria Novosti/Science Photo Library **749** Hulton Archive/Getty Images **751** Kent Gavin/Keystone/

Hulton Archive/Getty Images **752** David Parker/Science Photo Library **753** Manchester Daily Express **754** Courtesy Pilkington United Kingdom Limited **755** Science Museum **756** Science Photo Library **758** Mary Evans Picture Library/Alamy **760** Amy Trustram Eve/Science Photo Library **761** George Steinmetz/Corbis **763** NASA **764** Giphotostock/Science Photo Library **765** Ferranti Electronics/A. Sternberg/Science Photo Library **766** Science Museum **767** Image Source/Corbis **768** Istock **769** The Coca-Cola Company/The Advertising Archives **770** Hank Morgan/Science Photo Library **772** Steve Shott © Dorling Kindersley **774** JupiterImages/Comstock Images/Alamy **775** Car Culture/ Corbis **776** Hulton Archive/Getty Images **777** Bettmann Corbis **778** Muntz/Getty Images **785** General Electric **786** Justin S. Osborne/U.S. Navy/Reuters/Corbis **787** Jack Pritchard/Getty Images **789** Peter Ruck/Bips/Getty Images **790** Science Museum **790** Science Museum **792** Courtesy Leonard Herman **793** Bootstrap Alliance **794** Roger Ressmeyer/Corbis **795** Heinz Kluetmeier/Time & Life Pictures/Getty Images **797** Shutterstock **798** Phototake Inc./Alamy **801** David Scharf/Science Photo Library **802–803** Andy Piatt **805** Science Museum **807** NASA **808** Mediscan/Corbis **810** Science Museum **811** Motorola, Inc. **812** Courtesy of InPhase Technologies **813** Vintage Calculators **815** Vo Trung Dung/Corbis Sygma **817** Maximilian Stock Ltd/ Science Photo Library **818** Roger Ressmeyer/Corbis **819** Science Museum **821** © Magimix **822** Wellcome Images **823** Hulton-Deutsch Collection/Corbis **824** Science Photo Library **825** NASA/ Science Photo Library **826** Bettmann Corbis **827** Roger Ressmeyer/ Corbis **828** Gregor Schuster/zefa/Corbis **829** Science Museum **830** Roger Ressmeyer/Corbis **833** Volker Steger/Science Photo Library **834** Shutterstock **835** Science Museum **836** Philippe Plailly/ Eurelios/Science Photo Library **837** Image Source/Getty Images **838** Institute Archives **839** Kim Kulish/Corbis **840** Photodisc/Alamy **841** Curtesy Kodak **842** Roger Ressmeyer/Corbis **843** © Sony **845** Charles O'Rear/Corbis **846** Courtesy Porter Novelli **847** © Sony **849** Hank Morgan/Science Photo Library **850** Dyson **851** The Advertising Archives **854** TRW Corporation/Science Photo Library **855** NASA/Science Photo Library **856** Garo/Phanie/Rex Features **857** Philippe Psaila/Science Photo Library **859** Qilai Shen/epa/ Corbis **861** LWA/Getty Images **863** Mauro Fermariello/Science Photo Library **864** www.protoolerblog.com **865** Science Museum **866** Matt Stroshane/Getty Images **867** National Aeronautics & Space Adm/Science Museum **868** Science Museum **870** Science Museum **871** Science Museum **873** Getty Images **875** Science Museum **876** Science Museum **877** James King-Holmes/Science Photo Library **879** Peter Menzel/Science Photo Library **881** Science Museum **882** Cordelia Molloy/Science Photo Library **882** Wellcome Images **884** Damien Lovegrove/Science Photo Library **885** Victor De Schwanberg/Science Photo Library **886** Eurelios/Science Photo Library **889** Simon Fraser/Science Photo Library **891** Don Farrall/ Getty Images **892** O. Louis Mazzatenta/National Geographic/Getty Images **895** Patrick Dumas/Eurelios/Science Photo Library **896** CERN/science Photo Library **897** Jung Yeon-Je/AFP/Getty Images **898** Andrea Matone/Alamy **900** Corbis **902** Science Museum **903** D. Roberts/Science Photo Library **905** Zephyr/Science Photo Library **907** Jung Yeon-Je/AFP/Getty Images **908** Courtesy Trevor Baylis **910** bobo/Alamy **912** Gideon Mendel/Corbis **915** Volker Steger/Science Photo Library **916** Pasquale Sorrentino/ science Photo Library **918** Vo Trung Dung/corbis Sygma **919** Marijan Murat/dpa/Corbis **920** Fa Changguo/Xinhua Press/Corbis **922** Yoshikazu Tsuno/AFP/Getty Images **924** NASA **927** Linna Photographics, Inc. **929** Courtesy of *What Hi-Fi? Sound and Vision* **930** James King-Holmes/Science Photo Library **931** Brian Jackson/Alamy Stock Photo **932** Thomas Deerinck, NCMIR/Science Photo Library **933** Justin Sullivan/Getty Images **935** Arnaldur Halldorsson/Bloomberg/Getty Images **937** Waverly Labs **938** Guy Corbishley/Alamy Stock Photo **939** Misha Friedman/ Getty Images

Acknowledgments

Quintessence would like to thank the following picture libraries, and in particular the individuals named:

Vera Silvani/SCALA Picture Library

Tom Vine/Science & Society Picture Library

Nick Dunmore/The Bridgeman Art Library

John Moelwyn-Hughes/Corbis

Hayley Newman/Getty Images

Jessica Talmage/Mary Evans Picture Library

Anna Mosley/The Art Archive

Quintessence would also like to thank the following individuals for their assistance in producing this book:

Helena Baser

Rebecca Gee

Neil Lockley

Ann Marangos

Bruce Nicholson

Fiona Plowman

Tobias Selin